# TOOLEY'S DICTIONARY OF MAPMAKERS

The travels of Abraham Ortelius and Joris Hoefnagel in the 1570's were sometimes depicted in Braun and Hogenberg's Civitates Orbis Terrarum. Here they can be seen scrambling down a mountainside on the view of Tivoli.

# TOOLEY'S DICTIONARY OF MAPMAKERS

Revised Edition

E - J

Editor
Josephine French

Consulting Editors
Valerie Scott, Mary Alice Lowenthal

Picture research
Valerie Scott

EARLY WORLD PRESS
2001

© EARLY WORLD PRESS, 2001
1111 East Putnam Avenue
Riverside, CT. 06878


Designed by Roy Walker MCSD

Printed in England by Bookcraft, Midsomer Norton, Somerset


All rights reserved. No part of this publication may be reproduced, stored in a retrieval system, or transmitted in any form or by any means electronic, mechanical, photocopying, recording or otherwise, without the prior permission of the copyright owner.


ISBN 0-906430-19-4

# List of contributors

*Thanks are due to the following for their assistance:-*

Abbott, Jeffrey. London, UK
Ala'i, Cyrus. London, UK
Astengo, Corradino. Genova, Italy
Babinski, Mark. Garwood, New Jersey, USA
Bartholomew, John. Edinburgh, Scotland
Baskes, Roger. Chicago, IL, USA
Baynton-Williams, Ashley. London, UK
Bely, Alexander. Minsk, Republic of Belarus
Bendall, Sarah. Oxford, UK
Burden, Eugene. Ascot, Berks, UK
Burden, Philip. Herts, UK
Burgess, R.A. Edenbridge, Kent, UK
Calhoun, Zachary. California, USA
Carpine-Lancre, Jacqueline. Monaco
Carroll, Raymond. Spalding, Lincs, UK
Cartwright, M.F. Cape Town, South Africa
Conzen, Michael. Chicago, IL, USA
Cook, Andrew. London, UK
Crampton, Jeremy. Fairfax, VA, USA
Dahl, Bjorn. Copenhagen, Denmark
Dilke, Oswald. Leeds, UK (deceased)
Dörflinger, Johannes. Vienna, Austria
Dorothy Sloan, USA
Erling, Paul. Chicago, Illinois, USA
Fisher, Susanna. Southampton, UK
Galera i Monegal, Montserrat. Barcelona, Spain
Gasset-Argemi, Josep. Barcelona, Spain
Goffart, Walter. Toronto, Canada
Goss, John. Bletchley, Bucks, UK
Gròf, Làszlò. Oxford, UK
Heckrotte, Warren. Oakland, CA, USA
Herbert, Francis. London, UK
Hodson, Donald and Yolande. Herts, UK
Hofmann, Catherine. Paris, France
Hooker, Brian. Orewa, New Zealand
Hyde, Ralph. London, UK
Jones, Ieuan E. Birmingham, UK
Junko, Suzuki. Tokyo, Japan
Kapp, Kit S. Osprey, FL, USA
Kelly, James. Oxford, UK
King, Geoffrey. Eccleshall, Staffs, UK
Lane, Christopher. Pennsylvania, USA
Layland, Michael. Vancouver, BC, Canada
Loupis, Dimitris. Athens, Greece
Lourie, Ira S. Rockville, MA, USA
Maura O'Connor, Canberra, Australia
Montillo, Alberto. The Phillippines
Moore, John. Glasgow, Scotland
Oehrli, Marcus. Wabern, Switzerland
Oehrli, Markus. Wabern, Switzerland
Paas, John Roger. Northfield, MN, USA
Pastoureau, Mireille. Paris, France
Pedley, Mary. Ann Arbor, MI, USA
Peile, A.J. Bracknell, Berks, UK
Pelletier, Monique. Paris, France
Ristow, Walter. Mitchellville, MA, USA
Ritchie, G.S. Aberdeen, UK
Ritter, Michael. Sielenbach, Germany
Roberts, Iolo and Menai. Newcastle, Staffs, UK
Ross, Robert. Calabasas, CA, USA
Rossetti, Keith. Menorca, Spain
Scharfe, Wolfgang. Berlin, Germany
Steward, Harry. Worcester, MA, USA
Tallis, E. Farnham, Surrey, UK
Tyner, Judith. Long Beach, CA, USA
Worms, Laurence. London, UK

# How to use the Dictionary

**Order of entries**
Entries appear in alphabetical order except in the case of entries under the same name. These are headed by families of that name, irrespective of the names of the individuals.
Generations are indicated by bullet points [¥].

After family entries, individuals are then listed in alphabetical order by their given names. Where an initial is known only, that individual precedes those whose given names begin with that initial. Where two or more individuals have the same given names, they are listed chronologically.

Companies follow individuals with the same name. Successions are indicated by bullet points [¥].

NB. The examples on this page are fictitious.

**Williams**, family. Engravers of London.
¥ **Williams**, John (1720-1772).
¥¥ **Williams**, Michael George (1743-1791).
¥¥ **Williams**, John & Son (*fl*.1761-1769). Partnership of John Williams and his son Michael.
¥¥¥ **Williams**, George (1764-1835). Son of Michael.
¥¥ **Williams**, Adam James (1747-1810). Son of John and brother of Michael.

**Williams**, Adam (1826-1911).

**Williams**, G. (1803-1856).

**Williams**, George (1780-1850).

**Williams**, George (1820-1880).

**Williams & Co.** (*fl*.1760-1776).
¥ **Williams, Smith & Co.** (*fl*.1777-1781).

---

**Content of entries**
The mapmaker's name is followed by birth and death dates where known, and a small amount of biographical information if available.

This is followed by a few examples of work in chronological order, together with any available information about others involved.

Where possible references have been included. These give details of where further information on this mapmaker or his works can be found.

**Williamson[Williamsun][1]**, George (1780-1850)[2]. Engraver at the sign of the burin, Cornhill[3] London (1800-1820); 22 Fleet Street, London[4] (1821-1845[5]). John Rose's *The Road from London to Birmingham*[6] (3 sheets[7]), 1810[8], 1812[9]; *Atlas of England* (28 maps[10]), London[11], J. Smith[12] 1815, Oxford[13], 1817[13]; *Map of London, Westminster and Southwark*, London, J. Smith 1822; map of Ireland[14], 1823; *London*, 1840 (with D. Beech[15]). **Ref.** SMITH, J. *George Williamson: Engraver* (London 1996); WILLS, C. *London engravers* (1987) pp.10-16.[16]

**Key to example**
1. Alternative names, spellings, mis-spellings or misreadings.
2. Birth and death dates unless otherwise indicated.
3. Address as shown on works at given date(s).
4. Supplied address, not necessarily on works.
5. Dates at this address.
6. Title of work.
7. Number of sheets for a single map.
8. Earliest known date for the work.
9. Later issue or edition (not always the same title).
10. Number of maps in a publication.
11. Place of publication.
12. Publisher.
13. Details of later issue or other edition.
14. Unverified title or description of an untitled map.
15. Collaborator.
16. Useful references in reverse chronological order.

---

**The following abbreviations have been used**

| | | | |
|---|---|---|---|
| *b.* | born | *q.v.* | which see |
| *d.* | died | Ref. | reference |
| *c.* | circa | SJ | Society of Jesus |
| ed. | editor | RE | Royal Engineers |
| *fl.* | flourished | RN | Royal Navy |
| MS | manuscript | USN | US Navy |

**E., A.W.** Survey of Woodford Row, 1863.

**E.V.** See **Vico**, Enea.

**Eachard** [Echard], Laurence (1670-1730). Historian and geographer. *Compleat compendium of geography*, 1691; *Gazetteer*, 1692 (16th edition, 1744).

**Eager**, E.N. County surveyor in the U.S.A. County of Solano, California, 1890.

**Eagleton**, Thomas. Surveyor. Map of an estate in Buckinghamshire, England, 1777 MS.

**Eami** Bokushin. Japanese artist and cartographer. *Arima-yama no zu* [plan of Arima], c.1737 (after Shunboku).

**Earhart**, J.S. *Map of the City of Hamilton, Ohio*, 1864.

**Earl**, George Windsor. Arafura Sea, Admiralty chart 1837, reduced version 1840.

**Earl**, Thomas. Lagos River, Admiralty chart 1857.

**Earle**, Augustine (*fl*.1726-1754). Survey of Cawston Warren [Norfolk, England], 1732.

**Earle**, J.H. *Atlas of Knox County, Illinois*, Chicago 1891.

**Early**, Eleazer. Publisher of Savannah, Georgia. *Map of the State of Georgia*, 1818 (compiled by Daniel Sturges).

**Earnshaw**, J. *Plan of Bridlington Quay*, 1891.

**Easburn** [Eastburn], Benjamin. Surveyor General of Pennsylvania, 1733-1741. Parts of Pennsylvania and Maryland, 1740; surveys used for *A Plan of the City of Philadelphia...*, London, Andrew Dury 1776 (engraved by P. Andre).

**East**, Thomas (1540-1608). Printer of London. *The Post of the World*, 1576.

**Eastburn**, James [and Company]. Publisher of New York City. J.H. Eddy's *The State of New York*, 1818.

**Eastgate**, John. Engraver of Islington Road, and of 1 Fletcher Road, Clerkenwell, London (1811). E.J. Johnson's *Chart of the coasts of Northumberland and Durham*, 1819.

**Eastin**, Lucien Johnston. *Emigrant's Guide to Pike's Peak* [Colorado], Leavenworth City 1859 (with a map of the gold mines).

**Eastman**, *Captain* S. Indian tribes of the U.S.A. for Henry Rowe Schoolcraft's *Historical and Statistical Information... Indian Tribes of the United States* (6 volumes), Philadelphia, Lippincott 1851-1853, re-issued 1854-1857; *Map of Nebraska & Kansas Territories*, Philadelphia, Lippincott, Grambo & Co. 1854.

**Eastoe**, —. Engraver. Texel, Vlieter Roads &c. [Netherlands], 1812-1813.

**Easton**, Alexander. Civil engineer. Mailroads Milford to Caermarthen, 1824-1827; *Map shewing the route of the North Pennsylvania Rail Road from Philadelphia to Bethlehem*, Philadelphia, Crissy & Markley 1857 (lithographed by P.S. Duval & Co.).

**Easton**, E. *London Water Supply 1892*, Judd & Co. 1893.

**Easton**, John. Surveyor. *London and Liverpool Mail Road*, 1826 (with T. Telford and T. Casebourne, drawn by L. Hebert).

**Easton**, Captain L.C. Fort Laramie [Wyoming] to Fort Leavenworth [Kansas], 1849.

**Eastwood**, —. See **Eschuid**, John.

**Eaton**, —. See **Hunt** & Eaton.

**Eaton**, Amos (1776-1842). Geologist and botanist. Survey for Erie Canal, 1825; *This colour'd map exhibits a general view of the economical geology of New York and part of the adjoining states...*, Albany 1830 (printed by Websters and Skinners).

**Eaton**, Charles Y. Surveyor. Assisted with George N. Colby's *Atlas of Somerset County, Maine*, Houlton 1883.

**Eaton**, George C. Surveyor of Delaware, Ohio. *Muskingum County* [Ohio], Zanesville, W.P. Bennett 1852.

**Eaton**, J.H. *Sketch of the Sabine River Lake and Pass from Camp Sabine to the Gulf*, Washington D.C. 1838.

**Eaton**, S. Dwight. American engraver. W.E. Ferguson's *Map of the Old Colony Rail Road with its branches & Connecting roads* [E. Massachusetts], Boston, J.H. Bufford & Co's Lith. 1850.

**Eaton**, William C. American surveyor. *Map of the Town of Canaan, Grafton Co. N.H.*, Philadelphia, E.M. Woodford 1855; Town of Enfield, New Hampshire, 1855; *Map of Windham County Connecticut*, Philadelphia, E.M. Woodford 1856 (with others); *Map of Tolland County, Connecticut*, Philadelphia, Woodford & Bartlett 1857 (with H.C. Osborn).

**Eayre**, T. See **Eyre**, Thomas.

**Ebano**, Eleodoro. Portuguese mapmaker. Map of the bay of Paranaguá [Brazil], 1653.

**Ebbesen**, —. Danish engraver. *Holstein und Lauenburg* (6 sheets), 1861-1863 (with others); *Generalstabens topographiske Kaart over Danmark*, 1862.

**Ebden**, William (fl.1811-1856). *The Country twenty-five miles round London*, 1815; *Spanish America*, 1820; *The Pedestrians Companion fifteen miles round London* (engraved Sidney Hall), 1822; *Twelve Miles round London*, 1823 (engraved by Hoare & Reeves); from 1824-1828 Ebden produced a series of loose county maps, engraved by Hoare & Reeves and published initially by Hodgson and Company with titles such as *Hodgson's New Map of the County of Somersetshire*, 1824 (when W. Cole took over, titles were amended to begin *Ebden's New Map of...* for example *Ebden's New Map of the County of Buckinghamshire...*, 1825 and later); All the maps appeared in several editions in different publications including *A New Atlas of England and Wales*, London, James Duncan [and others], 1833 and later editions, also *A Complete County Atlas of England & Wales*, 1833 and later, also *The New British Atlas*, H.G. Collins 1855; they were also revised as a series of loose county maps with titles beginning *Collins' Railway and Telegraph map of...*, 1855 and later. **Ref.** CARROLL, R.A. *The printed maps of Lincolnshire 1576-1900* (1996) pp.216-219; SMITH, D. *Imago Mundi* 43 (1991) pp.48-58; KINGSLEY, D. *Printed maps of Sussex 1575-1900* (1982) pp.170-172.

**Ebdy**, T.G. Surveyor. *Plan of the City of Durham*, 1865.

**Ebel**, J.A. *Charte der Elbmündungen*, 1846.

**Ebel**, Dr. med. Johann Gottfried (1764-1830). Natural scientist and travel writer from Züllichau, Prussia [now Sulechów, Poland], worked in Frankfurt-am-Main, Paris and Zurich. Wrote a guide for travellers through Switzerland, c.1793 (included a panorama of the Alps drawn by Ebel, an edition of 1804-1805 included a more scientific panorama by H. Keller). **Ref.** GERMANN, Thomas *Cartographica Helvetica* 13 (1996) pp.23-30 [in German].

*Christoph Daniel Ebeling (1741-1817). Portrait by P. Suhr (By courtesy of the Library of Congress, Washington DC)*

**Ebeling, Christoph Daniel** (1741-1817). German historian, studied at Göttingen from 1763, taught in Hamburg from 1769, rising to become Professor of history and classical language at Hamburg Gymnasium. Known for his major work on the history and geography of the United States. *Erdbeschreibung und Geschichte von Amerika...* (7 volumes, no maps), Hamburg, Carl Ernst Bohn 1793 and later (maps were made by Sotzmann to accompany the work, published as *Atlas von Nord-Amerika*, 1796-1799, though only 10 sheets of the planned 18 were completed).
**Ref.** RISTOW, W.W. *American maps and mapmakers* (Detroit 1986) pp.169-178; RISTOW, W.W. *The Map Collector* 14 (1981) pp.2-9.

**Ebeling, E.** Plan Stadt Braunschweig, 1860.

**Ebenreyter, Philipp** (*fl.*16th century). German artist and mapmaker. Painted maps for the bishop of Würzburg, 1538.

**Eberhard,** Dupré. Madagascar, 1667.

**Eberhard, H.** See **Eberhart, H.**

**Eberhard, Philipp** (1563-1627). Swiss mathematician, stonemason and topographer of Zurich. One of the founders of the Zurich surveying school (with J. Praetorius).

**Eberhardt,** —. Surveyor. Plan of Tübingen, 1876.

**Eberhardt von Romunde,** Johann. Karte von Ellenser Deichwerk, 1619.

**Eberhart, G.A.** Illustrated Historical Atlas of Carroll County, Ohio…, Chicago, H.H. Hardesty 1874; Illustrated Historical Atlas of Hancock County, Ohio, Chicago, H.H. Hardesty 1875.

**Eberhart, H.** Engraver for H. Stein, 1852; for Heinrich Berghaus, 1863; engraved text for *Spanien und Portugal* (4 sheets), Justus Perthes 1871 (with E. Kühn).

**Eberle,** Joseph Meinrad. Swiss surveyor. Appenzell, 1846.

**Ebersberger** [Ebersperger], family.
• **Ebersberger,** Johann Georg (1695-1760). Engraver and publisher. Son-in-law of Johann Baptist Homann, continued business with Johann Michael Franz as Homännische Erben [Homann Heirs] from 1730. **N.B.** See **Homann Heirs**.
•• **Ebersberger,** Barbara Dorothea. Daughter of Johann Georg. Inherited her father's interests in Homann Heirs on his death in 1760. When she married in 1761, control passed to her husband G.P. Monath.

**Ebert,** Frederick J. Colorado Territory, 1862.

**Ebert von Ehrentreu,** Karl (1754-1813). Austrian cartographer.

**Ebi** Gaishi [Gensui]. Japanese surveyor. *Kaihin shūkō no zu* [Coastal navigation], 1680 MS.

**Ebner,** Ferdinand (1786-1859). Publisher in Augsburg. *Umgebung von Augsburg*, c.1838.

**Ebner,** Georg. Publisher in Stuttgart, Germany. *Neueste Karte von dem Koenigreich Wuertemberg* (4 sheets), c.1834; *Post-und Reisekarte von Deutschland* (2 sheets), 1844.

**Ebray,** —. *Carte géologique du Département de la Nièvre*, 1864 (with Bertera).

**Ebstorf Map.** Anonymous 356-cm circular mappamundi, c.1235. It was rediscovered at Ebstorf Benedictine Convent in 1830, but destroyed during an air raid in 1943. It is now known only from facsimile. **Ref.** KUGLER, H. & ECKHARD, M. (Eds.) 'Ein Weltbild vor Columbus: Die Ebstorfer Weltkarte' in: *Interdisziplinäres Colloquium 1988* (Weinheim 1991) [in German]; WOLF, A. 'Die Ebstorfer Weltkarte…' in: *Cartographica Helvetica* No.3 (1991) pp.28-32 [in German]; HARVEY, P.D.A. *Medieval Maps* (London 1991) p.30; WOLF, A. 'Nieuws over de Ebstorfer Wereldkaarte' in: *Caert-Thresoor* Vol.7 No.2 (1988) pp.21-20 [in Dutch]; KRETSCHMER, DÖRFLINGER & WAWRIK *Lexicon zur Geschichte der Kartographie* (1986) vol.C/1, pp.184-185 [in German] (citing numerous references).

**Ecckebrecht, P.** See **Eckebrecht,** Philip.

**Echard, L.** See **Eachard,** Laurence.

**Echave de Assu,** Francisco. Lima, Antwerp 1688; La Estrella de Lima, 1688.

**Echegaray,** Martín de. Pilot. *Mapa de las costas del Golfo de Mexico y de la America Septentrional*, 1686 MS.

**Echenique,** Santiago. *Mapa de la Provincia de Córdoba*, 1866.

**Echigoya** Jihei. Japanese publisher. Joint publisher (with Hiranoya Mohei) of Shikata Shunsui's map of the Nishikawa Canal, 1863.

**Eck,** C.W.A. von. Historical notes to the maps in *Gegenden von deutschen Städten*, Berlin 1828 (small atlas with circular maps of the environs of German cities).

**Eck,** *Doctor* Heinrich A. (1837-1925). *Karte und Profile zur geologischen Beschreibung von Rüdersdorf und Emgegend*, Berlin, 1872; Schwarzwaldbahn, 1884.

**Eck**, Johannes (1486-1543). Theologian, mathematician, geographer and cosmographer at Ingoldstadt. Wrote theoretical works on geography and cartography, as well as a treatise on the construction and use of astrolabes.

**Eckard**, —. Engraver for Karl Friedrich Rudolf von Wiebeking's *Hydrographisch und militaerische Karte von dem Nieder Rhein von Lintz bis unter Arnheim...*, 1796.

**Eckberg**, *Captain* C.G. Plan and view of Cocos-Keeling Island, 1749, published Dalrymple 1792; chart of the south-east coast of Hainan, Dalrymple 1792.

**Eckebrecht** [Ecckebrecht], Philip (1594-1667). Mathematician and astronomer of Nuremberg. Friend of Johann Kepler for whose *Tabulae Rudolphinae...* he drew a map of the World embraced by a double headed eagle entitled *Nova Orbis Terrarum Delineatio...*, 1630 [all known copies date from after 1658] (map engraved by J.P. Walch of Nuremberg). **Ref.** SHIRLEY, R. *The mapping of the World* (1983) N° 335 pp.358-359.

**Eckener**, F. Lithographer and printer. *Karte von Schleswig-Holstein*, 1867.

**Ecker**, Johann Anton (1755-1820?). Artist and cartographer of Graz, worked in Vienna. Nördliche and Südliche Halbkugel der Erde (2 sheets), 1794, re-issued 1800. **N.B.** Compare with John Alexander **Ecker**, below.

**Ecker**, John Alexander (1766-1829). Bohemian physician. Südliche Halbkugel, 1800. **N.B.** Compare with Johann Anton **Ecker**, above.

**Eckersberg**, Christoph Wilhelm (1783-1853). German painter and engraver. Town plans.

**Eckert-Greifendorf**, *Professor* Max (1868-1938). German geographer and cartographer from Chemnitz. Wrote extensively on cartographic history and theory. Primary school atlas, 1898 with editions to 1914; *Wirtschaftsatlas der deutschen Kolonien*, Berlin, D. Reimer 1912. **Ref.** SCHARFE, W. *Imago Mundi* 38 (1986) pp.61-66; KRETSCHMER, DÖRFLINGER & WAWRIK *Lexicon zur Geschichte der Kartographie* (1986) vol.C/1, pp.185-186 [in German].

**Eckhardt**, *Doctor* Christian Leonhard [Christoph Ludwig] Philipp (1784-1866). Land surveyor. *Fortsetzung der topographischen Karte von Schwaben* (4 sheets), Cotta Verlag 1820-1823; Hessen und Nassau, 1823; Stern-Karte, 1853; also various globes.

**Eckhoff**, Emil A. *Official Map of the Territory of Arizona*, New York 1880 (with P. Riecker).

**Eckstein**, C.A. (1840-1925). Dutch military cartographer. *Kaart van het gedeelte Java en Sumatra...*, 1883. **N.B. Compare with Charles Eckstein, below.**

**Eckstein**, Charles F. Director of the Topographical Department of the Ministry of Defence in the Netherlands in the 1860's. Developed the 'typoautography' method of map lettering, and flat colour map reproduction through the use of lithography.

**Ecosse**, Jean d'. Topographical painter and mapmaker, worked from horseback. Travelled with the notaire Antoine Actuhier conducting a survey of the settlements, lands and castles of the Comtés of Valentinois and Diois [Drôme and Isère, France] from which he prepared a map, 1423 MS (now lost). **Ref.** DAINVILLE, F. de SJ *Imago Mundi* XXIV (1970) p.102 [in French].

**Eddison**, John. English surveyor. Roundhay Park Estate, Leeds, 1871.

**Eddy**, James (b.1806). Artist and draughtsman working for Pendleton's Lithography Co. of Boston, U.S.A. Maps of Lake St Clair, River St Clair and Detroit for a Congress report, 1827; *Map of the Country Embracing the Several Routes Examined with a View to a National Road from Zanesville to Florence*, 1828; United States-Canadian boundary in Lake Huron & Lake Erie (8 maps to accompany government report), 1828; *Map of Lynn and Saugus* [Massachusetts], 1829 (surveyed by Alonzo Lewis); *A Map of the Bounty Lands in Illinois...*, after 1830 (compiled by Zophar Case); *A Plan of the Public Lands in the State of Maine...*, 1835 (from surveys by G.W. Coffin). **Ref.** RISTOW, W. *The Map Collector* 21 (1982) pp.26-31.

**Eddy**, John Henry (1784-1817). Geographer and map publisher of New York. *Map of the Western Part of the State of New-York Showing the route of a Proposed Canal from Lake Erie to Hudson's River*, 1811;

draughtsman for *Map of the Straights of Niagara from Lake Erie to Lake Ontario 1813*, New York, Prior & Dunning 1813 (engraved by Robbinstone & Son); 30 miles round New York City, 1814; *Map illustrative of a communication between the Great Lakes and the Atlantic Ocean, by means of a canal...*, 1816; *The State of New York*, New York, James Eastburn 1818 (published posthumously). **Ref.** RISTOW, W. *The Map Collector 67 (1994) pp.21-23.*

**Eddy**, R.A. Publisher of San Francisco. W.M. Eddy's *The official map of the State of California*, 1853 (engraved by Schedler and Liebler).

**Eddy**, William M. City surveyor for San Francisco, Surveyor General of California. *Official map of San Francisco...*, 1849; *Official Map of the City of San Francisco...*, San Francisco, Benjamin F. Butler 1851 (lithographed by B.F. Butler, another edition was lithographed by Britton & Rey); ...*The Offical Map of the State of California*, R.A. Eddy 1853 (engraved by Schedler and Liebler); *Official map of the State of California*, 1854 (engraved and printed by J.H. Colton).

**Edelgestein**, Regnier. See **Frisius**, Gemma.

**Edeling**, A.C.J. *Kaart van de Eilanden beoosten Java*, 1842 (with W.T. Baars, C.W.M. van de Velde and L.J. de Vriese).

**Edelmuller**, Friedrich (1840-1905). Austrian military cartographer.

**Eden**, P. Surveyor of Iver, Buckinghamshire. Part of Iver, 1850 MS.

**Eden**, Richard (c.1521-1576). Scholar and translator with a particular interest in the New World and the Northwest and Northeast Passages. Published an English translation of the section on America from Sebastian Münster's *Cosmographia*, 1553; *The Decades of the Newe Worlde* (with a map of the New World derived from Jean Bellère), 1555 (thought to be the earliest collection of voyages in English); *The Arte of Navigation*, 1561 and editions to 1615 (with a map titled *The Newe Worlde*, derived from the work of Pedro de Medina, 1545); *The History of Travayle in the West and East Indies...*, London, Richard Jugge 1577 (augmented edition of *The Decades*, completed by Richard Willes). **Ref.** BURDEN, P. *The mapping of North America (1996) pp.28-29, 33.*

**Eden**, William. *Windsor Park and part of the Forest*, 1800.

**Eden**, W. [& Co.]. Printer and publisher of 14 Basinghall St., London. Captain Edward Smith's Coast of Syria, 1840.

**Éder**, Franciscus Xavier, SJ (1727-1773). Hungarian cartographer. Spent 20 years surveying in Peru, 1749-1769, surveys published posthumously. *Provincia Moxos*, Buda 1791. **Ref.** RAUM, F. *Magyar térképészek Délamerikában* [Hungarian cartographers in South America] in: *Geodézia és Kartográfia* [Geodesy and cartography] (1983/3) [in Hungarian].

**Eder**, Johann. *Prospekt der Stadt Warschau*, 1700.

**Eder**, Joseph (b.c.1750). b. Vienna. Prepared a map of Jerusalem *Neuer und wahrer Abriss der Stadt Ierusalem...*, [n.d.]; *Hand oder Schulatlas der Oesterreichischen Monarchie*, 1813.

**Eder**, W. *Handbuch der Allgemeinen Erdkunde der Länder-und Staatenkunde*, Darmstadt 1860 (supplement to L. Ewald's *Hand-Atlas der Allgemeinen Erdkunde der Länder- und Staatenkunde... in 80 Karten*).

**Eder**, Wolfgang. Publisher of Ingolstadt. Johannes Myritius's *Universalis orbis descriptio*, in *Opusculum geographicum rarum*, 1590.

**Edgar**, John (fl.1719-1735). English surveyor, worked in Dorset, Somerset and Wiltshire. *Manor of Strickland, Dorset*, 1735.

**Edgar**, *Lieutenant* Thomas, RN. Marine surveyor. Survey of West Falkland, 1786-1787, published as *A Chart of West Falkland Island*, Aaron Arrowsmith Sr. 1797 (engraved by T. Foot).

**Edgar**, William (fl.1717-d.1746). Scottish geographer, surveyor and architect. Apprenticed to George Riddell of Edinburgh in 1717, burgess and guild brother in Edinburgh, 1736, served as surveyor with the army of the Duke of Cumberland 1745-1746.

*A map of the Firth and River of Forth...*, 1740 MS; *A New and Correct Map of the Shire of Peebles or Tweeddale*, 1741 (engraved by Richard Cooper); *The Plan of the City and Castle of Edinburgh*, 1742, reprinted 1765 and later (engraved on copper by P. Fourdrinier, used in William Maitland's *History of Edinburgh*); *A new and correct map of Loch-Lomund with the country circumjacent*, 1743 MS; *A new and correct map of Stirling-Shire and Clackmannan-Shire*, 1743 MS; *The course of the King's Road makeing betwixt Dumbarton and Inveraray...*, 1745 MS; *Old and New Aberdeen...*, c.1746; *Description of the River Forth above Stirling* [Perthshire], 1746; *An exact map of the country about Perth...*, 1746 MS; *Inverness with the adjacent country including to Nairn with the field of battle near Culloden...* (2 sheets), 1746 MS; *A map of the roads from Forfar and Montrose to Aberdeen*, 1746 MS; map of Kinloch Rannoch, c.1746 MS; and others. **Ref.** MOIR, D.G. The early maps of Scotland to 1850 (2 volumes, 1973, 1983) passim.

**Edgcome**, W.H. *British Burma, Pegu Division* (4 sheets), 1885 (with F. Fitzroy).

**Edgell**, *Captain* Harry Fowkes, *RN*. Surveys of Paquet Harbour, Newfoundland, 1810-1815.

**Edgerton**, H.H. Arlin Gold Fields [Canada], 1899.

**Edgeworth**, father and son. Surveyors.
• **Edgeworth**, Richard Lovell (1744-1817). Surveyor from Somerset, worked for the Bogs Commission in Ireland, 1809-1814. *Bogs of County Longford & Westmeath* (4 sheets), 1811; *Bogs on River Shannon* (2 sheets), 1814. **Ref.** Memoirs of Richard Lovell Edgeworth, Esquire. (London 1820).

•• **Edgeworth**, William (1794-1829). Surveyor, assisted his father in the work for the Bogs Commission in Ireland, 1809-1814. Produced manuscript maps of areas in Ireland and Yorkshire. County Longford (4 sheets), 1814 (with his father); Co. Roscommon, 1817 (with R.J. Griffith).

**Edkins**, Joseph. *Eastern and Western Hemispheres*, 1864.

**Edkins**, S.S. Globemaker and seller, son-in-law of globemaker T.M. Bardin. One of those who sold later editions of Bardin's *New British Globes*. Pair of globes, 1799-1800 (terrestrial globe dedicated to Sir Joseph Banks).

**Edler**, Anton. *Würm- oder Starnberger-See*, 1866.

**Edler**, E.G. (fl.1816-1855). Engraver of maps for Adolf Stieler from 1816, and for later editions of Stieler atlases up to 1850; for Johann Ludwig Spruner von Merz, 1855.

**Edmands**, Benjamin Franklin. *The Boston School Atlas...*, Boston, Lincoln & Edmands 1830, 1831, 1832, 1833.

**Edmonds**, A.M. Draughtsman for the Canadian Pacific Railway. *Map of the Dominion of Canada* (6 sheets), 1882 (lithographed by The Burland Lithographic Co., Montreal); *Map Shewing the Railways of Canada* (3 sheets), Montreal, The Sabiston Litho. & Publishing Company 1891.

**Edmonds**, Alfred. *Plan of the City of Pekin*, 1900.

**Edmonds**, Christopher. *Plan of the Parishes St. Mary Lambeth, St. Mary Newington &c.*, 1833.

**Edmonds**, *Captain* Joseph, *RN*. Marmorice Harbour, 1802; Laurie & Whittle's *New Chart of Greece*, 1812.

**Edmonston**, *Doctor* Arthur. *The Zetland Islands* [Shetland Islands], 1809.

**Edmunds**, William. Surveyor in Kent, England. The Taddy Estate [Thanet] (22 maps), 1828-1830; *A Plan of the Town of Margate*, 1821 (used in *The New Margate, Ramsgate and Broadstairs Guide*, 1823 and Charles Kadwell's *Kentish Topography*, 1846); *Map of the Harbour, Pier and Bay of Margate...*, 1839 (lithographed by G.F. Cheffins).

**Edmunds**, *Captain* William H. After service in the Boer War, he was appointed draughtsman in the Department of Lands in Adelaide, becoming a pioneer of the mapping of South Australia. Sketch map of Adelaide, 1913-1914; *Road Plan of the Adelaide Hills*, 1938.

**Edoya** Kichiemon. Japanese publisher. Joint publisher (with Shimuraya Yohachi) of Hashimoto Tokuhei's *Fujimi On-Edo Ezu* [plan of Edo with a view of Mount Fuji], 1818.

**Edward**, Charles. *Plan of the Town of Dundee with the improvements now in progress*, 1846, 1847 (engraved by W. & A.K. Johnston).

**Edward**, Robert. See **Edwards**, Robert.

**Edwardes**, *Major* —. Country around Shanghai, 1865.

**Edwards**, brothers. American surveyors, employed by D. Jackson Lake in the early 1870's on surveys of Missouri and Kansas, both also worked on an atlas of Lorain County, Ohio, 1874 (with D.J. Lake and others). They also published atlases.
- **Edwards Brothers** (*fl*.1875-1885). Publishers of Missouri and '209 S. Fifth Street', Philadelphia. From 1875 they published many atlases of counties in Missouri and Kansas, later atlases were published by John P. Edwards. *An Illustrated Historical Atlas of Boone County, Missouri*, Missouri, Edwards Brothers 1875; *Atlas of Bourbon County, Kansas*, 1878; *Labette County, Kansas*, 1880; Sumner County, Kansas, 1883; Cloud County, Kansas, 1885. **Ref.** RISTOW, W.W. American maps and mapmakers (Detroit 1986) p.419.
- **Edwards**, Charles H.
- **Edwards**, John P. Worked with Charles as Edwards Brothers, and later published under his own name from Quincy, Illinois. County atlases of Missouri and Kansas including *Edwards' Map of Adams County, Illinois*, Quincy 1889.

**Edwards**, *Lieutenant-Colonel* —. *Map of the Shire Highlands*, 1897.

**Edwards**, A. *Map of the United States Road from Ohio to Detroit*, 1825 (with P.E. Judd and S. Vance, drawn by J. Farmer).

**Edwards**, A.F. Chief engineer of the Sachett's Harbor & Saratoga Railroad Company. *Report of the different routes and estimates of the Sacket's Harbor & Saratoga rail-road, October 1853* [Lake Ontario], Saratoga Springs, Steam Press of G.M. Davison 1853.

**Edwards**, Bryan (1743-1800). West India merchant. *The History, Civil and Commercial of the British Colonies in the West Indies* (with maps of the islands), London, J. Stockdale 1794, 1801, 1807, London 1818, the maps were re-engraved (mostly by Tanner) for *A New Atlas of the British West Indies...*, Charleston, E. Norford, Willington & Co. 1810, re-issued Philadelphia, I. Riley 1818 (the *Atlas* was prepared to accompany the Philadelphia edition of Edwards's *History*).

**Edwards**, Charles (1628-c.1691). Welsh puritan author and theologian. *Darluniad y Ddaear* [A view of the World], 1677 (engraved by Richard Palmer, said to be the first published Welsh language map, the map was included in the 3rd & 4th editions of Edwards's *Y Ffydd Ddiffuant* [The unfeigned faith], Oxford 1677, 1722).

**Edwards**, D. Publisher of '12 Ave Mary Lane', London. Publisher of *Pinnock's Guide to Knowledge* 1834 (included woodblock town plans by J. Archer).

**Edwards**, D.C. Publisher of Chicago. Republished Warner & Beers's *Atlas of the State of Illinois*, 1879 edition (first published Warner & Higgins 1869).

**Edwards**, Dennis [& Co.]. Printer and publisher of Cape Town. Cape Town, 1888 (in *General directory and guide-book to the Cape of Good Hope and its dependencies*, 1888); *A Plan of Cape Town*, 1900.

**Edwards**, George (1694-1773). Amateur scientist and ornithologist, library keeper of the Royal College of Physicians, London. Map of Britain and coast of Europe [known as the 'beetle map'] from *A Natural History of Uncommon Birds*, *c*.1746. **Ref.** SHIRLEY, R. Printed maps of the British Isles 1650-1750 (1988) p.55; SHIRLEY, R. The Map Collector 21 (1982) p.46.

**Edwards**, J. Engraver and publisher of Betchworth, Surrey. *Plan of Grove Hill, Camberwell*, 1793. **N.B.** Compare with **Edwards**, James E.

**Edwards**, James E. Surveyor, engraver and publisher from Surrey, England, later worked from Nº·23 Belvidere Place, Southwark, London (1799); Lee Place, Old Brompton

*George Edwards (1694-1773). (By courtesy of the National Portrait Gallery, London)*

(1816); Cromwell's Lane, Kensington (1817). Undertook surveys of the southern Home Counties, 1787 to 1800, which were used as a basis for his *A Companion from London to Brighthelmston* (9 maps of the route and town plans), 1796-c.1800, later republished as *Edwards's topographical surveys through Surrey, Sussex and Kent*, 1817-1820; *Trigonometrical Land Chart*, 1800; *Edwards's General Map* [Sussex], 1800, 1816, 1817; View of Brighton, 1817. **Ref.** KINGSLEY, D. *Printed maps of Sussex 1575-1900* (Lewes 1982) p.384.

**Edwards**, John P. See **Edwards** brothers, above.

**Edwards**, *Reverend* Joseph. *A New Map of the Diocese of Lichfield*, 1873.

**Edwards**, Langley (d.1774). A plan of the Low Fell Lands and Marshes... [the Fens], 1761.

**Edwards**, Leicester. *Chart of the Boom Kittam* [Sierra Leone], 1878-1879.

**Edwards** [Edward], *Reverend* Robert. *b*. Dundee, Minister of Murroes, Scotland, from 1648. Description of the County of Angus, together with a map *Angusia provincia Scotiae...*, Amsterdam, J. Janssonius van Waesbergen, M. Pitt, S. Schwart *c*.1678, re-issued by G.Valk & P. Schenk *c*.1715 (possibly intended originally for Moses Pitt's *English Atlas*; used as a source for *The Shire of Angus or Forfar* in H. Moll's *A Set of Thirty Six New and Correct Maps of Scotland*, 1725). **Ref.** MARTIN, A.I. 'A study of Edward's map of Angus, 1678' in: *Scottish Geographical Magazine* 96 (Edinburgh, 1980) pp.39-45.

**Edwards**, Robert. Engraver. *A Mapp of ye Orkney Ilands*, 1711.

**Edwards**, Robert. Hydrographer, *RN*. *Harbour of Kittie, Caroline Is.* [Pacific Ocean], 1839.

**Edwards**, Talbot. Military engineer. *Plan showing land, ...at Chatham*, 1706; *The harbour, with the town, Gosport, Spithead and Porchester Castle* [Portsmouth], 1716; *Isle of Wight, Portsmouth and Langston Harbours...*, 1717.

**Edwards**, Thomas. American engraver. Worked for Pendleton's Lithography at Boston, Massachusetts, then co-founded (with Moses Swett and Annin & Smith) the Senefelder Lithographic Company, 1828. Senefelder became part of Pendleton's Lithography Co., Boston, Massachusetts in 1831.

**Edwards**, William. Printer of the 4th edition of John Rocque's *A New and Accurate Survey of the Cities of London and Westminster, the Borough of Southwark...* (16 sheets in one volume), 1748 (first published 1746 as *An Exact Survey...*).

**Edwardson**, *Captain* William Laurence *RN*. Marine surveyor, undertook surveys of the New Zealand coast, 1822-1823 including Foveaux Strait, 1822; Ruapuke Island, 1823 MSS published by the Hydrographic Office in 1840. His surveys were used as the basis of at least two of the charts prepared by Jules de Blosseville and published in Duperrey's *Atlas: Hydrographie*, 1827 (*Carte de la Côte Méridionale de l'ile de Tawaï-Poénammou (Nouvelle-Zélande)*, 1827 and *Plan du Havre Chalky (Ile Tawaï-Poénammo)*, 1827), the surveys were also used (uncredited) by Beaufort, 1833. **Ref.** HOOKER, B. *The Map Collector* 44 (1988) pp.30-31, 33-34.

**Eek**, P. van. Polder near Dordrecht, 1811.

**Eekhoff**, Wopke (1809-1880). *Nieuwe Atlas van de Provincie Friesland* (32 maps), 1849-1859.

**Effendi**, Ferid. Officer in the Turkish Army. One of the Turkish representatives in the international team which draughted *...la frontière Serbo-Turque selon article 36 du traité de Berlin... 1878* (11 sheets plus title), 1879.

**Effendi**, *Lt*. Halid. Officer in the Turkish Army. One of the Turkish representatives in the international team which draughted *...la frontière Serbo-Turque selon article 36 du traité de Berlin... 1878* (11 sheets plus title), 1879.

**Effner**, Joseph Xaver. Area map of Hochstift Eichstätt [Franconia], 1806.

**Egan**, Frederick William (1836-1901). Geologist working for the Geological Survey of Ireland, 1871-1888.

**Egede**, family.
• **Egede**, Hans Poulsen (1686-1758). Norwegian missionary and natural scientist in Danish Greenland, 1721-1736. *Des alten Grönlandes...*, 1742 (with map of Greenland).
•• **Egede**, Poul Hansen (1708-1789). Clergyman and author, worked with his father Hans on the maps of Greenland. *Nachrichten von Grönland...*, 1790 (with map).
••• **Egede**, Christian Thestrup (1761-1803). Naval officer, traveller and cartographer, son of Poul. *Reisebeskrivelse til Øster-Grønlands...*, Copenhagen 1789 (with map).
• **Egede**, Niels. *Tredie Continuation Af Relationerne Betreffende Den Gølandske Missions...*, Copenhagen 1744 (with map).

**Egerton**, *Commander* F.W., *RN*. *Gorringe or Gettysburg Bank*, Admiralty 1877.

**Egerton**, *Lieutenant* I.F. *Map of the Districts of Midnapur and Hijellee*, 1849.

**Eggenberger**, Kiadja. Hungarian cartographer. *Ausztralia*, 1870.

**Eggenstein**, Stephen. Cordiform map of the World, 1664 (derived from Ortelius). **Ref.** The Map Collector 64 (1993) front cover & p.6.

**Eggers**, August. *Münz-Weltkarte* [distribution of the currencies of the World], Vienna 1873.

**Eggers**, Christian Ulrich Detlev von. Map of Iceland in his *Physikalische und statistische Beschreibung von Island*, Copenhagen 1786.

**Eggers**, Baron H. *Croquis des Weges von Jarabacoa, St. Domingo*, 1888.

**Egglofstein**. See **Egloffstein**, Friedrich W. von.

**Egidius Boleavus Bulionus**. See **Boileau** de Bouillon, Gilles.

**Eglesfield** [Egglesfeild], Francis (*fl.*1644-*c.*1676). Printer, publisher and mapseller 'at the Marigould in Paules Churchyard', London. *A New & Exact Mappe of England...*, 1644 (etched onto copper by W. Hollar).

**Egli**, *Professor* Johan Jakob (1825-1896). Teacher and geographer from Zurich, worked in Flaach, Winterthur and Zurich. Specialist in place names. *Nomina Geographica*, Leipzig 1872.

**Eglin**, brothers [G.; Gebrüder]. Lithographers of Lucerne, Switzerland. Kanton Luzern, 1838.
• **Eglin**, Karl Martin (1787-1850). Engraver and lithographer in Lucerne, worked with brother Bernhard.
• **Eglin**, Bernhard (1798-1875). Draughtsman and lithographer of Lucerne and Mulhouse.

**Egloff**, Georg. ...*Post... Karte des Königreichs Polen...*, 1856.

**Egloffstein** [Egglofstein] Baron Friedrich W. von (1824-1885). Artist, cartographer and topographer from Bavaria. Served in the Prussian army, went to the USA in 1848 where he was involved in exploratory expeditions west of the Mississippi and was described as 'Topographer for the Route' [Pacific Rail Road]. Associated with the Geographical Institute, N°·164 Broadway, New York. Patented a half tone screen allowing the printing of shaded relief on maps, 1865. Egloffstein died in Dresden. *Explorations and Surveys for a Rail Road from the Mississippi River to the Pacific Ocean*, Washington, Selmar Sieberts Engraving and Printing Establishment 1855 (with E.G. Beckwith); *Map of the Territory of the United States from the Mississippi to the Pacific Ocean*, 1857 (with E. Freyhold); *New style of topographical drawing, derived from late experiments with the photograph and daguerrotype, from mountain models* (monograph), Washington 1857; *Map of Explorations and Surveys in New Mexico and Utah*, New York 1864 (from the work of J.N. Macomb and C.H. Dimmock).

**Egmond**, A. van. *Kaart van den Haarlemmermeerpolder*, 1867.

**Egmont**, —. Madagascar, 1773.

**Egon**, F. Buddhist priest and mapmaker. Map of Jambudvipa (woodcut), 1845.

**Ehesbroush**, J.C. Engineer, one of those who assisted in the surveys laying out the line illustrated in P. Anderson's *Map Exhibiting the Experimental and Located Lines for the New-York and New-Haven Rail-Road*, New York 1845 (lithographed by Snyder & Black).

**Ehmann**, Andreas. Engraver, worked in Augsburg. *Universal-Historie auf der Land-Karten*, after 1733.

**Ehnlich**, I.F.C. Engraver for Adam Christian Gaspari, 1808.

**Ehrenberg**, Christian Gottfried (1795-1876). Egypt, 1826; *Carte de l'Arabie Petrée*, 1828-1834.

**Ehrenberg**, Herman. Civil engineer in California. *La Paz and its environs showing the positions occupied by the U.S. troops and the Mexicans...1847*, c.1847; *Sacramento City*, 1849; *Klamath Gold Region*, 1850; *Plan of Sacramento City, State of California...*, New York 1850 (copied from the survey by W.H. Warner, lithographed by W. Endicott & Co.); *Silver regions of Tubac*, 1857; *Map of the Gadsden Purchase, Sonora and portions of New Mexico, Chihuahua & California*, Cincinnati, Middleton, Strobridge

& Co. 1858. **Ref.** NORTH, Diane T. *The maps, plans and sketches of Herman Ehrenberg: a cartobibliography* (California Map Society 1988).

**Ehrenberg**, Peter Schubert von. *Nova Hungariae*, Vienna 1696 (engraved by Johann Franz de Langgrafen).

**Ehrenreich**, *Doctor* P. Ethnologische Karte von Brasilien, 1891.

**Ehrenschantz**, Gerh. Officer in the Swedish Army. Plan of the battle of Narva, 1700; plan of the crossing of the River Düna, 1701.

**Ehrenstein**, Heinrich Wilhelm von (1811-1874). Cartographer of Dresden. *Königreich Sachsen*, 1856.

**Ehrgott, Forbriger & Co.** Lithographers, engravers and printers of 'Carlisle Building, S.W. cor. of 4th and Walnut Street, Cincinnati'. A. Alberti's *Map of the Saginaw Valley, Michigan*, 1860; Birds eye view of Camp Chase near Columbus, Ohio, Cincinnati 1860s (drawn by A. Ruger); Union Pacific Railroad, 1866.

**Ehrhardt**, —. Umgebung von Ruppin, 1888; Umgebung von Marburg, 1896.

**Ehrhardt**, Ludwig. Palestine, 1834.

**Ehricht**, C. Engraver for Joseph Meyer, 1830-1849.

**Ehrmann**, Theodor Friedrich (1762-1811). Geographer and printseller in Weimar. Guinea, 1793; Kafferland, 1797; Senegal, 1804.

**Eian**, Yamazaki. See **Yamazaki** Eian.

**Eiberger**, F. *Gewässer und Höhenkarte Württemberg*, 1893.

**Eichbaum**, C.M. Draughtsman for G.W. Bacon's *Historical Picture Map of England and Wales*, 1875.

**Eichen**, Frederick W. French inventor. Patented the *Navisphère* (a celestial globe and mount used for navigation), 1881.

**Eichhorn**, A. Cartographic depiction of the duration of sunshine across the German Empire, 1903.

**Eichler**, [Johann] Gottfried (1715-1770). Draughtsman and engraver in Augsburg. Cartouches for Matthäus Seutter, e.g. *Palestinae*, c.1750; title-page for T. Lobeck's *Atlas geographicis portatilis*, T.C. Lotter c.1760.

**Eichler**, Matthias Gottfried (1748-1821). German draughtsman and engraver working in Basle, Bern and Augsburg. *Grundriss von Bern*, 1790; engraver for J.E. Müller and J.H. Weiss's *Atlas Suisse* (16 sheets), Aarau 1796-1802.

**Eichman**, William. Partner in Zadok Cramer's print and bookshop in Pittsburgh, from 1811. **N.B.** See Zadok **Cramer.**

**Eichmann**, J. See **Dryander**, Johannes.

**Eichoff** [Eichov, Eichovius], Cyprian. Probably a pseudonym for Matthias Quad.

**Eicholz**, D. *Plan von Königsberg* [Kaliningrad], c.1863.

**Eichovius**, C. See **Eichoff**, Cyprian.

**Eichstrom**, F. *Lauf der Planeten*, 1847.

**Eichwald**, Eduard (1795-1876). Russian naturalist and traveller. *Karte des Kaspischen Meers*, 1834; *Géographie ancienne de la mer Caspienne*, 1838.

**Eick**, Albert. *Reise Karte für Mittel Deutschland*, 1869.

**Eicken**, Franz. Lithographer. Vienna, 1848.

**Eigenbrodt**, G. Geographer, engraver and publisher. *Voies de communication de la Belgique*, 1870; *Bruxelles*, 1874.

**Eikelenberg**, S. *Westfriesland, Kennemerland en Waterland*, c.1720.

**Eillarts**, Johannes (1568-1612?). Engraver.

**Eimmart**, Georg Christoph (1638-1705). Astronomer, geographer, copper engraver and globemaker from Regensburg. Worked in Nuremberg from c.1660, where he built his own observatory and many of the instruments which equipped it. d. Nuremberg. *Planisphærum Cæleste*, Homann c.1690; gores for 30-cm celestial and terrestrial globes, Nuremberg 1705 (these are often found in Homann atlases). **Ref.** Van der KROGT, P. *Old globes in the Netherlands* (1984) pp.122-123

**Eiriksson**, Leif. Son of Eirik the Red. Norse explorer, described in *Grænlendinga saga* as the first [non-native American] to set foot on the North American continent *c.*1000. **Ref.** SEAVER, K.A. *The frozen echo: Greenland and the exploration of North America c.1000-1500* (1996).

**Eisen**, C.F. *Chemins de Fer de Prusse*, 1846.

**Eisenmenger**, S. See **Isenmenger**, Samuel.

**Eisenschmid**, Johann Caspar (1656-1712). Cartographer and mathematician of Strasbourg. *Tabula novissima totius Germaniae* (4 sheets), J.B. Homann *c.*1720, and later issues.

**Eitel**, Edward E. *County atlas of California*, San Francisco, D.S. Stanley 1894; *County atlas of Oregon and Washington*, San Francisco, D.S. Stanley & Co. 1894; *County atlas of California and Nevada*, San Francisco 1909.

**Eitner**, F. One of the engravers of *Reymann's Special-Karte* [Schleswig-Holstein] (8 sheets), Glogau, C. Flemming *c.*1860.

**Eitoku**, Kano. See **Kanō** Eitoku.

**Eitzinger** [Eitzing], M. See **Aitzing**, Michael.

**Ekeberg** [Ekberg], *Captain* Carl Gustav Eric (1716-1784). Plan and view of the Keeling or Cocos Is., 1749 published Alexander Dalrymple *c.*1792; Southeast coast Hainan, published Dalrymple *c.*1792; *Charta üfver Cap Godt Hop*, Stockholm, H. Fougt 1773.

**Ekebom**, Karl Gustav. *Karta öfver Finlands Södra Skärgård*, Helsinki, 1886.

**Ekel**, Friedrich Christian. Architect and draughtsman. *Plan de la ville de Reinsberg*, 1773.

**Ekholm**, N. Karta öfver Amsterdamön, 1896-1897.

**Ekholm**, Nils (1730-1778). Swedish land surveyor of Vasa.

**Eklund**, A.W. (1796-1885). Furstendömet Finland, 1857.

**Ekmansson**, Frederik. *Erichsens och Schönnings Charta öfver Island... 1772*, published in Uno von Troil's description of Iceland published Uppsala 1777.

**Eland**[t], Hendrik (d.1705). Draughtsman and engraver of Amsterdam. Designed and engraved the cartouche for François Halma's edition of Ptolemy's *Geographia*, Utrecht 1704.

**Elandt**[s] [Elands], Cornelis (fl.1660-1670). Artist, surveyor and engraver of The Hague. *s'Graven-Hage*, 1666.

**Elberts**, W.A. *Kaart van Overijssel*, *c.*1866.

**Elcano**, J.S. See **Cano**, Juan Sebastian del.

**Eldberg**, Carl (*fl.*late-15th century). Swedish chartmaker.

**Elder**, John. Cleric and scholar from Caithness, Scotland. Travelled widely throughout Scotland and the Western Isles and studied at the universities of Glasgow, Aberdeen and St. Andrews. Compiled a manuscript map of Scotland which he presented to Henry VIII *c.*1543 (no longer extant); it has been suggested that he may have supplied the source material for Gerard Mercator's *Angliae, Scotiae & Hiberniae*, Duisburg 1564. **Ref.** DELANO-SMITH, C. & KAIN, R.J.P. *English maps: A history* (1999) p.63; MOIR, D.G. *The early maps of Scotland* Vol.I (1973) pp.12-16.

**Eldred**, Edward (*fl.c.*1597-1603). English estate surveyor. Shalford [Essex], 1603 MS.

**Eldridge**, George, father and son.
• **Eldridge**, George (1821-1900). Mariner, hydrographer and surveyor. Charts of the coast of North America including: *Chatham Lights to South West Part of Handkerchief*, 1851; *Chart of the Vineyard Sound and Nantucket Shoals*, Boston 1865; *Eldridges Coast Pilot*, 1880-1893. **Ref.** McGLAMERY, P. 'George Eldridge, the Chatham chartmaker' *Meridian* No.3 (1990) pp.27-28; GUTHORN, P.J. *United States coastal charts 1783-1861* (1984) p.12; GUTHORN, P.J. *The Map Collector* 23 (1983) pp.18-19.
•• **Eldridge**, George Washington (1845-1914). Son of George, above. Sold his father's charts through a ship's chandlery at Vineyard Haven. *George W. Eldridge's Harbour Chart Book, Boston to Bar Harbour*, 1925, published posthumously by Wilfrid O. White (George Washington

Eldridge's son-in-law) under the name Eldridge. **Ref.** GUTHORN, P.J. *Imago Mundi* 43 (1991) pp.72-80.

**Eldridge.** See **Davies** & Eldridge.

**Electric Telegraph Company.** *Telegraph lines of central Europe*, 1852 (lithographed by Day & Son); *The Electric Telegraph Company's map of the telegraph lines of Europe*, Day & Son 1854.

**Elekes,** Ferenc [Franz von]. Hungarian military cartographer and lithographer working in Vienna. Designed a set of terrestrial globe gores for the Viennese publisher F.L. Schöninger, 1844; *Grundriss Residenzstadt Wien*, 1857.

**Elena,** Felix. Professor of drawing in Uruguay. *Map of the Railways of South America...*, 1893 (with Juan José Castro; engraved by P. Ludwig; printed by the South American Banknote Company).

**Elfert,** Paul (1861-1898). Geographer of Leipzig, worked for Heinrich Wagner and Ernst Debes.

**Elford,** James M. Conducted astronomical observations in South Carolina, *c.*1820, his results were used as basic data by John Wilson for his *Map of South Carolina*, 1822.

**Elías,** José Antonio. *Atlas histórico, geográfico y estadístico, de España...*, Barcelona 1848, 1853 (engraved by R. Alabern and E. Mabon).

**Elias,** Ney. Plan of Tsien-tang river, 1867 MS; Yellow River, 1871 MS; Ortos Country, 1876 MS; *Sketch of a journey in Pamir & Upper Oxus regions*, 1886.

**Elie de Beaumont,** J.B. See **Beaumont**, Jean Baptiste Armand Louis Léon Elie de.

**Eligerus** [Algeri], Johannes (*fl.c.*1350). Made improvements to the design of the quadrant.

**Eliot,** J.B. American engineer, aide-de-camp to General George Washington. *Carte du Théâtre de la Guerre Actuel Entre les Anglais et les Treize Colonies Unies de l'Amerique Septentrionale*, Paris, Mondhare 1778, 1781; *Carte générale des Etats Unis de l'Amérique Septentrionale, avec les limites de chacun des dits Etats*, Paris, Mondhare 1783. **Ref.** RISTOW, W.W. *American maps and mapmakers* (Detroit 1986) pp.61-62.

**Eliseo,** P. *Calabria Ulteriore*, 1780.

**Eliza,** Francisco. Spanish explorer based at San Blas in Mexico. Undertook two voyages exploring and charting the north-west coast of America, including the vicinity of Vancouver Island, 1790-1791; Monterey and San Diego, also Trinidad Bay and Bodega Bay, 1793. **N.B.** Manuscript charts from the voyages, not necessarily by Eliza himself, are in the Library of Congress. **Ref.** WAGNER, H.R. *The cartography of the northwest coast of America to the year 1800* Vol.I (1968) pp.223-224, 236-237.

**Elkan,** C. Bookseller. Grundriss von Harburg, *c.*1859.

**Ellacott,** C.H. Rossland & its mines [British Columbia?], 1897.

**Ellery,** Robert Louis John (1827-1908). Government Astronomer in Australia, President of the Royal Society of Victoria, Melbourne Department of Lands and Surveys. Directed the trigonometrical surveys of Victoria from 1858. *Map shewing the rainfall over south eastern Australia and Tasmania*, 1884.

**Ellet,** Charles *junior* (1810-1862). American civil engineer. *Map of the City and County of Philadelphia*, 1839; *Map of the western railroads tributary to Philadelphia*, 1851 (engraved by Wellington Williams).

**Ellicott,** *Major* Andrew (1754-1820). Maryland surveyor. In 1791 he was appointed surveyor of Federal Territory in Virginia, and completed the planning of Washington D.C. With David Rittenhouse he joined the commission set up to establish the boundaries between New York and Pennsylvania, 1786-1787; survey of the land ceded by Virginia and Maryland for the Territory of Columbia, 1791; *Plan of the City of Washington* (8.5" x 10.25"), 1792 (engraved by Thackara and Vallance; published in *The Universal Asylum and Columbian Magazine*, March 1792; another version of this small plan was engraved by S. Hill and published in *Massachusetts Magazine*); *Plan of the City of*

*Major Andrew Ellicott (1754 - 1820). (By courtesy of Cartographic Associates, Fulton, Maryland, USA)*

*Washington in the Territory of Columbia* (20.25" x 27"), 1792 (engraved in Philadelphia by John Vallance and James Thackara; described as the first official engraved plan of the city; another version of the same date was engraved by Samuel Hill of Boston; Ellicott's plan was also issued as a 'handkerchief map'); *Territory of Columbia*, 1794 (printed by J.T. Scott); map of the Mississippi River, before 1803; *Map of Morris's Purchas or West Geneseo in the State of New York*, 1804. **Ref.** RISTOW W.W. *A la Carte* (1972) pp.135-138, 146-148; VERNER, C. *Imago Mundi* 23 (1969) pp.60-66; MATHEWS, C.V. *Andrew Ellicott: his life and letters* (New York 1908).

**Elliger**, Ottmar (1666-1735). German painter and engraver from Hamburg, moved to Amsterdam 1679. Appointed court artist at Mainz, 1716, moved to St Petersburg and became engraver to the Academy, 1726. *Map of France*, N. Visscher *c*.1704; also plates in François Valentijn's *Oud- en Nieuw-Oost-Indien*, 1726.

**Elliot**, —. Printer of 14 Holywell Street, Strand. *Map shewing the Tracks of the Arctic Expedition by Sea & Land in Search of Captn. Ross*, 1833 (lithographed by C. Ingrey).

**Elliot**, *Captain* —. See **Elliott**, *Captain* William.

**Elliot**, S.A. *Plan of the city of Washington...*, 1822, 1827 (published in William Elliot's *The Washington Guide*, Washington, S.A. Elliot 1822).

**Elliott**, Charles L. *Atlas of the city of Newport, Rhode Island*, Springfield, Mass., L.J. Richards 1893 (with Thomas Flynn).

**Elliott**, *Lieutenant* Charles P. *Map of Mount St Helens*, 1897.

**Elliott**, George, *RN. Entrance to South Harbour Balembangan*, 1845.

**Elliott**, Henry C. *Valley of the Amazon*, to accompany Lt. Herndon's Report, 1854.

**Elliott**, Henry W., Surveyor. Yellowstone Lake, 1871; St. Paul, Bering Sea, 1880; Bering Sea, 1881.

**Elliott**, J.E., *RN*, of HMS *Assistance*. Baffin's Bay, 1851.

**Elliott**, John (1759-1834). Artist and surveyor from Hemsly, near York. Served on Cook's second voyage and produced several charts, surveys and views, including a chart of the track of the *Resolution*. Others include: Vanuatu, 1774 (with Joseph Gilbert); a chart of New Caledonia, 1774 (with Joseph Gilbert); Savage Island, 1774; Leper island, 1774; chart of South Georgia, 1775 (with Joseph Gilbert); chart of South Sandwich Islands, 1775 (with Joseph Gilbert); and others. **Ref.** DAVID, Andrew (Ed.) *The charts and coastal views of Captain Cook's voyages* Vol.II.

**Elliott**, S.G. Civil engineer. *Map of Central California showing the different railroad lines completed and projected*, Nevada 1860; *Map of the battlefield of Gettysburg...*, New York, H.H. Lloyd & Co. 1864.

**Elliott** [Elliot], *Captain* William, *RN*. Bay & Town Kingston, St. Vincent, 1817 MS (with James Langley), Admiralty 1820; Courland Bay, Tobago, Admiralty 1820.

**Ellis**, A.D. American surveyor. *Beers, Ellis & Soule's Map of Venango Co. Penn. from actual surveys*, F.W. Beers & Company 1865 (with Frederick Beers and G.C. Soule). **N.B.** See also under **Beers** family.

**Ellis**, A.G. Town of Astor, Wisconsin, 1835; Plat of Navarino, Wisconsin, 1836.

**Ellis**, Charles H. *Detroit & its environs*, 1876.

**Ellis**, D.W. *The D.W. Ellis Map of Ogden City, Utah*, Pueblo, Colorado, Cactus Printing Co. 1891.

**Ellis**, George. Publisher of 5 Smith's Square, Westminster. *Ellis's New and Correct Atlas of England and Wales*, 1819 (a reprint of *Wallis's New British Atlas*, the maps for which were engraved by James Wallis and first published by S.A. Oddy dated 1812 or 1813).

**Ellis**, Henry (1721-1806). Traveller and hydrographer. Voyage to Hudson's Bay, 1746-1747; *Chart of the Coast where a North West Passage was attempted in...1746*, London 1748 (in Ellis's *A voyage to Hudson's-Bay*).

Ellis, Henry T. Manila, 1859.

Ellis, J. Engraver. *Carte de L'Isle de Corse...*, after 1741.

Ellis, J. Bay South West of Suez, Dalrymple 1801.

Ellis, J. Lithographer of Toronto. *Plan Shewing the Region explored by S.J. Dawson and his Party between Fort William, Lake Superior and the Great Saskatchewan River... 1857 to... 1858*, 1858; W. Gibbard's *Chart of Collingwood Harbor...*, 1858.

Ellis, James (fl.1755-d.1800). Estate surveyor in Essex, England. Produced manuscript maps of parts of eastern England. Havering-atte-Bower, 1755 MS.

Ellis, John. See Ellis, Joseph, below.

Ellis, Joseph (*d*.1800). Engraver and publisher of Clerkenwell, London. Apprenticed to Richard William Seale in 1749. *The New English Atlas*, 1765 re-issued as *Ellis's English Atlas*, 1766, 1768, 1773, 1777, (French edition published as *Atlas Britannique, ou Chorographie Complette de l'Angleterre*, 1766); *The Kentish Traveller's Companion*, 1776; Bernard Scalé's *Hibernian Atlas*, 1776; Bowles's *Pocket Plan of London and Westminster*, 1780; *Bowles's New Pocket Map of England and Wales...*, 1780; P.D. Lesaux's *A Parochial Map of the Diocese of Canterbury*, 1782; one of the engravers for P. Crosthwaite's *Maps of the Lake District*, 1783 (with H. Ashby and S.J. Neele); *Corsica*, c.1790; Daniel Paterson's *Paterson's Twenty Four Miles round London...*, Bowles & Carver 1797 (with B. Baker). **N.B.** Sometimes incorrectly referred to as John Ellis. **Ref.** HODSON, D. *County atlases of the British Isles published after 1703* Vol.III (1997) pp.26-51; TOOLEY, R.V. *The Map Collector 8* (1979) pp.55-57.

Ellis, R.G. Part New Brunswick Geology (3 sheets), 1880.

Ellis, Robert. Java, 1811 MS; District of Boondelkhund [India], 1813 MS.

Ellis, T.J. Lithographer. *Bucks., Berks., London and Windsor Railway...* (6 sheets and title), 1833 (surveyed by Cruickshank & Gilbert).
**N.B.** Compare with Thomas Joseph Ellis, below.

Ellis, Thomas Joseph. Surveyor, Vauxhall Bridge Rd., Westminster. *Map of the County of Huntingdon* (4 sheets), surveyed 1823-1824, published 1824, 1829 (engraved Thomas Foot); *Nottinghamshire* (4 sheets), surveyed 1824-1825, published 1825, 1827, 1831 (engraved by T. Foot); England, 1832.

Ellis, William (1792-1872). Served in Polynesia for the London Missionary Society, arrived Hawaii 1822. *Narrative of a Tour through Hawaii*, 1826, 1827.

Ellison, —. Survey of the Borough of Bradford [England?], used by Thomas Dixon, 1856.

Ellison, James. Soondurbuns, 1891.

Ellison, Richard (*fl*.1735-1738). Survey of the Rivers Swale & Ouze [Yorkshire], 1735.

Ellobet, Francisco. *Sevilla Regnum*, for Homann's Heirs, 1781.

Ellung, I. Kort över Jylland, 1820.

Elmes, James (1782-1862). Surveyor of the Port of London. Contributed 6 charts to *A Guide to the Port of London*, 1842 (engraved by James Wyld).

Elmore, *Captain* —, RN. Straits of Singapore, 1799.

Elmore, Publius Virgillius (1798-1857). Surveyor in Canada. *Adolphustown*, 1825; *Map of the District of Prince Edward*, 1835; survey of Ameliasburgh, 1839; *Chart of Picton Bay*, 1841; Survey of Beverly, 1843; *Plan of a Survey...of the Settlement in the rear of the Township of Madoc*, 1848; *Plan of the Town of Norwood*, 1856; and many other township surveys.

Elmore, William R. *Map of the County of Hastings and Adjoining Townships*, New York, Snyder, Black & Sturn 1856

Elmpt, *General* Phillip, *Freiherr* von (1724-1795). Military cartographer, Austrian army. In charge of 33 engineers in survey of Temes Banat [Southern Hungary] (208 sheets), 1769-1773.

Elmsley [Elmsly], Peter (1736-1802). Bookseller of London. One of the sellers of the

*Adam Elsheimer [Elzheimer] (c.1574 - 1620). (By courtesy of Rodney Shirley)*

4th edition of Gibson's version of William Camden's *Britannia*, 1772 (last edition with Robert Morden's maps); H. Skrine's *Rivers of Note in Great Britain*, 1801. **Ref.** HODSON, D. *County atlases of the British Isles* Vol.I (1984) pp.109-112.

Elorriaga, Miguel. Palaos [Philippines], 1709.

Eloy de Almeida, Romão. Portuguese engraver. *Carta militar... de Portugal*, 1808.

Elphinston, Mountstuart (1779-1859). Governor of Bombay. Afghanistan, 1808-1809.

Elphinstone, John (1706-1753). Scottish military engineer and surveyor, Corps of Royal Engineers (1744). British Isles, 1743; plan of Burntisland, 1743; *A New & Correct Map of the Lothians from Mr Adair's observations*, 1744; *A New & Correct Mercator's Map of North Britain*, London, A. Millar 1745 (engraved T. Kitchin; apparently used by both sides during Jacobite Rising); *Plan of the field of battle near Preston*, 1745; *...plan of the ground adjacent to Fort William...*, 1748. **Ref.** MOIR, D.G *The early maps of Scotland* Vol.I (1973) pp.86-88.

Elphinstone, *Captain* John, *RN*. Saldanha Bay [South Africa], surveyed 1796, published Admiralty 1813.

Elsden, James Vincent. *Map of Herts shewing the superficial geology* in *Transactions of Hertfordshire Natural History Society*, 1883.

Elsenwanger, Anton. Bookseller and publisher of Prague. District maps of Bohemia, *c*.1771, 1794.

Elsevier. Publishing company of Rotterdam. *De Elsevier Globe*, 1881 (collapsible globe in a wooden case). **Ref.** KROGT, P. van der *The Map Collector* 33 (1985) pp.28-29.

Elsevier, Isaac. See Elzevier, Isaac.

Elsheimer [Elshaimer; Elzheimer], Adam (*c*.1574-1620). Artist and engraver of Frankfurt. *d*. Rome. Engraved the decorative double-hemispherical world map showing the 1595-1598 voyage of Cornelis Houtmann to the East, published in Sebastian Brenner's *Continuator temporis quinquennalis*, Frankfurt am Main 1598, 1599. **Ref.** SHIRLEY, R. *The mapping of the World* (1983) N°.208 p.227.

Elson, *Captain* Thomas, *RN*. Admiralty charts. *Chart of the coast of the Morea...* [Greece], Hydrographic Office 1825 (with Smyth and Slater, engraved by J. Walker); Port Mandri, 1829-1843; *The Graham Shoal* [inset on chart of Mediterranean Sicily], surveyed 1841, Hydrographic Office 1851 (engraved by J. & C. Walker).

Elstob [Elstobb], William (1737-1793). Engineer and teacher of mathematics, of King's Lynn and Cambridge, England. Undertook many specialist surveys, mostly in Cambridgeshire and Norfolk. Sutton & Mepall Levels, [n.d.]; Isle of Ely, 1750; Wisbech Channel, 1773-1775, MS; Bottom River Ouse (4 sheets), 1776; Great Level of Fens, 1793.

Elstracke, Reynold [Renier; Renold] (1571-1630). Flemish engraver born in London. For the English edition of Jan Huyghen van Linschoten, 1598; Giovanni Battista Boazio's *Irelande*, 1599; John Speed's historical and military map of England, Ireland and Wales (4 sheets), 1603-1604; William Baffin's East India, 1619; engraved replacement plate for the map of Norfolk in Speed's *The Theatre of the Empire of Great Britaine...*, 1623 edition (and perhaps revised other maps for this edition); engraver for Henry Briggs's *The North part of America...* [with California as an island], in *Purchas His Pilgrimes*, 1625 (the map was possibly issued earlier). **Ref.** SHIRLEY, R. *Early printed maps of the British Isles 1477-1650* (1991) pp.106, 107.

Elsworth, Richard. Ultima Aethiopum, 1739.

Elton, Andrew. Calabria & Naples, 1780; Livorno, Salonicki & others, *c*.1780.

Elton, *Lieutenant* Isaac Marmaduke *RE*. *Plan of Charleroi*, 1815 MS.

Elton, *Captain* James Frederic, *RN* (1840-1877). British hydrographer and explorer, died of malaria in Ugogo in 1877. Route Tati... Delagoa Bay, 1870 MS; *The Slave Caravan Route from Dar es Salaam to Kilwa*, 1873 (with T.F. Pullen, lithographed by Harrison & Sons); Route from North of Lake [Malawi] to Ujiji..., 1877 MS; *Travels and researches among the lakes and mountains of Eastern and Central Africa*, 1879 (edited by H.B. Cotterill). **Ref.** BRENCHLEY, S. *The Map Collector* 42 (1988) pp.30-32.

*Captain James Frederic Elton, RN (1840 - 1877).*

**Elton**, John. English naval captain in Russian service. Caspian Sea, 1753.

**Eltzner**, Adolf. *Plan von Leipzig*, 1847; *Das biblische Jerusalem aus der Vogelschau*, Leipzig 1852; *Saechsische Schweiz*, 1867; *Plan von Breslau*, 1870; *Plan von Bremen*, c.1871.

**Elwe**, Jan Barend [Barent] *(fl.1785-1809)*. Publisher and bookseller of Amsterdam, worked with D.M. Langeveld. *Compleete Zak-Atlas, van de Zeventien Nederlandsche Provinciën*, 1785, 1786; *Volkomen Reisatlas van geheel Duitschland*, 1791. **Ref.** KOEMAN, C. *Atlantes Neerlandici (1967-1971)* Vol.II pp.104-108.

**Elwon**, *Captain* T. Indian Navy. Red Sea, 1830-1804; Jiddah, 1858; Musawwa Channel, 1876.

**Elwood**, S.D. Stationer and lithographer of Detroit. *Map of the Port Huron and Milwaukee and the Detroit and Milwaukee Railways...*, c.1854 (drawn by J.T. Baker, lithographed W. Felt & Co.); *Map of the Detroit and Milwaukee Railway*, c.1854 (drawn by T. Baker, lithographed by W. Felt & Co.); lithographers for *Saint Mary's Falls Ship Canal...*, 1858; *Map of the Upper Peninsula of Michigan*, 1858; *Map of the Lower peninsula of Michigan*, 1858.

**Ely**, A.E.M. *City of Palmyra, Marion County, Missouri*, c.1860.

**Ely**, Chas. E. Engraver of Philadelphia. Thomas R. Tanner's *A New & Authentic Map of the State of Michigan and Territory of Wisconsin...*, Philadelphia, H.S. Tanner 1839; *A New Sectional Map of the State of Michigan and Territory of Wisconsin*, New York, H.S. Tanner 1844.

**Ely**, W.A. Spanish Texas, 1835.

**Ely**, W.W. New York Wilderness, 1868, 1874.

**Elzevier** [Elsevier; Elzevir], Isaac (1596-1651). Printer of Leiden. Petrus Bertius's *Theatrum geographiae veteris* (47 copperplate maps), Amsterdam, Hondius 1618-1619 (maps by Mercator and Ortelius); Philip Clüver's *Italia Antiqua* (2 volumes, 15 maps), 1624, 1629; Clüver's *Introductionis in Universam Geographicam*, 1624; Johannes de Laet's *Nieuwe Wereldt* (10 maps), 1625.

**Elzheimer**, A. See **Elsheimer**, Adam.

**Emakimono** Dochu. Picture scroll of the sea and land routes from Edo to Nagasaki, mid-17th century.

**Emdre**, S. van. *Nieuwe gemaakte afteekening der stad Jerusalem*, Utrecht 1707.

**Emery**, Henry (*b.c.*1766/7-*after* 1841). Essex estate surveyor, produced manuscript surveys including: *Freehold Estate called High Hall* [Tolleshunt d'Arcy], 1805; *Estate called High-Fields* [Great Coggeshall], 1806; *Manor of Great and Little Birch*, 1811; *Estate called Spicers* [Wethersfield], 1816; *Estate called Headboroughs* [Aldham], 1820; *Estate called Crisps* [Great Totham], 1824; and others.

**Emery**, Louis. *Carte de Suisse*, 1798.

**Emiya** Kichiemon. Japanese publisher. Kondō Kiyoharu's woodblock plan of the city gates of Edo, [early 18th-century].

**Emmerich**, N. Geographer and publisher. *Regierungs-Bezirk Arnsberg*, l845, 1860.

**Emmert**, Ludwig. Lithographer in Munich. Lithographic engraver for some map sheets of forestry maps including: *Forst-Wirthschafts-Karte der Koeniglichen Salinen Waldungen in dem Forstant e Berchtesgaden* (25 sheet map with 2 text sheets), 1819; *... Marquartstein...*, c.1821; *Karte der Umgebung Münchens*, c.1826; *... Marquartstein...*, c.1829.

**Emmett**, S.B. *Sketch Map of the North West Portion of Tasmania shewing the portion of the Hellyer River reported as being auriferous by Mr. S.B. Emmett*, c.1865.

**Emminger**, Eberhard (1808-1885). Artist and lithographer of Munich. *Erinnerung an den Rhein*, c.1870.

**Emmius** [Emmio], Ubbo[ne] (1547-1625). Cartographer and historian, *d.* Groningen. *Typus Frisiae Orientalis*, 1590-*c.*1595; maps of Groningen and Friesland, 1616, used by Hondius, Blaeu and others; town plans used in J. Blaeu's 'Townbooks' from *c.*1649.

**Emmons**, Samuel Franklin (1841-1911). American geologist and mining engineer. *Geological Atlas of Leadville, Colorado*, 1882, 1883; *Anthracite-Crested Butte folio, Colorado*, Washington D.C., U.S. Geological Survey 1894; *Map of Alaska, showing known gold-bearing rocks with descriptive text*, U.S. Geological Survey 1898; *Ten mile district special folio, Colorado*, Washington D.C., U.S. Geological Survey 1898; *The Downtown district of Leadville* [Colorado], Washington D.C. 1907 (with J.D. Irving).

**Emmoser**, Gerhard (*fl.*1573-1579). Clockmaker of Vienna. Silver celestial globe, Vienna 1579 (with clockwork mechanism).

**Emory**, *Lieutenant* (later *Lieutenant-Colonel*) William Hemsley (1811-1887). Topographical Engineer in the US Army, principal assistant on the North Eastern Boundary survey, between the U.S. & Canada, 1844-1846. Later became Mexican Boundary Commissioner, surveys used in the planning of the Pacific Railroad. Reduced J.N. Nicollet's Upper Mississippi River, 1843; *Map of Texas and the Countries Adjacent*, Washington D.C., War Department 1844; *Rio Grande and lower Colorado River*, 1846; *Military Reconnaissance of the Arkansas, Rio del Norte, and Rio Gila*, 1847; *Military Reconnaissance from Fort Leavenworth, in Missouri to San Diego, in California*, 1847 (assisted by others), published 1848 (drawn by Joseph Welch; lithographed by Edward Weber); United States, 1854-*c.*1857; Pacific Railroad surveys, 1857 (with G.K. Warren);

USA-Mexico Boundary, 1857. **Ref.** REBERT, P. 'Mapping the United States-Mexico boundary: co-operation and controversy' in: *Terrae Incognitae* 28 (1996) pp.58-71; THROWER, N.J.W. 'William Emory and the mapping of the American southwest borderlands' in: *Terrae Incognitae* 22 (1990) pp.41-91.

**Emperger**, Josef Edler von (*d*.1818). Cartographer and lawyer of Klagenfurt.

**Emphinger**, —. *Geognostische Charte Sachsen Schlesien*, 1836.

**Emrik & Binger**. Lithographers. Dr. G.J. Dozy's *Schoolatlas der geheele aarde*, Arnhem, J. Voltelen 1877; Dozy's *Historische atlas der Algemeene Geschiedenis*, Zutphen, W.J. Thieme & Cie 1909 edition (possibly also an edition of 1874); Dozy and Brugmans's *Historische Atlas, ten gebruike bij het onderwijs in Algemeene en Vaderlansche Geschiedenis*, Zutphen, W.J. Thieme & Cie. *c*.1918 edition.

**Emslie**, family.
• **Emslie**, John (1813-1875). Draughtsman, illustrator and engraver of London. [James] Reynolds's *Travelling Atlas of England*, 1848; *The Chief Objects of Interest in London*, 1851; John Airey's *Railway Diagram of London and its Suburbs*, 1869.
• **Emslie**, John & Sons. Engravers and lithographers of London.
•• **Emslie**, John Philipps. Engraver, son of John, above.
•• **Emslie**, William R. Brother of John Philipps, above.
•• **Emslie**, J.P. & W.R. (*fl*.1876-1899). Partnership of the Emslie brothers, engravers and lithographers of London. England & Wales (2 sheets), 1877, (4 sheets), 1877, (4 sheets), 1896; railway maps of Cumberland & Westmorland, 1897; Edinburgh & Glasgow, 1898; *Official Railway Map of London and Its Environs*, 1899.

**Enackel von Hoheneck**, C.A. See **Enenckel von Hoheneck**, *Baron* Georg Acacius.

**Enagrius**, C.E. *Charta ofver Swerige med Tilgränsande Länder...*, 1797 (in Hermelin's atlas of Sweden).

**Enchin**, — von (814-891). *Dai-Tō Koku Zu* [map of China in the Tang Dynasty].

**Enaro**, I. de. Publisher of Madrid. Pellicer de Touar's Cataluna, 1643.

**Enciso**, Martín Fernández de. See **Fernández de Enciso**, Martín.

**Encke**, Johann Franz (1791-1865). German astronomer. Led a group of astronomers who together compiled a celestial atlas entitled *Akademische Sternkarten*, 1830-1859.

**Endasian**, Elias. Armenian mapmaker. Maps of America, Asia, Europe, published Venice 1787.

**Ende** [Endenus], Joost [Joos] van den (1576-*c*.1618). Military cartographer and engraver of Zierikzee. Zirizea, Blaeu 1649.

**Ende** [Eynde], Josua van den (*c*.1584-1634). Engraver of Amsterdam. *Nova et exacta terrarum orbis tabula geographica ac hydrographica* [world map] (12 sheets), published by Willem Blaeu, *c*.1604, after Pieter Plancius's 1592 map, possibly re-issued 1617-1618; Willem Blaeu's single-sheet *Nova totius terrarum orbis...*, 1606; 1607; H. Hondius's Europa, 1617; *Gallia vetus*, 1636 (used by J. Janssonius in *Novus Atlas*, *c*.1645); *La Principaute de Dombes*, 1638 (in *Novus Atlas*). **Ref.** SHIRLEY, R. *The mapping of the World* (1983) N°243 pp.255-259.

**Enderlin**, Jakob. Publisher of Augsburg. *Theatrum Adriaticum* (14 maps), 1685; *Trinum marinum*, 1693 (with 53 maps after Vincenzo Maria Coronelli).

**Endersch**, Johann Friedrich (1705-1769). Doctor of medicine. Prussia, for Homann Heirs 1753; for Franz Anton Schrämbl and Franz Johann Joseph von Reilly; map of Weichseldelta, 1753; map of Episcopatum Warmiensem in Prussia, 1755; diocesan map of Ermland, 1755; *Mappa Geographica Borussiam Orientalem æque Occidentalem...*, 1758.

**Endert**, V.A. Lithographer of Gotha. *Australien*, Justus Perthes 1850.

**Enderunlu Ressam Mustafa**. Map of Russia, Poland, Bogdan, Wallachia, and part of Hungary, *c*.1768; Map of the Regions North of the Black Sea, *c*.1768 (in Ottoman Turkish). **Ref.** GOODRICH, T.D. 'Old maps in the library of Topkapi Palace in Istanbul' in *Imago Mundi* 45 (1993) p.125; KARAMUSTAFA, A.T. 'Military, administrative, and scholarly maps and plans' in HARLEY & WOODWARD (eds.) *The history of cartography* vol.2 book 1 *Cartography in the traditional Islamic and south Asian societies* (Chicago and London 1992) pp.217, 226.

**Endicott**, company and family. Lithographers of New York City, active for up to 40 years and noted at 59 Beekman Street, N.Y.C.
• **Endicott, G. & W.** *Map and profile of the proposed Paterson and Dover Rail Road and Paterson and Ramapo Rail Road*, c.1847 (by J.W. Allen).
• **Endicott, Francis.** Involved in the firm of Endicott & Company.
• **Endicott, William [& Co.].** *Map of the Northwestern States...*, 1850; *Skeleton Map. Shewing the position and connections of the Michigan Southern Rail Road...*, 1850; *Map of the Village of Eagle Harbor, Houghton County, Michigan*, 1851; *Atlantic to Mississippi*, 1854.
• **Endicott & Company.** *Map of the Croton Water Pipes with the Stop Cocks* [map of the new fresh water supply to Manhattan], c.1842; *Plan of Sacramento City, State of California*, 1849; *A Map Showing the route of the proposed Rail Road from the Copper and Iron Mining District of Lake Superior... to the State of Wisconsin*, c.1855; *Map of the Chicago, St. Paul & Fond du Lac Railroad*, 1855.

**Endlicher**, Stephan Ladislaus (1804-1849). *Atlas von China nach der Aufnahme der Jesuiten-Missionare* (4 sheets), Vienna 1843.

**Endner**, Gustav Georg (1754-1824). Engraver of Leipzig. *Grundriss der Stadt Cadix*, 1810; *Heligoland*, [n.d.]; *Grundriss der Stadt Flensburg*, [n.d.].

**Endter**, Wolfgang Moritz. Bookseller of Nuremberg. *Carinthia*, 1688.

**Enenckel** [Enackel; Ennenckel] **von Hoheneck**, *Baron* Georg Acacius. Austrian historian and cartographer. *Disputatio astronomica vel geographica de zonis*, Tübingen 1592; *Greece*, in 1614 edition of Thucydides.

**Enfant**, P.C. See **L'Enfant**, Pierre Charles.

**Engall**, J. Sherwin. Belgian Congo, 1890.

**Engel**, B.F. Mecklenburg-Schwerin, 1846; Mecklenburg-Strelitz, 1850.

**Engel**, Ernst (1821-1896). German geographer and statistician.

**Engel**, Samuel (1702-1784). Swiss geographer and librarian of Bern, worked also in Aarberg. *Mémoires et observations géographiques... sur les pays septentrionaux de l'Asie et de l'Amérique*, Lausanne 1765 (maps of the American Northwest); *Extraits raisonnés des voyages faits dans les parties septentrionales de l'Asie et de l'Amérique*, Lausanne, J.H. Pott et Comp. 1779. **Ref.** PULVER, P. *Samuel Engel. Ein Berner Patrizier aus dem Zeitalter der Aufklärung (1702-1784)* Bern (1937) [in German].

**Engelbrecht** brothers.
• **Engelbrecht, Christian** (1672-1735). Engraver, publisher and printseller in Vienna and Augsburg, pupil of Johann Jakob von Sandrart; brother of Martin Engelbrecht (below). *Theatrum Belli Italica*, c.1701; maps for W. Purgstall's *Germania Austriaca*, Vienna 1701; engraver of L. Anguissola and J. Marinoni's map of the city of Vienna, 1706 (with J.A. Pfeffel); *Mappa Regni Hungariae* (4 sheets), 1709.
• **Engelbrecht, Martin** (1684-1756). Publisher and engraver in Augsburg, worked alone and with brother Christian (above). Published a vast number of prints and views of European towns. Basel, 1700; Strasbourg, 1720; Berlin, c.1730; Mainz, c.1730; Palermo, c.1730; *Plan von Maienfeld*, 1730.

**Engelbrecht**, Karl von. Member of the international Turco-Greek Boundary Commission, worked under J.C. Ardagh in the preparation of *Carte de la Nouvelle Frontière Turco-Grecque...* (13 sheets), 1881.

**Engelbrecht**, Thieß Heinrich (1853-1934). *Die Landbauzonen der außereurop. Länder*, Berlin 1899; *Die Feldfrüchte Indiens in ihrer geograph. Verbreitung*, Hamburg 1914; *Das landwirtschaftl. Nebengewerbe in Rußland*, St Petersburg 1914.

**Engeldue**, J. *Master, RN.* Additions to Plan of Maldonado Bay [Uruguay], 1820.

**Engelhard-Reyher**. Printers of Gotha. *Spruner-Menke Hand-Atlas für die Gesichte des Mittelalters und der neuren Zeit* (90 maps),

Gotha, Justus Perthes 1880 (T. Menke's edition of Dr. K. von Spruner's *Hand-Atlas*).

**Engelhardt,** *Lieutenant* Friedrich Bernhard (1768-1854). Topographer and surveyor. Worked with D. Gilly on a map of Prussia entitled: *Karte von den Königl. Preußischen Provinzen Pommerellen un dem Netzedistrict,* 1791-1795 (the map was unpublished); Plan of Danzig, 1813; Umgebung von Berlin (10 sheets), 1816-1819; Deutschland, 1824; Preussischen Staate, 1827; *Karte von dem Königreich Pohlen, Gross-Herzogthum Posen und den angrenzenden Staaten...,* [n.d.], corrected edition published Berlin, Simon Schropp & Co. 1862 (engraved by Carl Mare); and others.

**Engelhardt,** Friedrich Wilhelm (1824-1889). Historian and scholar of Luxembourg. *Plan von Luxemburg,* 1850 (engraved and printed by Conrad Rosbach for Engelhardt's *Geschichte der Stadt und Festung Luxemburg,* Luxembourg, F. Rehm and V. Bück 1850).

**Engelhardt,** Moritz von. Worked with F. Parrot on *Reise in die Krym und den Kaukasus,* Berlin 1815 (maps engraved by C. Mare).

**Engelhardt,** *Dr* Paul. *Karte von Central-Ost-Afrika...,* Berlin, J.H. Neumann 1886 (with I. von Wensierski); *Eisenbahn-Karte Mittel Europa,* 1887-1888.

**Engelman,** J. (*fl.*1784-1802). Rhynstroom, 1793; Land von Heusden en Altena, 1798 (with F.W. Conrad).

**Engelmann,** father and son.
• **Engelmann,** Johann Wenzel I (1713-1762).
•• **Engelmann,** Johann Wenzel II (1748-1803). Engraver of Vienna. Worked on maps for F.A. Schrämbl's *Allgemeiner Grosser Atlas,* 1786-1800; Joseph Oehler's *Oestreichische Niederlande,* 1793.

**Engelmann,** father and son. Lithographers from Alsace, worked in Mulhouse and Paris. One of the foremost lithographic companies of the early-19th century. They were responsible for the introduction of dry-paper lithography, a technique which was adopted by the Imprimerie Nationale. Engelmann C$^{ie.}$ was succeeded by Thierry Frères. **Ref.** COOK, K. *Imago Mundi* 47 (1995) pp.160-161; WALLIS, H & ROBINSON, A. *Cartographical innovations* (1987) p.292.
• **Engelmann,** Godefroy [Gottfried] (1788-1839). Worked as a lithographer in Mulhouse from 1814, and Montmartre, Paris, from 1816 (where he worked at some time with P. Thierry). He returned to Mulhouse when the company was dissolved during the revolution of 1830.
•• **Engelmann,** Jean (1816-1875). Son of Godefroy, with whom he worked on the development of colour lithography. Jean restored the company to Paris in 1837.
•• **Engelmann & C**$^{ie.}$ Printers of Paris. *Portefeuille géographique et ethnographique* (32 lithographed maps), 1820 (G. Engelmann and G. Berger); *Carte géologique du Département du Haut Rhin,* 1831; *Carte minéralogique des Pyrénées,* 1831; U.J. Walker's *Carte des Cantons Solothurn* (4 sheets), 1832; credited with the printing of several maps of parts of Greece and the Islands which had been lithographed by F. Rivier of the Dépôt de la Guerre, *c.*1832.

**Engelmann,** Julius Bernhard. *Post- und Reise-Karte von Deutschland,* Frankfurt 1835.

**Engelmann,** Wilhelm (1808-1878). Publisher of Leipzig. *Bibliotheca geographica* (2 volumes), Leipzig 1857 (contains list of maps printed in Germany).

**Engelvaart,** P. Dutch cartographer. Voorne, 1675.

**Enguidanos,** Tomás López. Spanish engraver. *Mapa del Reyno de Valencia,* 1795 (with script engraver J. Asensio).

**Ennenckel,** G.A. See **Enenckel** von Hoheneck.

**Enouy,** Joseph Christopher. English geographer and map engraver. Invasions of England, 1797; *Egypt with part of Arabia and Palestine,* London, Laurie & Whittle 1801; *A New Map of Ireland...,* London, Laurie & Whittle 1802, 1807, 1808 (in *A New and Elegant Imperial Sheet Atlas...*); *A new map of Scotland,* 1803, 1807, Laurie & Whittle 1810 (engraved by Joseph Bye); *New Map of England and Wales,* 1805, 1811; *Europe,* 1809; *The United Kingdom of Great Britain & Ireland with the Adjacent Parts of the Continent...,* 1828?.

**Enrile,** Nicolás. Mindanao [Philippines], 1826 MS; Burias, 1832 MS.

**Enriques,** Luis. Rio Grande de la Magdalene, 1601.

**Ens,** Gaspar. Signed the dedication on the reverse of the map of Britain in 4th edition of P. Keschedt's pirate version of G.A. Magini's edition of Ptolemy's *Geographia*, Arnhem, Johannes Janssonius 1617.

**Enschede,** Isaac & Johannes. Engravers and publishers of Haarlem. *Kaart der Nieuwe Ontdekkingen Benoorden de Zuyd Zee...* [North America], 1754 (reduced from J.N. Delisle, 1752).

**Ensenius,** G. See **Buondelmonti**, Christoforo.

**Ensign,** D.W. [& Co.]. Publishers of Chiacago and Philadelphia. Black Hawk County, Iowa, 1869; *History of Colmbiana County, Ohio*, 1879; *History of Cuyahoga County, Ohio*, 1879; *History of Berrien and Van Buren counties, Michigan*, 1880; *History of Allegan and Barry Counties, Michigan...*, Philadelphia 1880; J.S. Schenck's *History of Ionia and Montcalm counties, Michigan*, 1881; *Plat Book of DeKalb County, Illinois*, Chicago 1892; *Waco, Texas 1892*, 1892; *Atlas of Kane County, Illinois*, Chicago 1892; *Plat Book of Carroll County, Illinois*, Chicago 1893; *Plat book of Cass County, North Dakota*, Philadelphia 1893; *Plat book of Grand Forks, Walsh and Pembina counties, North Dakota*, 1893; *Index Map of Polk County, Minn.*, 1896.

**Ensign,** companies. Engravers and publishers, 25 Park Row, N.Y. & 10 Main St., Buffalo; 36 Ann St., N.Y. (1848); 50 Ann Street (1849). Probably successors to Phelps, Ensigns & Thayer.
• **Ensign, T. & E.H.** Booksellers of New York. Noted on the imprint as sellers of James H. Young's *Map of the United States*, 1844 (first published 1831); *Ensign's Traveller's Guide and Map of the United States...*, 1846.
• **Ensign's & Thayer** (*fl.*1848). Publisher of '36 Ann Street', New York (1848); '50 Ann Street' (1849). Map of New England, 1847; J.M. Atwood's *Map of Canada East and West*, 1848, 1849 (also published by D. Needham in Buffalo).
• **Ensign & Thayer** (*fl.*1850). Publisher of '50 Ann Street & 12 Exchange St. Buffalo, N.Y.' (1850). *Map of the Seat of War* [Mexico], 1847; re-issue of Humphrey Phelps's *Ornamental Map of the United States*, 1848, 1851 (first published 1847); *Map of Massachusetts, Rhode-Island & Connecticut*, 1848; *Map of the City of New York, with Adjacent Cities of Brooklyn & Jersey City*, 1849, 1851 and later (street directory); Phelps's *National Map of the United States*, 1849, 1851; *Map of the Gold Regions of California*, 1849; republished Atwood's *Map of Canada...*, 1850; *Map of Canada East and West*, 1850 (smaller, folding edition). Many of these maps were later republished by Ensign, Bridgman & Fanning.
• **Ensign, Thayer & Co.** *Pictorial Map of the United States*, 1852.
• [Horace Thayer & Co.] (*fl.*1852). See **Thayer**.
• [Thayer, Bridgman & Fanning] (*fl.*1853). See **Thayer**.
• **Ensign, Bridgman & Fanning.** Publishers of '156 William St. & 8 Exchange St. Buffalo, N.Y.' (1854); '156 William St. corner of St. Ann, New York' (1855); successors to Thayer, Bridgman & Fanning, 1854. *Map of the United States with its Territories along*

*Lóránd Eötvös (1848 - 1919). (By courtesy of László Gróf)*

with Mexico and the West Indies, 1854; *Ensign, Bridgman & Fanning's Travellers' Guide through the States of Ohio, Michigan, Indiana, Illinois, Missouri, Iowa, and Wisconsin*, 1854, 1855, 1856, (accompanied by J.M. Atwood's *Map of the Western States...*, 1849); re-published J.M. Atwood's *Map of Canada East and West*, New York 1854 (first published 1848); republished *Map of Canada East and West*, 1854, 1855, 1862, 1865 (smaller, folding edition, first published 1850); *Township Map of Michigan*, 1855 (engraved Beveridge); *Ensign, Bridgman & Fanning's Rail Road Map of the Eastern States*, 1856; *Ensign, Bridgman & Fanning's Raolroad Map of the United States*, 1859; *A New Township Map of the State of New York*, 1860; *Map of the Present Theater of Events*, 1861.

• **Ensign, Everts & Everts.** One of the imprints used by Louis H. Everts and his partners for a series of county atlases in 1876. **N.B.** See also under **Everts**.

**Ensinck**, François Jan (1806-c.1856). Lithographer of Breda. Antwerpen, 1856.

**Enthoffer**, J. Compiler (with A. Bastert, assisted by P.H. Donegan) and co-publisher (with A. Petersen) of *Map of the City of Washington* (10 sheets), Boston, A. Petersen & J. Enthoffer 1872; *Map of the city of Washington showing the progress of buildings...*, 1873.

**Entick**, John (c.1703-1773). Schoolmaster. *A Plan of the Harbour and Town of Louisbourg in the Island of Cape Breton*, Dublin, Exshaw 1758 (in Exshaw's Magazine); *A New and Accurate Map of the Island of Martinico*, 1763, 1766; *A New and Accurate Map of the Seat of the Late War on the Coast of Choromandel in the East Indies*, 1763; *A New and Accurate Map of North America*, 1763; *New and Accurate Map of Minorca...*, 1764; *A New and Accurate Plan of the River St. Lawrence from the Falls of Montmorenci to Sillery...*, 1766; *Entick's Description of the British Empire*, London, W. Lane 1770; *View of the Taking of Quebec...1759*, Dublin, J. Exshaw 1774. **N.B.** Some of these maps appeared in Entick's The General History of the Late war Containing It's Rise, progress, And Event, In Europe, Asia, Africa And America..., London, Edward Dilly and John Millan 1763-1764, they were also copied by Exshaw for the The Gentleman's And London Magazine.

**Entrecasteaux.** See **Bruny** d'Entrecasteaux, Joseph Antoine.

**Entresz**, K.A.W. Provinz Posen, 1842.

**Entwistle**, J.C. Publisher of Washington. *Potomac river from Washington to Chesapeake Bay*, 1887 (engraved by W.M. Dougal).

**Eosander**, — (*fl*.18th century). North German cartographer.

**Eötvös**, Lóránd (1848-1919). Hungarian scientist, inventor of the Eötvös pendulum, designed to demonstrate the rotation of the Earth. **Ref.** VAJDA, P. 'Creative Hungarians' in: Technika Történeti Szemle [Technical History Review] (1979); RENNER, J. 'In memory of Roland Eötvös' in: Bulletin Géodesique (1949/12).

**Ephorus** of Cyme (c.405-330 BC). Greek historian. Works lost, but his maps were used as a source by Cosmas Indicopleustes, c.540 AD.

**Epler**, —. US Army Corps of Engineers. *Topographical Map of Humboldt County State of Nevada*, New York, G.W. & C.B. Colton 1866 (with Parkinson).

**Epner**, Gustavus. *Gold Regions in British Columbia*, 1862.

**Epworth**, Christopher (1737/8-1824). Surveyor, produced manuscript maps of parts of Leicestershire, Lincolnshire and Rutland. *Chart of part of the River Humber*, 1820.

**Erasmy**, Mathias (b.1833). Lithographer and publisher of Luxembourg and Woippy-lez-Metz, Lorraine. Produced maps, plans, guides and views of his region, including *Plan der Stadt & Festung - de la ville & forteresse de Luxembourg*, 1861, 1867, published with other plans in his own *Guide du voyageur dans le Grand-Duché de Luxembourg*, Luxembourg, V. Bück 1861 (the map was also published separately, and in the works of others).

**Eraso y Prados**, Modesto (*fl*.1890-1900). *Mapa ilustrado de España y sus posesiones para la Guardia Civil*, 1895.

**Eratosthenes** (c.276-196 BC). Mathematician, polymath and geographer of Cyrene, Director of the library at Alexandria. Using observations taken at Alexandria and Aswan, he calculated the circumference of the earth at 252,000 *stades* (of disputed length); said to have been the first to construct a map of the known world scientifically. **Ref.** KRETSCHMER, DÖRFLINGER & WAWRIK *Lexicon zur Geschichte der Kartographie* (1986) Vol.C/1 pp.197-198 [in German]; DILKE, O.A.W. *Greek and Roman maps* (1985) pp.32-35 et passim.

**Erault**, —, sieur Desparées. *Carte marine de l'Isle de Ré*, c.1680.

**Erbe**, Ludwig. Relief maps. Palestine, 1842; Europe, 1844-1850; Germany, 1850.

**Erben**, Josef (1830-1910). Geographer and globemaker of Prague. Drew the gores for Czech and Russian language globes, 1860s; map of Bohemia, 1873.

**Erber**, Bernhard (1718-1773). *Notitia illustris Regni Bohemiae scriptorum, Geographica et Chorographica*, Vienna 1760 (with a 12-sheet map based on J.C. Müller's, 1712-1717).

**Erdélyi**, M. See **Transylvanicus**, Maximilianus.

**Erdeswicke**, Sampson (d.1603). A Survey of Staffordshire, 1593 MS, printed 1717.

**Erdinger**, G. Printer. W. Müller's *Plan der Residenzstadt Hannover*, 1822 (script engraved by Rower, map engraved by Wagner).

**Erdmann**, A.J. (1814-1865). Swedish geologist. Carte géologique de la Suede, 1860-1865.

**Erdmann**, J.A. Die Schweiz, 1748.

**Erédia** [Héredia], Manuel Godinho de (1563-1623) SJ. Portuguese mathematician and cosmographer from Malacca. Charts of south east Asia, 1601-1622; *Declaraçam de Malaca* (39 maps), 1613; Goa, c.1616; *Tratado Ophirico* (11 maps), 1616; *Livro de Plantaforma das Fortalezas da India* (22 maps), c.1620; and others. **Ref.** CORTESÃO & TEIXEIRA DA MOTA (eds.) *Portugaliae monumenta cartographica* (Lisbon 1960) Vol.IV, 39-46.

**Erhard**, family and companies. Known variously as Erhard, Erhard Schièble, Erhard Frères and similar.
• **Erhard**, —. Mapmakers, engravers and lithographers noted at 'rue St André-des-Arts' (1852, 1862); '42 rue Bonaparte' (1854, 1862), Paris. Founded by Georges Erhard Schièble. Engravers for *Carta Geográfica Plana del curso del Rio Magdalena...* [Colombia], c.1849; *Carte de la Grèce*, Paris, Dépôt de la Guerre 1852; *Carte de la Presqu'île de Gallipoli...*, Paris, Dépôt de la Guerre 1854; credited as engraver on 3 maps in Lavallée's *Atlas de géographie militaire adopté pour l'Ecole impériale militaire de Saint Cyr*, Paris, 1861 (printed by Lemercier); engraver of F. Bazin and F. Cadet's *Atlas Spécial de la Géographie Physique, Politique et Historique de la France*, Paris, c.1862; *Cartes des Parties Centrales du Sahara...*, 1862; V.A. Malte-Brun's *Carte des Régions Semiretschinsk et Tranilienne dans l'Asie Centrale*, Paris, Société de Géographie 1864 ('après les derniers travaux de MMrs. Sémenof, Veniukop et Golubef'); *Carte de France...* (95 maps), Paris, Librairie Hachette 1870 and later; *Lagos et Sés Environs*, 1873; engravers for E.G. Rey's *Carte de la Montagne des Ansariés et du Pachalik d'Alep* [Syria], Paris Société de Géographie 1873; *Nouvelle carte générale de l'Egypte*, 1879 (engraved by Erhard Schièble, printed by Erhard); engravers and printers for A. Bertrand's *Carte de la Gaule*, Paris, Société de Géographie 1879 ('V.A. Malte-Brun, Script.'); *Environs de Bizerte...1880* [Tunisia], Dépôt de la Guerre 1880 (printed by Lemercier); 16 coloured maps for A. Vuillemin's *Atlas de Europe physique - Bassins des grands fleuves de la France et de l'Europe*, Paris 1882; *Cannes et environs*, 1895.
• **Erhard Schièble**, Georges (1821-1880). Geographer and engraver at 'rue St. André-des-Arts 14' (1851); '12 r. Duguay-Trouin, Près le Luxembourg', Paris, member of the 'Société de Géographie', elected to 'Cercle de la Librairie', 1866. He devised an electro-chemical process for the transfer of engraved maps from stone to copper. He was succeeded in the Erhard business by his three sons, the 'Erhard Frères'. *Carte de la Mer d'Aral et du Khanat de Khiva*, Paris, Société de Géographie 1851 (compiled by Jacques de Khanikoff, maps engraved on stone for Malte-Brun's *Atlas de la France illustré*, 1852-1855); *Le Paris de Napoléon III*, Paris, Lanée 1869 (printed by Lemercier et Cie.); *Nouveau Plan de Paris*, Paris, Lanée 1870 and later (printed by Monrocq); *Nouvelle carte générale de l'Egypte*, 1879 (printed by Erhard); *Alpes Maritime*, 1880.
•• **Erhard**, Georges *fils*. One of the 'Erhard Frères', son of Georges Erhard Schièble.

Elected to 'Cercle de la Librairie', 1879.

•• **Erhard**, Henri. One of the Erhard Frères, son of Georges Erhard Schièble.

•• **Erhard**, Eugène. One of the Erhard Frères, son of Georges Erhard Schièble.

•• **Erhard** F<sup>res.</sup> (fl.to 1916). Engravers and printers of Paris. Partnership of Georges (fils), Henri and Eugène Erhard. Succeeded their father in the 'Etablissement de Erhard, rue Duguay-Trouin à Paris (Près Luxembourg)'. *Itinéraires en Asie...*, c.1885; A.A. d'Oliveira's *Carta da Africa Meridional Portugueza*, 1886; *Carta de Moçambique*, 1889, 1894; E. Giffault's *Carte du Soudan Occidental...*, 1890; *Carta das Possessões Portuguezas da Africa...*, 1891; engravers for F. Schrader's *Atlas de Poche*, Paris, Librairie Hachette et C<sup>ie.</sup> 1894; *Carte du Maroc......*, 1897; *Mapa Del Peru* (35 map sheets), Paris 1898; *Carte de la frontière nord-est de la France...*, Paris, Chapelot & Cie. 1914; and many others.

**Erhard**, Eugen Leo (1865-1915). Cartographer of Paris. **N.B.** Compare with Eugène Erhard, above.

**Erhard**, Georg (1854-1898). Cartographer from Baden, worked in Paris. **N.B.** Compare with Georges Erhard (fils), above.

**Erhardt**, —. Engraver of 'rue Bonaparte 42', Paris. For J. Andriveau-Goujon, 1841; Bonange, 1878; Société de Géographie, 1879. **N.B.** Compare with Erhard family and companies, above.

**Erhardt**, Karl. Geographer of Stuttgart.

**Erich[ius]**, Adolar[ius] (1559/1561-1634). Priest, chronicler and cartographer. *Thuringia landgraviatus* (24 sheets), Erfurt, P. Wittel 1625 (the legend contained useful historical information on the area) the map was later used by Willem Blaeu 1634.

**Erichsen**, Hr. G.R.S. Norge, 1785.

**Erichsen** [Eriksen; Eriksson], *Professor* John [Jön] (1728-1787). *Nyt Carte over Island*, c.1771 (the first map to show Reykjavik); *Det Sydlige Norge...*, 1785; *Udtog af Christian Lunds Indberetning...*, 1787.

**Ericsson**, Emil. Sverige, 1876.

**Erikson**, Leif. See **Eiriksson**, Leif.

**Eriksson** [Ericsson], Harald. *Frövi-Falu Jernbanan*, 1867.

**Eriksson**, Jön. See **Erichsen**, *Professor* John.

**Erizzo**, *Conte* F.M. Terre Polare, 1853.

**Erkel**, M. van. Architect. *Plattegrond der Stad Arnhem*, 1853.

**Erkert**, Rodrich von. *Atlas ethnographique... des polonais*, 1863.

**Erlinger**, Georg (c.1485-1541). Woodblock cutter and printer of Augsburg, later worked in Bamberg (from 1519). Produced an *Erklärung* (explanation) for a revised version of Etzlaub's road map of the Holy Roman Empire *Das heilig Römisch Reich mit allen Landtstraßen*, c.1515; *Gelegenheit Teutscher Lanndt* (pilgrim routes to Rome), 1524, 1530; *.. dem jungen Wandersmann fast nützlich*, Nuremberg [before 1530].

**Erman**, Georg Adolf (1806-1877). Ural-Gebirge, 1837; Kamtchatka, 1838; *Geognostische Skizze von Nord-Asien*, 1842.

**Ermil**, Antonio [e filio]. French bookseller in the 'Strada di Toledo', Naples. Seller of S. Giraud's *Nuova Pianta di Napoli*, 1767.

**Ermirio**, Girolamo. *Contorni di Firenze*, c.1840.

**Erns**, E. *Herzogthum Württemberg*, 1842; railway maps, 1844; *Karte von Württemberg*, c.1853.

**Ernst**, Hans (fl.late-16th century). Globemaker.

**Ernst**, K. German cartographer. World [wall map], 1830; Atlas of Biblical Geography, 1852.

**Ernst & Co**. Illustrated plan of Manchester & Salford, 1857.

**Ernst & Korn**. Publishers of Berlin. *Der Rheinstrom und seine wichtigsten Nebenflüsse von den Quellen bis zum Austritt des Stromes aus dem Deutschen Reich* (22 sheets), 1889 (compiled by the Central Meteorological and Hydrographic Office, Baden, printed by Giesecke & Devrient of Leipzig).

**Eroedi** [Eródi]-Harrach, Bela (1846-1936). Hungarian geographer.

**Eroedi** [Eródi], Kalman (*fl.*19th-century). Hungarian geographer, specialised in school atlases.

**Erp**, Theodore van (*b.*1874). Military surveyor and engineer in Dutch East Indies.

**Errard de Bar-le-Duc**, Jean. Ingénieur du Roi. *Plan de la Ville de Montreuil*, 1605.

**Erskine**, A. Vincent. Government surveyor. *General Plan of all the farms taken up by the Government of The New Republic, Zululand*, 1885 MS. **N.B.** Compare with Saint Vincent **Erskine**, below.

**Erskine**, J.F. *General View of the Agriculture of the County of Clackmannan*, 1795 (with a map of the county engraved by H. Gavin).

**Erskine**, Robert. Cartographer. Plans of Iberian ports, 1727-1734; *Plan of Gibraltar Bay with the enemies attack...*, London, W.H. Toms 1744 (with C. Knowles).

**Erskine**, Robert (1735-1780). *b.* Dunfermline. Land surveyor and engineer 'next to the Crown, Scotland Yard', London. Fellow of the Royal Society, 1771; emigrated to North America 1771 to manage the American Iron Works. Appointed geographer and surveyor general to Washington's army, 1777, and took Simeon De Witt as one of his assistants in 1778. Simeon DeWitt took over as geographer when Erskine died suddenly in 1780. Most of Erskine's maps are held by the New York Historical Society. *A Map of part of the States of New York and New Jersey*, 1777; *Surveys in New York & Connecticut State for His Excellency General Washington*, 1778; Roads in the United States, 1778-1779; map of the Highlands of the New York, 1779 MS; manuscript maps of the road to Yorktown, 1781 (completed by DeWitt). **Ref.** RISTOW, W.W. *American maps and mapmakers* (Detroit 1986) pp.37-38; GUTHORN, P. *The Map Collector* 2 (1978) pp.11-13; GUTHORN, P. *American maps and mapmakers of the Revolution* (1966) pp.17-22.

**Erskine**, Saint Vincent. Gold Fields in South East Africa, 1876; Transvaal, 1877; Kaap Gold Fields, 1887.

**Erslev**, Edvard (1824-1892). Danish geographer and naturalist, founder Geografiske Selskab, Copenhagen. *Skoleatlas*, 1870; published various manuals of geography, 1873-1876.

**Erstein**, Hans (*fl.*16th century). Alsatian globe maker.

**Ertinger**, François (1640-*c.*1710). Draughtsman and engraver. *Plan de la ville et citadelle de Palamos*; *Bataille d'Ensheim, près Strasbourg...1674*; *La bataille de Liorens en Catalogne...*: and other maps, battle plans and views published in Beaulieu's *Les glorieuses conquêtes de Louis le Grand*, Paris 1694.

**Ertinger**, P. French engraver. Volan's *Habitation des Hollandois au Cap de Bonne Esperance*, Paris, Coignard 1691 (in S. La Loubère's *Description du Royaume de Siam*).

**Ertl**, Anton Wilhelm (1654-*c.*1715). Bavarian jurist and historian. *Des Chur-Bayerischer Atlantis*, Nuremberg, 1705.

**Erwin**, John W. *Elkhart County, Indiana*, 1871.

**Erzey**, Jan (1780-*c.*1842). Cartographer of Amsterdam.

el-**Erzincani**, es-Seyyid Abdu'laziz bin Abdu'lgani. Map of Europe, Asia and North Africa, 1813 [in Ottoman Turkish]. **Ref.** GOODRICH, T.D. 'Old maps in the library of the Topkapi Palace in Istanbul' in *Imago Mundi* 45 (1993) p.126.

**Escalante**, V. de. See **Velez** de Escalante.

**Escande**, Léon. *Haut Fleuve Rouge*, 1895.

**Escandon**, José de, *Conde de la Sierra Gorda. Mapa de la Sierra Gorda y Costa del Sero Mexicano*, 1747 MSS; Nuevo Santander, 1748 MSS.

**Eschauzier**, Brand. *Encyklopaedischer Atlas*, 1838.

**Eschels-Kroon**, Adolf. *Beschreibung der Insel Sumatra*, Hamburg, Carl Ernst Bohn 1781.

**Eschenard**, F. See **Eschinardi**, Francisco.

**Escher**, A. See **Escher von der Linth**, Arnold.

**Escher**, Lieutenant B.G. *Vaarwaters naar de Reede van Batavia*, 1840-1857.

**Escher von der Linth**, father and son.
- **Escher von der Linth**, Hans Conrad (1767-1823). Swiss geologist and engineer of Zurich. Father of Arnold. Produced many maps and panoramas of Swiss regions including: *Plan des Ausflusses des Wallensees und des Laufs der Linth*, 1804, 1807. **Ref.** SOLAR, G. *Hans Conrad Escher von der Linth. Ansichten und Panoramen der Schweiz. Die Ansichten 1780-1822* (Zurich 1974) [in German].
- • **Escher von der Linth**, *Professor* Arnold (1807-1872). Swiss geologist of Zurich. *Geologische Karte des Sentis*, 1837-1872, published 1873; *Carte géologique de la Suisse* (4 sheets), Winterthur 1853, 1867 (with B. Studer).

**Eschinardi** [Eschenard], Francisco, *SJ* (1623-c.1700). Mathematician of Rome. *Imperii Abassini tabula geographica* (4 sheets), Paris 1674, 1684.

**Eschmann**, *Major* Johannes (1808-1852). Swiss astronomer, engineer and cartographer, born Wädenswil, worked in Zürich; assisted Guillaume-Henri Dufour in *Carte de la Suisse* series. *Topographische Karte des Cantons St. Gallen mit Einschluß des Cantons Appenzell* (16 sheets), 1841-1854 (with others).

**Eschuid** [Eastwood; Estwood; Eschenden], John (*d.*1380). English physician and astrologer. *Summa astrologiae*, Venice 1489 (with world map after Ambrosius Macrobius). **Ref.** CAMPBELL, T. *The earliest printed maps 1472-1500* (1987) pp.115-117.

**Eschwege**, Wilhelm Ludwig von (1777-1855). Cartographer of Weimar. *Petrographische und orographische Karte...*, 1811 (map showing the route from Rio de Janeiro to Villa Rica, with a geological profile); America, 1818; Rio de Janeiro, 1819; Brazil, 1821.

**Escluse**, Charles de l'. See **Clusius**, Carolus.

**Escudé Bartolí**, Manuel. *Atlas geográfico iberoamericano. España: description geográfica y estadistica de las provincias españolas*, Barcelona, after 1898.

**Escudero**, Ramón. Plan de Yquique, Chile, 1861.

**Eseitz**, D.F. van. See **Fabricius**, David.

**Esenev**, Ya. Geodesist and surveyor of the Moscow province of Russia, 1733.

**Esenius**, G. See **Buondelmonti**, Christoforo.

**Esents**, D.F. van. See **Fabricius**, David.

**Eskes**, H.P. *Platte Grond der Stad Amsterdam*, 1842.

**Eskrich**, Pierre. Artist and woodcutter, born in Paris, worked in Lyon (1545-1551) and Geneva (1552-1564). *S. Quinten...1557*, [n.d.]; *Mappemonde nouvelle papistique*, 1566-1567; his Bible maps include Canaan, c.1566; Eden, c.1566; Holy Land and Eastern Mediterranean, c.1573; Exodus, c.1580.

**Eslick**, —. *Eslick's Patent Dissected Map of England & Wales* (jigsaw puzzle), mid-19th century (engraved by W. Hughes).

**Esnauts**, companies.
- **Esnauts et Rapilly** (1777-1791). Jacques Esnauts and Michel Rapilly at 'rue St. Jacques à la Ville de Coutances', Paris. Joint publishers of various works. R. Phelipeau's *Carte detaillé des Possessions Angloises dans l'Amérique Septentrionale*, 1777 (engraved by E. Voysard); Brion de la Tour's *Carte du Théatre de la Guerre entre les Anglais et les Américains...*, 1778; *Nouveau plan des environs de Paris...*, 1778 (later published by Esnauts alone); *Nouveau plan routier de la ville et faubourgs de Paris*, 1778 (engraved by C.B. Glot and E. Voysard); *l'Amérique Septentrionale*, 1779; Brion de la Tour's *Presqu'isle des Indes Orientales...*, 1780; N. Chalmandrier's *Plan de Gibraltar...*, 1782; Louis Brion de la Tour's *Tableau Général de l'Amérique*, 1783; *Carte de la partie de la Virginie... 1781*, 1784; M. Poirson's *Carte des Départements de Paris, de l'Oise, de la Seine et l'Oise, de la Seine et Marne...*, 1791.
- **Esnauts** [Esnault], Jacques (*d.c.*1813). Publisher of 'Boulevard de Richelieu dit des Panoramas au coin de la Rue de la Loi Terrasse Frascati au Pavillon de la Paix N[o.]7' (1811); 'Boulevard Montmartre Terrasse Frascati au Pavillon de la Paix N[o.]7 près la rue de Richelieu' (1811), Paris. *Nouveau plan des environs de Paris...*, 1811 (first published Esnauts & Rapilly 1778); *Plan Routier de la Ville et Faubourgs de Paris. Divisé en Douze Mairies*, 1811.

**España**, Casildo. Engraver for M. Rivero Maestre's *Atlas Guatemalteco*, 1832.

**Espedic**, J. *Carte des vignobles du Médoc*, 1868.

**Espejo**, J. Draughtsman at the Dirección de Hidrografía, Madrid. *Plano de la Ria de Corcubion*, Madrid 1837 (surveyed by I. Fernández Flórez, engraved by J. Carrafa); *Plano de la Ria de Muras y de Noya*, Madrid 1838 (surveyed by I. Fernández Flórez, engraved by C. Noguera and J. Hermoso); *Plano de la Ria de Camariñas*, Madrid 1838 (surveyed by I. Fernández Flórez, engraved by C. Noguera and N. Gangoiti).

**Espin**, Thomas Espinelle. *A New Star Atlas*, 1895 and later (revised and corrected edition of Proctor's *Small Star Atlas* of 1872).

**Espin**, W.M. Surveyor. *Fletcher & Espin's Map of Matabeleland* (4 sheets), London, Stanford's Geographical Establishment 1896, 1897; *Plan of Bulawayo Township*, 1898 (both with P. Fletcher).

**Espinalt y García**, Bernardo. *Atlante español*, 1778-1795; *Postas de España*, 1794.

**Espinha**, Joseph d', SJ (1722-1788). Member of the team which surveyed Xinjiang province of China 1755-1759 (with F. da Rocha); surveys used by Michel Benoist for *Qianlong neifu yutu* [104 woodblock maps], *c.*1770, copperplate version published 1775.

**Espinosa**, Antonio Vásquez (*d.*1630). Andalusian geographer.

**Espinosa**, Luis & Manuel. *Plano de la ciudad de México*, 1867.

**Espinosa de los Monteros**, Antonio. *Plano topográfico de la Villa... Madrid*, 1769.

**Espinosa y Tello**, José de (1753-1815). Astronomer and surveyor of Madrid. In the 1790's he sailed with the Malaspina expedition to the west coast of America, making observations on latitude and longitude, founded Depósito Hidrográfico, 1797. *Relacion del Viage hecho por las goletas Sutil y Mexicana...* [California and the west coast of north America] (9 maps), 1802-1806; *Memoria sobre observaciones astronomicas*, 1809; *Carta esferica America Meridional*, 1810; *Antillas Mayores*, 1811; *Carta General para les Navegaciones a la India Orientale...* (6 sheets), 1812. **Ref.** TOOLEY, R.V. *The Map Collector* 2 (1978) p.53.

**Espinoza**, Enrique (*b.*1848). Atlas de Chile, 1897.

**Esquemeling** [Exquemelin: Oexmelin], Alexandre Oliver (*c.*1645-*c.*1707). Privateer and navigator. *Bucaniers of America*, 1684 (from Dutch edition of 1678); *Histoire des Avanturiers qui sont Signalez dans les Indes...*(2 volumes), 1688 (includes charts of central America).

**Esquivel**, Pedro de (*fl.*late-16th century). Iberian cartographer, commissioned by Philip II of Spain to lead a group of surveyors in a detailed topographical survey of Spain, from 1570. The manuscript surveys were bound together *c.*1585 and became known collectively as the 'Escorial Atlas' (the key map is thought to be by J.L. de Velasco). **Ref.** MUNDY, B. *The mapping of New Spain* (1996) pp.2-8.

**Ess**, Jay [*pseudonym*]. *Ess's American guide to London and its suburbs*, 1873 (taken from *Waltham Bros. Pocket Map of London*, 1870).

**Essen**, Dirk van (1829-1880). Lithographer of Rotterdam. Plans of Rotterdam, [n.d.].

**Essen**, John van. Manuscript charts of the East Indies, *c.*1700.

**Esser**, Johann Georg. German cartographer. *Grund-Riß der Reichs Stadt Kempten*, 1729; *Grundriß des Ihler-Flusses*, 1730.

**Essex**, Frank B. Draughtsman in United States War Department. *Map of Cuba* (2 sheets), Washington 1911 (with A.B. Williams).

**Essex**, James (1722-1784). Architect and antiquarian. Plan of Stonehenge and the Ancient Sites, [n.d.].

**Establecimiento Geográfico**, Madrid. *Plano de la plaza de Cadiz*, 1836; *Mapa itinerario de la parte Nord-Oeste de España y parte de Portugal*, 1838.

**Estancelin**, *Lieutenant-General* —. Woods and forests of Europe (40 sheets), 1768.

**Este World Map**. Circular Catalan world

map, c.1450 (deposited in the Biblioteca Estense, Modena by Cesare d'Este in 1598). **Ref.** DILKE, O.A.W. & M.S. *The Map Collector* 53 (1990) pp. 15-19.

**Esteller**, Eduardo Moreno. *Teatro de la guerra de Oriente*, c.1876.

**Estévanez**, Nicolás (1838-1914). *Atlas geográfico de América...* [Central and South America], 1885, 1896.

**Esteve**, R. Engraver. Buenos Ayres, 1812-1828.

**Estienne**, Charles (1504-1564). Writer and publisher of Paris. Produced what is thought to be the first French road book: *La Guide des chemins de France*, 1552 and editions to 1668 (compiled without maps, but some later editions included a route chart); *Les Voyages de plusieurs endroits de France et encores de Terre Saincte, d'Espaigne, d'Italie et autres pays...*, 1552.

**Estienne**, François. Publisher of Geneva. Bible with maps of the Holy Land and eastern Mediterranean, Geneva 1567.

**Estienne**, father and son.
• **Estienne**, Robert (1503-1559). Printer, of the 'rue Jean de Beauvais', Paris.
•• **Estienne**, Henri. Son of Robert. *La Bible*, Geneva 1565 (containing woodcut maps of Eden, Exodus, and Canaan).

**Estienne**, *veuve et fils*. 'Rue St Jacques', Paris. Worked for J.B.B. d'Anville, c.1743. Co-publisher (with d'Anville and Jaillot) of d'Anville's *L'Italie...*, 1743.

**Estorff-Veerssen**, A. von. *Karte des Kreises Uelzen*, 1890.

**Estorgo y Gallegos**, Francisco Xavier. *Quiapo*, Philippines, 1746; *Plano de Manila*, 1770.

**Estrabou**, —. *État-Major du Soudan Français*, c.1890.

**Estrabou**, G. Draughtsman. *Frontière des Alpes*, published in L. Grégoire's *Atlas Universel de Géographie Physique & Politique*, 1892.

**Estrada**, Angel [& Cia.]. Publisher of Buenos Aires. *Atlas general de la república Argentina*, Buenos Aires 1910; E.A. Bavio's *Atlas escolar de la república Argentina...*, Buenos Aires, A. Estrada & Cia. c.1910.

**Estrada**, Bartolomé Ruiz de. See **Ruiz** de Estrada.

**Estrémont de Maucroix**, — d'. *Banc au sud-ouest des Îles de Farroë*, 1846.

**Estridge**, A.W. Survey Wanstead Flats, 1862 MS.

**Estruc y Jordan** [Estruch i Jordan], Domènec (1786-1851). Spanish engraver from Muro de Alcoy, near Alicante, worked in Mallorca, Havana and Barcelona, died in Madrid. *Gran carta de la Isla de Cuba*, c.1820s; *Nueva descripción del principado de Cataluña*, Barcelona 1824; *Germania*, 1832; *Carta Geogr.º Topográfica de la Isla de Cuba...Años de 1824 á 1831*, Barcelona, c.1835; *Posesiones de America: Isla de Puerto Rico*, Madrid 1851 (in Francisco Coello de Portugal y Quesada's *Diccionario geografico-estadistico-historico: Atlas de España y sus posesiones de ultramar*; Estruc possibly also engraved the 2-sheet map of Cuba in the same atlas).

**Estwood**, J. See **Eschuid**, John.

**Esveldt**, Steven van (d.1774). One of the Amsterdam publishers of Antoine-François Prévost's *Historische Beschryving de Reizen...* 1758 (maps by Bellin); Esveldt copied at least one map (*De Eilanden of het Keizerryk van Japan*), the second edition of which was published in A.J. Roustan's *Geschiedensis van alle Volken* (9 volumes), Amsterdam, Willem Holtrop 1787-1788.
• **Esveldt**, *widow* —. Continued the business after the death of Steven, entering into partnership with Willem Holtrop in 1778, who took over in 1784.
•• **Esveldt-Holtrop**, J.S. van. Publisher of Amsterdam. *Een blik op Holland of Schilderij van dat Koninkrijk in 1806* (includes a 'Leo Belgicus'), 1807; *Atlas van Saxen*, 1810.

**Eszler** [Aeszler; Aeschler], Jacob[us]. Co-publisher (with Georg Übelin) of the 1513 Strasbourg edition of Ptolemy's *Geographia* (edited by Martin Waldseemüller).

**Eszterházy**, *Prince* Pál [Paul] (1635-1713). Hungarian military cartographer. In campaign against the Turks, 1664, mapped Babóca,

Türbék, Eszék [now Osijek, Croatia]. **Ref.** VÉRTES, J. *A nagy palatinus* [The great palatine] Sopron, 1939.

**Etablissement Géographique de Bruxelles.** Founded by Phillippe Vandermaelen. *Namur & Liége & Mons & Manage Railways. Comparative Map of distances between Paris & Cologne and likewise between Dover & Cologne,* 19th-century; *Nouvelle Carte Générale de la Belgique,* c.1840 (drawn by J.F. Keyser, engraved by P.J. Doms); *Carte speciale des chemins de fer Belges,* 1843; *Traité entre Sa Majesté le Roi des Pays-Bas, Grand Duc de Luxemburg, et Sa Majesté le Roi des Belges..* 1842 (8 maps with text), c.1845; Vandermaelen's *Carte Topographique de la Belgique...* (250 sheets), c.1854; *Carte des Chemins de Fer de Belgique,* 1868.

**Etanduère, l'.** See **L'Etanduère.**

**Ethersey,** *Lieutenant* Richard. Surveyor in the Indian Navy. India, 1836; *Chart of Paumben Pass,* 1838; *Gulf of Cambay,* 1845; *Chart of the Malacca Banks,* 1845.

**Ethicus** [Aethicus of Isteria] (*fl.*6th century). Roman geographer. *Cosmographia,* first printed Venice 1513. **N.B.** In the past he has been credited (wrongly) with authorship of the Antonine Itinerary.

**Etienne,** Jean d' (1725-1798). *b.* Cernay. Cartographer in Bückeburg.

**Etienne-Gayet,** J. Draughtsman at the Reale Corpo di Stato Maggiore, Italy. Worked with Mudotti on *Carta topografica degli stati in terraferma di S.M. il Re di Sardegna...* (68 sheets), 1851-1858 (lithographed by Biasoli).

**Ettling,** Theodor (*b.*1823). Engraver, lithographer and draughtsman, initially in Amsterdam, later traded at '3 Red Lion Square, Holborn', London. Lithographed maps and charts of South Africa for Jacob Swart, 1847; *Goud-Districten van Opper Californie,* 1849; *Ettling's Drawing-room Atlas of Europe...,* London, Longman, Brown, Green and Longmans 1855; maps for Day & Son's *Weekly Dispatch Atlas,* London 1855-1863; *Map of the United States of North America, Upper & Lower Canada, New Brunswick, Nova Scotia & British Columbia...,* 1861 (supplement to *Illustrated London News,* printed by Panicographie de Gillot, Paris); *Times Map of North America,* 1861.

**Etzel,** Franz August von. See **Oetzel.**

**Etzlaub,** Erhard (*c.*1460-1532). Physician, cartographer, mathematician, instrument and compass maker of Nuremberg. Noted for making the first printed road map of Central Europe. Etzlaub is considered to be the originator of a form of map projection, later refined by Mercator and known as Mercator's Projection. It was first used in an engraving on the cover of a sundial c.1511-1513. Germany, 1492; road map of the environs of Nuremberg (218 mm. diameter), 1492; *Das ist der Rom-Weg von meylen zu meylen mit puncten verzeychnet...,* c.1500 (first printed European road map, 6 copies recorded from 3 different printings); *Das sein dy Lantstrassen durch das Romisch Reych* (central Europe), 1501; World maps on compass lids, 1511-1513 (attributed); *Nurnberg im Reichswald,* 1516. **Ref.** ENGLISCH, B. *Imago Mundi* 48 (1997) pp.103-123; DELANO SMITH, C. *The Map Collector* 26 (1984) pp.38-41; KRETSCHMER, DÖRFLINGER & WAWRIK *Lexikon zur Geschichte der Kartographie* (1986) vol.C/l, pp.205-206 (numerous references cited) [in German]; CAMPBELL, T. *Imago Mundi* 30 (1978) pp.79-91; SCHNELBÖGL, F. *Imago Mundi* 20 (1966) pp.11-26; KRÜGER, H. *Imago Mundi* 8 (1951) pp.17-26.

**Eudoxos** of Cnidus (*c.*390-*c.*340 BC). Greek philosopher and astronomer. Made earliest recorded celestial globe.

**Eufredutius.** See **Freducci.**

**Eugene,** Carl Ludwig. Hypsographic map of Sweden, 1858.

**Eugéniéwitch,** G. *Carte des Communications ... de Serbie,* 1893.

**Eulenstein,** F. Engraver for H. Kiepert, 1857.

**Euler,** Leonhard (1707-1783). Swiss cartographer, astronomer and mathematician, of Basle, worked also in St. Petersburg and Berlin. Served in Russian navy from 1727, became professor of mathematics at the Academy (founded 1725), went to the Berlin Academy, 1741. Supervised the preparation of *Atlas Russicus,* St Petersburg 1745 (with J.-N. Delisle and M.V. Lomonosov); *Mappa Geographica Regni Hiberniae...* [Ireland], 1750; *Tabula Geographica Regni Hungariae,* Berlin, Michaelia 1753, 1760, 1777; *Atlas*

*Geographicus*, Berlin 1753, 1756, also published as *Geographischer Atlas*, 1760; devised, and wrote about, map projections, 1753, 1777, 1779. **Ref.** KRETSCHMER, DÖRFLINGER & WAWRIK Lexikon zur Geschichte der Kartographie (1986) vol.C/1 pp.206-207 (cites many references) [in German]; KRETSCHMER, I. in: SCHARFE, W. & JÄGER, E. (Eds) Kartographiehistorisches Colloquium Lüneburg '84. Vorträge Berlin (1985) pp.29-38 [in German].

**Eulitz**, Oskar. Publisher in Lissa [Vis, Dalmatia] and Posen [Poznan, Poland]. *Kreis Heydekrug* [Lithuania], 1914; *Kreis Tilsit* [Russia], 1915.

**Eulzofen**, T.L. *Guyane Française*, c.1881.

**Eumenius** of Autun (*b.c.*264). Rhetorician and teacher. Described a large wall map of the World intended for his school at Autun, France. **Ref.** DILKE, O.A.W. Greek and Roman maps (1985) pp.53-54.

**Eunson**, George. *A chart of the islands of Orkney with the... coast of Scotland*, c.1795.

**Eussen**, Jacob (*fl.*1745). Engraver in Amsterdam.

**Eustis**, Henry Lawrence (1819-1885). American engineer and professor. *Plan exhibiting the ravages of the Tornado of 22nd Aug. 1851*, Boston 1853.

**Euting**, Julius (1839-1913). *Odilienberg und Umgebung*, 1874.

**Evans**, *Corporal* —. British member of the international team which draughted frontier plans for the Servian Boundary Commission according to the Treaty of Berlin, 1878.

**Evans**, A.E. [& Sons.]. Publishers of 403 Strand, London. Plan of the cities of London and Westminster, 1857 (engraved by George Jarman as a copy of the Faithorne and Newcourt plan of 1658).

**Evans**, Albert S. *White Pine, Nevada*, 1869.

**Evans**, Sir Arthur John (1851-1941). Archaeologist and Balkanologist. *Diagrammatic map of Slav territories east of the Adriatic* (scale 1:1 500 000), London, Sifton, Praed & Co. (for The Balkan Committee) 1915; *Diagrammatic map of Slav territories east of the Adriatic* (scale 1:2 000 000), London, Royal Geographical Society 1916 (published in *The Geographical Journal* to accompany an article by Evans).

**Evans**, Cadwalader (1762-1841). Surveyor. Draught of a Tract of Land in Bucks County, Pennsylvania, 1786. **N.B.** Compare with Cadwallader **Evans,** below.

**Evans**, Cadwallader. *A Map of Delawar Bay and River with the Islands therein*, 1756 MS.

**Evans**, Charles. English surveyor, produced manuscript maps and plans including many associated with property of the Crown. Richard Crowle's estate adjoining the Castle [Windsor], 1754; Site of old palace [Richmond], 1756; Buckingham House and grounds, 1760; Lands in the parish of St. Giles in the Fields, 1764; Spring Gardens [Westminster], 1765; Frogmore [Windsor], 1766; and others.

**Evans**, E. *The Pictorial Missionary Map of the World*, c.1860.

**Evans**, E.J., *RN*. North East coast of Australia, Admiralty 1855. **N.B.** Compare with F.J.O. **Evans,** below.

**Evans**, Edward. Rivers of Ireland, 1887.

**Evans**, *Master* (later *Captain Sir*) Frederick John Owen (1815-1885). Hydrographer of the Navy 1874-1884. Survey of Central America, 1833-1836; North East Australia, 1841-1846; Isle of Man surveys, 1847 (with G. Williams); surveys of New Zealand, c.1848-1855 (with others) published 1856 and later; *Tasmania formerly Van Dieman Land from various authorities...*, 1860, 1893; Atlantic Ocean Pilot, 1868.

**Evans**, G. *Topographical map of Madison County, New York*, 1853.

**Evans**, *Commander* George, *RN. Port Louis, Mauritius*, 1819.

**Evans**, *Captain* George *RN*. Plan of Birkenhead Docks, 1847 (zincographed by J. & C. Walker); *Ulverston and Lancaster Railway*, Ordnance Survey 1851.

**Evans**, George William (1778-1852). English engineer, worked as a surveyor in New South Wales from 1802, appointed acting surveyor,

*Sir Arthur John Evans (1851 - 1941). Portrait by Francis Dodd, 1935. (By courtesy of the National Portrait Gallery, London)*

1803, settled in Hobart, Tasmania. Port Dalrymple, Tasmania, 1809; Hobart Town, 1812; *Chart of Van Diemens Land*, 1821, published J. Souter 1822 (engraved by T. Tyrer); *Van Diemen Island. MacQuarie Harbour*, 1822 published 1845 (engraved by J. & C. Walker).

**Evans**, Henry Smith, *FRGS*. World, 1847, 1851; *Geology Made Easy*, 1851; *The Crystal Palace Game, A Voyage Round the World*, c.1854.

**Evans**, John. **N.B.** Confusingly, several mapmakers of this name were working at the end of the eighteenth century, they are clarified below. **Ref.** ROBERTS, I. & M. *The Map Collector* 46 (1989) pp.18-23.

**Evans**, John. Father and son. **Ref.** ROBERTS, I. & M. *The Map Collector* 46 (1989) p.18.
• **Evans**, John [of Llwynygroes, **John Evans 1**] (1723-1795). Amateur cartographer. *Map of the Six Counties of North Wales* (9 sheets), 1795; reduced version published 1797 and 1802 by his son, below.
•• **Evans**, John [**John Evans 2**] (1756-1846). Doctor of Medicine, also of Llwynygroes, son of John Evans 1. Published a reduced version of his father's *Map of the Six Counties of North Wales*, 1797, 1802.

**Evans**, *Reverend* John [of Islington, **John Evans 3**] (1767-1827). Baptist minister in Islington, north London from 1792. *The Juvenile Tourist*, London 1804; *A New Royal Atlas* [the maps for *The New Geographical Grammar*], London, J. Cundee 1810. **N.B.** Not to be confused with Reverend John **Evans** (1768-c.1812), topographical writer and editor. **Ref.** ROBERTS, I. & M. *The Map Collector* 46 (1989) pp.18-19.

**Evans**, John Thomas [**John Evans 4**] (1770-1799). Explorer who left his native North Wales in 1792 to go in search of 'Welsh Indians' on the Missouri River. Chart of the Missouri River from Fort Charles [Iowa] to the Mandan and Hidatsu villages in North Dakota, c.1792 (a copy on 7 sheets was used by Lewis & Clark). **Ref.** ROBERTS, I. & M. *The Map Collector* 46 (1989) pp.19-20; WOOD, W.R. 'John Thomas Evans and William Clark: Two early western explorers' maps re-examined' in *We Proceeded On: trans-Mississippi-west exploration* Vol.9 N°.1 (Lewis & Clark Heritage Heritage Foundation, Great Falls, MT 1983) pp.10-16; WILLIAMS, G.A. 'John Evans's Mission to the Madogwys, 1792-1799' in *Bulletin of the Board of Celtic Studies* 27 (1978) pp. 569-601; WILLIAMS, D. 'John Evans' Strange Journey' in *American Historical Review* 54 (1949) pp. 277-295, 507-529.

**Evans**, John [of West Smithfield, **John Evans 5**] (*fl.*1794-1799). Publisher of 41-42 Long Lane, West Smithfield, London. *A New Map of England and Wales...* (2 sheets), 1794; *A New & Accurate Plan of the Cities of London & Westminster & Borough of Southwark...*, 1796, 1798, 1799; *A General Map of France*, 1796. **Ref.** ROBERTS, I. & M. *The Map Collector* 46 (1989) p.20-22.

**Evans**, John [of Penygraig, **John Evans 6**]. *A New Map of the Vicinity of Aberystwyth*, 1842. **Ref.** ROBERTS, I. & M. *The Map Collector* 46 (1989) p.22.

**Evans**, Lewis (c.1700-1756). Surveyor and draughtsman. Born in Wales and later emigrated to Philadelphia, USA. *A Map of that Part of Bucks County released by the Indians to the Proprietors of Pensilvania in September 1737* [the Indian Walking Purchase], 1738 MS; possibly designed map of Cape Hatteras to Boston for James Turner, 1747; *A map of Pensilvania, New-Jersey, New York and the Three Delaware Counties*, 1749 (engraved by L. Hebert), re-issued 1752; *A General Map of the Middle British Colonies in America*, 1755 (engraved by James Turner) and numerous later editions (used as a base map by many other cartographers until c.1814). **Ref.** GIPSON, L.H. *Lewis Evans* (Philadelphia 1939); STEVENS, H.N. *Lewis Evans, his map of the Middle British colonies in America. A comparative account of 18 different editions...1755-1814* (London 1924).

**Evans**, R.M. Surveyor. Washoe Mines, California, 1860; Gold Hill, American Flat & the Divide [Nevada Territory], 1864.

**Evans**, Thomas (1739-1803). Bookseller at 54 Paternoster Row, London. P. Russell and O. Price's *England Displayed*, 1769 (maps by J. Rocque and T. Kitchin); one of the sellers of E. Gibson's edition of Camden's *Britannia*, 1772 (with maps by R. Morden); Evan Jones's *A New and Universal Geographical Grammar* (2 volumes), 1772 (with G. Robinson).

**Evans**, *Lieutenant* Thomas, *RN*. Hydrographer from Fishguard in South Wales. Straits of Sincapore [?], 1804; survey of Liverpool & Chester, 1812-1828; Bay & Harbour Fishguard, 1817; Barrow, 1832.

*John Evelyn (1620 - 1706). Portrait by Robert Walker, 1648. (By courtesy of the National Portrait Gallery, London)*

**Evans,** W.H. Engraver of Chester. *New Illustrated Plan of the Ancient City of Chester*, 1869.

**Evans & Bartle.** Engravers and publishers of Washington. Charts for the Coast and Geodetic Survey, 1892 including *District of Columbia* (48 sheets), 1892 (incomplete).

**Evelyn,** John (1620-1706). English gentleman and diarist from Wotton, Surrey. Founder member of the Royal Society. Evelyn's broad range of interests led him to write significant works on topics varying from pollution in London to forest trees, he also played an active role in the rebuilding of St Paul's Cathedral. Made additions to an earlier map of Deptford, 1650; produced a scheme for the replanning and rebuilding of London after the Great Fire of 1666 entitled *London Redivivum*, 1666.

**Everaert,** Martin. Teacher of navigation. Translated Medina's *Arte de Navegar* into Dutch as *De Zeevaert oft Conste van ter zee...*, Antwerp c.1580; contributed to Peter Apian's *Cosmographia*, 1584; translated Lucas Jansz. Waghenaer's *De Spieghel der Zeevaerdt* from Dutch into Latin (45 charts), Plantin 1586.

**Everest,** Sir George (1790-1866). English military engineer and surveyor in India until his retirement in 1843. Mount Everest is named after him. Successor to William Lambton as Superintendent of the Great Trigonometrical Survey [of India], from 1823; Surveyor-General of India 1830-1843; Meridian arc of India. **Ref.** GILL, B. 'The Big Man: Surveying Sir George Everest' in *Mercator's World* Vol.5 N°·4 (July/August 2000) pp.24-27.

**Everett,** Arthur (*fl.*1868-1887). Indigo, Chiltern and Wahgunjah, Australia, 1868; Victoria, 1869; draughtsman and colourist for R. Brough Smyth's *First Sketch of a Geological Map of Australia including Tasmania*, 1875 and later [earliest geological map of Australia]; Continental Australia (6 sheets), 1887.

**Everett,** Edward. *City of Quincy, Adams County, Illinois*, 1857.

**Everett,** James. Printer, bookseller and publisher of Market Street, Manchester. *A New Plan of Manchester and Salford with their vicinities... in 1832*, J. Everett 1834.

**Everett,** Joseph David (1831-1904). *b.* Ipswich, Suffolk. Worked as a globeseller and geographer in Belfast.

**Eversmann,** A.E.F. Produced industrial and economic maps. *Darstellung derjenigen Niederrheinisch-Westphälischen Gegenden...* (4 sheets), [n.d.]; *Geschäfts-Karte der Königl. Preußischen Rheinprovinz und Westphalen*, 1840.

**Everson,** A.H. Engraver. Railway, postal and telegraph map of South Australia, 1888.

**Everson,** Charles. *Map of Devon & Cornwall*, c.1895.

**Evert,** E. *Plan der Stadt Posen*, 1896.

**Everts,** Louis H. (*b.*1836). After army service he formed a partnership with a former colleague Thomas H. Thompson at Geneva, Illinois (c.1866). They published illustrated 'combination' atlases of counties in northern Illinois and Iowa. When the partnership dissolved c.1872-1873, he worked alone and with others in a confusing range of concurrent companies associated with the name of Everts. **Ref.** RISTOW, W.W. *American maps and mapmakers* (Detroit 1985) pp.413-416.
• **Thompson & Everts** (*fl.c.*1866-c.1872). (*q.v.*)
• **Everts, Baskin & Stewart.** Partnership of Louis Everts, O.L. Baskin and David Stewart, trading as draughtsmen and publishers of Chicago. *Combination Atlas Map of Lee County, Illinois*, Chicago 1872; *Combination Atlas Map of McHenry County, Illinois* (24 maps), Chicago 1872; *Combination Atlas Map of Ogle County, Illinois* (32 maps), 1872; Rock County, Wisconsin, 1873; county atlases of Michigan, 1874; New Jersey, 1874-1876; *Evert's & Stewart's Illustrated Historical Atlas*, 1876.
• **Everts, Stewart & Company.** Atlas of Sandusky County, Ohio, 1874.
• **Everts & Stewart.** Combination atlas maps of Lenawee County and Washtenaw County, Ohio, 1874.
• **Everts, L.H. [& Co.].** Produced combination atlases for the northern, western and central counties of Ohio, 1874-1875; *The Official State Atlas of Kansas*, 1887.
• **Everts, Ensign & Everts** [Ensign, Everts & Everts]. Atlases of counties in New York, Pennsylvania, New Jersey, Michigan, 1876 including Genesee County, and Yates County, New York, 1876.

*Sir George Everest (1790 - 1866). Surveyor General of India 1830 - 1843. Portrait by William Tayler, 1843. (By courtesy of the National Portrait Gallery, London)*

• **Everts & Kirk.** *The Official State Atlas of Nebraska*, Philadelphia 1885.

**Everts & Richards.** Bristol County, Massachusetts, 1895; Providence County, Rhode Island, 1895.

**Everwijn,** R. *...Zuid Ost Kust van Sinkep*, 1872.

**Eveau de Fleurieu d'.** See **Claret** de Fleurieu, Charles Pierre.

**Evia,** José de. Produced maps of the great arc of the Gulf coast from Florida to Vera Cruz. *Costa de la Florida*, 1783.

**Evia,** Simon de. *Plano y descricion* [sic] *de la Probincia* [sic] *de la Luociana* [sic], 1736 MS.

**Evreinov** [Yevreinov], Ivan Mikhailovic (*d*.1724). Russian cartographer and geodesist. Worked with F. Luoin on surveys of Siberia and Russia. First map of Kurile lslands, 1719-1723.

**Ewald,** Julius W. (1801-1891). German geologist. Geological maps of Germany including: *Geologische Karte der Provinz Sachsen*, 1865.

**Ewald,** Ludwig Wilhelm (1813-1881). German geographer and statistician. *Hand-Atlas der Allgemeinen Erdkunde*, 1852-1858; *Erdkarte*, 1853; *Europaeischen Staaten*, 1858; *Hand-Atlas der Allgemeinen Erdkunde der Länder- und Staatenkunde... in 80 Karten*, 1860-1862.

**Eward,** J. Mapseller 'at the Beehive against Northumberland St., Strand', London. William Edgar's *Plan of the City and Castle of Edinburgh*, 1742.

**Ewart,** Major J.H. *Map showing roads and trade routes from the River Prah, north through Kumasi...* [Ghana], *c*.1889; Lagos survey [Nigeria], 1890-1892 MS (with A.G. Fowler and M. de Fesigny); *Map of the Pokra Kingdom* [Nigeria], 1891 (lithographed by Judd & Co.); *Route survey in the colony of Lagos*, 1893.

**Ewart,** Major John A. *Military Sketch of the S.W. part of the Crimea...*, 1854-1856 MS (with others); *Position of the British Army before Sevastopol Jan$^y$ 1855...*, 1855 MS; *Yeni-Kali shewing the position of the Allied Force...*, 1855 MS.

**Ewart,** Lieutenant John Spencer. Cameron Highlanders. Kassassin & Tel EI-Kebir [Egypt], 1882-1887; *Sketch showing east Bank of Nile, with Position of the Enemy around Giniss* [Sudan], 1885.

**Ewen,** Daniel. Surveyor in the city of New York from 1817. Produced many coloured manuscript surveys of the New York waterfront (6 volumes), from 1827 [Manhattan Borough President's Office].

**Ewich,** Hermann. Historian of Wesel. *Patriae Antiquae Inter Iulii Et Caroli Magna Caesarum Romanorum Tempora Descriptio* [Low Countries], Johannes Janssonius *c*.1652 (engraved by S. Saevery).

**Ewing,** Maskell C. Surveyor. *Plan of the town of Alexandria, D.C.*, Philadelphia 1845 (lithography by T. Sinclair).

**Ewing,** Thomas. Cartographer. *New General Atlas*, Edinburgh 1817, editions to 1862.

**Ewing,** Thomas. *Gold Region of West Kansas*, 1860.

**Ewing,** W.C. *Columbia River*, 1837; *St. Louis Harbour*, 1837.

**Ewliya,** Celebi (1611-1684). Wrote a description of the Nile entitled *Seyahat-name*, [n.d.].

**Ewoutsz.,** Jan. Woodcut engraver of Amsterdam. Cut Cornelis Anthonisz.'s 12-sheet Amsterdam, 1544; Sailing directions, 1560-1561.

**Ewyk** [Ewijk], Nicholas van. *Hémisphère meridional*, Amsterdam 1752 (after Guillaume Delisle's map of 1714).

**Exchaquet,** Charles-François (1746-1792). Swiss mineralogist. *Carte pétrographique du St. Gothard*, 1791. **Ref.** SEYLAZ, L. Die Alpen II, 5 (1935) pp.187-195 [in French].

**Expilly,** abbé Jean Joseph Georges d' (1719-1793). French diplomat, traveller and geographer. *La Géographie Manuel*, 1757 &c.; *Topographie de l'univers*, 1757-1758; *Dictionnaire géographique des gaules*, 1762-1770; *Mappe Monde represente en Hemispheres Oriental et Occidental*, 1765, used in his *La Géographie Manuel...*, 1771.

**Exquemelin**, A.O. See **Esquemeling**, Alexandre Oliver.

**Exshaw**. Family of publishers '...at the Bible on Cork-Hill, over against the Old-Exchange', Dublin. Publishers of the Dublin editions of the monthly *The London Magazine*, 1741-1754 (copied from the London original). From 1751 the Exshaws began to draw from a wider range of sources, and in 1755 the title of their journal was changed to *The Gentleman's and London Magazine*.... Some 120 maps were published in the magazine. **Ref.** JOLLY, D. *Maps in British periodicals Part II* (1991) pp.155-173 (list of maps published); HODSON, D. *County atlases of the British Isles published after 1703* Vol.2 (1989) pp.175-179 (list of British county maps published).
• **Exshaw**, Edward (d.1748). Published the first Dublin edition of *The London Magazine* in May 1741. He was joined in partnership by John Exshaw in 1745. On Edward's death he was succeeded in the partnership by his widow Sarah.
• **Exshaw**, John. Joined Edward Exshaw in 1745.
• **Exshaw**, Sarah. Second wife of Edward Exshaw. Joined John Exshaw in partnership on the death of her husband in 1748.

**Eyb**, Otto Franz von (fl.1858-1882). Engraver for H. Kiepert, 1860.

**Eyckelberch**, Gillis van. See **Hooftman**, Gilles Egidius.

**Eyes**, family. Surveyors of Liverpool.
• **Eyes**, John (fl.1725-d.1773). Land surveyor and chart publisher of Liverpool. Liverpool Bay, 1725; surveyed Lancashire coast, 1736-1737; collaborated with Samuel Fearon on *A Description of the Sea Coast of England and Wales*, 1736-1738 published 1738 (thought to be the first charts to show Greenwich as the Prime Meridian); chart of Furness and Anglesea, 1738; Plan of Liverpool Docks, 1742; surveys used by Bowen & Kitchin in the *Large English Atlas*, 1749-1760; Survey of Dee, 1755 (with Sumner); resurveyed Liverpool Bay, 1764; Plan of Liverpool, 1765 (engraved by Thomas Kitchin).
•• **Eyes**, Charles (1754-1803). Surveyor of Liverpool, nephew of John Eyes. Plan of Liverpool, 1785.

**Eygird**, Krzysztof (fl.17th century). Produced town plans and maps of battles in Europe, associated with the wars between Sweden, Russia, Turkey and the Cossacks.

**Eyler**, —. *Topographical Map of Humboldt County, Nevada*, 1865.

**Eynde**, J. van. See **Ende**, Josua van den.

**Eynon**, I. 'Print and Mapseller, corner of Castle Alley, Threadneedle Street, Royal Exchange', London. Second edition of Robert Withy's *A Pocket Map of London Westminster and Southwark...*, 1760 (first published Withy 1759).

**Eynon**, R. Publisher and mapseller behind the Royal Exchange, London. Staten Island, 1776.

**Eyre**, Edward John (fl.1751-1769). Estate surveyor, produced manuscript maps of various areas in England and Wales, including The manor of East Greenwich, 1754; New Park, Richmond, 1754; Essex estate plans, 1758-1767; plans Minterne, 1767 MS.

**Eyre**, Edward John (1815-1901). Led an expedition into the northern and western parts of New South Wales, 1840. Lieutenant-Governor of New Zealand, 1847. Mount Remarkable, South Australia, 1857; Australasia, 1863.

**Eyre**, John. The exact surveyor, 1654.

**Eyre**, *Captain* John. Plans of fortifications London, 1643 MS.

**Eyre**, Joseph. *A true mapp or plott of the division of the wastes and commons within the hamlets of Hope, Aston, Thornehill and Nether Ashopp...* [Derbyshire], 1675.

**Eyre**, Mary. Commissioned a tapestry map of Nottinghamshire. *Nottinghamshire described*, 1632 (based on the maps of Saxton and Speed).

**Eyre** [Eayre], Thomas (1691-1757). Surveyor, bell founder and clockmaker of Kettering, England. Produced surveys of parts of Northamptonshire and Bedfordshire. Plan of Kettering, Northamptonshire, [n.d.]; *The County of Northampton*, 1779, 1780, 1791 (revised by Thomas Jefferys, engraved by William Faden).

**Eyre**, William (d.1765). Engineer. Fort Edward [New York State], 1755; Plan of Fort William-Henry, c.1755 and c.1756.

**Eyring**, Hans. Publisher in Wroclaw, Poland. Re-issued Martin Helwig's map of Silesia, 1627 (with Johann Perferts).

**Eysbroek**, H. Pilot. *Wind-Kaart van den Noord Atlantische Ocean*, 1856.

**Eysenbroek**, Pieter. Instrument maker in Haarlem, The Netherlands. Terrestrial globe, 1760.

**Eytzinger**, M. See **Aitzing**, Michael.

**Eyzinger**, M. See **Aitzing**, Michael.

**Ezhevsky**, F.K. (*fl.*early-18th century). Russian cartographer and geodesist.

**Ezuya** Shōhachi (*fl.*from 1837). Publisher based in Nara, Japan. Specialised in city plans and travel maps. *Washū Nara no Zu* [plan of Nara city], 1844. **Ref.** YAMASHITA Kazumasa *Japanese maps of the Edo Period* (Japan 1998) p.157.

F., A. Milano, *c*.1580.

F., B. Celestial globe, 1600.

F., D.N. Plan des Armées Autrichiennes et Françaises aux deux Côtés du Rhin, *c*.1744.

F., F. Vitenbergo, *c*.1565.

F., G.W. Indian country to the East and West of Mississippi, 1836?.

F., H. Woodblock cutter. Philipp Apian's *Chorographia Bavaria*, 1568.

F., I.G. Town of Appanoose, Hancock County, Illinois, 1857.

F., M. See **Florimi**, Matteo.

F., T. Chart of the coasts of England, *c*.1603 MS.

F., W. See **Fairbank**, William.

**Faber**, *Lieutenant* — von. Siege of Badajoz 1812 [Spain], 1850.

**Faber** [Fabre], A.F. Engraver for a late edition of work by Christophorus Cellarius, published as *Tabulas geographicas orbis veteris*, 1774.

**Faber**, Ernst (1839-1899). Missionary, translator and writer in China. Credited with various documents on geography and natural history.

**Faber**, J.H. Crown surveyor. Manuscript survey of British Honduras, 1858 (this survey was used as the basis for several later maps including that by E.C. Abbs, 1867).

**Faber**, Konrad (*c*.1500-1552 or 1553). Artist in Frankfurt am Main. Plan of Frankfurt am Main, 1552, used by Braun and Hogenberg in volume I of *Civitates*, 1572.

**Faber**, Martin (1587-1648). Architect, painter, and cartographer of Emden, Germany. Pascaerte Riviere Oost ende West Eems, 1642. **Ref.** LANG, W. 'Martin Faber's map of the Ems mouth' in: *Imago Mundi* IX (1952).

**Faber**, Samuel (1657-1716). Theologian and geographer, *b*. Altdorf, *d*. Nuremberg. Faber's name appears on some of the globes prepared by J.H. Andreae (contribution unknown); *Atlas scholastico-hodeoporicus*, Nuremberg 1712, *c*.1720, 1740 (engraved by C.W. Weigel the leder).

**Fabert**, Abraham. French magistrate in Metz. His description of the diocese and area of Metz or Pays Messin was used by J. Leclerc in *Théâtre géographique du Royaume de France*, 1619, 1622 and later, by Willem Blaeu in *Atlantis Appendix*, 1630, M. Tavernier 1632, Henricus Hondius 1633, J. Boisseau 1642, Jansson 1652.

**Fabius**, Nicasius. Ancient Flanders, 1641.

**Fabré**, *Lieutenant* —. Plan de la rade de l'île St. Pierre, Dépôt Général de la Marine 1824; charts for H.-Y.-P.-F. Bougainville, 1828.

*William Faden (1749-1836). Portrait by John Russell. (By courtesy of The Paul Mellon Centre for Studies in British Art, London)*

Fabre, A.F. See Faber, A.F.

Fabre, Jean. French cartographer. *Carte des pais reconquis... grises* [Valteline], *c.*1626; *Carte des Sevenes*, 1629.

Fabregat, J. Joaquín (1748-1807). Engraver for Domingo de Aguirre, 1775; for Vincente Tofiño de San Miguel, *Isla de Menorca*, 1786.

Fabretto [Fabretio], Raffaello (1619-1700). Antiquarian, artist and secretary. *De Aquis et Aqueductibus Romae* (3 folding maps), Rome 1680; Rome, *c.*1695; *Roma Vetus*, *c.*1710.

Fabri, François. Printer and publisher of Douai. Published Cornelis van Wytfliet's *Descriptionis Ptolemaicae Augmentum*, 1603 (a 'supplement' to Ptolemy's *Geographia* covering America and first published in Louvain, 1597), French translation published as *Histoire Universelle des Indes, Orientales et Occidentales* (3 volumes), 1605, 1607, 1611 (the two additional volumes prepared for this French edition were entitled *Histoire Universelle des Indes Orientales* and *La Suite de l'Histoire des Indes Orientales*).

Fabri, Johann Ernst (1755-1825). German geographer. *Handbuch der neuesten Geographie*, Halle 1797.

Fabricius, Antonius Bleyrianus. Cartographer. France, 1624.

Fabricius [Eseitz; Van Esents], David (1564-1617). Cartographer, astronomer and pastor of Essen. *Frisiae Orientalis*, Emden 1589; *Ost-Friesland*, 1598; Oldenburg, 1592; *Oost en West Vrieslant*, *c.*1610; Plan von Emden, 1619. **Ref.** MEURER, P. Fontes cartographici Orteliani... (1991) pp.141-142 [in German].

Fabricius, Herman. See Fabronius, Hermann.

Fabricius [Fabritius], Paulus (1529-1588). Astronomer, physician and cartographer, professor of Mathematics, Vienna. *Marchionatus Moraviae...* (6 sheets), Vienna 1568 (used by Ortelius, 1573); Austria (now lost), [n.d.]. **Ref.** MEURER, P. Fontes cartographici Orteliani... (1991) pp.142-143 [in German].

Fabricius, Dr. Wilhelm (1861-1920). Professor of ancient history and historical cartographer. *Atlas der Rheinprovinz*, 1894.

Fabris, Dominik Tomiotti de (1725-1789). Austrian general and cartographer.

Fabritius, P. See Fabricius, Paulus.

Fabronius, Hermann [*pseud.* Erasmus Sabinus Hofnerus; Harmin de Mesa; Herman Fabricius; Mosemann] (1570-1634). Theologian and poet. *Newe Summarische Welt Historia...* (with woodcut world map), Schmalkalden 1612.

Facius, brothers.
• Facius, Georg Siegmund (b.1750). Artist and engraver of Frankfurt am Main, in London from 1776. Worked for Johann Wilhelm Abraham Jaeger, 1789.
• Facius, Johann Gottlieb. Engraver, worked with brother Georg. *Carte Topographique d'Allemagne...*, J.W. Jaeger 1778.

Fackenhofen, Georg Carl Adam Joseph, *Freiherr* von (1745-1804). Cartographer of Würzburg. Regional map of the diocese of Würzburg, 1788-1791, engraved 1804.

Faden, William (1749-1836). Publisher, engraver and cartographer, 'at the Corner of St. Martins Lane, N°·5 Charing Cross, London'; 'Geographer to His Majesty & to his R.H. the Prince of Wales'. Born in London, son of Fleet Street printer William Faden [Mackfaden] and his wife Hannah. Faden was apprenticed to engraver James Wigley in 1764, becoming a Freeman of the Clothworkers' Company in 1771. He joined the family of the late Thomas Jefferys in partnership sometime before 1773 and took control in 1783. His contribution to the development of cartography was considerable, commissioning new surveys and publishing the work of mapmakers throughout Europe including J.F. Dessiou and V. Tofiño de San Miguel. On his retirement in 1823, one of his own former apprentices, James Wyld, took over the company. Engraver for *A Plan of the continuation of a proposed Collateral Cut or Branch to communicate with the intended Canal from Stourbridge to the Canal from the Trent to the Severn... in County of Stafford, to... Netherton in... County of Worcester*, *c.*1771; John Mitchell's *A Map of the British and French Dominions in North America* [still dated 1755], Jefferys & Faden *c.*1773 and later; *A New Map of the German Empire...*, London, Faden & Jefferys 1773; World,

1775; Bernard Ratzer's *Plan of the City of New York*, 1776; *The North American Atlas...*, 1776, 1777, 1778; *General Atlas*, 1778 and later editions; *A New Plan of the Island of Grenada*, 1780; *Plan of the Bay, Rock and Town of Gibraltar...*, 1781; *A Map of the Mediterranean Sea with the Adjacent Regions and seas in Europe, Asia and Africa*, c.1785; *A Map of the Frontiers of the Emperor and the Dutch in Flanders and Brabant...*, 1789; *A map of the Seven United Provinces with the Land of Drent and the Generality Lands*, 1789, 1794; *A Map of the Austrian Possessions In The Netherlands or Low Countries, with the Principalities of Liege and Stavel, &c.*, 1789; T. Jefferys's *Mappa ou Carta Geographica das Reinos de Portugal e Algarve*, 1790 (engraved by M. Rodrigues); *Atlas of the Battles of the American revolution...*, 1793; Dessiou's *Le Petit Neptune Français*, 1793, 1805; *A Map of the Post Roads of Germany and the Adjacent States*, 1795 (compiled by F.J. Heger); L.S. Delarochette's *A General Chart of the West India Islands...*, 1796; *Atlas minimus*, 1798 and later; *Nouvelle Carte de la Suisse*, 1799; A.F.G. de Palmeus's *A Topographical map describing the Sovereign Principality of the Islands of Malta and Goza...*, 1799; L.S. Delarochette's *Italy with the Addition of the Southern Parts of Germany...*, 1800; *Chart of the Straits between Denmark and Sweden...*, 1801 (from the surveys of C.C. Lous and J. Nordenankar); *An Entirely New & Accurate Survey of the County of Kent*, 1801 (the first published Ordnance Survey map); *A Topographical Map of the County of Essex*, 1804; *A Map of Europe* (3 sheets), 1804; *Plan of Cadiz*, 1805 (surveyed by V. Tofiño de San Miguel, engraved by Cooper); *A Correct Map of France...* (2 sheets), 1806 (reduced from Cassini); *A New General Chart of the Mediterranean Sea...*, 1806; *Chart of the Baltic Straits...*, 1807; *Plan of the City of Copenhagen...*, 1807; *A general chart of the West Indies and Gulf of Mexico*, 1808; *A Military Plan of the District of Colchester*, 1809; *A new Map of Spain and Portugal...*, 1810; *The Bay of Algeciras or Gibraltar...*, 1812 (from the surveys of V. Tofiño, engraved by Thomson); *Europe, Describing all the Changes of Territory... Confirmed By The Definitive Treaty of Paris... 1815* (2 sheets), 1816; *Map of the Central States of Europe... describing their New Limits... 1815*, 1816 (engraved by S. Hall); *Ottoman Dominions*, 1822; and many others. **Ref.** WORMS, L. in: *Dictionary of National Biography: Missing Persons* (Oxford 1993) p.218; CAMPBELL, T. *The Map Collector* 32 (1985) p.37 (note).

**Faehtz,** E.F.M. Worked with F.W. Pratt on the compilation and publication of *Real estate directory of the city of Washington, D.C. ...Containing a separate plat of each square in the city* (3 volumes), Washington D.C. 1874 (for Fitch & Fox).

**Fagan,** J. [& Son.]. Electrotypers of Philadelphia. Printers of H.F. Walling and O.W. Gray's *New Topographical Atlas of the State of Pennsylvania*, Philadelphia, Stedman, Brown & Lyon 1872.

**Fagan,** Lawrence (b.c.1825). Irish artist and surveyor active in the United States from c.1850. Produced township plans and county maps. Worked on county wall maps of New York State, Connecticut, Massachusetts and New Hampshire, 1852-1868; worked with James C. Sidney and James Neff on surveys for a map of Orange and Onondaga counties, New York, published by Sidney 1852, 1855; *Map of the Town of New Hartford, Litchfield Co. Conn.*, Philadelphia, Richard Clark 1852; *Tompkins County New York*, Philadelphia, Robert Pearsall Smith 1853; *Map of New Milford, Litchfield Co. Connecticut*, Philadelphia, Richard Clark 1853; *Map of the Town of Cornwall, Litchfield County, Connecticut*, Philadelphia, Richard Clark 1854; *Map of the Town of Newtown, Fairfield Co. Conn.*, Philadelphia, Richard Clark 1854; *Washington* [Connecticut], Clark 1854; map of Chenango County, New York, 1855 (with others); Schenectady County, New York, 1856 (with others); *Cheshire County, New Hampshire*, 1858; *Map of Berks County, Pennsylvania...* (Atlas), Philadelphia, Henry F. Bridgens 1861 (from a wall map of 1860), republished as *Township Map of Berks County, Pennsylvania*, 1862; and others.

**Fage,** Lieutenant (later *Major-General*) Edward, Royal Artillery (d.1809). *Plan of Rhode-Island*, 1777; *Plan of the adjacent coast to the northern part of Rhode Island*, 1778; *...line of defence for the Town of New-Port in Rhode Island*, 1778; *Original Plan of Brentons Neck*, 1779; *A plan of York and*

*Gloucester* [Virginia], London, J.F.W. Des Barres 1782; Woolwich, 1797 MS.

**Fage**, Robert. *A Description of the whole World*, 1658 re-issued as *Cosmography*, 1663, 1667, J. Overton 1671 (1671 edition sometimes contains John Overton's *A New and most Exact map of America*, 1668).

**Fahlberg**, Samuel (1758-1834). *Charta öfwer St. Barthelemy* [Leeward Islands], 1795 (used as the basis for *Chart of the Islands and Channels of St. Bartholomew, St. Martin, Anguilla, Dog & Prickly Pear, &c.*, London, Jas. Whittle and Richd. Holmes Laurie 1814).

**Faiano**, J. See **Fayen**, Antoine Jean du.

**Faidherbe**, Louis Léon César (1818-1889). Soldier, geographer and explorer in Senegal and French Sudan. Governor of Senegal 1854-1861, 1863-1865, in which capacity he organised the exploration of the colony and founded 'La Commission de la Carte de Sénégambie'. **Ref.** BROC, N. *Dictionnaire illustré des explorateurs Français du XIXe siècle: I Afrique* (1988) pp.137-138.

**Faille**, P. de. See **La Feuille**.

**Faindrick**, *sieur* —. *Carte générale de haute et basse Allemagne*, 1669.

**Fairbairn**, J. *Traveller's Guide through Scotland and its Islands* (with a general map of Scotland), Edinburgh 1798.

**Fairbairn**, James. Rector of the Academy, Bathgate. *Schoolroom map of North America*, Edinburgh, William and Robert Chambers 1861 (also published in London by W.W. Orr & Co.).

**Fairbairn**, William. Surveyors.
• **Fairbairn**, William (*fl*.1784-1807). Schoolmaster, thought to be the uncle of William (below) to whom he taught land surveying. *Plan of the lands and Barony of Galashiels*, 1795.
•• **Fairbairn**, Sir William (1789-1874). Nephew and pupil of William, above. Plan improvements Manchester, 1836.

**Fairbank** family. Quaker land surveyors of Sheffield, flourished over four generations. They produced canal plans, as well as numerous inclosure, tithe, estate, turnpike, insurance and poor rate surveys and maps. **N.B.** The Fairbank Collection, comprising maps, plans, surveyors' books, letters & papers is in Sheffield City Archives.
• **Fairbank**, William I (*c*.1688-1759). Schoolmaster and surveyor of Sheffield. Produced manuscript maps of parts of Derbyshire, Lincolnshire, Staffordshire and Yorkshire.
•• **Fairbank**, William Fairbank II (1730-1801). Schoolmaster and surveyor, son of William (above), father of William III and Josiah (below). Produced manuscript surveys of areas throughout England. Estate at Thurcroft, 1767 MS; plan of Sheffield, 1771; plan of Broxted, Essex, 1779 MS; *A Plan of the Dearn & Dove Canal* [proposed], 1793 MS; *Stainforth Cut to River Trent* [intended canal], 1793; *Parish of Sheffield*, 1795 (with William III) published by Wm Fairbank & Son 1796; *A Plan of the Town of Sheffield*, Sheffield, John Robinson 1797 (engraved by Thomas Harris, published in *Directory of Sheffield*); *Dun: Course of River from Sheffield to confluence with Ouse*, 1801 MS; *Dun: intended alterations in navigation of River*, 1803 **N.B.** This last work is either incorrectly dated or should be credited to son William III.
••• **Fairbank**, William III (1771-1846). Land surveyor, son of William Fairbank II. Between 1801-1828, worked in partnership with brother Josiah I (below) on manuscript surveys of parts of Derbyshire, Lincolnshire, Staffordshire and Yorkshire. Manor of Eckington, Derbyshire, 1803 MS (with Josiah); Plan of Sheffield, 1808 (with Josiah).
••• **Fairbank**, Josiah I (1778-1844). Land surveyor. Son of William Fairbank II, father of William IV and Josiah II (below). Worked with brother William III (above). Manor of Eckington, Derbyshire, 1803 MS (with William III); *Plan of Sheffield*, 1808 (with William III); inclosure map of Eckington, 1822 MS.
•••• **Fairbank**, William IV (1805-1848). Son of Josiah I. Produced manuscript maps of Derbyshire.
•••• **Fairbank**, Josiah Forster II (1822-1899). Land surveyor, son of Josiah I.

**Fairburn**, John. Publisher, geographer and map seller at 146 Minories, also noted at N$^{o.}$2 Broadway, Ludgate Hill, London (1820). *A New Map of France divided into Metropolitan Circles and Departments agreeable to the Decree of the National Assembly 1790*, 1794; *London and Westminster*, 1795 (engraved by E. Mogg), 1796 (inset 'Proposed Wet Docks' added), 1797

('Rules of the Fleet Prison' added), re-issued 1800, 1801 (docks added), 7th edition with further additions published 1802, 1803, 1806; *Fairburn's Map of the Country Twelve Miles Round London*, 1798 (engraved by E. Bourne), boundary of the Penny Post added 1800 (also published on calico as *The Travelling Handkerchief*, 1831); North America, 1798; *The Junction Of The Counties In England and Wales*, 1798 (instructional card game for children incorporating county maps copied from *Cary's Traveller's Companion*, 1789); *Fairburns Atlas in Miniature*, c.1800; *Fairburn's Map of the Scene of Action on the German, French, Dutch, Swiss and Italian Frontiers... Exhibiting at One View the whole Theatre of War...*, 1805; Spain and Portugal, 1808; *Fairburn's Circular Plan of London*, c.1820, and others.

**Fairchild**, John F. Atlas of Mount Vernon & Pelham, New York, 1899.

**Fairchild**, W. (*fl.*1743-1784). Surveyor, prepared manuscript maps of areas in all parts of England including Shalford, Essex, 1760 MS; Manors of Hillesden, and Cowley Buckinghamshire 1763 MSS.

**Fairfax**, *Captain* Henry (1837-1900) RN. Admiralty Charts of Kerguelen Islands [Indian Ocean], 1874-1875.

**Fairfax**, Wilson M.C. Engineer and draughtsman. *Map of such portions of the Boundary Lines between the States of Maine and New Hampshire and the British Provinces in North America*, 1841 MS (with R.D. Cutts); *Map of the Boundary Lines between the United States and the adjacent British Provinces from the mouth of the River St. Croix to... the River St. Lawrence near St. Regis...*, 1843 (with T.J. Lee; engraved by W.J. Stone); draughtsman for *Patapsco River and the approaches...*, 1859 (with M.J. McClery and A. Palmstein).

**Fairfield**, G.A. Surveyor for the U.S. Coast Survey. Triangulations for charts of St. John's River, Florida, 1853-1856.

**Fairfowl**, George. Surgeon, HMS *Dromedary*. Sketch of the Coast of New Zealand, 1820.

**Fairholme**, *Captain* William Ernest. Map to illustrate notes on Races and Religions in *Turkey-in-Asia* [sic], 1897 (manuscript additions to Stanford's *The States and Provinces of the Balkan Peninsula*).

**Fairlove**, Ich$^d$ [Ichabad; Ichabod]. Surveyor. *A True Plan of the City of Excester* [Exeter], Edward Score 1709 (engraved by Joseph Coles).

**Fairholme**, *Captain* William Ernest. One of the boundary commissioners for the Turco-Greek frontier. *Map to illustrate notes on Races and Religions in Turkey-in-Asia*, 1897.

**Fairman** brothers. Engravers from Newtown, Connecticut. **Ref.** RISTOW, W.W. The Map Collector 36 (1980) p.22.
• **Fairman**, Gideon (1774-1827). Engraver of maps, charts, bank notes and vignettes, apprenticed to Isaac and George Hutton of Albany (1794), set up on his own in Albany in 1796. Moved to Newburyport, Massachusetts and worked with William Hooker until 1810. Amongst other maps they engraved charts for Edmund March Blunt (Hooker's father-in-law). In 1810 Fairman moved to Philadelphia, where he joined Murray & Draper, the bank note engravers. He was also a friend and business associate of engraver and inventor Jacob Perkins. *A Map of the Oneida Reservation*, 1796; *A Map of the Middle States...*, 1798; Ontario and Steuben Counties, 1798; De Witt's *A Map of the State of New York*, 1802; *Map of the State of New York* (smaller version), 1804; N. Bowditch's *Chart of the Harbours of Salem...*, 1806 (with Hooker, revised edition published 1834).
•• **Fairman, Draper, Underwood & Company**. Engravers of Philadelphia specialising in bank notes. One of the names assumed by the partnership of Gideon Fairman and Murray & Draper and others. **Ref.** RISTOW, W.W. American maps and mapmakers (Detroit 1986) pp.307-308; **N.B.** See also **Murray, Draper, Fairman & Company**.
• **Fairman**, David (1782-1815). Brother of Gideon. Artist and engraver specialising in bank notes. One of the engravers of maps for Samuel Lewis and Aaron Arrowsmith's *A New and Elegant General Atlas*, Philadelphia, Conrad 1804 and later editions (atlas to accompany J. Morse's *Geography*).
• **Fairman**, Richard (1788-1821). Engraver working in G. Fairman's company c.1820. Possibly his brother.

**Fairman**, Francis. English estate surveyor. *A true mapp of several pieces of Woodland and*

*William Faithorne the elder (1616-1691) in 1650. Portrait attributed to Robert Walker. (By courtesy of the National Portrait Gallery, London)*

*Arable land lying in the parishes of Wrotham, Leybourne and West Peckham* [Kent], 1726.

**Fairman**, George. Photolithographer of Queensland, Australia. Darling Downs 40 chain map series, 1884-1890; Moreton 20 chain map series, 1885-1886.

**Fairweather**, *Captain* Patrick. Chart entrance Old Calabar, Dalrymple 1790.

**Fairweather**, W.G. Surveyor to Northern Rhodesia government. Surveyed farms and topographically mapped Rhodesia for first time. **Ref.** *A Colonial Surveyor at Work: The Field Diary of W.G. Fairweather...1913-1914.*

**Faistenauer** [Feistenhauer], Johann (1610-1636). Draughtsman. *Das Landt und... Stifft Berchtolsgaden* [Tyrol] (woodcut, 4 sheets), 1628.

**Faithorne**, William *the elder* (1616-1691). Engraver of prints, titlepages, bookplates, portraits and maps 'at the Sign of the Drake against Palsgrave's Head Tavern', London (1650-*c*.1680), later Printing House Yard, Blackfriars. Exiled to Paris as a Royalist during the Interregnum. Richard Newcourt's *An Exact Delineation of the Cities of London and Westminster* (12 sheets, woodcut), 1658 (only one complete copy known, Bibliothèque Nationale); *The Art of Graving and Etching*, 1662; River System between Mediterranean and Toulouse, 1670; Augustine Herrman's *Virginia and Maryland As it is Planted and Inhabited this present Year 1670...* (4 sheets), London, A. Herrman & T. Withinbrook 1673.

**Fakymolano**. Brother of Sultan of Magendanao. Map of Magendanao, *c*.1774 MS.

**Falaki-al Mahmud**, *pasha* (1815-1885). Egyptian geographer and astronomer. Directed the team which (on the orders of the Khedive Said) mapped Egypt, 1860-1885.

**Falbe**, *Lieutenant* —. *Côtes de Sicile et Tunis*, 1840-1856; *Régence de Tunis*, 1842, 1857; *Grönland Vestkyst*, 1863-1866.

**Falcão**, Luís de Figueiredo. Codex of Portugal (43 maps), 1617 MS.

**Falckenstein**, Albert de. Ingeniero del Estado. *Atlas géographique de la République de Pérou*, 1862-1865.

**Falckenstein**, E.V. von. See **Vogel** von Falckenstein.

**Falckenstein** [Falkenstein], Johann Heinrich von [de] (1682-1760). *Delineatio Nordgoviae veteris*, 1733; numerous maps for Homann's Heirs 1753 and later.

**Falda**, Giovanni Battista [da Valduggia] (1648-1678). Italian architect and engraver working in Rome. *Provincie Unite*, 1672; Città di Roma, Giovanni Domenico de' Rossi 1676; *La Scandinavia*, 1678.

**Faleiro** [Falero], Francisco (*fl*.early-16th century). Portuguese mathematician and cartographer.

**Faleiro** [Falero], Rui (mentioned 1523). Portuguese navigator and cartographer.

**Faleleeff** [Faleleyev], —. One of the calligraphy engravers for Pydyscheff's *Atlas géographique de l'Empire de Russie, du Royaume de Pologne et du Grand Duché de Finlande...* (83 sheets), 1820-1827, 1829, 1834.

**Falero**, F. See **Faleiro**, Francisco.

**Falero**, R. See **Faleiro**, Rui.

**Faleti**, Bartolomeo. Publisher in Rome. G.A. Dosio's map of Rome, 1564 (engraved by S. de Ré).

**Falger**, Anton (1791-*c*.1855). Draughtsman, engraver and lithographic artist from Tyrol, worked as a draughtsman in Munich and Weimar from *c*.1820, then returned to Tyrol. Lithographic engraver for some map sheets of *Gefürstete Grafschaft Tyrol und Vorarlberg*, Innsbruck, F. Unterberger 1826 (includes town plans and vignettes showing local scenes and costumes).

**Falk**, —. *Königsberger Land Kreise*, 1841.

**Falk**, Miksa (1828-1908). Hungarian historian. Revised Antal Vállas's *Új kézi és iskolai atlasz*, 1855 (first published 1845).

**Falk**, R. Printer and publisher. *Stadt von Paris*, 1870; *Karte von Frankreich*, 1871.

**Falka**, Sámuel (1766-1826). Hungarian map engraver, born Fogaras [now Fagaras, Transylvania, Romania], died in Buda.

Worked with Ferenc Karacs on the maps of János Lipszky, 1803 and later.

**Falkema**, Jacob. Engraver of Amsterdam. Cornelis Danckerts's *Nova totius terrarum orbis tabula*, *c*.1695.

**Falkenstein**, E.V. von. See **Vogel** von Falckenstein.

**Falkenstein**, J.H. de. See **Falckenstein**, Johann Heinrich von.

**Falkner**, George [& Sons]. Lithographic printers of Manchester. Manchester Ship Canal, 1888.

**Falkner**, Thomas (1708-1784). Jesuit missionary. *Description of Patagonia*, 1774 (contains 2 maps engraved by T. Kitchin).

**Falle**, P.J. Madras Roadstead, 1876.

**Falledo y Rivera**, Vicente. Colombia, *c*.1805 MS.

**Fallex**, Maurice. *Maroc*, Paris, Delagrave *c*.1913; *Allemagne, royaume de Prusse et empire d'Autriche...*, Paris, Forest *c*.1917.

**Fallen**, O. Oesterreich (3 sheets), Vienna 1822.

**Fallon**, Ludwig August, *Freiherr* von (1776-1828). Austrian general and cartographer, chief of Austrian Topographical Bureau. *Das Oesterreichische Kaiserthum mit beträchtlichen Theilen der angrenzenden Staaten* (9 sheets), Vienna 1822.

**Fallowes**, Benjamin (*fl.*1714-1720). Quaker farmer and estate surveyor in Maldon, Essex. His numerous manuscript estate plans include Dengie, 1714; Rivenhall Place estates, Essex (*c*.26 maps for the Western family), 1715-1717; Abington Hall, Cambridgeshire, 1716; surveys of the estates of the Earl of Abergavenny [Monmouth, Hereford and Worcester], from 1718. **Ref.** MASON, A.S. Essex on the map (1990) pp.48-50.

**Falmar**, —. *Road from Chagres to Panama*, 1851.

**Fan** Cho (*fl.*9th century). Chinese geographer.

**Faneau** [Faveau] **Quesada**, *Don* Antonio. California, 1768 MS; *Chart of Balabac & East coast of Palawan*, 1775.

**Fanelli**, Francesco. Plan of Athens, 1707 (after Verneda and Spon).

**Fangel**, H. Lithographer. *Plan von Hadersleben*, 1862.

**Fannin**, N.W. Eustace. Land surveyor in South Africa. *Boundary Survey...* [Orange Free State and Natal] (9 sheets), 1884 (with A.H. Hime, '...showing surveys by Orpen... [and] G. Tatham...').

**Fannin**, Peter, *Master*, RN. Sailed on the *Adventure*, producing a number of surveys and coastal views. Retired from the Navy in 1775 to found a school of navigation on the Isle of Man. Best known for his *Plan of the Isle of Man*, 1789.

**Fanning**, —. *Map of New York City*, 1857.

**Fanning**, J.T. City of Manchester, New Hampshire, 1873.

**Fanning**. See **Ensign**, Bridgman & Fanning, also **Thayer**, Bridgman & Fanning.

**Faqih**, Ahmad bin-Muhammed (*fl.*1st century). Arab cartographer. Atlas of Islam, 903.

**Farabi** [Alpharabius], Abu Nasr, Muhammad, Ibn-Muhammad (872-950). Called also the Second Teacher (Aristotle being the first). One of the greatest Islamic philosophers, mathematicians and musicians of Persian or Turkish origin. Responsible for some 100 books and treatises including *Book of latitudes and longitudes*.

**Farey**, John, father and son.
• **Farey**, John, *the elder* (1766-1826). Mineral surveyor, geologist and estate surveyor of London, noted at 12 Upper Crown Street, Westminster (1809). Produced estate and mineral maps of areas throughout England and Scotland. Section of the Principal Strata of England, 1808 (earliest geological section across country); Great Clacton, 1809 MS; Survey of Derbyshire, 1811-1813.
•• **Farey**, John *the younger* (1791-1851). Civil engineer and draughtsman, son of John, *the elder*. Drew Derbyshire for his father's Survey of Derbyshire, 1811-1813.

**Farey**, William. South West District St. Pancras, 1820.

**Farghani** [Alfraganus], Abu al-Abbas, Ahmad, Ibn-Muhammad, Ibn Kathir (*fl.*9th century). Mathematician and astronomer. *Usul Ilm al-Nujum* [The Principles of Astronomy]; *Fusul al-Thalathin*, both published in Latin in Ferrara, 1493.

**Fargwarson**, Andrei Danilovich (1675-1739). Scottish born mathematician, worked in Russia from 1699. Taught at the Moscow School of Mathematics and Navigation, and the St Petersburg Naval Academy. Surveys for the Moscow-St Petersburg road, 1709.

**Fargola**, F. (*fl.*19th-century). Engraver of Naples.

**Faria**, J.C. de Sá e. See **Sá e Faria**, José Costodio de.

**Faribault**, Eugène Rudolphe (*fl.*1898-1908). Surveyor. Forest Hill Gold District, 1898; *Gold Fields of Nova Scotia*, 1906 (with a map of the province of Nova Scotia).

**Faries**, Robert. Engineer. *Map of the Sunbury and Erie Rail Road and its Connections*, 1870.

**Farin**, Joaquin Gonzalo. Engineer, National Mining Corps. *Carta Geografico-Minera de la Provincia de Huelva* [Spain], 1870 (lithographed by Alemana, Seville).

**Farini**, G.A. Routes in Kalahari Desert, London 1886.

**Farler**, *Reverend* J.P. Maps of the Tanganyika Territory. *The country of the Wabondei, Wasambara and Wakalindi*, 1877; *Map of the southern Usambara*, 1878; *Native routes to the Masai Country and to the Victoria Nyanza*, 1882.

**Farley**, Minard H. *Farley's Map of the Newly discovered Tramontane Silver Mines in Southern California and Western New Mexico...*, San Francisco, W. Holt 1861.

**Farley**, William *RN. Chart of The Rabbit Islands* [Dardanelles], 1820, 1830.

**Farmer**, family (*fl.*1825-1915). Publishers of Detroit. Produced numerous maps of Michigan, Wisconsin and states in the upper Mississippi and Great Lakes. **Ref.** RISTOW, W.W. *American maps and mapmakers* (Detroit 1985) pp.273-277.
• **Farmer**, John (1795-1859). Surveyor and teacher from Saratoga County, New York. After compiling a map of Michigan for Orange Risdon in 1824 he established his own map making company in Detroit in 1825. He sold the copyright in his maps to J.H. Colton in 1835, but returned to copper engraving and map publishing in 1844. In the ensuing years he built up a successful map engraving and publishing company, which on his death was continued by his widow and three children. *Map of the Surveyed Part of the Territory of Michigan...*, Orange Risdon 1825 (engraved by Rawdon Clark & Company, J. Farmer also published a similar version of his map at the same date and with the same title, but engraved by V. Balch and S. Stiles); draughtsman for *Map of the United States Road from Ohio to Detroit*, 1825 (by P.E. Judd, S. Vance and A. Edwards); *Map of the Territories of Michigan and Ouisconsin*, 1830, 1835 (engraved by Rawdon Clark & Co.); *Emigrants Guide, or Pocket Gazetteer of the Surveyed Part of Michigan*, Albany 1830, 1831 (with a new edition of Farmer's 1825 map of Michigan); *Plat of the City of Detroit as laid out by the Govr. and Judges*, 1831 (lithography by Bowen & Company); *Map of the Surveyed Part of Michigan*, New York, J.H. Colton 1843 (engraved by S. Stiles & Co.); *Map of the State of Michigan and the Surrounding Country...*, Detroit 1844 and many later editions to 1873 (edition of 1873 published by Silas Farmer); *A geological map of the Mineral Region...*, 1847 (from the chart by Lieutenant Bayfield); *Map of the States of Michigan & Wisconsin embracing a great part of Iowa and Illinois... with a Chart of the Lakes*, 1848 and later editions; *Map of the Southern Part of Michigan*, 1855 (engraved by S. Stiles & Co.); and many others.
• **Farmer**, Roxanna [R. Farmer and Company] (1800-1890). Widow of, and joint successor (with her three children) to, John.
•• **R. Farmer & Company** (*fl.*1859-1863). Publisher of 85 Monroe Avenue, Detroit, 'successors to the late John Farmer' (1862). *Map of the States of Michigan and Wisconsin... projected and engraved by John Farmer...*, Detroit, R. Farmer & Co. 1861; *Farmer's Rail*

*Road & Township Map of Michigan and Chart of the Lakes*, 1862 (drawn and engraved by John Farmer, completed and published by R. Farmer & Co.).

•• **Farmer**, Esther. Daughter of John and Roxanna, briefly held an interest in the Farmer publishing company from the death of her father until she sold her share to Silas in 1866.

•• **Farmer**, John H. Dentist, son of John and Roxanna. Inherited an interest in the Farmer publishing company on the death of his father. Silas bought his share in 1866.

•• **Farmer**, Silas [S. Farmer & Co.] (1839-1902). Publisher and historian of 35 Monroe Avenue, Detroit (1863-1866); 'Corner of Monroe Avenue and Farmer Street' (1867); '31 Monroe Avenue. Corner of Farmer Street' (1872, 1873), successors to John Farmer and R. Farmer & Co. Silas took over the company in 1863, and later bought the interests of his mother, brother and sister. *Guide Map of the City of Detroit*, 1863; republished *Farmer's Rail Road & Township Map of Michigan and Chart of the Lakes*, 1863, 1864, 1866 (new plate), 1867, 1868 (new plate), 1871, 1872, 1873 (first published by R. Farmer 1862); *Farmer's Map of Lake Superior and the Mining Regions*, 1867, 1868, 1872; *Map of Lake Superior*, 1867; *Rail Road and Township Map of Wisconsin*, 1867; *The New Official State Map of Michigan...*, 1873 (engraved by Calvert Lithography Company).

••• **Farmer**, Arthur John (*b*.1876). Son of, and successor to, Silas Farmer. He re-issued existing Farmer maps, and also prepared road maps and atlases of Michigan. Map plates were sold to C.M. Burton, *c*.1915.

**Farmer**, *Captain* George (1732-1779). Information for chart of Bay of Bengal, 1794.

**Farnau**, J. (*fl*.1680-1698). Heligoland, [n.d.].

**Farnham**, Thomas Jefferson (1804-1848). Publisher and geographer of New York. *Travels in the Great Western Prairies...*, 1841, 1843; *History of Oregon Territory...* (with an untitled woodcut map of Oregon Territory), New York 1844, 1845; *Map of the Californias*, New York 1845 (published in *Travels in the Californias, and Scenes in the Pacific Ocean*, and in S.E. Morse and S. Breese's *The Cerographic Atlas of the United States*); map of Mexico, Texas and California in *Mexico: its geography - its people - and its institutions*, 1846.

**Farquhar**, Alexander. Road surveys in Aberdeenshire. *Plan... of a proposed... turnpike from Bogniebrae to Milton of Rothiemay*, 1841 (with A. Ogg, lithographed by J. Henderson); *Plan... of a proposed turnpike... from Rothimay to Braco*, 1841 (with A. Ogg, lithographed by Henderson); *Plan... of a proposed turnpike from Port Elphinstone to Kemnay*, 1841 MS (with A. Ogg); Plan of Huntly, 1845 (with A. Ogg).

**Farquharson**, John. Surveyor of Invercald. *A map of the Forrest of Mar*, 1703 MS.

**Farquharson**, John. North side of Loch Tay [Perthshire] (18 sheets), 1769 MS (John McArthur produced 24 sheets of the south side).

**Farquharson**, *Colonel* John [born Iain Cosmo MacPherson]. Director General of the Ordnance Survey, 1894-1899. **Ref.** OWEN, T. & PILBEAM, E. *Ordnance Survey - Mapmakers to Britain since 1791* (Southampton 1992) pp.80-81, 84.

**Farrand**, William. Donnabruoke & Tannee [Down Survey], 1655-1657; Barony of Rathdown, Dublin 1656-1657 (with William Storck).

**Farrant**, R.G. Gosha [Somalia], 1896 MS.

**Farrell**, *Lance Corporal* A.J. RE. Manuscript plan of Gibraltar, 1894.

**Farren**, R. *The Thames, the Isis, the Cam*, 1884.

**Farrer** [Farrar; Ferrar; Ferrer], father and daughter.

• **Farrer**, John (*d.c*.1651). Resident of Little Gidding, Huntingdonshire. Involved with the Virginia Company. Manuscript map of Virginia, 1650 (New York Public Library); *A mapp of Virginia discouered to ye Falls...*, London, J. Stephenson 1651, (worked up from the manuscript, engraved by J. Goddard) various states of the map appeared in the 3rd edition of Edward Williams's *Virgo triumphans: or, Virginia richly and truly valued*, London, John Stephenson 1651, and in Edward Bland's *The Discovery of New Brittaine*, J. Stephenson 1651. **Ref.** Cumming, W.P. *The southeast in early maps* (1998 edition revised by L. De Vorsey) pp. 148-151; BURDEN, P. *The mapping of North America* (1996) pp.387-398.

•• **Farrer**, Virginia (1628-1668). Daughter of John above. Published the 3rd and 4th states of her father's map (the title was amended to read 'Hills' rather than 'Falls' for the 4th state, c.1652), the plate passed to Peter Stent (who may have issued it c.1658), then to John Overton who issued it c.1665-1673. **Ref.** RITZLIN, M. 'Women's contributions to North American cartography' in Meridian N°2 (1989) pp.5-16.

**Farrington**, J. Plan Makasar, 1815 MS.

**Farris**, J.J. Publishers of T.M. Fowler's bird's-eye view of High Point, North Carolina, 1913.

**al-Farsi.** See **Istakhri**.

**Farsky**, — (c.1864-1912). Lithographer and printer of Prague. Umgebung von Teplitz [now Teplice, Bohemia], c.1875.

**Fasbender**, Karl. *Hölzels Automobilkarten* [Austria-Hungary], Vienna, E. Hölzel 1913-1915 (with Theodor Bradaczek).

**Fasching**, Antal (1879-1931). Hungarian engineer who in 1906 suggested use of an oblique cylindrical map projection. **Ref.** 'Fasching Antal' Bulletin géodesique XXXI (1931) [in Hungarian].

**Fassbinder**, P. *Palästina zur Zeit Jesu Christi*, 1899.

**Fatio de Duillier**, brothers.
• **Fatio de Duillier**, Jean-Christophe (1656-1720). Swiss engineer born at La Forge, France, worked in Geneva. One of the sources for Antoine Chopy's *Carte Du Lac De geneve...*, 1730.
• **Fatio de Duillier**, Nicholas (1664-1753). Mathematician and astronomer, born in Basle, worked in Geneva and England.

**Fatout**, —. Publisher of 'Bvd. Poissonniere 17', Paris. Alexandre A. Vuillemin's *Atlas illustré de géographie commercial...*, 1860.

**Fatout**, Henry B. Atlas of Indianapolis and Marion County, Indiana, 1889.

**Fauche**, Pierre François. Bookseller in Hamburg. *Carte de Russie...*, 1794.

**Fauche**, Samuel (1732-1802). Swiss printer and publisher of Neuchâtel.

**Faucit**, T. See **Fawcet**, Thomas.

**Fauck**, A. *Map of Idaho Territory*, 1865.

**Faucou**, Lucien. *Plans historiques de Paris*, 1889.

**Faujas de Saint-Fond**, Barthelemi. (1741-1819). French geologist. *Volcans... du Vivarois*, 1778.

**Faul**, August. *Map and profile of the Orange and Alexandria Rail Road*, New York, Ackerman Lith. c.1854; Swann Lake and Aqueduct [Baltimore], 1862.

**Faulhaber**, Albrecht. Map of Spandau, 1685.

**Faulhaber**, Johann Matthias, *the younger* (1670-1742). Cartographer of Ulm.

**Faulkner**, George (c.1699-1775). Printer, publisher and bookseller, worked with James Hoey Sr. at Christchurch Lane, Skinner's Row, Dublin, (1728); Essex Street, Dublin (from 1730-1765); 15 Parliament Street, Dublin (1766-1775). Published the works of Dean Swift. Sold Thomas Kitchin and Thomas Jefferys's *Small English Atlas*, 1751; C. O'Conor's *Dissertation on the History of Ireland* (map based on Ptolemy), 1766 (first printed by James Hoey Sr., and published by Michael Reilly 1753).
• **Faulkner & Smith**. Partnership of George Faulkner and his nephew Samuel Smith. Mapsellers of Dublin. Sold Thomas Jefferys's *A New and Accurate Map of the Kingdom of Ireland...*, 1759 edition (also sold by Thomas Jefferys in London).

**Faulkner**, Thomas. *An Historical and Topographical Description of Chelsea and Its Environs*, 1810 (contains *A Map of Chelsea Surveyed in 1664 By James Hamilton Continued To The Year 1717*).

**Faunthorpe**, Reverend John Pincher FRGS (c.1844-1924). Priest, teacher and geographer. *Elementary Physical Atlas*, 1867; produced a series of county geographies for schools which were published by George Philip & Son c.1872-1873 including *The geography of Sussex, for use in schools*, London & Liverpool, George Philip & Son 1872; *The geography of Bedfordshire, for use in schools*, London & Liverpool, George Philip & Son

1873; and others similarly titled including *Lincolnshire*, 1873, and *Hertfordshire*, 1873 (the maps, which were by John Bartholomew, were used for *Philips' County Atlas*); *Outline Atlas*, 1874; *Projection Atlas*, 1874.

**Faure,** —. Hydrographer on the Freycinet expedition to Australia, 1801-1803. Contributed to charts which were published in Louis Claude de Saulces de Freycinet's *Voyage de découvertes aux Terres Australes*, Paris 1811-1812, 1824.

**Faure,** A. Carte géologique des parties de la Savoie, du Piedmont et de la Suisse, 1862.

**Faure,** *Captain* Antonio de la. Caminos... Provincia de la Guayana, 1783 MS.

**Faure,** Guillaume Stanislas (1765-1826). Printer, administrator and politician of Le Havre. Associated with Nouveau flambeau de la mer, 1822-1824. **N.B.** Compare with **Faure**, Pierre and **Faure**, P.J.D.G., below.

**Faure,** P. Surveyor. *Plan de la paroisse royale de St. Germain l'Auxerrois*, 1739.

**Faure,** P.J.D.G. Printer, chart and bookseller of Le Havre. Publisher of some editions of R. Bougard's *Le petit flambeau de la mer...*, 1763, 1770, 1789 (first published J. Gruchet, 1684, then by Pierre Faure). **Ref.** PASTOUREAU, M. *Les atlas français XVIe-XVIIe siècles XVIe-XVIIe siècles* (1984) pp.75-79 [in French].

**Faure,** Pierre. Printer and bookseller of Le Havre. In partnership with the widow of G. Gruchet as publisher of some editions of R. Bougard's *Le petit flambeau de la mer...*, 1742, 1752 (first published J. Gruchet, 1684). **Ref.** PASTOUREAU, M. *Les atlas français XVIe-XVIIe siècles* (1984) pp.75-79 [in French].

**Fausel,** C. Bird's-eye views of U.S. cities. *Graniteville* [Massachusetts], Troy, New York, L.R. Burleigh 1886; *Antwerp* [New York], 1888; *Plattsburg* [New York], 1899.

**Fauth,** Philipp Johann Heinrich (1867-1941). Created a large scale chart of the moon, 1932 (published by his son Hermann in 1964).

**Fauvel,** Albert-Auguste (1851-1909). French colonial officer, economic geographer and naturalist based in China. He published several local and geographical studies including *The province of Shantung, its geography and natural history*, 1876 (in the *China Mail*). **Ref.** BROC, N. *Dictionnaire illustré des explorateurs Français du XIXe siècle: II Asie* (1992) pp.142-143.

**Fauvel,** Maurice (1851-1908). French geographer.

**Fauvet,** —. *Nouvelle mappemonde*, 1805.

**Fava,** Duarte Jozé. Captain of Engineers. Plano do Lisboa, 1807-1826.

**Favanne,** — de. Engraver for J. de Beaurain's *...Ducheé de Mantoue*, 1734.

**Favarger,** H.F. Golfo Adriatico, 1854.

**Favart,** Jean-Baptiste. Manuscript plans including: *Plan de L'Isle d'Aix*, 1714; *Plan du Fouras*, 1714.

**Faveau Quesada,** A. See **Faneau** Quesada, Don Antonio.

**Favenc,** Ernest. Australian journalist, explorer and historian. Map of Tasman's Track, 1885, in *The History of Australian Exploration from 1788 to 1888*, 1888; *Map showing the distribution of Australian Aborigines*, 1892 MS.

**Favoli** [Favolius], Hugo [Ugo] (1523-1585). Dutch poet, doctor and translator, *b*. Middelburg, *d*. Antwerp. Latin translation of Ortelius-Galle miniature atlas, 1585.

**Favre,** Jean Alphonse. Geologist in Geneva. *Carte géologique de la Savoie*, 1862; *...du canton de Genève*, 1878; *Glaciers des Alpes suisses*, 1884.

**Favre,** Maximin. *Chart of the Mediterranean Sea*, 1772-1774.

**Favrot,** —. French cartographer. Cours du Rhin, 1696-1697.

**Fawcet** [Faucit], Thomas. Published Cornelis Vermyden's *A Discourse Touching the Drayning The Great Fennes* (with map), 1642.

**Fawcett,** John (*fl*.1818-1828). Estate surveyor, New Ormond St., London, produced canal and estate maps of Essex and London. Corringham, 1818 MS.

**Fawcett**, T. Gold Region, Yukon, 1898.

**Fawekner**, W.H., *Second Master*, HMS *Phoenix*. Draughtsman for E.A. Inglefield's *Chart Shewing the North West Passage...*, Hydrographic Office 1853 (2 versions, one lithographed by J. Walker, the other by Day & Son).

**Fawkner**, John Pascoe (1792-1869). Australian pioneer of Victoria. Plan of the town of Melbourne, 1841.

**Fay**, *Général* Charles Alexandre (*b*.1827). Writer. *Théâtre des opérations en Bohème*, 1867; *Campagne de 1870*, *c*.1889.

**Fay**, Theodore Sedgwick (1807-1898). Diplomat and printseller of New York. *Outline of geography*, 1867; *Atlas of Universal Geography*, 1869-1871; *First Steps in Geography*, 1873.

**Fayano**, J. See **Fayen**, Antoine Jean du.

**Fayard de la Brugère**, Jean Arthème (*b*.1836). *Atlas universel*, 1877 (with Alphonse Baralle).

**Faye**, Hervé Auguste Etienne Albans (*b*.1814). French astronomer. *Leçons de cosmographie sur l'origine du monde*, [n.d.].

**Fayen** [Faiano; Fayano; Fayanus], Antoine Jean du (*c*.1530-*c*.1616). Physician at Limoges. Produced the first known map of the Limousin, published as *Totius Lemovici et confinium provinciarum* in Maurice Bouguereau's *Le Théâtre François*, 1594 (plate re-issued by J. Leclerc, 1619, copied by Ortelius, Blaeu, Tavernier and others).

**Fayolle**, —. *Plan de la rade de Dantzig*, 1815.

**Fayram**, Francis. One of the publishers of Camden's *Britannia*, 1730 edition (with maps by Robert Morden).

**Fayram**, J. *Stockholm ab Oriente*, 1725 (engraved by J. Mynde).

**Fazari**, Ibrahim, Ibn-Habib, Abu-Ishaq (*fl*.8th century). Mathematician. Translated the Indian astronomy into Arabic, entitled *al-Sind Hind al-Kahir*.

**Fazello** [Fazelli], *Padre* Tomaso (1490-1570). Sicilian priest and historian. *Isola... di Sicilia*, 1628; *Historia di Sicilia*, Palermo 1628.

**Fea**, *Captain* Peter. Manuscript charts of the west coast of Madagascar, 1765, 1784 published London, Dalrymple *c*.1792.

**Fearnside**, T. Manuscript plan of the Citadel at Hull, 1745 MS.

**Fearon**, Samuel. Shipwright, hydrographer and publisher of Liverpool, collaborated with John Eyes on *A Description of the Sea Coast of England and Wales*, 1736-1737, published 1738 (thought to be the first printed charts to show Greenwich as Prime Meridian); Furness & Anglesea, 1738; corrected chart Liverpool Bay, 1757.

**Featherstone**, Henry. See **Fetherstone**, Henry.

**Featherstone**, Joseph. Deeping Fen [Lincolnshire], 1763 (from the surveys of Vincent Grant).

**Featherstonhaugh**, George William. British Boundary Commissioner in North America. Surveys in New Brunswick, Quebec and Maine, *c*.1839 (with R. Mudge).

**Featherstonhaugh**, James D. Surveyor for the Boundary Commissioners. Worked with W.E. Delves-Broughton on surveys of the Quebec/Maine border including *A Map of... the Country in dispute with the United States...*, 1842, 1848 (engraved James Wyld).

**Febiger**, *Rear-Admiral* John Carson, US Navy. Yellow Sea, 1868.

**Febure**, Le. See **Le Febure**, Valentin.

**Febvre de la Barre**. See **Barre**, Jean François Le Febvre de la.

**Fedchenko**, A.P. Russian cartographer. Maps of Asiatic Russia, 1870-1872.

**Feddes** [Harliensis; Harlingensis], Pieter (1586-1634). Poet, artist, engraver and publisher of Harlingen [Friesland].

**Feer** [Fehr], Johannes (1763-1823). Fortifications engineer, astronomer and cartographer from Rheineck, Switzerland, worked in Zurich and Meiningen. Also carried out some of the surveys for the Dufour map of Switzerland. *Geometrischer Plan des* *ganzen Gemeindebahns von Wollishofen*, 1788; *Specialcharte des Rheinthals*, 1796; *Canton Zürich, Rheinthal*, 1797.

**Fees**, Theodor. Worked as a cartographer for the Vienna publisher E. Hölzel. School wall map of the region of Styria (6 sheets), 1900.

**Fegraeus**, Ludvig. *Karta öfver Wisby*, 1879.

**Fehlbaum**, Eduard. Lithographer. *Plan der Stadt Bern*, c.1876.

**Fehse**, C. *Gemeente-Atlas van de Provincie Groningen*, Groningen, J. Oomkens 1862.

**Feichtmayr** [Feuchtmayr], Joseph Anton (1696-1770). Engraver and sculptor. Plan von Madach, [n.d.].

**Feige**, *Lieutenant* —. Worked with others on a map of the environs of Berlin (10 sheets), 1816-1819.

**Feignet**, Johan Carl de (1740-1816). Cartographer and surveyor.

**Feild** [Field], John. Surveyor. Manuscript estate plans of Buckinghamshire: Dorney, 1781-1782; Boveney, 1790; Burnham, 1796.

**Feilden**, *Colonel* Henry Wemyss (b.1838). Routes Spitzbergen, 1898.

**Feind**, Bartholomäus. Cosmographia, Erfurt 1692.

**Feingroit**, Peretz. Lithographer of Warsaw. *Rabbi* Dov ben Joseph Yospa Baer's's *Eretz Israel*, 1881.

**Feistenhauer**, J. See **Faistenauer**, Johann.

**Feistmantel**, Ottoker. (1848-1891). Czech geologist, secretary to Geological Survey, Calcutta.

**Feiwel**, Jehoshua ben Israel. Map of Eretz Israel, in Kitzwei Eretz, Zolkiew 1772, revised by Arieh Leib ben Isaac, Grodno 1813.

**Felberthann**, E. Engraver for H.C. Kiepert, 1860.

**Felbiger**, Johann Ignatz von (1724-1788). Teacher and cartographer. *Carte du Diocese de Breslau*, 1751.

**Felbinger**, Kaspar (fl.1564-1576). Engraver of C. Henneberger's *Prussiae* (9 sheets), 1576.

**Felch**, *Reverend* Cheever (b.1787). Began his working life as a bookseller, printer, publisher and methodist preacher, then joined the navy as chaplain in 1813. Served as instructor in mathematics and navigation, as well as chaplain to American naval officers at Sacketts Harbour on Lake Ontario. Chart of *Cape Ann Harbour* [Massachusetts], 1819 (with M.T. Malbone), published in Blunt's *The American Coast Pilot*, 1827; survey of Georges Bank, c.1820 (with E. & E.M. Blunt). **Ref.** GUTHORN, P. United States Coastal Charts 1783-1861 (1984) pp.13-14, 33.

**Felden van Hinderstein**, G.F.D. van de. See **Derfelden** van Hinderstein, Gijsbert Franco von.

**Felgate**, Robert (fl.1672-1675). Estate surveyor of Gravesend, worked with Gregory King and John Ogilby. Surveys of Maldon and Ipswich, c.1674 (with Gregory King; Ipswich was engraved by Thomas Steward for publication in 1698); surveyed Westminster with Gregory King, 1674-1675; Aldham &c., 1675 MS; Essex, J. Ogilby & W. Morgan 1678 (surveyed with Gregory King, engraved by F. Lamb).

**Felkin**, Robert Felkin (b.1853). Reiseroute Ladó bis Dara, 1881.

**Felkis** [Felky], Antal (fl.late-18th century). Hungarian cartographer. Official surveyor to Esztergom County, numerous maps, e.g. area of Buda and Hidegkút now in Budapest. **Ref.** HRENKO, P., 'A Fehéregyháza kutatás...' in: Geodézia és Kartográfia [Geodesy and Cartography], (1976/1).

**Felkl**, father and son. Globemakers of Prague.
• **Felkl**, Jan (1817-1887). Bohemian printer and prolific globemaker based in Prague. Worked with several geographers of the time, including Otto Delitsch. From the 1850s onwards Felkl made thousands of terrestrial, celestial, and moon globes in a number of languages for export throughout Europe and to the United States. Designed for schoolroom and institutional use, many of them incorporated a clockwork mechanism to demonstrate the rotation of the earth. **Ref.** KRETSCHMER, DÖRFLINGER & WAWRIK Lexikon zur Geschichte der Kartographie (1986) vol.C/1, p.218 (with references) [in German].
•• **Felkl**, Kryštof Zikmund (1855-1894). Youngest son of, and successor to, Jan. Joined his father as a globemaker in 1870.

•• **Felkl & Sohn** (*fl.*1870-1950s). Jan was joined by his son Kryštof in 1870, when production was moved to Roztok. Sales outlets were maintained in Prague and Vienna. Lithography allowed faster production and in 1873 the firm produced approximately 15,000 globes in at least 8 sizes and 17 languages. After the deaths of Jan and Krystof, the firm remained in the control of the Felkl family.

**Fell**, Thomas M. *Australian Gold Fields*, *c.*1854.

**Fell**, *Lieutenant* (later *Commander*) William, Indian Navy. St. Bees Head to Duddon, 1824; surveyed coasts of India, 1841-1848; Pegu, 1852; North coast of Sumatra, 1853.

**Fellechner, Müller & Hisse**. Mosquito Küste, Berlin 1845.

**Fellenberg**, Edmund von (1838-1902). *Doldenhorn und Weiße Frau* [Switzerland], 1863 (with A. Roth, included a reproduction of a watercolour map by J.R. Stengel of 1850).

**Fellows**, Alfred. Lithographer. Colony of Victoria from the latest surveys, *c.*1859-1862.

**Fellows**, John (*fl.*1773-1825). Surveyor, produced manuscript maps of parts of Midland counties. Estate Radclive, Buckinghamshire, 1773 MS.

**Fellows**, William (*fl.*1793-*d.*1816). Surveyor. *Public Foreign Sufferance Wharfs* [London], 1803.

**Fellweck**, Johann Georg (*fl.*1772). Globemaker of Würzburg, Germany.

**Felsecker**, A. J. See **Felssecker**, Adam Jonathan.

**Felsing**, father and son. Engravers of Darmstadt.
• **Felsing**, Conrad (1766-1819). Royal copper engraver of Darmstadt, specialised in town maps of the Rhine-Main area. Trier und Saarburg, 1793-1794; Frankfurt, 1811; Situations-carte Darmstadt (24 sheets), *c.*1820.
•• **Felsing**, Johann Heinrich (1800-1875). Engraver of Darmstadt, son of C. Felsing (above). Situations Karte v. Rhein, 1822.

**Felssecker**, Adam Jonathan (1683-1729). Printer, publisher and copper engraver of Nuremberg. *Des neu eröffnenden Kriegs Theatri in Ungarn...*, 1737; *Kayserlich-Türkischen Friedens-Theatrum in Ungarn*, 1739.

**Felt**, W. [& Co.]. Lithographers of New York. J.T. Baker's *Map of the Port Huron and Milwaukee, and the Detroit and Milwaukee Railways...*, 1854.

**Felterus** [Feltero], Kjetill. See **Claesson**, Kjetil.

**Feltham**, M.*ris* [Mrs.] —. Mapseller in Westminster Hall, London. One of the sellers of Christopher Browne's *A Newe Map of England* (2 sheets), *c.*1693 edition.

**Felton**, John. Printer, colourist and publisher of Oswestry. William Williams's map of Denbigh & Flint (4 sheets), *c.*1720 (engraved by J. Senex).

**Felton**, Matthew. Map of the Harbour of Waitemata, New Zealand, 1841; The City of Wellington, 1841.

**Felton**, S.M. American surveyor. Beckford Estate in Charles Town, Massachusetts, 1837.

**Felton & Parker**. American surveyors. *Plan of the City of Charlestown...*, 1848 (with Ebenezer Barker).

**Fembo**, father & son.
• **Fembo**, Georg Christoph Franz (1781-1848). Printer, publisher and geographer of Nuremberg. Bought Georg Christoph Franz's interest in Homann's Heirs in 1813. *Atlas Silesiae...*, Nuremberg, Homann Heirs 1813; Laustiz, 1815; Nassau, 1818; Neuer Handatlas im kleinen Format von 24 Blättern, 1818; Great Britain and Ireland, 1818; Australien, 1820; *Neuer Handatlas* (24 sheets), 1821; Baiern, Württemberg, Baden und Hohenzollern, 1823; *Karte von den Grossherzogthuemern Mecklenburg*, 1826; Kur-Hessen, 1836.
•• **Fembo**, Christoph Melchior (1805-1876). Publisher and mapseller of Nuremberg, son of Georg, above. Took over the firm on the death of his father in 1848, but closed the business in 1852. On Christoph's death the stock and materials were sold at auction, present locations unknown. *Charte von Ungarn, Siebenbürgen, Slavonien und der Bukovina*, Nuremberg 1815 (copper engraving); *Charte von Australien*, 1830.

**Fening**, —. Plan of Drumsheugh, 1822.

**Fennel**, Otto (1826-1891). Instrument maker of Kassel, worked on the development of surveying equipment.

**Fennema**, R. *Geologische Beschrijving van Java en Madoera* (26 sheets), Amsterdam, Joh. G. Stemler Cz. 1886-1894 (with R.D.M. Verbeek). **Ref.** KOEMAN, C. Atlantes Neerlandici (1967-1971) Vol.VI pp.232-233.

**Fenner**, Rest. Engraver of Paternoster Row, London. *Pocket Atlas, Classical*, London, R. Jennings 1828; *Pocket Atlas of Modern Geography*, London, R. Jennings 1830; *Fenner's Pocket Atlas of Modern and Antient Geography*, Robert Jennings 1836 (by Rowland Bond; includes maps from the earlier publications).

**Fenner**, W.A. Trigonometric survey of tidal islands in Indus, 1852.

**Fenner**, Walter. Island of Carriacou, 1784.

**Fenner, Sears & Co.** Engravers and printers. Maps and views for J.H. Hinton, 1831, used in Hinton's *An Atlas of the United States of North America*, 1832 and later.

**Fenning**, Daniel. Engraver, geographer, globe-seller and publisher in London. *A New System of Geography*, 1765, 1773 (with J. Collyer); *A New and Easy Guide to the use of Globes...*, [n.d.], 6th edition published Dublin, J. Rice 1796, 7th edition, Dublin, P. Wogan 1797.

**Fenton**, James (1804-1875). Land surveyor of Chelmsford. Great Burstead, Essex, 1831 MS.

**Fenwick**, *Lieutenant* Thomas Howard *RE* (*fl.*1821-1849). Surveyor credited with military maps in County Durham and Kent. Manuscript maps of Chatham, 1821.

**Fenwick**, *Captain* W. *RE* Military maps and plans of sites in Halifax, Nova Scotia, 1800-1805 MSS. **Ref.** Compare with William **Fenwick**, below.

**Fenwick**, *Lieutenant-Colonel* William *RE* Endorsed several plans of the coast of County Cork, 1809-1812 (surveyed and drawn by others).

**Fényes**, Elek [Alexander] (1807-1876). Hungarian cartographer and publisher of Pest. *Közönséges Kézi- és Oskolai Atlasz...* [school atlas with 8 maps], 1843; transport map of Hungary, 1868. **Ref.** BORBÉLY, A., 'Fényes Elek térképei' [The maps of Elek Fényes], in: *Földrajzi Értesítö* [Geographical Bulletin], (1958/2).

**Feodoseyev**, V. (fl.17th century). Russian scribe and draughtsman. Map Sevsk, 1650.

**Fer, de**, family. **Ref.** PASTOUREAU, M. *Les atlas français XVIe-XVIIe siècles* pp.167-223 [in French].

• **Fer**, Antoine de (*fl.*1644-1672, *d.*1673). Mapseller, engraver and colourist, 'quay de l'isle du Pallais, pres le pont de bois' (1633); 'demeurant sur le Quay qui regarde la Mégisserie près l'Orloge du Palais à l'Age de Fer'; 'à la Sphère royale' (1652), Paris. Bought the plates of N. Tassin (with N. Berey). Re-issued Turquet's azimuthal north polar map, with rules for a travel game, 1648; C. Tassin's *Plans et profilz des principales villes & lieux considerables de France*, 1652 (first published M. Tavernier 1634); editor of P. Duval's *Cartes de géographie...*, 1657 (maps re-used in *l'Europe*, 1686); Isle de Madagascar (2 sheets), 1666, used by Jean de Beaurain 1751; *Carte générale de France avec la route de postes*, 1668; geographical games, 1670-1672.

• [Hourlier], Geneviève (*d.*1690). Widow of Antoine, ran the business from his death until her retirement in 1687, when it was taken over by Nicolas.

•• **Fer**, Nicolas de (1646-1720). Geographer of Paris. 'Géographe ordinaire de Sa Majesté catholique' noted at 'l'Isle du Palais sur le Quay de l'Orloge à la Sphère Royale'. Son of Antoine, above, apprenticed to Paris copper engraver Louis Spirinx from 1659. In 1687 he took over the family business from his mother and built it into a flourishing map publishing company. He was succeeded by sons-in-law Jacques-François Bénard and Guillaume Danet. *France écclésiastique*, 1674; *Franche-Comté*, 1674; *Plan de l'Isle de Goreé*, 1677; *Haute Lombardie*, 1682; decorative map of the course of the Rhine (3 sheets), 1689 (engraved by Liébaux and N. Guérard); *Les Costes de France Sur l'Ocean et sur la Mer Mediterranée...*, 1690 (from the atlas of C. Tassin, 1634, the plates of which had been bought by A. de Fer in 1644; *Les forces de l'Europe...* (8 parts), 1690-1695 and later (the first part, published in 1690 was entitled *Introduction à la fortification*), 2 volume edition published in Amsterdam, P. Mortier 1702; *Les Frontiers de France et d'Espagne*, 1694; *Atlas royal...*, 1695, 1697, 1699 (used the maps of Sanson, Jaillot and others); *Le theatre de la guerre ou representation des principales villes du Pais Bas catholiques...* (25 plans), 1696 (plans re-used from *Forces de l'Europe*); Carel Allard's *Le theatre de la Guerre, dans les Pays-Bas* (53 maps and plans in 2 volumes), 1696-1697 (many of the town plans were taken from *Forces de l'Europe*); *Petit et Nouveau Atlas* (19 maps), 1697, 1705, 1723; untitled wall map of the Americas [the original 'Beaver Map'], 1698; *Le Théâtre de la guerre en Allemagne* (26 maps), 1698 (many of the town plans were taken from *Forces de l'Europe*); *Postes de France et d'Italie*, 1700; *L'Atlas Curieux...* (6 parts), 1700-1705 and later (the maps re-appeared in several later works); *Cartes et descriptions... au sujet de la Succession de la Couronne d'Espagne...* (18 maps), 1701; *Cartes nouvelles et particulieres pour la guerre d'Italie* (29 maps), 1702 (used some maps from *Atlas Curieux*); *Le theatre de la guerre dessus et aux environs du Rhein...* (29 maps), 1702, (28 maps), 1705 (re-used maps from *Atlas Curieux* and *Forces de l'Europe*); *Les Jonctions des deux grandes Rivieres de Loire et de Seine par le Nouveau Canal d'Orleans et celuy de Briare*, 1705 (engraved by C. Inselin), a smaller version (engraved by P. Starckman) was published in 1716; *Cartes nouvelles et Particulieres Pour la Guerre d'Allemagne* (62 maps), 1705 and later (re-used maps from *Atlas Curieux* and *Forces de l'Europe*); *Les Beautés de la France*, 1708 and later (re-used maps from *Atlas Curieux*); *Frontieres de France et des Païs Bas* (24 sheets), 1708-1710 (engraved by P. Starckman, republished as *La Flandre en 23 feuilles...*, 1722); *Atlas ou Recüeil de cartes géographiques...*, 1709-1728, Desbois 1746; *Suite de l'Atlas Curieux*, 1714, 1716; *Introduction à la Géographie* (6 maps), 1717 (first published with no maps in 1708). **Ref.** SHIRLEY, R. *Printed maps of the British Isles 1650-1750* (1988) pp.57-59; KRETSCHMER, DÖRFLINGER & WAWRIK *Lexicon der Geschichte der Kartographie* (1986) Vol.C/1 pp.220-221 [in German]; PASTOUREAU, M. *Les atlas français XVIe-XVIIe siècles* (1984) pp.167-223 (for collations of individual atlases) [in French]; DAHL, E.H. *The Map Collector 29* (1984) pp.22-26; M. PASTOUREAU *Imago Mundi 32* (1980), pp.45-72; *The Map Collector 9* (1979) pp.52-58 (collation of *L'Atlas Curieux*).

**Fer**, Edward W. Railway map of India, 1878.

**Fer**, F. See **Ojea**, Hernando.

**Fer de la Nouerre**, de. See **De Fer** de la Nouerre.

**Ferber**, J.J. (1743-1790). Swedish geologist. Mineral geschichte von Böhmen, [n.d.].

**Ferck**, Th. Draughtsman. *Schleswig Holstein. Der Nord-Ostsee-Kanal*, Berlin, Dietrich Reimer 1895 (lithographed by C.L. Keller).

**Ferdinando**, S.M. Contorni di Napoli, 1819.

**Ferdinandus**, Filips. Artist of Antwerp. Plan Papa [Hungary], 1566, for Braun & Hogenberg.

**Fergola**, Filippo (fl.19th century). Engraver and calligrapher of Naples.

**Fergus**, G.H. Publisher of Chicago. Nehemiah Matson's *Map of Bureau County, Illinois* (atlas with 26 maps), 1867.

**Fergus**, John (fl.1753-1762). Surveyor. Plans of Edinburgh, 1759 (with Robert Robinson).

**Ferguson**, —. Engraver of *Africa 1792* ('for Jacksons's edition of *Payne's New System of Universal Geography*', 1792).

**Ferguson**, —. Handkerchief map entitled *A New Map of the Roads of England and Scotland*, c.1810.

**Ferguson**, A.M. Ceylon, Colombo 1875.

**Ferguson**, Charles. Entrance Gipps Land Lakes, 1866.

**Ferguson**, G.R. Isthmus of Tehuantepec, James Wyld 1851.

**Ferguson**, George E. Draughtsman in West Africa. *A Portion of the Leeward part of the District of Accra* [Ghana], 1881; Map of the territories of Katani [Kétonou], Porto Novo, Appah and Pokra in Dahomey..., 1883 (lithographed by Dangerfield of Covent Garden); maps of parts of Ghana and the Ivory Coast, 1883-1884; *Sketch Map of the Divisions of the Gold Coast Protectorate*, Stanford's Geographical Establishment 1884; *Map of a portion of the Gold Coast*, London, Stanford's Geographical Establishment 1890; *Map showing Routes in Awuna and Krepi traversed by the Anglo-German Boundary Commission...*, 1892, 1893; *Map of the Hinterland of the Gold Coast Colony...*, War

Office 1893 (larger scale version on 4 sheets published by the Ordnance Survey, 1897); *Sketch map of the principal trade roads* [Gold Coast], 1894; *Sketch Map of the Junction of the River Daka, with the Volta River...*, Wyman & Sons 1895; and others.

**Ferguson**, J.H.D. Rossland, British Columbia, 1898.

**Ferguson**, James (1710-1776). Scottish philosopher and astronomer from Banffshire. Bought most of John Senex's copper plates, globe moulds and tools from his widow Mary Senex in 1755. Made his own globes and republished those of Senex. His copper plates passed to Benjamin Martin in 1757, but many later globes by B. Martin, W. Bardin and G. Wright continued to be titled 'Ferguson's' well after this date. A pair of celestial and terrestrial globes, *c*.1750 (engraved James Mynde); *Astronomy Explained upon Sir Isaac Newton's Principles...* (description only), 1756 (later editions, which ran up to 13th edition, 1811 included increasing numbers of charts); "Papers for covering a terrestrial globe", 1773 (MS globe gores 4" diameter). **Ref.** DEKKER & VAN DER KROGT *Globes from the Western World* (1993) pp.112-115; MILBURN, J.R., KING, H.C. *Wheelwright of the Heavens: The life and work of James Ferguson FRS* (1988).

**Ferguson**, James (1797-1867). American surveyor to the Board of Boundary Commissioners under the Treaty of Ghent [U.S.A./Canada border], later worked for the United States Coast Survey. Numerous surveys of the boundary lands including: Drummond Island to St. Joseph Island, *c*.1820; Pigeon River to Lake Namekan, 1823-1824 (with G.W. Whistler); route from Kaministiquia River to Lac La Croix, 1824 (with G.W. Whistler); and many others (with G. Whistler, S. Thompson and D. Thompson); triangulation survey of New Haven Harbor, 1833 (with Edmund Blunt, directed by F.R. Hassler), published 1846, triangulation of New York Bay and Harbour, published 1844 (with Edmund Blunt); *The Harbor of Annapolis*, 1846 (with F.H. Gerdes, directed by G.M. Bache, engraved by J.H. Young); one of the surveyors for *Map of Delaware Bay and River*, published 1848 (surveys directed by F.R. Hassler and A.D. Bache); one of the surveyors for *Mouth of the Chester River*, published 1849 (with J.E. Johnston, directed by A.D. Bache); triangulation surveys for *Patapsco River and the Approaches*, published 1859; and other contributions to U.S. Coast Survey charts.

**Ferguson**, James. Lithographer at the Topographical Department, War Office, London. *Map of Montenegro...*, 1860; drew the hills on E.G. Ravenstein's *Map of Abyssinia*, 1868; plans of sites in the Sinai Peninsula, *c*.1870 (printed by Vincent Brooks, Day & Son).

**Ferguson**, Patrick (1744-1780). Born in Aberdeenshire, served in New Jersey as an officer of the Royal North British Dragoons. *Long Island Sound Shore...*, 1779 MS; *Hudson River from Peekskill to Slaughters Landing*, 1779 MS; *Verplanks Point*, 1779 MS; parts of Union and Bergen Counties, New Jersey, 1780; *Proposed fortifications of Savannah*, 1780 MS; and others now in Clements Library, Michigan, U.S.A.

**Ferguson**, W.E. *Map of the Old Colony Rail Road with its branches & Connecting roads* [E. Massachusetts], Boston, J.H. Bufford & Co's Lith. 1850 (engraved by S. Dwight Eaton).

**Fergusson**, John. *Plan of the minerals in the parish of Muirkirk and Shire of Ayr* [Scotland], 1841.

**Fergusson & Mitchell**. Lithographers and publishers. Provinces of Otago & Southland, New Zealand, 1866.

**Ferīdūn**, Ahmed [Ferīdūn Beg] (d.1583). Ottoman historian and illustrator. HIs work contains topographica; maps. *Nüzhetü 'l-ahbar der sefer-i Sigetvar* [Chronicle of the Sigetvar campaign], 1568-1569. **Ref.** ROGERS, J.M. 'Itineraries and town views in Ottoman histories' in HARLEY & WOODWARD *The history of cartography* vol.2, book 1 *Cartography in the traditional Islamic and South Asian societies* (1992) p.254.

**Ferimontanus**, Daniel. See **Cellarius** Ferimontanus, Daniel.

**Ferlettig**, A. *Pianta di Trieste*, 1871, 1892.

**Ferlin**, —. Plans of Toulon, *c*.1873.

**Fernandes**, Marcos (*fl*.late-16th century). Portuguese mapmaker.

**Fernandes** [Fernandez], Pero. Portuguese pilot and cartographer, father of Luís Teixera. Manuscript charts of the Atlantic, *c*.1525, 1528.

**Fernándes**, R.A. Navigational distances on Amazon, Manaos 1869.

**Fernandes** [Fernando], Simão [Simon]. Portuguese pilot and cartographer in the service of the English. Chart of the Atlantic, *c*.1580 [British Library].

**Fernandes**, Valentim. Portuguese mapmaker. Manuscript maps of Atlantic islands, *c*.1506-1510.

**Fernándes Pereira**, Francisco. See **Francisco** [Fernándes Pereira], *Father*.

**Fernández**, Antonio. Printer in Madrid. Bernardo Espinalt y García's *Atlante español*, 1778.

**Fernández** [Fernándex], P.S. South America, 1800.

**Fernández**, P. See **Fernandes**, Pero.

**Fernández**, Pascual. *Aspecio geográfico del mondo Hispanico*, 1761.

**Fernández**, R.G. See **González** Fernández, Ramón.

**Fernández**, S. See **Fernandes**, Simão.

**Fernández de Córdova**, Francisco. Added the Spanish text to the indigenous 'Relación Geográfica' logographic map of Misquiahuala [New Spain], *c*.1579 MS.

**Fernández de Enciso**, Martín. Spanish navigator and geographer. *Suma de geográfia*, Seville 1519.

**Fernández de Medrano**, Sebastian. *Geografía o moderna descripcion de el Mundo*, Antwerp 1709.

**Fernández de Palencia**, Diego. Historia del Peru, 1571.

**Fernández de Roxas**, Antonio. *Topografia de la ciudad de Manila*, 1717.

**Fernández Flórez**, Ignacio [Ygnacio]. Surveyor. *Plano de la Ria de Arosa*, Madrid, Dirección de Hidrografía 1833 (map drawn and engraved by C. Noguera, lettering engraved by N. Gangoiti); *Carta esférica de la costa de Galicia...* (2 maps), Madrid, Dirección de Hidrografía 1835-1836 (drawn and engraved by C. Noguera, lettering engraved by M. Giraldós); *Plano de la Ria de Corcubion*, Madrid, Dirección de Hidrografía 1837 (drawn by J. Espejo, engraved by J. Carrafa); *Plano de la Ria de Muros y de Noya*, 1838 (drawn by J. Espejo, engraved by C. Noguera, lettering engraved by J. Hermoso) amended edition published 1868; *Plano de la Ria de Camariñas*, Madrid, Dirección de Hidrografía 1838 (drawn by J. Espejo, engraved by C. Noguera, lettering engraved by N. Gangoiti).
**Ref.** MÉNDEZ MARTÍNEZ, G. *Cartografía antigua de Galicia* (1994) pp.132-133 *et passim*.

**Fernando**, Simon. See **Fernandes**, Simão.

**Fernel**, Jean. French physician. Measured the arc of the meridian [Paris to Amiens], 1625.

**Ferra**, J. *Carta esferica... Isla de Santo Domingo*, 1802.

**Ferrande**, Pierre Gracie. *Le grand routier et pilotage*, Poitiers 1520.

**Ferrar**, John & Virginia. See **Farrer**.

**Ferrara**, A. *Carta minerologica dell'Isola di Sicilia*, 1810; *Carta dell'Etna*, 1818.

**Ferrari** [Galateus Leccensis], Antonio (1444-1516). Physician, poet and geographer, born Galatina. Maps used by Leandro Alberti, *Isole Appartenenti alla Italia*, 1567.

**Ferrari**, Filippo (1551-1626). Mathematician and geographer of Pavia.

**Ferrari**, G.G. See **Giolito**, Gabriele.

**Ferrario**, F.F. *Epitome geographicus*, 1605.

**Ferraris**, Joseph-Johann-Franz [Joseph-Jean-François], Graf von [comte de] (1726-1814). Army officer and military cartographer from Lunéville, Lorraine, Director General of artillery in Austrian Netherlands. *d*. Vienna. *Pays-Bas Autrichiennes* (275 sheets), 1770-1774 MSS; *Carte chorographique des Pays-*

*Bas* (25 sheets), Mechelen 1777 (engraved by L.A. Dupuis), reduced versions of this map were published in London by W. Faden, 1789 and in Brussels by J. Coché, [n.d.] (engraved by J.C. Maillart). **Ref.** KRETSCHMER, DÖRFLINGER & WAWRIK *Lexicon der Geschichte der Kartographie* (1986) Vol.C/1 pp.221-222 [in German]; AUDENHOVE, Marcel van (editor) *La Cartographie au XVIIIe Siècle et l'Oeuvre du Comte de Ferraris 1726-1814* (Brussels 1978).

**Ferrarius**, —. Wrote some of the geographical descriptions published in Pieter van der Aa's *Le Nouveau Théatre du Monde*, Leiden 1713 (the text on Europe was by N. Gueudeville).

**Ferraz**, Guilherme Ivens. *Provincia de Moçambique*, 1894.

**Ferre**, —. de, 'Capt. du Regt. de la Marine et ing. ord. du Roy'. Military maps of Basse Alsace, 1692 MSS.

**Ferreira**, *Captain* Joseph Maria. Manuscript plans of Almeida, Portugal, 1808 (with J.J. da Cunha).

**Ferreira de Loureiro**, Adolpho. Atlas, 1886.

**Ferreiro**, Martin (1830-1901). Geographer of Madrid. *Los Alfaques*, 1858; *Atlas de España*, 1864; *España y Portugal*, c.1867; *Cartagena*, 1873; *Provincias Vascongadas y Navarra*, 1874; *Provincia de Zambales*, c.1880.

**Ferreiro Reimão**, Gaspar (*fl*.16th century). Published *Roteiro*, after 1612 (contained 15 maps by Luis Teixeira).

**Ferrel**, William (1817-1891). American meteorologist, worked for the U.S. Coast and Geodetic Survey. *Meteorologial researches* (with maps), Washington, Government Printing Office 1877-1882.

**Ferrer**, Fermin. Member of the American Geographical Society, one time Secretary of State of the Republic of Nicaragua. *Government map of Nicaragua...* [with inset *Map of Central America*], New York, A.H. Jocelyn 1856; *Colton's map of Central America by Fermin Ferrer...* (2 sheets), New York, J.H. Colton & Co. 1859.

**Ferrer**, J.S. y. See **Seyra** y Ferrer, Juan.

**Ferrer**, Jaume [Jayme]. Catalan cosmographer. World (in northern and southern hemispheres), 1495 MS (lost).

**Ferrer**, John. See **Farrer**, father & daughter.

**Ferrer**, Juan. Spanish hydrographic draughtsman. *Carta esferica del Golfo de Gascuña y Canales de la Mancha y Bristol*, Madrid, Déposito Hidrográfica 1803 (engraved by Fernando Selma).

**Ferrer**, *Professor Doctor* K. Karte von Palästina 1891.

**Ferrer**, V. See **Farrer**, father & daughter.

**Ferreri**, Giovanni Paolo. Worked in Paris. World, c.1554 MS.

**Ferreri**, Giovanni Paolo (*fl*.1600-1624). Globe and instrument maker in Rome.

**Ferrerios**, Rainaldo Bartolomeo de. Majorcan chart maker. Chart, 1592 (with Matteo Prunes).

**Ferrero**, *General* A. *Collegamento geodetico dell'isola di Malta alla Sicilia*, 1900 MS (Public Record Office, Kew). **N.B.** Compare with Annibale **Ferrero**, below.

**Ferrero**, Annibale (1839-1902). Mathematician and geographer in Florence, Director Istituto Geografico Militare.

**Ferrero della Mamora**, Alberto. *Carte della Sardegna*, 1839; Carte géologique d'Italie, 1846.

**Ferret**, Pierre Victor Adolphe. *Carte d'Acir-Hedjaz*, 1840; *Abyssinie*, 1843; *Carte géologique du Tigre*, 1848; *Voyage en Abyssinie*, 1847-1848 (with Joseph Germain Galinier).

**Ferretti**, *Captain* Francesco. *Diporti notturni...* (28 maps of European islands), Ancona 1580.

**Ferri**, *Father* Francesco Maria. Engraver for Vincenzo Maria Coronelli, 1691-1696.

**Ferris**, George Titus (*b*.1840). Associated with Daniel Appleton & Co's *Appleton's Atlas of the United States*, 1888.

**Ferris**, Warren Angus. Rocky Mountain Fur Region, 1836 MS.

**Ferron**, —. Carte des Chemins de Fer Français, 1874?.

**Ferry**, A.C. Atlas of Lorrain County, Ohio, 1874 (with D.J. Lake).

**Ferry**, Hypolite. Nouvelle Californie, 1850.

**Ferslew**, Martin Wilhelm (1801-1852). Lithographer in Copenhagen. Maps of Slesvig.

**Ferslew**, W. Eugene. Geographer and publisher. City of Richmond, Virginia, 1859.

**Féry**, Jules H. Cariboo Gold Region, 1861.

**Fesca**, A. Engraver in Kiel. *Trieste*, 1865.

**Fessard**, —. French surveyor. Worked with Langelay and others on at least eight sheets of the Cassini *Carte de France* (sheets 154 & 168 in Provence, 129, 130, 157, 159, 171 & 172 in Brittany), 1778-1787.

**Fessler**, Ignaz Aurel. Cartographer and publisher of Leipzig. *Die Gesichte der Ungern und ihre Landsassen...* [Hungary], 1825 (maps engraved by T. Goetz).

**Fest**, Matthias (1736-1788). Austrian cartographer of Eisenstadt.

**Festa**, Felice. Lithographer. *Atlante Itinerario delle Provincie di terraferma di S. M. il Re di Sardegna*, 1820.

**Fester**, Diderich Christian (1732-1811). Mathematician and cartographer of Trondheim, Norway. Collaborated with Erik Pontoppidan on the preparation of the seven-part *Den Danske Atlas*, 1763-1781 and later; *Generalum et prorsus Novam Siælandiæ Tabulam Geographicum...* [Denmark], 1773 (engraved by H. Quist).

**Festing**, *Major* Augustus Morton. Assistant colonial secretary and treasurer in Freetown, Sierra Leone. Compiled and authorised the publication of maps of West Africa 1885-1887. Map approximately showing the Territory of Port Lokoh, 1885 (lithographed by Dangerfield & Co.); *Plan of Quiah Territory...*, Dangerfield & Co. 1886; *Map showing the country to the north and south of the River Gambia under the sovereignty or influence of the French*, 1887; coloured manuscript map of Sierra Leone, 1887.

**Feterus**, Olasson (*fl.*late-17th century). Scandinavian cartographer.

**Fetherstone** [Featherstone], Henry [Henrie]. Printer and publisher 'in Paul's Church-yard at the Signe of the Rose', London. Bought 55 plates of the Mercator-Hondius *Atlas Minor*, first published 1607. Samuel Purchas's *Purchas his Pilgrimage or relacons of the World...*, 1613, 1614, 1616, 1626; publisher of *Purchas his pilgrimes* (4 volumes), 1625 (printed by William Stansby, incorporated the Hondius plates of 1607).

**Feuchtmayr**, J.A. See **Feichtmayr**, Joseph Anton.

**Feuersfeld**, Franz Schönfelder von. See **Schoenfelder** von Feuersfeld, Franz.

**Feuerstein**, W. Engraver for F.A. Meyer, 1867.

**Feuille**, family. See **La Feuille**.

**Feuillée**, Louis (1660-1732). Botanist and astronomer. Measured longitude between the Isle of Fero and Paris Observatory, 1724.

**Fèvre**, — sieur le. Map of the Low Countries, 1672.

**Fewson**, *Captain* —. Chart for John William Norie, 1840.

**Feyhel**, Martin (*fl.*16th century). Maker of measuring instruments. Worked in Naumburg, Germany.

**Fhert**, A.-J. de (*b.*1723). French engraver. Worked on at least 6 sheets of the Cassini *Carte de France*, 1757-*c.*1760.

**Fiala**, John [János] (1822-1911). Hungarian cartographer. Army officer, emigrated to USA after failure of 1848-1849 War of Independence. Made maps for Union Pacific Railroad Co. *General map of the United States & their territory between the Mississippi & the Pacific ocean* (with post routes, railroads and ore discoveries), 1859 (lithographed by A. McLean, engraved on stone by A. Janicke); Missouri, 1860-1865.

**Fiamengo**, A. di A. See **Arnoldi**, Arnoldo di.

**Fickler** [Weylensis], Johann Baptiste. Translated from Latin the history of Scandinavia by Olaus Magnus as *Olai Magni Historien der mittnächtigen Inder*, Basle, Henricpetri 1567 (with a woodcut map of Scandinavia copied from Olaus's *Carta Marina* of 1539, and thought to have been cut by Thomas Weber).

**Fidalgo**, Joaquin Francisco (1755-1820). Charted the coast of Venezuela, 1792-1803; *Carta esférica de las Islas Antillas*, 1802 (with C. de Churruca); *Plan Porto Cabello*, 1804; Costa del Darien, 1817; charts published in *Atlas geográfico de la América Meridional y Pacífico*, Madrid 1840.

**Fiddes**, *Lieutenant* James, RE. *Plan of James Is., River Gambia*, 1783 MS.

**Fidler**, Peter (1769-1822). Explorer and surveyor in American Northwest, worked for the Hudson's Bay Company. Travelled alongside the local native tribes, whose knowledge and maps assisted him with his study of the trading routes and topography of the Rocky Mountains. He contributed to the accurate updating of the Arrowsmith maps. Collected and copied native maps including those of the Upper Missouri by the Blackfoot chief Ac ko mok ki, 1801-1802 MSS (for the Hudson's Bay Company). **Ref.** BELYEA, B. in LEWIS, G.M. (Ed.) *Cartographic encounters: Perspectives on native American mapmaking and map use* (1998) pp.138-139, 143-145, 148-149; *The Map Collector* 39 (1987) pp.16-20.

**Fidling**, Theodor Christoph. *Plan vom Dorfe Babic bei Warschau*, c.1750.

**Fiedler**, F. Kreis Westprignitz, 1899.

**Fiedler**, Ferdinand Ambrosius (d.1756). Surveyor of Magdeburg, Germany.

**Fiedler**, Karl Gustav. Geologist of Saxony. *Geognostisch-Bergmännische Karte von dem Königreich Griechenland* [Geological and mining map of Greece], 1840.

**Fief**, J.B.A.J. du. See **Du Fief**, Jean-Baptiste-Antoine-Joseph.

**Field**, *Lieutenant Commander* Arthur Mostyn, RN. Hydrographer of the Navy, 1904-1909. Chart of Tasmania approaches to Hobart, 1886-1888, published Admiralty 1889; Vavau Group [Tonga], Admiralty 1898.

**Field**, Barnum (1796-1851). *The American School Geography*, Boston, W. Hyde 1832.

**Field**, Barron (1786-1846). Lawyer and writer. Editor of *Geographical Memoirs on New South Wales by Various Hands* (4 maps), London, John Murray 1825.

**Field**, John. See **Feild**, John.

**Field**, John (fl.1806-1829). Draughtsman, produced manuscript copies of Ordnance plans, 1811-1816 including: a plan of Limerick, 1811 MS (originally compiled by F. Birch, 1803); ordnance plan of Clonmel, Tipperary, 1811; ordnance plan of Omagh, Co. Tyrone 1811; town and castle of Scarborough, 1812; Bahama Islands, 1813; frontier fortresses of France, 1816.

**Field**, Richard. Publisher of London. *A Summarie and true discourse*, 1589 (incorporated Giovanni Battista Boazio's map of Drake's voyage, *The Famouse West Indian voyadge made by the Englishe fleet of 23 shippes and Barkes...*).

**Field**, Samuel. *Draught of the Windward Coast of the Mosquito Shore* [Central America], Mount and Page 1740.

**Field Publishing Co.** Sangamon County, Illinois, 1894.

**Fielding**, John. Publisher in partnership with Walker until c.1781, when he began trading alone.
• **Fielding & Walker.** Publishers of 'Pater Noster Row', London. *A General View of the Roads of England and Wales*, 1777 (engraved by J. Lodge); W. Russell's *History of America* (2 volumes), 1778; maps for *Westminster Magazine* including the following: *Plan of the Sea Engagement Fought on 27th of July 1778, between the English and French fleets off Ushant*, 1779 (engraved by J. Lodge); *A Chart of the British Channel...*, 1779 (engraved by J. Lodge); *A New Chart of the Straits of Gibraltar*, 1780 (engraved by J. Lodge); R. Cowley's *Charles Town, South Carolina*, 1780 (engraved by J. Lodge); *A Chart of the Coast of France with the Isles of Guernsey, Jersey, Alderney &c.*, 1781

(engraved by J. Cary); *The Seven United Provinces*, 1781 (engraved by J. Cary).
• **Fielding**, John. Publisher of 'N°·23 Paternoster Row', London. As well as other works, he published maps for *European Magazine*, 1782-1783 (some of these were published jointly with J. Sewell and J. Debrett). *An accurate Map of the Country twenty Miles round London...* 1782 (engraved by J. Cary, published in *Fielding's London Guide*); *A New Pocket Plan of the Cities of London, Westminster and Borough of Southwark...*, 1782 (engraved by J. Cary, published in *The Stranger's guide through & round London*, 1872, 1786); *An Accurate Map of the Islands of St. Christophers and Nevis*, 1782 (engraved by J. Cary, published by J. Fielding, J. Debrett & J. Sewell for *European Magazine* in 1782); *An Accurate Map of Ireland*, 1782 (engraved by J. Cary for *European Magazine*); *Plan of the Bay and Roads of Cadiz*, 1783 (engraved by J. Cary for *European Magazine*); *The British Channel*, 1783 (engraved by J. Cary for *European Magazine*); *Turky in Europe*, 1783 (engraved by J. Cary for *European Magazine*); *A Map of the United States of America, as Settled by the Peace of 1783*, 1783 (published by Fielding for *European Magazine*); John Andrews's *History of the War with America*, 1785; and others.

**Fielding**, Lucas Jr. See **Lucas**, Fielding.

**Fielding & Walker**. See **Fielding**, John.

**Fielitz** [Filitz], Friedrich August von (*fl.*1821-1825). Engraver in Berlin. A view of Berlin, *c.*1835.

**Figari**, Antonio. Etudes géographiques, 1866.

**Figg**, William, father and son. Surveyors and draughtsmen of Lewes, Sussex. Between them the Figgs produced at least 150 plans. William Figg was succeeded by George Fuller. **Ref.** BAYNTON-WILLIAMS, R. *The Map Collector* 46 (1989) p.15.
• **Figg**, William (1770-1833). Land surveyor. Lands in East Hoathly, 1822 MS.
•• **Figg**, William, *the younger* (1799-1866). Surveyor, son of above, took over father's business before 1825. Corrected Christopher and John Greenwood's *Map of the County of Sussex* (6 sheets), 1861. **Ref.** KINGSLEY, D. *Printed maps of Sussex* (1982) pp.177-178.

**Figueiredo**, Manoel de (1558-*c.*1630). Portuguese cartographer, examiner of pilots and 'master of the navigation instruments and keeper of the charts' for the Council of India. Chronographia, 1603; *Hydrografia. Exame de pilotos...*, Lisbon 1608; Roteiro e navegâcao das Indias, [n.d.].

**Figueroa**, C. de la R. y. See **Rocha** y Figueroa.

**Figueroa**, José Vazquez. 'Secret[ario] de Estado y del Despacho Univ[ersal] de Marina'. Estrecho de San Bernardino [Philippines], Madrid 1816.

**Figueroa**, Rodrigo de. Porto Rico, 1519 MS.

**Figuier**, Louis. *Terre & mers: description physique du globe*, Paris, Hachette 1884.

**Fijnje** [Fynje], H.F. (*fl.*19th century). Dutch surveyor for Topographische Dienst. Survey maps of the Netherlands provinces.

**Filaeus**, Petrus [Pehr]. Swedish cartographer, mid-18th century.

**Filastre**, G. See **Fillastre**, Guillaume.

**Fileti**, Giovanni. Author and director of navigation school at Palermo. *Carta Piana del Mare Mediterraneo* (11 sheets), 1802.

**Filiberto**, Emanuele. Worked in Rome. Thought to have made a terrestrial globe, *c.*1570.

**Filisov**, Yakov Kondratyevich (1698/1697-after 1753). Early eighteenth century Russian geodesist and cartographer. Surveys in Novaya Zemlya from 1720; geographer heading the team surveying Enisey and Irkutsk provinces in Siberia, 1738-1750.

**Fillastre** [Filastre; Fillâtre], *Cardinal* Guillaume (1348-1428). French prelate, geographer and classicist. Made a manuscript copy of the Latin version of Ptolemy (the Nancy codex *c.*1427), which for the first time incorporated 'modern' maps, including a world map, and a map of the Nordics based on the work of Claudius Clavus. **N.B.** Some scholars date the world map c.1418.

**Fillatreau**, —. Artist and engraver. Worked with Soudain on an inset view which appeared on J. Migeon's *Afrique Physique*, 1891 (the map itself was engraved by G. Lorsignol).

**Fillian**, John. Engraver. Title for Peter Heylin's *Cosmographie*, 1669.

**Fillingham**, William (1734-1795). English surveyor, produced large scale manuscript maps. Inclosure map of Beckingham [Lincolnshire], c.1770; plan to inclosure award at Westborough cum Doddington, 1771; Crown estate at Barnby-in-the-Willows, Nottinghamshire, 1806.

**Fillmore**, Millard (1800-1874). Lawyer, later to become 13th President of the United States. Earned a living as a surveyor for his employer, Judge Walter Wood, in whose office he was studying law. In later years his interest in maps and geography was reflected in his collection of maps and atlases. Although the collection was largely dispersed in 1890, many of the maps are now held in the Library of Congress.

**Fillot**, H. Plan de Mojanga [Madagascar], 1895.

**Filosi**, Giuseppe. *Pianta di Verona*, 1737.

**Fils**, *Lieutenant* August Wilhelm von (1799-1878). Artillery officer, cartographer in Berlin and Schleusingen. Glatz, 1832; Sachsen, 1836; Rudolfstadt, 1848; *Thüringer Wald*, 1860-1867; *Gegend von Ilmenau*, c.1876.

**Filson**, John (c.1747-1788). Surveyor and historian of Pennsylvania, moved to Kentucky c.1782-1783, where he acquired land holdings. Compiled a book about Kentucky with a map entitled *This Map of Kentucke, Drawn from Actual Observations...*, Philadelphia, Tenoor Rook 1784 (engraved by H.D. Pursell), and many other editions . **Ref.** NEBENZAHL, K. The Map Collector 56 (1991) pp. 40-45; WHEAT & BRUN Maps and charts published in America before 1800 (1978) entry N° 631 p.138.

**Finaeus**, Orontius. See **Fine**, Oronce.

**Finagenov** [Finaghenof], —. One of the calligraphy engravers for Pyadyschev's *Atlas géographique de l'Empire de Russie, du Royaume de Pologne et du Grand Duché de Finlande...*(83 sheets), 1820-1827, 1829, 1834.

**Finch**, Heneage. Surveyor. Upper Castlereagh Street, Sydney, New South Wales (1827). *The estate of Lewinsbrook... Co. Durham*, 1841.

**Finch**, J.A. *Alaska and Pacific coast map*, San Francisco, Hicks-Judd Co. 1898.

**Finck**, *Lieutenant Colonel* — von. *Königreich Württemberg*, 1895.

**Finckh**, Georg Philipp (c.1608-1679). Administrative official. *S. Rom. imp. Circuli et electoratus Bavariae Geographica Descriptio...* (with 28 sheet map and an index map), issued by his son, 1684 (based on P. Apian's map).
• **Finckh**, Georg Philipp. Son and successor to Georg Philipp, above.

**Finden**, brothers. English engravers.
• **Finden**, Edward Francis (1792-1857). Engraver, worked with brother William. Best known for numerous Victorian steel-engraved illustrated books. *Road Book from London to Naples*, 1835.
• **Finden**, William (1787-1852). Engraver. Worked with brother Edward.

**Findlater**, *Reverend* Charles (1754-1838). *General View of the Agriculture of the County of Peebles*, 1802 (with a map entitled *Map of the County of Peebles or Tweedale*).

**Findlay**, Alexander, father and son. **Ref.** FISHER, Susanna. The makers of the 'Blue Back' charts; a history of Imray, Laurie, Norie & Wilson (forthcoming).
• **Findlay**, Alexander (1790-1870). Draughtsman and engraver of Scottish descent, worked at Pentonville, London, then Quality Court, Chancery Lane, London. One of the founding Fellows of the Royal Geographical Society. Engraved charts, 1816-1865 (many compiled by his son, Alexander George Findlay); Abraham Rees's *Cyclopaedia ancient & modern*, 1820 (London editions); R.H. Laurie's *Survey of the Country Around London to the Distance of Thirty-Two Miles from St. Paul's*, 1829 and later; *A Survey of the Estuary of the Thames and Medway*, London, R.H. Laurie 1831, 1839; maps for Thomas Kelly, 1834-1836; maps for *A New and Comprehensive Gazetteer* (5 volumes), London, T. Kelly, 1835 (maps re-used in James Barclay's *A New & Universal English Dictionary*, T. Kelly, 1843); map of Ireland, London, T. Tegg 1845; maps for the *Journal of the Royal Geographical Society*.

*Alexander Findlay FRGS (1790-1870). Portrait bequeathed to the RGS by W.R. Kettle, 1918. (By courtesy of the Royal Geographical Society, London)*

•• **Findlay**, Alexander George (1812-1875). Geographer, chartmaker, draughtsman and engraver, Fellow of the Royal Geographical Society. Son of Alexander (above), with whom he worked in Pentonville, then at 4 Quality Court, Chancery Lane, London (c.1844-1858). Worked with John Purdy on charts and sailing directions for R.H. Laurie, and took charge of the firm's hydrographic work when Purdy died in 1843. In the following years he revised much of Purdy's work. He became executor and successor to R.H. Laurie, first as manager then as proprietor of the firm of R.H. Laurie at 53 Fleet Street (1858-1875). He was succeeded by his nephews Daniel and Henry Kettle. *New South Wales*, London, T. Kelly 1836, 1843 (engraved by A. Findlay); *Chart of the Mediterranean*, 1839; *Australia*, London, R.H. Laurie 1841 (engraved by A. Findlay); *New Zealand*, London, R.H. Laurie 1841 (engraved by A. Findlay); completed Purdy's *New American Navigator*, 1843; *Modern Atlas*, 1843, W. Tegg 1850; *Classical Atlas*, 1847, W. Tegg 1854; *The Bermuda or Somers' Islands...*, London, R.H. Laurie 1847, 1852, 1868 (engraved by A. Findlay); *A Directory for the Navigation of the Pacific Ocean*, R.H. Laurie 1851 (later divided into separate north / south directories); *Map of the routes between Zanzibar and the Great Lakes in Eastern Africa*, 1860 (from the surveys of J.H. Speke, engraved by E. Weller for Richard Burton); *Lighthouses of the World*, 1861 (an annual guide for navigators); *Laurie's General Chart of the Mediterranean Sea, and a portion of The Euxine or Black Sea...*, R.H. Laurie 1866 (engraved by Alexander Findlay); *A Directory for the Navigation of the Indian Ocean*, 1866; *A Directory for the Navigation of the Indian Archipelago, and the Coast of China*, 1869 (and other nautical Directories); contributed maps and charts for the *Journal of the Royal Geographical Society*. **Ref.** HERBERT, F. *The Map Collector* 35 (1986) pp.24-25.

**Findlay**, John. Tahiti, 1877 MS.

**Findlay**, *Lieutenant* J.K. Surveys of the Kennebec River [Maine], 1826 (with others, drawn up by H.A. Wilson).

**Findorf**, Dietrich (1722-1772). Painter and engraver. Lauenburg, [n.d.].

**Findorf**, Friedrich Christoph (*fl.*1784-1805). General Charte von Bremen, 1795.

**Fine**, E.E. Surveyor. Humboldt Mining Region, San Francisco, Gensoul 1864.

**Fine** [Finaeus; Finnaeus], Oronce [Orontius Delphinas] (1494-1555). Renaissance astronomer, mathematician, teacher, cartographer, illustrator and wood engraver from the Dauphinais area of France (his work often incorporates a dolphin to represent his origins). Wrote theoretical works on many subjects including astronomy, astrology, mathematics and music. *La cité de Iherusalem*, 1517 (published in B. Breydenbach's *Le Grant Voyage de Iherusalem...*, 1517, map copied from the work of E. Reuwich); *Recens et integra orbis descriptio*, 1519 MS (world map on a heart-shaped projection, no longer extant), engraved and published 1534-1536 (woodcut, now known in only two examples, served as model for Haci Ahmed's Ottoman Turkish version, Venice 1559, for Giovanni Cimerlini's engraved map of 1566, and a later derivative by Giacomo Franco 1586-1587); *Nova Totius Galliae Descriptio* (6 wood blocks), Paris, Simon de Colines 1525, Paris, Jérome Gourmont 1538, 1546, 1553 (a reduced version was published Venice, Vavassore 1536, also copied by S. Münster, 1540 and others); *Nova, et integra universi orbis descriptio* (woodcut, double-cordiform world map), 1531 and at least 6 states to 1555 (first published in Johann Huttich and Simon Grynaeus's *Novus Orbis Regionum*, Paris 1532); *Chorographiae gallicae in aliarum exemplum descriptio* [outline map of France], 1532 (in Fine's *De Cosmographia, sive Mundi sphaera*); map of the travels of St. Paul in Europe, *c*.1536; *Un partie de la Gaule de huit degrez de longitude, cinq degrez de latitude* [S.E. France], 1551 (woodcut map in his *Sphaera mundi...*, 1551); survey and map of 'le Pré aux Clercs', Paris, 1551 MSS; and other works, including unpublished maps. **Ref.** KARROW, R.W. *Mapmakers of the sixteenth century and their maps* (1993) pp.168-190; KRETSCHMER, DÖRFLINGER and WAWRIK *Lexikon zur Geschichte der Kartographie* (1986) vol.C/1, pp.222-223 (numerous references cited) [in German]; SHIRLEY, R. *The mapping of the World* (1983) N°.66 p.73; *Ibid.* N°.69 p.77; DAINVILLE, F. de. *Imago Mundi* XXIV (1970) pp.49-55; GALLOIS, L. 'La Grande Carte de France d'Oronce Fine' in *Annales de Géographie* (1935) pp.337-348.

**Fineman**, Johann. Swedish surveyor. Wisby [Visby, Sweden], 1797 MS.

*Alexander George Findlay FRGS (1812-1875). (By courtesy of the Royal Geographical Society, London)*

**Finger**, Friedrich August (1808-1888). German geographer.

**Finiels**, Nicolas de. Cours de Mississippi, 1797-1798.

**Fink**, *Major* —. *Plan von Lübeck*, 1872.

**Fink**, Albert (1827-1897). American draughtsman. *Map & profile of the location of the Baltimore & Ohio Rail Road...*, Baltimore, A. Hoen & Co. 1850.

**Finlay**, Anthony. See **Finley**, Anthony.

**Finlay**, *Lieutenant* John, 83rd Regiment. Manuscript map of the island of Guernsey, 1782.

**Finlay**, *Captain* John, *RE*. Plans of the Royal Gunpowder Works at Waltham Abbey, 1800-1801; gunpowder mills at Faversham, Kent (10 plans), 1801 MSS.

**Finlayson**, J. *Mexico*, c.1840.

**Finlayson**, James. Draughtsman working in North America. *Geographical, Statistical and Historical Map of America*, 1822 (engraved by Young & Delleker); *Geographical, Statistical, and Historical Map of North America*, 1822 (larger map on larger scale than above, engraved by J. Yeager); *Geographical, statistical and and historical map of Michigan Territory*, 1822 (in *A complete historical, chronological and geographical American atlas*, Philadelphia, H.C. Carey & I. Lea 1822, 1823, 1827). **N.B.** Compare with James **Finlayson**, publisher, below.

**Finlayson**, James. Publisher of Philadelphia, described as 'Successor to John Melish'. Republished John Melish's *Map of the United States*, 1823 (first published 1816).

**Finlayson**, John. *A Plan of the Battle of Culloden* [Inverness-shire, Scotland], 1746.

**Finley** [Finlay], Anthony (c.1790-1840). Engraver, map and atlas publisher in Philadelphia. *A New General Atlas...* (60 plates), Philadelphia 1824, 1826, and editions to 1834 (maps engraved by Young & Delleker); *A New American Atlas...*, 1826 (maps drawn by D.H. Vance and engraved by J.H. Young, the plates were re-used by S. Augustus Mitchell for his *A New American Atlas*, 1831); *Atlas Classica* (10 maps), 1829 (this work was also bound in with the *New General Atlas* in its 1829 and 1834 editions); revised edition of R. Mills's *Atlas of the State of South Carolina*, 1838 (printed by J. & W. Kite); also produced a series of state maps. **Ref.** RISTOW, W.W. in: WOLTER, J.A. and GRIM, R.E. *Images of the World* Washington (1997) pp.318-320; RISTOW, W.W. *American maps and mapmakers* (1986) pp.268-270.

**Finley**, Josiah (*fl.*1747-*c.*1765). Estate surveyor of Billericay, and later Chelmsford, Essex. Ramsden Bellhouse &c., Essex, 1747 MS; Little Waltham, c.1760 MS. **Ref.** MASON, S. *Essex on the map* (1990) pp.79-80.

**Finley**, Julia Vilett [J.V.F.]. *The tourist's guide to the city of Washington...*, Washington 1889.

**Finne**, *Captain* E. *Omegnen of Mjösen*, 1845; *Palaestina til Skolebrug*, c.1876.

**Finnie**, James. Printer of Melbourne, worked with George Lusty. R. Brough Smyth's *First Sketch of a Geological Map of Australia including Tasmania*, Melbourne, Department of Mines 1875 (drawn and coloured by A. Everett, lithographed by Richard Shepherd).

**Finsch**, *Doctor* Otto (1839-1917). Traveller, zoologist and ethnographer. Brandenburger Küste, 1894; Krakau, 1896.

**Finsler**, *General* Hans Conrad (1765-1839). Cartographer of Zürich, also worked in Bern. *Die Schweizerischen Landvogteyen Lauis und Mendris*, Zurich 1786 (reduced version of P. Neuroni's map of Sottoceneri, 1780).

**Finsterwalder**, Sebastian (1862-1951). Mathematician, cartographer and expert in glaciation. Surveyed the glaciers of the eastern Alps for which purpose he devised specialised surveying instruments, 1887 and later; used a balloon to take aerial surveys and photographs from which he constructed a relief map of Gars am Inn (scale 1:10 000), 1902. **Ref.** KRETSCHMER, DÖRFLINGER & WAWRIK *Lexikon zur Geschichte der Kartographie* (1986) Vol.C/1 pp.223-224 [in German].

**Fiorini**, M. See **Florimi**, Matteo.

**Fiorini, Matteo** (1827-1901). Italian geographer. *Le sfere cosmografiche e... terrestri*, 1894.

**Fiorini, N.** See **Florino**, Nicolo.

**Fire Insurance Maps.** See under C.E. **Goad**; G.T. **Hope**; W.H. **Martin**; and D.A. **Sanborn**.

**Firks, G.** [& Sons]. Publishers of Plymouth. *Westward Ho!* [Devon & Cornwall], *c*.1895.

**Firmin-Didot.** See **Didot**, family.

**Firminger, R.E.** Sketch surveys Kricor, 1886.

**Firth, C.** Lithographer. Proposed new London Bridge, with its approaches on both banks of the river (14 plans) [City of London, Bermondsey and Southwark], 1828-1834. **N.B.** Compare with C.M. **Firth**, below.

**Firth, C.M.** Zincographer of 7 St. Michael's Alley, Cornhill, London. Plan of the City of London showing ward boundaries, Corporation of London, 1839; *A General Plan of the Public Sewers Within the City of London and the Liberties Thereof*, 1841, 1847 (surveyed by R. Kelsey).

**Fisch, J.C.** *Plan de Lyon*, 1875.

**Fischbein, Georg.** Engraver of Georg Sigismund Otto Lasius's *Harz Gebirge*, 1789.

**Fischer, —.** Involved in the production of maps of Turkey, possibly as printer, all published by Simon Schropp & Co. of Berlin. *Plan von Afium-Kara-Hissar*, 1838; *Situations Plan der Stadt Brussa*, 1838; *Plan der Stadt Karaman und der Umgebung*, 1838; *Plan der Umgegend von Kiutachia*, 1838.

**Fischer, A.** *Wand-Karte von Herzogthum Anhalt*, 1875.

**Fischer, Albert.** *Württemberg, c*.1844; *Umgebungen von Wildbad, c*.1846.

**Fischer, C.** One of the engravers of A. Papen's *Topographischer Atlas des Königreichs Hannover und des Herzogthums Braunschweig...* (80 sheets), Hanover 1832-1847. **N.B.** Compare with Carl **Fischer**, below.

**Fischer, Carl.** Saxony, 1836-1850.

**Fischer, Eduard.** *Bukowina* [West Ukraine], 1898, 1899.

**Fischer, Friedrich von** (1826-1907). Austrian general and cartographer.

**Fischer,** *Doctor* G.A. Maps of East Africa, 1891-1895.

**Fischer, Georg Peter.** Draughtsman of Munich. Plans of German towns for Matthäus Merian, 1644 and later.

**Fischer, H.** [& Co.]. Publishers. Africa, 1825; Northwest Germany, 1829.

**Fischer, Hans.** Hauran, 1889; Palästina, 1890.

**Fischer, Hermann von** (*b*.1851). *Geographie der Schwäbischen Mundart (mit Atlas)*, 1895.

**Fischer, J.** Engineer. *Plan von Constantinopel*, 1877.

**Fischer, J.B.** *Das Herzogthum Nassau*, 1828.

**Fischer, Johann Eberhard.** *Sibiriæ Veteris* (2 maps), St. Petersburg 1768.

**Fischer, Johann Christoph** (1729-1801). Cartographer in Darmstadt.

**Fischer, Johann Jakob** (1709-1753). Surveyor of Bern, Switzerland.

**Fischer, Josef,** SJ (1858-1941). Priest and historian, specialist in geography of the 14th-16th centuries. Best known for his works on Ptolemy's *Geographia* and for attempts to demonstrate early cartographical evidence of Norse voyages to North America. **N.B.** It has been suggested that it was Josef Fischer who created the controversial 'Vinland Map'. The map, which shows Norse explorations in Greenland, Iceland and North America, has been dated to *c*.1440, and is now held in Yale University Library. **Ref.** SEAVER, K.A. 'The "Vinland Map": who made it, and why? New light on an old controversy' in *The Map Collector* 70 (1995) pp.32-40.

**Fischer, Max.** *Landeshauptstadt Brünn* [Brno], *c*.1871.

**Fischer,** *Major* N. Klein-Asien, 1844; Plan-Atlas von Klein Asien, 1854.

**Fischer, Theobald** (1846-1910). Geographer, publisher and map historian in Marburg, his publications included reproductions of medieval world maps and sea charts. *Raccolta di mappamondi*, 1871-1881; H. Moehl and W. Keil's *Orohydrographische und Eisenbahn-Wandkarte von Deutschland* (coloured wall chart), Kassel 1879.

**Fischer, Theodor.** Publisher in Kassel. *Handkarte des heiligen Landes*, 1843; *Deutschland's Eisenbahnen*, 1844.

**Fischer, Wilhelm.** Cartographer. Historische und geographische Atlas, Berlin 1834-1837 (with Friedrich Wilhelm Streit).

**Fishbourne, R.W.** (fl.1851-1858). Lithographer of 529 Clay St., San Francisco. G.H. Baker's *Map of the Mining Region of California*, Barber and Baker 1855, 1856; for E.E. Fine, 1864. **N.B.** See also **Nagel**, Fishbourne & Kuchel.

**Fisher**, family and companies. English printers and publishers also known as The Caxton Press. **Ref.** SMITH, D. *Victorian maps of the British Isles* (1985) pp.136-137; GARDINER, R.A. in: WALLIS, H. & TYACKE, S. (Eds.) *My head is a map* (1973) pp.59-60.
• **Fisher, Henry** (d.1837). English publisher and printer of Liverpool. First recorded in 1816 as active in the partnerships of Nuttall, Fisher & Co. q.v., then Nuttall, Fisher & Dixon q.v.. Founded the Caxton Press at Liverpool c.1818; moved to Owens Row, Clerkenwell, and Newgate Street, London (1821); Angel Street, St. Martin le Grand (1823). Map of the States of America, 1823.
•• **Fisher, Robert.** Son of Henry, joined the family business in 1825.
•• **Fisher, Son and Co.** (fl.1825-c.1845). Publishers of London and Paris. The imprint originated in 1825 when Robert joined his father in business. Known as 'Fisher, Son. & Co., Caxton Press, Angel Street, St. Martin's-le-Grand, London; Post-Office Place, Liverpool; Piccadilly, Manchester'. In 1833, the firm was joined by Peter Jackson. Map of Africa, 1825; J.& C. Walker's map of Lancashire, 1831 (used in E. Baines's *History of the County Palatinate and Duchy of Lancaster*, 1836); in 1842 Fisher took over James Gilbert's proposal for a county atlas, and published it as *Fisher's County Atlas of England and Wales*, 1842-1845 (first maps were engraved by Joshua Archer, the remainder were by F.P. Becker & Co.).

••• **Fisher and Jackson.** An imprint of Fisher, Son & Co. used occasionally after the death of Henry Fisher. Succeeded by Peter Jackson c.1843.

**Fisher**, *Lieutenant* Arthur à Court *RE*. Draughtsman. Manuscript plans of Gibraltar, 1852-1854; *Plan of Sevastopol Shewing the Attacks of the Allied Armies in 1854-1855...*, 1856 MS.

**Fisher**, Charles H. *Lake Ontario Shore Rail Road*, 1868.

**Fisher**, I. Engraver. Faeroe Islands, 1781.

**Fisher**, J. Engraver. I. and R. Dewe's *Pictorial Plan of Oxford*, 1850.

**Fisher**, James (fl.1841-1847). Surveyor. Plan of Hurstpierpoint [Sussex], 1841.

**Fisher**, John (fl.c.1670-1703). English surveyor. Survey of Whitehall Palace, 1670, engraved by G. Vertue, 1747; Survey plot of Whitehall, 1680; Oak End, Gerrard's Cross [Buckinghamshire], 1680 MS; Bulstrode Estate [Buckinghamshire], 1686 MS.

**Fisher**, Jonathan. *Waterford* [Ireland], 1772.

**Fisher**, Joshua (1707-1783). Cartographer, conducted surveys of the Delaware Bay and River. *To The Merchants & Insurers of the City of Philadelphia This Chart of Delaware Bay... is Dedicated*, Philadelphia 1756 (engraved by J. Turner, printed by J. Davis), the chart was suppressed to keep the information from the French and was re-engraved at a smaller scale, probably by Henry Dawkins c.1760/1776 (copies were made by William Faden, 1776, Sayer & Bennett in *North American Pilot* part II, 1776, Andrew Dury, 1776, Mount & Page, 1776, George Le Rouge & Dépôt de la Marine). **Ref.** WHEAT & BRUN *Maps and charts published in America before 1800* (1969) N$^{os}$.475-476 pp.102-103.

**Fisher**, Ralph. Plan River Sherbro, Sierra Leone, 1773-1794.

**Fisher**, *Doctor* Richard Swainson. Editor of the Journal of the American Geographical and Statistical Society. Specialised in the descriptive texts and statistical data often found in Colton atlases. Text for Colton's *Atlas of America*,

1854, 1856; *Colton's Atlas of the World... Accompanied by Descriptions Geographical, Statistical and Historical...*, 1855, 1856; *Colton's General Atlas*, 1857 and later; *Johnson's New Illustrated (Steel Plate) Family Atlas*, 1864 and later (Fisher was not credited in the earlier editions of the work).

**Fisher**, Robert. See **Fisher**, family and companies (above).

**Fisher**, Samuel B. *Coal Fields in Pennsylvania*, 1849.

**Fisher**, T. *R.M.S.D.* Ordnance draughtsman. Town and castle of Scarborough, 1812; Hull, 1813; Plymouth, 1814 MSS.

**Fisher**, Thomas J. [& Co.]. Publisher of Washington D.C.. *Map of the City of Washington*, 1884; *Plat and survey and subdivision of 'Washington Heights'...*, 1888 (with Wm. P. Young and Wm. Forsyth).

**Fisher**, Thomas [W.T.] (1741-1810). Part of Pennsylvania and Maryland, 1771, 1789.

**Fisher**, William (fl.1656-1691; d.1692). Publisher and chartmaker 'at the Postern-Gate on Tower Hill', London, member of the Stationers' Company. In 1677 joined John Seller in publication of the *Coasting Pilot*. Seller's interests in *The English Pilot* and the *Atlas Maritimus* were transferred to Fisher in 1679. Later entered into association with John Thornton, Fisher supplying the finance and Thornton the charts, to publish the *English Pilot the Fourth Book* [included the West Indies], 1689. Richard Mount, apprenticed to William Fisher in 1670, married Fisher's daughter Sarah in 1682, was taken into partnership and succeeded him. **Ref.** TYACKE, S.J. London Map-sellers 1660-1720 (1978) p.114; VERNER, C. 'John Seller and the chart trade in seventeenth century England' in THROWER, N.J.W. Ed. The Compleat Plattmaker (1978) pp.149-150.

**Fisher**, Son & Co. See **Fisher**, family and companies (above).

**Fisk**, W.W. Publisher of Boston. *Part of Alaska and Northwest Territory*, 1898.

**Fisk & Russell** (fl.1850s-1869). Engravers of New York. Gaylord Watson's *The American Republic and Rail-Road Map of the United States...*, New York 1867; Chicago, 1869; Hannibal & St. Joseph Rail Road, 1869.

**Fisquet**, Honoré Jean Pierre (1818-1883). *France...l'Algérie et des colonies*, Paris 1878; *Grand atlas départmentale de la France*, 1878.

**Fisquet**, Théodore Auguste. *Chart of the coast of California*, 1851.

**Fisscher**, J.W. van Overmeer. Manuscript plans of West African forts, 1786.

**Fitch**, Asa. Practising physician. Only one cartographic work is known - a survey of Washington County, New York, 1850 (engraved and printed by J.E. Gavit, based on the work of D. Burr, 1829).

**Fitch**, Charles H. *Indian Territory*, 1895-1899, 1902.

**Fitch**, George W. Worked with G.W. Colton on the preparation of atlases, all of which were published in many editions. *Introductory Lessons in Geography*, New York, George Savage 1853; *Colton & Fitch's Introductory School Geography*, New York, J.H. Colton 1856, (maps by G. Woolworth Colton) and other editions including, Ivison & Phinney 1859, Ivison Phinney & Co. 1863, Sheldon & Company 1876; *Colton & Fitch's Modern School Geography*, New York, J.H. Colton & Company 1856 (maps by G. Woolworth Colton) and other editions published by Ivison & Phinney 1857, 1859, Ivison, Phinney & Co. 1860, 1863, Ivison, Phinney, Blakeman & Co. 1867, 1868, 1871 (1867+ editions were revised by Charles Carroll Morgan); *Colton & Fitch's Primer of Geography*, 1866 (maps by G. Woolworth Colton); *Colton & Fitch's School Geography*, 1868, 1870; *Primer of Geography*, 1870.

**Fitch**, Herbert [& Co.]. Lithographers of Bury Street, London E.C. *Pictorial Chart of London And Guide to the Principal Streets, Railways, Places of Interest and Amusement, &c.*, 1890.

**Fitch**, John (1743-1798). American metal craftsman and inventor. *A Map of the North West Parts of the United States of America*, 1785. **Ref.** PHILLIPS, P.L. The rare map of the Northwest 1785 by John Fitch inventor of the steamboat (Washington

D.C. 1916); WINSOR, J. (ed.) [Remarks on John Fitch map], in Massachusetts Historical Society, *Proceedings* 2d Ser., 7 (Boston, 1891-1892) pp.364-366.

**Fitch**, John. Draughtsman of Golden Square, London. Manuscript plans of Church Passage, Piccadilly (4 sheets), 1834.

**Fitch & Fox**. Real estate brokers of Washington. Commissioned *Real estate directory of the City of Washington D.C.* (3 volumes), Washington D.C., E.F.M. Faetz & F.W. Pratt 1874.
• **Fitch, Fox & Brown**. Real estate brokers of Washington D.C. Published *Map of the city of Washington and environs*, c.1885 (copyright A.G. Gedney).

**Fitchatt**, John (1729-1776). English estate surveyor and headmaster in Brentwood, Essex. Prepared manuscript maps and taught surveying. Glebe lands of Great Warley, 1750; Romford & Noak Hill, 1768; farm at Horndon, 1772; Brentwood, 1772.

**Fitton**, H.G. Surveys in Sudan, 1894-1896; Routes to Kuror, 1897.

**Fitz**, Ellen Eliza (*b*.1836). American governess from New Brunswick. Patented a mount for a terrestrial globe, designed to show the passage of the Sun, 1875. *Handbook of the Terrestrial Globe; or, Guide to Fitz'z New method of Mounting and Operating Globes*, Boston 1876, 1878, 1880, 1884; *Fitz Globe*, Boston, Ginn & Heath 1879; *Fitz's New Terrestrial 12-inch Globe*, 1881. **Ref.** DECKER, E. & Van der KROGT, P. *Globes from the Western World* (1993) pp.128-129; WARNER, D.J. *Rittenhouse* Vol.2 No.2 (1987) p.62.

**Fitzer**, Wilhelm (*fl*.1650). Born in London, worked in Frankfurt for Matthäus Merian.

**Fitzgerald**, E.A. Sketch map Southern Alps, 1896.

**Fitzgerald**, G. Projected Canal in Nicaragua, James Wyld 1850-1851.

**Fitzgerald**, J.W. Surveyor. Pringle Parry Sound, 1878.

**Fitzgerald**, *Lieutenant* R. *Plan of the battle of Meanee*, c.1843.

**FitzHenry**, *Captain* Claude Brittain. *Map of the Country around Pietermaritzburg from a sketch by Captain C.B. FitzHenry, 7th Hussars, 1897*, Southampton, Ordnance Survey Office 1907.

**Fitzherbert**, Anthony. Produced what might be described as the first book on surveying in English. It covered estate management and valuation rather than measurement. *Here Begynneth a right Frutefull Mater: and hath to name the boke of Surveyenge and Improvements*, 1523. **N.B.** Some sources give the date 1528 for this work, it is also sometimes credited to Anthony's elder brother John. **Ref.** WOODWARD, D.A. in THROWER, N.J.W. *The compleat plattmaker* (1978) p.160; RICHESON, A.W. *English Land Measuring to 1800: Instruments and practice* (Cambridge, Mass. 1966) pp.31-33.

**Fitzhugh**, Augustine [Anthony] (*fl*.1683-1697). Chartmaker, apprenticed to John Thornton of the Drapers' Company in 1675. 'Living next doare to the Shippe in Virgine Street' and 'at the corner of the Minnories neare little Tower Hill', credited with 9 charts of North America and Asia. *Anglia Nova... Cannada...*, 1693; Coasts of Newfoundland, 1693; draughtsman for Cyprian Southack's chart of Boston harbour, 1694.

**Fitzmaurice**, L.R. Hydrographer, *RN*. Bird Island, 1816; Gough's Island, 1816; River Zaire, 1817; North Sea, 1817-1818; England, 1819-1824; Port Talbot, 1846.

**FitzOsborne** [FitzOsberne], James (*fl*.1629-1656). Estate surveyor of Sussex, England. Manor of Hawards Heth & Trubweeke, 1638.

**Fitzpatrick**, —. See **Henecy & Fitzpatrick**.

**Fitzroy**, Alexander. The Kentuckie Country, 1786 (brochure to accompany edition of John Filson's *This Map of Kentucke*, c.1784 and later).

**Fitzroy**, F. *British Burma, Pegu Division* (4 sheets), 1885 (with W.H. Edgcome).

**FitzRoy**, *Commander* (later *Vice-Admiral*) Robert, *RN* (1805-1865). Navigator and marine surveyor, commander HMS *Beagle* 1828-1836; Governor of New Zealand 1843-1845. Commanded voyage of HMS *Adventure* (1826); and of HMS *Beagle* to Tierra del Fuego (1831-1836), the voyage which included Charles Darwin's visit to the Galapagos Islands in 1835. Surveys of the Straits of Magellan, 1826-1830, used on Admiralty Chart 554 *The Strait of Magalhaens commonly called Magellan*, 1879; South America, 1828-1830; surveys of Tierra del Fuego, 1830-1834 used in Admiralty Chart 1373 entitled *The South-Eastern part of Tierra del Fuego...*, 1841, 1880 (engraved by J. & C. Walker); surveys of the Falkland Islands, 1834 (used on many maps including L. Hebert senior's *Map of the East and West Falkland and Adjacent Islands*, 1836, Admiralty Charts, and maps published by Waterlow & Sons, 1839 and Edward Stanford 1870); *Patagonia. Rio Negro to Cape Three Points*, 1834, published as Admiralty Chart 1288 Sheet X, 1840 (engraved by J. & C. Walker); charts of the Galapagos Islands, 1835 published as *Pacific Ocean / Galapagos Islands*, 1836; *Part of the Isthmus of Darien* [Panama], 1853 (engraved by J. & C. Walker); Cocos Islands, 1856.

**Fitzwilliam**, Brian. Plan of Portsmouth, c.1585 MS.

**Fitzwilliams**, E.C.L. Aberporth Bay, 1860 (with Owen Thomas).

**Fitzwilliams**, Michael. Thought to have drawn a map of Belfast Lough [County Antrim, Ireland], before 1569 (copied by Christopher Saxton and others).

**Fix**, W. *Wandkarte Preussischen Staats*, 1855; *Rheinland u. Westfalen*, 1864.

**Fixlmillner**, Placidus (1721-1791). Made two large quadrants for the Observatory of Kremsmünster, which incorporated telescopes, 1767, 1777.

**Flachat**, Eugène. *Chemin de fer de Paris à Meaux*, 1839.

**Flacourt**, Etienne de (1607-1660). French traveller and administrator in Madagascar. Madagascar, 1656.

**Flahaut**, —. French engraver. [A.E.?] Lapie's *Carte génerale de la Turquie d'Europe en XV feuilles...*, Paris, C. Picquet 1822; Lapie's *Carte Physique, Historique & Routière de la Grèce...*, Paris, C. Picquet 1826; [A.E.?] Lapie's *Carte du pourtour septentrional de la Mer Noire...*, Paris 1829; M. Lapie's *Carte de l'Egypte*, 1829; M. Lapie's *Carte de la Turkie d'Europe...*, 1829; *Carte du Bosphore et des environs de Constantinople...*, c.1829; one of the engravers for *Carte topographique de la France* (12 sheets), Dépôt de la Guerre 1832-1847. **N.B.** Compare with **Flahaut**, Mlle and **Flahaut**, P. (below).

**Flahaut**, Mlle —. Engraver, for Dépôt de la Marine, 1818; for Jean Antoine Letronne, 1827; for A.-H. Dufour, 1836; for J. Andriveau-Goujon, 1836; *Océanie*, Paris, J. Andriveau-Goujon 1854 ('Gravé par Mlle. Flahaut et Smith. Les Ecritures par P. Rousset, Les Eaux par Mme. Fontaine').

**Flahaut**, P. Engraver. Europe, 1841.

**Flamand**, George[s] Barthélemy (1861-1919). French explorer and geologist in North Africa; director of the Service Géologique des Territoires du Sud; taught at the Faculté des Sciences d'Alger from 1905. Worked on a geological map of Algeria, from 1880; *Recherches géologiques et géographiques sur le haut pays de l'Oranie et sur le Sahara*, 1911.

**Flamichon**, —. French surveyor. Worked in the Basses Pyrénées on the Cassini *Carte de France*, sheets 107, 108, 139, 140 with Moyset, 1766-1773, and sheet 138 with Capitaine, 1773-1784.

**Flaminius**, Leo. Itinerarium per Palaestinam (with 7 maps), Rothenburg 1682 (pirated edition of Rauwolff's Travels).

**Flammarion**, Camille (1842-1925). French astronomer. *Atlas astronomique*, 1875; *Atlas céleste*, 1877; *Les étoiles et les curiosités du ciel*, 1882.

**Flammarion**, E. Publishers of Paris. *Guides Flammarion* (15 volumes), 1898-1899 (French route guides prepared by A. Sauvert).

**Flamsteed** [Flamstead], John (1646-1719). English astronomer from Derbyshire, in 1675 he became the first Astronomer Royal, a post which he held until his death (he was succeeded by Edmund Halley). From 1676 he began regular and reliable telescope observations from the Royal Observatory at Greenwich. His star catalogues formed the basis of many of the star positions as depicted on subsequent celestial globes, and his placing of the fixed stars helped in the development of a method for calculating longitude at sea. Published the first of several catalogues of eclipses of Jupiter's satellites in *Philosophical Transactions*, 1683; Catalogue of Observations of Stars, 1707; *Historia Coelestis Britannica* (3 volumes), London 1725 (published posthumously); *Atlas coelestis* (27 star maps to illustrate the *Historia*), London, J. Hodgson 1729 (and other editions).

**Flanagan**, F.W. Cartographer active in New Zealand. Port Adventure, Stewart Island, 1872; Orepuki Town, 1873.

**Flandrus**, G.C. See **Carolus**, Joris.

**Flannigan**, M. Surveyor and draughtsman. Tourists' map of Forestier's and Tasman Peninsulas, 1894-1897.

**Flatters**, *Colonel* Paul François Xavier (1832-1881). French explorer charged with establishing routeways through the Sahara. On his second mission, 1880-1881, he and his companions were murdered by Tuareg tribesmen.

**Flatman** [Flattman], R. Surveyor of Canterbury, worked mostly in Kent. Map of Margate, 1681 MS (now lost); *Littlebourne and Wickham Brookes...*, 1695.

**Flaxman**, Charles. Surveys of the sources of the Rhine, 1839-1843.

**Flèche**, J. le L. de la. See **Le Loyer** de la Flèche.

**Flecheux**, M. (1738-1793). French astronomer and mathematician. *Planétaire ou planisphère nouveau*, 1778.

**Fleck**, *Doctor* E. *Deutsch Süd-West Afrika*, 1899.

**Fleck**, G. *Agricultural lands at Kyneton, Victoria*, 1854.

**Fleck**, Matthew (*fl.*1822-1829). Estate surveyor, produced manuscript maps of parts of Durham and Northumberland. Ryton Inclosures [County Durham] (10 plans), 1829 MSS.

**Fleetwood**, *Sir* William (*fl.*1592-1629). Survey of the Crown Lands, 1610-1629 MSS.

**Flegel**, Edward Robert (1852-1886). German explorer and colonist in Cameroon and Nigeria, 1880-1885.

**Flegel**, K. *Die Erzlagerstätten Oberschlesiens...* [topographical map showing the mines of Upper Silesia], Breslau 1913.

**Fleischer**, Johann Friedrich (1690-1765). Publisher, bookseller and mapseller of Frankfurt am Main.

**Fleischer**, Johann Georg. Publisher in Frankfurt and Leipzig. *Geschichte der See-Reisen nach dem Südmeere*, 1775 (from John Hawkesworth's account of Cook voyages, published 1773); *Der neuen Entdeckungen der Russen zwischen Asien und America...*, 1783 (from the work of William Coxe, 1780).

**Fleischhauer**, W. Hildesheim [Germany], *c.*1880.

**Fleischman**, C.L. *Der Nord amerikanische Landwirth...*, New York 1848.

**Fleischmann**, Andreas (1811-1878). Engraver in Munich.

**Fleischmann**, August Christian. German engraver. *Erster Abriss der Stadt Jerusalem*, *c.*1708 (in a German Bible); *Vorstellung der Stadt Jerusalem*, 1736

**Fleischmann**, F. Worked with Knecht, Rebmann and Sommer on the lithography for *Karte von dem Königreiche Wurtemberg...* (*c.*58 sheets), 1827-1838.

**Fleischmann**, Friedrich (1791-1834). Engraver in Munich. Heligoland, 1814-1815; Battle plans, 1818.

**Fleischmann**, Johannes Joseph. Printer in Nuremberg. For Homann Heirs, 1748.

**Fleming**, George H.F. 'Engineer-in-Charge of Survey of Nigerian Eastern Railway'. *Alternative Routes for Eastern Railway*, 1912; *Nigerian Eastern Railway, Northern Section*, 1914.

*John Flamsteed (1646-1719). A portrait made about 1680 by an unknown artist. (By courtesy of the National Portrait Gallery, London)*

**Fleming**, James. Draughtsman in the Surveyor General's Office, Vryburg, British Bechuanaland. *Sketch Map of the British Protectorate* [Botswana], *c*.1890; *Map of the Surveyed portion of British Bechuanaland*, 1892; surveyed part of British Bechuanaland Protectorate (7 sheets), 1894; Trade routes Transkei, 1899.

**Fleming**, John (1785-1857). Scottish naturalist and mineralogist. Mineralogy of Orkney & Shetland Islands, 1808.

**Fleming**, Peter. *Map of the City of Glasgow and suburbs* (6 sheets), 1807 (engraved by R. Scott), re-issued A. Findlay & W. Turnbull 1820 (reduced version published 1808); Plan of different lines of road between Hamilton and Elvan-foot, 1813.

**Fleming**, *Sir* Sandford Arnot (1827-1915). Scottish civil engineer and land surveyor, arrived in Canada 1845. Appointed Provincial Land Surveyor, later Engineer-in-Chief to the Canadian Pacific Railway. Engraved and sold plans of Ontario towns, harbours etc. Noted at 67 Yonge Street, Toronto (1851). *Plan of the Town of Peterborough Canada West*, Toronto, Scobie & Balfour Lithog. 1846; *Plan of the Town of Cobourg*, Toronto, Scobie & Balfour 1847; *Newcastle & Colborne Districts*, Toronto, Scobie & Balfour 1848; manuscript plans of Toronto Harbour, 1850-1851; *Topographical Plan of the City of Toronto...*, 1851; surveys of urban landholdings, 1855-1857; *Sketch of part of Canada West... Northwest or Saugeen Railway*, Toronto, Maclear & Co. Lith. 1857, 1867; *Sketch shewing the position of the flowing wells at Enniskillen*, 1863; maps to accompany *Report on the Intercolonial Railway Exploratory Survey*, 1864; *Map of the Country to be traversed by the Canadian Pacific Railway...*, 1876 (lithographed by Burland Desbarats); surveys for the Canadian Pacific Railway, 1877, 1879 (lithographed by Burland Desbarats Co., Montreal).

**Flemming**, family and company. **Ref.** KRETSCHMER, DÖRFLINGER & WAWRIK *Lexikon zur Geschichte der Kartographie* (1986) Vol.C/1, pp.226-227 [in German].
• Flemming, Carl von (1806-1878). Founder of the Carl Flemming publishing house. He was succeeded by sons Carl Martin & Georg.
• Flemming, Carl. Publishing company of Glogau and Berlin, run by Carl, then by sons Carl Martin and Georg. The firm was sold to C. Dünnhaupt and H. Müller in 1888. The company closed in 1932. F.H. Handtke's *Schul-Atlas der neueren Erdbeschreibung*, 1840 and later editions; *Vollstandiger Hand-Atlas* (80 sheets), Glogau, 1841 and later (a 30 sheet supplement was added in 1845 to make the *Vollständiger Universal-Atlas*); Rudolph von Wedell's *Historisch-geographischer Hand-Atlas*, Berlin 1843, Glogau 1856; *Reymann's Special-Karte* [Schleswig Holstein], *c*.1860; *Raab's Eisenbahn-Karte von Russland*, 1873; *General-Karte der Europäischen Türkei und des Vladikats Montenegro*, Glogau *c*.1875; *F. Handtke's Special-Karte der Europäischen Türkei*, Glogau *c*.1876; *Karte der Eisenbahnen Mittel-Europa's mit Angabe sämtlicher Bahnstationen Hauptpost- u. Dampfschiff-Verbindungen*, Glogau 1878 (compiled by H. Müller); *F.H. Handtke's Handatlas vom Preuß. Staat*, Glogau 1878; *Dislokationskarte des Deutschen Heeres und seiner Grenznachbarn*, *c*.1900 (compiled by A. Herrich, edited by H. Müller); *Flemmings topographische Karte der Balkanstaaten und Länder Massstab* (4 sheets), Glogau-Berlin, C. Flemming 1908; *Offizielle Automobil-Tourenkarten des kaiserlichen Automobil-Clubs*, 1908; *Flemmings Karte von British Indien*, Berlin and Glogau, C. Flemming 1913; *Der Weg vom Continent nach England*, Berlin & Glogau, C. Flemming 1914; *Flemmings Karte für das türkishe Interessen-Gebiet*, 1914; *Grosse Spezialkarte vom belgischen und französischen...*, 1914; *Flemmings Karte der Grenzgebiete zwischen Russland, Österreich-Ungarn und Deutschland*, Berlin & Glogau 1914; *Flemmings Kriegskarten* (16 maps), 1914.
**N.B.** If the later works above were published by C. Dünnhaupt and H. Müller, this does not seem to be indicated on the maps. The subject matter suggests new material, rather than re-publications of earlier work.
•• Flemming, Carl Martin (1835-1891).
•• Flemming, Georg (1843-1893). Sons of Carl, continued the business from 1878-1888, when the firm was sold to C. Dünnhaupt and H. Müller.

Flemming, M. *Grosser Atlas der Eisenbahnen v. Mittel-Europa*, 1892.

Flender, R. *Schlesien*, 1854; *Special Karte der Krim*, 1855.

Flesher [Fletcher], Miles. Printer of London. John Speed's *Prospect*, William Humble 1646 (miniature edition).

Fletcher, —. Surveyor. *Plan of Birkenhead & Claughton-cum-Grange*, 1858.

Fletcher, Edward Green (*fl*.1775-d.1790). Manuscript maps of parts of Essex, Derbyshire and Nottinghamshire, including plan of Warley Camp, Essex, 1779; Intended canal from Cromford to Langley-Bridge, 1789.

Fletcher, Edward Taylor (1817-1897). Map of the Peninsula of Niagara, 1866.

Fletcher, F. Map of east and west Kootenay [British Columbia], 1897.

Fletcher, Hugh. Geological map of Cape Breton [Canada], 1879.

Fletcher, James. Bookseller of The Turl, Oxford. Re-issued Benjamin Cole's *Twenty miles round Oxford*, *c*.1735 (first published 1706).

**Fletcher**, Joseph (1813-1852). Barrister and statistician. Twelve thematic maps of England and Wales to illustrate his *Moral Statistics of England and Wales*, 1849. **Ref.** SMITH, D. *The Map Collector* 30 (1985) pp.4-6; ROBINSON, A.H. *Early thematic mapping in the history of cartography* (1982) pp.164-166.

**Fletcher**, L. *Isle of Wight*, 1806 (re-issue of T. Baker's 1785 map).

**Fletcher**, Miles. See **Flesher**, Miles.

**Fletcher**, P. *Fletcher & Espin's Map of Matabeleland* (4 sheets), London, Stanford's Geographical Establishment 1896, 1897; *Plan of Bulawayo Township*, 1898 (all with W.M. Espin).

**Fletcher**, Robert H. *Map of the Nez Percé Indian Campaign*, 1877.

**Fletcher**, William. Surveyor for the East India Company. *Plan of Prince of Wales' Island* [Penang], 1820.

**Flett**, J.H. Bird's-eye view of Denver, Colorado, 1881.

**Fleuriais**, George Ernest. *Plan du Mouillage de Campèche* [Mexico], 1865.

**Fleurieu**, C.P.C. de. See **Claret** de Fleurieu, Charles Pierre.

**Fleurieu de Bellevue**, L.B. (1761-1852). French geologist. Works on geology, mineralogy and meteorology, 1790-1847.

**Fleuriot de Langle**, Alphonse Jean René (1809-1881). French navigator. Spitsberg, 1838-1840; Cap de Monte, 1842-1852; Bumbalda, 1843-1845; Baie de Santa Isabel, 1846-1851, all for Dépôt de la Marine.

**Fleury**, E. de. *Nueva Mapa de Sonora &c.* [Mexico], 1864.

**Fleury** [Flury], François Louis Teisseidre [sometimes misnamed Henri] de (fl.1777-1792). French military engineer, served in the American army 1777-1779, he later returned to America as an officer in the French Army under Jean Baptiste de Rochambeau. *A Sketch of the Siege of Fort Schuyler*, 1777 MS; *Mudd Island*, c.1777; *Figure Approximatif du Fort Mifflin*, 1777; *View of the Enemy Fleet before Philadelphia*, 1778; and others. **Ref.** GUTHORN, P. *American maps and mapmakers of the Revolution* (1966) pp.22-23.

**Fleury**. See also **Flury**.

**Flexney**, William. Bookseller of London. One of the many sellers of E. Gibson's edition of William Camden's *Britannia*, 1772 (last edition to contain Robert Morden's maps).

**Fliegner**, Ferdinand. Cartographer and lithographer in Breslau [Wroclaw]. Poland, 1848; Plan of Vienna, c.1848.

**Flierbaum**, G. Harzkarte, c.1896.

**Fligely**, *General* August von (1810-1879). Austrian general and cartographer, worked for the Militärgeographisches Institut Wien.

**Flinders**, *Lieutenant* (later *Captain*) Matthew (1774-1814). Navigator and hydrographer from Donnington, Lincolnshire. Surveyed the coast of New South Wales with George Bass, 1795-1800; circumnavigated Tasmania, 1798-1799 (with Bass); surveyed the west coast of Australia, 1801-1803. Charts of Australia, 1789-1829 later published as an atlas of Australia in *A Voyage to Terra Australis... in the years 1801, 1802 and 1803* (16 charts), London, G. & W. Nichol 1814 (later formed basis of Admiralty Charts for Australian waters); chart of the Bass Strait, 1798 (survey used in *Neptune François*, 1802, and later editions of *Hydrographie Françoise*). **Ref.** SCURFIELD, G. & J.M. *The Map Collector* 62 (1993) pp.14-20; *The Map Collector* 64 (1993) pp.54 (letters).

**Flint**, A.R. American army engineer. *Chart of Owls Head Harbour (Maine) and Projection for a Breakwater...*, 1837 (with T.A. Barton).

**Flint**, J. Map of Louisville, Kentucky, and Vicinity, 1824.

**Flint**, James. *Report on the proposed new road from Perth to Elgin* [Scotland], 1832 (with a map of the line of the road).

**Flint**, Ole Nielsen (1739-1808). Engraver of Copenhagen. Grund-Tegning Kiøbenhavn, 1784.

*Captain Matthew Flinders (1774-1814). (By courtesy of the Bridgeman Art Library, London)*

**Flintoff,** John (*fl*.1751-*d*.1773/1778). Estate surveyor, produced manuscript maps of estates in northern England. Property of the Savoy Hospital at Sutton Grange [Yorkshire], 1751.

**Flinzer,** Theodor (*fl*.1850). Lithographer.

**Flitcroft,** H. (*fl*.1730-1744). Manuscript surveys of Buckingham House and grounds [Westminster], 1744.

**Flocquet,** J.A. See **Floquet,** Jean André.

**Floder,** A. Lithographer and publisher of Vienna. *Strassen-Karte des Koenigreiches Ungarn,* 1832, 1853; *Strassen-Karte der Oesterreichischen Monarchie,* 1835; *F. von Freisauff's Ektypographischer Schul-Atlas für Blinde,* Vienna *c*.1837.

**Floquet** [Flocquet], Jean André. Hydraulic engineer. Planned an irrigation canal to run from the River Durance in France. Work on the 'Canal de Cadenet' began in 1752. *Carte du Cours du Canal de Provence ou Canal d'Aix et de Marseille. c*.1750s.

**Flore,** *Captain* —. Faeroe Islands, 1781.

**Floreda,** Pascual. Zifur [Philippines], 1768 MS.

**Florentinus,** S. See **Buonsignori,** Stefano.

**Florentius**, J. See **Langren** family.

**Flores**, Manuel Antonio. Guyana, 1777 MS.

**Florez**, *Don* I.F. Ferrol Harbour [Spain], 1846; Cape Finisterre to Vigo Bay [Atlantic coast of France & Spain], 1846.

**Flórez de Setién y Huidobro**, *padre* Enrique [Henrique] (1702-1773). Contributed to the monumental religious history of Spain *España Sagrada. Theatro Geographico Historico de la Iglesia de España* (76 volumes), Madrid, Antonio Marin *c.*1764. **Ref.** MÉNDEZ MARTÍNEZ, G. *Cartografía antigua de Galicia* (1994) pp.63-70 *et passim*.

**Floriano** [Floriani, Florianus], Antonio [Antonius]. Italian painter, architect and cartographer, born Udine, worked in Venice. Untitled world map in the form of globe gores (2 sheet copperplate), *c.*1555 (after Gerard Mercator, 1538, possibly engraved by Paolo Cimerlini). **Ref.** GALLO, R. 'Antonio Florian and his mappemonde' in *Imago Mundi* 6 (1949).

**Floriantschitsch** [Florjančič] **von Grienfeld**, Johann Dismas (1691-*c.*1757). Cartographer of Krain. *Ducatus Carnioliae Tabula Chorographica* (12 sheets), Ljubljana 1744 (engraved by A. Kaltschmidt).

**Florianus** [Blommaerts], Johannes [Jan] (1522-1585). Born at Antwerp, teacher at Bergen op Zoom. *Frisiae Orientalis descriptio*, used in the Ortelius *Theatrum Orbis Terrarum*, 1579; *Frisia orientalis*, 1595 (replaced the above map in the *Theatrum*). **Ref.** MEURER, P. *Fontes cartographici Orteliani...* (1991) p.145-146 [in German].

**Florimi** [Fiorimi], Matteo (*fl.c.*1570-1612). Publisher in Siena. *Chorographia Tusciae*, *c.*1570 (one of many copies of Girolamo Bellarmati's map of 1536); *Hungaria*, Siena *c.*1595 (from Mercator's map); *Nova descrittion della Lombardia...*, *c.*1602 (copied from G. Gastaldi's map dated 1559); *Geographia Moderna De Tutta La Italia*, [n.d.] (copied from Gastaldi's map of 1561); Orvieto, 1600; Totius Terre Promissionis, 1600; many other undated maps in so-called 'Lafreri' atlases. **Ref.** VAN DER HEIJDEN, H. in: KÖHL, P.H. & MEURER, P.H. (Eds.) *Florilegium Cartographicum* Leipzig, Bad Neustadt a.d.Saal. (1993) pp.117-130 [in German].

**Florin**, Johan (1739-1796). Swedish land surveyor.

**Florino** [Fiorini], Nicolo. Portolan maker of Venice. Chart of the Mediterranean (8 charts on 2 sheets), 1462; chart, 1489.

**Floris**. See **Berckenrode**, Floris Balthasar van.

**Flosie**, Michael (1724-1794). Engraver. Flensburg, [n.d.].

**Flotte-Roquevaire**, René de. French surveyor in Morocco. *Carte du Maroc* (2 sheets), 1897. **Ref.** BROC, N. *Dictionnaire illustré des explorateurs Français du XIXe siècle: I Afrique* (1988) pp.142-143 [in French].

**Flower**, W.L. *Map of Cook County, Illinois*, Chicago, S.H. Burhans & J. Van Vechten 1862; Chicago, drawn for Davie's *Atlas*, 1863.

**Floyd**, Charles. *Siège de Maestricht 1632*, 1633.

**Floyd**, Thomas. Midshipman aboard the *Racehorse*. Manuscript maps and views for Constantine John Phipps's *A voyage towards the North Pole*, including *The track of the Racehorse and Carcass on a voyage of discovery towards the North Pole in the year 1773*, 1774; *Part of Spitsbergen*, 1773; *Part of the North East Land*, 1773.

**Floyd**, William H. Jr. [& Co.]. Atlas of St. Joseph [Missouri], 1884.

**Floyer**, Ernest A.S. Persia, 1882; Egypt, Eastern Desert (3 sheets), 1891.

**Fludd**, Robert. Disertatio cosmographicam, 1621.

**Fluddes** [Fluddus; Fludus]. See **Lluyd**, H.

**Flrul**, Mathias von (1756-1823). Bavarian mining officer. *Gebürgs-Karte von Baiern*, 1792.

**Flury** [Fleury], —. Pérou, 1824 (with Joseph Lartigue).

**Flury**, F. [or Henri]. See **Fleury**, François Louis Teisseidre.

**Flury**, L. Sketch of the siege Ft. Schuyler, New York, *c.*1850.

**Flyn,** J. Engraver. *A New and Correct Plan of London Westminster and Southwark...*, 1770 (for H. Chamberlain's *A new and complete history and survey of the cities of London and Westminster*).

**Flynn,** John. Florence, Missouri, 1857.

**Flynn,** Thomas. Atlas of Newport, Rhode Island, 1893 (with C.L. Elliott); Atlas of the Suburbs of Cleveland, Ohio, 1898.

**Foa,** Edouard (1862-1901). French explorer from Algeria. Travelled the Zambezi, Tanganyika and the Congo, 1893-1897. On his return to France in 1897 he was awarded a gold medal by the Société de Géographie in Paris. Amongst his contributions were maps of the Zambesi River, 1893; *Pays entre Zambèze et Chiré*, 1893; Tanganyika, 1894. **Ref.** BROC, N. Dictionnaire illustré des explorateurs Français du XIXe siècle: I Afrique (1988) pp.143-145.

**Focht,** *Colonel* —. *Carte générale du Gouvernement de Voronège* (5 sheets), 1780 MS.

**Fock,** H. (1766-1822). Dutch engraver. Haarlem, 1801.

**Focken,** Hendrik. See **Fokken,** Hendrik.

**Fodor,** Ferenc (1887-1962). Cartographer and historian from Tenke, Hungary. Worked in the Kogutowicz Institute where he made many maps. d.Budapest. Researched the history of Hungarian cartography, *A magyar térképirás* [Hungarian cartography], Budapest 1952-1954.

**Foeh** [Föh], —. Hafen von Marca [East Africa], 1875.

**Foeltz-Eberle** [Föltz-Eberle], E. Drew and lithographed *Karte von Texas*, 1839; *Grundriss von Frankfurt*, c.1868.

**Foerster** [Förster], *Doctor* Fr. *Neuester Plan von Wien*, 1872.

**Foerster** [Förster], H.G. *Plan von Leipzig*, 1878.

**Foerster** [Förster], L. Plan of Trieste, c.1860.

**Foerstner,** E. *Carta geologica dell'isola di Pantelleria* [Sicilian Channel], 1881.

**Foetterle** [Fötterle], Franz (1823-1876). Geological maps including: *Geologische Übersichts-Karte von Süd-Amerika*, 1856; *Geologische Karte der Markgrafschaft Mähren und des Herzogthumes Schlesien* (2 sheets), Vienna 1866; *Übersichtskarte des Vorkommens, der Production und Circulation des mineralischen Brennstoffes* [Austrian Empire], 1868.

**Foin,** Augustin (*b.*1726). French engraver, worked on 4 sheets of the Cassini *Carte de France*, published from 1756-c.1763.

**Foisse,** Jacques de. Engraver in Hamburg, for Hess's Topographie, 1785.

**Fokken** [Focken], Hendrik. Publisher 'inde Molsteegh, Amsterdam'. *Genève*, 1650; Gibraltar, c.1650; Batavia, c.1680.

**Folderbach,** A. (*d.*1656). Land surveyor of Friesland.

**Foley,** *Captain* Thomas. Soundings for a plan of the Haven of Ferrol, Spain, 1751; manuscript plans in the Bahamas including *A Plan of the Islands of Ilathera, Abbacco and Providence &c...*, 1765; *A Plan and View of the Harbour and Town of Providence...* [Bahamas], 1765.

**Folger,** L.B. Engraver. *Utah*, for W.H. DuPuy's *Atlas*, New York, Phillips & Hunt 1886; another map of Utah in *The Columbian Atlas of the World We Live In*, New York, Hunt & Eaton 1893.

**Folger,** *Captain* Timothy. Mariner and whale fisherman from Nantucket, cousin and friend of Benjamin Franklin. His experience, advice and sketch map formed the basis of Franklin's Gulf Stream charts including *A Chart of the Gulf Stream* (with accompanying *Remarks* compiled by Folger), London, Mount & Page 1769 and other editions including a French edition, Paris, Le Rouge early 1780s. **Ref.** COHN, E.R. Imago Mundi 52 (2000) pp.124-142; DE VORSEY, L. The Map Collector 15 (1981) pp.6-7.

**Folie,** A.P. French geographer and draughtsman active in the United States from 1791. *Plan of the Town of Baltimore and Its Environs...*, Philadelphia 1792, Ghequiere & Holmes 1794 (engraved by James Poupard); *To*

*Thomas Mifflin Governor... This Plan of the City and Suburbs of Philadelphia...*, 1794 (engraved by R. Scot & S. Allardice).

**Foligne**, *Lieutenant* — de. Plan de Cap François [San Domingo], 1781.

**Folino** [Folin, Follino], Bartolomeo [Bartlomiej] (1730-*c*.1808). 'Capitaine du Corps d'Artillerie de la Couronne de Pologne'. Engraver of *Carte générale et nouvelle de toutte la Pologne...*, Warsaw, Michael Groell 1770, 1772, re-issued F.A. Schrämbl 1788, 1801.

**Folkard**, Thomas (*fl*.1720-1727). East Anglian estate surveyor. Mendham [Suffolk], 1721; Stokesby Common [Norfolk], 1721 MSS.

**Folkema**, T. Engraver for A. Allard, also for Schotanus à Sterringa, 1717.

**Folkingham**, William (*fl*.1605-1622). Land surveyor of Lincolnshire. *Feudigraphia: The Synopsis or Epitome of Surveying Methodized*, London, 1610, [an early guide to the use of colour on land-use surveys].

**Follenweider**, R. (1779-1847). German engraver. Plans of Swiss and German towns.

**Follin**, O.W. *Plan of the Isthmus of Tehuantepec* [Mexico], 1851.

**Folque**, *General* Filippe. Director-General of 'Trabalhos Geodesicos do Reino', Portugal. Map of Portugal, *c*.1865; *Carta topographica da Cidade de Lisboa...*, 1871 (engraved by Mesquita); Plano hydrografico do Lisboa, 1878.

**Folsom**, Charles J. Publisher of 40 Fulton St., New York. Mexico & Texas, 1842.

**Folsom**, P. *Map of McLean County, Illinois*, Hance and Taylor 1856.

**Fomine**, Alexander (1713-1802). Russian historian and geographer. Description of the White Sea, 1797.

**Fonbonne** [Fombonne; Fonbone], Quirin [Qt.] (*b.c*.1680). Engraver, worked in Amsterdam *c*.1705, in Paris 1714-1734. Engraved three maps for A.F. Frézier's *Relation du Voyage de la Mer du Sud aux Côtes du Chily et du Perou...1712, 1713 et 1714 par M. Frézier*, Paris, J.-C. Nyon 1716; L.P. Daudet's *Carte des environs de la Ville de Reims*, Paris, G. Demortain 1722, *c*.1740; engraver for André Desquinemare's *Plan générale de la Forest de Fontainebleau...* (2 sheets), Paris, Desquinemare 1727 (script engraved by Jenvilliers); *Carte générale des duchez de Lorraine et de Bar*, published 1746.

**Foncet**, —. France and Savoie (14 sheets), Paris 1760 (with Bourset). **N.B.** The Bourset credited on this map could well be **Bourcet** q.v., though this has not been assumed.

**Foncin**, Pierre (1841-1916). French historian and geographer from Limoges. Géographie historique, 1888; *Géographie générale*, Paris, A. Colin 1889.

**Fonné**, E. *Plan et carte des environs de Toulon*, Lyon, G. Toursier 1909.

**Fonseca**, J.G. da. See **Gonsalves** da Fonseca.

**Font**, *Fr*. Pedro (*d*.1781). Franciscan missionary and cartographer. Manuscript maps drawn into his diary 1776-1777, including *Plan o mapa del viage que hicimos desde Monterey al puerto de San Francisco*, 1777 MS. **Ref.** BOLTON, H.E. *Font's Complete Diary Translated...* (Berkeley 1933); RICHMAN, I.B. *Three maps...reproduced in facsimile...* (Providence 1911).

**Fontaine**. See also **La Fontaine**.

**Fontaine**, —. Yazoo County, Missouri, 1874 (with J.M. Mercer).

**Fontaine**, *Mme* —. Engraver who specialised in portraying areas of water. Océanie, Paris, J. Andriveau-Goujon 1854 ('Gravé par Mlle. Flahaut et Smith. Les Ecritures par P. Rousset, Les Eaux par Mme. Fontaine'); engraved the water on *Chemins de fer d'Espagne et de Portugal*, Paris, E. Andriveau-Goujon 1863 (railways by Alfred Potiquet, outline engraved by Smith, calligraphy by P. Rousset, mountains by Gérin); one of the engravers for *Carte orographique hydrographique et routière de l'Empire français...*, Paris, E. Andriveau-Goujon 1870; Suisse, 1880; Europe, 1882; one of the engravers for *Algérie....*, Paris, Andriveau-Goujon 1890 (with Gérin, Dalmont & Smith).

**Fontaine**, G. French surveyor. Worked with

others on at least 6 sheets of the Cassini *Carte de France*, 1759-1780.

**Fontaine** [Fontanus], Jacques [Jacobus] de la. *Territorium Bergense*, used by Antonius Sanderus in *Flandria illustrata*, 1641-1644, also by Joan Blaeu, 1662; *Berga S.ti Winoci*, used in Blaeu's *Toonneel der Steden* [Townbooks of the Netherlands], 1649 (also used by Sanderus).

**Fontaine**, P.F.L. *Plans de plusieurs châteaux*, 1831.

**Fontaine**, *Reverend* Peter. Chaplain and surveyor to Virginian Commissioners, 1749. Virginia and North Carolina, 1752 MS.

**Fontán Rodríguez**, *Doctor* Domingo (1788-1866). Spanish mathematician and geographer. Two maps of parts of Galicia, *c.*1823; *Plano del País adyacente a las tres rias de la Coruña, Betanzos y Ferrol*, Madrid 1826 (used in S. de Miñano's *Diccionario Geografico-Estadistico de España y Portugal*); *Plano del corrigimiento de La Cañiza...*, *c.*1826 MS; *Plano de la Ría de Arosa*, Madrid 1828 (lithographed by José Santiró, used in Miñano's *Diccionario*); *Carta Geométrica de Galicia* (12 sheets), Paris 1845 (engraved by L. Bouffard, lithographed by Lemercier, used as a source for later maps).

**Fontana**, Francesco (*c.*1580-*c.*1656). Italian mathematician and astronomer. *Novae Coelestium Terrestriumque Rerum Observationes* [atlas of the phases of the moon] (29 maps), Naples 1646.

**Fontana**, Giovanni Baptista (1525-1587). Artist and engraver of Verona, worked in Innsbruck, court painter to Archduke Ferdinand. Palestine, for Bolognini Zaltieri, 1569.

**Fontana**, Jacomo. Cartographer of Ancona. Ancona, 1569 (used by Braun and Hogenberg).

**Fontaney**, Jean de, *SJ*. Mathematician and missionary in south east Asia and China. *Nagasaki Appellé par le Chinois Tchangki*, 1704 (published in volume 8 of *Lettres édifiantes et curieuses*, Paris 1708), later used by Joseph Stöcklein, 1726, and Emanuel Bowen, 1743.

**Fontanus**, Jacobus. See **Fontaine** Jacques de la.

**Fonte**, *Admiral* Bartholome de. A track of his apocryphal discovery of a north-west passage on a voyage in 1640, appeared on various maps of the mid-18th century, including those of Delisle and Le Rouge. The reports of his voyage, on which these maps were based, were later discounted. **Ref.** HECKROTTE, W. & DAHL, E.H. *The Map Collector* 64 (Autumn 1993) pp.18-23.

**Fonten**, G.G. (*fl.*early-19th century). Swedish surveyor.

**Fontenelle**, —. (*fl.c.*1670s). French engraver. 'L'Empire de la poésie', fantasy map published in *Mercure Galant*, 1678.

**Fonton**, Feliks Petrovich (1801-1862). *Plan du siège et de l'assault d'Akhaltsikh*, 1830 (engraved by C. Avril); *Atlas Russie dans l'Asie-Mineure*, 1840 (engraved by C. Avril).

**Fonville**, — de. New France, Québec 1699 MS.

**Fooks**, C.E. Architect, surveyor and publisher of Lichfield St., Christchurch, New Zealand. *Christchurch, Canterbury, New Zealand*, 1862.

**Foord**, William (*fl.*1767-1771). Estate surveyor, worked in Monmouthshire. Caldicot, 1771 MS; the township of Shirenewton, 1771 MS.

**Foot**, Peter. Surveyor of Dean Street, Soho, London. Middlesex, 1794.

**Foot**, Thomas. Principal engraver to the Board of Ordnance, 'Weston Place, St. Pancras', also 'Weston Place, Battle Bridge', London. Engraved for John Meares's *Voyage...*, 1790; *Geometrical Survey of the Gulf of Naples... communicated by Alexander Dalrymple Esqr.*, and other plates in J.F. Dessiou's *Le Petit Neptune Français*, W. Faden 1793, 1805 (3rd edition); for Aaron Arrowsmith, 1794-1798; *Wallis's New and Correct Plan of London and Westminster...*, John Wallis 1795; William Faden's St. Eustatius, 1795; W. Gardner, T. Yeakell and T. Gream's *A Topographical Map of the County of Sussex...*, 1795; *A Chart of West Falkland Island...*, London, A. Arrowsmith 1797 (surveyed by Lt. Thomas Edgar); *A Chart shewing part of the Coast of N.W. America...* (2 sheets - Alaska and California), London, J. Edwards and G.

Robinson 1798; *French Republic*, 1798; William Yates's County of Stafford, W. Faden 1798; William Mudge's *Kent*, 1801 (the first Ordnance Survey map); Thomas Joseph Ellis's *Map of the County of Huntingdon* (4 sheets), 1824, 1829; Nottinghamshire (4 sheets), T.J. Ellis 1825, 1827.

**Foote**, family and companies. **Ref.** RISTOW, W.W. *American maps and mapmakers* (Detroit 1985) pp.410-412.
• **Foote**, Charles M. (*b*.1849). Worked in partnership with George E. Warner as Warner & Foote [*q.v.*], from *c*.1876 until 1886, then worked alone and in association with Edwin C. Hood and John W. Henion. He was succeeded by his son Ernest in 1899. Plat books of Brown, Dunn and Wanpaca counties, Wisconsin, 1889-1890 (with W.S. Brown).
• **Charles M. Foote and Company** (*fl*.1886-1899). American publisher of Minneapolis, successor to Warner & Foote. Specialised in maps and atlases of counties in the states of Wisconsin, Iowa, Michigan and Minnesota.
•• **Foote**, Ernest B. (*fl*.1899-1903). Succeeded his father as publisher of U.S. county atlases.

**Foote**, J. Chart of the bay of Tetuan [Morocco], 1802 (inset on a chart of the Mediterranean in *The Marine Atlas*, London, William Heather 1808).

**Foote & Davies Co.** Bird's-eye view of Atlanta, Georgia, 1919.

**Foppens**, François. Publisher of Brussels. Published Jean-Baptiste Chrystin's *Les Délices des Pais-Bas*, 1711, 1713.

**Foppens**, J.F. Bibliotheca Belgica, 1739.

**Foppiani**, Celestino Luigi. *Genova pianta topografica*, Genoa 1846, *c*.1860.

**Foquett**, William. *A map of the Isle of Wight...*, Newport, Isle of Wight 1818.

**Fora**, — de la. See **Lafora**, Nicolás de.

**Forbes**, Alexander. Topographical draughtsman and surveyor. *Galloway*, 1690; *Ground Plott of Coventre*, 1691; *Newcastle under Lyne*, 1691 MSS; *Walthamstow*, 1699; plan of Barcelona, Anna Beeck *c*.1705.

**Forbes**, Archibald. *General gazetteer or geographical dictionary*, London, T. Tegg 1823.

**Forbes**, Edward. Co-operated with Alexander Keith Johnston. *Physical Atlas*, 1848; Palaeontological map of the British Isles, 1850. **N.B.** Compare with Dr. James D. **Forbes** (below).

**Forbes**, H.O. Timor Laut, 1882; South East New Guinea, Brisbane 1889.

**Forbes**, James (1749-1819). Explorer and geographer. Oriental Memoirs, 1813.

**Forbes**, *Doctor* James D. (1809-1868). Scottish geologist. Assisted Alexander Keith Johnston in *Physical Atlas*, 1848. **N.B.** Compare with Edward **Forbes** (above).

**Forbes**, W.H. Civil War Maps of Richmond, Virginia, 1863-1864.

**Forbes**, William. Chart of Blackwood Bay, 1836 (inset on a chart of part of the northeast coast of Australia in *The Complete East India Pilot*, London, J.W. Norie 1838).

**Forbiger**, Albert (*fl*.19th century). Attempted the reconstruction of a classical world map, based on the work of Strabo (*c*.64 BC-AD 21). *Orbis terrarum antiquus*, 1842 and later.

**Forbriger**. See **Ehrgott**, Forbriger & Co.

**Force**, Jean-Aymar Piganiol de la. See **Piganiol** de la Force, Jean-Aymar.

**Forchhammer**, brothers.
• **Forchhammer**, Johann Georg (1794-1865). Danish geologist. Works on geology and prehistoric Scandinavia. *Geognostische Karte der Herzogthümer Schleswig und Holstein*, 1847.
• **Forchhammer**, P.W. (1803-1894). Topographer, brother of Johann. *Topographie von Athen*, 1841; *Tenedos u. Festlande*, 1856; *Die Ebere von Troja*, 1850.

**Ford**, *Captain* Augustus. Officer in the U.S. Navy. *Chart of Lake Ontario*, New York 1836 (lithographed by Curriers Lithography), similar smaller map published Oswego, Henry Adriance 1836 (lithographed by Sarony & Major); surveys also used as one of the sources for *Chart of Lake Ontario*, Toronto, W.C. Chewett 1863.

**Ford** [Forde], *Sir* Edward (1605-1670). *A map of the Colne with all its branches* [Essex], 1641, 1720; *A Designe for bringing a Navigable river from Rickmansworth in Hertfordshire to St Gyles in the Fields*, London, J. Clarke 1641, 1720.

**Ford**, James (*d.c.*1812). Engraver of Dublin. Lake of Killarney, *c.*1786; Ancient Ireland (2 maps on one sheet), 1800; D. Cahill's Queens County [Ireland] (4 sheets), 1806.

**Ford**, Richard. Joint publisher of Daniel Neal's *History of New England*, London, J. Clark, R. Ford and R. Cruttenden 1720 edition.

**Ford**, Reuben W. City of Austin, Texas, 1872.

**Ford** [Forde], Richard. Surveyor. *A New Map of the Island of Barbadoes*, London, John Overton, Robert Morden, William Berry and Joseph Pask, *c.*1681, London, Philip Lea and John Seller, 1681-1682, London, P. Lea 1685, London, George Willdey *c.*1730; Ford's map was also used as a source for maps by J. Seller, 1682, 1686, 1690 and H. Moll, 1717.

**Ford**, Richard (1796-1858). *Travelling map of Spain*, 1845.

**Ford**, *Major* W.H. Endorsed manuscript plans of Dover, 1811, 1817.

**Ford & West**. Lithographers of 54 Hatton Garden, London. Oregon, 1855.

**Forde**, Edward. See **Ford**, *Sir* Edward.

**Forde**, Richard. See **Ford**, Richard.

**Fordyce**, *General* Charles Francis. Maps of the Ukraine. Plan of the Heights of Inkermann, 1854-1855; Battlefield of Alma, *c.*1855.

**Fordyce**, John (*fl.*1794-1808). Surveyor General of Land Revenue, worked in London, Dorset and Lincolnshire. Directed a survey of Crown leases within the cities of London and Westminster and the Borough of Southwark, 1804 MS; improvements Hamilton Place & Piccadilly, 1805-1812.

**Fordyce**, *Captain* (later *Lieutenant-General Sir*) John. District of Agra, Calcutta 1837-1846.

**Forel**, François-Alphonse (*b.*1841). Swiss physician and glacier scientist. Lac Leman, 1892, 1895.

**Fores**, D. See **Dumas** de Fores, Isaac.

**Fores**, Samuel W. (*fl.*1789-1836). Publisher of Piccadilly, London, at various addresses: N°.3 (1789), N°.50 (1807), N°.41 (1829-1836). Specialised in maps of London, satirical cartoon maps and sporting prints. *A New Plan of London...*, 1789, 1807; *Fores's New Plan of London Including the New Improvements*, 1822 (engraved by N.R. Hewitt); *Fores's New Plan of London including all the late Improvements*, 1829, 1836.

**Fores**, W. Travelling Companion London to the Isle of Wight, 1819.

**Forest**, F. de Belle. See **Belleforest**, François de.

**Forest**, Jules. Publisher of Paris. *États des Balkans et Roumanie*, 1917: M. Fallex's *Allemagne, royaume de Prusse et empire d'Autriche...*, *c.*1917.

**Forestier**, Jacques de (*fl.*16th century). Printer of Rouen. Routier de la mer.

**Foret**, Auguste. Sénégal, 1888.

**Forlani** [Furlani], Paolo (*fl.*1560-1571). Copper plate engraver and publisher from Verona, worked in Venice 'al segno della colonna in merzaria' and also 'in Merzaria alla libreria della Nave'. Worked with G.F. Camocio, F. Bertelli and B. Zaltieri. Prolific output includes: *Mapa de España y Portugal*, 1559; world map, G.F. Camocio 1560; *Navigationi dil mondo nouo*, Venice, G.F. Camocio 1560, 1563, *c.*1575; *Graeciae chorographiae...*, 1562; *Descrittione di tutta la Toscana*, 1563; *Il Disegno del discoperto della noua Franza...*, Venice 1565, re-issued with the imprint of Bolognino Zaltieri 1566 (possibly the first printed map exclusively devoted to North America); *La Discrittone di Tutto il Peru*, 1565; *Venetia...*, 1565; *Africa a veteribus...*, 1566; *Vera et ultima discrittione di tutta l'Austria, Ungheria, Transilvania, Dalmatia, et altri paesi come nel disegna apare*, 1566; *Totius Galliae Exactissima Descriptio*, 1566; *Il Primo Libro Delle Citta, et Fortezze Principali del Mondo* [town book], 1567; *...descrittione del Regno di*

*Polonia*, 1568; Diogo Homem's portolan chart of the Mediterranean Sea, Venice 1569, re-published Rome, A. Lafreri 1572; *Al Mag.co Sig.or Antonio Tognale Sig.or Mio Osser.mo ... tutte le navigationi del Mondo Nuouo* [North and South America] (2 sheets), 1570, Simon Pinargenti 1574 (based on the maps of G. Gastaldi); Cyprus, 1570; maps for Camocio's atlas of the wars between Venice and the Turks, *c.*1572. **Ref.** WOODWARD, D. *Imago Mundi* 46 (1994) pp.29-40; WOODWARD, D. *Imago Mundi* 44 (1992) pp.45-64; WOODWARD, D. *The maps and prints of Paolo Forlani. A descriptive bibliography* (1990); TOOLEY, R.V. 'Maps in Italian Atlases...', *Imago Mundi* 3 (1939) pp.12-47.

**Forling**, Gustaf Johann (*fl*.mid-19th century). Swedish land surveyor.

**Formaleoni**, Ab.V. Script engraver. Antonio Zatta's *L'Africa divisa Ne'Suoi Principali Stati Di Nuova Projezione*, Venice, A. Zatta 1776.

**Formaleoni**, Vincenzio Antonio (1752-1797). Historian, geographer and poet. *Teatro della guerra...*, Venice 1781; *Essai Sur La Marine Ancienne Des Vénitiens*, Venice 1788 (with 2 maps reproduced from Andrea Bianco).

**Forman**, J.E. Civil engineer, Dubuque, Iowa. Nebraska Railroads, 1857.

**Formentus**, —. Engraver. *Moncalieri*, Joan Blaeu 1682.

**Formiggini**, —. Publisher of Milan. *Atlante ad Uso del Nuovo Dizionario Geografico* (6 maps), 1814 (maps engraved by the Bordiga brothers).

**Fornari**, Mauro. *Provincia di Lodi*, 1789.

**Fornaseri**, Jacques de. Italian engraver. *Siege of Bricherasio*, 1594.

**Förnebohn**, A.E. *Geologiska översigs-karta öfver Sveriges ...*, 1873.

**Fornel** [Fournel], Louis (1698-1745). *Coste des Eskimaux* [Labrador], 1748 MS.

**Forrest**, *Lieutenant* C.R. Chart Lake Borgne [Haiti?], 1815.

**Forrest**, John (later *Sir* John, *Baron* Forrest of Bunbury) (1847-1918). Explorer and statesman, from Bunbury, Western Australia. Explored Western Australia, 1869-1874; Surveyor-General and Commissioner for Crown Lands; Member, later Premier of the Legislative Council in Western Australia (1890-1901). Strong advocate of the mapping of Australia. *Map Showing the route of the West Australian Exploring Expedition... commanded by John Forrest F.R.G.S.*, 1874.

**Forrest**, *Captain* Thomas (1735-1802). English navigator, discovered the Forrest Strait. Voyage to New Guinea and the Moluccas, 1774-1776, published 1779, French edition with 24 maps and charts published 1780; charts of East Indies harbours and islands, 1781-1786, surveys published by Alexander Dalrymple 1792, in Laurie & Whittle's *East India Pilot*, 1800, and by John Norie, 1833; Keysers Bay, 1774; *Journal... Bengal to Quedah*, 1783-1789.

**Forrest**, William (*fl*.1799-1832). Surveyor in Scotland. Drumpellier & Coats, 1801 MS; Haddingtonshire [East Lothian] (4 sheets), 1799-1801 (engraved by James and Robert Kirkwood); Plan of the estate of North Berwick, 1804; The County of Lanark (8 sheets), 1813 published 1816, 1818 (engraved by J. & G. Menzies); Linlithgowshire [West Lothian] (2 sheets), 1817 published 1818.

**Forrester**, A. Lithographer of Edinburgh. William Bald's *Plan of the Island of Benbecula*, 1805; Bald's *Plan of the Island of South Uist*, 1805; Plan of the town of Inverary, 1825 (for J. Wood's *Town Atlas of Scotland*).

**Forrester**, Joseph James (1809-1861). Wine grower in the Alto Douro, Portugal. Won the Oliveira prize for an essay on Portugal which included a map. *Map of the Wine District of the Alto-Douro*, London, Royston & Brown? 1843, also published Edinburgh, J. Menzies, London, John Weale 1853, 1854 (engraved by J. Wyld); *The Portuguese Douro and the Adjacent Country...* (2 sheets), 1848, 1852 (engraved by W. Hughes).

**Forrester**, W. Lithographer. *Map of the County of Edinburgh...*, 1850.

**Forresters & Co.** Lithographers of Edinburgh. J. Boulton's plan of estate at Strichen [Aberdeen, Scotland], *c.*1860. **N.B.** Compare with A. **Forrester** and W. **Forrester**, above.

*Captain Thomas Forrest (1735-1802). Engraving by I.K. Sherwin, artist William Sharp. (By courtesy of the National Library of Australia)*

**Forsell** [Forssell], Carl Gustaf (1783-1848). Mathematician and cartographer of Stockholm. From 1806-1808 he worked on maps for S.G. Hermelin's *Geographiske Chartor öfver Swerige*, including *Karta Öfver Göteborgs och Bohus Läns Höfdingdöme*, 1806 (engraved by S. Andersson); *Karte öfver Halmstads Höfingdöme*, 1807 (engraved by S. Anderson); *Karta öfver Skaraborgs*, 1807 (engraved by S. Andersson); *Karta öfver Elfsborgs...*, 1808 (engraved by S. Andersson); Sweden and Norway (9 sheets), 1815-1826; Göta Canal, 1823 (with Jacob Forsell, below); map of the southern part of Scandinavian peninsula, 1830 (early use of colour to show altitude zones).

**Forsell**, Jacob. Göta Canal, 1823 (with Carl Gustaf Forsell).

**Forsell**, Lars. Finnish surveyor. Plan of Helsinki, 1696.

**Forshaw**, W. Lithographer of Liverpool. R.B. Hughes's *Map of the Rivers Parana and Paraguay...* (4 sheets), 1842.

**Forsmann**, M. Transvaal, 1868.

**Forssell**. See **Forsell**, Carl Gustaf.

**Forster**. See also **Foerster**.

**Forster**, father and son. Ref. DAVID, A (ed.) The charts and coastal views of Captain Cook's voyage Vol. 2 (1992) pp.xlvi-liv, lviii, and lxxx-lxxxiii.
• **Forster**, Johann Reinhold (1729-1798). German priest, botanist and naturalist from West Prussia. Moved to London with his son George in 1766; worked with George on translations into English including Bougainville's *Voyage*, 1771; sailed on *Resolution* as part of Captain James Cook's Second Voyage 1772-1775; returned to Germany in 1780 and died at Halle in 1798. Surveys on the Volga River, 1765, published London 1768; *Geschichte der Entdeckungen und Schiffahrten im Norden...* (3 maps), Frankfurt, C.G. Strauss 1784.
•• **Forster**, George [Johann Georg Adam] (1754-1794). Artist, botanist and translator, son of Johann Reinhold Forster with whom he worked. Sailed with his father on Captain James Cook's Second Voyage, returned to Europe in 1778 and died in Paris. *A Chart of the Southern Hemisphere...*, 1777 (engraved by W. Whitchurch) in his *A Voyage round the World*, 1777 (based on his own account and on his father's Journal).

**Forster**, A. Engraver. Grèce, Athens 1838.

**Forster**, Charles. Pologne, Paris 1811.

**Forster**, D.I.R. *Carte von der Südlichen Spitze von Africa*, 1797 inset on *Vergrösserte Charte der Gegend um die Capstadt* [Cape of Good Hope], Nuremberg 1797.

**Forster**, H. *Plan of the attack of the Mountain of Faron* [Toulon], 1793 MS.

**Forster**, Jerome. Ephemerides meteorographicae ad annum, 1575.

**Forster**, J.H. American surveyor, assistant to Lt. Col. J. Kearney of the Corps of Topographical Engineers. One of the surveyors for *West End of Lake Erie and Detroit River*, 1849, published Washington 1852; worked with W.H. Hearding and I.L. Beghlin on surveys for *Preliminary Chart, Lower Reach of Saginaw River...*, surveyed 1856, published 1867 (engraved by W.H. Dougal); *Chart of East Neebish Rapids River St. Mary...*, 1858 (drawn by J.T. Baker, lithographed by J. Bien); *Field Chart of middle channel Lake George St. Mary's River...*, 1858 (drawn by J.T. Baker, lithographed by T.S. Wagner); *Channel Cut of Middle Chanel, Lake George St. Mary's River*, Washington 1858 (drawn by J.T. Baker, lithographed by C.B. Graham).

**Forster**, John. Surveys in Nicaragua, 1847, published Berlin 1849.

**Forster**, R.P. East India Islands, 1818.

**Forster**, T. ...*Plan of the City of Durham...*, 1754. N.B. Compare with Thomas **Foster**, below, but note that Durham City Library has no record of the **Foster** map. The **Forster** map has vignettes, but no views nor any mention of the engraver.

**Forster**, Thomas. South Shore of Lake Erie, 1830 (with James Maurice).

**Forster**, Westgarth. *A manor at Tynehead* [Cumberland], 1823 MS.

**Forster**. See **Baskin**, Forster & Co.

*Johann Reinhold Forster (1729-1798) and George Forster (1754-1794), father and son. Portrait by Daniel Beyel. (By courtesy of the National Library of Australia. Rex Nan Kivell Collection)*

**Forster-Heddle**, M. Geological Map of the Shetland Is., 1879; Sutherland, 1881.

**Forstman**, G. See **Fosman**, Gregorio.

**Forstman**, G.A. (1773-1830). Engraver in Hamburg.

**Forsyth**, Charles. Geological map of West Lothian, 1846 (engraved by W.H. Lizars).

**Forsyth**, *Lieutenant* (later *Commander*) Charles C. *Waterloo Bay* [Cape Province], 1846 published London 1849; mouth of the Buffalo River [South Africa], 1847; Port Michael Seymour, 1856.

**Forsyth**, *Sir* Thomas Douglas. Turkestan, Calcutta 1875; Central Asia, 1878.

**Forsyth**, William. American surveyor. *Plats of subdivisions of the city of Washington*, Washington, R.A. Waters 1856; *Todd & Brown's subdivision of part of "Pleasant Plains & Mt.Pleasant", suburbs of Washington D.C.*, Washington, J.F. Gedney 1868; *Plan of the city of Washington, D.C.*, 1870; *Plat and survey and subdivision of 'Washington Heights'...*, 1888 (with Wm. P. Young and Thomas J. Fisher).

**Fort Dearborn Publishing Co.** Publisher of Chicago. *National Standard Family & Business Atlas of the World*, 1896 and later; *International Office & Family Atlas*, 1897.

**Forten**, M. *Master*, HMS *Heroïne*. Improvements to J.B.N.D. d'Après de Mannevillette's chart of Junkseilon Island [Thailand], 1798.

**Fortier**, Claude François (1775-1835). Engraver and etcher of Paris. Maps for *Voyage en Espagne*, 1803.

**Fortin**, Jean. French engineer. *Plan de l'île de St. Pierre*, Versailles, J.N. Bellin 1763, (Fortin's survey of St. Peter's Island was also inset on a chart of the south coast of Newfoundland, in Sayer & Bennett's *North American Pilot*, 1775-1776); *Carte particulière des îles de St Pierre et Miquelon*, in *Neptune François*, Dépôt Général de la Marine 1782.

**Fortin**, Jean Baptiste (1740-1817). Publisher & globe maker, 'Ingénieur mécanicien du Roy, Rue de la Harpe, près la Rue du Foin', Paris, bought the Vaugondy stock from Didier Robert de Vaugondy in 1778, he was himself succeeded by C.F. Delamarche. Pair of globes, 1770; published the French edition of Flamsteed's Star Catalogue as *Atlas Céleste de Flamstéed*, 1776; re-issued Didier Robert de Vaugondy's *Nouvel atlas portatif*, 1778; *Carte du Canada et des Etats-Unis de l'Amérique Septentrionale*, 1778 (third state of Vaugondy's *Carte de Pays... de Canada*); terrestrial globe, 1780; sold Charles Messier's celestial globe, 1780; armillary sphere, 1780 (with Charles Messier). **Ref.** PEDLEY, M.S. *Bel et utile: The work of the Robert de Vaugondy family of mapmakers* (Tring 1992) pp.117-119.

**Fortun**, J.M. See **Martínez** Fortún, Jacques.

**Fortune**, Joseph. Surveyor in Canada. Survey of Lancaster Township, 1806; Caledonia Township, 1807; surveys of Alfred and Plantagenet, 1816-c.1820; *Plan of the Village of Richmond* [Johnstown province, Upper Canada], 1818 (copied by W.A. Austin, 1859); Longueuil, 1833 (with D. McDonell).

**Fortune**, William (d.1804?). Canadian surveyor. *...Six Townships on River Radeau and Part of the Waters of River Thames*, 1794; survey of Alfred Township, 1797; surveys in Hawkesbury, 1797-1798; surveys in Plantagenet, 1797; and others.

**Fosdick**, Nicoll. Surveyor, mariner and trader of New London. Surveys of Long Island Sound used for *Blunt's New, And Correct Chart Of Long Island Sound From Montauk Point to Frog's Point...*, 1805 (J. Cahoone's surveys were also used, engraved by Peter Maverick), New Edition 1813, 1819, 1825 (engraved by W. Hooker).

**Fosman** [Forstman], Gregorio. Spanish engraver, worked for Francisco de Aefferden, 1696.

**Fossard**, —. Engraver for J.B.B. d'Anville, 1756.

**Fosse**, Chatry de la. See **La Fosse**, Chatry de.

**Fossé**, J.B. de la. See **Delafosse**, Jean-Baptiste.

**Fossé**, N. du. See **Du Foss**, Nicolas.

**Fosses**, Chaumette des. See **Chaumette** Des Fosses, Amadée.

**Fosset**, —. Printer, at 'Rue du Faubourg St. Jacques N°·19), Paris. Oregon, 1854.

**Foster**, family.
• **Foster**, George. Publisher, printer & mapseller at the 'White Horse opposite the north gate in St. Paul's Church Yard, London'. *A New and Correct Map of the World*, 1737; *A New & Correct Map of England and Wales*, 1737; *A New & exact Plan of the Cities of London and Westminster...*, 1738 (engraved by E. Bowen), later republished by others, E. Foster 1752, F. Bull 1754, R. Sayer c.1761, 1768, 1775; *A Pocket Map of London, Westminster & Southwark*, 1739; Norfolk, 1739 (engraved by W. Roades); *The Seat of the War in the West Indies...* [relating to the so-called War of Jenkins' Ear] (5 maps on 1 copper plate), 1740 (engraved by Emanuel Bowen); seller of Emanuel Bowen's *A Sequel to the Seat of War in the West Indies...*, 1740; *A New Map of Island of Jamaica*, 1740.
• **Foster**, Elizabeth. Publisher at the White Horse, Ludgate Hill. Wife of George (above); republished his plan of London and Westminster, 1752 (first published 1738).

**Foster**, —. Bay of Colonia [Uruguay], 1819.
**N.B.** Compare with Commander Henry **Foster**, below.

**Foster**, G.E. *West Ontonagon group of mines, Lake Superior*, 1864.

**Foster**, Henry. Surveys in Santa Fe, Argentina, 1889-1891 (with F. Wiggin).

**Foster**, *Commander* Henry (1796-1831), *RN*. Admiralty surveyor and astronomer; Fellow of Royal Society, 1824, sailed as astronomer with William Edward Parry, 1824-1825, 1827; led expedition to study ellipticity of the earth, 1827; Copley Medal of Royal Society, 1827. Survey La Plata, 1819; South Shetlands, 1820; East coast Greenland, 1823; Chart of Atlantic, 1827-1831; River Para, 1831; and others.

**Foster**, J.G. Cyclists' & pocket maps of Ontario and Toronto, 1895-1900.

**Foster**, J.T. *Map of the towns of Hyde Park, Lake, Calumet and the east half of Worth*, 1871 (with J. Van Vechten).

**Foster**, J.W. Geologist in the United States. Worked with others on *Geological Map of the Lake Superior Land District in the State of Michigan*, 1847 (with J.W. Foster, lithographed by J. Ackerman), used (with other surveys and maps), to accompany C.T. Jackson's *Report on the geological and mineralogical survey of the mineral lands of the United States in the State of Michigan...*, 1849-1850 and other U.S. Congress reports (surveys also used by John Farmer 1849, 1853, 1861 and others); *Route of the Peninsula Railroad in the State of Michigan*, c.1852.

**Foster**, John (1648-1681). Massachusetts printer and engraver. Established the first printing house in Boston, 'over against the sign of the Dove', 1675. Produced the first known map to have been printed in English America: *A Map of New-England, being the first that was ever here cut...* (woodblock) [the so-called 'White Hills map', based on a survey by William Reed, 1665], published in William Hubbard's *A Narrative of the Troubles with the Indians in New-England*, Boston c.1677 (another version, taken from a different block and known as the '*Wine* Hills map' because of differences in spelling and style, was used for the London issue of Hubbard's work.) **Ref.** WHEAT, J.C. & BRUN, C.F. *Maps and charts published in America before 1800* (1978) N°s·144 & 145 p.29; WOODWARD, W. *Imago Mundi* 21 (1967) pp.52-61.

**Foster**, John. Essex estate surveyor. Survey of a farm at Wix, Essex, 1712 MS; farm at Wrabness, Essex, 1712 MS.

**Foster**, John, *junior* (1758-1827). Surveyor of Liverpool, son of a builder, also called John. North West part of Liverpool (3 plans), 1793 MSS; Ordnance plan in Liverpool, 1803 MS.

**Foster**, N.G. Cranford, New Jersey, 1870.

**Foster**, *Lieutenant* T. Manuscript map and plans of Government land and buildings on New Providence Island [Bahamas], 1831.

**Foster**, Thomas. Town plan of Durham (with 4 views), 1754 (engraved by James Mynde).
**N.B.** Note that Durham City Library has no record of this map. Compare with T. **Forster**, above, whose map of the same date has vignettes, but no views nor any mention of the engraver.

**Foster**, W.H. [& Co.]. Lithographers and

publishers of 86 Treville Street, Plymouth. *New Map of Plymouth, Devonport, Stonehouse and Neighbourhood*, 1897.

**Foster & Hauck.** Publishers of Houston, Texas. Texas, 1834.

**Foster & Marion.** Denver, Auraria & Highland, c.1860.

**Fotherby**, John. Lincolnshire surveyor specialising in estate, drainage and fen maps. Level lying upon the River Ancholme, 1640 (with F. Wilkinson).

**Fothergill**, J. Engraver of Market Street, Manchester. T. Tinker's *A Plan of the Towns of Manchester and Salford... in the County Palatine of Lancaster*, 1772, 1822 and later copies; *A New Plan of Manchester and Salford with their vicinities... in 1832*, J. Everett 1834.

**Fotheringham**, T. East Coast of England, 1803; survey of Yarmouth, David Steel 1804.

**Fouard**, Moises. French engraver. *Plan de la ville de Mons en Hainault*, 1691; *Plan de la Ville de Tripoli en Barbarie...*, 1694, both published in S. de Beaulieu's *Les Glorieuses Conquêtes...*, 1694.

**Fouché**, —. *Seine Inférieure*, 1857; *Arrondissement de Dieppe*, 1858; *St. Valéry*, 1867; and others.

**Foucherot**, —. Ingénieur des ponts et chaussées. *Mer de Marmara*, for J.D. Barbié du Bocage, 1788 and later.

**Foucquet**, H. Surveyor in Kent. Defence works at Sheerness, 1732 MS; plans of Deal Castle, 1733 MSS; Dover Castle, 1736 MSS; Plan of the town, harbour and fortifications of Dover, 1737 MS.

**Foudrinier**, P. See **Fourdrinier**, Paul.

**Fougasse**, Thomas de. *History of Venice* (with map), 1612.

**Fougeu**, Jacques.(*fl.*1595-1606, *d.*1646). Member of the army of King Henry IV of France, responsible for lodgings, camps and supplies. Prepared several manuscript maps of the Champagne region for military planning purposes. **Ref.** DESBRIERE, M. 'L'oeuvre de Jacques Fougeu' in: BOUSQUET-BRESSOLIER, C. (Ed.) L'oeil du cartographe (Paris 1992) pp.233-244 [in French]; SAVARY, B. 'Les cartes de Jacques Fougeu, Maréchal des Logis des armées d'Henri IV' in Bulletin du Comité Français de cartographie 130 (Paris, décembre 1991) pp.29-34 [in French].

**Fouilliand**, Francisco. Provincia de Corrientes, Argentina, 1891.

**Foulerton**, *Captain* John. Royal Sovereign Shoals, 1813; Dartmouth to Start Point, 1813; *Breakwater at Portland*, c.1825.

**Foulkes**, A. Llandudno & environs, c.1886.

**Foulkes**, *Lieutenant* Charles Howard. Elmina [Ghana], 1897; Cape Coast Castle, before 1897; Accra and neighbourhood [Ghana], before 1898.

**Foullon-Norbeck**, H. de (1850-1896). Austrian geologist. Scientific mission to Australia, 1893.

**Fouqué**, P.A. (*b.*1828). French geologist. Collaborated in production of Carte géologique de la France, [n.d.].

**Fouquet**, P., *the younger*. Print seller and publisher. Amsterdam 1783.

**Fourcade**, Henry Georges (1865-1948). Silviculturist, pioneer in stereo-photogrammetry. Plan of Devil's Peak [Africa] (33 sheets), 1905. **Ref.** STORRAR, C.D. 'Fourcade's map of Devil's Peak' in Maps of Africa: prodeedings of the Symposium on maps, South African Library, Cape Town, 24th-26th November 1988 (Cape Town 1989) pp.29-38.

**Fourcault**, Jean Baptiste. *Département de Maine et Loire*, 1860.

**Fourcy**, Eugène de. *Carte géologique Côtes du Nord*, 1843; *Atlas souterrain de Paris*, 1855.

**Fourdrinier** [Foudrinier, Fourdrinière], Paul (*fl.*1720-1760). French engraver, publisher & mapseller; came to London 1720, corner of Crag's Ct., Charing Cross. Peter Gordon's *A View of Savanah as it stood the 29th March 1734*, 1734; T. Lediard's plan of Westminster, 1740; *The South prospect of Tetuan*, London c.1740; P. Durell's *A plan of the Harbour Town and Castles of Carthagena*, 1741; William Edgar's plan of Edinburgh, 1742 (published in Maitland's

*History of Edinburgh*); S.R. Spalart's *An exact plan of the battle of Dettingen*, 1743; plan of the City of London, 1744; Plan of the Ancient City of Westminister, *c*.1761.

**Foureau**, Fernand (1850-1914). Geographer and explorer in North Africa. *Carte d'une partie du Sahara septentrionale*, 1908. **Ref.** BROC, N. *Dictionnaire illustré des explorateurs Français du XIXe siècle: I Afrique* (1988) pp.147-150 [in French].

**Fourier**, *baron* Jean Baptiste Joseph de (1768-1830). French physician and surveyor. Wrote on the geography of Egypt.

**Fournel**, Henri (1799-1876). French engineer & explorer in Algeria 1840-1844. *Carte géologique du Bocage Vendéen*, 1835; *Richesses minérales de l'Algérie...* (2 volumes), 1850-1854.

**Fournet**, Jean Joseph Baptiste Xavier (1801-1869). French geologist and meteorologist.

**Fournier**, Adam Gabriel. *Plan of that part of the New Fort at Placentia...* [Newfoundland], 1745 MS; *A Plan of Fort William in St. John's, Newfoundland*, 1749 MS.

**Fournier**, Daniel (*c*.1710-1766). 'A la mode Beefseller, shoemaker & engraver. Drawing master & teacher of perspective'. Produced maps of Jamaica from the surveys of Thomas Craskell and James Simpson, including *The Map of the Island of Jamaica...*, London 1763 (surveyed 1756-1761); *...Map of the County of Surry* [Jamaica], 1763; *...Map of the County of Cornwall* [Jamaica], 1763; *...Map of the County of Middlesex* [Jamaica], 1765.

**Fournier**, Georges (1595-1652). French priest and almoner for the French navy. Wrote an encyclopaedia on seafaring entitled *Hydrographie contenant la theorie et la pratique de toutes les parties de la navigation*, Paris, Michel Soly 1643, 1667, 1679 (chapter 4 in book 14 is an evaluation of charts, maps and globes). **Ref.** KROGT, P. van der *Globi Neerlandici* (Utrecht 1993) pp.232-233.

**Fournier**, H. Printer, Rue de Seine, Paris. Atlas Ceran, Lemonnier 1837.

**Fournier**, *Lieutenant* Joseph Marie. Served aboard *Héroine* on Jean-Baptiste Thomas Cécille's voyages in New Zealand in the 1830s. Manuscript chart of Port Cooper [Lyttleton Harbour] and Port Levy, 1838 (used as a source for later maps); *Plan de la Rivière Kawa-Kawa*, 1838 (with L. Durand-Dubraye), published Dépôt-général de la Marine 1840 (engraved by Ambroise Tardieu and J.M. Hacq); *Plan du Port Akaroa*, 1838 (with L. Durand-Dubraye), published Dépôt-général de la Marine 1840 (engraved by A. Tardieu and J.M. Hacq); *Plan des Iles Chatam* (3 maps), 1838 (with L. Durand-Dubraye), published Dépôt-général de la Marine 1840 (engraved by Tardieu & Hacq).

**Fournier d'Albe**, E.E. Statistical maps of Ireland, 1891-*c*.1894.

**Fournier de Saint-Martin**, Désiré. *Tableau géographique des Pays Bas*, *c*.1825.

**Fournier des Ormes**, G. (1777-1850). French engraver. Plan Battle of Waterloo, after 1815.

**Fouzy**, *Lieutenant* H. Route de E. Obeiyad à El Facher, Cairo 1876.

**Fowkes**, Francis. Convict. *Settlement at Sydney Cove*, 1789 (thought to be the earliest printed plan of the colony).

**Fowler**, A.G. Surveyor. *Lagos Survey* [Nigeria], 1890-1892 MS (with J.H. Ewart and M. de Fesigny); *Map of Route... from Lagos to Ilorin...*, 1893; *Lagos. Proposed Ship Canal*, 1895; and others.

**Fowler**, Charles (*fl*.1819-1853). Surveyor of Leeds. *Plan of the Town of Leeds*, 1821; *Plan of the Several Turnpike Roads between Leeds & Doncaster...*, 1822; East Lothian, 1825; Fife, Midlothian, 1828; Hungerford Market, 1829; Yorkshire, 1836, 1859 (engraved by J. Neele).

**Fowler**, *Captain* Charles J., RE. Map of the Isle of Wight (6 sheets), 1859 MS.

**Fowler**, Frank. Acting Crown Surveyor and Commissioner of Mines in Guyana. *Chart of the sea coast of the colony of British Guiana*, London, Stanford's Geographical Establishment 1901; *Map of the Northern Portion of British Guiana*, 1906.

**Fowler**, H. Plan River Hondo [Central America], 1886-1887 MS.

*Thaddeus Mortimer Fowler, left (1842-1922). (By courtesy of the Library of Congress)*

**Fowler**, John. Engineer. Improvements proposed by Great Grimsby and Sheffield Junction Railway Company, 1845 (engraved by J.W. Lowry).

**Fowler**, John. Plan of London, 1860; Wady Halfa-Shendy, 1871-1872; Plan of the River Tees, 1879; Ambukol-Shendy, 1884.

**Fowler**, John. Surveys of Middle and Upper Egypt used for *Middle Egypt*, London, Edward Stanford 1882; and *Map of portions of Middle and Upper Egypt*, London, Edward Stanford 1883.1908. **N.B. Compare with John Fowler, above.**

**Fowler**, L.D. Assessment map of Jersey City, 1870-1873.

**Fowler**, Richard. Apprenticed to the Drapers' Company of London or 'Thames School' of chartmakers, 1719.

**Fowler**, Thaddeus Mortimer (1842-1922). Native of Lowell, Massachusetts. After an injury sustained during military service with the New York Volunteers during the Civil War, he began working with his uncle, a photographer in Madison, Wisconsin. Fowler established his own firm specialising in panoramic maps or bird's-eye views. His first work, a view of Omro, Wisconsin, was published in 1870. After several moves he settled in Morrisville, Pennsylvania in 1885. He is known to have produced and published over 300 bird's-eye views including towns and cities in 18 states and in Canada, many of them now in the Library of Congress. He was also involved in a variety of creative and publishing partnerships including:
• **Fowler & Evans**. Publishers of Asbury Park, New Jersey. T.M. Fowler's *Somerville* [New Jersey], 1882 (printed by Beck & Pauli, Milwaukee).
• **Fowler & Downs** (*fl.*1888-1914). Boston based partnership of T.M. Fowler and Albert E. Downs. Draughtsmen and publishers for *Gettysburg* [Pennsylvania], 1888 (lithographed by A.E. Downs); draughtsmen for *Scranton* [Pennsylvania], Fowler & Moyer 1890 (lithographed by A.E. Downs); publishers for A.E. Downs's *Grafton* [West Virginia], 1898; publishers of A.E. Downs's *Canal Dover* [Ohio], 1899; publishers of A.E. Downs's *Cordele* [Georgia], Morrisville 1908; draughtsmen for *Haverhill* [Massachusetts], Hughes & Bailey 1914.
• **Fowler & Henry**. Partnership of T.M. Fowler and F.P. Henry, publisher of T.M. Fowler's *Hamburg* [Pennsylvania], 1889.
• **Fowler, Downs & Moyer**. Draughtsmen and publishers for *Wilkes-Barre* [Pennsylvania], 1889 (lithographed by A.E. Downs).
• **Fowler & Moyer** (*fl.*1889-1902). Partnership of T.M. Fowler and James B. Moyer. Published over 100 of T.M. Fowler's views of Pennsylvania alone, as well as others for towns and cities in Ohio (8), Oklahoma (2), Texas (5) and West Virginia (17), also Michigan and Missouri.
• **Fowler & Bailey**. Partnership of T.M. Fowler and Oakley H. Bailey. Publishers of Boston. T.M. Fowler's *Dover* [New Jersey], 1903.
• **Fowler & Basham**. Publishers of Flemington, New Jersey. T.M. Fowler's *Bluefield* [West Virginia], 1911.
• **Fowler & Hughes**. Draughtsmen for *Peekskill* [New York], New York, Hughes & Bailey 1911.
• **Fowler & Browning** (*fl.*1912-1913). Publishers of Asheville, North Carolina. *Black Mountain*, [North Carolina] 1912; *Hendersonville* [North Carolina] 1913 (both by T.M. Fowler).
• **Fowler & Kelly** (*fl.*1905-1907, 1917-1918). Publishers of Morrisville, Pennsylvania (1906-1907) and Passaic, New Jersey (1917-1918). Published views by T.M. Fowler including 2 in West Virginia, 1905; 7 in Maryland, 1906-1907; 7 in Pennsylvania, 1906-1907; 1 in Virginia, 1907; 2 in Oklahoma, 1918. **Ref.** HEBERT, J.R. *Panoramic maps of Anglo-American Cities* (Washington 1974) *passim*.

**Fowler**, William. Engraver. John Gibson's Oxfordshire, 1765; engraver of *Worcester Shire* for Joseph Ellis's *The New English Atlas*, 1765.

**Fowler**, William. Maps of Bridgewater estates, Shropshire, 1650-1651.

**Fowler**, William. Publisher from Wakefield, published county maps in association with Christopher and John Greenwood and T. Sharp. *The County Palatine* [Lancashire] (6 sheets), Wakefield and London, W. Fowler and C. Greenwood 1818 (drawn by R. Creighton from the survey by C. Greenwood, engraved by S.J. Neele & Son); C. Greenwood's *Cheshire* (4 sheets), W. Fowler & C. Greenwood 1819; C. Greenwood's

*Staffordshire*, W. Fowler, C. Greenwood & Co. 1820; *Map of the County of Haddington...* [East Lothian] (2 sheets), Thos. Sharp, C. Greenwood & Wm. Fowler 1825, republished W. Fowler 1844 (engraved by J. Dower); *Map of the County of Berwick...* (2 sheets) (2 sheets), Thos. Sharp, C. Greenwood & Wm. Fowler 1826, W. Fowler 1826, 1845 (engraved J. Dower); *Map of the Counties of Fife and Kinross...* (4 sheets), Thos. Sharp, C. Greenwood & Ww. Fowler 1828 (engraved by J. Dower) and other editions; *Map of the County of Edinburgh...* (2 sheets), Thos. Sharp, C. Greenwood & Wm. Fowler 1828 (engraved by John Dower), Wm. Fowler 1840.

**Fowles**, Arthur William. 'Marine Artist' of Ryde, Isle of Wight. *The Isle of Wight*, Ryde, J. Briddon *c.*1880, 1889 (mainland railways added), Ventnor, J. Briddon 1897 (engraved by Lewis Becker).

**Fownes**, —, RN. Improvements to Captain Joseph Huddart's False Bay, 1798.

**Fox**, —. See **Martin** & Fox.

**Fox**, A.W. *Map of Queensland* (4 sheets), Brisbane 1878 (lithographed by H.W. Fox).

**Fox**, *Sir* Douglas. Proposed Channel Tunnel, 1883.

**Fox**, G. Worked on maps for John Pinkerton's *Modern Geography*, Philadelphia 1804.

**Fox**, H.I. Plan of estates *...in the Parishes of Abington, Littlington and Steeple Morden* [Cambridgeshire], 1804.

**Fox**, H.W. Drew and lithographed A.W. Fox's *Map of Queensland*, 1878.

**Fox**, Jesse W. Surveyor of Salt Lake City. *Map of Salt Lake City and environs*, 1890.

**Fox**, Justus. *Die Meer-Enge der Dardanellen* [Straits of the Dardanelles] (woodcut), 1772.

**Fox**, Richard. Surveyor appointed by the Governor of Kentucky to survey the Kentucky/Tennessee boundary, 1818-1819 (with W. Steele).

**Fox**, Samuel. *An Improv'd Map of the County of Derby*, 1760.

**Fox**, *Major* Walter Reginald. *Map of the Nile Provinces*, 1884.

**Fox & Otley**. Sectional Map of Lee County, Iowa, 1861.

**Foxe** [Fox], *Captain* Luke (1586-1635). English navigator, explored Hudson Bay in search for northwest passage, 1631-1632. Untitled map of his voyage, 1635. **Ref.** Kenyon, W. Arctic argonauts (1974) pp.81-87.

**Foy**, *Général* —. *Histoire de la guerre de la Peninsule sous Napoléon*, Paris 1827.

**Fraas**, Oscar (1824-1897). German geologist and explorer. *Geognostiche Karte v. Württemberg, Baden u. Hohenzollern*, 1870; *Drei Monate im Libanon*, 1876.

**Fracastoro** [Fracastro], Girolamo (1483-1553). Italian cosmographer, geographer and mathematician. Friend of Giovanni Battista Ramusio and dedicatee of his *Navigationi et Viaggi*, 1553. **N.B.** There seems to be no evidence for Tooley's suggestion that Fracastoro was responsible for the map of America in Ramusio's Delle navigationi, 1556. Fracastoro seems merely to have advised Ramusio and passed on to him a map by Gonzalo Orviedo.

**Fracanzano da Montalboddo**. See **Montalboddo**, Fracan Antonio.

**Frachus**, G. See **Franco**, Giacomo.

**Fraitteur**, W. Plan der Stadt Mannheim, 1813.

**Frambottus**, Paulus. *Verona fidelis*, *c.*1660.

**Frame**, Edward H. *Flushing*, 1871.

**France**, J. la. See **La France**, Joseph.

**Franceschi**, Domenico de. Edited Venice (6 sheets), 1565.

**Franceschi**, Girolamo. Publisher in Florence. Stefano Buonsignori's Florence, 1594.

**Franceschini**, F. *Città de Bologna*, 1822.

**Francesco**, Jean. See **Della** Gatta, Giovanni Francesco.

**Francesco Padovani**, Patrizio. Manuscript atlas (27 maps), 1540.

**Francese**, Stefano. See **Tabourot**, Etienne.

**Francheville**, — de. Côte occidentale d'Afrique, 1828; Baie de Santo Antonio, 1828; Côte d'Or, all 1828.

**Franchi**, S. *Carta geologica dei d'intorni di Ventimiglia*, 1894.

**Franchus**, G. or J. See **Franco**, Giacomo.

**Francia y Ponce de Léon**, Benito. Isla de Luzon [Philippines], 1889.

**Francini**, Alessandro (d.1648). Florentine sculptor working in France. *Maison Royale de Fontaine Belleau*, 1614; *Chasteaux Royaux de Sainct Germain en Laye*, 1614.

**Francis**, Absalom. Map of the Mines of Cardiganshire, 1878.

**Francis**, Charles. *Chart of the sea coast of the colony of British Guiana*, London, Stanford's Geographical Establishment 1901; *Map of British Guiana*, London, Stanford's Geographical Establishment 1902.

**Francis**, Edward J. Published a facsimile of Ranulph Agas's *Civitas Londinum*, 1874 (first published c.1563).

**Francis**, Richard. Plot of Killbeg (County Wicklow), 1656 MS.

**Francis**, *Reverend* William F. Diocesan maps, 1864; Lichfeld, c.1867.

**Francisco**, Diego. See **Francoso**, Diogo.

**Francisco** [Fernándes Pereira], *Fr.*—. *Diccionario geografico de Portugal*, 1862.

**Franciscus Monachus Mechliniensis**. See **Monachus**, Franciscus.

**Franck**, Matthäus (d.c.1568). Mapmaker and printer of Augsburg. *Warhafftige Beschreibung der Porto von Malta...*, 1565 (siege map taken from a similar one by Zenoi of the same year). **Ref.** GANADO, A. & AGIUS-VADALÀ, M. *A study in depth of 143 maps representing the Great Siege of Malta 1565* (1994) p.159 et seq..

**Francke**, R. *Carlshafen u. Umgegend*, 1896.

**Franco** [Frachus; Franchus; Francus], Giacomo [Jacomo; Jacobus] (1550?-1620). Engraver of Venice. Map of Cyprus, 1570 (used as a source by Ortelius, 1573); *Cosmographia Universalis...*, c.1586-1587 (taken from the map of Oronce Fine); frontispiece to Livio Sanuto's *Geographia*, 1588; Rome, 1589; with Giuseppe Rosaccio engraved maps and plans for a travel guide *Viaggio da Venetia*, Venice, Franco 1598, 1606, later editions published by Marco Sadeler and Stefano Scolari.

**Franco**, Isaac. See **François**, Isaac.

**François**, Auguste (1857-1935). French diplomat in the Far East. Geographical works with maps including *Le Lieou-Kiang et la rivière de King-Yuan-Fou, au Kouang-Si*, 1904. **Ref.** BROC, N. *Dictionnaire illustré des explorateurs Français du XIXe siècle: II Asie* (1992) pp.193-195 [In French].

**François**, *Lieutenant* Curt von. Belgian Congo, 1885-1886.

**François** [Franco; Frank; Francus], Isaac [Ysaaco] (1566-1650). Architect of Tours, Royal overseer at Touraine. Map of Touraine, 1592 (used by Maurice Bouguereau in *Théâtre françois*, 1594, by Ortelius, 1598, by J. Leclerc, 1619, J. Boisseau 1642 and others).

**Francoso** [Francisco; Froncoso], Diogo. Engraver of J. Serra's *Caifornias: Antigua y Nueva*, Mexico City, F. de Zúñiga y Ontiveros 1787 (used in Fr. F. Palou's *Relation Historica de de la Vida... del... Padre Fray Juniero Serra*).

**Francq**, B. le. See **Lefrancq**, B.

**Francus**, Jacobus. See **Franco**, Giacomo.

**Francus** [Frank], Ysaacus. See **François**, Isaac.

**Frank**, Martin (d.1666). Mapmaker for the Fürstentum Kulmbach-Bayreuth.

**Franke**, A. *Geognostiche Karte der Grafschaft Schaumburg*, 1867.

**Franke**, Alcuin Rudolf. Environs Leipzig, 1874; *Sächsisches Vogtland*, c.1891.

**Franke**, C. *Schul-Atlas*, c.1866.

*Benjamin Franklin (1706-1790). (By courtesy of the National Portrait Gallery, London)*

**Franke**, Julius. [Karte der] *Königlichen Preussichen Provinz Sachsen*, 1858; *Planiglob in zwei Wandkarten*, 1861.

**Franken**, J.H. *Historische Notitie Atlas*, 1924.

**Frankendaal**, Nicolaas van. Draughtsman and engraver of Amsterdam. Naumburg, 1749.

**Frankland**, George (1800-1838). Assistant Surveyor Van Diemen's Land, 1826; Surveyor General & Commissioner of Crown Lands, Tasmania. Explorations along Derwent, Gordon, Huon & Nive Rivers, 1828-1835. Military operations against the Aboriginals of Van Diemen's Land, 1831; *Chart of Forestier's and Tasman's Peninsulas*, London, 1834 (lithographed by James Basire, printed by James & Luke Hansard) 1837 edition (lithographed by L. Schönberg, printed by James & Luke Hansard); *This Map of Van Diemen's Land...*, 1837, J. Cross 1839.

**Franklin**, Captain —. Plan of the British Settlement at Singapore, 1828.

**Franklin**, Benjamin (1706-1790). American printer, author, scientist, diplomat, statesman, Deputy Postmaster General for the North American Colonies, pioneer of Gulf Stream charting. Map in Articles of Agreement between Maryland and Pennsylvania, Philadelphia, c.1733; *A Chart of the Gulf Stream*, 1786 (after Captain Timothy Folger), re-issued in *Remarks upon the Navigation*, 1789. **Ref.** COHN, E. *Imago Mundi* 52 (2000) pp.124-142; DeVORSEY, L. *The Map Collector* 15 (1981) pp.6-10; WHEAT & BRUN *Maps and charts published in America before 1800* (1978) N°·474 p.102, N°·721 p.157, N°·723 p.157; *Imago Mundi* 28 (1974) pp.105-120.

**Franklin**, J.J., *RN*. Chart of Palk Strait [India / Sri Lanka], 1838-1845; Tuticorin [S. India], 1842; Tinnevelly [S. India], 1846.

**Franklin**, John (*fl.*1721-1724). Estate surveyor. Melksham, Wiltshire, 1724.

**Franklin**, John (*fl.*1767-1770). Surveyed estates in Bedfordshire and Buckinghamshire including Hawns, Bedfordshire, 1767 MS.

**Franklin**, *Lieutenant* (later *Rear-Admiral Sir*) John, *RN* (1786-1847). English explorer and hydrographer from Spilsby, Lincolnshire. Took part in expeditions to Spitzbergen and the Canadian Arctic coast, 1818-1822 and 1825-1827; Governor of Van Diemen's Land [Tasmania], 1837-1843; commanded an expedition in search of the Northwest Passage, 1845-1847 from which he did not return. *Narrative of a Journey to the shores of the Polar Sea* [1819-1822], published John Murray 1823, 1824; surveys in Arctic America, 1825 published Admiralty 1845; *Narrative of a second expedition...* [that of 1825-1827], London, John Murray 1828, 1829.

**Franklin**, Robert. Land surveyor, Thaxted, Essex. Langley, *c.*1851 MS.

**Franklin**, William Buel (1823-1903). U.S. Corps of Topographical Engineers. Rocky Mountains, 1845.

**Franklin Globes**. Globe-making company, possibly originated by Franklin Field of Troy. A series of terrestrial and celestial globes between 6 inches and 30 inches in diameter. They were issued by Merriam Moore and others, published Troy, New York between 1851-1896. **Ref.** WARNER, D. *Rittenhouse* Vol.2 N°2 pp.63-64; *ibid.* Vol.2 N°3 pp.88-89.

**Franks**, J.H. Draughtsman and engraver of Commutation Row, Liverpool. J. Sherriff's map of the district of Liverpool, 1817, 1823; *Canada*, 1820 (in Daniel Blowe's *A Geographical, Commercial, and Agricultural View of the United States of America*, Liverpool, Henry Fisher *c.*1820); Plan of Liverpool, 1821; C. Greenwood's map of Lancashire (6 sheets), G.F. Cruchley 1836.

**Franks**, Theodore. Public Land States & Territories, 1865; United States, 1866.

**Franks & Johnson**. Engravers of Leeds. *Manchester and its environs engraved from an actual Survey made in 1824 by William Swire*, Liverpool, W. Wales & Company *c.*1824. (published in *History, Directory and Gazetteer of Lancashire* by Edward Baines).

**Franquelin**, Jean Baptiste Louis *SJ* (1653-*c.*1725). French cartographer, 'hydrographe du roi', official cartographer for New France. Many manuscript maps including: *Carte pour servir à l'éclaircissement du Papier Terrier de la Nouvelle France*, 1678; *Carte de la mine dargent*, *c.*1680; *Carte du Fort St Louis de Québec*, 1683; *Carte de la Louisiane ou des voyages du Sr de La Salle... les années 1679, 80, 81 & 82*, 1684; *Carte du Grand Fleuve St. Laurens...*, 1685 (after Jolliet); *Carte de l'Amerique Septentrionalle... Québec*, 1688; *Ville de Manathe ou Nouvelle-Yorc*, *c.*1693 (later published in: *Le Petit Atlas Maritime*, Paris, J.-N. Bellin 1763); *Partie de l'Amerique Septentrionalle ou est compris la Nouvelle France*, 1699; *Carte de la Nouvele* [sic] *France où est compris Nouvelle Angleterre, Nouvelle Yorc, Nouvelle Albanie, Nouvelle Suede...*, *c.*1702; several of these maps were later published by Nicolas de Fer, Coronelli and Delisle. **Ref.** COHEN, P.E. & AUGUSTYN, R.T. *Manhattan in maps 1527-1995* (1997) pp.50-51.

**Franquet**, Louis. (1697-1768). *Plan du fort du Sault de S. Louis et du village des sauvages Iroquois*, 1752.

**Franseckij**, Ernest von. *Umgegend von Düsseldorf*, 1842.

**Frantzius**, A. von. Costa Rica, Gotha, Justus Perthes *c.*1860.

**Frantzl**, Augusto. Plan of Trieste, 1843, 1844 (with Pacher).

**Franz**, family. Publishers of Nuremberg, associated with the company of Homann Heirs.
• **Franz**, Johann Michael (1700-1761). Geographer and cartographer, later publisher of Nuremberg. Fellow student and friend of J.C. Homann. Professor of Geography in Göttingen, wrote several treatises on the practice of cartography. Founder of the short lived Cosmographical Society of Nuremberg. In 1730 he was joint successor (with Johann Georg Ebersberger) to J.C. Homann, and together they formed Homännische Erben

*Sir John Franklin (1786-1847). (By courtesy of the National Library of Australia)*

[Homann Heirs]. Franz worked on improving the cartographic quality of Homann Heirs' publications, including the addition of descriptive information, the name of the cartographer and the date of publication. He relinquished control of the company in 1755. **N.B.** See also **Homann Heirs**. **Ref.** HEINZ, M. *Imago Mundi* (1997); KRETSCHMER, DÖRFLINGER & WAWRIK *Lexikon zur Geschichte der Kartographie* (1986) Vol.C/1, p.237 [in German].

• **Franz**, Jacob Heinrich (*b*.1714). Brother of Johann Michael, bought his brother's share of Homann's Heirs in 1759 and succeeded him as Director of Homann's Heirs, 1761.

•• **Franz**, Georg Christoph (1747-1823). Son of Jacob. Sold his share in Homann's Heirs to Georg Christoph Franz Fembo in 1813.

**Franz**, — (*fl*.1795-1820). Copper engraver. Prussia, 1802-1809; for Johann Marias Friedrich Schmidt, 1820.

**Franz**, H. *Schlesien*, *c*.1850.

**Franz**, J. Railway maps of Europe, 1867-1891.

**Franz**, Johann Georg (*c*.1776-1836). Globemaker, antiquary and printseller of Nuremberg. Constructed celestial and terrestrial globes, 1790-1810 (based on the work of D.F. Sotzmann and J.E. Bode).

**Franz**, Leopold. District of Kennedy, Queensland, 1860.

**Franza**, Peter. Mapsellers and publishers of Prague.
• **Franza**, Peter (1767-1830). New editions of Anton Elsenwanger's 12 maps of the Districts of Bohemia.
•• **Franza**, Peter (1797-1834). Son of, and successor to, Peter above.

**Franzini**, *Major* Marino Miguel, Corps of Engineers. Charts of Portugal, 1811-1816; *Carta Reduzida da Costa de Portugal...*, A. Arrowsmith 1811; *Plan de la Barre de Lisbone*, 1816. **Ref.** DIAS, M.H. & ALEGRIA, M.F. 'Na transição para a moderna cartografia: as cartas nauticas da região de Lisboa segundo Tofiño e Franzini in: *Finisterra: revista Portuguesa de Geografia* 29, 58 (Lisboa: Centro de Estudos Geograficos, Univ. de Lisboa, 1994) pp.231-265 [Portuguese with summaries in French & English].

**Frary**, E. *Atlas of Herkimer County, New York*, 1868.

**Fraser**, A. Manuscript plans of Quebec with the fortifications, before 1790.

**Fraser**, *Colonel* A. Central Ceylon, 1845.

**Fraser**, *Captain* A., Bengal Engineers. Isthmus of Kraw [Burma and Thailand], 1861; Route Bengal-Siam, 1862 MSS.

**Fraser**, *Lieutenant* Alexander. South Carolina Regiment. *Battle of Savannah* [Georgia], 1778 MS.

**Fraser**, F.A. Mackenzie. *Mauritius*, 1835.

**Fraser**, G.A. See **Frazer**, George Alexander.

**Fraser**, *Lieutenant* G.J. Worked with the Survey of India. *Map of the District of Mozufurnugur*, 1827-1832 (with W. Brown), published Agra 1858; *District of Bareilly*, 1833-1837 (with J. Abbott), published Calcutta 1858; District of Budaon, Agra 1857.

**Fraser**, J. Plymouth Sound, 1788.

**Fraser**, J. Engraver. Bering Straits, 1790; Palestine, *c*.1790.

**Fraser**, James. *Map of the Counties of Fife & Kinross*, Scotland (4 sheets), 1841, 1846, 1847 (engraved by W. & A.K. Johnston, after T. Sharp, C. Greenwood and W. Fowler, 1828).

**Fraser**, James (*d*.1841). Publisher of 215 Regent Street, London. *Fraser's Panoramic Plan of London*, 1831, *c*.1835, *c*.1837 (engraved by Josiah Neele).

**Fraser**, James. *A New Map of Ireland...*, Dublin, William Curry Jun$^r$ & Company 1837 (used in Fraser's *Guide Through Ireland*, Dublin, Wm. Curry Jnr. & Co., London, S. Holdsworth, Edinburgh, Fraser & Co. 1838 with editions to 1854); *Travelling Map of Ireland*, 1852, 1884; Dublin, 1860.

**Fraser**, *Major-General* John. *Map of the Island of Ceylon*, 1862.

**Fraser**, Malcolm. Route of the Western Australian Exploring Expedition, 1875.

**Fraser**, Robert. Soil of Devonshire, 1794.

**Fraser**, R.H. Lithographer. William Lord's chart of the Mersey River from Rock Lighthouse and Bootle Bay to Oglet Point and Ellesmere Port, 1852.

**Fraser**, Simon. Plan of the general attack on Fort Mifflin, 1777 MS (Clements Library, MI).

**Fraser**, *Captain* William. Lands around Dampier's Strait, 1782.

**Fraser**, William. Plan of the proposed Tay & Loch Earn Canal [Scotland], 1807.

**Fraser**, William H. Worked with Augustus, Charles and L.C. Warner on surveys in Indiana and Ohio, 1862-1867.

**Fraslin**, Pedro (*d.c.*1766). *Carta reducida tersera parte de la navegacion de Philipinas al Puerto de Acapulco* [coast of California], *c.*1765 MS.

**Frattino**, Giulio Carlo. Cartographer and publisher of Milan. Stato di Milano, 1703.

**Frauenberger**, George. Oil Territories, Pennsylvania, *c.*1860.

**Frauendorff**, Carl von. Member of the Academy of Sciences, St Petersburg. Theatrum belli Crimeae, 1737-1738, published *c.*1740; Crimeae conspectus, *c.*1740.

**Frazer**, *Captain* —. Chart of Holy Island [east coast of England and Scotland], 1798, published William Heather 1808.

**Frazer**, A. Brassa Sound, 1783; Fortifications at Leith, 1785 MS; Firth of Forth, 1785 MS.

**Frazer**, *Lieutenant* Daniel. Officer in the Royal Staff Corps. Plans to illustrate military tactical lectures, *c.*1800; one of those involved in a military survey of northern France: *British Lines of Occupation. N. of France (after 1815)...* (10 sheets), 1818-1823 MSS.

**Frazer** [Fraser], *Lieutenant* (later *Captain*) George Alexander, *RN*. Admiralty surveyor. Carlingford Lough [Louth, Ireland], 1831 MS; surveys of Ireland and St George's Channel, 1837-1852; Scotland, 1841-1844.

**Frazer**, Robert. Map for Lewis & Clark Expeditions, 1807 MS.

**Frazer**, William. Recorded at Shadwell Water Works, London. *A correct ground plan of the dreadful fire at Ratcliff on July 23rd 1794*, 1794.

**Frazier**, Walter S. *Illinois*, 1864 (lithographed by C. Shober).

**Frear**, Thomas. *City of Philadelphia*, 1869.

**Frearson** brothers [Frearson & Brother; Frearson's Printing House]. Printers and publishers of Adelaide, South Australia, issued a periodical entitled *The Pictorial Australian*, 1885-1895. Plan of the City of Adelaide, *c.*1876; Map of the City of Adelaide, 1877; *Frearson's Map of Australia*, 1879; Settlements in Australia, *c.*1879; *Frearson's plan of the mineral leases at Kalgoorlie (Hannan's), Western Australian goldfields*, *c.*1895.
• **Frearson**, Samuel (*d.*1887).
• **Frearson**, Septimus.

**Frech**, F. Geologische Skizze der Kärnischen Alpen, 1887.

**Frécinet**. See **Freycinet**, brothers.

**Frederic**, J.A. Insula Anticosti, 1758-[1760].

**Frederici** [Fredrici], *Lieutenant* Johann Christian. Surveys of the south coast of Africa, 1789-1801. St. Helena Bay [west coast of Africa], 1791 (used by Dalrymple 1797); *Map of the course of the Oliphant's River...* [Cape Province], 1801 MS.

**Frederici**, M. Dutch surveyor. Collaborated with B. Schotanus à Sterringa on his atlas of Friesland, 1698.

**Frederick**, J.L. State Capitol Harrisburg, Pennsylvania, 1820.

**Frederick**, W. Publisher and mapseller. W. Hibbart's *Five miles round Bath*, Bath 1773; W. Hibbart's *Plan of the City of Bath*, 1780.

**Fredonyer**, —. North East California and North West Nevada, 1865 MS.

**Fredricci**, J.C. See **Frederici**, Johann Christian.

**Freducci**, family. Portolan makers of Ancona.
• **Freducci**, *Conte* Hoctomanno [Conte di Ottomano]. Fourteen charts and atlases

known. Portolan chart of Europe, 1497; manuscript chart of the Mediterranean Sea and Black Sea, Ancona 1524; portolan atlas, 1533; manuscript atlas of five charts, Ancona 1537; Chart of Mediterranean, 1538. **N.B.** E.L. Stevenson, in *Maps, charts, globes: Five centuries of exploration* (New York 1992 edition) pp.12-13, notes an earlier portolan maker of the same name (fl.c.1460); Tony Campbell in Harley and Woodward's (eds.) *History of cartography* Vol.I (1987) p.411 note 299, credits a certain Conte Hectomano Freducci with a chart thought to date from 1424.

•• **Freducci**, Angelo. Son of Hoctomanno, above. Atlas, 1555; Atlas of Asia, Europe and part of America, 1556 MS.

**Freebairn**, Alfred Robert (1794-1846). Engraver who used the 'anaglyptograph' method patented by John Bate. Atlas to accompany William Siborne's *History of the War in France and Belgium in 1815*, London, Boone 1844 (with relief shaded battle maps produced by anaglyptography from Siborne's 3D model of the battlefields). **Ref.** WALLIS & ROBINSON *Cartographical innovations* (Tring 1987) entry 7.011 p.287.

**Freed**, I.G. Surveyor for Geil & Jones and Geil & Harley. *Map of Wayne Co., Michigan*, Philadelphia, Geil, Harley & Siverd 1840; *Topographical map of the Counties of Ingham & Livingston, Michigan*, Geil, Harley & Siverd 1859 (with others); and other county surveys.

**Freeling**, Arthur. *Grand Junction Railway Company*, 1837; *London & Birmingham Railway Company*, c.1838; *London & Southampton Railway Companion*, 1839.

**Freeling**, Arthur Henry (1820-1885). Surveyor-General of South Australia.

**Freeling**, *Sir* Francis (1764-1836). Secretary to the General Post Office. Revised Daniel Paterson's *Roads*, 1811.

**Freeman**, —. Engraver for F. Lucas, 1823.

**Freeman**, Edward Augustine (1823-1892). English historian and geographer. *Historical Geography of Europe* (2 volumes), London, 1881, London, Longman, Green & Co. 1882, and later.

**Freeman**, G.L.B. Historical map Anglo-Saxon and Roman Britain, 1838.

**Freeman**, J.D. Land surveyor. Texas, 1836.

**Freeman**, Joseph (1734/5-1799). Estate surveyor of Cambrdgeshire, produced maps of estates in various parts of England. Great Dunmow, 1768; *A particular of several farms belonging to St Thomas' Hospital in Essex, Cambridgeshire and Buckinghamshire*, 1781; Aveley, 1782 (copied).

**Freeman**, Thomas (d.1821). *Map of the market-squares and that part of the Canal and Mall which is north of the Tiber-Creek* [Washington D.C.], 1794 MS.

**Freeman** [Freman], W. Surveyor General, New South Wales. *Co. of King* [New South Wales] (5 sheets), Sydney, Surveyor General's Office 1867; Georgiana, New South Wales, 1877.

**Freeman**, W.J. Printer of London. 'Steam-Lith$^{o.}$ 2 Old Swan Lane, Upper Thames S$^{t.}$ E.C.'. Lithographic transfer of Thomas Dix's map of Bedfordshire under the title *Official Map of Bedford*, London, Simpkin, Marshall and Co. 1877 (first published Darton, 1818); *Official Map of Lincoln*, London, Simpkin, Marshall and Co. 1877 (from Thomas Dix's map, first published Darton 1820); *Official Map of Warwick*, London, Simpkin, Marshall and Co., c.1878 (from Thomas Dix's map of 1820); and others.

**Freeman**, William. Publisher 'near Temple Bar', London. Published an English edition of Commelin's collection of voyages (first published 1644) under the title *A Collection of Voyages Undertaken by the Dutch East India Company...*, 1703 (with T. Newborough, J. Nicholson, R. Parker and J. Walthoe).

**Freeman, Hunt & Company**. See **Hunt**, Freeman [& Co.].

**Freher**, Marquard (1565-1614). Jurist, politician and historian in Heidelberg. *De Lupoduno*, 1618 (with an archaeological map of Kurpfalz).

**Freibe**, W.C. Atlas von Liefland, Riga 1798 (with Ludwig August von Mellin).

**Freire** [Freiral], João. Portuguese cartographer. Manuscript atlas of the North Atlantic and Mediterranean, 1546.

**Freire**, Olavo. One of the draughtsmen for Homem de Mello's *Atlas do Brazil*, 1908 (maps engraved by A. Simon).

**Freisauff**, F. von. *Ektypographischer Schul-Atlas für Blinde*, Vienna *c.*1837 (lithographed by A. Floder).

**Freitag** [Freytag], Gerard. *Frisia Occidentalis*, (with A. Metio, used by Blaeu, 1635 and J. Janssonius, *c.*1645).

**Freiwill**, —. Umgegend von Berlin, 1875.

**Frellon**, Johann. Eastern Mediterranean, Lyons 1568.

**Freman**, W. See **Freeman**, William.

**Frémin**, A.R. Geographer and publisher of 'Rue des Fossés St. Jacques N$^{o.}$34', Paris; pupil of Jean Baptiste Poirson, attached to the Dépôt Général de la Guerre. *États Unis*, 1820; *Atlas de la France*, 1844 (with Alexis Donnet); *Carte du Chemin de Fer de Paris à Rouen et au Havre*, Paris, Auguste Logerot [*c.*1850s] (engraved by C. Dyonnet); corrected Eustache Hérisson's *Océanie*, 1854; *Carte physique et routière de la France*, 1859, editions to 1878; *Mappemonde*, 1868.

**Frémont**, —. Cartographer of Dieppe. *Diocèse de Rouen* (6 sheets), 1715.

**Frémont**, *Lieutenant* (later *General*) John Charles (1813-1890). U.S. Corps of Topographical Engineers. Worked on surveys of the Upper Mississippi River, 1838-1839 (with J.N. Nicollet), later expeditions were charted in his *Map of an exploring expedition to the Rocky Mountains in the year 1842 and to Oregon and North California in the years 1843-44*, 1845, published in the *Report* of the expedition, Washington, Gales & Seaton 1845 (lithographed by E. Weber and Company, surveys used by other mapmakers); *Map of the City of St. Louis*, St. Louis, J. Hutawa 1846 (with J.N. Nicollet); *Geographical Memoir on Upper California*, 1848 (incorporated *Map of Oregon and Upper California*, drawn by Charles Preuss, lithographed by E. Weber & Co., Baltimore); *Military Reconnaissance from Fort Leavenworth, in Missouri to San Diego, in California*, 1848 (with W.H. Emory, lithographed by E. Weber).

**Frémont d'Ablancourt**, Nicolas (1625-1693). French diplomat, worked in Spain and Portugal 1659-1668, where he acquired maps which were later used in the preparation of *Suite du Neptune François*, Amsterdam, Pieter Mortier 1700.

**French**, brothers.
- **French**, John Homer (*b.*1824). Teacher, surveyor and mathematician from New York State, headmaster of Newtown Academy [Connecticut], 1852-1855, began producing and publishing township plans in his spare time. In 1855 he resigned his teaching post to work with Robert Pearsall Smith in Syracuse, co-ordinating surveys for a state map, gazetteer and county maps of New York State. Amongst his team of surveyors he employed several former pupils, including members of the Beers family and D. Jackson Lake, to undertake an ambitious programme of new and updated surveys, 1857-1859. These included a survey of Oneida County, 1857 (with Silas Beers, F.W. Beers and D. Jackson Lake) published as *Gillette's Map of Oneida Co. New York...*, Philadelphia 1858; the individual surveys culminated in the main state works: *The State of New York from New and Original Surveys under the Direction of J.H. French*, Syracuse, Robert Pearsall Smith 1859, 1860, New York, H.H. Lloyd 1865; *Gazetteer of the State of New York Embracing a Comprehensive View of the Geography, Geology, and General History of the State, and a Complete History and Description of Every County, City, Town, Village and Locality*, Syracuse, R.P. Smith 1859 and many editions including *Historical and Statistical Gazetteer of New York State*, Syracuse, R.P. Smith 1860. Ref. RISTOW, W.W. American maps and mapmakers (Detroit 1985) pp.363-375; RISTOW, W.W. The Map Collector 36 (1980) p.22-23.
- **French**, Frank F. Surveyor, brother of John Homer French with whom he worked on the New York survey, 1856-1858, mostly on the main State map under the supervision of F. Mahler. Town of Milo, New York, 1856; Albion, Orleans County, New York, 1857; Ovid, New York, 1858; *Map of Orange and Rockland Cos. New York*, Philadelphia, Corey & Bachman 1859 (with W.E. Wood and S.N. Beers).

**French**, E.R. Printer of Brattleboro, Vermont. Jeremiah Greenleaf's *New Universal Atlas*, 1840, 1842.

**French**, Frederick. American surveyor. Plan of the Town of Dracut, Massachusetts, 1791.

**French**, *Captain* George. Coast of Sumatra, 1784-1785, published Alexander Dalrymple 1786.

**French**, J.O. Province of La Rioja, 1839.

**French**, W. Publisher of 67 Paternoster Row, London. *Payne's Illustrated Plan of London*, 1853 edition (first published *c*.1851).

**French & Bryant**. Atlas of Brookline, Massachusetts, 1897.

**Frend**, A.B. *Approaches to the proposed Ranelagh suspension bridge*, 1843.

**Frentzel**, Georg Friedrich Jonas (*c*.1754-1799). Copper engraver of Leipzig. Produced maps and town plans of Saxony. Saxony and Bohemia (20 sheets), 1780.

**Frenzel**, Johann (1788-1858). Engraver of Dresden. Dresden, 1808.

**Frere**, Israel. Liberties of Oswestrie, 1602 (with John Norden).

**Frère de Montizon**, Armand-Joseph (*b.c*.1788, *fl*.1818-1848). French science teacher and mapmaker. Produced educational pamphlets and maps for which he ran his own 'imprimerie-lithographie' from 1821. Specialised in the graphic representation of statistics on maps. *Carte Philosophique figurant la Population de la France*, 1830 (the first printed dot map to show population distribution); produced the first known French electoral maps, 1834-1835. **Ref**. PALSKY, G. 'Les developpements de la cartographie statistique au XIXe siècle' in *La cartografia francesca* (Barcelona 1996) pp.153-155 [in French].

**Frérot d'Abancourt**, Charles. See **Abancourt**, Charles-François Frérot d'.

**Frese**, D. See **Friese**, Daniel.

**Frese**, G.W. de. *Carlstads Stift*, 1871.

**Freshfield**, Douglas W. *The Western Atlas*, 1886.

**Fresne**, Marc Macé Marion du (*d*.1772). In command of a French expedition to the Pacific in 1772. Manuscript charts, including *Partie de la Nouvelle Zélande*, and *Plan du Port Marion* (New Zealand), later used by Julien Crozet in *Nouveau Voyage à la Mer du Sud*, 1783. **Ref**. HOOKER, B. *The Map Collector* 43 (1988) pp.18-19.

**Fresnoy**, N.L. du. See **Lenglet**-Dufresnoy, *abbé* Nicolas.

**Fresnoy**, *sieur* du D. de T. See **Templeux**, Damien de.

**Freud**, Alexander. Railway & Post Map Austria-Hungary, 1899.

**Freudenfeldt**, H. Der Preussische Staat, 1867; Preussen u. Deutschland, 1872.

**Freudenham[m]er**, *Doctor* Georg [Jerzy]. Physician and cartographer of Poznan. *Palatinatus Posnaniensis...*, 1645, first used by J. Blaeu 1662 (engraved by G. Coeck, descriptive text by Szimon Starowolski), a different version was published by Covens & Mortier, *c*.1740.

**Freudhoffer von Steinbach**, brothers. Hungarian cartographers.
• **Freudhoffer von Steinbach**, Antal (*fl*.1764-1773). Hungarian cartographer of German descent. From 1767 worked on mapping northeast Carpathians, published as Der Höchste Carpathus; map of Tisza River, 1773.
• **Freudhoffer von Steinbach**, Ferenc. Brother of Antal and József, also became a cartographer and worked in southern Hungary and Dobsina.
• **Freudhoffer von Steinbach**, József. Like his brother Ferenc, worked as a cartographer in southern Hungary and Dobsina.

**Freund**, —. Plan von Wittenberg, 1890.

**Freusberg**, Marquard Rudolph von (*fl*.early-18th century). Cartographer of Württemberg.

**Frey**, J.B. Historice ab orbe, 1677.

**Frey**, J.J. *Plan über der Stadt Bezirk der August Rauracorum*, *c*.1830.

**Frey**, Johann Jakob (1783-1849). Engineer of Knonau. Worked on the triangulation of the canton of Bern, Switzerland, 1824-1825. *Trigonometrischen Höhenbestimmungen in Grindelwald, Lauterbrunnen & Frutigen*, 1816. **N.B. Compare with Frey**, J.J., above.

*John Homer French (b.1824). (By courtesy of the Library of Congress, Washington DC)*

**Frey**, Johannes Michael (1750-after 1818). Draughtsman and engraver in Augsburg. Views of *Augsburg*, 1795; *Babenhausen*, c.1800; *Kirchheim*, c.1800.

**Frey**, Julius (1872-1915). Cartographer and later publisher of Bern. In 1898 he joined G. Kümmerly to form the Swiss company Kümmerly & Frey *q.v.*

**Frey & Nell**. Publishers of 79 Nassau St., New York. *Topographical Railroad & County Map of the States of California and Nevada...*, 1868 (from information supplied by L. Nell, lithographed by F. Mayer & Co.).

**Freycinet** [Frécinet], brothers. French hydrographers who in 1800 sailed under the command of Nicolas Baudin on his voyage of discovery to southern Australia; Louis sailed on *Le Naturaliste*, and Henri aboard *Le Géographe*. Both individually and together they charted parts of the coasts of Tasmania and Australia, 1800-1804. **Ref.** TOOLEY, R.V. *The mapping of Australia* (1979) pp.82-88.
• **Freycinet**, Henri (1777-1840). Charts were published in Louis-Claude's *Voyage*, from 1807.
• **Freycinet** [Desaulces de Freycinet], Lieutenant Louis-Claude de Saulces de (1779-1842). French navigator and hydrographer, when *Le Naturaliste* returned to France, Louis took command of the schooner *Casuarina* on its explorations of the Tasmanian and South Australian coasts; circumnavigation aboard *Uranie*, 1817-1820. Over 30 maps and charts published in *Voyage de Découvertes aux Terres Australes* (12 volumes), 1811-1812, (4 volumes) Paris, A. Bertrand 1824 (with M. Boullanger and others, printed by Langlois); *Carte générale de la Nouvelle Hollande*, 1812; *Voyage Autour du Monde... Sur les Corvettes l'Uranie et la Physicienne. Atlas de Navigation et Hydrographie* [1817-1820] (22 maps), Paris 1824 and later; *Carte de la Province de Rio de Janeiro...*, 1824.
• **Freycinet**, Rose Marie de [née Pinon] (1784-1832). Dressed as a sailor she accompanied her husband Louis Claude on the *Uranie*, 1817-1820, and gave her name to Rose Island, Samoa. Her journal was published in 1927.

**Freycinet**, Louis de. *Plan du Havre de Balade*, 1854-1856.

**Freydenberg**, Henri (1876-1975). French colonial officer. *Explorations dans le bassin du Tchad* (2 maps), 1907; *De Nguigni à Bilma*, 1908. **Ref.** BROC, N. Dictionnaire illustré des explorateurs Français du XIXe siècle: I Afrique (1988) pp.151-152 [in French].

**Freyer**, Heinrich (1802-1866). *Special-Karte des Herzogthums Krain* (16 sheets, chromolithograph), Vienna 1846.

**Freyhold**, Alexander von (1813-1871). *Methodischer Netz-Atlas*, 1846; *Vollständiger Atlas*, 1850; *Neue Karte von Deutschland, zugleich historisch-geographische Karte von Preussen*, Berlin, Deitrich Reimer 1853 (drawn by A. von Schmidt, lithographed by L. Kraatz).

**Freyhold**, Edward (fl.1855-1879). Topographer and draughtsman. Worked on Gouverneur Warren's *Map of the Territory of the United States from the Mississippi River to the Pacific Ocean*, 1858 (with F.W. von Eglofstein, printed by Julius Bien to accompany G.K. Warren's *Memoir* and included in *Pacific Railroad Reports*); draughtsman for *Sketch Exhibiting the Routes between Fort Laramie and the Great Salt Lake*, 1858; contributed detailed maps for the Boundary Commission survey of the north-west boundary of the United States, 1866; revised and redrew *Territory of the United States from the Mississippi River to the Pacific Ocean*, New York, J. Bien 1868; topographical sheets for Clarence King's *Geological and Topographical Atlas Accompanying the Report of the Geological Exploration of the Fortieth Parallel*, New York, Julius Bien, 1876; *Map of the Territory of the United States, West of the Mississippi River*, 1879.

**Freytag**, Adolf. *Mitteleuropa*, c.1889.

**Freytag**, Gerard. See **Freitag**, Gerard.

**Freytag**, Gustav (1852-1938). Cartographer and publisher of Vienna. Trained as a lithographer with his uncle Friedrich Köke, and between 1872-1878 he worked as a cartographer with F.A. Brockhaus, also worked in London and Berlin. With Wilhelm Berndt founded G. Freytag & Berndt which published his cartographic work. *Eisenbahnen Russlands*, 1884; *Afghanistan*, c.1885; tourist maps of eastern Alps, 1888 onwards; *Reise- und Verkehrs-Atlas von Österreich-Ungarn*, 1896; *Völker-*

*und Sprachenkarte von Österreichisch-Ungarn*, Vienna after 1900; *Hand-Atlas für den politischen und gerichtlichen Verwaltungsdienst in der Östrreichisch-ungarischen Monarchie*, Vienna 1901; *Export-Atlas*, 1901; *G. Freytag's Touristen-wanderkarten...*(13 maps), G. Freytag & Berndt 1907; and many others. **Ref.** ESPENHORST, J. Andree, Stieler Meyer & Co: Handatlanten des deutschen Sprachraums (1800-1945) (1994) pp.338-344 [in German]; KRETSCHMER, DÖRFLINGER & WAWRIK Lexikon zur Geschichte der Kartographie (1986) Vol.C/I pp.241-242 (numerous references cited) [in German].

• **Freytag & Berndt** [Kartographische Anstalt G. Freytag & Berndt] (*fl.*1885-1920). Cartographic publishing company formed by G. Freytag and W. Berndt in 1885. Until the First World War they were one of the foremost European publishers of cartographic works. *Térkép az 1905-évi országgy. képviselóvalasztások eredményéról* [map showing the results of Hungarian elections in 1905], 1905; made the first *Skikarten* (ski-ing maps), starting with the areas Lilienfeld, Reisalpe and Stuhleck-Pretul in the Alps, 1910. **N.B.** See also **Berndt**, Wilhelm.

•• **Freytag-Berndt & Artaria.** Freytag & Berndt merged with Viennese publishers Artaria & Co. in 1920.

**Frézier**, Amédée François (1682-1773). French navigator and 'Ingénieur ordinaire du Roy'. *Relation du Voyage de la Mer du Sud aux Côtes du Chily et du Perou... 1712, 1713 & 1714* (24 sheets of maps, plans, charts and views), Paris, Jean-Geoffroy Nyon 1716, (2 volumes) Amsterdam, Pierre Humbert 1717, English editions published London, Jonah Bowyer 1717, London, John Senex 1722 (used as a source by J. van Keulen [n.d.], Delisle 1722 and later, d'Anville 1730, J. Homann 1733, Prévost 1746, J. de Beaurain 1751 etc.).

**Fricx** [Frickx; Friex; Frix], father and son.
• **Fricx**, Eugène Henri (1644-1730). Bookseller, publisher and 'imprimeur du Roi' (from 1689) of 'Rue de la Madeleine', Brussels (1706-1711). Issued maps from 1703 onwards. Best known for his maps of the military campaigns of the War of the Spanish Succession, 1701-1714 which included *Carte Particuliere des Environs de Bruxelles...* (2 sheets), 1706 republished G. Fricx 1746 (engraved by J. Harrewyn); *Carte Particuliere des Environs de Louvain...* (2 sheets), 1706, G. Fricx 1746; *Carte Particuliere de Mons d'Ath de Charleroy...* (2 sheets), 1706, G. Fricx 1745; *Carte Particuliere des Environs de Dunkerque, Bergues, Furnes, Gravelines, Calais et Autres*, 1707 (engraved by Harrewyn); *Carte Particuliere des Environs de Bruges, Ostende, Damme, l'Escluse et autres...* (2 sheets), 1707, G. Fricx 1744; *Plan de la Bataille d'Oudenaerde du 11 Juillet 1708*, 1708 (drawn by G.L. Mosburger, engraved by Harrewyn); *Carte Particuliere des Environs de Liege, Limbourg, et Partie de Luxembourg* (2 sheets), 1708, republished 1746; *Carte Particuliere des Environs d'Anvers, Gand, Hulst, et de tout le Pays de Waes*, 1708; *Carte Particuliere des Environs de Maestricht, Partie de Liege, Faucquemont, et Pays d'Outre-Meuse*, 1708, G. Fricx 1745; *Carte Particuliere des Environs de Roermonde, Venlo...* (2 sheets), 1709; *Partie de l'Angleterre* [Kent], 1709, 1744 (engraved by Harrewyn); *Bouchain Ville Forte du Comte de Hainaut Située sur la Riviere d'Escaut*, 1711 (engraved by P. Devel); *Carte Particuliere des Environs d'Avesnes, Landrecy, la Capelle, Guise, etc*, 1712; the campaign maps were drawn together as *Table des Cartes des Pays-Bas et des Frontières de France* (26 sheets), Brussels, E.H. Fricx 1712 (the same maps were also published under the title *Recueil des cartes des Provinces Meridionales des Pais Bas*). **Ref.** KRETSCHMER, DÖRFLINGER & WAWRIK Lexikon zur Geschichte der Kartographie vol.C/I (1986) pp.242-243 [in German]; KOEMAN, C. Atlantes Neerlandici (1967-1971) vol.2 pp.109-110.

•• **Fricx**, Georges. Son of Eugène Henri (above); 'Imprimeur de Sa Majesté, rue de la Madeleine', Brussels. Republished many of the maps of E.H. Fricx 1744-1748. **Ref.** KOEMAN, C. Atlantes Neerlandici (1967-1971) Vol.VI p.XIII.

**Frid**, Ottone. See **Friedrichs**, Otto.

**Fridelli**. See **Friedel**, E.

**Friderich**, Johannes. Augsburg, *c.*1624 MS.

**Friderichsen**, L. Karte des westlichen Theiles der Südsee, Hamburg 1885.

**Friderici**, Jan Christiaan. Dutch surveyor. South coast of Africa, Cape Aquilles to Swart Kop's River Bay, 1789-1790 MS. **N.B.** Compare with **Frederici**, Johann Christian, above.

*Louis-Claude de Saulces de Freycinet (1779-1842). (By courtesy of the National Library of Australia. Rex Nan Kivell Collection)*

**Fridericus von St. Emmeran,** —. Monk at Stift Klosterneuburg, near Vienna. Drew a manuscript 'T-O' map known as the 'Fridericus-Karte', c.1421-1422 or c.1440 (now lost), said to have been the first modern and comprehensive map of middle Europe. A contemporary manuscript description of the map was so detailed that many people have since been able to reconstruct it. **Ref.** KRETSCHMER, DÖRFLINGER & WAWRIK Lexikon zur Geschichte der Kartographie (1986) p.243 [in German].

**Fridrich,** Johann Gottlieb (1742-1809). Engraver of Regensburg and Copenhagen. J. Erichsen and G. Schönning's map of Iceland, c.1771; A.H. Godiche's Slesvig, 1781; Christian Jochum Pontoppidan's *Det Nordlige Norge...*, Copenhagen, I.G. Blankensteiner 1795.

**Fried,** Franz (fl.1811-1868). Austrian geographer of Vienna, many of his maps were published by Artaria & Compagnie. *Karte der Oesterreichische Monarchie...*, 1817; *Amérique* (4 sheets), 1818, 1841; *Atlas der neuesten Geographie für Jedermann und jede Schulanstalt...* (25 sheets), Vienna, Artaria & Co 1825 and later editions; *Karte des europäisch-osmanischen Reiches... Moldau, Besarabien, Wallachey, Bulgarien und Rumelien* (6 sheets), 1828; *Neueste General-Post-und Strassen Karte der Oesterreichischen Monarchie* (4 sheets), Vienna, Artaria 1829 (with M. de Traux); *General-Karte des Erzherzogthums Oesterreich...*, Artaria 1832; *Deutschland*, 1868.

**Fried,** J. Serbia & Bosnia, 1812.

**Friedberg,** Emanuel von. *General Karte von Serbien*, 1853.

**Friedel** [Fridelli], Ehrenbert Xavier [Xaver], SJ (b.1673). Austrian Jesuit. Worked with J.B. Régis, P. Jartoux, J. Bouvet and de Mailla on the preparation of a map and description of China: *Allgemeinen Reichskarte unter der gegenwärtigen Dynastie* (woodcut, 28 sheets), 1717, (copper etching), 1719, (woodcut, 32 sheets), 1721.

**Friedemann,** Hugo. *Schul-Wandkarte Sachsen*, c.1877.

**Friedenreich,** P.C. *Danmark... Slesvig... Faeröerne*, 1861.

**Friederich,** Charles [& Company]. Lithographers, printers and publishers of St Louis. Edward & Julius Hutawa's *Plan of the City of St. Louis*, 1838, 1842.

**Friederichs,** Joachim. Publisher of 5 Nassau Street, Soho Square, London. *The Circuiteer or Distance map of London* (designed for checking cab fares), c.1847, c.1850, 1851, c.1862.

**Friederichsen,** Ludwig F. [& Co.] (1841-1915). Cartographer and publisher of Hamburg, founder Cartographische Institut Hamburg. Published a range of hiking maps. Worked for Adolf Stieler, 1863; Costa Rica, 1876; *Karte des Samoa*, 1879; *Sir Walter Ralegh's Karte von Guayana mit dem Lauf des Orinoco und des Maranon order Amazonas um 1595. Facsimile...*, 1892; *Concessionsgebietes Süd-West Afrika Compagnie*, 1897; C. Waeber's *Map of north eastern China*, 1900; and many others.

**Friederichsen,** Max H. (b.1874). *Methodischer Atlas zur Länderkunde von Europa*, Hannover & Leipzig, Hahns 1914, 1917.

**Friederichsen,** Peter (fl.1830-1865). Cartographer for Justus Perthes.

**Friedlein,** D.E. Bookseller of Cracow. *Plan okolic Krakówa*, 1830.

**Friedlein,** G.H. Leipzig, 1846.

**Friedrich,** Doctor Ernst (1867-1937). German geographer. *Kleinasien*, Halle 1898; *Produkten und Verkehrsrate von Afrika* (3 sheets), Velhagen & Klasing 1903; *Handatlas für das Deutsches Volk*, Leipzig 1926, and later editions with title changes.

**Friedrich,** Jacob Andreas *the elder* (1684-1751). Engraver and publisher in Augsburg. *Carte de la Haute et Basse Alsace* (3 sheets), 1727.

**Friedrich,** Jacob Andreas *the younger* (1714-1779). Engraver in Stuttgart and Nuremberg. *Ladenburg und Neuenheim* (military map on 2 sheets), c.1745.

**Friedrich,** L. Route maps of central Europe, 1859-1866.

**Friedrichs** [Frid], Otto[ne]. Engraver for Ubbo Emmius's *Typus Frisiae Orientalis*, c.1595.

**Friend,** John (*c*.1665-*c*.1720). Chartmaker of East Lane, Rotherhithe. As an apprentice of Joel Gascoyne of the Drapers' Company he became one of the 'Thames School' of chartmakers, gaining his freedom in 1689. Friend seems to have been the only one who was not based in Wapping. Coasts of Asia (15 charts), 1703-1708; estate map at Laindon Hall, Essex, 1705 MS; chart Whitby Flamborough Head, 1707; chart of the Persian Gulf, 1709 MS; chart of Newfoundland, *c*.1713; and others. **Ref.** PAYNE, A. *The Map Collector* 53 (1990) pp.38-39; SMITH, T.R. 'Manuscript and printed sea charts in seventeenth-century London: the case of the Thames School' in THROWER, N.J. *The Compleat Plattmaker* (1978) pp.45-100; CAMPBELL, T. 'The Drapers' Company and its school of seventeenth century chart-makers', in WALLIS & TYACKE (eds.) *My Head is a Map* (1973) pp.85-86, 100-102; WALLIS H.M. and CUMMING W.P. 'Charts by John Friend preserved at Chatsworth House, Derbyshire, England' in *Imago Mundi* 25 (1971), p.81 (with list).

**Friend,** N.M. *Sheboygan, Wisconsin*, 1838.

**Friend,** Norman (*fl*.1847-1887). Lithographer of 332 Walnut Street, Philadelphia. Briefly in partnership with Jacob Aub as Friend & Aub, below. Involved in the production of many U.S. county maps and atlases, including the publications of Robert Pearsall Smith and Thompson & Everts. J.C. Sidney's *Map of the Circuit of Ten Miles Around the City of Philadelphia*, Philadelphia, R.P. Smith 1847 (lithographed by Friend, printed by Peter Duval); Plan of Wilmington, Delaware, 1850; Edward Roberts's map of Philadelphia, 1868; Jackson County, Michigan, 1874; and others.
• **Friend & Aub** (*fl*.1860-1863). Partnership of Norman Friend and Jacob Aub, trading as lithographers and engravers at 330 [80] Walnut Street, Philadelphia. Henry Macintyre's Boston, 1852; *Map of the Town of Litchfield*, 1852; *Map of the Town of Torrington*, 1852; *Map of the Town of Norfolk*, 1853; R.P. Bridgens's *Map of the City of San Francisco...*, M. Bixby 1854 (printed by Wagner & McGuigan); H.F. Bridgens's map of Cumberland County, Pennsylvania, 1858 (printed by Wagner & McGuigan); *Clark & Tackabury's new Topographical Map of the State of Connecticut*, 1859.

**Friend,** Robert (*fl*.1711-1739). Apprenticed to John Friend (above) of the Drapers' Company or 'Thames School' of chartmakers, for eight years from 1711 (there is no evidence that he was related to his master). Later 'to be heard of at the Jerusalem Coffee House in Change Alley'. East Indies, 1719 MS; Chart of part of Malaya, Tanasam &c., 1738 MS; 'Gerritsz' chart, 1739. **Ref.** CAMPBELL, T. 'The Drapers' Company and its school of seventeenth century chart-makers', in WALLIS & TYACKE (eds.) *My Head is a Map* pp.100-101.

**Friend,** William. Straits of Jubal, 1802; Tor Harbour, 1804.

**Fries** [Friess; Frisius; Phrysius; Phryes; Phrijsen], Lorenz [Laurens; Laurentius] (*c*.1490-*c*.1531). Physician, astrologer and geographer from Colmar, Alsace. Studied medicine at several European universities, and worked in Strasbourg *c*.1519-1527. Produced a number of scientific instruments and revived a method of projection recorded by an Andalusian astronomer. Reworked most of the maps of Martin Waldseemüller. Worked with Peter Apian on the preparation of *Tipus orbis universalis...*, 1520 (based on Waldseemüller's world map of 1507, published in J. Camers's *Solini Enerrationes*, Vienna 1520, also in Pomponius Mela's *De situ orbis*, Basle 1522); worked with the Strasbourg printer J. Grüninger on a new edition of Ptolemy's *Geographia* (with woodcut maps reduced from M. Waldseemüller), Strasbourg 1522 (the maps were re-used in versions edited by W. Pirkheimer, published Strasbourg, J. Grüninger, Nuremberg, J. Koberger 1525, Lyons, M. & G. Trechsel 1535, Lyons, G. Trechsel 1541 etc.); revised M. Waldseemüller's *Carta marina* (12 sheets), 1525, 1527, 1530, Latin edition, C. Grüninger 1531 (first published 1516); *Uslegung der Mercarthen oder Carta Marina* (commentary on 32 leaves to accompany *Carta Marina*), Strasbourg 1525, (26 leaves) 1527, (22 leaves) 1530, Latin edition 1530. **Ref.** KARROW, R.W. *Mapmakers of the sixteenth century and their maps* (1993) pp.191-204; MEURER, P. *Fontes cartographici Orteliani...* (1991) pp.146-148 [in German]; KRETSCHMER, DÖRFLINGER and WAWRIK *Lexikon zur Geschichte der Kartographie* (1986) vol.C/1 pp.243-244 (numerous references cited) [in German]; SHIRLEY, R. *The mapping of the World* (1983) N°45 p.51; *ibid*. N°47 p.53; *ibid*. N°56 p.60.

**Friese** [Frese], Daniel (1540-1611). Artist and cartographer, worked in Hamburg and Lüneburg. Worked with H. Rantzau on town plans for Vol.5 of Braun & Hogenberg's *Civitates Orbis Terrarum*, including *Bardewick*, 1588; *Heide*, 1596; *Meldorp*, l596.

**Frieseman**, H. Publisher of Amsterdam. *Carte de la Crimée*, 1787; *Mer de Marmora* (4 sheets), 1791.

**Friesen**, Karl Friedrich (1785-1814). Mathematician, architect and cartographer of Magdeburg.

**Friess**, L. See **Fries**, Lorenz.

**Friex**, E.H. See **Fricx**, Eugène Henri.

**Friis**, Jens Andreas (1821-1896). Professor at the University of Christiania, Oslo. Made maps for the Norwegian Geological Institute. *Kart over Russisk Lapland*, before 1884 MS; *Ethnografisk Kart over Finmarkens Amt...* (4 sheets), before 1888; *Russisk Lapland* (2 sheets), c.1888; Tromsø, 1890.

**Frijlink**, Hendrik (1800-1886). Publisher and bookseller of Amsterdam. His copperplates passed to Noothoven van Goor of Leiden in 1869. *Kleine School-Atlas*, 1843 and twelve editions to 1873; *Nieuwe Hand-Atlas der Aarde* (24 copperplate maps issued in 4 instalments), Amsterdam 1851-1855 and five editions to 1868, Leiden, Noothoven van Goor 1869 and three editions to 1881. **Ref.** *Caert-Thresoor* 19 (2000) N°2 pp.37-44 [in Dutch with English summary].

**Frijman**, Jonas Andersson (*d.c*.1704). Swedish surveyor of Gotland.

**Frintzel**, —. Engraved Christlieb Benedict Funk's *Südliche Erd-Oberflaeche auf Aequatorflaeche entworfen...* [Australia], 1781.

**Friquegnon**, *Capitaine* (later *Commandant*) Jean-Baptiste (1859-1934). French military geographer in Indochina, founded the Service géographique de l'Indochine in 1900, of which he was Director 1904-1909. *Carte du Tonkin*, 1884; *Carte de la Cochinchine* (4 sheets), 1890; *Carte du Tonkin* (4 sheets), 1890; *Carte de l'Annam* (6 sheets), 1890; *La Grande Carte de L'Indo-Chine*, 1893, 1894, 1899, 1907 (with Cupet and De Malglaive, printed by Dufrénoy), reduced edition published Paris, Augustin Challamel 1895, 1909, 1914; *Plan de la Ville de Haphong*, 1898; *Chine Méridionale et Tonkin*, 1899; *Province de Yunnan*, 1900; *Carte de la Cochinchine Française*, 1901 (revised version of the map by A. Koch); *Service Géographique des Colonies... Tonkin et Haut Laos* (4 sheets), Paris, A. Challamel 1902 (engraved by R. Hausermann); *Carte de la Chine Orientale* (9 sheets), 1908. **Ref.** BROC, N. *Dictionnaire illustré des explorateurs Français du XIXe siècle: II Asie* (1992) pp.196-197 [in French].

**Friquegnon**, *Captain* Nicolas. Sometimes credited with the maps of Indo-China listed under Jean-Baptiste Friquegnon, above. **N.B.** Neither Broc nor the Bibliothèque Nationale mentions Nicolas, and both credit Jean-Baptiste with maps of the region.

**Frisak**, *Captain* H. von. *Skagestrands Bugt i Island*, 1808-1810, published 1818; *Sydlige Kyst af Island*, 1823.

**Frisch**, F. (*fl.c*.1800-1815). Worked in Augsburg. Town views after L.A.G. Bacler d'Albe and P. Marchioretto.

**Frisius** [Phrysium; Reineri], Gemma [Jemme] (1508-1555). Mathematician, physician, cosmographer and cartographer from Dokkum in Friesland. Studied and worked in Louvain from c.1525, wrote many influential theoretical works on mathematics, surveying and cosmography. Gemma was succeeded by his son Cornelius as professor of mathematics and medicine at the University of Louvain. Made corrections to Peter Apian's *Cosmographicus Liber*, Antwerp, R. Bollaert 1529 and many later editions (first published in 1524); thought to have made a cosmographic globe, 1529; *De Principiis Astronomiae & Cosmographiae*, Louvain and Antwerp 1530 and later; essay on the use of triangulation in surveying, 1533 (published in *Cosmographics Liber* and *Principiis Astronomiae*); 37-cm terrestrial globe, 1536 (engraved by Gaspard van der Heyden and Gerard Mercator); 37-cm celestial globe, 1537 (with Gaspard van der Heyden and Gerard Mercator); world map, Louvain 1540 (no surviving examples known); reduced woodcut world map, first published as *Carte Cosmographique, ou universelle description du Monde...* in the first French edition of Apian's *Cosmography*, 1544 and other editions incuding A. Du Pinet, 1564.
**N.B.** The name Reineri describes him as son of Reinier,

*Gemma Frisius (1508-1555). (By courtesy of Rodney Shirley)*

though Gemma himself chose to adopt the name Frisius after the place of his birth. Some scholars have tried to translate his given name, Gemma, resulting in confusing alternatives such as Edelgestein and Van der Steen. **Ref.** KARROW, R.W. *Mapmakers of the sixteenth century and their maps* (1993) pp.205-215; KROGT, P. van der *Globi Neerlandici* (1993) pp.48-57, 75-77, 410-412; KRETSCHMER, DÖRFLINGER & WAWRIK *Lexikon zur Geschichte der Kartographie* (1986) Vol.C/1 p.255 (numerous references); KISH, G. *Medicina, Mensura, Mathematica: The Life and Works of Gemma Frisius 1508-1555* (Minneapolis 1967); HAARDT, R. 'The globe of Gemma Frisius' in *Imago Mundi* 9 (1952) pp.109-111; ORTROY, F. van *Bibliographie de Gemma Frisius, fondateur de l'école belge de géographie, de son fils et des neveux les Arsenius* (Brussels 1920).

Frisius, L. See **Fries**, Lorenz.

**Frits** [Fritsch], András [Andräa] Erik [Erich] (1715-1778). Hungarian cartographer, worked with Sámuel Mikoviny on county map series. *Tabula Nova Inclyti Regni Hungariae*, Pozsony 1753 (engraved by Sebestyén Zeller, in J. Tomka-Szászky's *Compendium Hungariae Geographicum...*, 1753); numerous manuscript maps extant.

**Fritsch**, Caspar. Printer of Amsterdam. Orbis antiqui sive geographiae plenibus, 1706.

**Fritsch**, Gustav Theodor (1838-1927). German naturalist and explorer. Drei Jahre S. Afrika, 1868; Eingeborenen S. Afrikas, 1872.

**Fritsch**, *Doctor* I.H. Charte vom Harz, Magdeburg, Heinrichschafen 1833.

**Fritsch**, Johann Theobald. *Zweybrücken*, 1774, 1794.

**Fritsch**, Karl von (*b*.1838). German geologist and explorer; visited Canaries and Morocco. *Kanar Inseln*, 1867; *Tenerife geologisch topographische dargestellt...*, Winterthur, J. Wurster & Co 1867 (with G. Hartung and W. Reiss); *Tenerife, nach Vorhandenen Materialien und eigenen...*, Winterthur c.1869 (with G. Hartung and W. Reiss); *Sanct Gotthard*, 1873.

**Fritsche**, Hermann P.H. (*b*.1839). Russian explorer of German origin. *Carte de la Chine septentrionale, de la Mongolie, de la Mantchourie, du pays de l'Amour et de l'Oussori...*, St Petersburg 1874.

Fritsche, W.H. See **Fritzsche**, Wilhelm Heinrich.

**Fritschen**, Thomas. Krieg in Italien, 1702.

**Fritschi**, J.N. *Umgebungen v. Baden*, c.1855; *Umgebung v. Heidelberg*, c.1877.

**Fritsen**, P.E. *Noord-Braband*, 1841.

**Fritz**, A. *Plan von Carlsruhe*, c.1872.

**Fritz**, H. Nordlichtes, Gotha 1874.

**Fritz**, Samuel (1656-1728). Bohemian Jesuit missionary and traveller. Contributed to the exploration and mapping of the Amazon Basin. *El Gran Rio Marañón o Amazonas* (2 sheets), 1691 published Quito 1707, English edition by Herman Moll 1717.

**Fritzsch**, C. Engraver of Hamburg. *Landt zu Ditmers...*, 1733; Jonas Hanway's *Baltic M. Die levant*, 1754 edition.

**Fritzsche**, Guglielmo Enrico. See **Fritzsche**, Wilhelm Heinrich.

**Fritzsche** [Fritsche], Wilhelm Heinrich [Guglielmo Enrico] (1859-1894). German cartographer, also worked in Italy. One of the founders of the Istituto Cartografico Italiano at Rome. Sudan, 1885; *Nuovo atlante geografico*, Rome 1887; *Karawanstrasse*, Gotha 1890; *Carta Generale della Sicilia...*, Rome, Istituto Cartografico Italiano 1891 (engraved by T. Peuckert); *La rappresentazione orographica a luce doppia nella cartografia moderna*, 1892; *Regno d'Italia* (20 sheets), 1893. **Ref.** CERRETI, C. 'Un industria che da lungo tempo non fioriva più nell'Italia'; W.H. Fritzsche e l'Istituto Cartografico Italiano in: *Notiziario del Centro Italiano per gli Studi Storico-Geografici* (agosto-dicembre 1996) An.4, 2-3, 21-27 [in Italian].

Frizell, family.
• **Frizell**, Charles (*fl*.1747-1803). Estate surveyor working in England and Ireland.
•• **Frizell**, Charles. Son of Charles, above, also worked as an estate surveyor. Their work cannot be separated.
•• **Frizell**, Richard (*fl*.1750-1797). Son of the elder Charles (above), surveyed estates in England and Ireland, including Stoke Hammond, 1774 MS.

Frizon. See **Frisius**, Gemma.

*Sir Martin Frobisher (c.1535-1594). [Note mis-spelling of his name on engraving].*
*(By courtesy of the National Library of Australia, Rex Nan Kivell Collection)*

**Froben**, father and son. Printers and publishers of Basle.
• **Froben**, Johannes (1460-1527). Born in Hammelburg, Germany, worked as a printer and publisher in Basle. **Ref.** HERNANDEZ, A. *Johannes Froben und der Basler Buchdruck des 16. Jahrhunderts* Basle (1960) [in German].
•• **Froben**, Hieronymus (1501-1563). Born in Basle, son of Johannes Froben. Printer of Erasmus's edition of Ptolemy (no maps), Basle 1533.

**Frobisher**, *Sir* Martin (*c.*1535-1594). English navigator and hydrographer from West Yorkshire; made first English attempts to find the Northwest Passage in 1576 and 1577; third voyage to Greenland, 1578; Vice-Admiral on Francis Drake's expedition to the West Indies 1586. Annotated William Borough's chart of the Atlantic, 1578; untitled world map in George Best's *A True Discourse of the Late Voyages... of Martin Frobisher*, London 1578; Plot of Croyzon 1594 MS. **N.B.** There is no record of the birth of Frobisher in the registers of his home parish. He himself implied that he was born in 1538 or 1539. The years 1535 or 1536 pre-date the parish records and would match the known facts of his life better. **Ref.** SYMONS, T.H.B. *et al* (Eds) *Meta Incognita: a discourse of discovery: Martin Frobisher's Arctic expeditions, 1576-1578* (Canadian Museum of Civilization 1999).

**Froebel** [Fröbel], J. (1805-1893). German mineralogist. *Aus Amerika*, 1857-1858.

**Froger**, François (1676-c.1715). French engineer and traveller. *A Relation of a voyage made in the years 1695, 1696, 1697, on the coasts of Africa, Streights of Magellan, Brazil, Cayenna, and the Antilles...*, London 1698 and other editions.

**Froggatt**, Walter Wilson (1858-1937). b. Victoria. Maps of the Australian coast & New Guinea, 1885-1892.

**Froggett**, John. Engraver of London, for Robert Wilkinson, c.1808-1821 including *New South Wales New Zealand New Hebrides*, London, Rob.t Wilkinson 1808.

**Froggett**, John Walter. Engraver and publisher, No.3 West Sq., London. *Rape of Arundel*, 1819; engraved World, 1825; *Froggett's Survey of the Country 30 Miles round London*, 1831, 1832; *Froggett's Map of the Country Fifteen Miles round London*, London, G.F. Cruchley, c.1842, after 1853. **N.B.** Compare with John **Froggett**, above.

**Frogley**, Arthur (fl.c.1709-1724). English estate surveyor, possibly from Cambridgeshire. Surveyed estates in Cambridgeshire, Essex and London including land at Hatfield Broadoak, Essex, 1714.

**Froiseth**, Bernard Arnold Martin (1839-1922). Civil engineer from Trondheim, Norway, lived in Minnesota from childhood, trained at Montreal. Moved to Salt Lake City in 1869, where he worked as a surveyor, 1870-1900. *Map of the Territory of Utah Territory*; *Great Salt Lake Valley*; *Plat of Salt Lake City* (3 maps on one sheet), 1869, 1870; *Map of the Territory of Utah*, 1870 and other editions; *Crofutt's New Map of Salt Lake City*, New York 1871 (engraved by the Actinic Engraving Co.); *New Mining Map of Utah*, New York, American photo lith. co. 1871 (with H.R. Durkee); *Froiseth's New Sectional and Mineral Map of Utah...*, 1871, 1875, 1878, 1879, 1898 (re-engraved 1875, printed by A.L. Bancroft & Co., San Francisco); *Froiseth's Map of Little Cottonwood Mining District...*[Utah], 1873; *Map of the Territory of Utah*, 1874; *Salt Lake City, prepared expressly for Crofutt's Salt Lake City Directory 1885*, 1885 (engraved by Hilpert and Chandler); *Froiseth's New Sectional & Mineral Map of Utah*, 1898 (drawn and printed by W.B. Walkup). **Ref.** RISTOW, W.W. *American maps and mapmakers* (Detroit 1986) pp.462-463; MOFFAT, Riley Moore *Maps of Utah to 1900* (Western Association of Map Libraries, 1981).

**Froloff** [Frolov], —. Plan of Moscow, 1819; one of the calligraphy engravers for Pyadyschev's *Atlas géographique de l'Empire de Russie, du Royaume de Pologne et du Grand Duché de Finlande...*(83 sheets), 1820-1827, 1829, 1834.

**Frolov**, K. (fl.1770-1783). Engraver of St Petersburg, worked for the Imperial Academy of Sciences. Engraved maps of areas of Russia surveyed by I. Islenyev, J.F. Schmid and J. Truscott, 1770-1783 including J. Truscott's *Mappa Gubernii Astrachanensis*, [n.d.]; J. Truscott's *Mappa Gubernii Sibiriensis*, [n.d.].

**Fromann** [Frommann], August Bernhard (1737-1817). Map of the Coburg region, 1783.

**Fromann**, Gustav. *Special-Karte des Odenwaldes*, 1867.

**Frome**, *Captain* [later *Colonel*] Edward Charles, RE (1802-1890). Surveyor-General of South Australia, 1839-1849. *Report on the country to the eastward of Flinders' Range, South Australia*, 1844 (based on his explorations of 1843; map by John Arrowsmith).

**Fromman**, A.B. See **Fromann**, August Bernhard.

**Frommann**, Maximilien. Lithographer for Carl Glaser's *Vollständiger Atlas über alle Theile der Erde*, Darmstadt, L. Pabst 1836, 1840; Hessen, 1867.

**Frommel**, Gustaf F. (fl.late-18th century). Cartographer of Baden.

**Frommel**, Karl Ludwig (1789-1863). Engraver and artist in Karlsruhe; founded an engraving workshop in England with Henry Winkles in 1824.

**Frommhold,** Tobias. Decorative wall map of a private estate called Flur Benndorf, Saxony, 1773.

**Froncoso,** D. See **Francoso,** Diogo.

**Frontinus,** Sextus Julius (*c*.30-104). Roman governor of Britain, wrote on stratagems, aqueducts and surveying. Illustrated excerpts on surveying in *Corpus agrimensorum romanorum*; mention of maps in work on aqueducts of Rome.

**Frontpertius,** Adrien Front de (*b*.1825). Geographer and publisher. Canada, 1867; États Unis, 1873; États latins de l'Amérique, 1883.

**Froriep,** family.
- **Froriep,** *Doctor* Ludwig Friedrich von (1779-1847). Doctor of medicine, took control of the Geographische Institut Weimar, 1822, which later passed to his son.
- **Froriep,** Robert (1804-1861). Undertook the management of his father's business when bankruptcy threatened. In 1845 he was joined by H. Kiepert of Berlin, who did the cartographic work. *Physikalischer Erdglobus*, 1846; *Der noerdliche Sternenhimmel*, 1848; school atlas of ancient history, 1848.

**Froschauer,** family. Printers and publishers of Zurich. **Ref.** KARROW, R.W. *Mapmakers of the sixteenth century and their maps* (1993) pp.310; ibid. 511-515; LEEMANN-VAN ELCK, P. *Die Offizin Froschauer. Zürichs berühmte Druckerei im 16. Jahrhundert. Ein Beitrag zur Geschichte der Buchdruckerkunst anlässlich der Halbjahrtausendfeier ihrer Erfindung* Zurich (1940); RYCHNER, M. *Rückblick auf vier Jahrhunderte Entwicklung des Art. Institut Orell Füssli in Zürich* Zurich (1925) [in German].
- **Froschauer,** Christoph [Christoffel] *the elder* (*c*.1490-1564). German born printer and publisher of Zurich, founder of the publishing company Orell Füssli [*q.v.*]. Map of the Holy Land for the Zurich Bible, 1525, 1579; published Joachim Vadianus's *Typus cosmographicus universalis*, 1534; Johannes Honter's *Rudimenta Cosmographica* [school atlas of Switzerland], 1546, 15 later editions to 1602 (14 woodblock maps cut by H. Vogtherr); Johann Stumpf's chronicle of Switzerland, 1548 and later, including reduced edition 1554; Stumpf's *Landtafeln... XII*, 1548, 1556, 1562 and later editions (re-used maps from Stumpf's Chronicle).
- **Froschauer,** Christoph *the younger* (1532-1585). Nephew of Christoph, above. Printed woodcut maps by Jos Murer, including *Statt Zürych*, 1576 (woodblock cut by L. Fryg).

**Froschauer,** Eustachius (*d*.1552). Publisher of Zurich, presumably related to the Froschauers above, and also credited with the publication of a work by Stumpf. *Verzeichnung der loblichen Eydgnoschafft* (poem with a map of Switzerland), *c*.1538.

**Frosne,** I. Worked as an engraver of portraits and decorations which appeared on battle plans and profiles in S. de Beaulieu's *Les glorieuses conquestes de Louis le Grand*, *c*.1694.

**Frost,** Geo. S. *The town and harbor of Frankfort* [Michigan], 1870.

**Frost,** H. Engraver. Christopher Greenwood's *Map of the county of Sussex...*, 1829; Dorset, 1829.

**Frost,** H.J. *Kalcaska County, Mich.*, 1878.

**Frost & McLennan.** *Map of Lake County, Illinois*, Chicago 1873.

**Frugoni,** Juan. Estancia de San Jorge Uruguay, London 1880.

**Fruin,** H.J. *The new map of China*, 1916 (with E.J. Dingle).

**Fry,** L. See **Fryg,** Ludwig.

**Fry,** *Colonel* Joshua (*c*.1700-1754). American surveyor and mathematician, Commander of Virginian Regiment. Surveyed the boundary between Virginia and North Carolina, 1749-1751 (with Peter Jefferson). Worked with Jefferson on the preparation of *A Map of the Inhabited part of Virginia Containing the Whole Province of Maryland with Part of Pensilvania, New Jersey And North Carolina drawn by Joshua Fry and Peter Jefferson in 1751*, London, Thomas Jefferys *c*.1753, 1754 (*most* is added to the title), 1755, Rob$^{t.}$ Sayer & Tho$^{s.}$ Jefferys 1768, 1775, Rob$^{t.}$ Sayer 1782, 1794 (used by Robert de Vaugondy for *Carte de la Virginie et du Maryland*, 1755 and later (in *Atlas Universel*), and by George Louis Le Rouge for *Virginie, Maryland en 2 feuilles...*, 1777). **Ref.** VERNER, C. *Imago Mundi* 21 (1967) pp.70-94; VERNER, C. *The Fry and Jefferson map* (Princeton University Press 1950).

*Captain Edward Charles Frome (1802-1890). (By courtesy of the National Library of Australia. Rex Nan Kivell Collection)*

**Fry,** W. Ellerton. Sketches used for *Map of the Route taken by the B.S.A. Co's expedition to Mashonaland 1890* [Zimbabwe], 1890; *Victoria Falls*, 1892.

**Fry & Son.** Surveyors of Grays Inn, London. Estate plan at Calverley, Yorkshire, 1852 MS.

**Frye,** Alexis Everett (1859-1936). *Home & School Atlas*, Boston, Ginn & Company 1895, 1896; *Complete Geography*, Boston & London, Ginn & Company 1895; *Grammar School Geography*, Boston & London, Ginn & Company 1902, 1903, 1911, 1913, 1920; *Geography Manual*, Boston 1903; and many other school geographies.

**Fryer,** C.E. Fishery Districts England & Wales, 1884-1885.

**Fryer,** George. Plan of Borough of Kingston-upon-Hull, 1885; Plan of the City of Hull, 1898.

**Fryer,** father and sons. Surveyors of Newcastle. Produced manuscript maps of areas in northern England.
• **Fryer,** John (1744/5-1825). River Tyne, 1773; Castle Garth, Newcastle, 1777; inclosure maps at Warden, Northumberland, 1779; Alston Moor, Cumberland, 1797 (with John Bell); surveys in Hexamshire, Northumberland, 1800; land at Ryton, Durham, 1801; Lanercost, Cumberland, 1804, 1806.
•• **Fryer,** Joseph Harrison (1778-1855). Surveyor, son of John (above). Produced manuscript maps in northern England. Park at Ennerdale, Cumberland, 1805; Ulverston, 1805; Skeldon Moor, Lancashire, 1805; Torver Common, Lancashire 1805.
•• **Fryer,** William (*fl.c.*1809-1833). Son of John. Map of Northumberland, 1833
• **John Fryer & Sons.** The above, working together. Parish of Simonburn, Northumberland, 1809 MS; Northumberland, 1820 (engraved by M.W. Lambert).

**Fryg** [Frei; Frey; Frig; Fry], Ludwig (*fl.*1559-1586, *d.c.*1600). Woodblock cutter of Zurich. J. Murer's plan of Zürich, 1576 (used by Sebastian Münster).

**Fuca** [Apostolos Valerianos], Juan de la (*d.*1602). Greek-born navigator, in the service of Spain. His claim to have sailed through a 'Northwest Passage' from Labrador to the Pacific in 1592 contributed to the idea that California was an island (an account of his voyage appears in *Purchas his Pilrimes*).

**Fuchs,** A. Charte von der Insel Bandt, 1743.

**Fuchs,** *Doctor* C.W.C. *Carta geologica dell' isola d'Ischia, c.*1872.

**Fuchs,** Caspar Friedrich (1803-1874). Engraver in Hamburg. Schmalkalden, 1848 (with C.F. Danz); Helgoland, *c.*1850.

**Fuchs,** Charles. Lithographer of Hamburg. *Das Terrain der Danewerk-Stellung, c.*1864; *Holsteinischer Kanal*, 1866.

**Fuchs,** F. Draughtsman. Bird's-eye view of Boston, Philadelphia, John Weik 1870.

**Fuchs,** Johann Conrad. German military cartographer. *Karten Einiger an dem Ufer deß Rheins ligender Festungen*, 1707.

**Fuchs,** Károly Henrik (1851-1916). Hungarian cartographer and scholar, from Pozsony [now Bratislava]; taught in Pozsony, Brassó [now Brasov, Transylvania, Romania], and Sopron; worked on development of photogrammetry in cartography. Died in Pozsony. Manuscripts kept at Budapest Academy of Sciences.

**Fuchs,** *Dr.* W. Die Venetianer Alpen, Vienna 1844.

**Füchsel,** Georg Christian (1722-1773). German physician. *Historia terrae et Maris, ex historia Thuringiae...*, 1762 (contained a geological map of part of Germany).

**Fucini,** Alberto (*b.*1864). University professor of geology. Carta geologica del cicondario di Rossano, [n.d.].

**Fucker,** András (*fl.*1733-1749). Hungarian cartographer, studied geometry and law at Jena from 1717. Sáros County, 1733; Tokaj region, 1749.

**Fuehrer** [Führer], *Lieutenant* Carl. Plan attack on Fort Washington, New York, 1776.

**Fuente,** F. Charts of various islands of the Galapagos, including Carenero Island; Esperanza y S. Marcos...; Los Hermanos; Quito Sueño; San Clemente, all dated 1748.

**Fuenzalida,** Jose del C. *Nueva mapa de Chile* (2 sheets), Valparaiso, Tornero & Bertini *c.*1912.

**Fuerst,** Paul. See **Fürst,** Paul.

**Fuerstaller,** J. See **Fürstaller,** Josef.

**Fuerstenhoff,** J.G.M. von. See **Fürstenhoff,** Johann Georg Maximilian von.

**Füesslin,** J.C. *Staats- und Erdbeschreibung der Schweizerischen Eidgenosschaft,* 1770.

**Fuhrmann,** Matthias (*c.*1690-1773). Engraver of Vienna.

**Fuji** Fuchiomi. Chart of Ise Bay [Japan], 1868.

**Fujii** Hanchi. See **Ochikochi** Doin.

**Fujita** Junsai [Ryo]. Japanese surveyor. *Ezo Kokyo yochi zenzu* [map of Ezo], 1854 (draughted by Hashimoto Sadahide).

**Fuko,** Mineta. See **Mineta** Fuko.

**Fukuzumi** Kinrindo. Japanese publisher. Kageyama Noritaka's map of Mutsu and Dewa Provinces [Japan], 1868.

**Fulgosio,** Fernando. Islas Filipinas, 1871.

**Fullarton,** Archibald [& Co.] (*fl.*1833-1872). Publishers and engravers of Glasgow (1833-1842), Edinburgh and London, also Dublin from 1845. 34 Hutcheson Street, Glasgow (1833); 31 South Bridge, Edinburgh (1833); 110 Brunswick Street, Glasgow (1838, 1840); 12 King's Square, Goswell Street Road, London (1838, 1840); 6 Roxburgh Place, Edinburgh (1838, 1840). James Bell's *A New and Comprehensive Gazetteer of England and Wales...* (partwork in 4 volumes), 1833-1837 and later (from *c.*1840 entitled *The Parliamentary Gazetteer*); James Bell's *A System of Geography...* (3 volumes), 1838; republished J. & C. Walker's *British Atlas...,* 1846 (first published 1841); *Gazetteer of the World* (with atlas), 1850-1856 and later; Rev. J.M. Wilson's *Imperial Gazetteer of Scotland,* 1854 (some plans engraved by G.H. Swanston, also published as *The county atlas of Scotland); Royal illustrated atlas,* 1860 and later; *Royal Illustrated Atlas of Modern Geography,* 1864 (previously published in 27 parts, 1854-1862); J. Bartholomew's *Imperial Gazetteer of England and Wales,* 1868. **Ref.** SMITH, D. *Victorian maps of the British Isles* (1985) pp.137-139.

• **Fullarton, Macnab & Co.** Publishers. New York publisher of later editions of J.M. Wilson's *The Imperial Gazetteer of Scotland.*

**Fuller,** [Richard] Buckminster (1895-1983). American engineer. A minor creation of his was a world map which folded into a self-assembly globe with flat triangular faces: *Fuller Projection Dymaxion Globe,* Buckminster Fuller Institute 1938, and later (with Shoji Sadao).

**Fuller,** C.A., U.S. Corps of Engineers. Buffalo, 1850; Island of Montreal, 1850; Memphis 1850; Sketch Map of the Red River, 1858.

**Fuller,** Corydon Eustathius (*b.*1831). Teacher from Indiana, kept a diary of his adventures and misfortunes while working as the travelling salesman for J.H. Colton in Arkansas (now in the William L. Clements Library, University of Michigan). Sold *Colton's General Atlas,* and *Colton's Atlas of the World,* 1857-1858. **Ref.** BOSSE, D. 'A canvasser's tale' in *The Map Collector* 57 (1991) pp.22-26.

**Fuller,** Edward. Inclosure plans. Exmoor Forrest, 1817; Delamere Forrest, 1817.

**Fuller,** Edward Bostock (*fl.*1687-1696). Surveyor of estates in Berkshire and London including the manor of Cholsey, Berkshire, 1695-1696.

**Fuller,** Francis (*fl.*1832-1864). English estate surveyor, worked in London and the Home Counties. Plan of Muswell Hill Park 1860 MS.

**Fuller,** George. Successor to William Figg, surveyor of Lewes, Sussex. Lewes, 1877; Eastbourne, 1879.

**Fuller,** John F. Whitman County, Washington, 1895 (with W.J. Roberts).

**Fuller,** S.P. Boston, 1838.

**Fuller,** T.W. *Map of Rock Island County, Illinois,* Dixon, Illinois, Clifford L. Hubbard 1915; *Map of Henry County, Illinois,* 1915; *Map of Carroll County, Illinois,* 1916.

**Fuller,** Doctor Thomas (1608-1661). *Cantabrigia* (bird's-eye view of Cambridge), 1634; maps to History of the Holy War,

*Corydon Eustathius Fuller (b.1831). (By courtesy of the Clements Library, University of Michigan, USA)*

1639; *A Pisgah-Sight of Palestine*, London, John Williams 1650 [or earlier]. **Ref.** The Map Collector 3 (1978) pp.50-51.

**Fuller**, *Captain* William. Charts of the entrance to St Mary's River, and the mouth of the Nassau River, 1769 published as insets on Thomas Jefferys's *Plan of Amelia Island in East Florida*, London, Thomas Jefferys 1770 (later used in Faden's *American Atlas*, 1781).

**Fullerton**, *Lieutenant* J.D. Plan of the City of Harar (4 sheets), 1885; Somaliland, 1885.

**Fulljames** [Fulliames], Thomas (*fl*.1789-d.1847). Estate surveyor, produced manuscript surveys of estates in various parts of England. Maps of the properties of the Duke of Dorset in Ashdown Forest, 1791 MSS; Stead Quarter Farm, 1792 MS; Desborough Hundred, 1796; Dean Forest, 1804.

**Fulmer**, F.S. Rutland County, Vermont, 1869 (with F.W. Beers).

**Fulneck Academy.** Moravian Atlas, 1853.

**Fulton**, Hamilton, father and son.
• **Fulton**, Hamilton (*fl.*1810-*d.*1834). Surveyor and civil engineer of Newman Street, London (1813), father of Hamilton Henry, (below). Surveys in several English counties, also in the USA from 1819-1828. Worked for Thomas Telford, 1810, state engineer for North Carolina, from 1819, chief engineer of Georgia, 1826-1828. *Plan... Stamford Junction Navigation*, 1810; thought to have contributed to R.H.B. Brazier's *A New Map of the State of North Carolina...*, John MacRae 1833. **Ref.** RISTOW, W.W. *American maps and mapmakers* (Detroit 1986) pp.124-126.
•• **Fulton**, Hamilton Henry (1813-1886). Surveyor and engineer, trained by his father. Produced civil engineering maps for sites in England and Wales. Kings Lynn, 1846.

**Fulton**, David. Arkansas, 1839.

**Fulton**, Henry. Farm line map of the City of Brooklyn, 1874.

**Fulton**, J.A. Surveyor. Boundary line Ohio, 1818.

**Funck**, Carl Oscar. *Karta öfver Stockholm*, 1846.

**Funck** [Funk; Funke], David (1642-*c.*1705). Engraver and publisher in Nuremberg, commissioned maps from Johann Baptist Homann. *Des Schutz-reichen...*, 1690; *Terra Sancta*, *c.*1695; *Saxonia Inferioris*, 1690; *Cracau*, 1695; *Circulus Suevicus*, *c.*1700; *Lusatiae Superior*, *c.*1700; *Insula & Regnum Candia*, *c.*1700; *Poliometria Germaniae*, *c.*1707; and others.

**Fundanus**, Nicander Philippinus. Publisher of Rome. *Nova et Exactissima Totius Ungariae Descriptio...* (3 copper plates), 1595 (copied from Alexander Mair's 1594 map of Hungary).

**Funes de Pavia**, Juan Batista. Derrotero general, Guayaquil, 1700.

**Fungairiño**, E. Spanish engraver. *Mar Mediterràneo. Costa Se. de España. Plano del puerto y arsenal de cartagena con la Ensenada de Escombrera y las Algamecas*, Madrid, Hydrographic Department 1875-1876 (with S. Bregante y Martinez); topographic engraver for *Carta Generale... del Archipiélago Filipino* (2 sheets), Madrid 1875 (surveys by C. Montero i Gay, script engraved by J. de Gangoiti).

**Funk**, Christlieb Benedict (1734-1814). Professor of natural history at Leipzig. *Südliche Erd-Oberflaeche auf Aequatorflaeche entworfen...* [Australia], 1781 (engraved by Frintzel).

**Funk**, D. *Umgebung von Ingoldstadt*, 1865.

**Funk**, David. See **Funcke**, David.

**Funk**, *Captain* James de. *General map of the Malabar Coast*, Dalrymple 1755, 1792.

**Funk & Wagnall's Co.** Publishers of New York. *Standard Atlas of the World*, 1896; *Funk & Wagnalls new comprehensive atlas of the world*, New York and London 1916.

**Funke**, Carl Philipp (1752-1807). *Atlas der alten Welt*, Weimar 1800 and editions to 1884 (with G.U.A. Vieth).

**Funke**, Christian B. (1736-1786). Professor of Physics at Leipzig. *Anweisung zur Kenntniss der Gestirne* 1770 (with 4 star charts).

**Funke**, D. See **Funcke**, David.

**Funkenstein**, Judah. Land of Canaan (4 sheets), 1876.

**Funnell**, William. Mate to Captain Dampier. Accounts of Funnell's voyages of 1703-1706 appear in collections by J. Harris, 1764, J. Callandar, 1766-1768 and J. Burney, 1805-1806. Plan Bays of Le Grand, Brazil, 1703; *A Voyage Round the World* [an account of William Dampier's voyage in the South Seas 1703-1704], 1729 (volume IV of Dampier's *A Collection of voyages*).

**Funter**, *Captain* Robert. Raft Cove, Alexander Dalrymple 1789; charts and Port Cox in Captain John Meares's Voyage, 1790.

**Furck**, Sebastian. San Salvador, 1650.

**Furlanetto**, Ludovico. 'Ponte de Bareteri', Venice. *Laguna Veneta*, *c.*1780, also published as *Carte des lagunes de Vénise*, 1780; *Nuova Carta Topografica della Provincia*

Captain *Tobias Furneaux (1735-1781)*. *(By courtesy of the National Library of Australia. Rex Nan Kivell Collection)*

*di Dalmazia*, 1787 (surveyed by Melchiori and Zavoreo); *Nuova Carta Geografica Della Guerra Presente tra i Due Imperj e la Porta Ottomania*, Venice, Marten 1788; *Territorio di Friul*, 1793.

**Furlani**, P. de. See **Forlani**, Paolo.

**Furlong**, *Captain* Lawrence. *American Coast Pilot*, Newburyport, E.M. Blunt 1796 and later.

**Furnas**, Boyd Edwin (1848-1897). Atlas directory of Miami County, Ohio, 1883.

**Furne & Cie.** Publishers and booksellers 'Rue St. André des Arts N⁰·55', Paris. Louis-Philippe Ségur's *Atlas pour l'histoire universelle*, 1842.

**Furneaux**, *Captain* Tobias (1735-1781). Circumnavigator, sailed with Captain Samuel Wallis, 1766-1768, later sailed with Captain James Cook as commander of the *Adventure* and explored the coast of Van Diemen's Land, 1772-1775. A chart of the Southern Hemisphere, 1772-1773; *Part of Van Diemens Land...*, 1773 (published in an account of the voyages of Captain Cook, London, Wm Strahan 1777); *A Mercator's Chart* [the track of the *Adventure* east of New Zealand], 1773 **Ref.** DAVID, A. (Ed.) *The Charts and Coastal Views of Captain Cook's Voyages* Vol.2 pp.lix.

**Furnis**, B.T. Chart of Encounter Bay, 1838 (with Colonel W. Light) published on *The maritime Portion of South Australia...*, London, J. Arrowsmith 1839.

**Furnival**, E. Engraver of a reduced edition of Robert Baugh's map of Shropshire (1 sheet), R. Baugh & J. Furnival 1811 (from the 9-sheet edition published in 1808); Llanymynech near Oswestry, 1814 (with J. Furnival).

**Furnival**, J. Publisher, worked with R. Baugh and E. Furnival (above).

**Furrer**, Konrad. Swiss pastor and traveller. Palestina, 1863 (with Heinrich Lange).

**Fürst** [Fuerst], Paul (*c*.1605-1666). Publisher and printseller of Nuremberg. In the 1660s he re-issued Oterschaden's globes (first published c.1600), and the globe gores and planispheres of Isaac Habrecht II (first published 1620s). *Bourges*, 1638; *Tours*, 1838; *Koppenhagen*, c.1640; *Abbildung des Königreich Ungarn, Durch Türckey Bis Nach Constantinopel*, c.1660, 1665; *Nurnberg*, 1664; *Regensburg*, 1666; *Cölln*, 1667.

**Fürstaller**, Joseph Jakob (1730-1775). *Atlas Salisburgensis* (34 sheets), 1765 (now lost).

**Fürstenhoff**, Johann Georg Maximilian von (1686-1753). *Land-Charte des Chur-Fürstentums Sachsen...*, 1741.

**Furtado**, L.C.C.P. See **Pinheiro** Furtado.

**Furtenbach**, Joseph von (1591-1667). Architect in Ulm. *Newes Itinerarium Italiae*, Augsburg 1627 (with road map of Italy).

**Fuser** [Fusier], *Lieutenant* Lewis V. (*d*.1780). Royal American Regiment. *Plan of Canada or the Province of Quebec...*, 1760; maps of the Saint Lawrence Valley, 1761-1763 MSS.

**Fushimiya** —. Japanese publisher. *Shimpan Zoho Osaka no Zu* [Plan of Osaka], 1657, 1661, 1671, 1678 (one of the oldest printed plans of the city).

**Fussell**, James. Publisher, Sydney, New South Wales. Stanford's New Map of Australia, 1864; Fussell's New Map of Australia, 1868.

**Füssli & Co.** Part of the Orell Füssli publishing company of Zurich, founded by Christopher Froschauer in 1519. Became Orell Füssli Kartographie AG, which continued trading into the 20th century. Adrien Hubert Brué's *Amérique*, 1826. **N.B.** See also **Orell** Füssli.

**Fustinoni**, A. *Provincia di Como*, 1884.

**Fu Tsê-Hung.** Panoramic maps of rivers and lakes in *Hsing-shui chin-chien* (Golden Mirror of the Flowing Waters), 1725.

**Fyers**, *Lieutenant* William (*fl*.1773-1812). Engineer. *Plan of the post of Portsmouth* [USA], 1781 MS (Clements Library).

**Fynje**, H.F. See **Fijnje**, H.F.

**G., C.F.** *Plan of Milford Haven*, 1785-1790.

**G., D.E.** *Der Hochlöblich Schwäbische Circul...*, c.1680.

**G., E.** *Karte von Europa für Schulen*, c.1877.

**G., F.** (possibly Philippe [Filips] Galle or Leonardo Gaultier). Signed dedication of the map of the Americas entitled *Novus Orbis* in Richard Hakluyt's edition of Peter Martyr d'Anghiera's *Decades*, Paris 1587. **N.B.** According to Philip BURDEN (*The mapping of North America* p.78), previous opinion that F.G. was Galle is now disputed. There is evidence to suggest that Leonard Gaultier [Galter] might have been the engraver of this map.

**G., F.** *Topographische Karte... Schleswig*, 1863.

**G., F.L.** See **Güssefeld**, Franz Ludwig.

**G., G.B.V.** *Nuova Carta del Mare Adriatico*, 1815.

**G., G.W.** Draughtsman for [George] Higgie's map of the Isle of Bute, 1886.

**G., J.F.** *Provincia de Guipuzcoa*, 1875.

**G., L.** American engraver. Patrick May's map of Sacketts Harbour, 1815.

**Gabato.** See **Cabot**, Sebastian.

**Gabb**, William More (1839-1878). Paleontologist from Philadelphia. Californische Halbinsel, Gotha 1868; *Mapa de la Isla de Santo Domingo* (geological map), 1872; Costa Rica, 1877.

**Gabinetto Litterario.** Publishing firm in Naples. Atlante universale, 1813, 1817.

**Gäbler, F.E.** See **Gaebler**, Friedrich Eduard.

**Gaboriaud**, *Captain* —. *Carte du Sahara Algérien*, 1845 (based on information gathered by E. Daumas, engraved by J. Schwaerzlé); *Partie septentrionale de l'Afrique*, 1845.

**Gabriello de Sanctis**, —. Atlante corografico dell Due Sicilie, Naples 1840, 1843 and 1856.

**Gabrielli**, Ferruccio. Worked with others on *Region Oriental del Perú...*, Lima, Sociedad Geografica 1906.

**Gábriely**, Johann. *Finanz- und Handels-Karte des oesterreichischen Kaiserstaates* (4 sheets), Vienna 1857 (with A. Doležal).

**Gabrys**, J. *Carte Ethnographique de l'Europe*, Institut Géographique Kummerly & Frey 1918.

**Gadea**, Joseph. Austrian cartographer. *Tabula Geographica...Regnum Sclavoniae*, Vienna 1718.

**Gädertz, A.** See **Gaedertz**, A.

**Gadner** [Gadnerus; Gadmer], Georg (1522-1605). Swabian surveyor. Map of Württemberg, 1572 MS, used by A. Ortelius

1579, 1584, 1592, M. Quad 1596, complete version (20 sheets), 1596; Stuttgarter Amt, 1587. **Ref.** MEURER, P. *Fontes cartographici Orteliani...* (1991) p.148 [in German].

**Gadolin**, Jacob (1719-1802). Swedish bishop, mathematician, astronomer and surveyor.

**Gaebler** [Gäbler], Friedrich Eduard (1842-1911). Cartographer and publisher in Leipzig. *Special Atlas der berühmtesten und Städte Deutschlands und der Alpen*, Leipzig 1882; *Schul-Atlas über alle Teile der Erde*, 1883 (with Carl Diercke); *Taschen-Atlas des deutschen Reiches...*, Leipzig, 1886 and later editions; *Australien und Oceanien*, 1894; *Umgegend von Chemnitz*, 1898; *Diercke Schul-Atlas für höhere Lehran stalten*, Braunschweig, G. Westermann 1899 (with C. Diercke); and many others.

**Gaedertz** [Gädertz], A. *Provinz Schan-Tung*, 1898.

**Gaertig**, H. *Karte zur Apostolengeschichte*, c.1870.

**Gaetano**, Palma. *Khartis tis Europaikis Turkias, palai men Ellados*, Trieste 1811 (map in Greek and French); Italian map of the Provinces, Trieste 1812; Map of European Turkey, Trieste 1814.

**Gaetschy**, Gaspar. Mapa de los ferrocarriles mexicanos, Mexico 1888 (lithographed by Emanuel Moreau).

**Gaffarel**, Paul J.L. (1843-1920). French historian and geographer from Moulins, taught at Dijon. Histoire des colonies français, 1875; *Histoire de la Floride française*, Paris, Firmin Didot 1875.

**Gage**, Michael Alexander (*fl*.1836-1852). *Plan of the Town and Port of Liverpool...* (3 sheets), 1836 (engraved by Thomas Starling).

**Gage**, Thomas (c.1600-1656). English traveller in Central America, returned to England in 1637, d. Jamaica. *A New Survey of the West-Indies*, 1648, 2nd edition (4 maps) 1655, French language edition, Amsterdam, Paul Marret 1694, 1697 (maps after N. Sanson).

**Gage**, W.J. Publisher of Toronto. *Gage's new excelsior map of the Dominion of Canada...*(4 sheets), 1912 (made by G.W. Bacon).

**Gage**, *Reverend* William Leonard (1832-1889). Scholar, geographer and translator. *A Modern Historical Atlas...*, New York 1869.

**Gagen**, G.R. Surveyor of 14 Stracey St., Stepney, London. Ground plan of Hampton Court village, 1835 MS.

**Gagenhart**, Peter (*fl*.1490-1492). Globemaker of Nuremberg.

**Gagnebin à Neuchâtel**. Publisher. Lithograph map of Australia, New Zealand, and the East Indies, 1850.

**Gagnières des Granges**, *abbé* Claude François (1722-1792). French globeseller.

**Gail**, Jean Baptiste (1755-1829). French hellenist. Atlas à la géographie d'Hérodote, 1823.

**Gail**, Jörg. Schoolmaster and notary in Augsburg. *Raißbüchlin*, Augsburg 1563 (thought to be the first printed travel guide in German, formed the basis of several maps and atlases).

**Gaillard**, Louis, SJ. *Plan de Nankin*, 1898.

**Gaillard**, Tacitus. Merchant and surveyor. Worked with James Cook on a survey of the province of South Carolina, 1770 MS.

**Gaillardot**, Charles. (1814-1883). French doctor, naturalist, geographer and explorer in North Africa and the Middle East. Prepared the first French map of the area from Hauran to Damascus, c.1855.

**Gailloüe**, Pierre. Mapseller and publisher 'dans la Cour du Palais, Rouen'. Guillaume Le Vasseur de Beauplan's Normandie, 1667.

**Gaio**, Matheo da [Mathei]. Oval map of the World, Venice 1516 MS. **N.B.** David Woodward has raised doubts as to the authenticity of this and other manuscript world maps of the period. WOODWARD, D. *The Map Collector* 67 (1994) pp.2-10.

**Gair**, Robert (1839-1927). Globemaker of New York. Globes of 3, 4, 6, 9, 12, and 20-inch diameters from 1890 onwards. **Ref.** WARNER, D. *Rittenhouse* Vol.2 No.2 pp.63-64; ibid. Vol.2 No.3 pp.89-90.

**Gaishi** [Gensui] Ebi (*fl*.17th century). Japanese surveyor. Coastal navigation chart, 1680.

**Gaitte**, A.J. French engraver, pupil of P.C. de la Gardette. Engraver for volume II of Le Gentil de la Galaisière's *Voyage Dans Les Mers De L'Inde...*, Paris, 1781.

**Gajetius**, L. See **Guyet**, Lézin.

**Galabert**, Louis. *Carte minéralogique des Pyrénées*, 1831.

**Galaisière**, G.J.H.J.B. Le G. de la. See **Le Gentil** de la Galaisière.

**Galanti**, A. *Carta della Alpi Giulie*, Turin 1864 (with E. Cagli, A. Dardano and G. Bonatti).

**Galanti**, Luigi. Italian geographer. *Atlante di geografia moderna*, Naples 1834-1836.

**Galateus Leccensis**. See **Ferrari**, Antonio.

**Galaup**, J.F. de. See **La Pérouse**, Jean François Galaup, *Comte* de.

**Galbraith**, Frank H. Chicago railway mail clerk. *Galbraith's railway mail service maps* (8 state maps) [Routes and post offices of the Railway Mail Service in the mid-western states - Illinois (8 sheets), Indiana (4 sheets), Iowa (8 sheets), Kansas (8 sheets), Michigan (4 sheets), Minnesota (8 sheets), Missouri (8 sheets), Nebraska (8 sheets)], Chicago, McEwen Map Co. 1897.

**Gale**, A.C. Land agent of Winchester, Hampshire. Glenbervie and Willow's Green inclosures, Alice Holt Forest, Hampshire, 1850 MS.

**Gale**, G.N. Engineer. Proposed Line of Rail Station Buildings at Waimakarin Gorge, 1865.

**Gale**, J. Engraver. Contributed to I. Hargrave's map of Kingston upon Hull, 1791.

**Gale**, Richard Christopher (*fl*.1830-1855). Land surveyor of Winchester, produced large scale manuscript surveys in many parts of England. *Compton*, 1833; *Plan of Winchester*, 1836; Survey of Durrington, 1843 MS.

**Gale**, Samuel. *...Plan of part of the Province of Lower Canada*, 1795 MS (with J.B. Duberger). **Ref.** SEBERT, L.M. 'The first maps of the eastern townships', *Association of Canadian Map Libraries and Archives Bulletin* 77 (December 1990) pp.1-5.

**Gale**, *Doctor* Thomas (1635/6-1702). English Hellenist, Dean of York and High Master of St Paul's. Assisted in preparation of Moses Pitt's *English Atlas* project, 1680; Britannia romana, Christoph Weigel 1720.

**Gale & Butler**. Engravers of Crooked Lane, London. Plan of part of London, 1800; Daniel Janvrin's map of Guernsey, 1810.

**Galeatius**, Carolus. *Carta topografica della Stato di Milano*, 1777.

**Galego**, João (*fl*. late-16th century). Portuguese chartmaker.

**Galen**, Antony van (*fl*.1715-1738). Dutch surveyor and cartographer, worked in Brabant, styled 'landmeter van de Prins en over de stad en Baronie van Breda'.

**Galiani**, —. Engraver of calligraphy. *Contorni di Napoli*, c.1820 (drawn by G. Riesso, map engraved by Bartoli).

**Galiano**, *Lieutenant* (later *Brigadier de Marina*) Dionisio Alcalá. Straits of Magellan (5 charts), 1785-1786 MSS; *Atlas para el viage de las goletas Sutil y Mexicana al reconocimiento del Estrecho de Juan de Fuca en 1792*, Madrid 1802 (with Cayetano Valdes); Carta esférica Nord-Oeste America, 1795; *Atlas... del Estrecho de Juan de Fuca*, Madrid 1802; *Archipelago de Graecia*, 1806; *Carta Esférica... del Mar de Marmara*, 1806.

**Galignani**, family. Printers and publishers of Venice and Padua. **N.B.** The relationships between them have been assumed.
• **Galignani de Karera**, Simon. Publisher in Venice. Tomaso Porcacchi da Castiglione's *Isole più famose del mondo*, 1572, 1576 (with maps engraved by Girolamo Porro).
•• **Galignani Heirs** [Heredi di Simon Galignani]. Publishers of Tomaso Porcacchi *Isole più famose del mondo*, 1590, 1605 editions; Giovanni Magini's *Geographiae Universae tum veteris tum novae...*, 1596 (Latin edition of Ptolemy's *Geographia*, with maps engraved by G. Porro).
•• **Galignani** [Galignani Fratelli], Giovanni Battista & Giorgio. Printers and publishers of Venice. L. Cernotti's *Geografia Cioè Descrittione Universale Della Terra*, 1597-1598 (Italian translation of Magini's 1596 Latin edition of Ptolemy's *Geographia*, with the same Porro maps as the 1596 edition).

••• **Galignani** [Galignani Fratelli], Paolo & Francesco. Printers and publishers of Padua. Tomaso Porcacchi *Isole più famose del mondo*, 1620 edition; republished Cernotti's *Geografia...*, 1621 edition (Porro maps from 1596 edition).

**Galignani**, A. *Plan of Paris*, 1850.

**Galileo Galilei** (1564-1642). Astronomer and mathematician from Pisa, Italy. One of the pioneers of the use of the telescope for astronomy. He proposed the movements of the satellites of Jupiter as a determinant in the measurement of longitude. Sketch map of the moon, 1610.

**Galindo**, *Colonel Don* Juan. *Usumasinta River* [Mexico], 1833; *Sketch of the State of Costarica in Central America*, London, John Murray 1836 (for Journal of the Royal Geographical Society).

**Galinée**, René de Bréhant de. See **Bréhant** de Galinée, René de.

**Galinier**, Joseph Germain. *Carte d'Acir*, 1840; *Carte de l'Abyssinie*, 1843; *Voyage en Abyssinie*, 1847-1848 (with Pierre Victor Adolphe Ferret); *Carte géologique du Tigré*, 1848.

**Galissonière**, *Marquis* de la. See **La Galissonnière**.

**Gall**, Carl. Publisher of Berlin. Plan Gegend Danzig, 1813.

**Gall** [Inglis] family, and firm Gall & Inglis.
• **Gall**, James *senior* (1783-1874). Printer of Potterrow, Edinburgh. In partnership as Hay, Gall & Co., then established his own business at 24 Niddry Street in 1810, specialising in religious works and aids for the blind. Joined by his son James *junior* in 1838-1847 to form James Gall & Son; James *junior* was replaced in the partnership by Robert Inglis c.1848.
• **James Gall & Son** (*fl.*1838-1847). Partnership of James Gall senior and James Gall junior.
•• **Gall**, *Reverend* James *junior* (1808-1895). Mapmaker, publisher and missionary, son of James Gall. Worked with his father as James Gall & Son for about ten years before resigning in 1847 to become a minister in the Church of Scotland. His work was published by the family firm. *An Easy Guide to Constellations*, 1866 and later editions; *The People's Atlas of the Stars*, Gall & Inglis 1857, 1862; the Orthographic Projection, published in the first issue of the *Scottish Geographical Journal*, 1885. **Ref.** CRAMPTON, J. 'Cartography's defining moment: The Peters Projection controversy 1974-1990', Cartographica 31 (1995) pp.16-32; SNYDER, J.P. Flattening the Earth (Chicago 1993).

•• **Gall & Inglis** (*fl.* from 1848). Publishers of Edinburgh and London, formed when Robert Inglis joined James Gall's company c.1848. Addresses in Edinburgh included: 24 Niddry Street (1810-1847); 38 North Bridge (1849-1856); 6 & 13 George Street (1857-1877); 20 Bernard Terrace (1878-1923); 12 Newington Road (1924-1980). London addresses included: 30 Paternoster Row, London (1872-1874); 25 Paternoster Square, London (1875-1909); 13 Henrietta Street, London (1909-1930); 12-13 Henrietta Street (1931-1973). In 1877 the company acquired at auction part of the stock of George Frederick Cruchley (including some Cary plates). These were still in use in 1910. Many of their maps were used by regional publishers for their local maps. By the turn of the century, Gall & Inglis had become well known for their mass production of cheap, small scale maps for cyclists etc.. *Edinburgh Imperial Atlas*, 1850; *Australian Colonies and New Zealand*, 1851; *Gall & Inglis' Map of United States*, Edinburgh 1852; *Gall & Inglis' Map of the Seat of War in the Danubian Provinces and Greece, the Black Sea and the Caucasus*, 1854; *Gall & Inglis' New One Shilling Atlas of Modern Geography*, 1871; *School Atlas of Modern and Ancient Geography*, 1871; *Sixpenny Atlas of Modern Geography*, 1872; *Cruchley's Railway and Telegraphic County Atlas of England and Wales*, c.1878, c.1887; *Reduced Ordnance Map of London*, c.1878 (first published by Cruchley 1868); *Gall & Inglis' Imperial Globe Atlas of Modern and Ancient Geography* (33 maps), c.1887; Hunting maps, c.1888 (lithographed by Gall & Inglis for A.H. Swiss of Devonport); *Cruchley's County Maps of England* ['county maps for cyclists, tourists &c.'], Gall & Inglis c.1890 (sheets taken from maps first published by George & John Cary, 1820-1832); *Gall & Inglis's Map of the Environs of London. For Cyclists, Tourists, &c. showing the country 25 miles on each side of St Pauls*, c.1895; *Gall & Inglis' Large Scale 'Five-Inch' Map of London*, 1898; Harry Inglis's 'Contour Road' books for cyclists, 1899; *Kent, for cyclists, tourists etc....*, 1901; *Short Spins round London*, 1906 (with Arthur C. Armstrong); and many others.

•• [**Inglis**], Robert (d.1887). Printer, in partnership with James Gall (his father-in-law) from 1848 as Gall & Inglis (see above). His two sons joined him and later succeeded him.

••• [**Inglis**], James Gall (d.1939). Joined his father Robert in partnership in 1880.

•••• [**Inglis**], Robert Morton Gall. Son of James Gall Inglis, entered the family partnership in 1939.

••• [**Inglis**], Harry Robert Gall (d.1939). Younger son of Robert Inglis, brother of James Gall Inglis, joined the family partnership in 1887. Wrote many works on the history of the cartography of Scotland, published in the *Scottish Geographical Magazine* and elsewhere. Led the expansion of the firm into cartographic and astronomical productions. Road maps, 1894; cycling maps, 1902; *Short Spins around London*, 1903, 1906 (with Arthur C. Armstrong).

**Gall von Gallenstein**, Josef, *Freiherr (fl.*19th century). Cartographer of Graz. *Special Karten* [Styria] (5 maps), 1831 and later.

**Gallaeus**. See **Galle** family.

**Gallaham**, Wilhelm von (1751-1788). Austrian military cartographer.

**Gallaher**. See **White**, Gallaher & White.

**Galland**, John *(fl.*1796-1817). Engraver of Philadelphia. Plan of Baltimore, George Keatinge 1797 (no known example).

**Gallardo**, Ygnacio P. *Plano de la Ciudad de Mexico*, 1867.

**Gallatin**, [Abraham Alphonse] Albert (1761-1849). Native of Switzerland, became United States Secretary of the Treasury, diplomat. *Map of the Indian Tribes of North America...* (ethnographic map), 1836 (lithographed by Pendleton); worked with H. Hale and K. Andree on *Ethnographische Karte von Nord-Amerika*, 1854 (in H. Lange's *Atlas von Nord-Amerika*, Braunschweig, G. Westermann 1854).

**Galle** family. Engravers publishers and printsellers of Antwerp.

• **Galle** [Gallaeus; Gaule], Philippe [Filips] (1537-1612). Editor, engraver and printseller from Haarlem, active in Antwerp. Co-publisher of G. Braun and F. Hogenberg's *Civitates Orbis Terrarum*, 1572 (the first part of their 6 volume series of town plans); *Spieghel der Werelt* (later known as *Epitome*), 1577 (miniature edition of Ortelius; text by P. Heyns), numerous editions in different languages up to 1602 (the rights passed to J.B. Vrients in 1601); map of the seventeen provinces (12 sheets), 1578; engraved portrait of Ortelius in *Theatrum orbis terrarum*, 1579 (and later). **Ref.** BURDEN, P. *The mapping of North America* (1996) pp.61-64, 151-153. **N.B.** On the basis of a signature 'F.G.', Galle has been credited with the engraving of the map of the Americas in Richard Hakluyt's edition of Peter Martyr d'Anghiera's *Decades*, Paris 1587. According to Philip BURDEN (*The Mapping of North America* p.78) the assumption that 'F.G.' was Galle is now disputed. There is evidence to suggest that Leonard Gaultier [Galter] might have been the engraver of that work.

•• **Galle** [Gallaeus], Theodore [Theodorus] (1571-1633). Engraver and printseller of Antwerp, son of Philippe. Republished Caspar Vopel's *Recens et Germana Bicornis ac widi Rheni omnium Germnaiae amnium celeberrimi descriptio*, 1595 edition (published G. de Jode 1569); Artois, *c.*1600; credited with engraving some maps for the Vrients edition of Ortelius, *c.*1602; Palatinatus Rheni, *c.*1620.

••• **Galle**, Joannes (1600-1675). Engraver and publisher of Antwerp, son of Theodore. Re-issued maps published by Philippe and Theodore.

•• **Galle**, Cornelis (1576-1650). Engraver of Antwerp, son of Philippe. Credited with engraving *Chorographia terrae sanctae...*, Antwerp, M. Nutius 1632 (copied from a map by C. Adrichom); three hemispherical world maps in *Imago primi saeculi Societatis Iesu...*, Plantin, Antwerp 1640; Sicilia et Magna Graecia, Antwerp 1644.

**Galle**, Jean [Giovanni]. French cartographer. Picardie, 1540.

**Gallego**, Hernán (*b.*1508 or 1517). Pilot to Alvaro de Mendaña de Neyra. Chart of the Solomon Islands, 1568 (now lost).

**Gallegos**, F.J.E. y. See **Estorgo** y Gallegos, Francisco Xavier.

*Theodore Galle (1571-1633). From a painting by Anthony van Dyck. (By courtesy of Rijksprentenkabinet, Amsterdam)*

**Galler,** Hieronymus. Printer for the De Bry *Voyages*, 1619 edition.

**Gallet,** —. Engraver. *Nouveau Plan Itinéraire de la Ville de Paris...*, Paris, Jean Goujon 1824 (with Perrier, calligraphy engraved by Lale).

**Gallet,** Georges. Printer and bookseller in Amsterdam from *c.*1691, where he supervised the Huguetan printing house, many of whose publications bore his name. He also worked in London 1704-*c.*1724. A.P. De la Croix's *Géographie Universelle*, 1693; *Nouvelle Introduction à la géographie*, 1695 (Jaillot's *Atlas Nouveau*); *Le Theatre de la guerre en Allemagne...*, 1698; *Las Fuerzas de la Europa...*, 1700 (Spanish edition of N. de Fer's *Les Forces de l'Europe*).

**Galletti,** Johann Georg August (1750-1828). Cartographer and historian from Altenburg. *Allgemeine Weltkunde*, Leipzig 1807-1810, 1818.

**Galli,** Fiorenzi. Engraver. *Texas*, Mexico City, Linati 1826.

**Galli,** Marc. Miniature woodcut map showing the Atlantic and the coasts of Nova Spagna and Africa, published in *Miscellaneo matematico*, 1694.

**Gallo,** Agostino. *Le Dieci Giornate Della Vera Agricoltura, e Piaceri Della Villa* (with a map of Brescia), Brescia, L. de Sabbio and G.B. Bozzola 1564.

**Gallois,** —. *Plan von Hamburg und Altona*, 1868.

**Gallois,** Lucien (1857-1941). Geographer, cartographer and historian from Metz. Took over editorship of *Géographie universelle* in 1918 (after death of Pierre Vidal de la Blache); wrote several works on the history of cartography. He died in Paris. *Carte murale du théâtre de la guerre*, Paris, A. Colin *c.*1914. **Ref.** KRETSCHMER, DÖRFLINGER & WAWRIK *Lexikon zur Geschichte der Kartographie* (1986) p.245 [in German].

**Gallucci** [Galluci], Giovanni Paolo (fl.1569-1597). Italian astronomer from Salo. *Figura Mundum Novum Continens* [the Americas], and *Figurae Europae, Asiae, Africae* [eastern hemisphere], both woodcut maps in *Theatrum Mundi et Temporis*, Venice 1588 (considered to be the first printed star atlas), also in *Margarita Philosophica*, Venice 1599; *Mappa Mundi Meridionale* and *Mappa Mundi Settentrionale* [two woodcut hemispheres of the World] in: *Della Fabrica et uso...*, Venice, 1597. **N.B.** Tooley listed a 1586 edition of the map of America but this has not been found. BURDEN, P. *The mapping of North America* (1996) p.83.

**Gallwey,** *Captain* H.L. *RN*. Manuscript maps of areas in Nigeria. *Sketch map of the Benin River*, 1892; *Sketch Map of the Benin and Warri Districts*, 1893; Gwato Creek, 1893.

**Galpin,** Thomas Dixon. Seaman from Dorset, came to London *c.*1851 and entered into partnership with George William Petter to form the printing firm of Petter and Galpin [*q.v.*] in 1852. They took over Cassell's in 1855 to form Cassell, Petter & Galpin [*q.v.*].

**Galpin,** *Colonel* George. Plan of the Town of Alexandria, District of Columbia, 1798.

**Galt,** J.L. [& Co.]. *Ware* [Massachusetts], 1878 (printed by Beck & Pauli of Milwaukee).

**Galt & Hoy.** Publishers of New York. Produced and published a bird's-eye view of *Newport* [Rhode Island], 1878; published William I. Taylor's *The City of New York* [bird's-eye plan], 1879.

**Galter,** Leonardo. See **Gaultier,** Léonard.

**Galton,** Francis (1822-1911). Damar Land [Indonesia?], London 1854; *Meteorographica or methods of mapping weather* (33 plates), London, Macmillan 1863; Isochronic Passage Chart for Travellers, 1881.

**Galvão** [Galvaon], Antonio (*c.*1507-1557). Portuguese captain and geographer, *b.*India, *d.*Lisbon. Governor of the Moluccas. Information used for *D'Indiaanze Landschappen, Zeen en Eylanden, van Couchin Af, tot in de Moluccos...*, in volume II of *Zee-En Landreizen der Portugeezen*, Leiden, P. van der Aa before 1727; Tractado dos descubrièmentos antiguos e modernos, 1731.

**Galves,** *Don* Manuel. Lampun Bay, 1754; *Plan of Capa Luan (Luzon)*, A. Dalrymple 1774; *Port of Salomague*, A. Dalrymple 1781; and others.

**Gálvez**, Bernardo de. Plan of the Siege of Pensacola, 1781 MS (William Clements Library).

**Gálvez**, Juan. Provincia de Santa Fe, Rosario, 1888.

**Gálvez**, *Doctor* Mariano. Atlas Guatemalteco, 1832.

**Gama**, Luiz Philippe de Saldanha da. Plano da Guerra do Paraguay, 1869.

**Gamba**, Joseph. Publisher of Legorn [Livorno, Italy]. *Receuil de Plans des Principaux Ports et Rades de la Mediterranée*, 1800?, 1817 (included 40 maps by J.J. Allezard, first published 1795).

**Gambara**, Lorenzo. Poet of 16th century. Untitled map of a portion of the Atlantic Ocean from just below the Equator, north to the latitude of France in *De navigatione Christophori Columbi*, Rome 1583. **Ref.** BURDEN, P. The mapping of North America (1996) p.71.

**Gambarini**, Bernardo. Tiber, 1744 (with Andrea Chiesa).

**Gambillo**, Enrico. *Carta delle Strade Ferrate Italiane* (4 sheets), Bologna 1886, (2 sheets) 1888.

**Gambino**, Domenico. *Pianta topografica de la città di Palermo*, Palermo, 1862.

**Gambino**, *Professor* Giuseppe (1841-1913). *Grande carta murale della Sicilia* (6 sheets), Palermo 1886; Atlante scolastico muto, 1888; *La Sicilia itineraria*, 1888.

**Gamble**, *Colonel* Dominic Jacotin. Deputy Quarter-Master General. *Sketch of the Ground at Rangiriri, the scene of action between the Maoris and the Troops under the command of Lt. Genl. Cameron...1863*, London, War Office 1864; *Part of the North Island of New Zealand 1863-1864*, 1867.

**Gamble**, William Henry (*fl*.1867-1887). Draughtsman and engraver of Philadelphia. Prepared maps for the atlases of Samuel Augustus Mitchell, from *c*.1861, including: *Map of Kansas, Nebraska and Colorado*, Philadelphia, S.A. Mitchell jr. 1861; *County Map of Michigan and Wisconsin*, 1863 (used in *Mitchell's New General Atlas*, 1864); for M.W. White's *Atlas of the State of West Virginia*, 1873; for William M. Bradley, 1884; plan of Philadelphia and Camden, 1887.

**Gambola**, —. Draughtsman and engineer. *Plan Topographique des Fortifications de la Ville de Constantinople entre le lac de Tchekmidje et le lac de Derkos*, c.1877.

**Gamidge**, S. Mapseller and publisher at Prior's Head, Worcester. *A Plan of the City of Worcester*, 1764; *An exact Ground Plot of the City of Worcester as it stood fortified 3 Sep: 1651*, 1769; *View of Boscobel House... with the wood where King Charles II concealed himself after the Battle of Worcester [1651]*, 1769; *East and West Prospects of Worcester, with a Table of the Mayors*, 1775.

**Gamond**, A. Thomé de. *Carte d'étude... pour servir à l'avant project du Canal Interocéanique de Nicaragua*, Paris c.1858 (lithographed by Avril frères); *Carte d'étude pour le tracé et le profil du Canal de Nicaragua*, Paris, Dalmont & Dunod 1858; *Carte d'étude... d'un Tunnel sous-marin entre la France et l'Angleterre*, c.1870; *France (bassins hydrographique)*, c.1873.

**Ganahl**, Johann, *Ritter* von (1817-1879). Austrian military cartographer.

**Gandavo**, Pero de Magalhães de. Manuscript map of Brazil, c.1574 MS (included in a copy of *História da Provincia de Santa Cruz*).

**Gandini**, François. Route maps of Europe. *Itinéraire de l'Europe*, Milan 1821.

**Gandini**, G. Engraver and mapmaker. Atlas of the Volga (with 8 maps), 1767.

**Gandouin**, Julien Michel. Publisher of Paris. Charlevoix's *Histoire et description générale du Japon*, 1763 (with map by Bellin, engraved by Dheulland).

**Ganeau**, *veuve* —. Bookseller at 'Rue St. Jacques, près la rue du Plàtre', Paris. Pierre François Xavier de Charlevoix's *Histoire de la Nouvelle France*, 1744.

**Ganeparo**, *Professor* Salvino. *Città di Torino*, 1870.

**Gangoiti, J. de.** Script engraver. *Carta Generale... del Archipiélago Filipino* (2 sheets), Madrid, Direccion de Hidrografia 1875 (surveyed by C. Montero i Gay, topography engraved by E. Fungairiño).

**Gangoiti, N.** Spanish letter engraver. I. Fernández Flórez's *Plano de la Ria de Arosa*, Madrid, Dirección de Hidrografía 1833 (map drawn and engraved by C. Noguera); I. Fernández Flórez's *Plano de la Ria de Camariñas*, Madrid, Dirección de Hidrografía 1838 (drawn by J. Espejo, map engraved by C. Noguera).

**Gangoiti, Pedro Manuel** (1779-1830). Spanish artist, *b.*Bilbao, *d.* Madrid. Charts of Chile and Peru, 1798-1799; worked for Isidoro de Antillon y Marzo, 1801-1804.

**Ganière, Pierre** (1663-1721). Script engraver and geographer of Paris. Globes, 1695; Jesuit map of Mojos, Peru, *c.*1713.

**Gankei** Sanjin. Japanese cartographer. Woodblock map of Mikawa Province, *c.*1715-1735 and later.

**Gannett, Henry** (1846-1914). Chief Geographer of the United States Geological Survey, 1882, founder of the Association of American Geographers. Montana & Wyoming Territories, 1870; topographical maps for F.V. Hayden's *Geological and Geographical Atlas of Colorado*, New York, Julius Bien 1877; hypsometric map of the United States, Washington 1877; *Statistical Atlas of the United States* (10th census), New York City, Charles Scribner's 1883 (with F.W. Hewes); *United States* (9 sheets), 1891, 1899; worked with the Division of Geography and Forestry of the U.S. Geological Survey on a series of maps showing forest, pasture and cultivated land, 1897-1903; *Statistical Atlas of the United States, Based Upon the Results of the Eleventh Census*, New York, Julius Bien 1898; published a commercial geography, 1905; Topographic Maps of the United States, 1907.

**Gannett, S.S.** Surveyor with the U.S. Geological Survey. *Utah*, 1898 (with R.U. Goode and R.B. Marshall).

**Gannon, P.** Assistant surveyor, Victoria, Australia. *Allotments Ballarat East*, 1859.

**Ganocchi, Giovanni.** Canal du lac de Trasimene [Umbria, Italy] (2 sheets), 1788.

**Ganon, Abraham.** Jewish cartographer of Jerusalem. Prepared 7 maps for Itzhak Goldhar's *Adamath Kodesh* [Holy Country], Jerusalem 1907 and later (lithographed by Eizik Zvi Shild).

**Ganong, W.F.** Map of the Negoot or South Tobique Lakes compiled by W.F. Ganong from Garden's and McInnes' surveys and New Observations [Canada], 1900.

**Ganteff, —.** Set of twelve globe gores, 1697.

**Gantrel** [Cantrell], S. Engraver of an allegorical plate in volume II of Beaulieu's *Les glorieuses conquestes de Louis le Grand...*, 1694.

**Ganxales, T.** Engraver. Ocean Atlantico, 1813.

**Gapp, John.** Surveyor. Survey of Crown lands at Walsoken, West Walton and Tilney, Norfolk, 1837 MS.

**Garandolet, Giovanni.** Globemaker at Palermo, 1527. De orbis situ ac descriptione.

**Garay, José de.** Boca del Rio Coatzacualcos, 1843.

**Garbs, F.A.** German author. *Karte von Palaestina* (7-sheet wall map), 1854, 1856, *c.*1862.

**Garcaeus** [Gartze], **Johannes** (1530-1574). German theologian and astronomer. *Tertius tractatus de usu globi*, Wittenberg 1565.

**Garci-Aguirro, Pedro.** Baia del Salvador, 1798.

**García, A.C. y.** See **Centeno** y Garcia, José.

**García, B.E. y.** See **Espinalt** y García, Bernardo.

**García, R.E.** Mapa de Bolivia, 1897.

**García Conde, Diego** (1760-1822). Army officer from Barcelona, served in Mexico and adopted Mexican citizenship. Plan of Mexico City, 1793; *Nouvelle Espagne*, 1807 (from Miguel Costansó's surveys of 1769-1770). **N.B.** The entry for Pedro Garcia **Conde** in Tooley's dictionary of mapmakers Vol.I. (1999) is incorrect. It is not placed correctly, and conflates the infomation on two distinct individuals, Diego **García Conde**, and Pedro **García Conde**, below.

**García Conde**, Pedro José (1806-1851). Officer in the Mexican Army Corps of Engineers, teacher of mathematics at the Military Academy, and commissioner of the Mexican Boundary Commission from 1848. In addition to practical road and navigational engineering projects hs was responsible for surveys of the Gulf coast, Mexico City and parts of the interior. He died in his birthplace, Arizpe, Sonora. Reconnaissance of the southern coast, 1831-1833; completed a map of the state of Chihuahua, 1834 (begun by Estevan Stapples); surveys of the California boundary and the southern boundary of New Mexico, from 1848. **N.B.** See note at Diego **García Conde**, above. **Ref.** HEWITT, H.P. 'The Mexican boundary survey team: Pedro García Conde in California' in *Western Historical Quarterly 21* (May 1990).

**García de Cespedes.** See **Cespedes**, Andreas Garcia de.

**García de Leon y Pizarro**, Ramón. Governer and Captain General of the Spanish Province of Salta, Argentina. *Plan topografico del Valle de Centa*, 1794 MS; *Plan de la Ciudad de la Nueva Oran*, 1794 MS.

**García de Toreno**, Nuño (*fl.*early-16th century). Chart and instrument maker. Pilot and master cartographer in the Casa de la Contractación in Seville, 1519. 23 of his charts used by Magellan's voyage to the East Indies in 1519. Planisphere (updated with information from Magellan's voyage), 1522 [Royal Library, Turin].

**García Martinez**, *Don* José. 'Capitán é ingeniero'. *Mapa de la Isla de Iviza* [Balearic Islands], 1765, 1778.

**García y Cubas**, Antonio (1832-191?). *Atlas de la Republica Mexicano*, 1856-1858; *Arzobispado de Mexico*, 1872; *Atlas Pintoresco Estados Unidos Mexicanos*, Mexico, Debray 1885.

**García y García**, *Captain* Aurelio. *Puerto Mollendo*, 1871.

**García y García**, Julian. *Atlas general de planos de la 49 capitales de España*, 1905.

**García y González**, *Lieutenant* Emilio. Port of Valencia, 1867-1869.

**Garcie** [Gracie], Pierre [sometimes called Ferrande] (*c.*1430-*c.*1503). Born in France of Spanish extraction, considered the first French pilot and hydrographer to compile and publish sailing directions for mariners. *Routier de la Mer*, Rouen, Jacques le Forestier 1502, and later (translated into English by Robert Copland as *The Rutter of the See*, 1528 and later); *Le Grant Routtier*, Poitiers, Enguilbert de Marnef 1520 and other editions to the 1640s. **Ref.** WATERS, David W. *The rutters of the sea: The sailing directions of Pierre Garcie* (Yale University Press 1967).

**Garcin**, Etienne. *Celto Lygie ou la Provence*, 1847.

**Garde**, J. de la. See **De la Garde**, J.

**Garden**, *Major* —. Country round Kandahar, 1878; Country round Cabul, Calcutta 1879.

**Garden**, —. *Map of the Negoot or South Tobique Lakes compiled by W.F. Ganong from Garden's and McInnes' surveys and New Observations* [Canada], 1900.

**Garden**, Francis. Engraver of London. Maps in Thomas Salmon's *Universal Traveller*, London 1752-1753; *Plan of the town of Inverness*, 1754 (for Edward Burt's *Letters from a Gentleman...*); *Plan of the Town and Citadel of Bayonne*, 1760 (in *A New Military Dictionary...*, London, J. Cook 1760).

**Garden**, William (*fl.*1771-1806). Land surveyor of Laurencekirk, produced manuscript maps of various parts of Scotland. *A Map of Kincardineshire*, 1774, engraved on 2 sheets by P. Begbie, 1776, A. Arrowsmith 1797.

**Gardener**, T. & J. Engravers. Plan of Westminster, 1847.

**Gardette**, Pierre-Claude de la (*d.c.*1781). Engraver of Paris. Jean Chappé d'Auteroche's *Plan de la Ville de Mexico*, 1772 (published in *Voyage en Californie*, 1772); maps in Le Gentil de la Galaisière's *Voyage dans les Mers de l'Inde*, 1779-1781.
• **Gardette**, Marie Madeleine Folleville, *veuve* de la. Widow of Pierre-Claude de la Gardette. Printseller of 'Rue du Roule, Près de la rue St Honoré' (1781); 'Gallerie Palais-Royal No.141' (1785), Paris. *Baie de Chesapeak...*, 1781.

ALLEN GARDINER.

*Captain Allen Francis Gardiner (1794-1851). (By courtesy of the National Library of Australia)*

**Gardiner.** See **Smith** & Gardiner.

**Gardiner,** Alfonso. *Pupil Teacher's Year Book of Memory Maps*, Edward J. Arnold 1889.

**Gardiner,** *Captain* Allen Francis (1794-1851). Sailor and missionary in Tahiti, S. Africa and New Guinea. *Country of Natal, Proposed to be colonised*, 1836 (lithographed by C. Ingrey) published in Gardiner's *Narrative of a Journey to the Zoolu Country in South Africa undertaken in 1835*, 1836.

**Gardiner,** C.K. Surveyor General. *Map of the Oregon Territory west of the Cascade Mountains*, 1855.

**Gardiner,** J.T. See **Gardner,** James Terry.

**Gardiner,** Ralph. See **Gardner,** Ralph.

**Gardiner,** Samuel Rawson (1829-1902). *School Atlas of English History*, 1891.

**Gardiner,** William. See **Gardner,** William.

**Gardner,** —. Credited as a source on the title page of Le Rouge's *Atlas Amériquain Septentrional*, 1778.

**Gardner,** —. Associated with American globes of the early nineteenth century. *Gardner's Twelve Inch Celestial Globe*, Boston 1823; *Gardner's Twelve Inch Terrestrial Globe*, Boston 1823; *Gardner's Four Inch Celestial Globe*, 1825; *Gardner's Twelve Inch Celestial Globe*, 1829.

**Gardner,** —. Engraver. *Huntingdonshire*, 1822.

**Gardner,** Aston W. [& Co.]. *Plan of the City of Kingston and its suburbs...* [Jamaica], London 1889, 1907 (lithographed by Weller & Graham).

**Gardner,** B.H. Plan of the Parish of St. Sepulchre, 1823.

**Gardner,** C.J. Manuscript maps of the Chinese Empire, 1880.

**Gardner,** Frank A. *Soil Map. Salt Lake Sheet* [Utah], Washington 1899 (with John Stewart, lithographed by A. Hoen & Co.).

**Gardner,** H. *Plan of Nunnery... of St Helen's, Bishopgate St.* [London], 1817; *Parish of Christ Church, Surrey*, 1821. **N.B.** Compare with B.H. **Gardner**, above.

**Gardner,** H. Colchester-Harwich Railway, 1836 MS.

**Gardner,** James. Mapsellers and publishers of London.
• **Gardner,** James *senior*. Surveyor for the Ordnance Survey; in 1823 he was granted the sole agency for the supply of Ordnance Survey maps and became an engraver, mapseller and publisher based at 163 Regent St., London. He retired in 1840. *New Post Map of Central Europe, Exhibiting the Great and Secondary Routes...*, 1825 and later (engraved by W.R. Gardner); *New Plan of the Cities of London & Westminster with the Borough of Southwark*, 1827, and editions to 1846 (engraved by W.R. Gardner); William Marsden's *Map of the Island of Sumatra*, 1829; George Bradshaw's *...Canals... of England*, 1830; engraver for Robert K. Dawson's *Plans of the Cities and Boroughs of England and Wales* (2 volumes), London, James & Luke G. Hansard & Sons 1832; engraver of maps for *Atlas to Accompany the Second Report of the Railway Commissioners, Ireland* (6 maps including a population density map and 2 passenger flow maps by Henry Drury Harness), 1838; joint publisher of Richard Griffith's *A General Map of Ireland to accompany the Report of the Railway Commissioners* (6 sheets), London, J. Gardner and Dublin, Hodges & Smith 1839 (compiled 1836, engraved 1837-1838), and other editions including Philadelphia, Durkan, Beehan & Maher 1860.
•• **Gardner,** James *junior*. Mapseller at 33 Brewer St., Golden Sq., London also 129 Regent Street, London (from c.1846). Last 2 editions of *New Plan of the Cities of London & Westminster with the Borough of Southwark*, c.1845, 1846 (engraved by W.R. Gardner, first published 1827); *Geological map of England and Wales*, 1846; Blayney William Walsh's *Plan of the City of Bridgetown, Barbados*, 1847.

**Gardner** [Gardiner], James Terry (1842-1912). Surveyor and engineer. Worked with C.F. Hoffman on surveys of the Sierra Nevada, 1863-1867 (used by the Geological Survey of California for their map of 1868); *Map of the Yosemite Valley*, 1865; Geological Map Washoe Mining District, 1870-1880; Central Colorado, 1873; State of New York (2 sheets), 1879; Colorado, 1881.

**Gardner,** John. Seaman. Map of Louisbourg and its harbour, published in *New York Weekly Journal*, 24th December 1733.

**Gardner** [Gardiner], Ralph. *The River of Tyne*, 1650, 1653, 1660.

**Gardner,** Thomas. Engraver. *A Pocket-Guide to the English Traveller*, London, J. Tonson & J. Watts 1719 (strip road maps based on John Ogilby).

**Gardner,** Thomas (1690-1769). Antiquary and historian of Dunwich, Suffolk, Comptroller of the Port of Southwold, Suffolk. Owner of Ralph Agas's 16th-century map of Dunwich, which he had newly engraved by Joshua Kirby for inclusion in his *An Historical Account of Dunwich*, London 1754.

**Gardner,** W.B. *Sketch plan of Canterbury*, A. & C. Black 1907.

**Gardner, W.R.** Engraver and geographer of 17 Oxford St., London. engraved many of the maps in [Charles] *Smith's New English Atlas...*, 1822, 1825, 1828, 1833, 1844 and other editions; Charles Smith's *Map of the Country Twelve Miles Round London*, 1822 and editions to 1847; *New and Improved Map of England and Wales*, London, W. Darton 1823; *New Post Map of Central Europe, Exhibiting the Great and Secondary Routes...*, London, James Gardner 1825; Brighton (4 sheets), 1826; James Gardner's *New Plan of the Cities of London and Westminster...*, 1827 and editions to 1846; *Smith's Pocket Companion*, 1827; *Leigh's New Pocket Road-Book of Ireland...*, London, Samuel Leigh 1827 (also sold in Dublin by R. Milliken); London & Westminster, 1828; *New Improved Map of Scotland*, c.1830; Charles Smith's New Holland, 1837.

**Gardner [Gardiner], William** (*fl.*1725-1752). Land surveyor, produced manuscript maps of various places in England. Estate in Newport, Essex, 1727 MS.

**Gardner, William** (1739-1800). Surveyor, worked for the Duke of Richmond from c.1767, in partnership with Thomas Yeakell from 1770, joined the Board of Ordnance in 1784, succeeded Thomas Yeakell as Chief Draughtsman in 1787. *Plan of the City of Chichester* (2 sheets), 1769 (engraved by Yeakell); Goodwood & Halnaker, c.1770 (with T. Yeakell, engraved by Glot, 1776); undertook a survey of Sussex with Thomas Yeakell, 1778-1783, completed by Thomas Gream and published as *A Topographical Map of the County of Sussex...* (4 sheets), London, William Faden 1795 (engraved by T. Foot); *Plan of the Town of Brighthelmstone* [Brighton], 1779 (with T. Yeakell); detailed surveys of the Plymouth area, 1784-1786; *An Accurate Survey and measurement of the Island of Guernsey...* (2 sheets), 1787 (engraved by J. Warner); *An accurate survey and measurement of the Island of Jersey* (4 sheets), surveyed 1787 (with Thomas Cubitt, H. Lauzan, G. Pink, T. Yeakell the younger, and Thomas Gream), published 1795 (engraved by John Warner); drafted Ordnance Survey sheets for Kent, 1801.
**Ref.** BAYNTON-WILLIAMS, R. *The Map Collector* 71 (1995) pp.39-41; KINGSLEY, D. *Printed maps of Sussex 1575-1900* (1982) pp.91-95.

**Gardner, William Wells.** *Sixpenny Elementary Atlas*, c.1872.

**Garel, Avuray de.** French cartographer at the 'rue d'Enfer, vis à vis la petite porte de Luxembourg'. *La Mappe-monde ou representation de la terre en plan* (world map in two hemispheres), c.1680.

**Garella, Napoléon.** *Project d'un Canal... à travers l'Isthme de Panama*, 1845; *Central America*, 1850; Road Chagres to Panama, 1851.

**Garfield, Aquila.** Essex estate surveyor. Manuscript maps of Roxwell and West Ham (both Essex), 1682.

**Garfurth, W.** *Plan of fortifications of Carlisle*, c.1580 MS.

**Gariboldi,** *Hauptmann* —. *Plan der Umgebung von Cilli*, 1873.

**Garipuy, François** (1711-1782). Lawyer and member of the Société des Sciences at Toulouse. 'Directeur des travaux publics dans la senéchaussé de Toulouse, puis Carcassonne'. He organised works to improve the navigability of the Garonne River. *Carte Generale du Canal Royal de la Province de Languedoc* (3 sheets), 1771 (engraved by N. Chalmandrier); *Carte du Canal Royal de Languedoc* (15 sheets), 1774.

**Garma y Duran,** *Don* **Francisco Xavier de** (1708-1783). Catalan cartographer, archivist and genealogist, Director of the Arxiu de la Corona d'Aragó, Barcelona. Mapa del obispado de Barcelona, 1762; Mapa del Reino Balearico, 1765; *Mapa del principado de Cataluña, y condado del Rosellón*, c.1770, 1837, 1838 (used information which had first appeared on maps by Josep Aparici i Mercadal, 1720, and Oleguer Darnius, 1726).

**Garnett, John.** *Travelling maps of the Lake District*, 1855.

**Garnier,** *abbé* —. *Recueil de cartes de Paris*, Nyon 1787.

**Garnier,** —. J.B.L. Charle's *Nouvelle Carte de la France*, Paris, c.1866.

**Garnier,** —. See **Petrot**-Garnier.

**Garnier Frères** [Garnier Hermanos]. Publishers of '6, Rue des Saints-Pères', Paris. E. Liais's *Explorations scientifiques au Brésil*, 1865; *Atlas Géográfico de la República Argentina*, Paris, Garnier Hermanos 1877; L. Grégoire's *Géographie, physique, politique et économique de la France et ses colonies*, 1883; *Géographie de la France*, 1885; E. Zerolo's *Atlas Geográfico Universal*, Paris, Garnier Hermanos 1891; L. Grégoire's *Atlas Générale de Géographie Moderne*, 1891; L. Grégoire's *Atlas Universel de Géographie Physique & Politique*, c.1892; L. Poulmaire's *Carte du Royaume de Siam et des pays limitrophes...*, 1893; E. Nardin's *Maroc. Algérie. Tunisie. Carte politique et physique*, 1907; E. Nardin's *L'Afrique actuelle. Nouveau Carte politique et physique*, 1907 (drawn by E. Nardin and T. Platin); *Nouveau Plan d'Alger*, 1912 (engraved by L. Poulmaire); F. Ernest's *Plan d'Oran et ses Environs*, 1914; E. Nardin's *Plan de Toulon et ses Environs*, 1914.

**Garnier**, Adolphe. *Carte par Bassins du Department des Vosges*, 1872; *Carte Routière du Départment des Vosges*, 1874; *Carte Orographique du Départment des Vosges*, Paris 1874.

**Garnier**, B.L. Bookseller of 'Rua do Ouvidor 69', Rio de Janeiro. E. Liais's *Hydrographie du haut San-Francisco et du rio das Velhas* [Brazil] (20 maps), Paris & Rio de Janeiro, B.L. Garnier 1865. **N.B.** Compare with **Garnier Frères**, above.

**Garnier**, F.A. (1803-1863). French geographer, produced an unusual atlas of maps illustrating the globe by views centred on different continents. *Atlas Sphéroidal et Universel de Géographie*, Paris, J. Renouard 1860, 1861; *Tableau Synoptique et abrégé du Systeme Magnétique Terrestre...*, Paris 1860; School atlas, c.1885.

**Garnier**, Francis [Marie-Joseph-François] (1839-1873). French naval officer and colonial officer in Indochina. Governor of Saigon (1862). Took part in an expedition through the Mekong valley and Cambodia in search of trade routes, from 1866. They returned via the Yangtze and Shanghai in 1868 with a wealth of geographical information. *Voyage d'éxploration en Indochine*, 1873; *Itinéraire dans la Chine Centrale*, 1882. **Ref.** WALDMAN, C. & WEXLER, A. *Who was who in world exploration* (1992) pp.276-277; BROC, N. *Dictionnaire illustré des explorateurs Français du XIXe siècle: II Asie* (1992) pp.205-209 [in French].

**Garnier**, Hippolyte. *Plan de la Baie de Diego-Suarez* [Madagascar], 1837 (drawn by L. Bigeault in 1833); *Plan du Port de Tripoli de Barbaria*, 1837, 1856?; *Plan du Port de Santa Barbara*, 1847.

**Garnot**, Eugène Germain. *Atlas de l'expédition française de Formose 1884-1885*, published 1894.

**Garny**, —. *Plan... de la Ville de Paris*, 1818.

**Garofalo**, —. Engraver. *Città di Palermo*, c.1760; Giovanni Martinon's *Sicilia*, 1808?, 1812.

**Garran**, Andrew (1825-1901). Journalist and politician from London. Edited *Picturesque Atlas of Australasia*, Sydney & Melbourne, Picturesque Atlas Publishing Company 1886, 1888; *Australasia Illustrated*, London 1892.

**Garrard**, H.M. *Map of the Town...of Geelong*, 1848-c.1860.

**Garrard**, J. Jervis. Deputy Commissioner of Mines for Zululand. His *Report on the Mineral Resources of Zululand* was accompanied by *Map of Zululand showing the relative positions of various Gold and Coal Fields and the Routes of Approach...*, 1895.

**Garretson, Cox & Co.** *Columbian Atlas*, 1891 and later.

**Garrett**, G.H. *Map of Sulymah District, Sierra Leone*, 1888; *Sierra Leone*, London 1892.

**Garrett**, Henry A. *Bournemouth*, 1883, 1889.

**Garrett**, John (*fl*.1676-1718). Print and mapseller, member of the Merchant Taylors' Company, 'near the stairs of the Royal Exchange in Cornhill' or 'Exchange stayres in Cornhill' London. Brother-in-law of John Overton who married Garrett's sister Sara. Possibly the son of William Garrett, below. Took over Thomas Jenner's stock and shop at the Royal Exchange, c.1676. Republished Jenner and W. Hollar's map of England and Wales (the so-called 'Quarter-master's map', printed on silk), 1675, 1676, 1688 (first published by T. Jenner, 1644); *A Booke of Maps exactly describing Europe*, 1676; *A map of the land of Canaan*, 1676; Donald Lupton's *A most exact and accurate Map of the World*, 1676; *A New and exact map of America...*, engraved by Wenzel Hollar, 1676; republished J. van Langren's *A Direction for the English Traviller*, *c*.1677, *c*.1680 (maps first published 1635, previously republished T. Jenner, 1643); republished *A Book of Names of all Parishes in England and Wales*, 1677, 1680 (printed by Samuel Simmons, used maps from *A Direction...*, first published under this title by T. Jenner 1657); John Harris's *A View of the World in Divers Projections*, *c*.1697. **Ref.** CARROLL, R.A. The printed maps of Lincolnshire 1576-1900 (1996) pp.33-36; TYACKE, S.J. London map-sellers 1660-1720 (1978) pp.114-116.

**Garrett**, William (d.1674). Publisher and bookseller 'at the White Bear in Foster Lane, against Goldsmith's Hall'. Possibly the father of John Garrett, above. Bought John Speed's map plates from William Humble 1658-1659, but sold them soon after to Roger Rea.

**Garrigou**, T.E.A. (1802-1893). French geographer from Tarascon. Géographie de l'Aquitaine sous Caesar, 1863; Ibères, 1884.

**Garstin**, Symon. Castletowne, Co. Lowth, 1655 MS.

**Garti D'Alibrandi**, Giovanni. Engraver active in Tuscany. Some maps for Pazzini Carli's *Atlante Geografico*, 1788; Sardinia, 1793; Venetian States, 1794.

**Gartman**, Erven H. *Atlas der Oost-Indien* (38 maps), 1832.

**Gartman**, V.A. Telegraph map Russian Empire (4 sheets), St Petersburg 1870.

**Gartner**, Amandus. *Abriss der...Statt Wolffenbüttel*, 1627.

**Gartze**, Johannes. See **Garcaeus**.

**Garvan**, C.F. Surveyor of Federal Chambers, Kings Street, Sydney. Map of Lord's Paddock Estate, [n.d.]; *Sutton Forest villa and farm lots...*, Sydney, S.T. Leigh & Co. *c*.1887.

**Garvey**, E.A. *Map of the City of St. Louis*, 1874.

**Garvie**, Alex. Surveys round the town of Invercargill, South Island, New Zealand, 1856; sheep stations at Galloway and Moutere, 1858; *Reconnaissance Survey of the South Eastern Districts of Otago Province*, 1858; *Province of Otago, New Zealand*, *c*.1860.

**Gascaro**, Achille. Allegorical map *Le Pays de Tendre*, *c*.1700 MS.

**Gascoigne**, Joel. See **Gascoyne**, Joel.

**Gascoigne**, *Captain* John. Commander of H.M.S. *Alborough* from 1728. Surveys of the coast of South Carolina including Port Royal, and D'Awfoskée Sound, 1731 (used as a source by several later mapmakers including De Brahm in *A Map of South Carolina and a part of Georgia*, T. Jefferys 1757; T. Jefferys in *The North American Pilot...*, Sayer & Bennett 1776; J. Bellin in *Hydrographie Françoise*, 1778; Le Rouge in *Pilote Americain Septentrional...*, 1778 etc.). **Ref.** DE VORSEY, Louis The Map Collector 19 (1982) pp.30-31.

**Gascoyne**, John. *A true and exact draught of the Tower Liberties*, 1597 MS (with W. Hayward), published Society of Antiquaries 1742, copied by J. Heath 1752.

**Gascoyne** [Gascoin; Gascoigne], Joel (1650-1705). Surveyor, chart-maker, and engraver from Hull. Apprenticed to John Thornton of the London Drapers' Company in 1668. On gaining his freedom in 1676 he became one of the so-called 'Thames school of chartmakers' noted 'at ye Signe of ye Platt neare Wapping Old Stayres, 3 doares belowe ye Chappell'. In about 1689 began concentrating on land and estate surveys, and moved to the south west for the duration of his survey of Cornwall, 1693-1699. Charts in John Seller's *English Pilot*, 1677; Caribbean and east coast of North America, 1678 MS; North Atlantic, 1678 MS; *A New Map of the Country of Carolina* [so-called 'Second Lord Proprietors' Map'], 1682 (also sold by Robert Greene); Chart of the Atlantic, 1686; Hertfordshire estates of James Cecil, Earl of Salisbury, 1691; Sayes Court, Deptford, Kent (for John Evelyn), 1692; Manor of East Greenwich. 1693; *A Map of the County of Cornwall newly surveyed...*, 1699; Falmouth, 1700; Greenstead, Essex, after 1700; Enfield Chase, 1701 MS; Channel, 1702; Hamlet of Bethnal Green, Hamlet of Mile End, Limehouse,

Parish of St. Dunstan Stepney, Old Town in the Parish of Stepney, 1702-1703 MSS; and others. **Ref.** RAVENHILL, W. 'Joel Gascoyne', in NICHOLLS, C.S. (Ed.) *Dictionary of national biography: Missing persons* (1994) p.245; *A Map of the County of Cornwall Newly Surveyed: by Joel Gascoyne* Devon & Cornwall Record Society, Vol.34 (1991) facsimile map with introduction by W. RAVENHILL & O.J. PADEL.; RAVENHILL, W. *The Map Collector* 13 (1980) pp.34-35; CAMPBELL, T. 'The Drapers' Company and its school of seventeenth century chart-makers' in WALLIS & TYACKE, eds. *My head is a map* (1973) pp.81-106; RAVENHILL, W. 'Joel Gascoyne, a pioneer of large-scale county mapping' in *Imago Mundi* 26 (1972) pp.60-70.

**Gascoyne**, John. See **Gascoigne**, *Captain* John.

**Gaskell**, —. Atlas of the World, 1886; Family Atlas, 1887; Gaskell family and business atlas of the world, 1894, 1895.

**Gasp**, Raphaelo Fabretto. Mapmaker of Urbino, for Pieter van der Aa's *Galérie agréable du monde*, 1729.

**Gaspar**, Thomé. *Plan of Palaon Bay on Mindoro* [Philippines], 1761; *Port St Andres on the Island Marinduque*, 1761, both charts published together by Dalrymple *c.*1792.

**Gaspari**, Adam Christian (1752-1830). Geographer of Königsberg, worked for the Geographisches Institut, Weimar. *Vollständiges Handbuch der neuesten Erdbeschreibung*, 1797-1804 (from 1819 G. Hassel, J.G.Fr. Cannabich, J.C.F. GutsMuths and Fr.A. Ukert were also involved); *Allgemeiner Hand-Atlas der Ganzen Erde* (60 maps), Weimar 1797-1804, 1806; updated F.L. Güssefeld's *Neuer methodischer Schul-Atlas*, 1816 (first published 1792).

**Gaspary**, Nicolas. *Plan von Metz*, 1872.

**Gasselin**, E. *Regence de Tunis*, 1881.

**Gasser** [Gassarus], *Doctor* Achilles Pirmin (1505-1577). Physician, astronomer, geographer and topographer, from Lindau, active in Feldkirch and Augsburg (from 1546), where he died. Gasser adopted a series of symbols and conventional signs for use on maps. Manuscript map of the region of Allgäu, 1534 (sent to Münster but not used by him; now in the University Library, Basle); *Elementale Cosmographicum*, 1539, 1550; supplied material for Sebastian Münster's *Cosmographia*, 1544. **Ref.** BURMEISTER, K.H. 'Achilles Pirmin Gasser (1501-1577) as geographer and cartographer' in *Imago Mundi* 24 (1970) pp.57-62.

**Gasser**, Max (1872-1954). Specialised in layered maps for aeronautical navigation.

**Gasson**, John. Hampshire and Berkshire, *c.*1646.

**Gast**, A. *Plastischer Schul-Atlas*, *c.*1876.

**Gast**, brothers [Gast & Brothers]. Lithographers of St. Louis, Missouri. **N.B.** Assumed to be Leopold, John & Augustus.
• **Gast**, Leopold [& Bro.]. Lithographers of St Louis. *Part of the Mineral Region State of Missouri...*, 1853; Quin, Smith & Van Zandt's *Territory of Nebraska*, 1857; *Map of Routes to the Gold Region of Western Kansas*, 1859; *Map Exhibiting the Routes to Pike's Peak*, 1859; lithographers of D. McGowan and George H. Hildt's *Map of the United States West of the Mississippi...*, St. Louis, 1859; G. Hale and J.M. Truesdell's *Map of Lake County, Illinois*, 1861.
• **Gast**, John. Thought to be one of the Gast brothers.
• **Gast**, Augustus. Possibly one of the Gast brothers.
•• **Gast** (Aug) & Co. Publisher of St. Louis. *Map of the South Pacific Rail Road Co. of Missouri*, 1870; *Smith county* [Texas], 1880; *Edwards County, Illinois*, 1891.

**Gastaldi**, B. Schizzo di carta geologica di una parte degli Alpi, 1871.

**Gastaldi** [Gastaldo; Castaldi; Castaldo], Giacomo [Jacopo] (*c.*1500-1566). Italian astronomer, cartographer and engineer from Villafranca, Piedmont. By 1539 he was active in Venice, producing numerous general maps as well as engineering and waterways maps for the Venice area. By the 1540s he had developed his own distinctive style of copper engraving for his increasingly prolific output of maps. His maps were used as a source by many mapmakers including Camocio, Bertelli, Forlani, Ramusio, Cock, Luchini and Ortelius. Gastaldi died in Venice. His many works included: *...La Spaña...* [4 copperplates], 1544; *Isola della Sicilia*, 1545; *La vera descrittione di tutta la Vngheria...* [includes central Europe and part of Italy] (woodcut on 4 sheets), Venice, Matteo Pagano 1546; *Nova Totius Orbis Descriptio*, 1546, 1561 (lost), 1564; maps for Pietro Andrea Mattioli's Italian edition of Ptolemy's *Geographia* entitled *La Geografia di Claudio Ptolemeo Alessandrino...* (60 copperplate maps), Venice, Nicolo Bascarini 1548 (the maps were prepared from 1542 onwards); wall map of Africa for the Doge's Palace, Venice, 1549; *Dell'Vniversale. L'Vniversale Orbe della Terra...*, Venice c.1550; woodcut maps for Ramusio's *Delle Navigationi et Viaggi*, volume 1 (with 1 map) 1550, (4 maps) 1554, (edition with some copperplate maps), 1563, volume 3 (6 maps) 1556, 1606; *Il uero ritratto di tutta l'Alamagna*, G. Giolito 1552 and other editions; *Il Piamonte* (woodcut), 1555, and other editions; *Il Disegno Della Prima Parte Delasia* [Turkey and the Middle East], 1559 (the first part of a larger map); *Geographia particolare d'una gran parte dell'Europa...* (4 sheets), F. Licinio 1559-1560; *Il Disegno Della Geografia Moderna De Tutta La Provincia De La Italia* (2 sheets), 1561; *Il Disegno Della Seconda Parte Dell'Asia* (2 sheets), 1561, Olgiato 1570, De Jode 1578 (second part of the map of Asia begun in 1559); *Il Disegno Della Terza Parte Dell'Asia* (4 sheets), 1561, Olgiato 1570, De Jode 1578 (third part of the map of Asia); *Cosmographia Vniversalis Et Exactissima...* (woodcut, 9 sheets), 1561; *Il Disegno De Geografia Moderna Del Regno di Polonia..* (2 sheets), 1562; *Disegno dell'Italia...*, 1562; *Il desegno della geografia moderna de tutta la parte dell'Africa...* (8 sheets), 1564 (engraved by F. Licinio); *Il disegno d'Geografia moderna della provincia di Natolia, et Caramania...*, 1564, Camocio 1566, Zaltieri 1570; and many others. **Ref.** KARROW, R.W. *Mapmakers of the sixteenth century and their maps* (1993) pp.216-249; PICK, A. *The Map Collector 60* (1992) pp.30-31; MEURER, P. pp.148-154 *Fontes cartographici Orteliani...* (1991) [in German]; ASTENGO, C. 'I mappamondi di Giacomo Gastaldi e lo Stretto di Anian' in: *Annali di Ricerche e Studi di Geografia* 46 (Geneva 1990) pp.1-18 [in Italian]; KRETSCHMER, DÖRFLINGER & WAWRIK *Lexicon zur Geschichte der Kartographie* (1986) vol.C/1, pp.246-247 (numerous references cited) [in German]; ALMAGIÀ, R. Introduction to *La carta dei paesi danubiani e delle regioni contermini di Giacomo Gastaldi (1546)* (1939) [in Italian].

**Gastambide,** *Don* Pedro. *Plan of Pachiri town & fort... Dumaran Island*, 1762, published Alexander Dalrymple 1782.

**Gastel,** Johann *SJ* [*fl.*late-17th century]. Jesuit missionary in South America. Cartographical survey of the Marañon area of Peru.

**Gaston,** Samuel N. (*b.*1855). Draughtsman and publisher of New York. United States of America 1854; *Map of the State of Michigan*, New York, Gaston & Johnson 1856 (compiled and drawn by S.N. Gaston); publisher of Charles G. Colby's *The Diamond Atlas*, 1857; *The Campaign Atlas, for 1861...* (14 maps), New York, S.N. Gaston 1861

• **Gaston & Johnson.** Publisher of 115 & 117 Nassau Street, New York. *A new map of Our Country, present and prospective*, 1855 (engraved by L. Lipman); publisher of Samuel N. Gaston's *...Michigan*, 1856. **N.B.** See also **Morse & Gaston.**

**Gastrell,** *Captain* (later *Colonel*) J.E., *RE*. In charge of Indian cadastral surveys: *District of Jessore*, 1854-1856, published 1870; *City of Nagpoor*, 1865, 1870.

**Gates,** B.C. *Sullivan County, New York*, 1856; *Otsego County, New York*, 1856 (with C. Gates).

**Gates,** C. *Otsego County, New York*, 1856 (with B.C. Gates).

**Gates,** G.H. Drew Samuel Butler's *Dramatic Almanac Map for 1853*, 1853.

**Gáti** [Gáthy], István (1780-1859). Hungarian engineer and inventor, member of the Hungarian Academy of Sciences, 1836. Constructed a number of surveying and mapping instruments, 1835.

**Gatliff,** Charles. Map of London showing the locality of several blocks of Improved dwellings..., London, E. Stanford 1875.

**Gatonbe,** John. Iceland and Greenland, 1612.

**Gatta,** G.F. della. See **Della** Gatta, Giovanni Francesco.

**Gatti,** Aletinus (*fl.*16th century). Publisher of Rome. Geneva, [n.d.].

**Gatti,** Federico. School atlas, 1859 (with Bruno Colao).

*Abbé A.E.C. Gaultier (c.1745-1818). By Antoine Cardon. (By courtesy of the National Library of Australia)*

**Gaubil**, Antoine, SJ (1689-1759). French mathematician, astronomer and missionary from Gaillac. Worked in China from 1722, died in Peking. *Carte des Isles de Liou-Kiou*, 1752, published by Philippe Buache in *Considérations géographiques et physiques sur les nouvelles découvertes au nord de la Grande Mer*, Paris 1754.

**Gauché**, —. One of the engravers for *Carte topographique de la France* (12 sheets), Dépôt de la Guerre, 1832-1847; engraved parts of *Carte de la Russie...* (23 sheets), Dépôt de la Guerre 1854-1856 (printed by F. Chardon l'aîné).

**Gaudard de Chavannes**, Charles Philippe (1753-1780). French military cartographer.

**Gaudi**, J. See **Gaudy**, John.

**Gaudry**, Albert. Surveyed the island of Cyprus with Amédée Damour, his information was used by others including Edward Stanford (1878), and James Wyld (1878). *...Carte Agricole de l'Isle de Chypre*, 1854; *...Carte Géologique de l'Isle de Chypre*, 1860.

**Gaudy** [Gaudi], John. Marine surveyor. Barcelona, 1705; *Directions for Navigating from Cape Race to Cape St. Mary's...* [with] *Directions for sailing to Newfoundland*, 1718 published in *The English Pilot: The Fourth Book*, Mount & Page 1729 and later; *Levant*, c.1740.

**Gauld** [Gould], George (1732-1782). Scottish surveyor, studied at King's College, Aberdeen, appointed Royal Navy schoolmaster 1757, later described as 'Surveyor of the Coasts of Florida'. Many of his charts were used by Des Barres as a source for *The Atlantic Neptune*. Produced numerous manuscript charts of the Gulf Coast, 1764-1781; surveys of the coasts of Florida and Louisiana, 1764-1771 published as *An Accurate Chart of the Coast of West Florida and the Coast of Louisiana*, W. Faden 1803, 1820 (engraved by B. Baker); *A Survey of the Bay of Pensacola...*, 1766 MS; *A Plan of the Mouths of the Mississippi*, 1769 MS; *A General Plan of the Harbours of Port Royal and Kingston Jamaica*, 1772 MS, published Faden 1798; mauscript survey of East & West Florida, 1771-1772 (with Bernard Romans and David Taitt); *The West End of the Island of Cuba, and part of the Colorados*, 1773 MS, published W. Faden 1790; *The Island of Grand Cayman*, 1773 MS, published Captain T. Hurd 1815; *An Accurate Chart of the Tortugas and Florida Kays...*, 1773-1775 MS, published Faden 1790, 1820. **Ref.** WARE, John D. & REA, Robert R. *George Gauld, surveyor and cartographer of the Gulf Coast* (1982); GUTHORN, P. *British maps of the American Revolution* (1972) pp.22-23.

**Gaule**, — de. *Carte hydrographique de l'Embouchure de la Seine*, 1788.

**Gaulle**, F. See **Galle**, family.

**Gaultier**, *abbé* Aloisius Édouard Camille (c.1745-1818). Cartographer from Piedmont, worked in England during French Revolution, died in Paris. Devised a system of geographical games: *A Complete Course of Geography*, 1792 and later editions (the 1822 and 1825 editions included maps drawn by J. Aspin and engraved by N.R. Hewitt); *Atlas de géographie*, c.1810.

**Gaultier**, J. *Palestine*, c.1875; *Carte politique de la France*, c.1878; *...Cordillère des Andes*, 1880.

**Gaultier** [Galter], Léonard. Engraver of Paris. Said to have engraved the map of the Americas entitled *Novus Orbis* in Richard Hakluyt's edition of Peter Martyr d'Anghiera's *Decades*, Paris 1587 (previously attributed to Philippe Galle as engraver). **Ref.** BURDEN, P. *The Mapping of North America* (1996) p.78.

**Gaultier de Varennes et de la Vérendrye**, Pierre. See **La Vérendrye**.

**Gauss**, *Lieutenant* J. One of the draughtsmen for A. Papen's *Topographischer Atlas des Königreichs Hannover und Herzogthums Braunschweig...* (80 sheets), Hanover 1832-1837.

**Gauss** [Gauß], Karl [Carl] Friedrich (1777-1855). Mathematician from Braunschweig, also contributed to the fields of surveying, astronomy, and physics. Published a work on the earth's magnetic field, *Allgemeine Theorie des Erdmagnetismus*, 1833. Gauss died in Göttingen. Trigonometrical survey of the Kingdom of Hanover, 1816-c.1841 (his measurements were used as the basis of later maps including the work of F. Hartmann and C. Tomforde, 1827-1840 and A. Papen's

*Topographischer Atlas des Königreichs Hannover und Herzogthums Braunschweig...* (80 sheets), Hanover 1832-1837). **Ref.** BEECH, G. (Ed.) *Maps and plans in the Public Record Office 4. Europe and Turkey* (1998) N° 1702 pp.320-321 etc.; KRETSCHMER, DÖRFLINGER & WAWRIK *Lexikon zur Geschichte der Kartographie* (1986) vol.C/I, pp.247-248 [in German].

**Gaussin**, Pièrre Louis Jean Baptiste. *Îles Marquises* [French Polynesia], 1844-1849.

**Gauthey**, —. *Carte... des Montagnes de la France*, 1782.

**Gautier**, —. See **Gauttier**, Pierre Henri.

**Gautier**, — (*fl.*1758-1762). Mapseller of 'cul-de-sac Saint-Honoré' and 'Cloître Saint Honoré' (1760), Paris. Sold sheets of the Cassini *Carte de France*, from 1758; Brion de la Tour's map of France, 1762.
• **Gautier**, *veuve* — (*fl.c.*1763-1788). Publisher and print and mapseller of 'rue St. Jacques', Paris (1763); 'Cloître Saint-Honoré' (1765). Succeeded her husband in the sale of sheets of the Cassini map of France. *Tableau des 175 Feuilles de la Carte de France...*, 1765 (engraved by Aubin). **Ref.** PACHA, B. in: PACHA, B. & MIRAN, L. *Cartes et plans imprimés de 1564 à 1815* (Bibliothèque Nationale de France 1996) p.48 [in French].

**Gautier**, Antoine. Géographe du roi, 1629.

**Gautier**, Hubert (1660-1737). Ingénieur du Roi dans la Marine, *b.* Nîmes, *d.* Paris. *Nismes*, for Jean-Baptiste Nolin 1698; *Uzes*, *c.*1698; *Montagnes des Sevennes*, 1703.

**Gautier**, Jacques Fabien. *Carte des tremblements de terre arrivés en Europe l'année 1755*, Paris 1755.

**Gautier de Châtillon** [Gautier de Lille; Gualterus Philippus dictus Castillionensis]. 'T-O' mappamundi to accompany his poem *Alexandreidos*, *c.*1180 and later, varying copies. **Ref.** DESTOMBES, M. *Imago Mundi* 19 (1965) pp.10-12.

**Gautier de Metz** (*fl.*early-13th century). Poet from Lorraine, thought to have written *L'image du monde*, *c.*1245. The many known manuscript copies usually include 2 'T-O' world maps; also printed by William Caxton as *Myrrour of the worlde* (2 'T-O' maps), 1481. **Ref.** WOODWARD, D. in HARLEY & WOODWARD (Eds.) *The history of cartography Volume One: Cartography in prehistoric, ancient, and medieval Europe and the Mediterranean* (1987) pp.321 & 345.

**Gauttier**, *Captain* Pierre Henri. French naval officer. *Plan des Isles des Saintes*, 1803-1813; *Carte... de la Mer Noire*, 1820; surveys in the Aegean and Greek Islands, *c.*1825 (information used by Lapie, 1826); Carte de la Partie Méridionale de l'Archipel [Aegean Sea and Islands], Paris 1854-1863 (with F.A.E. Keller); one of several whose work was used in *Des Ports & Rades de la Mer Méditerranée*, Marseille 1855.

**Gauttier d'Arc**, Laurent Eduard (1799-1843). French historian and geographer.

**Gauvreau**, N.B. ...*Surveys in Northern British Columbia*, 1891.

**Gavard**, Ch. Engraver. *St. Petersbourg*, for A.-H. Dufour in Balbi's *Abrégé de Géographie*, *c.*1833; A.-H. Dufour's *Londres*, *c.*1838.

**Gavarrete**, Juan. *Carta de la Republica de Guatemala en la America Central trazado por J. Gavarette...*, Machado, Yrigoyen & Ca. *c.*1880 (engraved and printed by Erhard, Paris).

**Gavin**, H. *Virginia and Maryland*, *c.*1750.

**Gavin**, family. Engravers of Edinburgh.
• **Gavin**, Hector, *senior* (1738-1814). Native of Berwickshire, worked in Parliament Close, Edinburgh. *The Ancient City of Jerusalem and Places Adjacent*, 1759; Plan of the city of Edinburgh, 1763; *Scotland with a list of the royal burroughs...*, 1767 (in 10th edition of Thomas Salmon's *A New geographical and Historical Grammar*); Andrew Armstrong's map of Berwickshire, 1772; map of Scotland for W. Guthrie, 1772; Sketch Water of Druie, 1776; *Scotland from the best authorities*, 1780 (in *The Town and Country Almanac*); map of Oxnam Parish in: *The Statistical Account of Scotland*, 1794.
•• **Gavin**, Hector, *junior* (1784-1874). Apprenticed to the engraver Andrew Bell, later worked with his father as Gavin & Son.
•• **Gavin & Son**. Engravers. Thomas Brown's *A New and Accurate Map of Scotland with the Roads*, 1801 published by Brown in *A General Atlas*, 1801 and Brown's *Atlas of Scotland*.

**Gavit**, John E. Engraver and printer of Albany, New York. C. Brodhead's *A map of a tract of land in New York called Macomb's Purchase...*, 1850 (first published *c.*1796); for Asa Fitch's map of Washington County, New York, 1850.

**Gawemn**, A. *Karte des mittleren Ruhrgebietes*, *c.*1891.

**Gawler**, Colonel George (1795-1869). Second governor of South Australia, 1838-1841; South Australia, 1840 MS.

**Gawthorpe**, P.W., *RN*. Master HMS *Lion*. Manuscript charts of the Cape of Good Hope, South Africa, *c.*1812-1813, including Camps Bay, False Bay and Simon's Bay. His surveys were used by J.W. Norie on *A Chart of False Bay...*, 1819, 1827 and later (engraved by J. Stephenson, with an inset of Simon's Bay also surveyed by Gawthorpe); also used in Norie's *The Country Trade, or Free Mariners' Pilot*, 1833; also in *Complete East India Pilot*, 1838 and later.

**Gay**, Claudio. Atlas de Chile, 1854.

**Gay**, E.F. *Map... of the Canal from Columbia to Tide*, *c.*1845.

**Gayangos**, A.M. de. See **Martel** de Gayangos, Antonio.

**Gazis** [Gazes], Anthimos A. (1764-1823). Geographer and publisher. *Kharta tis Ellados* [Greece], Vienna 1800; *Stoikheia Geographias...* (with a map of the two hemispheres), Vienna 1804 (with N. Theotokis); publisher of the second edition of Meletius's *Geographia Palaia kai Nea*, Venice 1807; Map of Greece (6 sheets), Vienna 1810 (engraved by F.T. Müller).

**Gazü**, *Archimandrite* —. World map (4 sheets), Vienna, [n.d.].

**Gazulić** [Ghazulus Ragusinus], Ivan (1430-1476). Mathematician, astronomer and globemaker of Dubrovnik.

**Gebauer**, Emil. *Umgebungen von Zittau*, *c.*1895.

**Gebauer**, Johann Jacob (*d.*1819). German publisher in Halle. *Naturgraenzenkarte von Europa, Asien und Afrika*, 1787; J.C. Adelung's *Geschichte der Schiffahrten*, [n.d.] (with different maps of the northeast passage).

**Gebert**, Martin. Schwarzbildes, 1783.

**Gebhard**, H. *Plan von Nürnberg*, *c.*1868.

**Gedda**, father and son.
• **Gedda**, Johann Persson (*d.*1680). Surveyor of Uppland, father of Peter Gedda (below).
•• **Gedda**, Peter (1661-1697). Swedish surveyor, son of Johann Persson Gedda. Surveyor of Åbo and Bjöorneborg, Finland from 1680. From 1681 he worked as assistant to Werner von Rosenfeldt on surveys of the Baltic Sea, taking over from Rosenfeldt as Director of Pilotage in 1687. He continued the Baltic surveys, culminating in a manuscript atlas of charts, 1693 (Maritime Museum at Karlskrona), this was used as the basis for *General-Hydrographisk Chart-Book öfver Östersiön, och Katte-Gatt* (10 charts), 1695, 1696, (12 charts) 1699, 1764 (engraved by Anthoni de Winter, also published in Dutch and English editions), pirated maps were published by A. de Winter with J. Loots, 1697 and Hendrick Doncker, 1698. **Ref.** EHRENSVÄRD, U. 'Peter Gedda's maritime atlas of the Baltic' *Imago Mundi* 29 (1977) pp.75-77; KOEMAN, C. *Atlantes Neerlandici* (1967-1971) Vol.IV p.190.

**Gedde**, Christian. *Plans af de... Stad Kiöbenhafn*, 1757.

**Geddes**, George. *Salt Wells, Syracuse, New York*, 1868.

**Geddes**, James. Map and profile of Champlain Canal, 1825; Harbor of Oswego, 1825.

**Geddes**, John. Civil engineer from Edinburgh. *Plan of the Forth & Tay Railway*, Edinburgh, 1836.

**Geddes**, *Captain* William Loraine. Topographical map of the northeast part of the Niagara Peninsula, 1867.

**Geddy**, John. Bird's-eye view of St Andrews [Scotland], *c.*1580 MS (attributed).

**Gedney**, A.G. Lithographer and publisher of Washington. Produced many maps of the city of Washington, many of them for real estate brokers. *Hall and Elvans' subdivision of*

*Meridian Hill* [Washington D.C.], 1867; *Map of the City of Washington and environs*, 1884; R.O. Holtzman's *Map of the city of Washington and environs*, 1885; map of Washington for Fitch, Fox & Brown, 1885; B.H. Warner & Co's map showing a bird's eye view of the city of Washington, and suburbs, Washington, B.H. Warner 1886 (prepared by A.G. Gedney); A.J. Shipman's *Map of Fairfax County, Virginia*, 1886; *Map of the city of Washington*, 1887 (for James H. Marr); [J.R.D.] Morrison's *map of the country about Washington*, 1888; map of the city of Washington for the American University, 1893.

**Gedney**, Joseph F. Lithographer, engraver, plate printer and publisher of Washington, D.C. Ft. Abercrombie to Ft. Benton, 1863; W.J. Keeler's *Map of the Routes of the Union Pacific Rail Roads with their Eastern Connections*, 1867; Charles Du Bois's *A new sectional map of the state of Kansas showing the route of the Union Pacific railway*, 1867 (information compiled by W.J. Keeler); *National Map of the Territory of the United States from the Mississippi River to the Pacific Ocean*, Washington 1867 (compiled by W.J. Keeler and N. DuBois); William J. Palmer's *Map of the route of the Southern Continental R.R....*, 1868; *Todd & Brown's subdivision of part of "Pleasant Plains & Mt.Pleasant"*, suburbs of Washington D.C., 1868 (surveyed by William Forsyth); B.D. Carpenter's *Map of the roads in Washington county, D.C.*, 1870; published Joseph Henry's *Rain-Chart of the United States*, 1870, 1872; *Exhibit chart showing streets & avenues of the cities of Washington and Georgetown, Improved under the Board of Public Works, D.C...: gas mains*, 1873 (3 similar sheets of the same date cover pavements, sewers and water mains).

**Gedney**, *Lieutenant* Thomas R. Naval lieutenant working for the United States Coast Survey. Discoverer of 'Gedney's Channel'. *Chart of Narraganset Bay*, 1832 (with Charles Wilkes, G.S. Blake and Alexander Wadsworth); hydrography for F.R. Hassler's *Map of New York Bay and Harbor* (sheets 1-4), 1844, (sheets 5-6), 1845, (single sheet), 1845 (triangulation by J. Ferguson and E. Blunt, topography by C. Renard, T.A. Jenkins and B.F. Sands, hydrography engraved by F. Dankworth, lettering engraved by F. Dankworth and J. Knight, topography engraved by S. Siebert and A. Rolle, views engraved by O.A. Lawson).

**Gee**, father and son. English estate surveyors.
• **Gee**, Richard (*fl.*1777-*d.*1811). Surveyor, produced manuscript estate and inclosure plans of areas throughout England, especially southern and central counties. *Manor of Tarrant Rushton... Dorset*, 1789 MS; Reverend Hugg's Estate, Emberton, Bucks. 1799 MS.
•• **Gee**, Edward. Son of Richard, also a surveyor, specialising in estate, inclosure, road and tithe maps in central and northern England.

**Gee**, father and son. English estate surveyors.
• **Gee**, Thomas (*fl.*1790-*c.*1826). Produced manuscript estate and inclosure maps of various areas in central and northern English counties. Inclosure plan at Purston Jaglin, Featherstone, Yorkshire, 1811.
•• **Gee**, William (*d.c.*1825). Land agent and inclosure surveyor working in Lancashire, son of Thomas.

**Geel**, Joost van. See **Gelen**, Joost van.

**Geelkercken** [Geelkerken; Geilkerckius; Geylekerckius], family.
• **Geelkercken** [Geelkerken; Geerkerken; Geilenkerken; Geilkerckio; Geylekerck; Geylkercke], Arnoldus [Arnold; Arnoldo] van (*d.*1619). Dutch surveyor, elder brother of Nicolaes. *Terrae Sanctae seu Terrae Promissionis nova descriptio*, 1619, *c.*1670 (engraved by Nicolaes). **N.B.** It has been suggested that Arnold Geelkercken and Arnoldo di **Arnoldi** might be one and the same. **Ref.** HEIJDEN, H. van der 'Wie was Arnoldo di Arnoldi' in *Caert Thresoor* 18de jaargang 1999 nr.2 pp.37-40 [in Dutch with summary in English].
• **Geelkercken** [Geelkerken; Geilkerckius], Nicolaes [Nicolaus] van (*c.*1585-1656). Prolific publisher, cartographer, draughtsman, engraver and surveyor of Amsterdam and Friesland (1614-1616), Leiden (1616-1628) and thereafter Arnhem, where he died. His first published work is thought to be a plan of Nijmegen, 1610; *Universi Orbis Tabula De-Integro Delineata*, Davidt de Meine 1610; *Belegeringe van Bruijnswijck*, 1615; maps in Ubbo Emmius's *Frisia*, 1616; engraver for Philipp Clüver's *...Germaniae antiquae*, 1616; *Orbis Terrarum Descriptio Duobis Planis Hemisphaeriis Comprehesa*, Joannes Janssonius 1617; *America*, 1617; engraved for Eilhard Lubin's *Pomerania*, 1618; P. Clüver's *Rhetie*, 1618; Joris van Spilbergen's *Oost- ende West-Indische Spiegel*, Leiden 1619; map of the seige of Breda, 1625 (with

J. ten Berge); and Jacob Isaac Pontanus's *Historia Gelrica*, 1638. **Ref.** DEYS, H.P. 'De Gelderse kartograaf Nicolaas van Geelkercken', in *De Gelderse Vallei: geschiedenis in oude kaarten* (Utrecht 1988) [in Dutch].

•• **Geelkercken**, Isaac [Isaak] van (*fl*.1636-1672). Traveller, military engineer and surveyor of Arnhem; eldest son of Nicolaes. Worked in Nijmegen, 1636-1639; then in Norway as a military surveyor and engineer at Frederikshall, 1644-1657; succeeded his father as provincial surveyor of Gelderland, 1657. Usually signed work as 'J. van Geelkercken' (causing confusion with brother Jacob, below). Plan van Nijmegen, Frederick de Wit *c*.1680.

•• **Geelkercken** [Geelkerken], Jacob van (*fl*.1660-*c*.1677). Dutch surveyor and cartographer, *d*. Zutphen. Second son of Nicolaes, worked also as an apothecary at Zutphen (from 1660).

•• **Geelkercken**, Arnoldus [Arnold] van (*fl*.1649-1656). Dutch cartographer, youngest son of Nicolaes, active in Gelderland. Revised Thomas Witteroos's 1570 survey of Renkum and Ede, 1649-1656.

**Geelmuyden**, B. *Lomme-Atlas over Norge*, 1892.

**Geer**, Elihu. *Geer's Map of the City of Hartford* [Connecticut], 1847; *Geer's New Map of the City of Hartford*, 1858.

**Geerkerken.** See **Geelkercken**, family.

**Geerling**, W.J. Dutch cartographer. *Nieuwe Atlas voor Gymnasiën* (23 maps), 1859, 1870; *Geerling's Nieuwe Atlas der geheele aarde* (44 maps), 1872; *Gemeente Atlas van Nederland*, 's-Gravenhage, J. Smulders *c*.1882 (re-issued by Smulders as *Atlas der Provinciën*, 1886).

**Geernaert**, P.T. *Atlas ecclésiastique de la Belgique*, *c*.1850.

**Geerz**, Franz Heinrich Julius (1817-1888). Military cartographer of Berlin. Holstein u. Lauenburg, 1838-1845; worked on A.C. Gudme's *Kiel über die Gegend von Kiel...*, Kiel 1844; Schleswig, Holstein &c., 1859; Nordfriesische Inseln, 1888.

**Geest**, E. de. *Nederlandsch Oost-Indië*, 1871; *Het Koninkrijk der Nederlanden* (4 sheets), 1888-1897; *Platte Grond van Amsterdam*, 1890; *Sumatra* (12 sheets), 1892; *Atlas van Nederland* (7 maps), Amsterdam, Seyffart's Boekhandel 1902.

**Geerards**, M. See **Gerards**, Marcus.

**Gefferys**, T. See **Jefferys**, Thomas.

**Geiger** Hans Conrad. See **Gyger**, Hans Conrad.

**Geiger**, J.G. See **Gyger**, Hans Georg.

**Geiger**, Philipp. *Plan der Stadt Kaiserslautern*, 1883.

**Geiger**, W. *Karte der Erzdiözese Wien*, 1848.

**Geikie**, brothers. Geologists from Edinburgh.
• **Geikie**, *Sir* Archibald (1835-1924). Geologist from Edinburgh. Director-General of the Geological Survey of Scotland from 1867; he moved to London when he became Director General of the Geological Survey of Great Britain & Ireland, 1882-1901. Geological Map of Scotland, 1861; Edinburghshire, 1861-1864; Ayrshire, 1868-1872; Linlithgow, 1878; Geological Map of England and Wales (15 sheets), 1896.
• **Geikie**, *Professor* James Murdoch (1839-1915). Geologist from Edinburgh. Joined the Geological Survey [of Scotland] as an assistant in 1861, was promoted to geologist in 1867, and to distict surveyor in 1869. Between 1869 and 1875, he mapped Fife and the Lothians, the Southern Uplands, and the Ayrshire and Lanarkshire coalfields. He held the Murchison chair in geology at Edinburgh University from 1882-1914 and was president of the Scottish Geographical Society, 1904-1910. **Ref.** HERRIES DAVIES, G.L. 'James Murdoch Geikie', *Dictionary of national biography: Missing persons* (1994) p.246.

**Geil**, family and companies. Engineers, surveyors and publishers of Philadelphia. Both Samuel and John are credited as surveyors in the R.P. Smith *Gazetteer of the State of New York...*, 1859. **N.B.** See also **Harley**, D.S. & J.P.
• **Geil**, Samuel. Cartographer and publisher of 601 Chestnut Avenue (1864), also 602 Chestnut Street, Philadelphia. Worked alone and collaborated with others on numerous surveys of counties in North America, especially of New York State. Maps of counties in Indiana, Michigan, New York, Ohio,

Pennsylvania, 1852-1864; worked with Franklin Gifford on surveys of Niagara County, New York, published Wilson, N.Y., Franklin Gifford 1852 (the copyright was held by R.P. Smith); *Map of Orleans County, New York...*, Philadelphia, Lloyd van Derveer 1852 (with Jesse Lightfoot, copyright held by R.P. Smith); worked with Jesse Lightfoot on a map of Morris County, New Jersey, published J.B. Shields 1853; Cayuga County, New York, published Philadelphia, S. Geil 1853 (surveyed by S. Geil, F. Gifford and S.K. Godshalk); Montgomery County, New York, 1853 (with B.J. Hunter); Oswego County, New York, 1854 (with B.J. Hunter); map of Erie County, New York, 1854, 1855; *Map of the State of Michigan...*, 1864, 1865 (compiled by Geil & Harley, engraved by Worley & Bracher, printed by F. Bourquin, published by Samuel Geil - all of 602 Chestnut Street); geological maps, 1865; *The People's Map of the State of Michigan*, 1864 (compiled by Geil & Jones and Geil & Harley, engraved by Worley & Bracher, printed by F. Bourquin, published by Samuel Geil - all of 602 Chestnut Street); and many others. **N.B.** See also **Geil & Jones.**

• **Geil**, John F. Surveyor and partner in Geil, Harley & Siverd, publishers of Philadelphia. *Lorain County, Ohio*, Philadelphia, Matthews & Taintor 1857; *Medina County, Ohio*, Philadelphia, Matthews & Taintor 1857; *Laporte County, Indiana*, 1862.

• **Geil & Jones.** Topographical engineers and draughtsmen, Samuel Geil and S.L. Jones working together. *Map of Wayne Co., Michigan*, Philadelphia, Geil, Harley & Siverd 1840 (surveyed by I.G. Freed for Geil & Jones and Geil & Harley); *Map of the counties of Genesee & Shiawassee, Michigan*, Philadelphia, Geil & Jones 1859 (with C. Wilson, J.D. Nash, & J.W. Stout); draughtsmen for *Map of the Counties of Eaton and Barry, Michigan*, Philadelphia, Geil, Harley & Siverd 1860 (special survey by J.D. Nash, engraved by Worley & Bracher); *The People's Map of the State of Michigan*, Samuel Geil, 1864 (with Geil & Harley).

• **Geil & Harley.** Map compilers of 517, 519 & 521 Minor Street, Philadelphia (1861). *Map of Wayne Co., Michigan*, Philadelphia, Geil, Harley & Siverd 1840 (surveyed by I.G. Freed for Geil & Jones and Geil & Harley); *Map of Kalamazoo Co., Michigan*, Geil & Harley 1861 (surveys by I. Gross, drawn by S.L. Jones, engraved by Worley & Bracher); *Map of Lapeer Co., Michigan*, 1862 (by I. Gross and W.E. Doughty); compilers of Samuel Geil's *Map of the State of Michigan...*, 1864, 1864; *The People's Map of the State of Michigan*, Samuel Geil, 1864 (with Geil & Jones); directed the production of I.M. Gross's *Map of Allegan co., Michigan*, Philadelphia, Samuel Geil 1864.

• **Geil, Harley & Siverd.** Publishers of Philadelphia. Partnership of John F. Geil, David S. Harley and Eli F. Siverd. *Topographical map of the Counties of Ingham & Livingston, Michigan*, 1859; Geil & Jones' *Map of the counties of Eaton and Barry, Michigan*, 1860; *Map of the Counties of Cass, Van Buren, and Berrien, Michigan*, 1860.

**Geilenkerken.** See **Geelkercken** family.

**Geilkerckius.** See **Geelkercken** family.

**Geils**, *Lieutenant* Joseph Tucker. *Survey of the Route from Erzerum to Trebizond, made for the Topographical and Statistical Depôt* [Turkey] (3 sheets), 1855 MS.

**Geinitz**, F.E. Geological map of Mecklenburg [Germany], 1883.

**Geinitz**, Hans Bruno (1840-1900). *b.* Altenburg. *Atlas der Steinkohlen Deutschlands*, 1865.

**Geisendörfer**, J. Engraver of '12 r. de l'Abbaye', Paris. Luigi Pelion di Persana's *Plano topografico de La Habana, intra Extramuros y Cerro*, Paris, Goyer & Hermet 1865; *Côte de Syrie*, *c.*1866; E. Delpech's *Plan de la Ville de Bordeaux*, Bordeaux, Fret 1871; *Carte du bassin de l'Ebre pour servir l'intelligence de la Guerre Carliste*, Paris 1875; *Plan de la Ville d'Agen*, Paris, Lemercier 1875; *Carte topographique du Tonquin* 1883; *Guine Portugaise et possessions françaises voisines*, Lille, Danel 1887-1888; one of the engravers for L. Grégoire's *Atlas Universel de Géographie Physique & Politique*, published *c.*1892; *Carte Strategique de la frontière France - Allemande*, Paris, Lemercier 1893; *Plan des forts de St-Germain Marly et ses environs*, Paris, Lemercier 1896.

**Geisler**, A.D. Publisher of Bremen. E. Uhlenhuth's map of Europe and North America, 1849.

**Geisler, D.** Publisher of Bremen. *Charlston. Fünfte Ergänzungskarte zum Kriegsschauplatz in Amerika*, 1863.

**Geisler, J.T.** (*fl.c.*1718-1750). Swedish mine surveyor. *Falun grufwan* (9 maps), 1718 (for Fredrik I), another copy, 1750 (for Queen Christina).

**Geisler, L.** Engraver of Nuremberg. Sir John Barrow's *Küste von Africa von der Tafelbay am Cap der Guten Hofnung, bis zum Saldanhabay*, 1805; G. Bridges's *Militaerischer Plan von der Capschen Halbinsel*, 1806, (both maps were engraved for the German translation of Barrow's travels published Leipzig 1805 and 1806).

**Geispitz von Geispitzheim,** *Baron* Carl Heinrich (*d.*1787). Austrian military cartographer.

**Geissel,** Wilhelm. *Regierungs-Bezirk Wiesbaden, c.*1872.

**Geissendörfer, L.** Lithographer and printer. *Plan von Mannheim, c.*1866.

**Geissler, G.** One of the engravers for *Spruner-Menke Hand-Atlas ...*, maps prepared 1871-1879, published Gotha, Justus Perthes 1880; engraver of the South America plate in *Stanford's London Atlas of Universal Geography...* (folio edition), London 1884 (Geissler's name appears on only one map in this edition, printed for private circulation).

**Geistbeck,** *Dr.* Alois (1853-1925). *Die Seen der deutschen Alpen*, Leipzig 1885 (depth charts of the lakes of the German Alps); *Süddeutschland* (wall map), [n.d.].

**Geistbeck, Michael.** *Das Königreich Bayern in geographisch- statistischer Beziehung* (36 miniature maps for school use), 1878.

**Gelais,** Giovanni Vincenzo [Ioannis Vincentius]. Engraver. *Argonautica*, 1697; maps of Padua, 1699.

**Gelbke** [Gelbcke], Carl Heinrich von (*c.*1783-1840). Military cartographer and historian of Weimar. *Herrschaft Schmalkalden*, 1807; *Landvogtei Rothenberg, c.*1810; *Königreiche Württemberg* (4 sheets), 1813.

**Gelbrecht, H.** *Plan von Bremerhaven*, 1878.

**Gelder,** Jacob de (*fl.*1801-1848). Mathematician, map-copyist and cartographer of Rotterdam. Netherlands, 1809; Holland, 1811.

**Gelder, W. van.** *Atlas van Nederlandsch Oost-Indië*, Batavia [Jakarta], G. Kolff & Co. 1881; *Schoolatlas van Nederlandsch Oost-Indië*, Groningen, J.B. Wolters 1890 and later editons; *Atlas Sekolah Hindia-Nederland...*, Groningen, J.B. Wolters 1890 (later editions to 1919), the map was also published in a Malay version entitled *Atlas ketjil Hindia-Nederland*, 's-Gravenhage 1919 (and in other Indonesian languages); *Kleine Schoolatlas van Nederlandsch Oost-Indië*, 1925; and many others. **Ref.** KOEMAN, C. Atlantes Neerlandici (1967-1971) Vol.IV pp. 158-165.

**Gelen** [Geel], Joost van (1631-1698). Dutch cartographer, draughtsman, engraver and painter from Rotterdam. Engraved sheets for Jacob Quacq's *De Mond van de Maes*, 1665, a later issue was entitled *Afbeeldinge van de Maes van de Stadt Rotterdam tot in Zee...*, R. and J. Ottens 1740.

**Gelio,** *Captain d'État Major* —. *Nivellement de Jérusalem*, Paris 1863. **N.B.** Compare with Captain **Gelis**, below.

**Gelis,** *Captain* —. *Dead Sea, Jerusalem* 1863. **N.B.** Compare with Captain **Gelio**, above.

**Gelis,** *Capitaine* —. *Camp de Chalons sur marne. Plan du terrain affecté au cap et terrains environnans*, Paris, Dépôt de la Guerre 1865.

**Gell,** *Sir* William (1778-1836). British archaeologist, established the location of Troy. *Ilium or Troy*, 1804; *Geography and Antiquities of Ithaca*, 1807; Sketch map showing ruins of ancient Etruscan city ten miles north-west of Rome, 1822 MS; *Carta... di Roma*, 1827; *A Map... of the Peloponnesus* 1830.

**Gellatly,** John. (*c.*1802-1859). Scottish draughtsman, copper-plate printer and lithographer from Forfar. Worked in Edinburgh at 8 West Register Street (1826); 44 West Register Street; 1 George Street (1843); 26 George Street (1846). *Gellatly's new map of the country 12 miles round Edinburgh*, 1834, 1836, 1840, 1844 (by Wm. Johnson); published John Lothian's *Netherlands*, 1844; *Scotland*, 1845 (in *Atlas of Ancient and Modern Geography,*

Edinburgh, W. & R. Chambers 1845, and *Chambers's Atlas for the people*, Edinburgh, W. & R. Chambers 1846); worked for George Philip & Son, 1851-1852.

**Gellibrand**, Henry (1597-1636). English mathematician from London; Gresham professor of astronomy. Variation of the Magnetic Pole, 1635; also works on nautical geography, 1633-1635.

**Geltenhofer**, Stephanus. See **Keltenhofer**, Stephan.

**Gemelli-Careri**, John Francis. *Voyage autour du monde...* (containing maps), Paris 1719; hydrographical draught of Mexico, 1732.

**Gemert**, M.L. van. *Atlas behoorende bij "Het Firmament"*, Purmerend, J. Muuses 1910 (maps by J.F. Nuyens).

**Geminus** [Gemini; Geminy; Lambrechts; Lambert], Thomas (*c*.1500-1562). Publisher, engraver and instrument maker from eastern Flanders. He is thought to have learnt engraving at Louvain, and by 1540 he was working in 'Black Friers', London, as a copper engraver, mostly of medical and anatomical illustrations (including his own *Compendiosa totius anatomie delineatio...*, 1545 and editions to 1559). He is thought to have been one of the earliest engravers active in England, and some scholars believe he may have been the engraver of the so-called 'copper-plate plan' of London. Geminus also practiced as a surgeon under the patronage of Henry VIII, until barred by the College of Physicians in 1555. Astrolabe, 1552 (he is known to have made 6 others up to 1559); *Nova descriptio Hispaniae* (4 sheets), *c*.1553-1554, 1555 (copied from the work of Hieronymous Cock, 1553); *Britannia insulae...*, 1555 (re-issue of George Lily's 1546 map); published Leonard Digges's work on measurement entitled *A Boke named Tectonicon...*, 1556, 1561, 1562. **Ref.** KARROW, R.W. *Mapmakers of the sixteenth century and their maps* (1993) pp.250-254; O'MALLEY, C.D. Introduction to facsimile edition of Geminus's anatomical engravings: *Compendiosa totius anatomie delineatio. A Facsimile of the First English Edition of 1553 in the Version of Nicholas Udall* (1959); HIND, A.M. *Engraving in England in the sixteenth & seventeenth centuries* (1952-1955) pp.39-63.

**Gemma**, Cornelius (1535-1577). Astronomer at Leuven University. *De Prodigiosa Specie*, Antwerp 1578.

**Gemma Frisius**. See **Frisius**, Gemma.

**Gemmell**, J. American lithographer. *D.B. Cooke & Co's. miniature rail road map of Illinois*, 1857; *Map of the present and prospective rail-road connections of the city of Freeport*, 1857; *Franklin City* [Nebraska], 1857; *Chicago*, 1858; *Map of the Jacksonville and Savannah*, 1858; *Map of Rock County, Wisconsin*, 1858; *W.B. Horner's railway & route map to the gold regions in Kansas & Nebraska*, 1859; W.G. Wheaton's *Map of the Illinois River Railroad*, [n.d.]; *Sectional Nebraska & Kansas*, [n.d.]; *Chicago... for D.B. Cooke & Co's directory*, 1860.

**Gemmellaro**, Carlo. *Carta geologica della valle di Noto*, 1829; *Carta topografica dell'Etna*, 1860.

**Gemmellaro**, Giuseppe (1787-1866). Geologist from Catania, Sicily. *Map of the eruptions of Etna*, James Wyld 1828.

**Gemperlin**, Abraham. Publisher of Fribourg. Renwart Cysat's *Der... Japponischen Insel...*, 1586 (earliest Western printed map of Japan based on firm evidence).

**Gendebien**, Albert. *Coupe géologique du Bassin du Centre*, 1876.

**Gendre**. See **Le Gendre**-Decluy.

**Gendre**, A. Printer of A. Grandjean's *Carte du Nkomati Inférieur et du District Portugais de Lourenço Marques*, Neuchatel, Maurice Borel 1893.

**Gendrón**, Pedro. Spanish cartographer and publisher of Madrid. *La Alemania*, 1755; *Atlas*, 1756-1758.

**Genebelli**, Federico. Italian military engineer working in England, 1585-1602. Plans and maps of the fortifications of the port of Plymouth. Rye harbour, 1591, 1593 MSS.

**General Advertising & Circular Delivery Company**. Noted at 188 Strand, London. Published a plan of London '*Engraved Expressly for the Visitors Hand Book*', *c.*1857 (engraved by J. Dower, printed by J. Bradley).

**General Land Office.** *United States Including Territories and Insular Possessions...,* [n.d.].

**Genest,** P.M.A. *Carte de la Nouvelle France,* 1875.

**Genest,** François Xavier. Public Lands Surveyor. *Map of the Dominion of Canada and Part of the United States,* 1883 (with J. Tache).

**Gengen-do** family. Japanese artists and map-makers. **N.B.** Gengen-do is the professional or pen-name for members of the Matsumoto [Matsuda] family.
- **Gengen-do** [Matsumoto] Yasuoki [Gengendo I] (1786-1867). Copperplate engraver, late Edo period. *Chikyū bankoku zenzu* [World map], 1836; *Dōzen Nihonkoku yochi zenzu* [Map of Japan printed from a copperplate], [n.d.].
- ● **Gengen-dō** [Matsuda; Matsumoto] Rokūzan. Japanese cartographer and copperplate artist, printer and publisher. First son of Matsumoto Yasuoki, but changed the family name to Matsuda. Bird's-eye view of Edo on a folding fan, *c.*1856; bird's-eye view of Osaka city, 1856; map of the globe, 1856; copperplate map of Sanuki Province, Japan, published Murakami Kambei and others 1865; copperplate maps of the areas of Shikoku, Kantō and Chūgoku, *c.*1866; copperplate plan of the nobles' residential quarter of Kyoto, 1868. **Ref.** YAMASHITA Kazumasa *Japanese maps of the Edo Period* (Japan 1998) pp.36 and 84.
- ● **[Ranko-tei]** later [Seisen-do] [Gengen-do; Matsumoto] Ryūzan. Japanese cartographer, copperplate artist and publisher. Sixth son of Matsumoto Yasuoki, rather than the family professional name of Gengen-do, Ryuzan adopted the professional names of Ranko-tei and Seisen-do. *Hyōgo Saiken no Zu* [plan of Kobe], *c.*1870 (copperplate plan taken from an earlier coloured woodblock print).

**Genoi,** D. See **Zenoi,** Domenico.

**Genshu** Nagakubo. See **Nagakubo** Sekisui.

**Gensoul,** Adrien. Publisher at 'Pacific Map Depot', 511 Montgomery St., San Francisco. G. Woodman's *Map of the Mining Sections of Idaho & Oregon...,* 1864 (lithographed by B.F. Butler); E.E. Fine's *Humboldt Mining Region,* 1864; E. Fleury's *Nueva Mapa de los Estados de Sonora, Chihuhua, Sinaloa, Durango, y Territorio de la Baja California,* 1864; *Map de los Distritos Minerales de San Antonio, el Triunfo Las Cacachilas y Isla de Cormen Baja California...,* 1865; Richard Gird's *Official Map of the Territory of Arizona,* 1865; G. Owens's *New Map of the Mining Regions of Idaho and Montana...,* 1866.

**Gensui,** Ebi. See **Gaishi** Ebi.

**Gent,** Thomas (1693-1778). Printer and topographer, settled in York, 1724. *History of York,* 1730; *Ripon,* 1734; *Hull,* 1735; *Plan of the City of York,* 1771.

**Gent,** William (*fl.*1677-1685). English estate surveyor, worked mostly in London and the Home Counties. *Belsize & St. John's Wood* [London], 1679.

**Gentet,** Jaspar. *Golfe de Siam,* 1739.

**Genthe,** Herman. *Etruskische Tauschhandel,* *c.*1873.

**Gentil,** A. Printer of Paris. A.J. Bourdariot's *Carte de la Côte Ivoire...,* 1901.

**Gentil,** *Colonel* Jean-Baptiste-Joseph (1726-1799). French soldier in India from 1752, served as agent for the French government at the court of Oudh, 1765-1775. Gentil returned to France in 1777 with a large collection of Indian artefacts and paintings. Using the work of different Indian scholars he compiled a manuscript geographical and historical description of India, *c.*1769 (Bibliothèque Nationale); also an atlas entitled *Empire Mogol divisé en 21 soubas ou Gouvernements tiré de differens ecrivains du païs,* Faisabad 1770 MS (the maps are decorated with miniature paintings by Indian artists, now in the British Library). **Ref.** GOLE, S. *Maps of Mughal India drawn by... Gentil* (New Delhi 1988).

**Gentil,** L. Le. See **Le Gentil,** Labarinus.

**Gentil,** Louis (1868-1925). French traveller, naturalist and geologist. Geological map of Morocco, *c.*1910. **Ref.** BROC, N. *Dictionnaire illustré des explorateurs Français du XIXe siècle: I Afrique* (1988) pp.161-162 [in French].

*George, Prince of Wales later King George III (1738-1820). (By courtesy of Jo French)*

**Gentil**, Pierre. French author and traveller. *Deux véritables discours... sur la guerre de Malte* (with a plan), Paris 1567.

**Gentil de la Galaisière**, G.J.H.J.B. Le. See **Le Gentil** de la Galaisière.

**Gentleman's Magazine** (*fl*.1736-1858). English periodical containing numerous maps by E. Bowen, T. Jefferys, B. Cole, J. Gibson and many others. See Edward **Cave**.

**Genton**, Jean Louis Ambroise de, 'chevalier de Villefranche'. See **Villefranche**.

**Gentot**, —. Mapseller of 'Rue Mercière, Lyon', France. Sieur Delafosse's L'Amérique, 1771.

**Gentot**, —. *Carte de la Rade de Cherbourg*, *c*.1787.

**Gentry**, A.M. *Map of Texas showing the Sabine and Galveston Bay Rail Road...*, New York, Slote & Stone 1859; *Map of Texas showing the line of the Texas and New Orleans Rail Road*, 1860.

**Gentsch** [Gentzsch; Güntzsch], Andreas. Engraver in Augsburg and Saxony. *Prag* (view), 1620; *Abbildung der fürnembsten Städt*, c.1622.

**Genus**. See **Zeno**, Nicol.

**Geny-Gros**, —. Printer and lithographer of 'Rue de la Montagne, Ste. Geneviève 34', Paris. Lithographers for *Atlas du Guide Maritime et Stratégique dans la Mer Noire...*, Paris, J. Corréard 1854; for A.-H. Dufour, 1870.

**Geoffroy**, August. *Navigation par arcs de Grand Cercle*, 1867.

**Geoffroy**, Lislet. See **Lislet**-Geoffroy, Jean-Baptiste.

**Geographical Publishing Co.** Publishers of Chicago. *Imperial Royal Canadian world atlas: an atlas for Canadians*, 1935 (edited by Fred James, Lloyd Edwin Smith and Frederick K. Branom).

**Geographische Anstalt**. Publishers of Munich. *Georgien u. Hochland Armenien*, 1829.

**George** of Cyprus (*fl.*7th century). Byzantine geographer.

**George III** (1738-1820). King of Great Britain and Ireland 1760-1820. One of the great collectors of topographical material. His collections, which incorporated that of his uncle Prince William, Duke of Cumberland, eventually included over 50,000 atlases, maps, plans and views. With the exception of the military plans (which remain at the Royal Library, Windsor Castle), most of the material passed to the British Museum in 1828 and is now in the British Library. **Ref.** HODSON, Y. *The Map Collector* 44 (Autumn 1988) pp.2-12 [describes the Cumberland Map Collection]; WALLIS, H. *The Map Collector* 28 (Autumn 1984) pp.2-10 [British Library collections].

**George**, B. Lithographer of Cape Town. *Map showing the wine growing districts of Paarl, South Africa*, London, James F. Denman c.1890s.

**George**, Christopher. *Changes in mapping of Kennedy Channel* [Arctic], 1858 MS; *The Camaroon Mountains* for R.F. Burton's *Abeokuta and the Camaroons Mountains*, London, Tinsley 1863; Commander H.C. St John's *Island of Yesso* [Hokkaidō], 1871 MS ('projection by Cap$^{t.}$ C. George, Map Curator, RGS').

**George**, E. Engraved eight maps in V. Duruy's *Histoire de France*, Paris, Hachette 1862.

**George**, F.E. Engraver of 'Rue de la Harpe N$^{o.}$26' Paris. Worked with Charles Smith on some sheets of V. Levasseur's *Atlas National Illustré*, Paris, A. Combette 1845 and later.

**George**, Hans. *Der Fränckische Krayss...*, 1861.

**George**, S.A. Electrotype, 607 Sansome St., Philadelphia. Duane Rulison's map of the United States, 1860.

**George**, Thomas. English estate surveyor. Part of Iver, Buckinghamshire, 1856.

**George & Co.** Lithographers of 54 Huston [?] Gdn., London. Columbia, 1849.

**Georges**, —. One of the engravers for *Carte topographique de la France* (12 sheets), Dépôt de la Guerre 1832-1847.

**Georget**, —. 'Adjoint au Corps du Génie Militaire'. Drew maps 7 & 10 for the Atlas volume to *Mémoires du Maréchal Suchet, duc d'Albufera, sur les campagnes en Espagne, depuis 1808 jusqu'en 1814*, Paris, A. Bossange, Bossange père, F. Didot 1828.

**Georgi**, Johann Gottlieb. *Bemerkungen einer Reise im Russischen Reich...* (3 maps), St Petersburg 1775.

**Georgio**, Giovanni [Johannes]. See **Giovanni**, Giorgio.

**Georgio** [Georgius], Ludovicus. See **Barbuda**, Luís Jorge de.

**Georjio**, Giovanni. See **Giovanni**, Giorgio.

**Geörög**, D. See **Görög**, Demeter.

**Gephart**, Christian. Plan Oran, for Homann's Heirs 1732.

**Geppert,** Georg von (1774-1835). Austrian general and cartographer.

**Geppert,** Ludwig von (1777-1836). Austrian military cartographer.

**Geraghty,** Joseph P. (*fl.*1849-1856). Surveyor, produced estate and mineral maps in Gloucestershire and Lincolnshire. Iron Mines, Deans Meend [Forest of Dean], 1849 MS.

**Geraghty,** T.R. *Etheridge Goldfield, Queensland*, 1898.

**Gerald,** John. Draught Bellese [British Honduras], *c.*1750 MS.

**Gerald,** M.F. Costes de Terre Ferme, 1737.

**Gerald de Barri.** See **Giraldus** Cambrensis.

**Gerard** [of Cremona] (*c.*1114-1187). Born at Cremona in Italy but spent most of his life in Spain. Translated Ptolemy's *Almagest* and Al Farghani's *Elements of Astronomy*.

**Gérard,** *Captain* Alexander. *Karte des Hohen Himalaja*, 1832 (with Paul Gérard); *Koonawur*, 1841.

**Gérard,** Gilbert. *Monasterii B.ae Nuchariensis Scenographia* [Benedictine Monastery at Noyers], 1694.

**Gerard,** H. *Plan de la Bataille de Waterloo*, Brussels 1840.

**Gerard,** H. See **Gerritsz.**, Hessel.

**Gerard,** *Colonel* M.J. Sketch Gandamak to Tezin Valley (3 sheets), 1880; *Mesopotamia and Persia*, 1882-1886.

**Gérard,** Paul (1796-1866). Geographer from Ghent, worked for Philippe Marie Guillaume Vandermaelen. *Karte des Hohen Himalaja*, 1832 (with Alexander Gérard); *Carte topographique de la Belgique*, 1847-1859.

**Gerard,** Thomas. Survey of Dorset, 1633 MS.

**Gerardo,** Paulo. Portolano nuovo... del Levante et del Ponente, 1544; Portulano del Mare, Venice 1612.

**Gerards** [Geerards; Gerardus], Marcus [Marc] (1530-1590). Artist and engraver from Bruges. *d.* England. *Brugge Flandorum Urbs* (10 sheets), 1562.

**Gerards** W.J. *Teeken-Atlas van Ned. Oost-Indië*, 1898 and later; *Kleine Schoolatlas van Nederlandsch Oost-Indië*, 1927 and later.

**Gerardus,** H. See **Gerritsz.**, Hessel.

**Gerasimov,** Dmitrii [Demetrius] (*b.*1465). Muscovite Ambassador to the Pope, 1525; his information was used by Paolo Giovio as the basis for a book about Muscovy, and a map entitled *Moschoviae Tabula*, 1525 (one known example, attributed to Vavassore, copied by Battista Agnese, *c.*1540, and Giacomo Gastaldi, 1548). **Ref.** KARROW, R.W. *Mapmakers of the sixteenth century and their maps* (1993) pp.267-268.

**Geray,** Richard. English estate surveyor. Manuscript plan of an estate at Desford, Leicestershire, 1644.

**Gerbaud,** Georg. *Atlas der Völkerkunde*, 1886-1892.

**Gerbel** [Gerbelius], Nicolaus (1485-1560). Theologian, jurist and historian in Basle. Prepared a description of Greece to accompany the 8-sheet map of Greece by N. Sophianos, Basle 1545, 1550.

**Gerber,** E.B. American publisher. Sometimes worked in association with Charles S. Warner on the publication of maps by the Warner brothers. *Map of Noble Co., Ind.*, 1861.

**Gerber,** *Lieutenant* F.A.G. Manuscript copy of *Carte des Amtes Springe*, 1798 (from an original survey by Lasins, Kahle and Hase, 1782).

**Gerber,** Henrique. *Carta da Provincia de Minas Geraes* [Brazil] (4 sheets), Glogau 1862.

**Gerber,** Johann Gustav (*c.*1690-1734). Military cartographer from Brandenburg, in Russian service from 1710. Eastern Caucasus, St Petersburg 1735.

**Gerber,** W. Russian colonel. Chart of the Caspian West coast, 1735.

**Gerbert**, Martin (1720-1793). Abbot of St. Blasien Abbey in Germany. *Historia Nigrae Silvae*, 1788 (with a map of the Black Forest).

**Gerbier**, *Sir* Balthasar (*c.*1591-1667). Painter and architect from Middelburg, came to England in 1616. Works on navigation, cosmography, and geography, 1649; manuscript map of Gambia River, 1660 (British Library).

**Gerbig**, G. *Kreis Höxter* (4 sheets), 1892.

**Gerbillon**, Jean François, SJ (1654-1707). Priest from Verdun, France, travelled in China, died Peking. *Voyages dans la Tartarie Occidentale par l'ordre de l'empereur de la Chine en 1688 & 1698*, The Hague 1747-1780.

**Gerdes**, Ferdinand H. Surveyor, worked for the United States Coast Survey. Triangulation for *The Harbor of Annapolis...*, 1846 (with James Ferguson); triangulation of Mobile Bay, 1846; surveys for *Map of Delaware Bay and River*, 1848 (with several others); triangulation for *Cat and Ship Island Harbors*, 1850; *Rebecca Shoal Florida Reef*, 1851; *Reconnaissance of Channel N$^{o.}$IV Cedar-Keys Florida*, 1852; *Preliminary Reconnaissance of the Middle or Main, and West Entrances to St. George's Sound Florida*, 1853; *Preliminary Reconnaissance of the Entrance to Barataria Bay Louisiana*, 1853; topography for *Patapsco River and the Approaches*, 1859 (with R.D. Cutts, H.L. Whiting and J.B. Gluck); surveys of Chesapeake Bay for U.S. Coast Survey charts, 1862 (with several others).

**Gergens**, A. *Plan der Stadt Mainz*, 1795.

**Gerhard**, Carl August (d.1817). Cartographer of Baden.

**Gericke**, Johann Ernst (*fl.*1744-1769). Engraver in Berlin. *Landgrav. Thuringiae*, 1753; *Tab. Geogr. Africae*, *c.*1754.

**Gericke**, I.E. Engraver. *Borhanpor*, 1784.

**Gérin**, —. Engraver, for J. Andriveau-Goujon, 1841-1873; for Alexandre Vuillemin, 1857; engraved the mountains on *Chemins de fer d'Espagne et de Portugal*, Paris, E. Andriveau-Goujon 1863 (railways engraved by Alfred Potiquet, outline engraved by Smith, calligraphy by P. Rousset, water by Mme Fontaine); one of the engravers for *Carte orographique et routière de l'Empire français...*, Paris, E. Andriveau-Goujon 1870; *Algérie, d'après les cartes de l'Etat-Major...*, Paris, E. Andriveau-Goujon 1890 (with Dalmont, Smith & Fontaine, printed by Chardon).

**Gérin**, A. French engraver. *Tunisie-Côte Nord. Tabarca: plan levé en 1883 par M.L. Manen... et M.M.F. Hanusse...*, Dépôt des Cartes et Plans de la Marine 1885, 1887 (lettering by Vialard). **N.B. Compare with Gérin, above.**

**Geringius**, Erik (1707-1747). Engraver of Stockholm. *Södermanland*, *c.*1745.

**Gerini**, Giovanni. *Carta Corografica della divisione ecclesiastica, politica e pinanziaria dell'I.R. Litorale Austro-Illirico...* [Yugoslavia] (6 sheets), Vienna, I.R. Misurazione Catastale 1847.

**Gerlach**, C.W. *Plan der Stadt Leipzig*, 1814, 1845.

**Gerlach**, H. American lithographer in Buffalo, New York. Parkinson & Smith's *Chart of Lake Superior*, 1858 (printed by Sage & Sons).

**Gerlach**, H.D. 'Lieutenant du Genie et Aide-de-Camp du Prince'. *Plan du Siège de Cassel... 1762*, The Hague, Pierre Gosse & Daniel Pinet 1763; *Topographische Karte vom Herzogthum Braunschweig*, 1763-1769.

**Gerlach**, J.W.R. *Carte des Chemins de fer de l'Europe Centrale*, 1873.

**Gerlach** [Gurlach], *Captain* P. Deputy Quartermaster-General. *Plan of the action at Huberton* [Vermont, 1777], 1780 (engraved by W. Faden); *Prince Ann, Norfolk, Nansemond Counties, Virginia*, 1781 MS.

**Gerlachin**, Katherin. Publisher of Nuremberg. *Hist. zwolf Apostel*, 1584.

**Gerland**, Georg (1833-1919). Specialised in language maps. *Ethnologische Weltkarte*, Gotha 1871; *Polynesien*, 1872; *Atlas der Völkerkunde* (15 sheets), 1886-1891; made contributions to Heinrich Berghaus's *Physikalischer Atlas*, 1887-1892 (with O. Drude, J. Hann, W. Marshall, and G. Neumayer).

**Germain**, Adrien (1837-1895). French hydrographer. *Traité d'hydrographie*, 1882.

**Germain** [Germaini], Jean (*fl.*mid-15th century). Geographer.

**Germain**, Louis (1733-1770). French engraver, for abbé Jean Chappé d'Auteroche's *Carte du Kamchatka*, 1768.

**Germaine**, —. *Plan des Criques de Mascate et Khulbos*, 1862-1863; Chart of Madagascar, 1863-1864.

**German**, Marcin. *Miasto Wieliczka* [plan of the salt mining town of Wieliczka, Poland], 1645; *Delineatio Primae Saliafodinae Wielicensis Wizerunk Zupy Wielickzy Pierwszey* [The first underground level of the Wieliczka salt mines], 1645.

**German & Br.** [brother]. *Navarro County, Texas*, 1868.

**Germanus**, H.M. See **Metellus**, J.

**Germanus** [Germanico], *Donnus* Nicolaus [Nicolo]. Cosmographer and monk (possibly Benedictine) working in Italy. The name Germanus is indicative of German origins rather than being a family name. He produced several manuscript copies of Ptolemy's *Geographia*, 1460-1477 of which twelve are extant. These incorporated influential new maps, on which he used a distinctive trapezoidal projection. Three of the four 15th-century editions of Ptolemy (Bologna, 1477; Rome, 1478; and Ulm, 1482) are directly based upon the work of Germanus; he is also recorded as making a world map and a pair of globes for Pope Sixtus IV, 1477. **Ref.** KARROW, R.W. *Mapmakers of the sixteenth century and their maps* (1993) pp.255-265; KRETSCHMER, DÖRFLINGER & WAWRIK *Lexicon zur Geschichte der Kartographie* (1986) pp.522-523.

**Germond de Lavigne**, A. *Itinéraire descriptif... artistique de l'Espagne et du Portugal*, L. Hachette et C$^{ie.}$ 1859.

**Germontius**. See **Gourmont**, family.

**Gerner**, —. *Deutscher Zollverein*, 1869.

**Gerning**, Johann Isaac *Baron von*. *A Picturesque Tour along the Rhine*, London, R. Ackermann 1820.

**Gerold**, Carl. Austrian printer, studied lithographic techniques at Munich in 1816, then set himself up in Vienna as a lithographer specialising in the use of tints and colours.
• **C. Gerold's Sohn.** Publisher of Vienna. Baron F.A.H. von Hellwald's *Der Feldzug des Jahres 1809 in Süddeutschland*, 1864.

**Gerold**, Carl. *Situations-plan der Weltausstellung* [Vienna], 1873. **N.B.** Compare with **Gerold**, above.

**Gerolt**, Federico de. *Distritos minerales de Mexico*, 1827; Central Mexico, used by Heinrich C. Kiepert, 1858.

**Gerrich**, J.C. Engraver, for Carl Ferdinand Weiland, 1848.

**Gerrish**, E.P. American surveyor. *Map of Windham County Connecticut*, Philadelphia, E.M. Woodford 1856 (with W.C. Eaton, D.S. & H.C. Osborn, lithographed by W.H. Reese, printed by Wagner & McGuigan).

**Gerritsz.** [Gerrijtsz.; Gerritszoon], Adriaen (*c*.1525-1579). Navigator and chartmaker from Haarlem. *De Zeevaert*, 1588; *Generale Pascaerte* [Chart of Western Europe], 1591.

**Gerritsz.**, Cornelis. Native of Zuidland, Netherlands. *Sea Journal Unto Java*, Wolfe 1598 (with a map of Bali).

**Gerritsz.**, Dirck. Charts of Ireland, H. Gerritsz 1612. **N.B.** Compare with **Pomp**, Dirck Gerritsz.

**Gerritsz.** [Gerritszoon; Gerard; Gerardus; Gherritszoon van Assum], Hessel (1580-1632). Dutch engraver, cartographer, publisher and bookseller from Assum, Noordholland. Apprenticed to Willem Blaeu as an engraver, he set up on his own in Amsterdam, describing himself 'in de Paskaert' or 'sub signo Tabulae Nauticae'. Gerritsz. was recorded at various addresses in Amsterdam including: 'op't Water bij die oude Brug' (1609); 'by die Lienbaens Brugh' (1616); 'Nieuwe Zijds Voorburgwal' (1624); 'op den hoeck vande Doele-straat' (1627-1632). In 1617 he was appointed official mapmaker for the VOC (Dutch East India Company) in Amsterdam. He was responsible for compiling and supplying the company with accurate charts updated according to latest information from VOC ships' logs and annotated charts. He held and developed this role until his death, when he was succeeded by Willem Blaeu. Gerritsz. was also cartographer to the WIC (Dutch West India Company) from 1621. Engraved Netherlands, *c*.1608 (lost); Gulick Cleve, 1610; Spain, 1612; Beschrijvinghe van de Samoyeden Landt, 1612; *Tabula Nautica*, 1612 (published in the account of Hudson's 4th voyage); engraved Willem Blaeu's *Magni Ducatus Lithuaniae*, 1613 (later used in Atlantis Appendix to 1635, modified); *Tabula Russiae*, 1614; *Italia nuovamente piu perfetta* (4 sheets), 1617; *Landt van Eendracht* [Indonesian Archipelago and Australia], 1618, *c*.1628 and later; manuscript chart of East Asia, 1621; manuscript chart of the Pacific, 1622; Indian Ocean, 1622; maps in Johannes de Laet's *Nieuwe Wereldt*, 1625 and later; *Carte nautique des bords de Mer du Nort, et Norouest...* [North Atlantic], *c*.1628 MS; Rotario West Indies & South America, 1628-1631 MSS; *Nova Anglia, Novum Belgium Et Virginia*, Leyden 1630 (published in the second edition of de Laet's *De Nieuwe Wereldt ofte Beschrijvinghe van West-Indien*); *Indiae quae Orientalis dicitur...*, *c*.1631, and other maps published posthumously by Willem Blaeu in 1635; and many others. **N.B.** His son, confusingly called Gerrit **Hessels**, also worked for the VOC. **Ref.** ZANDVLIET, K. *Mapping for money* (1998) pp.86-100 et passim.; COHEN, P.E. & AUGUSTYN, R.T. *Manhattan in maps 1527-1995* (1997) pp.26-27; SCHILDER, G. in: KRETSCHMER, DÖRFLINGER & WAWRIK *Lexikon zur Geschichte der Kartographie* (1986) vol.C/1, pp.264-265 (numerous references cited) [in German]; KEUNING, J. *Imago Mundi* 6 (1947) pp.48-66.

• [Hessels], Gerrit. Son of Hessel **Gerritsz**. See **Hessels**, Gerrit.

**Gerritsz.**, Martin. See **Vries**, Maarten Gerritsz. de

**Gerritsz.**, Pieter (*fl*.1608-1624). Dutch geographer, bookseller and printer, worked in Haarlem and Rees. Noord-Holland, 1608.

**Gersdorff**, A. *Grundriss... Danzig*, 1822.

**Gerster**, B. *Canal de Corinthe*, 1882.

**Gerster**, Johann Sebastian (1833-1918). Swiss geographer, historian and teacher, born in St. Gallen, worked in Rorschach, Fribourg, Bern and Lucerne, producing thematic atlases, school wall maps and pocket maps. *Karte der Schweiz*, Frauenfeld, J. Huber [n.d.]; *Kanton Schaffhausen*, 1870; *Atlas Suisse*, Neuchâtel 1871; *Karte des Kantons Glarus*, 1877; *Vorarlberg und Liechtenstein*, 1895. **Ref.** SCHERTENLEIB, U. *Cartographica Helvetica* 20 (1999) pp.19-24 [in German]; SCHERTENLEIB, U. *Vermessung, Photogrammetrie, Kulturtechnik* 93, 10 (1995) pp.626-629 [in German].

**Gervais**, —. Publisher of 'Vieille rue du Temple No.47', Paris (1806). C.F. Delamarche's *Les usages de la sphère...*, 1799.

**Gervais**, Henry. Mapmaker of Paris. Devised a system of outline maps for use in schools. France physique, 1862; *France. Physique demi-muette* (5 maps), Paris 1867; *France Politique demi-muette*, Paris 1867; *Nouvelle Carte Physique de la France*, Paris *c*.1868; *Nouvelle Carte de la France*, Paris *c*.1874; *Nouveau Carte Politique de la France*, Paris *c*.1874; *Nouvelle Carte Physique de la France*, Paris, L. Hachette 1876.

**Gervais de la Rivière**, Marie Denise. See **Delahaye**, family.

**Gervais[e] de Palmeus**, A.F. See **Palmeus**, A.F. Gervais de.

**Gervaize**, —. Île Tsis, used by J.S.C. Dumont d'Urville, 1842-1848.

**Gervase of Canterbury**. Prepared a work which he called a *Mappa mundi*, *c*.1200. This was not a map, but a gazetteer of religious houses.

**Gervase of Tilbury** [Belmotus] (*c*.1160-*c*.1234). Chronicler and provost of Ebstorf. Associated, possibly as a source of information, with the 'Ebstorf' map, *c*.1235 (dates differ). The Ebstorf map was rediscovered in 1830, but destroyed in 1943.

**Gerwig**, R. Engineer and draughtsman of Karlsruhe, senior advisor on bridges and highways. *Chemin de fer par le St. Gothard. Plan de Fluelen à Biasca...* (4 sheets), *c*.1864 (with A. Beckh, lithographed by Wurster, Randegger & Co., Winterthur); *Chemin de fer par le St. Gothard. Plan général de Lucerne et Zug à Camerlata et Locarno...* (11 sheets), *c*.1864 (with A. Beckh).

**Gesner**, Conrad [Konrad] (1516-1565). Physician and scholar at Zürich. He prepared bibliographies describing in detail many contemporary maps, some of which are now lost and known only from this source, including maps by Oronce Fine, Gemma Frisius and other well known mapmakers. *Bibliotheca universalis*, Zurich, C. Froschauer 1545; contributed to Münster's *Cosmographia*, 1550; *Epitome bibliothecae Conradi Gesneri*, Zurich, Froschauer 1555. **Ref.** BURMEISTER, K.H. *Imago Mundi* 23 (1969) pp.75-76.

**Gessi**, Romulus. Manuscript maps of Uganda, 1876 used in the compilation of *Carte de Lac Albert...*, Cairo 1876; and *Carte du cours du Nil*, 1876.

**Gessner**, Abraham (1552-1613). Artist and goldsmith of Zürich, worked also in Stühlingen. Gilt silver goblet with 18.5-cm terrestrial globe, *c*.1580 [National Museum, Copenhagen]; other globes in form of drinking vessels, 1595 and 1600. **Ref.** KEJLBO, I.R. *Rare globes* (1995) pp.103, 105, 107, 138, 188.

**Getkant**, Friedrich [Fryderyk] (*d*.1666). Cartographer from the Rhineland in Polish service. Worked mainly on plans of town defences, military encampments and battlefields. *Delineatio situs Pucensis*, 1634 MS; *Tabula Topographica, demonstrans situs Pucensi...* [Baltic Sea], 1637; manuscript atlas *Topographica practica...*, 1638; Ukraine, 1638.

**Gette**, O. *Karte der Stadt Bremen*, 1875.

**Geus**, C.K. de. *Atlas à l'usage des voyageurs* (15 maps), 1858.

**Geus**, G.A. de. *Kaart van de... Haarlemmermeer*, 1857.

**Geusau**, *Lieutenant-General* Levin von (1725-1808). Cartographer of Berlin. *Topographische-Militaire Karte vom vormaligen Neu Altpreussen...*, Berlin, D.F. Sotzmann 1808 (engraved by Charles Mare).

**Geyer**, *Lieutenant* —. *Regensburg*, *c*.1866.

**Geyer**, F. *Karte der Zuckerfabriken Deutschlands*, 1879.

**Geyer**, Friedrich Wilhelm. Engraver in Nuremberg. *Plan von denen Gegenten am Rhein*, *c*.1744.

**Geyer**, G. *Hesse und Nassau*, 1868.

**Geyer**, V. *Die Schweiz*, 1864; *Karolinen, Palau und Marianen*, 1899.

**Geyger**, H.C. See **Gyger**, Hans Conrad.

**Geylekerckius** [Geylkercke]. See **Geelkercken** family.

**Gfug**, *Sek. Lt.* - von. *Umgebung von Graudenz*, 1874.

**Ghandia**, *Sargento Mayor* Antonio de. *Islas Philipinas*, 1727.

**Ghazulus Ragusinus**. See **Gazulić**, Ivan.

**Ghebellini** [Ghebellinus], Stefano [Stephanus]. Engraver of Brescia. *La carte du conte de Venayscin* 1574 (known from one example in Bibliothèque Nationale, Paris, used as a source by A. Ortelius, 1584). **Ref.** MEURER, P. *Fontes cartographici Orteliani...* (1991) p.156 [in German].

**Ghelen**, Johann Peter von (1673-1754). Printer, publisher and bookseller in Vienna. Matthias Roth's *Atlas Novus indicibus instructus* (26 maps), 1728; *Nova Maris Caspii... delineatio*, *c*.1735.

**Ghequiere & Holmes**. American publishers. A.P. Folie's *Plan of the Town of Baltimore and Its Environs...*, 1794 edition (first published 1792).

**Gherardi**, Giacomo (*d.c*.1593). Brother of Claudio Duchetti's wife Margherita. On the death of Claudio in 1585 he was appointed tutor to their children and until his own death carried on the Duchetti business as *haeredes Claudii Duchetti* which re-issued some of the Duchetti plates. **N.B.** See **Duchetti** family.

**Gherijtsz.**, Pieter (*fl.c*.1515). Dutch surveyor, working in Rijnland. Map of Rijnland, 1515 (now lost).

**Gherri[j]tszoon van Assun**, H. See **Gerritsz.**, Hessel.

**Gheyn**, Jacob de [Jacques II] (1565-1629). Artist, engraver and publisher from Antwerp, worked in Amsterdam.

**Ghigi** [Ghisius], Giovanni Baptista [Battista]. Geographer, painter and historian of Rome. Editor of Rome edition of C. Cellarius's

*Geographia Antiqua* (35 maps), Rome 1774; historical map of Rome, 1777; Paludi Pontine [south of Rome], 1778; Sicilia (4 sheets), Rome 1779.

**Ghijseman**, C. Map publisher of The Hague. Issued 7th edition of Johannes, Justus and David Vingboons's *'t Hooghe Heyraedtschap vande Landen van Woerden*, 1788.

**Ghillany**, F.W. (1807-1878). Published a copy of Johann Schöner's Nuremberg globe, 1853.

**Ghim** [Ghymmius], Walter [Gualterus] (1530-1611). Mayor of Duisburg and friend of Mercator. Wrote a biography of Mercator *Vita Mercatoris*, first published in the Duisburg edition of Mercator's *Atlas*, 1595.

**Ghisius**, Giovanni Baptista. See **Ghigi**, Giovanni Baptista.

**Ghisleri**, A. *Atlante d'Africa*, Bergamo 1909.

**Ghisolfi** [Gisolfo], Francesco. Genoese cartographer, copied Battista Agnese's and Giacomo Gastaldi's maps. 7 recorded manuscript atlases, 1546-1553; also possible globe gores, *c.*1550. **Ref.** ASTENGO, Corradino 'La produzione cartografica di Francesco Ghisolfi' [The cartographic production of Francesco Ghisolfi] in: *Annali di Ricerch e Studi di Geografia* (Genoa, 1993) [in Italian, with summaries in French, English & German].

**Ghisulfus** [Gisolfo], Jean [Giovanni]. L'Italie, 1757.

**Ghymmius**, Gualterus. See **Ghim**, Walter.

**Ghys**, F. Worked with F. Vandamme as printers of P. Vandermaelen's *Carte Topographique de la Belgique* (248 sheets), *c.*1854.

**Ghys**, Martin. Printer and engraver of charts in Antwerp. *Rade de Blankenberge...*, after 1872 (surveyed by A. Stessels); *Escaut. Depuis Flessingue jusqu'à Burght*, Belgian Hydrographic Department after 1881 (surveys by L. Petit); *Escaut...entre Moerzeke, Termonde et Kleyn Zand...*, Dutch Ministry of Marine before 1884 (surveys and soundings by L. Petit); *Mer du Nord... entre Ostende et Nieuport*, Belgian Hydrographic Department *c.*1884 (surveyed by L. Petit and E. Rochet); *Embouchure de l'Escaut*, Belgian Hydrographic Department *c.*1885 (surveyed and sounded by L. Petit and E. Rochet); *Escaut. Rade d'Anvers...*, 1886.

**Giambullari**, Pier Francesco (*c.*1494-1564). Italian philosopher and priest. *De'l Sito, Forma, & Misure dello Inferno di Dante*, Florence 1544 (with two hemisphere maps, one a north polar projection, the other showing the supposed location of Purgatory as referred to in Dante's *Inferno*).

**Giampiccoli**, Giuliano (1703-1759). Engraver from Belluno, active in Venice. Engraved the world map for G.B. Albrizzi's *Atlante Novissimo* (2 volumes), 1740.

**Gianelli** [Janellus], Giovanni [Joannes]. Globe and clockmaker in Cremona. Globe, Milan 1549.

**Giarattoni**, Giuseppe. Provincie e regni del universo, 1669 MS.

**Giarrè** brothers [Studio Giarrè] (18th-19th century). Engravers active in Florence under the name 'Studio Giarrè'. Engraved the lettering for *Atlas Historique*, Florence, Molini & Landi 1807; some maps for B. Borghi's *Atlante Generale*, 1816-1819 (including United States (4 sheets), 1818, and Tuscany, 1819); some maps in L. Cacciatore's *Nuovo Atlante Istorico*, Florence 1831 edition.
• **Giarrè**, Gaetano. Contorni di Firenze, *c.*1840.
• **Giarrè**, Raimondo.

**Giarré**, B. Calligraphy engraver. *Carta Topografica della Strada Ferrata Leopolda* [Florence], 1841 (map drawn by G. Piccioli and engraved by A. Verico).

**Gibb**, A. Plans of Aberdeen, 1861-1871. **N.B.** See also **Keith** & **Gibbs**.

**Gibb**, Andrew. *Road from Barglachan Coal Works to... near Carbello* [Ayrshire], 1826 MS.

**Gibbard**, William (1818-1863). Civil engineer and land surveyor in Canada. Plan of the Town Plot of Meaford, 1845; *Map of the County of Simcoe revised and improved...*, 1853 (lithographed by H. Scobie); *Creemore...*, 1853; *Plan of the Property of Messrs McMaster, Paterson, Hamilton & Robinson at Collingwood Lake Huron...*, New York, Miller's Lith. 1854; *Plan of the Building Lots* *in the Town of Barrie...*, Toronto, Scobie's Lith. 1854; *Cape Rich...*, Toronto, Maclear & Co. 1856; *Chart of Collingwood Harbor and its connections*, Collingwood 1858 (lithographed by J. Ellis); Track Chart of Lake Huron, 1862; Track Chart of Lake Superior, 1862; and many others.

**Gibbes**, Charles Drayton. Gold Region, California, 1851; *A new map of California*, Stockton, California, C.D. Gibbes 1852; Alpine County, 1866 MS; *Map of the States of California and Nevada*, San Francisco, Warren Holt 1869, reduced edition 1873, 1875 (with J.H. von Schmidt and A.W. Keddie, lithographed by S.B. Linton).

**Gibbings**, *Lieutenant* R. Malawa & adjoining countries, Calcutta 1845.

**Gibbins**, H. de B. *Atlas of Commercial Geography*, 1893.

**Gibbon**, John. Middlesbrough [Yorkshire], 1619 MS.

**Gibbon**, *Lieutenant* Lardner, *USN*. Maury's Wind and Current Chart. North Pacific..., 1852; *Exploration of the Valley of the Amazon* (3 volumes with 8 maps), 1854 (with W.L. Herndon).

**Gibbons**, *Captain* (later *Major*) Alfred St. Hill. East Yorkshire Regiment. Kingdom of Marutse [Rhodesia], 1897; Zambesi River, 1898 MS; *Map of the District lying between the Lakes Albert Edward and Kivo* [Zaire], 1900 MS.

**Gibbons**, Edward. Parish maps in Cambridgeshire & Huntingdonshire, 1800-1816.

**Gibbons**, *Sir* John, *Bart*. (1717-1776). Plan of the Manor of Stanwell [Middlesex], *c*.1760.

**Gibbons**, M.H.A., *RN*. One of the surveyors who assisted A.T.E. Vidal with *Survey of the Cape of Good Hope*, 1822 (with T. Boteler and C. Lechmere of HMS *Leven*), published London, Hydrographic Office of the Admiralty 1828 (engraved by J. & C. Walker); *A survey of Hout Bay, Cape of Good Hope*, 1822 (with T. Boteler), published London, Hydrographic Office of the Admiralty 1828 (engraved by J. & C. Walker).

**Gibbs**, Alexander (*fl*.1812-1818). Surveyor, produced manuscript maps of parts of Scotland, including a survey of the property holdings on the Hebridean island of Lewis, 1817 MS (National Library of Scotland).

**Gibbs**, Alexander. *Plan of the mouth of the River St. Pierre and adjacent islands in the River St. Lawrence*, 1825 MS.

**Gibbs**, George (1815-1873). Ethnologist and specialist in native American languages. *Sketch of the northwestern part of California...*, 1851 MS; *Sketch of the Wallamette [sic] Valley...*, 1851 (with E.A. Starling); *Map of the Western District of Washington Territory showing the position of the Indian Tribes...*, 1855 MS; *Map showing the Territory of the Blackfeet and the Common Hunting Ground of the Blackfeet and Western Indians*, 1855 MS; *Map of the Western Part of Washington Territory*, 1856 MS; assisted with the North Western Boundary Survey, 1857.

**Gibbs**, Joseph (1798-1864). Engineer of London, produced civil engineering maps of sites throughout England. *Bucks., Berks., London and Windsor Railway, proposed route from Paddington to Clewer*, 1833 (surveys by Cruickshank & Gilbert, lithographed by T.J. Ellis).

**Gibbs**, L. Drainage of Fenland, 1888.

**Gibbs**, S. Publishers of Bath. *A New and Correct Plan... of Bath*, 1835; *The City of Bath*, *c*.1858; *The Environs of Bath*, 1870; *Gibbs's New Plan of the City of Bath*, *c*.1870 (surveyed and drawn by Cotterell & Spackman); *A new and correct Plan of the City of Bath*, 1883.

**Gibbs**, Thomas Fraser. Provincial Land Surveyor, Ontario, Canada. *Plan of the City and Liberties of Kingston*, 1850 (lithographed by H. Scobie).

**Gibbs, Shallard & Co.** Publishers and printers of 108 Pitt Street (1869), later 70 Pitt Street (1883), Sydney, New South Wales. J. Jones's Road and Distance Map of New South Wales, 1871; *Sketch plan of Chamber's Creek mining leases*, 1872; Blackheath, 1881; *Gibbs, Shallard & Co.s Map of the City of Sydney and Suburbs*, 1884; *Captain Cook Estate*,

*Randwick*, 1885; *Map showing the railway systems of Australia*, 1888; *Liverpool*, 1892.

**Gibelli**, Giuseppe. Fortezzi... dello Stato Ecclesiastico, 1888.

**Gibson bros.** Publishers of Washington. *Roose's companion and guide to Washington and vicinity*, 1876, 1880, 1881, 1888 (with a map of Washington and suburbs).

**Gibson**, Arthur. Plat book Cottonwood County, Minnesota, 1896.

**Gibson**, Atkinson Francis (*fl*.1785-*d*.1829). Estate surveyor of Saffron Walden, worked in Essex. *Hatfield Peverel and Ulting*, 1786 MS.

**Gibson**, Edmund (1669-1748). Bishop of Lincoln, later London. Translated and revised William Camden's *Britannia*, 1695, 1722, 1730 and 1755 (with Robert Morden's maps).

**Gibson**, *Lieutenant* Francis (1753-1805). Customs officer of Whitby, Yorkshire. Plan of the natural and artificial defences of the port [Whitby], 1782 MS; Plan showing the purpose of the new east battery as a protection to shipping (with a view of the town and harbour of Whitby), 1794.

**Gibson**, George E. Omaha & South Omaha, Nebraska, 1887 (with William Gibson).

**Gibson**, John (*fl*.1750-1792). Geographer, engraver and draughtsman, N$^{o.}$18 George's Court, Clerkenwell, also noted at Red Lyon Street Clerkenwell (1753). *A new and accurate Map of America*, London *c*.1750; E. Bowen's *An Accurate Map of the West Riding of Yorkshire*, London, T. Kitchin 1750; engraved maps and charts for T. Salmon's *The Universal Traveller*, 1752-1753; maps for Jonas Hanway's *An Historical Account of the British Trade Over the Caspian Sea...*, London 1753; for Palairet's *Atlas Méthodique*, 1755; engraved a pirated copy of Lewis Evans's *A General Map of the Middle British Colonies in America*, T. Kitchin 1756; maps for *The Gentleman's Magazine*, 1758-1763; ... *This Map of the Post-Roads of Europe...*, London, J. Rocque 1758, French edition published R. Sayer 1771; *Atlas Minimus or a new set of Pocket Maps of... the known World*, London, J. Newbery 1758 (revised by E. Bowen), London, T. Carnan 1774, T. Carnan & F. Newbery 1779, Philadelphia, Mathew Carey 1798; *New and Accurate Maps of the Counties of England and Wales...*, J. Newbery 1759, post-1779; maps for W.A.B. Rider's *A New History of England*, 1761-*c*.1764; maps in *The American Gazetteer*, 1762; maps for *Universal Magazine*, 1763-1769; *An Accurate Map of the Island of Anglesey*, London, J. Knox 1765 (surveyed by Lewis Morris); *...Plan of the Cities and Suburbs of London & Westminster & Borough of Southwark...*, 1766; *South America*, *c*.1770; engraved maps for Joseph Smith Speer's *The West India Pilot*, 1771; *North America*, Sayer & Bennett 1776 (with T. Jefferys); *A Correct Map of Africa*, London *c*.1780; *A Map of the New Continent*, London *c*.1780; *Collieries on the Rivers Tyne & Wear*, 1787-1788; and others.

**Gibson**, Robert (*fl*.1731-*d*.1760). Irish estate surveyor. Treatise on Practical Surveying, 1802.

**Gibson**, T. Channel Islands, Emanuel Bowen 1750.

**Gibson**, William. Omaha & South Omaha, Nebraska 1887 (with G.E. Gibson).

**Gibson**, William T. Fold-out map to accompany John Delafield's *A General View and Agricultural Survey of the County of Seneca*, 1850, map also published separately as *Topographical Map of Seneca County, New York*, 1852.

**Giddings**, Amelia (*b.c*.1806). *Map of England* (silk embroidery), 1815 (aged 9).

**Gidmore**, J. Bath, 1694.

**Giegler**, J.P. Carte routière d'Italie, Milan 1818.

**Gielis Bouloigne.** See **Boileau** de Bouillon, Gilles.

**Gier**, H. *Plan von Bremen*, 1870; *Plan von Hannover*, 1895.

**Giesecke**, Karl Ludwig [Charles Lewis] (1761-1833). Bavarian mineralogist and musician, born Johann George Metzler, later changed his name to Karl Ludwig Giesecke. Worked in Copenhagen, the Faeroes, Greenland and Ireland, where he became professor of mineralogy at the Dublin Society. Mineralogy of Faeroes & Greenland, 1805-1807; Greenland, 1832. **Ref.** CRAIG, M. 'Charles Lewis Giesecke', in Dictionary of national biography: Missing persons pp.250-251.

**Giesecke & Devrient.** Printer of Leipzig. *Der Rheinstrom und seine wichtigsten Nebenflüsse von den Quellen bis zum Austritt des Stromes aus dem Deutschen Reich* (22 sheets), Berlin, Ernst & Korn 1889 (compiled by the Central Meteorological and Hydrographic Office of Baden).

**Giesemann.** See **Keller** & **Giesemann**.

**Giffart**, P. Engraver. *Partie inferieure occidentale de l'evesché du Mans*, 1706 (in Jaillot's *Atlas françois*).

**Giffart**, Pierre François. Publisher of 'Rue St. Jacques, à Sainte-Thérèse', Paris. Père Claude Buffier's *Géographie universelle*, *c*.1749, 1754, 1760.

**Giffault**, E. *Carte du Soudan Occidental et des Régions explorées par le Capitaine Binger de 1887 à 1889*, 1890 (engraved by Erhard Frères).

**Gifford**, C.B. Worked with W. Vallance Gray on bird's-eye views of U.S. cities. *San Francisco*, 1868 (drawn, printed and published by Gray & Gifford); *San José*, San José, G.H. Hare 1869 (drawn and printed by Gray & Gifford).

**Gifford**, Franklin. Surveyor and publisher of Wilson, New York. *Niagara County, New York*, Wilson, N.Y., Franklin Gifford 1852 (with Samuel Geil, copyright by R.P. Smith as part of the J.H. French - R.P. Smith series of surveys of New York State); survey of Cayuga County, New York, published Philadelphia, S. Geil 1853 (surveyed with S. Geil and S.K. Godshalk); *Broome County, New York*, 1855.

**Gifford**, John. *History of France*, 1798 (included John Gifford's *Sketch of the Harbour and Environs of Toulon*, engraved by Thomas Clarke).

**Gifford**, P.E. View of *Santa Barbara*, 1898 (printed by Los Angeles Litho. Co.).

**Gigas** [Gigante; Gigus], Johannes Michael (*c*.1580-1633). Physician, mathematician and geographer. *Prodromus geographicus* [Atlas of the Archbishopric of Cologne], 1620 (with so-called Pauluskarte, maps were later used by Blaeu, 1635, J. Janssonius, 1645, and Homann Heirs, 1672).

**Gignilliat**, T. Heyward. *Valley of the Orinoco River*, 1896.

**Gigus**, J.M. See **Gigas**, Johannes Michael.

**Gijsbertszoon** [Gyberlszoon], Evert. Chartmaker of Edam. East Indies, 1599; North Sea, 1601 MS.

**Gijseman**, Jan. Publisher of The Hague. Reissued Johannes Leupenius's *Het Hooge Heemraadschap vande Crimpenre Waard* (6 sheets), 1755 (with Geurt van Moelingen).

**Gilbert**, A.W. *Hamilton County, Ohio*, 1856; *Cincinnati & Vicinity*, 1861.

**Gilbert**, Frank Theodore (*b*.1846). Atlas of Yolo County, California, 1879.

**Gilbert**, George (*fl*.1828-1849). Estate surveyor of Colchester, Essex and Furnival's Inn, London. Estate, inclosure and tithe maps throughout England. Great Warley, Essex, *c*.1843 MS.
• **Gilbert & Tayspill** (*fl*.1837-1843). George Gilbert and Thomas Tayspill, practising together as land surveyors, of Colchester. Little Bromley, Essex, 1839 MS.

**Gilbert**, Giles. Estate surveyor in Ireland, work included the Down survey, 1655-1659; *Barony of Clonmoghan*, 1675 MS.

**Gilbert**, H. [& Co.]. Publishers of 16 Beekman St., New York. Pennsylvania Railway Company Atlas, 1875.

**Gilbert** [Gylbert], Sir Humphrey (*c*.1539-1583). English geographer, half-brother to Sir Walter Ralegh. After a failed attempt in 1578, he successfully established a new colony at St. John's, Newfoundland in 1583, and briefly became its governor. He died on the return voyage. *A General Map, made onelye for the Particuler Declaration of This Discoverie* (cordiform woodcut world map to show Northwest Passage), London 1576 (in Gilbert's *Discourse of A Discoverie for a*

*Sir Humphrey Gilbert (c.1539-1583). (By courtesy of the National Library of Australia)*

*New Passage to Cataia*). **Ref.** WALDMAN, C. & WEXLER, A. *Who was who in world exploration* (1992) pp.280-281; SHIRLEY, R. *The mapping of the World* (1983) N°.136 pp.158-160.

**Gilbert,** James. Map compiler and publisher of 49 Paternoster Row, London, also in partnership with E. Grattan as publisher Grattan and Gilbert. Guide to London, E. Grattan c.1824 (re-used W. Ebden's *The Pedestrian's Companion Fifteen Miles round London*, engraved by S. Hall, first published by M.J. Godwin, 1822); *The London Director*, Grattan & Gilbert 1840 (with a map by JoshuaArcher); *Modern Atlas of the Earth*, Grattan & Gilbert 1841; began a county atlas with maps engraved by J. Archer, published for M. Alleis either by J. Gilbert or Grattan & Gilbert, but after only 9 county maps were prepared the atlas was taken over by Fisher, Son and Co. (published as *Fisher's County Atlas of England and Wales*, 1842-1845); *Gilbert's Modern Atlas*, 1846, 1850; *Gilbert's College Atlas for Families and Schools*, 1847; *New Map of the World*, 1850; *Gilbert's Visitor's Guide to London*, 1851 (re-used map of London from *The London Director...*, 1840). **N.B.** See also **Grattan** & Gilbert.

**Gilbert,** *Captain* John. *Part of the Isl[an]d Cuba*, 1755.

**Gilbert,** John. *Pictorial Missionary Map of the World*, 1861.

**Gilbert,** *Captain* Joseph, RN (c.1733-c.1824). Master of the *Pearl*, and Master of *The Resolution* on James Cook's Second Voyage. Chart of the coast of Labrador, 1767, published London 1770 (engraved by Thomas Jefferys); South Coast of Newfoundland, 1768-1769; The head of the Bay Dispair and Conne River [Newfoundland], 1769 MS; survey of Plymouth Sound, 1769 MS; Vanuatu, 1774 (with J. Elliott); chart of South Georgia, 1775 (with J. Elliott); chart of South Sandwich Islands, 1775 (with J. Elliott); charts in the *North American Pilot*, 1775; Dusky Bay, New Zealand, 1775.

**Gilbert,** J.L. Manuscript plan of Regent's Park [London], 1824 MS.

**Gilbert,** Samuel A. Assistant surveyor for the U.S. Coast Survey. *Bull's Bay Harbor of Refuge Coast of South Carolina*, 1849, 1854; Plan of Charleston, 1849 (inset on map of South Carolina, 1854); triangulation for *Grand Island Pass, Mississippi*, 1852 published 1857; surveys for *Preliminary Chart of Charleston Harbor and its approaches*, U.S. Coast Survey 1855 (with C.O. Boutelle); surveyed the topography for *Plymouth Harbor Massachusetts*, 1857 (triangulation by A.D. Bache and T.J. Cram, hydrography by M. Woodhull and others, engraved by A. Sengteller, E.A. Madel & J.C. Kondrup); triangulation for *St. Louis Bay And Shieldsboro Harbor Mississippi*, 1857 (with J.E. Hilgard, topography by W.E. Greenwell, hydrography by B.F. Sands & others, drawn by S.B. Linton, lithographed by C.B. Graham); one of several triangulation surveyors for *Boston Harbor, Massachusetts*, 1857; triangulation for *Biloxi Bay Mississippi*, 1858 (topography by W.E. Greenwell, hydrography by B.F. Sands and others, engraved by R.T. Knight, A. Petersen & E.A. Madel); one of the surveyors for *Chesapeake Bay. Sheet No.4. From the Potomac River to the entrance of Pocomoke Sound*, 1859.

**Gilbert,** William (1540-1603). *De magnete*, 1600 (first great scientific work published in England); map of the moon, c.1600 MS, published posthumously in *De Mundo*, 1651.

**Gilbert,** William B. Chief engineer of the Rutland & Burlington Railroad (1848), of the Western Vermont Rail Road (1851) and of the N.Y, & Oswego Midland R.R. (1870). *Map & profile of the Rutland & Burlington Railroad*, Boston 1848 (lithographed by J.H. Bufford); *Map of the Western Vermont Rail Road and connecting lines*, New York 1851.

**Gilchrist,** *General* Charles A. *Atlas of Hancock County, Illinois*, 1874.

**Gilchrist,** James. *City of Wheeling &c. Ohio County, Virginia*, 1871.

**Gilde,** Adriaan Pieters 't. *Caerte Ende Figuere vanden Lande Ende Ambachten van den Axel...*, 1624; *Landt Caerte vande Schorren...*, 1636.

**Gildemeister,** J. *Karte des Gebiets der Reichs- und Hansestadt Bremen*, 1790-1793 (with C.A. Heineken).

**Giles**, family.
- **Giles**, Netlam (*c.*1780-1817). Solicitor and surveyor of Lincoln's Inn Fields (1849) London, produced manuscript maps of sites throughout England. Also worked with Francis Giles as Netlam and Francis Giles. *Plan of the town and harbour* [Dover], 1805 MS.
- **Giles**, Netlam and Francis. Surveyors and civil engineers of New Inn, London. Bishop's Stortford to Cambridge Canal, 1810 MS; plan for a canal from Dell Quay to Chichester, 1813; *Proposed Arundel and Portsmouth Canal...*, 1815; plan of Leeds, 1815; laid out a trigonometrical grid of Yorkshire which Christopher Greenwood used as the basis of his survey, 1815.
- **Giles**, Francis (1786-1847). Civil engineer and surveyor, worked initially in Yorkshire, later from New Inn, London (1815), and Salisbury St., London (1818-1819). Worked on civil engineering projects throughout England and Ireland, including surveys of rivers, harbours and railways. Also worked with Netlam Giles as Netlam and Francis Giles. *The Liffey from Carlisle Bridge to Dublin Lighthouse*, 1818-1819 MS; plans for Thames Haven Docks, [before 1840] MSS; Liverpool Harbour, 1821; proposals for the Portsmouth Junction Railway (3 plans), 1836 MSS; Whitechapel to Shell Haven, 1839.

**Giles**, Ernest (1835-1897). Explorer and surveyor from Bristol. Emigrated to Adelaide *c.*1850 and travelled extensively in Australia. Expedition to the Darling River, 1861-1865, made several attempts at an east-west transcontinental expedition from 1872, finally succeeding in 1875 by using camels. After further explorations in south-western Australia (1882), he settled in Coolgardie. *Australia twice traversed* (with maps), 1876 **Ref.** WALDMAN, C. & WEXLER, A. Who was who in world exploration (1992) pp.281-282.

**Giles**, George *Master, RN. Malooda Bay* [Borneo], 1845-1847 (Admiralty Chart).

**Giles**, George. Master of HMS *Arrogant. Position of the Allied Forces at the reduction of Bomarsund 16th August 1854* [Finland], 1854 MS; *Sketch of Sweaborg: shewing the Position of the Allied Forces. 17th August 1855* [Finland], 1855 MS.

**Giles**, H. *Map of the Hawaiian Islands*, 1876.

**Giles**, Joel. *Inner Harbour, Boston*, *c.*1850; *Plan of Back Bay, Boston*, 1852.

**Gilev**, Aleksey. Sergeant-Geodesist. Member of J. Billings's and G.A. Sarychev's expedition to Siberia, 1787-1788; described Arctic coast from East Cape to Kolyuchin Island; surveyed Avachinsk Bay, east coast of Kamchatka and Kurile Islands.

**Gilkison**, W.S. *City of Fort Wayne, Indiana*, 1866.

**Gilks**, Edward. Lithographer in Melbourne, Australia. *New England*, 1848; *Natal*, 1848; *Town Lots Ballarat, Melbourne*, 1859; *South Australia*, 1867; and others.

**Gill**, C.B. Roadstead of Salina Cruz [Mexico], 1871.

**Gill**, D. Surveys in South Africa, 1883-1907.

**Gill**, George *FRGS. Penny Atlas*, 1871.

**Gill**, George. County Geographical Cards, 1879.

**Gill**, J.K. Publisher of Portland, Oregon. *Habersham's Sectional and County Map of Oregon...*, 1874 (compiled by R.A. Habersham, engraved and printed by G.W. & C.B. Colton & Co.); *J.K. Gill & Co.'s map of Oregon & Washington Terr.*, 1878 (compiled by R.A. Habersham); Bird's-eye view of Portland, 1879 (printed by A.L. Bancroft).

**Gill**, Michael. Surveyed and published *Plaquemine, Louisiana*, 1854.

**Gill**, Reginald. *County of London*, 1900; *The Sphere atlas (elementary); political, physical and commercial*, London, Sphere Atlas Co. *c.*1900.

**Gill**, Valentine (*fl.c.*1798-1823). Surveyor, produced manuscript maps of Wexford. *New Map of the County of Wexford* (4 sheets), surveyed 1808, published W. Faden 1811, 1816 (engraved by J. Bailey and E. Ramsey). **N.B.** Compare with Valentine **Gill**, below.

**Gill**, Valentine. Provincial land surveyor in Canada from 1821. *Map of the First Division of the Roads Leading from Halifax to Truro and Windsor* [Nova Scotia], *c.*1816; plan of the town of Dundas, [n.d.], copied by J.G. Chewett, 1827. **N.B.** Compare with Valentine **Gill**, above.

**Gill**, *Lieutenant* (later *Captain*) William John, RE (1843-1882). *North East Persia*, 1873; *Western China & Eastern Tibet*, 1877.

**Gillberg**, I. Swedish surveyor. Gästrickland, 1789.

**Gilleland**, J.C. Ohio and Mississippi Pilot, Pittsburgh 1820; Plan of the Battle of Braddock's Defeat, 1838.

**Gillen**, —. Atlas of Mitchell County, Kansas, 1884 (with Davy).

**Gilleron**, —. *Carte géologique de la Suisse*, c.1867.

**Gillespie**, —. *Forest County, Pennsylvania*, c.1868.

**Gillespie**, W. American steel engraver working in Pittsburg, Pennsylvania. Map of Pittsburg and vicinity 'designating the portion destroyed by fire April 10th 1845', c.1845.

**Gillet**, —. Lithographer. *Plano de Valparaiso* [Chile], 1835.

**Gillet**, George (1771-1853). Surveyor general of Connecticut from 1813. *Connecticut, From Actual Survey, Made in 1811...*, Hartford, Hudson & Goodwin 1812-1813 (with Moses Warren, engraved by A. Reed & E. Windsor, reduced and/or enhanced editions of this map were published later by Hudson & Co. 1820 and A. Willard, 1820, 1829, 1833, 1842, 1847).

**Gillet**, N.J. *Prince Albert Gold District*, 1891.

**Gillet**, Thomas. Printer of Salisbury Square, Fleet Street, London (1799-1811). Captain Chauchard's *A General Map of the Empire of Germany, Holland, the Netherlands, Switzerland, the Grisons, Italy, Sicily, Corsica and Sardinia* (map in 23 sheets), London, John Stockdale 1800.

**Gillet and Son**. Printers of 'Crown-court, Fleet-street', London. One of the printers for Francis Grose's *Antiquities of England and Wales*, c.1810 edition. **N.B.** Compare with Thomas **Gillet**, above.

**Gillett**, A.G. Publisher of Addison, New York. P.J. Browne's *Map of Monroe County New York*, 1852.

**Gillette**, John E. One of the publishers registered at 517, 519, and 521 Minor Street, Philadelphia, the address also used by R.P. Smith and other publishers, such as D.J. Lake and W.O. Shearer in 1860, and A. Pomeroy in 1863, for the publication of Smith's maps of New York State. Map of Columbia County, New York, 1851 (surveyed by J.W. Otley and F.W. Keenan); *Gillette's Map of Oneida Co. New York*, 1858 (surveyed by S.N. Beers, D.J. Lake and F.W. Beers); Wayne County, New York, 1858; *Map of the Vicinity of Philadelphia*, John E. Gillette & C.K. Stone 1860 (surveyed by D.J. Lake, L.B. Lake, D.G. Beers, F.W. Beers, & S.N. Beers).

**Gilley**, *Lieutenant* William F., RE. Military plan of Devonport, 1854 MS.

**Gillier**, J. County Down (2 sheets), 1755.

**Gillig**, Charles Alvin. Compiled a *London Guide*, published by Gillig's United States Exchange, which incorporated maps of London 'specially prepared' by Edward Stanford, 1885, and James Wyld, c.1889.

**Gilling**, W. *A New General Atlas of the World*, London 1830.

**Gillingham**, Edwin. American engraver. John G. Hales's *Map of Boston and its Vicinity*, 1819, 1820, 1829 (also included in Hales's book *Survey of Boston and its Vicinity*, 1821); one of the engravers for H.S. Tanner's *New and Elegant Universal Atlas*, 1833-1836 (published as a partwork, when complete it was given the title *A New Universal Atlas...*, 1836).

**Gillion**, D.J. Mining Locations Seine River, 1899.

**Gillisse**, Pieter. *Caerte Ende Afteijckeninge Waer de Oude Slaecke...*, 1626.

**Gillman**, Henry (1833-1915). Surveyor and draughtsman from Kinsale, Ireland, worked with the United States Corps of Topographical Engineers. Assistant with Geodetic Survey of the Great Lakes. Contributed to charts of harbours on Lakes Superior and Huron, 1855-1864; *Preliminary Chart of Copper Harbor*, 1864.

**Gillone**, John, I & II. (*fl.*1767-1813). Scottish surveyors, County Surveyor of Galloway,

Scotland. Produced manuscript maps of various parts of Scotland including *Kirkcudbright Parish*, 1794; plan of the vicinity of Newton Stewart, 1807; Dee River, 1808. **N.B.** These works cannot be attributed accurately to either one.
- Gillone, John I (*d. before* 1809).
- Gillone, John II (*fl. to* 1813).

Gillot, Charles. French engraver. Louis Adolphe Thiers's Atlas de l'Empire, 1875.

Gillot, Firmin (1820-1872). French printer credited with 'gillotage, zincographie ou, au début, Paniconographie'. *Postal District Map of London* issued with the Illustrated Times ('...engraved in relief by Mr. Gillot of Paris by a new process of his own invention'), 1857; *Ettling's map of the United States*, 1861 (prepared by Gillot as a supplement to *Illustrated London News*); *The Visitor's Map of London*, 1862 edition (first published 1851).

Gillray, James (1757-1815). Artist and satirist, created political cartoons and satirical maps. *A new Map of England & France: The French Invasion*, London, Hannah Humphrey 1793.

Gilly, David (1748-1808). Topographical surveyor of Berlin. *Karte des Königl. Preuss. Herzogthums...*, Berlin 1789 (engraved by D.F. Sotzmann); worked with F.B. Engelhardt on a map of Prussia entitled *Karte von den Königl. Preußischen Provinzen Pommerellen und dem Netzedistrict*, 1791-1795 (the map was unpublished, but was later used by Schroetter); *Süd-Preussen*, Berlin 1802-1803.

Gilman, E. Draughtsman. *Table Showing the Estimated Surface of the territories of the United States*, c.1848 (lithographed by P.S.? Duval).

Gilmore, Joseph. *City of Bath*, 1697, 1715.

Gilmour, —. *National Atlas*, 1852 (with Dean).

Gilpin, —. *Hydrographic Map of North America*, c.1860; *Map Illustrating the System of Parcs, the Domestic Relations of the Great Plains, the North American Andes, and the Pacific Maritime Front*, 1873.

Gilpin, *Colonel* George. *Plan of the Town of Alexandria in the District of Columbia*, J.V. Thomas 1799 (engraved in New York by T. Clark).

Gilpin, Robert, *RN*. Surveyor. *Kagosima Harbour* [Japan], Admiralty 1863.

Gilpin, Thomas. Map of the proposed Chesapeake and Delaware Canal, 1769, published 1821.

Gilpin, William (1813-1894). First territorial governor of Colorado (1861). *Colorado Territory... Central Gold Region*, 1862.

Gilsemans, Isaack. Draughtsman employed by the VOC (Dutch East India Company) 1634-1646. Worked in the Moluccas, 1630s; sailed as draughtsman with A. Tasman's voyage to Australia and New Zealand, 1642-1643.

Gilson, Robert (*fl*.1779-1796). Surveyor, made manuscript maps of parts of Lincolnshire and Yorkshire. Survey of the River Foss, 1792 MS.

Gimbernat y Grassot, Carlos de (1768-1834). Catalan geologist worked in Barcelona and Switzerland, specialising in geological maps of Switzerland and the Swiss Alps. *Mapa Geognostico de la Suiza*, 1806 (said to be the first geological map of Switzerland); and others. **Ref.** PARRA DEL RÍO, M.D. *Los 'Planos Geognósticos de los Alpes, la Suiza y el Tirol' de Carlos de Gimbernat* (Aranjuez 1993) [in Spanish].

Gineste, *Captain* — de. *Karte der Insel Thera oder Santorin*, 1848.

Gingell, —. Engraver of Bath. Country round Cheltenham, 1810; *Plan of the city of Bath*, 1830.

Ginji, K.A. *Map of Yokohama*, 1871.

Ginn Bros. Publishers of Boston. Mary L. Hall's *Our World, No.II. A second series of lessons in geography*, 1872, 1875.

Ginn & Co. Publishers of Boston and London. Classical Atlas, 1882, 1895; A.E. Frye's *Home & School Atlas*, Boston 1895, 1896; Frye's *Complete Geography*, Boston & London, 1895; Frye's *Grammar School Geography*, Boston & London, 1902.

Ginnaro, Bernardino. *Nuova Descrittione del Giappone...* [Japan], in *Saverio Orientale*, Naples 1641. **Ref.** HUBBARD, J.C. *Imago Mundi* **46** (1994) pp.87-91.

**Ginver**, Nicholas. *A new chart of the Island of Scilly*, appeared in different editions of J. Seller's *The English Pilot* from 1772.

**Ginville**, Vincent de. French engraver. *Les Isles Britaniques*, 1701 (in N. de Fer's *L'Atlas curieux...*); *Les Costes aux Environs de la Riviere de Misisipi*, 1701 (in *L'Atlas curieux*); *Les Environs de Milan*, 1702 (in *L'Atlas curieux* and De Fer's *Cartes nouvelles et particulieres pour la guerre d'Italie*); *Les Environs de la ville de Naples*, 1702 (in *Cartes nouvelles et particulieres pour la guerre d'Italie*).

**Gioane de Bo da Venecia**. Now believed to be a misreading of the place name Caput de Bona Ventura. **Ref.** McINTOSH, G.C. *Imago Mundi* 52 (2000) pp.158-162.

**Giöding**, O.I. Swedish surveyor. Kungsholmen eller Stockholms, 1754.

**Giolito**, Gabriele [et fratelli] (*c*.1510-1578). Publisher, engraver, print and bookseller of Ferrara, worked in Venice 'al segno della Fenice'. G. Gastaldi's *Il uero ritratto di tutta l'Alamagna*, 1552; Z. Lilius's *Breve Descrittione del Mondo*, 1552 (2nd edition of F. Baldelli's Italian translation of Lilius's *Orbis breviarum*); publisher for G.T. Scandianese, 1556; *...descritto la regione dil piamonte* (copper-plate), 1556 (from Gastaldi's map of 1555).

**Giordano**, F. *Carta geologica del San Gottardo*, *c*.1872.

**Giorgi**, — de. *Nieuwe Atlas van het Koningrijk der Nederlanden*, 1845.

**Giorgi**, Giovanni. See **Giovanni**, Giorgio.

**Giorgius**, L. See **Barbuda**, Luís Jorge de.

**Giovanni**, A. (*fl*.19th-century). Italian engraver.

**Giovanni**, Giorgio. Portolan maker of Venice. Manuscript chart, 1494 (Biblioteca Palatina, Parma).

**Giovanni da Carignano**. See **Carignano**, Giovanni Maura da.

**Giovanni da Moncalerio**. See **Moncalerio**, Giovanni da.

**Giovanni da Udine**. See **Johannes** Utinensis.

**Giovio** [Iovii; Jovius], Paolo [Paulus] (1483-1552). Physician, historian, cartographer and collector from Como. Bishop of Nocera Inferiore, 1528-1548, *d*. Florence. Description of Lake Como (with map), *c*.1518 MS, published by Giordano Ziletti as *Pauli Jovii Descriptio Larii Lacus* (with a woodcut map), 1559 (copied by Ortelius, 1570); entertained Gerasimov, the Russian Ambassador to the Pope (1525), from whom he obtained information for a book about Russia and a map entitled *Moschoviae Tabula*, 1525 (one known example, attributed to Vavassore, copied by Agnese, 1540, and Gastaldi, *c*.1548); *Descriptio Britanniae, Scotiae, Hyberniae, et Orchadum...* (description, thought to have been accompanied by a map), Venice, M. Tramezini 1548. **Ref.** KARROW, R.W. *Mapmakers of the sixteenth century and their maps* (1993) pp.266-274.

**Giraldes**, M.J. Reino de Portugal, 1820.

**Giraldi[s]**, G. See **Giroldi**, Giacomo.

**Giraldon**, —. French script engraver and globemaker. French edition of James Rennell's Hindoostan, 1800; military map of northern Italy, Paris 1800; for L.-C. de Freycinet, 1807-1816; P.F. Tardieu, 1809; maps for Malte-Brun's *Précis de la géographie universelle*, Paris, Prudhomme 1810 (maps by P. Lapie and J.B. Poirson, printed by Prudhomme, coloured by Madame Diot); Pierre Lapie's *Atlas classique et `universel de géographie ancienne et moderne*, 1812 (with V. Adam); globe 1815 (with V. Adam).

• **Giraldon-Bovinet**, — [& Co.] (*fl*.1821-1840). French engraver and publisher of 'rue Pavée-Saint-André N⁰·5', Paris, successor to, and son-in-law of, Edme Bovinet. Atlas of route maps of France, n.d.]; historical plans of Paris, 1823; for Louis Vivien de St. Martin, 1823-1834; *Plan Topographique du Cimetière de l'Est...*, Paris, Emler Frères 1824, 1828; *Carte de l'Empire Ottoman Comprenant les Possessions de la Porte en Europe, en Asie, et en Afrique...*, Paris, Giraldon Bovinet & Co 1825 (compiled by Noël and Vivien, engraved by Giraldon Bovinet, printed by E. Jourdan, also published in London by J. Wyld and Cary, in Mannheim by Artaria and Fontaine, in Milan by P. & J. Vallardi, in Amsterdam by Boulton & Son, and in Glasgow by Brunin);

États Unis, 1825; L. Vivien de St. Martin's *Carte générale de l'Afrique Meridionale...*, Paris, 1826. **N.B.** In Volume I of this work, Giraldon-Bovinet appears incorrectly as **Bovinet**, Giraldon.

**Giraldós**, M. Spanish letter engraver. I. Fernández Flórez's *Carta esférica de la costa de Galicia...* (2 maps), Madrid, Dirección de Hidrografía 1835-1836 (drawn and engraved by C. Noguera).

**Giraldus Cambrensis** [Gerald de Barri; Gerald of Wales; Gerallt Cymru] (1146-1223). Norman-Welsh cleric who wrote descriptive works on Wales (*Descriptio Kambriae* and *Itinerarium Kambriae*) and Ireland (*Topographia Hiberniae* and *Expugnatio Hibernica*), c.1200. He prepared manuscript maps to illustrate these works - Ireland (one copy extant in the National Library of Ireland); the British Isles (found in several copies of his works); Wales (no surviving copies known). **Ref.** O'Loughlin, T. 'An early thirteenth-century map in Dublin: A window into the world of Giraldus Cambrensis' in *Imago Mundi* 51 (1999).

Girard, —. Morondava [Madagascar], 1725.

Girard, —. French surveyor. Worked with others in central France on 5 sheets (10, 11, 31, 49, 50) of the Cassini *Carte de France*, c.1754-1768.

Girard, *veuve* — [vueve de F. Girard]. Printer and publisher of Avignon. French translation of José Gumilla's study of the Orinoco. *Histoire Naturelle, Civile et Geographique de l'Orinoque*, 1758.

Girard, —. French cartographer. *Carte du Département de la Seine Inférieure...*, Paris 1830 (with Carbonnie, map engraved by F.P. Michel, letters by Aubert Junior, printed by Chardon).

Girard, —. Geologist. *Geologische Karte der Rheinprovinz und Westfalen*, 1855-1865.

Girard, Captain —. *Carte militaire et historique de la France*, Paris, J. Corréard 1860 (with A.M. Perrot).

Girard, —. *Carte Géographique, Physique et Politique des Royaumes d'Espagne et Portugal*, Paris c.1868, c.1878 (with J.B.L. Charle).

**Girard**, Charles (d.1866). French explorer in Africa. He compiled a map in 1866, used to illustrate *Exploration au Nouveau Calabar*, 1867.

**Girard**, E. *Perse: carte physique, politique, et economique...* [Iran], Paris, E. Girard 1933.

**Girard**, Xavier. Engraver and bookseller at 'Quai des Augustins n$^{o.}$25, maison du citoyen Voland', Paris. 'Géographe des postes' and 'Géographe de l'administration' (1841). *Plan de Paris*, 1820, 1840 revised by Achin and published as *Plan geometral de la ville de Paris*, Paris, Longuet 1854 (plan engraved by Thuillier oncle et neveu, text by P. Rousset); *Atlas Portatif et Complet du Royaume de France...*, Paris, Dondey-Dupré Père et Fils 1823 (with Roger l'aîné); *Carte générale Pour servir à l'Assemblage des 86 Départemens composant l'Atlas portatif du Voyaguer dans le Royaume de France*, 1823 (plan engraved by Vicq, text by M$^{elle}$ Vicq); *Carte des services de la Poste aux lettres...*, 1841.

**Girardet**, Abraham Louis (1772-1820). Swiss draughtsman and engraver, worked in France, Germany and the Netherlands. R.J. Bollin's *Plan der stadt und Gegend von Bern...*, Bern, J.J. Burgdorfer 1808.

**Giraud**, C. *Carte des Environs de Nice, de Monaco et de Menton*, Nice c.1860; *Carte des Environs de Nice*, 1870.

**Giraud**, Don Carlos A. *Plano de la Plaza de La Coruña con el Arrabel de la Pescadería, la peninsula; y el Puerto* [N.W. Spain], 1777 MS.

**Giraud**, Stefano [Étienne]. Engraver. *Nuova Pianta di Napoli / Nouveau Plan de Naples*, Naples, Antonio Ermil e filio 1767; *La Grande Golfe de Naples* (30 sheets), 1771; *Nouveau Plan de Naples / Nuova Pianta di Napoli...*, Naples 1790.

**Giraud**, Victor (1858-1898). French explorer, undertook an independent expedition to follow in the footsteps of David Livingstone, 1882-1885. His studies of the lakes and rivers contributed to the more accurate mapping of East Africa. *Itinéraire de Dar es Salam aux Lacs Bangouéolo et Moéro par Victor Giraud... 1882-1884* [Tanzania], Paris 1885 (drawn by J.A.A. Hansen, with inset *Carte d'ensemble des*

*Voyages aux Grand Lacs de l'Afrique Méridionale par V. Giraud)*; *Les lacs de l'Afrique équatoriale*, 1890. **Ref.** BROC, N. *Dictionnaire illustré des explorateurs Français du XIXe siècle: l Afrique* (1988) pp.163-164.

**Girault**, P.A. (1791-1855). French geographer from Saint Forgeau. *Dictionnaire de géographie physique et politique*, 1826. **N.B.** Compare with E. **Girault de Saint-Fargeau**, below.

**Girault**, Simon. *Le globe terrestre* (woodcut map of the World in two hemispheres), published in *Globe du monde...*, Langres 1592, 1598.

**Girault de Saint-Fargeau**, Eusèbe (1799-1855). *Petit Atlas National des Departements de la France*, Paris, Delaunay 1828 (maps engraved by Kardt & Hacq); *Guide Pittoresque du voyaguer en France* (6 volumes), Paris, Firmin-Didot frères 1838.

**Girava** [Giriva], Jeronimo [Hieronymus]. *b.* Tarragona, *d.* Milan. Cosmographer to Emperor Charles V. Portolan chart, 1552 (attributed); *Typo De La Carta Cosmographica* (woodcut cordiform world map based on Caspar Vopel's), 1556 (published in Girava's *La cosmographia, y geographia*, Milan 1556, second edition, Venice 1570), facsimiles issued during 1870s, blocks extant in Majorca. **Ref.** SHIRLEY, R. *The mapping of the World* (1983) N° 101 pp.114-116.

**Gird**, Richard. *Official Map of the Territory of Arizona*, San Francisco, A. Gensoul 1865.

**Girdlestone**, F.B. *Khor Rabbajy*, Admiralty 1872.

**Girelli**, Pietro Paolo. Engraver of Rome. Giovanni Battista Cingolani's Campagna di Roma (6 sheets), 1704.

**Giriva**, H. See **Girava**, Jeronimo.

**Girod**, J. *Plan of Port of Spain and Suburbs*, 1902 (lithographed by Waterlow & Sons), revised and published Port of Spain, Muir Marshall & Co., London, Waterlow & Sons 1912.

**Giroldi** [Ciroldis; Giraldis; Ziraldis; Zeroldis; Ziredis; Ziroldis], Giacomo [Jacobus de] (*fl.*1422-1452). Portolan maker of Venice. Chart of the Mediterranean, 1422; Portolan, 1425 (attributed); Atlas (6 charts), 1426; Atlas (6 charts), 1443; Atlas (6 charts), 1446; Atlas (3 charts), 1452.

**Giron**, *Vicomte* de. See **Grenier**, Jacques-Raymond.

**Giron**, Juan Manuel. *Atlas o compendio Geographico del Globo Terestre* (14 maps), Madrid 1756.

**Girona**, *Lieutenant* P. Draughtsman for the Ionian Engineers, Corfu. *Cefalonia* [Ionian islands], Corfu 1815 MS.

**Gironei**, Giorgio. Italian engineer, produced coloured manuscript surveys in the Ionian Islands including: *Isola di Paxo*, Corfu 1813; *Island of Cefalonia*, 1814; *Island of Ithaca*, 1815.

**Girongi**, Pietro. Topographer and draughtsman of Naples, prepared some maps for *Portulano delle coste della penisola de Spagna*, 1824 and later.

**Gironière**, P. de la. See **La Gironière**.

**Giroux**, —. One of the engravers of *Carte topographique de la France* (12 sheets), Dépôt de la Guerre 1832-1847; engraved two maps on steel for V. Duruy's *Histoire de France*, Paris, Hachette 1862.

**Giroux**, Alphonse [& Co.]. 'Boulevard des Capucines', Paris. L. Sagansan's *Carte des états de l'Italie...*, 1859.

**Girtin**, John. Engraver, 8 Broad St., Golden Square, London. Vicinity of Naples, 1815.

**Girtin**, T. *Barnard Castle in the County of Durham*, London, R. Ackermann 1800 (engraved by J. Hill).

**Gisborne**, Lionel. Civil engineer, prepared a proposal for the Darien Gap Canal. Isthmus of Darien [Panama], 1854 MSS.

**Gisevius**, Bogdan. Lithographer of Berlin. *Die Erzlagerstätten Oberschlesiens... von K. Flegel*, 1913; *Übersichtskarte der Besitz-Verhältnisse im Oberschlesischen Steinkohlenrevier und den Nachbarbezirken von Prof. Dr. R. Michael*, 1913 (drawn by Breitkopf); *Plan der Stadt Danzig. Angefertigt im Jahre 1908...*, Danzig,

A.W. Kafemann 1919?; *Plan der Stadt Danzig. Angefertigt im Jahre 1919...*, 1919.

**Gisolfo.** See **Ghisolfi**, Francesco.

**Gisolfo**, Giacomo. Atlas, *c.*1550 MS. **N.B.** Compare with F. **Ghisolfi**, above.

**Gisolfo**, Giovanni. See **Ghisulfus**, Jean.

**Gissing**, *Captain* C.E. Teita [Kenya], 1884.

**Gist**, Christopher (*c.*1706-1759). Explorer, trader and surveyor from Maryland. Employed by the Ohio Company to survey the Ohio Valley including parts of Pennsylvania, Kentucky and West Virginia, from 1750. His notes and surveys were handed over to the Ohio Company. He acted as guide to George Washington's expedition to Fort Duquesne (1753-1754), and to General Braddock in 1755. *Draught of Genl. Braddocks Route Towards Fort Du Quesne* [Pittsburgh], *c.*1755.

**Gittings & Dunnington.** *Clarksburg, Harrison County, West Virginia*, 1867.

**Giuli**, Carlo. Carta mineralogia utile de la Toscana, 1843.

**Giulietti**, G.M. *Carta originale delle Regioni Galla, Somali, Adal tra il Golfo di Tegiura e Harar... 1879...*, Turin, Guido Cora 1881.

**Giunti**, —. Publisher in Florence. *La Sfera di Messer Giovanni Sacrobosco*, 1572 (translated by Vicenzo Dante de Rinaldi, revised by Egnatio Dante).

**Giunti**, Lucantonio [and Heirs]. Publisher of Venice. Olaus Magnus's *Historia Delle Genti et Della Natura Delle Cose Settentrionali*, 1565 edition Giovanni Battista Ramusio's *Delle navigationi*, 1583-1606; republished G. Botero's *Relationi Universali*, Venice 1640 edition (with a new set of copperplate maps). **Ref.** BURDEN, P. *The mapping of North America* (1996) p.329-330.

**Giuntini**, Francesco. Doctor of Theology. Translated the work of J. de Sacrobosco into Italian, published as *La Sfera Del Mondo*, 1582.

**Giustiniani** [Justinianus], Agostino [Augustinus] (1470-1536). Historian and orientalist from Genoa, Dominican bishop of Nebbio, Corsica. Map of Corsica, *c.*1535 MS (used by Gastaldi, *c.*1560, by Leandro Alberti, 1567, by Ortelius, 1573, and Quad, 1596). **Ref.** KARROW, R.W. *Mapmakers of the sixteenth century and their maps* (1993) pp.275-279; MEURER, P. *Fontes cartographici Orteliani...* (1991) pp.157-158 [in German].

**Giustiniani**, Francisco. Italian cartographer. *El Atlas Abreviado*, Lyon 1739.

**Giustiniani**, Marc Antonio. Printer in Venice. Credited with the printing of *Cümle Cihan Nümenesi* [heart shaped world map], *c.*1559 (map thought to be by Hacci Ahmed).

**Givry**, A.P. 'Ingénieur Hydrographe de la Marine'. Baie de Cadiz, 1807; *Carte Réduite de la Côte Occidentale d'Afrique*, 1817; charts of Brazil for the Supplément to *Neptune François*, 1822-1824.

**Gjessing**, S.C. (1812-1897). Norwegian artillery officer, official surveyor for the government. *Kart over Christians Amt* [Norway] (s sheets), Stockholm 1845; *Kart over Buskerud Amt...* [Norway] (2 sheets), Copenhagen 1854; *Stavanger Amt*, 1866; and others.

**Glade**, Carlos. *Plano de Buenos Ayres*, 1867.

**Gladwin**, George. Engraver. Geometrical landscape of London and environs, showing the altitude of various buildings above the Thames, 1828 (surveyed by W. Moffat and F. Wood, aquatint by R. Havell junior).

**Glaeser.** See **Gläser**.

**Glareanus** [Gloria; Loriti; Loritus; Loritz], *Professor* Henricus [Heinrich] (1488-1563). Swiss humanist and geographer, born in Mollis [Glarus], professor of mathematics in Basle, Paris and Freiburg, where he died. Manuscript world and hemisphere maps (including two of the northern and southern hemispheres based on his own polar projection), *c.*1510-1520; *De Geographia*, Basle 1527, and editions to 1542 (with directions for the construction of globe gores). **Ref.** KRETSCHMER, DÖRFLINGER & WAWRIK *Lexikon zur Geschichte der Kartographie* (1986) vol.C/1, p.268 [in German]; HOHEISEL, K. *Geographers, Biobibliographical Studies* 5 (1981) pp.49-54; HEAWOOD, E. *Geographical Journal* 25 (1905) pp.647-654 (reproduced in *Acta Cartographica* 16 (1973) pp.209-216.

**Glas** [Glass], *Captain* George. Charts of El Rio harbour, Puerto de Naos and Puerto Cavallo, Canary Islands, *c*.1772. Used in *The West India Atlas*, London, Sayer & Bennett 1775, by William Faden in *The General Atlas*, 1778, and by Laurie & Whittle in *The Complete East India Pilot or Oriental Navigator*, 1800.

**Glas**, Gustav. *Deutsch-Tirol* 1855, *c*.1890; *Bayerischer Wald*, 1889.

**Glasbach**. See **Glassbach**, family.

**Glascott**, *Lieutenant* Adam G. *RN*. Route through Armenia, 1835 MS; Asia Minor, 1836; Southern Shore of the Black Sea, 1838 MS; *Entrance to the River Barima* [Guyana], 1841 (with R.H. Schomburgk); *Entrance to the River Waini or Guainia* [Guyana], 1841 (with R.H. Schomburgk); *Plan of the Port of Batoom in the Black Sea*, London, James Wyld *c*.1856; *Plan of Skutari*, 1856; and others.

**Glasener**, —. *Karte der Diocese Trier*, *c*.1869; *Trier und Coblenz*, *c*.1871.

**Gläser** [Glaeser], —. Geographer. *Thiergarten* [Berlin], 1822; *Gegend zwischen Berlin und Potsdam*, 1840.

**Gläser** [Glaeser], A. *Plan der Stadt Mainz*, 1852.

**Glaser** [Glaeser; Gläser], *Doctor* Carl. Deutschland, 1835; *Vollständiger Atlas über alle Theile der Erde*, Darmstadt, L. Pabst 1836 and later editions; *Atlas über alle Theile der Erde*, Mannheim, Hoff 1841-1842 with later editions including *Geographischer Hand-Atlas*, Stuttgart 1863; *Topisch-physikalischer Atlas*, Mannheim, Hoff 1844, Stuttgart, Krais und Hoffmann 1855; *Schul-Atlas*, 1846 with later editions; *Kreis Waldenburg*, 1892.

**Glaser**, Eduard (1855-1908). Undertook four trips to the Yemen, 1882-1894, resulting surveys were published in *Petermanns Geograph. Mitteilungen*, 1886.

**Gläser** [Glaeser], Friedrich Gottlob (1749-1804). German geographer and cartographer. *...Graffschaft Henneberg...*, 1774 (first use of colour to distinguish geological formations).

**Glaser**, Hans Wolff (*d*.1573). Engraver, printer and illuminator of 'die Judengasse' Nuremberg. *Zeitung auss der Insel Malta...*, 1565. **Ref**. GANADO, A. & AGIUS-VADALÀ, M. *A study in depth of 143 maps representing the Great Siege of Malta 1565* (1994) pp.262-266.

**Glaser**, Johann Heinrich. Publisher. *Vallistelina*, 1625.

**Glass**, Charles E. *Road guide to the Gold Fields*, *c*.1855.

**Glass**, George. See **Glas**, *Captain* George.

**Glass**. See **Judd** & **Glass**.

**Glassbach**, family. Prussian engravers.
• **Glassbach**, Christian Benjamin (1724-1779). Engraver, *b*. Magdeburg, *d*. Berlin. *Regni Poloniae Magni Ducatus Lithuaniae*, 1771.
•• **Glassbach**, Carl Christian (*b*.1751). Engraver from Berlin, son of Christian B. Glassbach (above), worked with brother Benjamin (below). Military maps, 1782-1800; *The Ganges...*, 1785 (copied from J. Rennell map of 1780); *The Burrampooter...*, 1786; *Die Nordwest America* for George Forster's *Geschichte der Reisen, die seit Cook an der Nordwest- und Nordost-Küste von Amerika...*, Berlin 1791.
•• **Glassbach** [Glasbach], Benjamin (*b*.1757). Engraver of Berlin, worked with his brother Carl Christian (above). *Hindoostan*, 1785 (copied from J. Rennell map of 1782); 6 maps for George Forster's *Geschichte der Reisen, die seit Cook an der Nordwest- und Nordost-Küste von Amerika...*, Berlin 1791; D.F. Sotzmann's Prussia, 1802.

**Glasser**, Johann Friedrich. Engraver in Augsburg. *Inter-valla Viaeque publicae Electoratus Saxoniae*, *c*.1760; *Episcopatus Augustanus*, *c*.1762.

**Glaudemans**, G.J. *Eenvoudige Atlas der geheele Wereld*, Den Haag, Johannes Ykema 1924 (with J.A. De Lint and J.S. Verburg, lithographed byM.V.J. Smulders & Co.).

**Glazier**, *Captain* —. *The Father of Waters* [Mississippi], 1881; *Lake Itasca*, 1881; *Lake Glazier*, 1894.

**Gledhill**, *Lieutenant Governor* Samuel. Chart of Placentia [Newfoundland], *c*.1740.

**Glegg**, John (1779/80-1845). Estate surveyor of Norfolk. *Plan of the Parish of Longham in the County of Norfolk*, 1816 MS.

**Glegg**, *Captain* J.B. Plan of Quebec and adjacent Country Shewing the principal Encampments & Works of the British & French Armies during the Siege by General Wolfe, 1759 MSS (with John Melish), reducedversion published 1813 (drawn by JOhn Melish, engraved by H.S. Tanner).

**Gliemann**, Johann Georg Theodor (1793-1828). Danish topographer, working in Copenhagen. *Post Kort over Danmark*, 1820; *Chart von Island*, 1824; *Amptskortatlas over Danmark*, 1824-1829.

**Glietsch**, G. *Communications télégraphiques... du monde*, 1871.

**Glimmerveen**, D.J. Dutch cartographer. Alblasserwaard, 1840.

**Glin**, R. See **Glynne**, Richard.

**Glindemann**, C. German cartographer. *Karte der Eisenbahnen Deutschlands*, 1848.

**Glinski**, G.V. Edited an atlas of Russia in Asia (2 volumes), St. Petersburg 1914.

**Globe Doré**. See **Paris Gilt Globe**.

**Globe Vert**. Anonymous globe dated 1515, now in the Bibliothèque Nationale, Paris.

**Globičz Búcina**, Samuel (*c*.1618-1693). Bohemian surveyor.

**Globuciarich**, J. See **Clobucciarich**, Johann.

**Glockendon** [Glogkendon; Glockenthon], father and son.
• **Glockendon**, Georg [Jorg] (1450-*c*.1514). Miniaturist, painter, engraver and publisher of Nuremberg. Woodcut Ptolemaic world map, *c*1490 (attributed); assisted Martin Behaim with globe, 1492; Erhard Etzlaub's regional map of Nuremberg, 1492; Etzlaub's *Rom-weg*, 1500; *Das sein dy lantstrassen durch das Romisch reych*, 1501.
•• **Glockendon**, Albrecht (*d*.1545). Son of Georg. Republished Etzlaub's *Romisch reych*, 1532.

**Glocker**, Adolf. *Karte von Korea*, *c*.l887.

**Glocksperger**, Jan (1678-1771). Czech surveyor and cartographer.

**Glöding**, O.I. Swedish surveyor. Chartan öfwer Kungsholmen eller Stockholms Wästre Malm, 1754.

**Glogkendon**, G. See **Glockendon**, Georg.

**Gloria**, H.L. de. See **Glareanus**, Henricus.

**Glot**, —. Engraver of Thomas Yeakell & William Gardner's map of Goodwood & Halnaker, *c*.1770; thought to be the engraver of sheets 1 & 4 of Yeakell and Gardner's incomplete map of Sussex, 1778-1783. **N.B.** Compare with C.B. **Glot**, below.

**Glot**, C.B. (*fl*.1777-1806). Engraver of Eustache Hérisson's *Plan de la Ville de Genève*, 1777; engraver (with E. Voysard) of Brion de la Tour's *Nouvelle Carte de la France*, Paris, J. Esnauts et M. Rapilly 1778; *Nouveau plan routier de la ville et faubourgs de Paris*, Paris, Esnauts et Rapilly 1778 (with E. Voysard); E. Hérisson's *Atlas de poche de géographie universelle*, 1799; engraver for Hérisson's maps in *Atlas du Dictionnaire de Géographie*, Paris, Desray 1809; *Nouvel atlas de la Bible*, Paris, Desray 1809; E. Hérisson's *L'Ireland*, Desray 1810 (from an unknown atlas).

**Glotsch**, Ludwig Christoph (*d*.1719). Engraver of Nuremberg. *Ducatis Tirolensis*, *c*.1725.

**Glover**, Eli S. American draughtsman and publisher. Produced many bird's-eye views of cities in North America, publishing some (but not all) of his own as well as those of others. Many of them were printed by A.L. Bancroft & Co. of San Francisco or Strobridge & Co. of Cincinnati. Published A. Ruger's *Grand Haven* [Michigan], 1868; A. Ruger's *Romeo* [Michigan], 1868 (and others of Michigan cities by Ruger); *London, Ontario*, 1872; *Central City & Blackhawk* [Colorado] 1873; *Bird's eye view of Logan City, Utah Territory*, Salt Lake City, E.S. Glover 1875 (lithographed by A.L. Bancroft & Co., San Francisco); *Bird's-eye view of Brigham City and Great Salt Lake, Utah*, Salt Lake City, E.S. Glover 1875 (lithographed by Strobridge & Co., Cincinnati); *Helena* [Montana],

Helena, C.K. Wells 1875 (printed by Bancroft); *Healdsburg and Russian River Valley* [California], Jordan Bros. 1876 (printed by A.L. Bancroft & Co.); *San Diego*, San Diego, Echneider & Kueppers 1876; *Walla Walla* [Washington], Walla Walla, Everts & Able 1876 (and other town views in Washington); *Salem* [Oregon], Salem, F.A. Smith 1876); *Los Angeles*, Los Angeles, Brooklyn Land & Building Co. 1877; *Los Angeles, Santa Monica & Wilmington*, Los Angeles, E.S. Glover 1877; *Victoria, British Columbia*, Victoria, M.W. Waitt & Co. 1878 (printed by A.L. Bancroft); *Port Townsend* [Washington], Portland, E.S. Glover 1878; *Portland* [Oregon], 1879 (printed by A.L. Bancroft & Co.); *Olympia, East Olympia & Tumwater* [Washington], Portland, Oregon, E.S. Glover 1879; *Anniston* [Alabama], 1888; *Muskegon* [Michigan], A.J. Little 1889; *Port Arthur* [Texas], Port Arthur Board of Trade 1912; and many others.

**Glover**, George. Architect and surveyor of Huntingdon. Produced maps of parts of Huntingdonshire and Suffolk. Plans for St. John's Hospital almshouses, Huntingdon, 1840 MSS.

**Glover**, *Lieutenant* (later *Sir*) John Hawley (1829-1885). Served in Royal Navy 1841-1877, later governor of Newfoundland, also Leeward Islands. *River Kwara* [Nigeria], 1857-1859 published on 2 sheets, London 1860; *Lagos & Central Africa Railway...*, 1858-1859; *Bight of Benin*, 1858-1862; *Inland Water communication between Lagos, Badagry, Porto Novo and Epè*, 1858 published 1863; *Lagos River*, 1859 published as Admiralty Chart 1861, 1879; River Niger, 1863-1899; surveys used for *Gold Coast Colony. Sketch Plan of the Town & Island of Lagos*, Stanford's Geographical Establishment 1883.

**Glover**, Moses (*fl.c.*1605-1640). Painter and architect of Isleworth, Middlesex. Plan of Petworth House, 1615 (attributed); *Isleworth Hundred being the Mannor of Sion...*, 1635 MS, facsimile published London, E. Stanford 1876.

**Glover**, Robert (1544-1588). Somerset Herald (1571), *b*. Ashford, Kent. *Kent*, 1570 MS; Survey of Herewood Castle, Yorkshire, 1584.

**Glover**, Stephen (*d*.1869). Compiled and published *A New Map of the County of Derbyshire*, London *c*.1850.

**Gluck**, J.B. Topographic draughtsman for the United States Coast Survey. *Minots Ledge off Boston Harbour*, 1853 (with others); *Boston Harbour, Massachusetts*, 1857 (with others); *Patapsco River and the Approaches*, 1859 (with others).

**Gluck**, Johann Paul. *Deliciae topographicae Norimbergensis* [regional maps of Nuremberg], 1733.

**Glümer** [Gluemer], Bodo von. *Estados Unidos Mexicanos*, 1882.

**Gluss**, I.R. *Chart of the Entrances of Goerhee & Quex Deep*, 1787-1795, 1801.

**Glynn**, *Lieutenant* James (1801-1871). United States naval officer. *Beaufort Harbour North Carolina*, 1839 (drawn by H.C. Flagg, lithographed by P. Haas); *A Chart of the Entrance of Cape Fear River* (4 sheets), 1839 (lithographed by P. Haas); *Cape Fear River North Caroline...* (3 sheets), 1839 (lithographed by P. Haas); *Continuation of the Survey of Cape Fear River...* (2 sheets), 1839 (lithographed by P. Haas).

**Glynne** [Glin], Richard (1681-1755). Clock and mathematical instrument maker, freeman of the Clockmakers' Company, 1705, later traded as a book and map seller. From *c*.1712-*c*.1725 he worked in partnership with his mother-in-law Anne Lea (widow of Philip Lea) at 'the Atlas & Hercules in Cheapside...', and in Fleet Street from *c*.1720. Pair of 12-inch globes, 1712; with A. Lea republished Robert Morden and Philip Lea's *London Westminster & Southwark*, *c*.1716 (editions to *c*.1725); armillary planetarium, 1730.

**Gmelin**, Johann Georg (1709-1755). German explorer and botanist from Tübingen. Served in Russia at the Academy of Sciences in St. Petersburg (1727), joined Bering's expedition to Siberia in 1733. In addition to his botanical studies, he undertook a survey of the depth of permafrost. ...*Reise durch Sibirien* (4 parts), Göttingen, A. Vandenhoecks 1751-1752.

**Gmelin**, Samuel Gottlieb. Maps and plans of Russia, including Astrakhan, Baku, etc. 'Reise-Beschreibung' (3 volumes), 1774.

*Charles Edward Goad (1848-1910). (Taken from Fire Insurance Plans in the National Map Collection, Public Archives of Canada 1977).*

**Gmunden**, Johannes von (c.1385-1442). Mathematician and astronomer of the Vienna school. Compiled a table of European cities and places, which may have incorporated a manuscript map of Central Europe. Thought to have made a plan of Vienna, upon which the so-called *Albertinischer Plan* was based.

**Ref.** KRETSCHMER, DÖRFLINGER & WAWRIK *Lexikon zur Geschichte der Kartographie* (1986) p.10; BERNLEITHNER, E. *Imago Mundi* 25 (1971) pp.65-67.

**Gnoli**, Bartolomeo. Artist and architect of Ferrara. Ducato di Ferrara, 1645; Valli di Comacchio, 1650.

**Gnudi**, Filippi de. *Città di Bologna*, 1702.

**Goad**, Charles Edward (1848-1910). English railway engineer, went to Canada in 1848, where he did not pursue engineering, but became a representative for the Sanborn Company which published fire insurance plans. From 1875 Goad began to compile and publish his own large-scale plans of Canadian towns from a base in Montreal, and soon became the largest private mapmaker in Canada. By 1885 Charles Goad had returned to Britain and established a branch at 4 Finsbury Circus, London E.C. (1886); 53 New Broad Street, London (1887); then Hatfield, Hertfordshire. Goad's highly detailed plans included information on property holdings, building construction, use and contents, and, like Sanborn's, functioned as Fire Insurance Plans. Once leased to an insurance company the map volumes were regularly updated by a team of workers who over-pasted sections with new information onto the existing maps, some of them building up to many layers. Goad's business rapidly expanded to cover the cities and industrial areas of Britain, Denmark, France, Egypt, Turkey, Venezuela, Bermuda, Mexico and South Africa. The insurance plan of London for example consisted of a key map and some 400 large-scale plans in 12 volumes, 1886-1892. **Ref.** ROWLEY, G. *The Map Collector* 29 (1984), pp.14-19; ROWLEY, G. *British fire insurance plans* (1984); HAYWARD, R.J. *Fire insurance plans in the National Map Collection* (Ottawa 1977); HAYWARD, R.J. 'Chas Goad and fire insurance cartography', in *Proceedings* (of the Association of Canadian Map Libraries, Eighth Annual Conference, Toronto June 9-13, 1974) pp.51-72.
• **Chas. E. Goad Company** (*fl.* from 1910). Successor company to Charles Goad, run by his three sons. Copyrights and assets of the Canadian branch of the company were sold to The Underwriter's Survey Bureau in 1931. The Goad Company in Britain continued producing fire insurance plans at Hatfield until the 1960s, when it began to specialise in shopping centre plans.
•• **Experian**. Map publishers of Hatfield. Successors to Goad as producers of shopping centre and town centre plans.

**Goad**, Thomas W. District New Mexico, Fort Leavenworth 1875; Costilla Estate, 1887.

**Goalen**, *Lieutenant* W.N., *RN*. Surveyor aboard H.M.S. *Swallow*. Survey of the east coast of China, province of Shan Tung and Kyan Chau bay, *c*.1863 (with G. Stanley, J. Hall, A. Hamilton & H.R. Harris); *Port Adelaide*, 1875 published as Admiralty Chart 1876, 1897, 1900; *Murray River*, 1876 published as Admiralty Chart 1879; *Australia. South Coast*, 1876 Admiralty Chart 1879.

**Gobanz**, *Doctor* Josef. *Hypsometrische Karte der Steiermark*, *c*.1864.

**Gobaut**, —. Draughtsman at the Dépôt de la Guerre. *Vue de Sébastopol...*, Paris *c*.1855 (with Jung, printed by Kaeppelin).

**Gobert**, Martin. Publisher, 'au Palais en la Gallerie des Prisonniers', Paris. One of the publishers of Christophe Tassin's *Cartes generales des provinces de France et d'espagne...*, 1633; C. Tassin's *Cartes générales des royaumes et provinces de la Haute et Basse Allemagne*, 1633; *Plan et profiles de Toutes les Principales Villes et Lieux considerable de France*, 1634.

**Gobert**, Th. Engraver 'Rue St. Jacques 171', Paris. L.E. Desbuissons's *Océanie*, Paris, Dufour, Mulat & Boulanger 1858 (lettering and colour by Beaurain); for Alexandre Vuillemin, 1873.

**Gobille**, Gédéon. Map seller of Paris, 'dans l'isle du Palais sur le quai du grand cour de l'eau, qui regarde la Megisserie, a l'Ache Royalle'. Untitled set of 12 terrestrial globe gores and a companion set of celestial globe gores, *c*.1650 (possibly a reprinting of an earlier work).

**Gobille**, J., SJ. French Jesuit. Mappa-Mundi, 1677; *Uranographie seu globi coelestis mappa*, *c*.1677.

**Gobin**, [Jean-Baptiste?; I.B.]. Engraver of Paris, apprenticed to painter Jean-Baptiste Belleville. Specialised in the design and engraving of cartouches, including many for Gilles Robert de Vaugondy, 1743-1753; designed and engraved 18-inch celestial globe gores for Gilles & Didier Robert de Vaugondy, *c*.1750.

**Goche**, Barnabe. Plan of the town of Galway, 1583 MS.

**Gochet**, A.M. (*b*.1835). Belgian geographer.

**Gôczel**, S. *Gold Region Western Australia*, 1896.

**Godalles**, Thomas Bonaventure. *Plan géométral du Port et Fort de Plaisance et de ces habitations en la coste occidentale de l'Isle de Terre Neuve*, 1712 MS.

**Godard**, —. One of the engravers of *Carte topographique de la France* (12 sheets), Dépôt de la Guerre 1832-1847.

**Godbid**, A. Printer and publisher of London. John Adams's *Index Villaris*, 1680 (with J. Playford); John Seller's *Atlas Maritimus* (pocket atlas), 1682 (with J. Playford).

**Godbid**, Anne. Printer of Jonas Moore's *A New System Of The Mathematicks: Containing I. Arithmetick... II. Practical Geometry... III. Trigonometry... IV. Cosmography... V. Navigation... VI. The Doctrine Of The Sphere... VII. Astronomical Tables... VIII. A New Geography...*, London, Robert Scott 1681 (with J. Playford).

**Godbid**, William. Printer and publisher of Thomas Philipott's *Villare Cantianum*, 1659 (with Philip Symonson's *A New Description of Kent*); printed the text for J. Speed's *The Theatre of the Empire of Great Britaine*, London, T. Bassett & R. Chiswell 1676.

**Goddard**, Charles. Engraver of 173 Pitt Street, Sydney, New South Wales (1858-1859). *Map of the county of Cumberland*, c.1851.

**Goddard**, George Henry (1817-1906). Architect and surveyor from Bristol, England, emigrated to California and worked as a government surveyor at Sacramento. He moved to San Francisco in 1862. Publisher of *Sonora*, 1852 (printed by Pollard & Brittons); surveyor for *Britton & Rey's Map of the State of California...*, San Francisco, Britton & Rey 1857 (engraved by H. Steinegger); panoramic map of San Francisco, Snow & May 1868 (printed by Britton & Rey); *San Francisco & Surrounding Country*, Snow & May 1876 (printed by Britton & Rey); and others.

**Goddard**, John *the elder*. Engraver. Some maps for Thomas Fuller's *Pisgah-Sight of Palestine*, 1650 and later; J. Farrer's *A Mapp of Virginia* in Edward Bland's *Discovery of New Brittaine*, London, J. Stephenson 1651; *Asia Descriptio Nova*, 1652 (for Heylyn's *Cosmographie*).

**Goddard**, John *the younger*. Engraver. Recut the Norfolk plate for Roger Rea's edition of John Speed's *Theatre of the Empire of Great Britaine*, 1665; map of the Cambridgeshire Fens (16 sheets), [n.d.].

**Goddard**, Thomas. Bookseller in Norwich. *A New and accurate Map of the County of Norfolk*, 1731 (with W. Chase) and 1740 (with R. Goodman).

**Goddard**, W. Hampshire estate surveyor. Bonham Estate, Hampshire, 1819 MS.

**Göde**, N. See **Goede**, Nikolaus.

**Godefin**, —. *Plan de la ville de St. Étienne*, St. Etienne 1866.

**Godefroy**, — (fl.1784). Publisher, Rue des Francs Bourgeois, Paris.

**Godet** [Godey], — *le jeune* (fl.1810-1825). Print and mapseller of 'Quai Voltaire N$^{o.}$21 près de la rue de Beaune', Paris (1810-1812). Designed and engraved *Nouveau Plan Routier de la Ville et Faubourgs de Paris*, 1825.

**Godfray**, Hugh. *Map of the Island of Jersey* (2 sheets), London 1849 (incorporated the surveys of E. Le Gros); Isle of Wight, 1849; *A Chart of South Latitudes*, London, J.D. Potter 1858 (drawn and engraved by E.J. Powell).

**Godfrey**, Jonathan (fl.1630-1658). Estate surveyor, produced manuscript maps of parts of Berkshire, Hampshire and Oxfordshire. *August 1654. Dinington Parke com Berks...*, 1654 MS.

**Godiche**, Andreas Hartvig (1714-1769). Printer and publisher of Copenhagen. *Dansk Atlas* (7 volumes), 1763-1781; *Den Kongal. Residentz Stad Kiöbenhafn í Grundtegning 1764*, published 1766; *Tabula generalis Jutiae Septentrionalis...*, 1767.
• **Godiche**, A.H. [Heirs]. Publishers. Sleswig, 1781 (engraved by J.G. Fridrich).

**Godin**, H.J. Engraver. C. Le Comte's *Plan de Spa... 1780* [Belgium], Liège, F.J. Desoer 1780.

**Godin**, Louis. French astronomer. Worked with P. Bouguer, C.M. de la Condamine and others on a survey of the arc of the meridian in Peru from 1735.

**Godinho,** Manoel *SJ* (1632-1712). Babylon, *c*.1660.

**Godinho de [H]erédia,** Manoel. See **Erédia,** Manuel Godinho de.

**Godman,** T. English surveyor. *Plan of the Site of the City of Verulam,* 1814; *Plan of the Town of St. Alban, Herts,* 1822.

**Godoy,** Alejandro. *Plano de la frontera del Salvador,* 1898; *Carta de la red telegráfica de la República de Guatemala,* 1898.

**Godreccius** [Godretius], W. See **Grodecki,** Wacław.

**Godshalk,** S.K. Surveyor. *Cayuga County, New York,* Philadelphia, S. Geil 1853 (with S. Geil and F. Gifford); *Ashtabula County, Ohio,* 1856; assisted S. Geil on surveys for *Cattaraugus County, New York,* 1856.

**Godson,** Richard. Surveyor. *A Plan of the proposed Line of navigable canal from the Warwick and Braunston Canal at the Foss Road in... Offchurch to the Oxford Canal in... Napton in the County of Warwick,* surveyed 1795 (engraved by B. Baker).

**Godson,** William. Draughtsman. *A new and correct Map of the World,* G. Willdey *c*.1715, T. Jefferys and W. Herbert *c*.1750.

**Godson,** William (*fl*.1717/32-*d.c*.1766). English surveyor. *Mannour of King-Sumborne,* 1734 MS; *An Accurate Survey of Odiam Park,* 1739 MS; *...City of Winchester...* (4 sheets), 1750 (engraved by R. Benning).

**Godunov,** Fëdor Borisovich (1589-1605). Russian cartographer. Map of Russia, used by Hessel Gerritsz. 1612.

**Godunov,** Pyotr [Peter] Ivanovich (*d*.1670). Governor of Siberia. *Tabula Russiae,* 1614; *Siberia,* 1667 (first map of the country).

**Godunov,** Simon. Russian mapmaker. Engraved map of Siberia published in a Bible, 1663, revised edition 1672; *Siberia,* 1667 MS (with Ulyan Remezov).

**Godwin,** C. Publisher of Bath. *A new and correct Plan of the City of Bath...,* 1810, 1816, 1825 (survey by Benjamin Donn).

**Godwin,** J. *Passage of the Douro, 12th May 1809* [Iberian Peninsula], *c*.1810; *Maps & plans of the operations, movements, battles & sieges of the British Army, during the campaigns in Spain, Portugal and the South of France, from 1808-1814,* London, J. Wyld *c*.1815.

**Godwin,** M.J. [& Co.]. Publishers of 41 Skinner St., London; later 195 Strand (1822). *A Guide through London...,* 1821, 1823 (engraved by S. Hall); W. Ebden's *The Pedestrian's Companion Fifteen Miles round London,* 1822 (engraved by S. Hall), re-issued by E. Grattan *c*.1824.

**Godwin,** R.H. *Part of the West Khási Hills* [Assam, India], Calcutta, Surveyor General's Office 1868.

**Godwin-Austen,** Robert Alfred Cloyne (1808-1884). Geologist. *On the possible Extension of the coal-measures beneath the South-Eastern part of England,* 1854; *A Physical and Geological Map of England and Wales,* 1865; *Map to illustrate the evidence in support of the continuity of productive Coal measures beneath the S.E. counties of England,* 1871.

**Goede** [Göde], Nikolaus (1561-1633). Pastor and geographer.

**Goedecke,** John Frederick. Draughtsman. *Annapolis Harbour Roads,* 1818 MS (surveyed by Jonathan Sherburne), used as an inset on *This Survey of the River Patapsco and part of Chesapeake Bay...,* Baltimore, F. Lucas Jr. 1819 (engraved by Cone & Freeman).

**Goedesbergh,** Gerrit van (*fl*.1651-1669. Publisher '...by de nieuwe-brugh inde Delfse Bybel', Amsterdam. Louis Vlasbloem's *Nieuwe Lees-Caert,* 1664 edition; *Kort Begrijp van de Nieuwe Lichtende Columne ofte Zee-Spieghel...,* *c*.1665.

**Goedesbergh,** Theodore van. *Atlas... Totius Orbis Tabulae* (3 volumes), 1646-1693 (composite atlas with differently dated maps by F. de Wit, H. Jaillot, N. Visscher, the Blaeu family, N. Sanson and others).

**Goedsche,** Bruno. Publisher of Chemnitz and Scheenberg. Plan of *London, c*.1838 (lithographed by H. Lehmann, printed Goedsche & Steinmetz, Meissen).

**Goedsche & Steinmetz.** See **Goedsche**, Bruno.

**Goeje,** M.J. de (b.1836). Dutch geographer and orientalist, translator of works of Arab geographers.

**Goepel,** Paul. Draughtsman for W.L. NIcholson's *Post Route Map of the States of Ohio and Indiana...*, 1870 (with A. Kilp, engraved by D. McLelland).

**Goerck,** Casimir Theodore (d.1798). American surveyor of New York. *A Map of the Ground & Different Routs from Newark to Paulas Hook* [New Jersey], published in *New York Magazine*, 1791; worked with J.F. Mangin on a large scale survey of the city, 1797-1798, it was completed by Mangin after Goerck's death, and published as: *A Plan and Regulation of the City of New York* (4 sheets), 1803 (engraved on copper by Peter Maverick). **Ref.** COHEN, P.E. & AUGUSTYN, R.T. *Manhattan in maps 1527-1995* (1997) pp.96-98.

**Goeree,** family.
• **Goeree,** Willem (1635-c.1706). Printer and publisher in Middelburg (1666-1677), then Amsterdam. Son-in-law of Johannes Janssonius van Waesberge. *Historische Landbeschryvinge van Groot-Brittanjen*, Middelburg 1666 (republished from a work first published by Valckenier, 1661); for Johannes van Loon, 1668; *Atlas ofte de geheele weerelt*, 1677.
•• **Goeree** [Goere], Jan (1670-1731). Dutch designer, artist, engraver and poet, son of Willem (above). Worked with A.H. Jaillot on *Hainaut* used by Nicolas de Fer in *Atlas royal...*, 1695, 1699; *Le Palatinat et electorat du Rhein*, Paris, A.H. Jaillot 1695; maps for Pieter van der Aa's *Le Nouveau Theatre du Monde*, 1713; Japan, 1715; *Nova et Exacta Chorographia Latii sive Territorii Romani*, Amsterdam c.1720; maps in Pieter van der Aa's *La Galérie Agréable Du Monde*, 1729.
•• **Goeree,** David. Son of Willem, above. Worked with the younger Willem.
•• **Goeree,** Willem. Son of Willem, above. Collaborated with his brother as publishers Willem en David Goeree.
•• **Willem en David Goeree** (fl.1711). Brothers David and Willem Goeree working together as publishers in Amsterdam. Cornelis le Bruyn's *Cornelis de Bruins Reizen over Moskovie, door Persie en Indie*, 1711.

**Goering,** Wilhelm. Cartographer. *Topogr. Karte v. Jerusalem und Umgebung*, Gütersloh, C. Bertelsmann 1929 (drawn by Walter Haucke to accompany Gustav Dalaman's *Jerusalem u. sein Gelände*).

**Goerög,** Demeter. See **Görög**, Demeter.

**Goerringer** [Görringer], —. Historische Atlas, c.1840.

**Goës,** Benedict [Bento] SJ (1562-1607). Portuguese lay brother with the Jesuit mission at Agra. Travelled overland from Agra round to the north of the Himalayas in search of Cathay. In 1605 he reached Suchow in China. Records of his travels were used as a source by Henry Yule in his *Cathay and the way thither...*, London c.1866 (this work included *Map of the Hindu Kush and the regions adjoining...*, the map, lithographed by Edward Weller, carries a credit to Goës but it was probably not compiled by him). **Ref.** WALDMAN, C. & WEXLER, A. *Who was who in world exploration* (1992) pp.285-286.

**Góes,** Damião de. Portuguese portolan maker and author. *Chronica da S. Principe D. João*, 1567.

**Goeschen,** Georg Joachim (1752-1828). Printer and publisher in Leipzig. *Atlas von Europa* (partwork), 1825-1830; *Atlas von Amerika*, 1830.

**Goessel,** Premier-Lieutenant — von. *Marsch-Routen-Karte für die Armee-Corps und Cavallerie-Divisionen der Oesterreichischen-Armee im Feldzuge 1866*, Berlin c.1867; *Marsch-Routen-Karte für die Divisionen der ersten, zweiten und Elb-Armee in Feldzuge 1866*, Berlin c.1867; *Marsch-Routen-Karte für die Divisionen resp. Brigaden der preussischen und süddeutschen Truppen im Main Feldzuge 1866*, Berlin c.1867 (3 above maps copied for *Campaign in Germany of MDCCCLXVI*, War Office c.1867); *Marsch-Routen-Karte für die Armee-Corps resp. Infanterie-u. Cavallerie-Divisionen der deutschen Armeen im Kriege gegen Frankreich 1870/71* (4 sheets), 1873.

**Goethals.** See **Algoet**, Liévin.

**Goethe,** Johann Wolfgang von (1749-1832). German writer, philosopher and thinker. *Die Höhen der alten und neuen Welt bildlich verglichen* (hypsometric map), Weimar 1813.

**Goethem**, G. van. *Polder van Oorderen*, 1723.

**Goetz** [Götz], —. *Plan de Moscou / Plan von Moskau*, Munich 1812 (with Lieutenant Dietrich).

**Goetz**, Lieutenant —. *Maroc au 100,000e. Environs de Fes*, Paris 1912.

**Goetz** [Götz; Goetzio], Andreas (1698-1780). German cartographer and philologist. Geographia antiqua, Nuremberg 1729; *Kurze Einleitung zur Alten Geographie* (10 maps), Nuremberg, J.C. Weigel 1729.

**Goetz**, Richard. *Karte der Umgegend von Aurich*, Aurich, D. Friemann 1901.

**Goetz**, T. Engraver for I.A. Fessler's *Die Gesichte der Ungern und ihre Landsassen...* [Hungary], 1825.

**Goetze** [Götze], A. Frederick. Geographer at Weimar Geographisches Institut. Maps for Adam Christian Gaspari, 1804-1811; revised Franz Ludwig Güssefeld's North America, 1812.

**Goetze** [Götze], Ferdinand. Naples, Weimar 1801; *Charte von Insel Corsica*, Weimar 1804.

**Goetzio**, A. See **Goetz**, Andreas.

**Goffart**, S. *Colonia Agrippina...* [Cologne], Cologne 1753 (drawn by J. Schott, engraved by Löffler Junior).

**Goffe**, J. Surveyor in Jamaica. Estates on Jamaica, 1688 MSS.

**Gogeard**, —. *Plan de la ville de Rouen*, Rouen 1894.

**Goggins**, Joseph. Political cartoon maps of Franco-Prussian War, 1870-1871, including *Novel Carte of Europe designed for 1870*, Dublin 1870 (and other editions in Danish, German and Swedish).

**Gogh**, *Luitenant ter Zee* J. van. *Kaart van het Vaarwater benoorden Makasser* [Macassar, Indonesia], 1849 published Amsterdam 1851.

**Goghe**, John. Irish cartographer. *Hibernia, insula non procul ab Anglia vulgare Hirlandia vocata*, 1567 MS.

**Gohlert**, Vincent (1823-1899). Austrian geographer and statistician.

**Going**, A.S. *Map showing mineral locations on Texada Island, Manaimo Mining District* [Canada], Victoria, B.C. 1897.

**Going**, Philip, *Master, RN. Port Davey, Tasmania*, 1850 published as Admiralty Chart 1852.

**Goizet**, F. *Plan de la Ville de Dijon*, Dijon, Damidot Frères 1900.

**Gold**, Joyce. Printer and publisher of 103 Shoe Lane, Fleet St., London, also the Naval Chronicle Office. The voyage of the young Anarcharsis [abbé Barthelemy], 1806 edition (first published Barbié du Bocage, 1788); maps in *Naval Chronicle*, 1812-1817 e.g. *Saldanha Bay*, 1812 and *Cape of Good Hope*, 1812; Robert Rowe's *English Atlas*, 1816; James Callander's *Chart of Infanta or Broad River*, 1817 (engraved by R. Rowe).

**Goldbach**, Christian Friedrich (1763-1811). German astronomer. *Neuester Himmels-Atlas*, Weimar 1799, 1803; *A Map of Africa for C.F. Damberger's Travels*, London, Longman & Rees 1800 (engraved by S.J. Neele, published in Damberger's *Travels in the Interior of Africa...* - a hoax travel book).

**Goldfrap**, *Ensign* John George. Manuscript plans for Magdelain &c., 1766; for part of the Province of Quebec, 1767.

**Goldhar**, Itzhak. Jewish geographer in Jerusalem. *Adamath Kodesh* [Holy Country] (7 maps), Jerusalem 1907 and later (maps drawn by Abraham Ganon; lithographed by Eizik Zvi Shild).

**Goldie**, A. New Guinea, 1878.

**Golding**, John (*fl.*1817-1820). Estate surveyor of Cambridgeshire and Essex. Finchingfield, 1820 MS.

**Goldingham**, John. *Survey of Pulicat Shoals* [S.E. India?], 1792, published Dalrymple 1794.

**Goldman**, B.F. *Full history of the war at a glance*, Cleveland, B.J. Golman c.1917

**Goldmann,** Charles Sidney. Witwatersrand, 1895; *Atlas of the Witwatersrand and other Goldfields in the South African Republic*, London, E. Stanford 1899 (with Baron A. von Mattzan).

**Goldschmidt,** A. The Berlin publisher of 'Grieben's guide books', which consisted of more than 160 volumes with maps and plans of Europe, 1909-1913 (they were also published in London by Williams & Norgate) including: *The Rhine a practical guide...*, 1910-1911; *Munich and the royal castles of Bavaria*, 1910-1911, 1913; *Nüremberg and Rothenburg on the Tauber*, 1910-1911; *Holland: a practical guide...*, 1910-1911; *Norway and Copenhagen*, 1910-1911; *Naples and environs, Mount Vesuvius, Pompeii, Sorrento...*, 1913; and many others.

**Goldschmidt,** Martinus Martini. *Statt Lucern*, 1597.

**Goldsmid,** *Sir* Frederick John. Engineer. Manuscript maps of the coast of Pakistan, c.1863-1865; *A Map of the Telegraphs to India*, London, Stanford 1874.

**Gold-Smith,** Eric Charles. *Tuhua or Mayor Id. Bay of Plenty* [New Zealand], 1884 MS.

**Goldsmith,** *Reverend* J. (pseud). See **Phillips,** *Sir* Richard.

**Goldthwait,** G.H. Engraver and publisher based in Boston. *Miniature County Map of the United States*, 1842.

**Goldthwait,** J.H. American draughtsman and engraver. *Map of Connecticut*, c.1838; *Massachusetts*, 1838; *Railroad map of New England & eastern New York*, 1849; *Goldthwait's map of the United States...*, 1861; insets on Colton's new Railroad map of the United States & Canada, 1861; *Pacific States*, 1865.

**Goldthwaite,** William W. Chicago globemaker. Collapsible silk globe, 1898, 1902; *Goldthwaite's Handy Hemispheres Map Globe of the World*, 1899. **Ref.** WARNER, D. *Rittenhouse* Vol.2 N°2 pp.63-64; *ibid.* Vol.2 N°2 pp.91-92.

**Goldtschmide,** Matthias. See **Petersen,** Matthias.

**Goldzweig,** Jacob. *Biblical Map of the Holy Land*, c.1893.

**Golemis,** And. *Kydoniai (Ayvali) kai Moskhonessoi (Eptanessoi)*, Athens 1925 (with Al. Khrysanthis).

**Golenishev-Kutuzov,** Login Ivanovich. Commissioned a series of maritime atlases for the Russian navy, the first volume of which was entitled *Morskoj atlas dlja plavanija iz Baltijskogo morja k Anglijskomu kanalu...* (23 sheets), 1798.

**Golescu,** Iordache (1768-1848). Romanian cartographer.

**Goliath,** Cornelis. See **Golyath,** Cornelis.

**Goll,** Johann Jakob (1809-1861). Cartographer and engraver of Zurich and Geneva, also for the Swiss Topographical Bureau, 1837-1860. Worked as a draughtsman and engraver with I.C. Wolfsberger on G.H. Dufour's *Carte Topographique du Canton de Genève levée par ordre du Gouvernement dans les années 1837 et 1838* (4 sheets), c.1838 (engraved by Bressanini); *Carte Physique de la Suisse*, 1850; *Thurgau*, c.1870.

**Golobardes,** Juan Bautista. Spanish naval officer. *Cordillera de los Pirineos*, 1817 MS; *Las 5 provincias maritimas de Cataluña*, 1828 MS.

**Golovnin,** Vasili Mikhailovich (1776-1831). Russian naval officer, also served with the British Navy (1801-1807). His numerous voyages included surveys of the coasts of Alaska and Kamchatka (1807-1810; 1817-1819), surveys of the Kuril Islands [between Kamchatka and Hokkaido] (1811), and a circumnavigation of the globe (1817). *Duché de Schleswig*, 1813; *Kurile Islands*, 1818; *Puteshestviye volkreg svieta po poreleniyu gosudara imperazhorskoye... Kamchatkie 1817-1819*, St Petersburg 1822. **Ref.** WALDMAN, C. & WEXLER, A. *Who was who in world exploration* (1992) p.286.

**Gollmes,** H. Engraver. Moore, Wilstach, Keys & Co.'s *World Chart*, Cincinnati 1850.

**Gollner,** *Colonel* E.G. Architect and civil engineer. Texas, Dallas 1876.

Hendrick Goltzius (1558-1617). (By courtesy of Rodney Shirley)

**Gollowin,** F. von. *Herzogthum Schleswig*, 1806.

**Goltz,** Colmar von der (1843-1916). German field-marshal, served in Turkey. *Karte der Eisenbahn von Salonik nach Monastir*, Berlin 1894; *Umgegend von Constantinopel*, 1897.

**Goltz,** Conrad. See **Goltzius**, brothers.

**Goltz,** Leonhard von der (1815-1901). *Provinz Pommern*, 1851.

**Goltzius,** brothers. Dutch engravers.
• **Goltzius,** Hendrick [Henricus] (1558-1617). Dutch artist and engraver in Haarlem. Engraved a portrait of G. Mercator, 1574, used by Mercator in his edition of Ptolemy's *Geographia*, 1584.
• **Goltzius** [Goltz], Conrad. Younger brother of Hendrick. Worked as an engraver in Cologne with Matthias Quad until 1596, when he left the city. *Angliae et Hyberniae compend: descriptio*, Peter Overadt, c.1594 (after Jodocus Hondius); *Germania*, c.1595 (copied from the work of C. Sgrooten); Map of France, C. Goltzius 1595 (after Hondius).

**Golyath** [Goliath; Golijath], Cornelis Bastiensz (d.1667/8). Surveyor and cartographer from Middelburg, worked as a clerk and cartographer for the Dutch West India Company in Brazil from 1634. Commander of the colony of Essequebo, 1657-1661. Portuguese fortresses in the Bahia de Todos los Sanctos, 1638; map of Dutch Brazil, c.1638-1640 MS; map of Recife and Mauritsstad, 1643 published C.J. Visscher 1648 (engraved by Pieter Hendriksz Schut); Siege Olinda de Pharnambuco, Nicolaes Visscher 1648.

**Gómara,** L. de. See **López de Gómara,** Francisco.

**Gombault,** —. One of the engravers for *Carte topographique de la France* (12 sheets), Dépôt de la Guerre 1832-1847.

**Gombault,** Louis. *Villehardouin*, c.1860.

**Gomboust,** Jacques (d.c.1665/1668). Engraver and 'Ingénieur du Roi'. Plan of Paris, 1652 (with Pierre Petit); collaborated with Caspar Merian on *Topographiae Galliae*, 1655-1661; Plan of Rouen, 1665.

**Gomes,** Estevão (*fl*.16-century). Portuguese navigator and cartographer in the service of Spain. Explored the northeast coast of North America from Florida to Newfoundland, c.1525. The information he brought back was used by later mapmakers including G. Ramusio, Venice 1534 and J. Bellère, Antwerp 1554.

**Gomes,** M. Publisher of Lisbon. *Africa Meridional. Mappa dos limites Portuguezes conforme aos ultimos Tratados...*, 1891

**Gomes,** M.A. See **Gómez,** Miguel Antonio.

**Gomez,** *Don* —. Cuba, William Heather 1809.

**Gómez** [Gomes], *Capitán* Miguel Antonio. 'Ingeniero ordinario'. *Plano Ychnographico de la plaza de Manila*, 1762, 1763 MSS; *Plano y perfil del fuerte nombrado nuestra Señora de la Concepción.. en la Isla de Mindanao...* 1754, 1765; *Copia de las perspectivas de las Islas Babayanes... del año 1772*, 1781; *Mapa de las pueblos que llaman de la Rinconada de Morong... en la Laguna de Bay*, 1773; and others.

**Gómez de Arteche y Moro,** José (*b*.1821). *Atlas de la guerra de la Independencia*, Madrid 1869-1901.

**Gómez Molenda,** Carmelo. 'Engineer director of the Commission of Roads to the south', Philippines. Manila, 1864 MS (painted on silk paper by Domingo Enriquez under the direction of D. Carmelo Gómez Molenda).

**Gómez y Parientos,** Moritz Georg (1744-1810). Austrian general and cartographer.

**Gomme,** *Sir* Bernard de (1620-1685). Engineer and surveyor from Lille, came to England with Prince Rupert, Quartermaster-General of Royalist army, 1642-1646, 'Engineer in chief of the King's Castles', 1661. Prepared military maps and plans including some for battles of the English Civil War. Plans for fortifications of Liverpool, 1644 MSS; *Portsmouth*, 1668; *Survey Citty of Dublin*, 1673 MS.

**Gönczy,** Pál (1817-1892). Hungarian cartographer. Produced a map for schools *Magyar Korona...*, Gotha 1866; school globes, 1869; Hungarian atlas, Pest 1890; *Atlasz*, 1897. **Ref.** HORVÁTH, G. 'Gönczy Pál kartográfiai munkássága'

['The cartographic work of Pál Gönczy'] in *Geodézia és Kartográfia* [Geodesy and Cartography], (1986/5) [in Hungarian].

**Góngora y Lujan**, Pedro Juan de, Duke of Almodóvar. See **Luján** y Góngora.

**Gonichon**, *sieur* —. *Fleuve Mississippi*, 1731 MS.

**Gonne**, —. Engraver. For Jedidiah Morse, 1792; *Ireland*, 1795.

**Gonneim**, — (*fl.*14th century). 3 manuscript planispheres; Imago Mundi (attributed).

**Gonsag**, Fernando. See **Koncság**, Ferdinando.

**Gonsalves da Fonseca**, J. *Governo de Pernambuco*, 1766.

**Gonzago**, Curzio (*c.*1536-1599). Italian printseller in Rome and Venice, possibly associated with a globe. **Ref.** WOODWARD, D. *The Holzheimer Venetian globe gores of the sixteenth century* (Madison, The Jupiter Press 1987).

**Gonzáles**, Alexandre. *Atlas maritimo de Peru, Chile, costa Patagonia* &c., 1797 MS.

**Gonzáles**, Andres. *Florida*, 1609 MS.

**Gonzales**, E. *Plan topographique complet et avec courbes de niveau de la ville d'Auxerre*, Auxerre, A. Lanier 1901.

**Gonzáles**, F. *Carta de las Nuevas Philipinas* [Palao archipelago], 1705 MS.

**Gonzales**, N. Lithographer of Madrid. *España y Portugal...*, 1862 (with Francisco Perez Banquero).

**Gonzáles**, T. Engraver for 'la Dirección Hidrográfica'. Maps for *Atlas marítimo de América y Oceanía*, 1750-1846.

**González**, A. y. See **Artero** y González, Juan de la Gloria.

**González**, E. Garcia y. See **García** y González, Emilio.

**González**, G. See **González** de las Peñas, German.

**González**, Joseph. Engraver. Miguel de Venegas' *Mar del Sur*, 1757. **N.B.** Compare with **GZ**, Joseph.

**González**, Juan de Dios. *Plano perfil y elevación del Fuerte de San Phelipe de Vacalar cituado en la Provincia de Yucatan...*, 1772; *Plano y elevación del actual estado en que se halla el Fuerte de San Phelipe de Bacalar cituado en la Provincia da Yucatan*, 1772; *Plan provincia de Yucatán*, 1774; and others.

**González**, M. *Carta Topografica de Uruguay*, Buenos Aires 1874.

**González**, Nicolas. *Peninsula Ibérica*, *c.*1871; Madrid, 1879.

**Gonzalez Canaveras**, Juan Antonio. *Planisferio ó Carta General de la tierra Segun los Ultimos Descubrimientos* (wall maps), Madrid 1800.

**González de Agueros**, Pedro, SJ. *Mapa de la Provincia y Archipelago de Chiloe en el reino de Chile*, Madrid *c.*1791.

**González de Carvajal**, Ciriaco. *Mapa del Archipielago Filipino, costas de China e islas de Sumatra, Java, Borneo, etc., mandado formar por el superintende de Filipinas D. Ciriaco González de Carvajal...*, 1787.

**González de las Peñas**, German. *Isla de Cuba*, Havana 1881.

**González de la Vega**, Rafael. *Cuba*, 1877.

**González de Mendoza**, Juan. *Regno della China*, 1589.

**Gonzáles Fernández**, Ramón. *Mapa del Archipielago Filipino*, 1875 (lithographed by Oppel & Co.); *Plano general de la ciudad de Manila y sus arrabales*, 1875 (both published in *Manuel del viajero en Filipinas*, Manila, P. Memije 1875); *Itinerario de la navegacion por vapor, de Cadiz a Manila y a HongKong 1877* (lithographed by Oppel & Co., published in *Annario Filipino...*, 1877).

**Good**, *Lieutenant* G.L., *RN*. *Soundings of Table Bay, 1894-1895*, published Admiralty 1898.

**Good**, J. *12 miles round Berwick*, 1806.

**Goodall**, Richard Wright. Surveyor and civil engineer of 137 King St., Sydney, New South Wales (1847). *Plan of Strathean*, 1840s; *Plan of Louth Park Estate in the parish of Maitland*, 1854.

**Goodbridge**, Charles Medyett. *Narrative of a voyage to the South Seas* (with 2 maps of Tasmania), London, Hamilton & Adams 1832.

**Goode**, J. Paul. American cartographer and geographer. Produced a series of political wall maps, Chicago and New York, Rand McNally c.1914.

**Goode**, R.U. Chief Topographer, United States Geological Survey, 1890-1893. *Utah*, 1898 (with S.S. Gannett and R.B. Marshall).

**Goodenough**, F.A. *Routes between India and China*, 1869.

**Goodenough**, *Commodore* James Graham (1830-1875). *Part of the South Pacific*, 1876.

**Goodhue**, J.H. *Concord, New Hampshire*, 1868; co-compiler, with H.B. Parsell, of *Atlas of the State of Rhode Island and Providence Plantations...*, Philadelphia, D.G. Beers & Company 1870 (lithographed by Worley & Bracher, printed by Bourquin).

**Goodman**, John. Engraver and printer of Frankfort, Kentucky. *Rapids of the Ohio River*, 1806.

**Goodman**, Robert. *A New and accurate Map of the County of Norfolk*, 1740 edition (with T. Goddard).

**Goodman**, William (fl.1599-1616). Surveyor, produced manuscript maps of Essex, London and Yorkshire. *Parish of St. Martin's Outwich, Essex*, 1599 MS; *Kelvedon & Little Coggeshall*, 1605 MS.

**Goodrich** brothers.
• **Goodrich**, *Reverend* Charles Augustus (1790-1862). Congregational clergyman, geographer and historian of Worcester, Massachusetts. *Atlas accompanying Rev. C.A. Goodrich's outlines of modern geography...*, Boston, S.G. Goodrich, M'Carty & Davis 1826.
• **Goodrich**, Samuel Griswold (1793-1860) [used pseud. Peter Parley]. Publisher and geographer of Boston, Massachusetts, brother of Charles Augustus Goodrich (above). *United States*, 1826 (engraved by J.W. Barber and A. Willard); Rev. C.A. Goodrich's *Atlas...*, Boston 1826 (with M'Carty & Davis); *Atlas*, 1830; *Atlas, designed to illustrate the Malte-Brun School Geography*, Hartford, F.J. Huntington 1832; *Peter Parley's Universal History on the Basis of Geography*, 1837; *A General Atlas of the World*, Boston, C.D. Strong 1841; *A Universal Illustrated Atlas*, Boston, Charles D. Strong 1842 (with Thomas Gamaliel Bradford, lithographed by B.W. Thayer); *A National Geography for Schools*, New York, Huntington & Savage 1845, 1846, 1848; *A Comprehensive Geography and History*, New York, Huntington & Savage, Mason & Law, 1850, later published by J.H. Colton 1855; *A National Geography for Schools...*, New York, George Savage 1852; *Shilling Atlas*, c.1859; *Outline Atlas*, c.1860.

**Goodrich**, Andrew T. [& Co.]. Stationer, map publishers and lending library at '124 Broadway, opposite the City Hotel, New York'. Bought John Melish's map and plate stock in 1822. *Hudson between Sandy Hook & Sandy Hill*, 1820; J. Melish's *Geographical Description of the United States*, 1826 edition; *A Map of the City of New York*, 1827 and later; *Stranger's Guide*, 1828. **Ref.** COHEN, P.E. & AUGUSTYN, R.T. *Manhattan in maps 1527-1995* (1997) pp.114-115.

**Goodridge**, John. Master Attendant in the Naval Office at the Cape of Good Hope. *Hout Bay...*, 1819.

**Goodwill**, —. Engraver. See **Consitt** & **Goodwill**.

**Goodwin**, John. Manuscript plans of Limehouse, Shadwell, Ratcliff and Wapping Wall [London docks] (4 plans), 1635 (with J. Marr).

**Goodwin**, M.P. Assisted Andrew Gray with *Topographical map of Guernsey...*, William Faden 1816; *Map of the Island of Guernsey*, J. Cochrane 1832.

**Goodwin**, N. *Plan of the City of Hartford, Connecticut*, c.1824 (with D. St. John).

**Goodworth**, W.G.W. *Germany*, 1899.

**Goodyear**, Charles (1800-1860). Inventor of New Haven, Connecticut. Developed vulcanised rubber and rubber tyres. Less well known are his 24-inch inflatable globes made with silk and rubber, and his 'India-rubber' maps, c.1850s.

**Goor**, D. Noothoven van. *Kaart van Zuid Holland*, Leiden 1866; *Nieuwe Kaart van het konigrijk der Nederlanden*, Leiden 1870; *Karta van Europa*, Leiden c.1872; Nieuwe School Atlas, 1875.

**Goor**, G.B. van, en Zonen [& Sons]. Publisher of Gouda. *Atlas van het Koningrijk der Nederlanden* (12 maps), 1882.

**Goos** family. **Ref.** KRETSCHMER, DÖRFLINGER & WAWRIK Lexikon zur Geschichte der Kartographie (1986) p.274 [in German].
• **Goos** [Goss], Abraham (c.1590-1643). Engraver, mapseller, cartographer and publisher, settled in Amsterdam c.1600. Son of Pieter Goos, diamond cutter from Antwerp, and Margareta van den Keere. This made him the nephew of Pieter van den Keere and of Collette van den Keere (who married Jodocus Hondius the elder). Worked with Pieter van den Keere, Jodocus Hondius the younger (Abraham's cousin), and Johannes Janssonius. Pair of 10-inch globes, 1612, published 1614 (with Pieter van den Keere); engraver of Pieter van den Keere's *Americae Nova Descriptio*, 1614; *Nieuw Nederlandtsch Caertboeck*, 1616, 1625; *Novus Tabularum Geographicarum Belgicae Liber*, 1619; 6-inch terrestrial globe, Johannes Janssonius 1621; draughtsman and engraver for a 17-inch terrestrial globe, published by Jodocus Hondius the younger and Johannes Janssonius 1623 (the celestial globe of the pair was by Adriaan Metius); *Tabula Magnae Britanniae...*, Visscher 1623; *West-Indische Spieghel*, Amsterdam 1624; for John Speed's Prospect, 1626; with Pieter van den Keere engraved *Atlas Minor Gerardi Mercatoris...*, Janssonius 1628 and editions to 1662 (pocket size maps known as miniature Mercators); Geldria, Henricus Hondius 1629; numerous maps in 1630 Mercator-Hondius Atlas and later editions; replaced the worn plate for 'The Kingdom of England' and other maps in Speed's *Theatre of the Empire of Great Britaine*, 1632 and later editions; Ancient Sicily, Johannes Janssonius 1636. **Ref.** Van der KROGT, P. Globi Neerlandici (1993) passim.; KOEMAN, C. Atlantes Neerlandici (1967-1971) Vol.II pp.120-122.

•• **Goos**, Pieter (1615-1675). Son of Abraham Goos and father of Hendrik. Cartographer, engraver, publisher, printer and printseller of Amsterdam, 'op't Water by de Nieuwe-brugh, inde Vergulde Zee-Spiegel'. In 1650 he bought the copper plates for A. Jacobsz.'s mariner's guide. *De Lichtende Columne ofte Zee-Spiegel* (2 volumes), 1650, 1654, 1656, 1657, 1666, 1670 (based on the work of Theunis Jacobsz., English editions 1667, 1668, 1669, 1670); *De Nieuwe Groote Zee-Spiegel*, 1662, 1664, 1668, 1671, 1674, 1675, 1676, H. Goos 1678 (English text editions *The Lightning Colomne or Sea-Mirrour*, 1658, 1660, 1662, 1667, 1669, 1670, 1675, French text editions *Le Grand & Nouveau Miroir ou Flambeau de la Mer*, 1662, 1667, 1671, 1672); *De Zee-Atlas ofte Water-Weereld...* (41 charts), 1666, 1667, 1668, 1669, 1670, 1672, 1673, 1675, 1676, H. Goos 1683 (French text edition *L'Atlas de la Mer, ou Monde Aquatique*, 1666, 1667, 1670, 1672, 1673, English text edition *The Sea-Atlas or the Watter-World*, 1667, 1668, 1669, 1670, Spanish text edition *El Atlas de la Mar, o Mundo de Agua*, 1668, 1676); republished Blaeu's *West Indische Paskaert*, c.1674. **Ref.** KOEMAN, C. Atlantes Neerlandici (1967-1971) Vol.IV pp.192-219.
•• **Goos**, *weduwe* [widow] — (d.1677). Continued the business after her husband Pieter's death in 1675. Re-issued *Nieuwe Groote Zee-Spiegel*, 1676. Her son Hendrik carried on the business.
••• **Goos**, Hendrik (fl.1675-1692). Publisher in Amsterdam, son of Pieter Goos. Collaborated with Hendrik Doncker and Casparus Lootsman. Re-issued editions of *De Nieuwe Groote Zee-Spiegel* and *De Zee-Atlas ofte Water-Wereld*.

**Goote**, G. *Atlas der Bijbelsche Aardrijks-en Oudheidkunde Kaarten geteekend*, 1912.

**Gopčevič**, Spiridon (1855-1936). Language map of Macedonia, 1889 (used by Petermann); *Etnografičeskaja karta Makedonije i Stare Srbije*, Berlin 1899.

**Gorcum** [Gorkum], Jan Egbert van (1780-1862). Military cartographer of Arnhem. *Oud-Nederland*, c.1831. **Ref.** CARMICHAEL-SMYTH, Sir James Memoir upon the topographical system of Colonel van Gorkum (London 1928).

**Gordeyev**, Konstantin. Map of the Demidov's metal works in the Kolyvan area, 1723.

Gordon, father and son. **Ref.** STONE, J.C. *The Pont manuscript maps of Scotland* (Tring 1989) pp.5-15; STONE, J.C. *Imago Mundi* 26 (1972) pp.18-26; SKELTON, R.A. *The county atlases of the British Isles 1579-1850* (1970) pp.97-110.

• Gordon, Robert [of Straloch] (1580-1661). Politician, geographer and surveyor from Kinmundy, Aberdeenshire. Bought the estate of Straloch in 1608. Edited, augmented and redraughted Timothy Pont's maps of Scotland for publication in Vol.V of *Atlas Novus* (49 maps), Amsterdam, Joan Blaeu 1654 (the first printed atlas of Scotland), and for Blaeu's *Atlas Maior*, 1662. **Ref.** STONE, J. *The Map Collector* 50 (1990) pp.12-16; STONE, J. 'The origins of three maps of Fife published by Blaeu in 1654' *Scottish Studies* 29 (University of Edinburgh 1989) pp.39-53; STONE, J. *The Map Collector* 10 (1980) pp.25-29; '...Scotland by Robert Gordon...', *Imago Mundi* 31 (1979) pp.84-87.

•• Gordon, *Reverend* James (1617-1686). Cartographer and pastor of Rothiemay, Aberdeenshire (1641), fifth son of Robert Gordon, above. Assisted his father with the revision of Timothy Pont's manuscript maps of Scotland. *Cowper of Fyff* [Cupar], 1642; plan of St. Andrews, 1642; Fife, 1645; plan of Edinburgh, 1647 published Blaeu 1649; *Abredoniae Novae et Veteris Descriptio. A Description of New and Old Aberdeen*, 1661 (facsimile published by J. Bartholomew, 1842).

Gordon, A. Lithographer of 66 Paternoster Row, London. G. Turnbull's map of the canals between Liverpool, Manchester, Birmingham and other principal towns in the west Midlands, 1831.

Gordon, Alexander. Historian, specialising in Roman Britain. Itinerarium septentrionale, 1726.

Gordon, *Lieutenant Colonel* Alexander RE. *Sketch of the routes from Kustenjeh to Chernavoda and Rassova with the Karasú Lakes* [Romania], 1854 (with T. Spratt and J. Desaint).

Gordon, Charles. *Ferro carriles de la Republica de Argentina*, 1889.

Gordon, *Lieutenant* (later *General*) Charles George (1833-1885) [Gordon of Khartoum]. Served in the Ukraine, later held important posts in China, Mauritus, and Africa. Governor Equatorial Provinces of Africa,

*A cartoon of General Charles George Gordon [Gordon of Khartoum] (1833-1885). A water-colour by Carlo Pellegrini, signed 'Ape', 1881. (By courtesy of the National Portrait Gallery, London).*

1874-1876; Governor-General of the Sudan, 1877 and 1884. *Intrenched Position of the Allied Army...* [Ukraine], 1855 MS (and others); *Sevastopol, shewing the French and English Attacks*, 1855 MS; *Sevastopol Russian Battery at the head of Dockyard Creek*, 1855 MS; *Sketch of Bolgrad & Tabac...* [Ukraine], 1856 MS; Road from Constantinople to Adrianople, 1856; Military plan of the country around Shanghai, 1865; Suakin to Khartoum, 1874; Contour map of Jerusalem, 1883. **Ref.** PLAUT, F. 'General Gordon's map of Paradise', *Encounter* (June/July 1982) pp.20-32.

*Robert Gordon of Straloch (1580-1661). (By courtesy of Map Collector Publications)*

**Gordon,** *Lieutenant* David McDowall, *RN*. Admiralty surveyor. Survey of the Coasts of China, 1845-1848; charts of Borneo and Brunei River, 1852; *Tam Sui Harbour*, 1855.

**Gordon,** *Major* Edward Charles Acheson *RE*. Credited with several manuscript plans and views. Active in the Ionian Islands, 1849-1851; contributed sketches for *Battle of the Alma... 1854* [Ukraine], London, J. Arrowsmith 1854; contributed views for *Map of Montenegro*, 1860; plan of the area south and east of Belgrade, 1863 (with others).

**Gordon,** *Captain* Harry [Henry]. Surveyor and engineer in the British army; Chief Engineer in North America. Fort Edward to Crown Point [New York], 1755; *Plan of Fort*

*Edward* [New York], *c.*1755 (with G. Bartman); travelled down, and surveyed, the Ohio River with T. Hutchins, 1766, surveys used for *A Plan of the Rapids in the River Ohio...*, 1766 MS published 1778; plans for a Citadel to improve the fortifications of Quebec, 1769.

**Gordon**, Henry [Harry]. Road from Stirling to Fort William [Scotland], 1749 MS; A plan of part of the new road from Stirling to Fort William, 1750, 1751; Plan of part of the road from Perth to Fort George, 1753, 1754.

**Gordon**, *Reverend* James. See **Gordon**, father and son, above.

**Gordon**, James Bentley (1750-1819). English geographer and historian. New System of Geography, [n.d.].

**Gordon**, Patrick (*d.c.*1702). Best known for one work of geography containing maps, the progress of which through 20 editions (most of them posthumous) over more than half a century, involved many of the eminent mapmakers, publishers and booksellers of the day. *Geography Anatomiz'd: or, The Geographical Grammar*, 1693 onwards (1st to 8th editions with maps from Robert Morden's *Geography Rectified*; 9th to 19th editions, 1722 onwards with maps by John Senex); 20th and last edition, 1754 (with maps by Emanuel Bowen). **Ref.** McCORKLE, B.B. *The Map Collector* 66 (1994) pp.10-15.

**Gordon**, Peter. First Bailiff of Savannah, 1732. *A View of Savanah as it stood the 29th of March 1734* [Georgia], 1734 (engraved by P. Fourdrinier).

**Gordon**, Peter. *Sketch of Reception Bay on the North Side of the Island of Tristan d'Acunha*, London 1814.

**Gordon**, Robert, of Straloch. See **Gordon**, father and son, above.

**Gordon**, Robert (1743-1795). Scottish explorer in South Africa. Produced many manuscript maps of the Cape Colony, 1777-1795.

**Gordon**, Robert. *Atlas of the Irrawaddy*, 1879-1880; Course of the Sanpo River, 1885; map of the ruby mines in Burma, 1888.

**Gordon**, Thomas (1778-1848). American lawyer, surveyor and publisher from Amwell, New Jersey. *A Map of the State of New Jersey*, 1828 (engraved by H.S. Tanner, E.B. Dawson and W. Allen), revised edition published T. Gordon & H.S. Tanner 1833, later editions revised by R.E. Horner published 1849, 1850, 1853, 1854; Gazetteer of the State of New York, Philadelphia 1836. **Ref.** RISTOW, W.W. *American maps and mapmakers* (1985) pp.115, 117.

**Gordon**, W. Commercial map of Scotland, 1785 (with J. Walter); *Traveller's Directory through Scotland*, 1792 (with R.N. Cheyne, re-used G. Taylor & A. Skinner's *A general map of the roads of Scotland*, first published 1776).

**Gordon**, William (*fl.*1730-1738). Surveyor and gentleman. *An Accurate Map of the County of Huntingdon* (6 sheets), surveyed 1730-1731 (assisted by A. Hammond), published 1731 (engraved by Emanuel Bowen, first separate map of the county); *An Accurate Map of the County of Bedford* (2 sheets; scale one inch to one mile), 1736 (engraved by J. Carwitham, still advertised by Carington Bowles in 1782).

**Gordon**, *Commodore* William Everard Alphonso, *RN*. Bird Islands [South Africa], 1853.

**Gordon**, *Lieutenant* William Stavely R.E.. Plan of Suakin, Sudan, [before 1887].

**Gordon**, *Sir* Willougby. Quarter Master General in 1811. Encouraged use of lithography for printing maps.

**Gordon**. See **Griffing**, Gordon & Co.

**Gordon & Gotch**. Publishers in Melbourne, Sydney & London. Map of Australia for the *Australian Handbook for 1877*, 1877 (shows the routes taken by explorers across Australia); *Newspaper Map of Queensland*, 1903; *Newspaper Map of New Zealand, North Island*, 1903.

**Gore**, *Captain* Arthur, *RN*. Turon Harbour, Cochin China, 1764, published Dalrymple; *Chart of the Islands of between St Johns and the Ladrones...*, 1786, published Dalrymple 1792.

**Gore**, *Major* Charles William. Paths round Newcastle, 1893.

**Gore**, *Lieutenant* George Corbet. Maps of Afghanistan and North-West Frontier Districts, 1880.

**Gore**, J. Publisher of Liverpool. Plan of Liverpool, 1821.

**Gore**, W.S. Surveyor General of British Columbia, *c*.1880.

**Gorges**, *Sir* Ferdinando (*c*.1566-1647). English colonist from Wraxall, Somerset. Governor of the fort of Plymouth, England, 1596. Founded two companies in Plymouth for planting lands in New England from 1606. He became part owner of the Colony of Maine in 1622, and was granted the proprietorship of Maine in 1639. Involved in the foundation of New Plymouth, 1628. *Platt of Plimowth* [Plymouth Fort, England], 1596.

**Gorio di Stagio**. See **Dati**, brothers.

**Gorjan**, August (*b*.1837). România, 1881; *Atlas-géografie România*, 1895 (with Ion Luncan).

**Gorkum**, J.E. van. See **Gorcum**, Jan Egbert van.

**Gorlinski**, Robert. U.S. Deputy Surveyor & Civil and Mining Engineer. *Saltair Beach* [Utah], Salt Lake Litho. Co. 1890.

**Gorlinsky**, Joseph. Draughtsman to General Land Office. *Map of the United States and Territories, Showing the extent of Public Surveys...*, 1867; *Great Rail Road Routes to the Pacific and their connections...*, 1869.

**Gormaz**, Francisco Vidal. *Plano de Puerto de Quintero*, Santiago 1866; *Plano del Rio Lebu*, *c*.1866 (with G. Peña and L. Señoret); *Plano del Puerto de Yañez*, Santiago 1866 (with G. Peña and L. Señoret); *Plano de la costa Araucana*, Santiago 1868; *Plano del rio Toltén i plaza militar...*, Santiago 1868; *Karte von Süd-Chile: Provinz Llanquihue und Theile von Valdivia und Chiloe*, in *Petermann's Geographische Mitteilungen*, Gotha 1880 (revised by C. Martin).

**Gormont**. See **Gourmont**, family.

**Görög** [Geörög], Demeter [Demetrius] (1760-1833). Hungarian cartographer. *Európának közönséges táblája* [Map of Europe], Vienna 1790; *Magyar Átlás* [Hungarian Atlas], 1802, 1848, 1860 (with Sámuel Kerekes and József Márton). **Ref.** MÁRTON, József *Görög Demeter életirása* [Biography of Demeter Görög], Vienna 1834 [in Hungarian]; NAGY, Júlia 'Görög Demeter', *Földrajzi Évkönyv* [Geographical Annual], 1977 [in Hungarian].

**Gorrell**, T.D. *Pleasants County, Virginia*, 1865.

**Gorries**, *Captain* Joannes. Swedish military surveyor in Bremen and Verden after 1648. *Ducatus Bremae & Ferdae*, J. Blaeu 1662.

**Görringer**, M. See **Goerringer**, M.

**Gorsuch**, Robert B. (*fl*.1850s). American railroad surveyor. Surveys in Mexico for US Army (with Andrew Talcott and M.E. Lyons).

**Gorton**, John (*d*.1835). *A New Topographical Dictionary of Great Britain and Ireland* [partwork], London, Chapman & Hall 1830-1832 (with county maps engraved by Sidney Hall and Selina Hall, re-used in later atlases).

**Gosch**, Christian Carl August (1829-1913). Editor. Republished Danish and English accounts of Captain James Hall's voyages to Greenland for the King of Denmark, 1605-1606, and on his own behalf in 1612. *The Danish expedition to Greenland...* (10 charts), London, Hakluyt Society 1897.

**Göschen**, Georg Joachim (1752-1828). Printer and publisher of Leipzig. W.E.A. von Schlieben's *Atlas von Amerika in 30 Charten*, 1830.

**Gosling**, Ralph (1693-1758). English topographer, produced the earliest known plan of Sheffield. *Plan of Sheffield*, 1732, 1736.

**Goslyng**, John. *Plott building in Nevills Alley, Fetter Lane* [London], 1670.

**Goss**, Abraham. See **Goos**, Abraham.

**Gosse**, Pierre. Publisher of The Hague. Joint publisher (with J. Neaulme) of *Carte de l'Abissinie*, 1728, and *Carte de l'Ethiopie Orientale*, 1728; joint publisher of E. Kaempfer's *De Beschryving van Japan...*, 1729 edition (with J. Neaulme and B. Lakeman), French edition, 1732 (with J. Neaulme).

**Gosse**, Pierre [Pieter], *junior* (1729-1765). Publisher of The Hague. Jean Palairet's *Atlas Méthodique*, 1755; *Atlas Portatif à l'usage de Messieurs les Officiers des Armées S.M. Britanniques en Allemagne...*, 1761 (also published in Münster by Vaudriancy Négociant); joint publisher, with Daniel Pinet, of battle plans published in F.W. von Bauer's *Plans pour servir à l'histoire de la guerre de Sept ans*, from 1763, and *Théâtre de la guerre en Allemagne entre la Grande-Bretagne et la France depuis l'an 1757 jusqu'à l'an 1762*, 1769.

**Gosse**, William Christie (1842-1881). English explorer from Hoddesden, Hertfordshire. He was taken by his family to Australia in 1850, where he became a government surveyor in South Australia. As leader of an overland expedition westwards in 1873, he discovered and named Ayers Rock. *Map showing the area of the Northern Territory and South Australia explored by William Gosse in 1873*, Adelaide, Surveyor General's Office 1874 (lithographed by Fraser S. Crawford).

**Gosselin**, *Colonel* —. Plans of part of the coast of Guernsey showing troop dispositions, 1795 MSS.

**Gosselin**, B. *Plan de Boulogne-sur-Mer*, 1863.

**Gosselin**, Charles. Publisher in Paris. *Oeuvres completes de Sir Walter Scott*, Paris, C. Gosselin & A. Sautelet 1826-1828, C. Gosselin 1829-1833 (maps by A.M. Perrot, engraved by P. Tardieu); *Carte des Etats-Unis d'Amerique Comprenant une partie des Districts de l'Ouest et de la Nouvelle Bretagne*, 1840; *Carte de la Prusse, de l'Autriche et de la Confederation Germanique*, 1866.

**Gosselin**, G. Carte des environs de Rome, 1868. **N.B.** Compare with Charles **Gosselin**, above.

**Gosselin**, Pascal François Joseph (1751-1830). French geographer from Lille, worked in Paris. *Recherches sur la Géographie Systématique et Positive des Anciens* (4 volumes), 1798-1813 (maps engraved by Chamouin); *Atlas ou recueil des cartes géographiques*, Paris 1814.

**Gosselmann**, *Captain* C.A. Swedish explorer in South America. Information was used by John Arrowsmith in the preparation of *Map of the Province of La Rioja, shewing the routes of Messrs. French, Gosselman & Hibbert*, published in *Journal of the Royal Geographical Society*, London 1839 (shows Gosselman's journey of 1837); *Resa in Södra Amerika*, 1842.

**Gosselo**, Edmund. Decorative, circular map of the Scilly Isles, showing where ships were lost, 1707-1710 MS.

**Gosset**, R. Plan of St. Helier, Jersey, 1850.

**Gosset**, *Major* William, *RE*. Tripoli, 1813; Algiers, 1816.

**Gosset**, *Captain* William Driscoll *RE*. Senior officer in the Ordnance Survey, one of the first to join the headquarters at Southampton in 1842, where together with W. Yolland he directed printing operations. He was sent to Paris in 1852 to study French map design and printing. He also experimented with ways of representing relief on maps.

**Gosson**, Henry. The Carriers Cosmographie, 1637.

**Gotch.** See **Gordon** & Gotch.

**Gotendorf**, —. *Rivière de Paris à Rouen*, 1878.

**Gotham**, William. Manor of West Harting, Sussex, 1632 MS.

**Gothus**, Olaus Joannis. See **Svart**, Olof Hansson.

**Gotofred**, Johann Ludwig. See **Gottfried**, Johann Ludwig.

**Gott**, *Captain* Reeve. Chart of the Gulf of Finland, 1785.

**Gotteberg**, E. de. Maps of Egypt, 1857-1868; *Ostaegyptischen Wüste*, 1859.

**Gottfried** [Gotofred] Johann Ludwig. Possibly a pseudonym for Johann Philipp Abelin. Geographer and author, worked with Matthäus Merian in Frankfurt. *Newe Welt und Americanische Historien* (45 maps), 1631, 1655, 1657; *Inventarium Sveciae* 1632; *Neuwe Archontologia Cosmica* (125 maps), Frankfurt 1638. **N.B.** Some sources claim that Gottfried and Abelin are one and the same, others, including Philip Burden in The Mapping of North America (1996) pp.273, 293, assert that they were not.

**Gotthard**, A. Cartographer and draughtsman. *Grossherzogthum Oldenburg und... Freien Stadt Bremen*, in L. Ravenstein's revised edition of *Meyer's Hand-Atlas der Neuesten Erdbeschreibung*, Hildburghausen, Verlag des Bibliographischen Instituts 1872.

**Gotthold**, August. *Plan von Kaiserslautern*, 1883.

**Gotthold** [Gottholdt], H.H. Denmark, 1808; Deutschland (55 sheets), 1808-1831; *Oesterreich vor und nach dem Wiener Frieden von 14ten October 1809*, Berlin c.1810; South Africa, 1810; Cape of Good Hope, 1815.

**Gottlieb**, August. See **Boehme**, August Gottlieb.

**Gottschalck**, Friedrich. *Plan von Dresden*, 1847.

**Gottschalk**, Adolf. Engraver for Joseph Meyer, 1830-1840.

**Gottwald**, Johann. *Österreichisch-Ungarische Monarchie*, c.1876.

**Gotz**, A. *Alten Geographie*, Nuremberg 1729.

**Gotz**, Marceli. *Mappa Królestwa Polskiego...*, Warsaw 1863, 1868 (with others).

**Götz**. See **Goetz**, Andreas.

**Götze**, Ferdinand. *Europa Nach den vorzüglichsten Hülfsmitteln neu entworfen und gezeichnet*, Weimar, im Verlage des Geograph. Institute 1815.

**Gotzmann**, W. *Greiz und Umgegend*, 1882.

**Gouault**, father and son. Papermakers of Troyes.
• **Gouault**, Jean [Jehan] (*d.*1603). Some of his paper, which carried the mark of a hand with the name 'Jehan Gouault', was exported to the Low Countries. It was used by Braun & Hogenberg for some sheets of *Civitates orbis terrarum*, 1572-1612, including some pulls of *Turo* [Tours] and *Monspessulanus* [Montpellier], Cologne, P. von Brachel 1612.
•• **Gouault**, Gilles (1601-1669). Son of Jean. A plan of the Abbey of Noyer is on paper bearing his mark [GG in a cartouche].

**Goubaud**, T. Lithographer of Ostend, Belgium. Theunissen's *Reis-kaart voor de binnelanden van Zuid-Afrika*, 1824.

**Goubaut**, —. *Plan de la Ville de Philipsbourg*, 1750.

**Goubet**, John (*fl.*1690-*d.*1695). Surveyor. Plans of fortifications in Ireland, 1690-1695 MSS; *Plan de la Ville de Portsmouth*, c.1692.

**Gouche**, Burnaby. Galway City, 1583.

**Goudriaan**, B.A. [B.H.] *Kaart der Hoofdrivieren op de Schaal*, 1830-1839.

**Gouge**, J. Publisher and mapseller in Westminster Hall. One of several sellers of Charles Price's *A Correct Map shewing all Towns, Villages, Roads... within 30 Miles of London...*, 1712.

**Gouge**, James (*fl.*1803-1834). Estate surveyor of Sittingbourne, Kent. Produced manuscript maps including *Boughton under Blean &c., Kent*, 1803; *Rodmersham, Tonge and Bapchild...*, 1815; *Bapchild and Tonge*, 1816; *A plan of the parish of Ashford...*, 1818; Glebe of Cuddington, Buckinghamshire, 1818; Panfield, Essex, 1819; *Hale Farm, situate in the Parish of Chatham in the County of Kent...*, 1821; *Cliffe at Hoo*, 1827; Lake estates in Kent, 1832 MSS.

**Gougeon**, H. *Prise de Chateaudun*, 1870.

**Gouget**, Jerome T. Engraver of Chicago. *Cabinet Map of the Western States and Territories*, Chicago, Rufus Blanchard 1869.

**Gough**, *Captain* Charles. Master of the *Richmond*. Three views of Gough Island [South Atlantic], 1732, published Dalrymple 1785, 1792.

**Gough**, *Captain* H. Chart of the Andaman Islands, 1708, published Dalrymple 1784.

**Gough**, John. Map of Ireland, 1567 MS.

**Gough**, Richard (1735-1809). English collector, historian and topographer, Director of the Society of Antiquaries, 1771-1797. Much of his personal collection was bequeathed to the Bodleian Library, Oxford. Compiled an inventory of British maps and topographical literature entitled *British Topography* (2 volumes), 1768, 1780, 1782; edited Camden's *Britannia* 1789 (with maps by John Cary). **Ref.** WALTERS, G. *The Map Collector* 2, pp.26-29; *Imago Mundi* 28 (1974) pp.124-128.

• **Gough map**. Map of Britain, *c*.1360, named after Richard Gough (above) who purchased it from the estate of Thomas Martin of Suffolk, *c*.1768; now in Bodleian Library, Oxford. **Ref.** HARVEY, P.D.A. *Medieval maps* (1991) p.73.

**Gouin**, Auguste-Jules (*b*.1850). French naval officer. Commissioned to prepare a map of Tonkin province in 1884. *Carte de Tonkin*, 1885. **Ref.** BROC, N. *Dictionnaire illustré des explorateurs Français du XIXe siècle: II Asie* (1992) p.220 [in French].

**Goujon**, Andriveau. See **Andriveau**-Goujon.

**Goujon**, Jean (*fl.c*.1793-1826). 'Marchand de cartes géographiques, Rue du Foin Saint-Jacques, chez Goujon, Maison-Egalité, ci-devant Palais Royal', Paris (1793); 'rue Fromenteau au N°·17' (1797); 'rue du Bacq N°·6' (from 1801). Worked for the Dépôt de la Guerre from 1801, he was also one of the sellers of the Cassini *Carte de France*, and maps by other contemporaries such as Frémin, Brué, Lapie, and Chanlaire & Dumez. His daughter married J. Andriveau, with whom he began working in 1825. Plan of Vienna, 1805; Thomas Lopez's *Carte des royaumes d'Espagne et de Portugal...*, 1808 (with others); *Nouveau tableau pour servir à l'assemblage des feuilles de la Carte de France de Cassini...*, 1808; P. Lapie's *Carte d'Europe*, 1812; P. Lapie's *Carte de l'Amérique méridionale...*, 1814 (engraved by P. Tardieu); *Carte détaillée des environs de Paris*, 1816; A.H. Brué's *Carte Physique, Administrative et Routière de la France...* (4 sheets), 1818; P. Tardieu fils aîné's *A Map of Louisiana and Mexico. Carte de la Louisiane et du Mexique*, 1820 (engraved by P.A.F. Tardieu père); North America, 1821; sold De Belleyme's *Carte itinéraire de la France*, 1824; *Nouveau Plan Itinéraire de la Ville de Paris...*, 1824 (engraved by Perrier & Gallet, calligraphy by Lale); *Carte des environs de Paris*, 1826. **Ref.** PACHA, B. in: PACHA, B. & MIRAN, L. *Cartes et plans imprimés de 1564 à 1815* (Bibliothèque Nationale de France 1996) pp.48-49 [in French]; See also **Andriveau Goujon**, J.

• **Goujon et Andriveau** (*fl*.1825-*c*.1830). Cartographers and publishers of Paris, partnership of Jean Goujon and his son-in-law J. Andriveau, succeeded by the Andriveau [Andriveau-Goujon] firm and family *q.v.*. Published C. Viard's *Carte speciale des postes de France*, 1829.

**Goulart** [Goulartio], Jacques [Jacobo] (1580-*c*.1622). Swiss theologian and cartographer, born Geneva, worked in Basle, Amsterdam, Nyon, Burtigny, Commugny, Aubonne and Arzier. *Chorographica tabula Lacus Lemanni...* , 1606 (used by Jodocus Hondius 1606, J.B. Vrients 1608, Willem Blaeu 1634, and Joannes Janssonius 1638).

**Gould**, Augustus Addison (1805-1866). Physician and conchologist. *New Ipswich, New Hampshire*, 1851.

**Gould**, Benjamin Apthorp (1824-1896). Astronomer, founded Observatory at Córdova, Argentina, 1870.

**Gould**, Charles. Government geologist. Expedition to Western Tasmania, 1860.

**Gould**, F.A. Atlas of Randolph and Wayne Counties, Indiana, 1847 (with D.J. Lake & G.P. Sanford).

**Gould**, Francis. Ordnance draughtsman. Manuscript copies of various plans, 1764-1781 including fortifications in the UK, and some of Bergen-op-Zoom, 1767-1768 (from originals dated 1751); *Plan of the peninsula and city of Gibraltar*, *c*.1779 MS (from an original of 1735).

**Gould**, G. See **Gauld**, George.

**Gould**, Hueston T. *Atlas of Franklin County, Ohio*, 1872 (with J.A. Caldwell); Atlas of Ross County & Chillicothe, Ohio, 1875.

**Gould**, James (*fl*.1718-*d*.1734. Made estate maps in Cambridgeshire, Essex and Surrey. Little Coggeshall, Essex, 1721 MS.

**Gould**, Jay [Jason] (1836-1892). Mathematician, surveyor and financier from Delaware County, New York. His county surveys were used by R.P. Smith in the compilation of the state map of New York. He abandoned cartographic work *c*.1856 to concentrate on business and finance. Worked with P.H. Brink and O.J. Tillson on the original survey for Brink & Tillson's *Map of Ulster County, New York...*, 1853

(Gould sold his interest in it before publication); *Map of Albany County New York...*, 1854 (with his cousin I.B. Moore); *Map of Cohoes New York*, Gould & Moore *c.*1854 (printed by Sarony & Major, New York); *Map of Delaware Co. New York...* Philadelphia, Collins G. Keeney 1856 (the copyright to the maps was bought by R.P. Smith). **Ref.** RISTOW, W.W. American maps and mapmakers (Detroit 1985) pp.379-386; RISTOW, W.W. The Map Collector 7 (1979) pp.2-10.

**Gould**, John. English surveyors in Kent, relationship unknown.
• **Gould**, John. English surveyor. *A Ground Plan of the Quay at Gravesend in 1829 before the Building of the Pier*, 1843; *A Plan of the present Quay and a Plan of the Pier at Gravesend*, 1845.
•• **Gould**, John *Junior*. Surveyor. *Plan of an estate situate in Stone Street and New Road Gravesend...*, 1852 (lithographed by Waterlow & Sons); *A Plan of the estate belonging to the Gravesend and Milton Waterworks Co., to be sold by auction...*, 1859; *A Plan of work to be done to Footpaths and road in the Grove, Milton*, 1876.

**Goulden**, William E. Publisher of Canterbury. *A Map of the County of Kent taken from the latest survey and corrected by John Dower*, 1875; *Plan of the City of Canterbury and the adjoining suburbs*, [n.d.] (in *Goulden's Canterbury guide*).

**Goumet**, J.M. *Plan et contour de la paroisse de Paudy* [Indre, France], 1757.

**Gourbeyre**, family. Papermakers of the Auvergne. Their papers were used by N. Sanson, from 1679; G. Delisle, 1701-1720; N. de Fer, from 1713; G. Robert, 1741; and others. **Ref.** MIRAN, L. in: PACHA, B. & MIRAN, L. Cartes et plans imprimés de 1564 à 1815 (Bibliothèque Nationale de France 1996) pp.66-67 [in French].
• **Gourbeyre**, Pierre [I] (*d.c.*1698). Worked with his father-in-law at a papermill at Noyras from 1660. He was succeeded there by his son Claude.
•• **Gourbeyre**, Claude (*d.*1733). Married Marie-Claudine, daughter of another papermaker, Thomas Dupuy. Succeeded his father at Noyras in 1698. Worked with his son Pierre II.
••• **Gourbeyre**, Pierre [II] (1702-1782). Son of Claude. Succeeded his father *c.*1732. Papers with his mark were used by Gilles Robert. The company failed in the mid-18th century.

**Gourdin**, —. Baie de St. Nicholas, 1838, and other plans for Jules Sébastian César Dumont d'Urville's *Voyage au pole sud...*, 1842-1848.

**Gourlay**, Robert Fleming (1778-1863). *Statistical Account of Upper Canada* (with maps), London, Simpkin & Marshall 1822; *Sketch of Proposed Improvements for Kingston now the Capital of Canada...*, 1841; Edinburgh, 1855.

**Gourmet**, —. Atlas, *c.*1770.

**Gourmont** [Germontius; Gurmontius] family. French printers and engravers of the 16th century. **Ref.** SHIRLEY, R. The Map Collector 18 (1982) pp.39-40; GRENACHER, F. Imago Mundi 14 (1960) pp.55-57.
• **Gourmont** [Gormont; Gurmontius], Hieronymus [Gilles; Hierosme; Jérome] de. Mapseller, publisher and printer 'devant le college de Cambray, aux trois Couronnes', Paris. Father of Jean, below. Imprint appears on the copy of Oronce Fine's *Recens integra orbis descriptio*, 1534-1536 (now held at Germanisches Nationalmuseum, Nuremberg); map of the Holy Land by Oronce Fine, *c.*1536 (now lost); anonymous map of Italy, 1537; Oronce Fine's *Nova totius Galliae descriptio*, 1538; *Vera Pulcherrimae Italiae Orbis...* [Italy], Paris 1544; *Universae Germaniae descriptio*, 1545; S. Münster's *Nouvelle description d'Angleterre*, Paris 1545, 1548; Champagne, 1546; Italy, 1548; *Islandia*, 1548 (from O. Magnus's *Carta marina*, 1539); G. Guéroult's *Nouvelle description du pays d'Almaigne*, *c.*1552; France, 1553; Guillaume Postel's Signorum coelestium, 1553.
•• **Gourmont** [Gurmontius; Germontius], Jean de (*fl.*1561-1585). Engraver and publisher, 'A Paris... demeurant rue Sainct Jean de Latran a l'Arbre sec'; succeeded his father and worked with brother François. Untitled woodcut world map framed by jester's head or 'foolscap' map, *c.*1575; cut blocks for Guillaume Postel's world map, *Polo aptata nova charta universi*, 1581 (sole copy known is dated 1621). **Ref.** SHIRLEY, R. The Map Collector 18 (1982) pp.39-40.
•• **Gourmont**, François de. Son of Hieronymus; worked with his brother Jean.

**Gournai**, Claude. Bookseller of 'quai de l'Horloge du Palais', Paris. Seller of the first version of *Le Neptune François* (29 charts of the coast of Europe from Norway to Gibraltar), 1693 (prepared by Sauveur, J.M de Chazelles, C. Pène and others). **N.B.** Compare with C. **Gournay**.

**Gournay**, C. Publisher and engraver 'à l'entré du Quay de l'orloge du côte du pont au change', Paris. One of the engravers for N. de Fer's *Les Costes de la mer mediterranée*, Paris 1690 (part of *Les Costes de France...*); Père Placide's *Le comté de Flandre*, veuve Du Val 1690; *Plan du Siege de Namur...*, Paris, C. Gournay 1692, 1708; maps used in de Fer's *Les Forces de L'Europe*, 1692, 1693; *Plan du champ bataille de la Marsaille* [Marsaglia, Italy], Paris, C. Gournay *c*.1693; *Plan du champ de bataille de Nerwinde*, Paris, Naudin & Gournay 1693 (drawn by Pennier; engraved by Naudin); *Saar-Louis Est...*, De Fer 1705; and others.

**Gourné**, Pierre Mathias, *abbé* de (1702-*c*.1770). Cartographer of Dieppe. Géographie méthodique, 1741; *Atlas abrégé et portatif*, 1763 (for Louis Charles Desnos); *L'Amérique. Table Géographique*, *c*.1765.

**Goussier**, —. Engraver for Diderot et d'Alembert's *Encyclopédie*, 1770-1779.

**Gouvion**, *Lieutenant-Colonel* Jean Baptiste de (1747-1792). French military surveyor and engineer from Toul. Went to America *c*.1777, involved with Duportail in the planning of West Point. Returned to France and service in the French Army in 1783. *Plan of the Attack of York in Virginia by the Allied Armies and France Commanded by His excellency General Washington...*, 1781 MS.

**Gouvion**, *Marshal* Laurent, *marquis* de Saint-Cyr (1764-1830). Atlas aux campagnes, 1828; Atlas à l'histoire militaire, 1831.

**Gouwen** [Gouwe], Gilliam [Gelliam; Willem; Wilhelm] van. Engraver, probably from Antwerp; designed and engraved title-pages, frontispieces, and illustrations for many books. Worked for the Allards in Amsterdam, *c*.late 1680s-*c*.1700, also in Haarlem. Engraved Joachim Bormeester's world map *Orbis terrarum nova accurata tabula* (4 sheets), *c*.1685; for Frederick de Wit, 1696, for François Halma, 1704; for Adriaan Braakman, 1706; cartouche for Pieter van der Aa's map of the World, 1713.

**Govantes**, Felipe Maria de. *Filipinas*, 1878.

**Gove**, F.W. *Rico Pioneer Mining District, Colorado*, 1880.

**Gover**, Edward. Draughtsman, engraver and publisher. *The Biblical Atlas...*, London, Religious Tract Society 1840; *Gover's General and elementary Physical Atlas*, London, E. Gover, Sen. 1845; *Gallia Transalpina* [Ancient France], London 1846 (drawn and engraved by Sidney Hall and Edward Gover); Carl Vogel's *The Illustrated general and elementary Physical Atlas with descriptive letterpress*, London, E. Gover, Sen. 1850; *Hand-Atlas of Physical Geography*, London 1850, London, Varty & Owen 1853 (from Berghaus's *Physikalischer Atlas*); *The Historic Geographical Atlas of the Middle and Modern Ages*, London, 1853 (translated from Spruner's *Historisch-Geographischer Hand-Atlas*); *Atlas of Universal Historical Geography*, London, E. Gover 1854; *The University Atlas, or Historical Maps of the Middle Ages*, London, T. Varty *c*.1854; Atlas of universal historical geography, 1854; Two-shilling physical atlas, 1854; and others.

**Govone**, *Major* T. *Plan of the Fortress of Silistria*, 1856.

**Gower**, Richard Hall (1787-1833). Naval architect and inventor, *b*. Chelmsford, *d*. Ipswich. *Chart of the Worcester's track over the Cape Bank...* 1791 [Table Bay to east of Algoa Bay, Southern Cape], London, A. Dalrymple 1793 (engraved by W. Harrison).

**Gowing**, John Sewell. Publisher of Swaffham, Norfolk. *Parish of Swaffham*, 1845.

**Gowland**, *Master* (later *Navigating Lieutenant*) John T., *RN*. Gulf of Patras [Greece], 1865-1866; survey of the east coast of Australia, 1866-1871 including *Gabo Island to Montague Island*, Admiralty Chart 1870 (engraved by Sharbau); *Botany Bay and Port Hacking*, London, Admiralty 1873 (with J.F. Loxton, engraved by E. Weller); *Newcastle Harbour* [New South Wales], London, Admiralty 1867 (with J.G. Boulton, engraved by E. Stanford).

**Goyder**, George Woodroffe (1826-1898). Surveyor General of South Australia. In

1869-1870 he led a survey expedition to lay out land (later to become the city of Darwin) on behalf of the Northern Territory Company. Mt. Denison Range, 1860 MS; South Australia, Adelaide 1880; *Plan of the southern portion of South Australia*, 1886.

**Gozzadino**, M A. Ritratto del antich. cella de Tivoli, 1622 (with J.A. de Paulis).

**Graaff** [Graaf; Graef; Graff], de. Father and son.
• **Graaff** [Graef], Abraham de. Teacher of navigation, and examiner of pilots. Wrote comments on the use of globes in: *Groote Zee-Vaert ofte de Konst der Stuer-Luyden*, Amsterdam, Pieter Goos 1657.
•• **Graaff** [Graaf, Graff], Isaac de (1667-1743). Clerk to the VOC [Dutch East India Company], from 1690, succeeded Joan Blaeu Jr as official VOC map supplier in 1705. Best known for the VOC manuscript atlas, c.1692 (incorporated copies of maps, plans and charts collected by VOC). Golfe du Bengale, 1709; Détroit de Malacca, 1710; La Sonde, 1711; Mer de Java, 1718; Atlantique, 1723; Océan Indien, 1728; chart of the Cape of Good Hope to Batavia, 1735. **Ref.** ZANDVLIET, K. *Mapping for money* (Amsterdam 1998) *passim*.

**Graaff** [Graaf], Cornelis Jacob van de (1734-1812). Cape administrator and engineer, Governor of the Cape Colony, 1785-2792. Commissioned many soundings and surveys of the Cape Coast. *Caart der Situatie van de Caap de Goede Hoop*, 1786. **Ref.** ZANDVLIET, K. *Caert-Thresoor* 6:1 (1987) pp.2-5 [in Dutch].

**Graaff**, Nicolaus [Nicolaas] de (1619-1688). Dutch naval surgeon, spy, draughtsman, mathematician and surveyor. Made many voyages in the service of the Dutch East India Company. Accounts of his travels accompanied by maps were published from 1701 onwards. Charts of the Malabar coast, from 1678 (used by Isaac de Graaff in his VOC atlas). **Ref.** BAREND-VAN HAEFTEN, M. *Caert-Thresoor* 9:1 (1990) pp.8-13 [in Dutch with English summary].

**Graah**, *Captain* Wilhelm August (1793-1863). Royal Danish Navy. *Vestlige kyst af Grönland*, 1823-1824; *Grönland*, 1832; Julianehaabs District, 1844.

**Graap**, H. Photo-lithographer. *Australien u. Neu-Seeland*, c.1876.

**Graberg**, Jakob. Mappamondo, 1802; *Imperio di Marocco*, 1834; *Maghrib ul Acsa*, 1834; *Kharesmia*, 1840.

**Grabowski**, Ambrozy. Cracow, 1830.

**Graça**, F.C. da. See **Calheiros** da Graça, Francisco.

**Grach**, C. German printer. *Karte des Bosporus mit Constantinopel...*, c.1870 (drawn by C. Stolpe, engraved by J. Straube).

**Gracher**, Johann Georg. Salzburg District, 1836 MS.

**Gracht**, Quentin van der (*b*.1534). Goldsmith of Antwerp. Plan of Béthune, used by Braun and Hogenberg, 1581 and later.

**Gracia**, —. Engraver of 'rue d'Iéna 9', Paris. *Siège et bataille de Turin, le 17 Septembre 1706*, c.1845 or 1850 (printed by Kaeppelin & Cie.).

**Gracie**, P. See **Garcie**, Pierre.

**Grack**, Carl. Lithographic printer in Berlin. *Das Schlachtfeld von Düppel Malerisch dargestellt von A. Meyer* [Denmark], Berlin, Julius Abelsdorff c.1864 (on the same sheet as a similarly titled map of Fredericia [Denmark], lithographed by R. Meinhardt); for C.A. Fay, 1871.

**Gracroft**, John. Greenland, Norway, Nova Zembla &c., 1740 MS.

**Grad**, Charles (1842-1890). Geographer and geologist, *b*. Turkheim. Alsace, 1889.

**Gradmann**, Robert (1865-1950). Population chart of Württemberg, 1914.

**Graef**, brothers. See **Gräf**.

**Graefé**, Carl. *Hippologische Karte von Preussen*, c.1855; *Bayern*, 1861.

**Graefe**, Th. Lithographer in Hof [Bavaria]. *Oberfranken*, 1838.

**Graessl**, Josef. German engraver, worked for Joseph Meyer. *Kurfürstenthum Hessen*, 1844; *Geognostische Karte von Central-und West-Europa*, 1844; Texas, 1852; *Wisconsin*,

1852; *Mississippi*, 1852; *Iowa*, 1853; maps for C. Arendts's *Vollständiger Hand-Atlas der neueren Erdkunde*, Regensburg, G.J. Manz 1858 and later.

**Graeve**, E. Publishers of Bucharest. J. Don's *Noul plan al capitalei Bucuresti...*, 1897.

**Gräf** [Graef], brothers.
- **Gräf** [Graef], Adolf (*fl*.1855-1878). Collaborated with Carl Gräf (below), Heinrich Kiepert, Carl Bruhns, C.F. Weiland and O. Delitsch on a variety of similar atlas titles (not necessarily different works). Hannover &c., 1857; *Hand-Atlas der Erde und des Himmels*, 1859 (with H. Kiepert, C. Gräf and C. Bruhns); *Atlas des Himmels und der Erde für Schule und Haus*, 1861 and later editions; *Africa*, [post 1867]; *Großer Hand-Atlas des Himmels und der Erde*, 1871 (with H. Kiepert, C.F. Weiland, C. Gräf, O. Delitsch, C. Bruhns); *Eisenbahn Karte von Deutschland*, *c.*1877; and others.
- **Gräf** [Graef], Carl (1822-*c*.1897). Collaborated with Adolf Gräf (above), Heinrich Kiepert, Carl Bruhns, C.F. Weiland and O. Delitsch. Sachsen, *c*.1855; draughtsman for *Das Europaeische Russland und die Statthalterschaft Kaukasus...*, Weimar, Geographisches Institute *c*.1878 (with A. Müller); *Die Schweiz*, *c*.1890; and others.

**Graf**, Charles. Lithographer of Great Castle Street, Oxford Street, London. *The Circuiteer or Distance Map of London...*, London, J. Friederichs 1850 edition.

**Graf**, Franz Josef (1811-1871). Engraver, worked with Johann Jakob Brack on *Topographische Karte des Kantons Zürich* (32 sheets), *c*.1850.

**Graf**, Fred. Draughtsman, printer and publisher of St. Louis, Missouri. *St. Louis* [bird's-eye view], Fred Graf 1893; *St. Louis*, Graf Eng. Co. 1896; *St. Louis*, Graf Eng. Co. 1907 (reduced edition).
- **Graf Eng. Co.** Publishers of St. Louis. Published the 1896 and 1907 versions of Fred Graf's bird's-eye view of St. Louis.

**Graf** [Grave], Hans. *Statt Franckenfurt*, 1552, used by Braun & Hogenberg, 1572 and later.

**Graf**, Hans. Swedish surveyor. Charta öfver Eskelstuna Stad, 1783 MS; Torshalla Stad, 1783 MS; Malmköping, 1798.

**Graf** [Graff], Urs, *the elder* (1485-1529). Swiss artist, engraver, block-cutter and goldsmith, born in Solothurn, worked in Basle. Credited with decorative work in some editions of Ptolemy's *Geographia*, Strasbourg 1525, Lyon 1535.

**Graf & Soret**. Plan of the Parish of St. Marylebone [London], 1833.

**Graff**, F. *Plan von Rostock*, *c*.1859.

**Graff**, I. de. See **Graaff**, Isaac de.

**Graff**, Kasimir (1878-1950). *Sternatlas*, Hamburg 1925 (with M. Beyer).

**Grafnetter**, Josef. *Plan von Prag*, *c*.1877.

**Grafton**, Richard. Printed 3 woodcut maps by William Patten, *The Expedicion into Scotla[n]d*, London 1548 (thought to be some of the first printed maps to be prepared in England).

**Graham**, —. Publisher of London. *Grahams new map of England and Wales with part of Scotland*, *c*.1838 (drawn by George Kemp).

**Graham**, A.W. Engraver of Montreal. Robert Barlow's *Geological Map of Canada*, Montreal, Dawson Brothers 1865 (printed in London at Stanford's Geographical Establishment).

**Graham**, Andrew B. Photo-Lithographer and publisher in Washington, D.C., and Baltimore (1903). F.L. Averill's *Map of Washington, D.C. and suburbs...*, 1892; *United States Coast and Geodetic Survey. General Chart of Alaska*, revised edition 1897; *Map of the District of Columbia*, Washington, Thos. J. Fisher & Co. 1899; for United States War Department, 1899; *Atlas accompanying the Counter Case of the United States before the tribunal convened at London under the provisions of the treaty between the United States and Great Britain* [Alaska], 1903 (maps from various surveys 1847-1903); *Alaska...*, 1906 (compiled by M. Hendges, revised and drawn by C. Helm).

**Graham**, *Lieutenant* C. Chart of Sandusky Bay [Lake Erie], 1826 (reduced version issued 1838).

**Graham**, C.A.E. Military plans. Dublin District, 1828 MS.

**Graham**, C.B. Worked with J.R. Graham as C.B. & J.R. Graham, later traded alone in Washington D.C.
- **C.B. & J.R. Graham** (*fl.*1835). Engravers and lithographers of 'N°·4 John St. New-York'. John Farmer's *Map of the City of Detroit in the State of Michigan*, 1835.
- **C.B. Graham's Lithography**. Active in Washington D.C. Produced many maps to accompany reports to Congress. *Map of that part of the Mineral Lands adjacent to Lake Superior ceded to the United States by the treaty of 1842 with the Chippewas*, 1845 (drawn by A.B. Gray and J. Seib); *Geological Diagram of the field notes of the surveys of township and subdivision lines, in the Northern Peninsula of Michigan*, 1846; *Isthmus of Nicaragua*, 1847; *New Mexico*, 1847; *Battles of Mexico*, Washington 1847; *South Pass to Great Salt Lake*, 1852; *Plat and subdivision of the Baltimore and Ohio rail road company's property, in... the city of Washington*, 1853; *Sketch of the Public Surveys in Michigan*, St. Paul, Surveyor General's Office 1857; *Preliminary Survey of Ocracoke Inlet North Carolina*, 1857; *St. Helena Sound South Carolina*, 1857; *St. Louis Bay and Shieldsboro Harbor Mississippi*, 1857; *Channel Cut of Middle Channel, Lake George St. Mary's River...*, 1858 (surveyed by J.H. Forster, drawn by J.T. Baker).

**Graham**, (later *Sir*) Cyril Clerke (*d.*1895). Elected to the Royal Geographical Society in 1858, Governor of Grenada in the mid 1870s. *Map to illustrate explorations in the desert east of the Haurán, and in the ancient land of Bashan... 1858* (engraved by J. Arrowsmith), in *Journal of the Royal Geographical Society* (Vol.28), London 1858.

**Graham**, G. (1675-1751). Instrument maker. Devised a form of 'planetarium' to demonstrate the movements of the planets. This was later developed by John Rowley, who gave it the name 'Orrery' in honour of Charles Boyle, Earl of Orrery.

**Graham**, George. American artist and draughtsman, drew the title cartouche and vignettes for Osgood Carleton's *Map of the District of Maine Massachusetts...*, 1801.

**Graham**, George. Waikato District, New Zealand, 1864.

**Graham**, H. Landscape painter. Corrected *A plan of the Lakes of Killarney* [Ireland], Dublin 1786 (engraved by J. Ford).

**Graham**, H. & C. Ltd. Color lithographers of S.E. London. Disease distribution maps in Prof. W.J. Simpson's *Report of Sanitary matters in East African Protectorate*, 1915.

**Graham**, *Lieutenant* Herman. *Rough Sketch of Routes through south-west Awunah* [Ghana], 1889.

**Graham**, J.E. Plan of Jeffersonville, 1817 MS.

**Graham**, J.R. Engraver and lithographer of New York. Worked with C.B. Graham *q.v.*

**Graham**, James (1827-1878). Surveyor, produced civil engineering maps of sites throughout England and Wales. Wicklands Level, Essex, 1858 MS.

**Graham**, *Major* James Duncan (1799-1865). Officer in the United States Corps of Topographical Engineers, directed and endorsed many surveys in USA, including maps of Canadian boundary lines (1840s), and later, as Superintending Engineer of Lakes Harbor Works directed the preparation of charts and plans of the Great Lakes harbours (1850s). One of the surveyors for *Charleston Harbour and the Adjacent Coast and Country, South Carolina* (4 sheets), 1823-1825 (engraved by W.J. Stone); *A Map of the Extremity of Cape Cod...*, surveyed 1833-1835 published 1836 (with others, engraved by W.J. Stone); surveys of the Sabine River and Pass, Texas, 1840-1841; *Map of the Boundary Lines between the United States and the Adjacent British Provinces*, 1843 (directed by J.D. Graham); Ports on Lake Michigan (12 sheets), Chicago 1854-1857 (directed by J.D. Graham); *Black Lake Harbor Michigan*, 1856 (surveyed by J.D. Graham).

**Graham**, John. *Map of Franklin Co.* [Ohio], Philadelphia, R.C. Foote 1856.

**Graham**, *Doctor* Patrick. Stirlingshire [Map of the Soils and Roads in Stirlingshire], 1812; Kinross & Clackmannan [soils, rivers and roads of Clackmannanshire], 1814.

**Graham**, W.J. Draughtsman for James White's *Topographical Map of the Rocky Mountains. Banff Sheet*, Ottawa 1903.

**Graham**, W.R.M. Compiled maps of the Provinces of East India & Bengal for the Surveyor General's Office in Calcutta, 1846-1852, including: *Map of the District of Ghazeepoor*, 1846 (surveyed by Lieutenant W. Maxwell 1839-1841, drawn by Mirza Mogulijann); *Map of the District of the Goruckpoor*, 1846 (lithographed by J. Wyld); *Map of the District of Bijnour*, 1847 (surveyed by Captain B. Brown 1833-1841).

**Grainger**, —. Engraver. Macao, *c.*1785.

**Grainger**, Thomas (1794-1852). Scottish civil engineer. Railway and civil engineering surveys in Scotland and England, including Plan of Riccarton, 1824 (used by James Horne for *Plan of the ground permanently occupied by the Union canal through Riccarton...*, 1833).

**Gramm**, Hans (1685-1748). Danish historian and mathematician, chief librarian to the Danish royal family. Founder of the Koneglige Danske Selskab [Danish Royal Society].

**Grambo**. See **Lippincott**, Grambo & Co.

**Grammaye**, J.B. (1579-1635). Belgian geographer and historian, *b.* Antwerp, *d.* Lübeck. Africae illustratae, 1622.

**Gramolin**, Alvise. Portolan maker of Venice. Portolans 1612, 1630.

**Granchain** [Grandchain?], *Captain* M. de. French Marine. *Plan de la Baye de St. Lunaire* [Newfoundland], 1784-1785.

**Grand**, — Le. See **Legrand**.

**Grand**, *bey* P. *Plan de la Ville du Caire* (4 sheets), 1874.

**Grandchamps**, —. *Département des Alpes Maritimes*, 1865.

**Grandi**, Francesco, *SJ* (*fl.*mid-18th century). Italian globemaker.

**Grandidier**, family. **Ref.** BROC, N. *Dictionnaire illustré des explorateurs Français du XIXe siècle: I Afrique* (1988) pp.165-168 [in French].

• **Grandidier**, Alfred (1836-1921). French traveller and naturalist, 1862-1864. Undertook a major survey of Madagascar (botanical, zoological, archaeological and geological as well as geographical), published under the title *Histoire physique, naturelle et politique de Madagascar* (30 volumes with maps), 1875-1917.

•• **Grandidier**, Guillaume (1873-1957). Continued work of his father Alfred on the *Histoire physique...*; also *Atlas des colonies françaises*, 1934.

• **Grandidier**, Ernest. Accompanied his brother Alfred on a journey to South America, 1857-1858.

**Grandis**, Alvise. *La Veneta Laguna*, 1799, 1820.

**Grandison**, H. Script engraver for Ordnance Survey. Waterford, Dublin 1842.

**Grandjean**, A. *Carte du Nkomati Inférieur et du District Portugais de Lourenço Marques*, Neuchatel, Maurice Borel 1893 (printed by A. Gendre).

**Grandpré**, A. de. Draughtsman in Canada. *Topographical Map of the Mount-Royal* [Montreal], 1898; *Around Mount Royal Park*, Montreal 1905; map of the city of Montreal, 1907; *Quebec, Montreal, and Eastern Townships*, Montreal 1907; Greater Montreal and vicinity, Montreal 1914; *Planiglobe of the Antipodes*, Montreal 1917.

**Grandpré**, Louis-Marie-Joseph O'Hier, *comte* de (1761-1846). French naval officer. *Plan de la citadelle du Cap de Bonne Espérance*, 1793, in *Voyage à la côte occidentale de l'Afrique*, Paris, Dentu 1801; *Dictionnaire universelle de géographie marine*, 1803.

**Grand-Voinet**, —. Geographer and engineer to the King Louis XV of France. *Carte géometrique de la châine des Pyrénées*, 1772.
**N.B.** Compare with **Grandvoynet**, below.

**Grandvoynet**, —. Worked with others in the Pyrenees and Languedoc-Roussillon on at least 4 sheets (20, 40, 58, 59, 59+) of the Cassini *Carte de France*, 1771-1778. **N.B.** Compare with **Grand-Voinet**, above.

• **Grandvoynet**, *père et fils*. Credited on sheet 20 (Mont-Louis) of the Cassini *Carte de France*, 1772-1777.

**Granelli,** Carlo, *SJ* (1671-1739). Jesuit theologian, cartographer and mathematician from Milan, worked in Vienna.

**Graner,** Fredrik. *St. Clair County, Illinois,* c.1854.

**Grange,** *Doctor* —. Geological map of Tasmania, 1847.

**Grange,** M.C. *Plan de la Ville de Poitiers,* Poitiers, E. Druinaud 1882 (used in *Atlas Générale de la Vienne,* L. Chanche 1882); *Plan Topographique de la ville de Chaumont,* Chaumont 1897.

**Granger,** —. *Map of Birmingham, c.1860.*

**Granges,** G. des. See **Gagnières** des Granges, Claude François.

**Granges,** — des. See **Desgranges**.

**Grangez,** Ernest. *Navigation de la France,* 1840; *Belgique & Hollande,* c.1877; and others.

**Granier,** —. *Montagnes de l'Arpette,* c.1760.

**Granier,** F. Painted a circular map of western France with Niort at its centre, 1636 MS (Public Record Office, Kew).

**Grant,** A.A. Standard indexed atlas, 1885; *Grant's Standard American Atlas of the World,* New York 1887; *Grant's Bankers & Brokers Railroad Atlas,* 1892.

**Grant,** Charles, *vicomte* de Vaux. *Nova Uranographia,* Brompton 1803 (star map).

**Grant,** Daniel. *Ecclesiastical map of England and Wales,* 1851.

**Grant,** Edward (*fl.*1736-1738). Estate surveyor of Somerset and Wiltshire. *Plan of Redlynch Park, Somerset,* 1738.

**Grant,** E. *Paris monumental et ses Environs,* Paris, c.1855.

**Grant,** E.S. [& Co.]. Publishers of Philadelphia. Thomas Gamaliel Bradford's *Illustrated Atlas... of the United States,* c.1838.

**Grant,** F.N. One of the surveyors of Nanaimo Harbour, Vancouver Island, 1899 (with B.T. Somerville, F.H. Walter and G.B.S. Simpson), published at the Admiralty 1901, 1906 (engraved by E. Weller).

**Grant,** *Lieutenant* J., *RN.* Australia, 1800-1802; *Chart of the North and West Parts of Bass's Straights,* 1803.

**Grant,** *Colonel* J.A. See **Grante,** James A.

**Grant,** J. Murray *F.R.G.S.*. *Sketch Map of Kaffraria...* [Cape Province], Cape Town, Saul Solomon & Co. 1872; *Map of part of Cape Province around East London...,* c.1875; military sketch of Transkei, 1875 (with G.P. Colley).

**Grant,** James. Surveyor and draughtsman, deputy surveyor of lands for the northern district of North America. Many of his surveys were used by Des Barres in *The Atlantic Neptune.* Surveyor for *A Plan of the River Connecticut...* (6 sheets), c.1760 MS (drawn by Thomas Wheeler); *A Plan of the Sea Coast from Ogunkett River to Cape Elizabeth....,* c.1765; worked with Thomas Wheeler on surveys of the Massachusetts coast, c.1770; Plan of Piscataqua Harbour, 1774; *A Plan of the Bay and Harbor of Boston,* 1775 (with J. Wheeler); *Plan of Perth Amboy,* c.1780 MS (used in John Hills's manuscript atlas of New Jersey); South East Coast of the Island of St. John, 1781 (for Des Barres); one of the surveyors for *A Topographical Map of the Province of New Hampshire,* London, William Faden 1784.

**Grant,** *Captain* (later *Lieutenant-Colonel*) James Augustus (1827-1892). British army officer from Nairnshire, Scotland. Accompanied John Hanning Speke on the journey of 1860-1863 which finally located the source of the Nile in Lake Victoria. Awarded the Gold Medal of the Royal Geographical Society, 1864. **Ref.** WALDMAN, C. & WEXLER, A. *Who was who in world exploration* (1992) p.291.

**Grant,** *Captain* P.W. Maps of Burma including Martaban, Ye, Tavoy & Marguie, 1870.

**Grant,** Robert (1814-1892). Scottish astronomer. A.K. Johnston's *Atlas of Astronomy... with an elementary Survey of the Heavens by R. Grant,* Edinburgh & London 1869 (also in *School Atlas of Astronomy*).

**Grant,** *Major* Samuel Charles Norton. Credited with the hillshading for H.H. Kitchener's *A Trigonometrical Survey of... Cyprus* (15 sheets), 1882 published London, Edward Stanford 1885; Anglo-Portuguese boundary, East Africa, 1893; *Sketch showing the proposed boundary of the Sultanate of Witu* [Kenya], 1897; Communications in Natal (18 sheets), 1897; *Atlas of the boundary between Guiana and Venezuela* (with Major-General Sir John Ardagh).

**Grant,** T.M. *The Western Lakes & Sounds, Middle Island NZ*, Wellington, New Zealand Survey 1888; *Mt Cook (Aorangi), its Glaciers & Lakes*, Wellington, New Zealand Survey 1888.

**Grant,** *General* Ulysses Simpson (1822-1885). President of the United States of America. *Field of Operations* [United States], 1865.

**Grant,** Vincent. *Survey of Deeping Fen* [Lincolnshire], *c.*1670 MS published 1763 (with additions by Joseph Featherstone).

**Grant,** W.P. Publisher of Cambridge. *Fifteen miles round Cambridge, c.*1865.

**Grant,** William. The Antiquity and Excellence of Globes, 1652.

**Grant & Co.,** Hudson Tunnel Railway, New York, 1898.

**Grant & Griffith.** Publisher of London. *The Complete War Map*, 1854.

**Grant-Wilson,** James S. *Bathymetrical Chart of Lochs Tay, Rannoch, Tummel & Earn,* published in *Scottish Geographical Magazine* 1888.

**Grante** [Grant], *Colonel* James A. [*baron* d'Iverque]. French officer of Scottish descent, joined Prince Charles Edward Stuart's campaigns of 1745-1746. France, 1745; *Carte où sont tracées toutes les différentes routes, que S.A.R. Charles Edward Prince de Galles, a suivies dans la grande Bretagne...* (9 sheets), Paris, Bernard-Antoine Jaillot 1747, Scottish edition (2 sheets), Edinburgh 1747 (engraved Alexander Baillie), single sheet edition published with the title *A Chart Wherein are mark'd all the different Routs of P[rince] Edward in Great Britain...*, Edinburgh 1749. **Ref.** SHIRLEY, R. *Printed maps of the British Isles 1650-1750* (Tring & London 1988) pp.61-64.

**Grantham,** Edward (*fl.*1688-1736). Estate and inclosure surveyor. Worked various parts of England including *Mannour of Seasonscot, Gloucestershire*, 1704 MS; Manor of Burston, Buckinghamshire, 1720 MS.

**Grantham,** *Captain* James. *Colony of Natal*, 1861; *Natal* (4 sheets), 1863.

**Grantham,** John. R. Shannon, 1830; *Plan of Belfast Harbour*, 1852.

**Grantham,** William. *Mannor of Testwood, Hampshire*, 1755.

**Granton,** G.R. *Midlothian*, 1795.

**Grantzow,** C. *Provinz Schlesien*, 1855.

**Graphaeus** [Grapheus; Schryver; Scribonius], Cornelius (1482-1558). Flemish poet from Aalst. Contributed verses to the reader in Gemma Frisius's *Arithmeticae practicae methodus facilis*, Antwerp 1540. **N.B.** Credited by R.V. Tooley with a map of Antwerp, 1565 (with Virgile de Bologne).

**Graphaeus** [Graphei], Johannes. Publisher of Antwerp. Latin editions of Peter Apian's *Cosmographicus Liber*, 1529, 1533.

**Grapow,** *Captain* —. Prussian naval captain. Surveys of the North Sea, including the Elbe, Jade, Weser & Eider rivers, 1868 (surveys used for an Admiralty Chart published 1869); Schleswig-Holstein (2 sheets), 1869.

**Gras,** Henri le. 'Libraire au 3e pilier de la grande salle du Palais, Paris'. Published playing card maps by Jean Desmarests under the title *Le Jeu de geographie*, 1644 and re-issues to 1698 (engraved by Stefano della Bella).

**Gras,** J. See **Honter,** Johannes.

**Gras,** Scipion. *Carte géologique du terraine anthracitère des Alpes de la France et de la Savoie*, 1855; *Carte Géologique du Départment de Vaucluse*, Avignon and Paris 1861-1862.

**Grasset,** F. *Plan de l'Île et des ports de Mombaze*, Dépôt de la Marine 1851; *Plan du Port d'Ambavaranou ou Baie Rigny* [Madagascar], Dépôt de la Marine 1851; *Carte de l'Île d'Abd-El-Kouri* [Gulf of Aden], 1847 (with L. Caraguel), published Paris 1851; *Côte de Syrie*, 1855.

**Grasset**, François (1722-1789). Printer, publisher and bookseller of Lausanne. *Carte dela Suisse ou sont les treize cantons, leurs Alliés, et leurs sujets*, 1769; republished Robert de Vaugondy's *Carte Nouvelle de l'Isle de Corse*, 1769 (first published Paris, 1756).

**Grasset de Saint-Sauveur**, Jacques (1757-1810). *b.* Montréal. Geographical works including Encyclopédie des voyages, 1796.

**Grassi**, Bartolomeo. Puteolis, 1584.

**Grassi**, Orazio. *Globe Maritime avec la cognoissance et pratique des longitudes...*, Paris, Jean Bessin 1622.

**Grassi**, Raniero. Engraver of Pisa. *Pianta della città di Pisa*, Pisa 1831.

**Grassmann**, R. Publisher of Szczecin, Poland. *Neuer Plan von Szczecin*, c.1845; Schul-Atlas, 1853.

**Grassmüller**, E. Great Streams of the World, 1834.

**Grassom**, John (*fl*.1817-1819). Scottish surveyor. *County of Stirling* (4 sheets), 1817 (engraved by J. & G. Menzies); *Town of Stirling*, 1819 (used as an inset on the map of Stirlingshire in John Thomson's *Atlas of Scotland*, 1820).

**Grataroli**, Guiliermo (1516-1568). Italian physician, *b.* Bergamo, *d.* Basle. Itinerary, Basel 1561.

**Gratorix**, Ralph. See **Greatorex**, Ralph.

**Gratia**, —. Engraver. *Environs de Mascara*, 1837; *Carte Générale des Chemins de Fer de France et d'une partie des Etats Limitrophes*, Paris, Napoléon Chaix & Co. mid-19th century.

**Gratiani** [Graziani], Paolo. Publisher in Rome; acquired some of the Lafreri-Duchetti plates. Sicily, 1582; Acquapendente, 1582.

**Gratiot**, C. Map of the Island of Michilimackinac, c.1843.

**Grattan**, Edward. Publisher of 51 Paternoster Row, London. Re-issued M.J. Godwin's *The Pedestrian's Companion, Fifteen Miles Round London*, c.1824 (engraved by S. Hall, first published by Godwin, 1822, 1824 edition issued with a guide by J. Gilbert); *A Guide through London and the Surrounding Boroughs*, 1836, 1837, 1840; *Porteus's Plan of London*, 1837; *The Pedestrians' Companion, Fifteen Miles round London*, 1837, 1838, 1840 (engraved by S. I. Russell).

• **Grattan & Gilbert**. Partnership of Edward Grattan and James Gilbert, traded as publishers and City of London agents for the Ordnance Survey, at 51 then 49 (1842) Paternoster Row, London. Published works by J. Gilbert and others. The company went bankrupt in 1842. J. Archer's *New Plan of London*, 1840 (in J. Gilbert's *The London Director...*, later used by Gilbert in his *Gilbert's Visitor's Guide to London*, 1851); *Modern Atlas of the Earth*, 1841; publishers of the first sheets of *Gilbert's County Atlas*, 1842 (engraved by Joshua Archer, after only 9 sheets the atlas was taken over by Fisher, Son & Co. to become *Fisher's County Atlas of England and Wales*).

**Grattan**, William. Map of San Antonio de Bexar, 1836.

**Gratton**, J. Sterland. *Plan of Ilkley*, 1885.

**Graurock**, — von. Prussian military cartographer. *Karte von Klein-Asien u. Syrien*, 1840.

**Gratz**, L.G. German cartographer. Collaborated with J.F. Allioli. Palestine (9 sheets), 1842-1844.

**Grave**, Hans. See **Graf**, Hans.

**Grave**, Heinrich. *Plan von Wien*, 1871.

**Gravelot** [Bourguignon], Hubert-François (1699-1773). French engraver and designer who worked both in London (1733), and Paris 'vis-à-vis les Pères de l'Oratoire, rue St-Honoré' (1754). As younger brother of Jean-Baptiste Bourguignon d'Anville, his family name was Bourguignon, but he worked under the name of Gravelot. Chart St Domingo, c.1730; charts for John Pine's Armada series, 1739; designed cartouches for J.-B.B. d'Anville's maps, 1746-1763 including *Amerique Septentrionale...*, 1746 (map engraved by G. Delahaye, cartouche designed by Gravelot, cartouche engraved by Major); cartouches for Buy de Mornas's *Atlas Méthodique et Elementaire de Géographie et d'Histoire...*, Paris, Buy de Mornas 1761 (also published by Desnos). **Ref.** PICK, A. in *The Map Collector* 42 (Spring 1998) p.50 (letter).

**Graves, —.** Geologist. *Carte géologique des Environs de Paris*, 1865.

**Graves, R.W.** *Flushing, Queen's County, New York*, 1871?.

**Graves,** *Reverend* Rosewell Hobart (1833-1912). Baptist Missionary in South China. Kwei Hong, 1866 MS.

**Graves,** *Lieutenant* (later *Captain*) Thomas (*d.*1856). *RN.* Admiralty surveyor. Succeeded Richard Copeland on the British Naval Aegean Survey aboard the *Beacon*, 1836, later captain of HMS *Volage*. Worked on charts of the west coast of Asia Minor and the Grecian Archipelago, his surveys forming the basis of many later Admiralty charts. Mediterranean Survey, 1832-1850; Gulf of Smyrna, 1836-1837; *Greece: Athens to Corinth...*, 1838, published 1843 (engraved by J. & C. Walker); Naxia, 1842; plan of Troy, 1842; *Cyprus...*, 1847-1850 (surveys formed the basis of several later maps); *Greece Talanta and Oreos Channels*, 1846-1847, published Hydrographic Office 1857 (engraved by J. & C. Walker); *The Harbour of Famagousta, with the ruins of Salamis*, 1850.
**Ref.** FISHER, S. *The Map Collector* 54 (1991) pp. 18-23.

**Graves, W.H.** American draughtsman and topographer. Worked for J.W. Powell on the U.S. Geographical and Geological Survey of the Rocky Mountain Region, 1878-1879, also assistant surveyor to D.L. Miller. Map of Utah Territory for C.E. Dutton's *Topographic and Geologic Atlas of the District of the High Plateaus of Utah*, 1879 (with others); county atlases for Massachusetts, 1895-1896.

**Graves & Hardy.** Eureka pocket atlas of the Red River Valley, 1894.

**Gravesande,** S.V. Volksplantingen in Essequebo [British Guiana], 1749; Rios Essequebe, 1750.

**Gravier,** Giovanni. *Pianta di Gibilterra*, 1762.

**Gravier,** Yves [Ivone]. Publisher and bookseller, '...sous la loge de Banchi', Genoa. Re-issued Jacques Nicolas Bellin's larger Mediterranean charts, 1790s; *Città di Genova*, 1789; *Guida per il viaggio d'Italia in Posta*, 1793; Joseph Roux's *Plans de ports et rades de la Méditerranée*, 1800; *Atlas maritime*, 1801.

**Gravius,** Nicholaes Theodorus. Publisher from Leeuwarden, worked in Amsterdam 1788-*c.*1800. Atlas van Duitsland, 1770; Nieuwe en Beknopt Kaart-boekje, vertoonende de XVII Nederlandsche Provinciën, 1770; G. Brender à Brandis's *Nieuwe... Zaken Reis-Atlas*, Amsterdam 1787; *Geographische en Historische Zak- en Reisatlas der Nederlanden*, 1789; *Nieuwe Reis- en Hand-atlas van Braband, Vlaanderen...*, 1789; *Zak-Atlas van Frankrijk*, *c.*1800.

**Gravius,** Sytse [Sixtus]. Dutch surveyor from Grouw, Friesland. Worked on Bernardus Schotanus à Sterringa's maps of Friesland, 1698.

**Grawert,** Gottfried (*c.*1670-1724). *Land-Charte des chur- und Fürstentums Sachsen...*, 1716.

**Grawert,** Julius August Reinhold von (1746-1821). Military cartographer of Königsberg [Kaliningrad], Prussia. Also worked on maps of Silesia. *...Pirmasens...*, 1793.

**Gray,** Alexander. *Ensign*, 40th Regiment of Foot. Fort Griswold [Connecticut], 1781 MS.

**Gray,** *Captain* Andrew, Quartermaster General's Department, Captain in Nova Scotia Fencibles. *Topographical map of Guernsey, Sark, Herm & Jethou*, William Faden 1816 (assisted by M.P. Goodwin).

**Gray,** Andrew B. United States surveyor and civil engineer. *River Sabine*, 1841; *Map of that part of the mineral lands adjacent to Lake Superior, ceded to the United States by the treaty of 1842 with the Chippewas*, 1845 (assisted by J. Seib, lithographed by C.B. Graham); drew John B. Weller's San Diego, 1849; United States-Mexican boundary, 1855.

**Gray,** *Captain* David. Ice chart Arctic, 1881; Arctic Ocean & Greenland, 1882.

**Gray,** Francis J. *RN.* Assisted with surveys in the Mediterranean around Corfu, Sicily and Malta, 1863-1866, published as Admiralty Charts, from 1865.

**Gray,** Frank Arnold. Mapmaker of Philadelphia, probably related to O.W. Gray (below). *Atlas of the United States*, 1874; *Maryland, Delaware and District of Columbia*, Philadelphia, O.W. Gray & Son 1875 (published in *Gray's atlas of the United*

States and *The National Atlas*); *New railroad map of the state of Maryland, Delaware, and the District of Columbia*, Philadelphia, O.W. Gray & Son 1876; New Map of the State of Texas, 1876; *Gray's new topographical map of Virginia and West Virginia*, O.W. Gray & Son 1877; *Gray's new map of Virginia*, 1879 (used in the Department of Agriculture's *Hand-book of Virginia*); *Gray's new map of Kansas*, Philadelphia, O.W. Gray & Son 1881; *Gray's New Map of Utah*, O.W. Gray & Son 1882; *Arizona and New Mexico*, 1884; contributed to M.R. Brown's *The continental atlas*, Philadelphia 1889.

**Gray**, *Honourable* Frederick. *Pianta del Porto di Tripoli*, 1842.

**Gray**, George Carrington. *New Book of Roads*, London 1824 (with maps by G.A. Cooke).

**Gray**, Henry. Surveys of Laurens District, South Carolina, 1820 used in Robert Mills's *Atlas of the State of South Carolina*, 1826.

**Gray**, Hugh. *Map of Canada &c.*, London, Longman 1809 (with inset *Bason of Québec*).

**Gray**, John. The Art of Land Measuring, 1757.

**Gray**, John. Part of Chopwell Royalty, 1846 MS.

**Gray**, John. *Marseilles, La Sable Co., Illinois*, 1868.

**Gray**, John C. *Atlas of the United States printed for the use of the Blind*, 1837.

**Gray**, Ormando Willis. Civil and topographical engineer of Danielsonville, Connecticut (1869); 55 North Sixth Street, Philadelphia (1872); 10 North Fifth Street, Philadelphia (1873-1876). Published his own atlases and also for a time collaborated with H.F. Walling on the production of US state atlases. County maps Canadian West and New York, 1859-1864; Atlas of Windham & Tolland Counties, Connecticut, 1869; *Official Topographic Atlas of Massachusetts*, Stedman, Brown & Lyon 1871 (with H.F. Walling); *New Topographical Atlas of the State of Pennsylvania*, Stedman, Brown & Lyon 1872 (with H.F. Walling); *New Topographical Atlas of the State of Ohio*, Cincinnati, Stedman, Brown & Lyon 1872 (with H.F. Walling); *New Topographical Atlas of the State of Maryland and District of Columbia*, Baltimore, Stedman, Brown & Lyon 1873 (with S.J. Martenet and H.F. Walling); [O.W.] *Gray's Atlas of the United States with general Maps of the World*, Philadelphia, Stedman, Brown & Lyon 1873, J.W. Lyon 1876 and editions to 1879 (some maps by Frank A. Gray); *Railroad Map of Michigan*, G.W. & C.B. Colton 1874; and others. **Ref.** RISTOW, W.W. *American maps and mapmakers* (Detroit 1985) pp.335, 427, 429-430; **N.B.** See also **Walling & Gray**.

• **Gray**, O.W. & Son. Publisher. *The National Atlas, Containing Elaborate Topographical Maps of the United States and the Dominion of Canada*, 1875 and editions to 1889 (with some maps by F.A. Gray); atlases of Dutchess and Essex counties, New York, 1876; Frank Gray's *Gray's New Topographical Map of Virginia and West Virginia*, 1877. **N.B.** Compare with Frank A. **Gray**, above.

**Gray**, W. Surveyor in North Carolina. *A Map of the Division Line between the Province of North & South Carolina...*, 1735 MS.

**Gray**, W. Vallance. Worked with C.B. Gifford on bird's-eye views of U.S. cities.
• **Gray & Gifford**. Draughtsmen, printers and publishers in California. *San Francisco*, 1868 (drawn, printed and published by Gray & Gifford); *San José*, San José, G.H. Hare 1869 (drawn and printed by Gray & Gifford).

**Gray**, Warren. *Map of LaFayette Co. [Wisconsin]*, 1860, 1866; *Grant County, Wisconsin*, 1868.

**Gray**, *Captain* William (d.1804). Pennsylvania Rifle Regiment. Prepared manuscript maps under the direction of Robert Erskine, geographer to Washington's army. These included a map of Colonel William Butler's line of march against the Iroquois at Unadilla, 1778; several maps covering *From Albany to Schoharie*, 1778.

**Gray & Son**. Steel engravers. Maps for James Bell's *A New and Comprehensive Gazetteer Of England and Wales*, Glasgow, A. Fullarton & Co. 1833 and later.

**Grayson**, John. Royal military surveyor and draughtsman. Plan buildings Royal Military Academy Woolwich, *c.*1810 MS.

**Grayston, —.** Series of Slate Cloth Wall Maps, 1893.

**Graziani, P.** See **Gratiani**, Paolo.

**Gream, Thomas** (*fl.*1782-1809). Surveyor with the Board of Ordnance, based at 7 Clifford's Inn, Fleet-street, London (1792) and Villiers St., Strand, London (1795). Assisted William Gardner and Thomas Cubitt with a survey of Jersey, 1787, published 1795 (engraved by J. Warner); *Environs of Brighthelmstone*, 1794; continued the work of William Gardner and Thomas Yeakell to produce *A Topographical Map of the County of Sussex* (4 sheets), London, W. Faden 1795 and later (engraved by Thomas Foot), reduced version, W. Faden 1799 (engraved by J. Palmer); Ordnance Survey of Kent, 1795-1799 published 1801 (with W. Gardner, T. Cubitt, G. Pink and T. Yeakell junior, thought to be the first Ordnance Survey map published); estate map Beckenham, 1809; plan of Brighton, 1817. **Ref.** BAYNTON-WILLIAMS, R. *The Map Collector* 71 (1995) pp.39-41.

**Greatheed, Samuel.** *Chart of the Islands within 2000 miles of Otaheite*, 1797.

**Greatorex, Albert D.** *Map of the Parish of Sutton, Surrey*, Sutton, W. Pile 1896 ('corrected by A.D. Greatorex'; published in *Pile's Commercial and General Sutton and District Directory*).

**Greatorex [Graterix], Ralph** (*c.*1625-1675). Surveyor, mathematician and instrument maker. Worked with Jonas Moore on an assessment of the destruction of the city of London following the Great Fire of 1666, used for John Leake's *An Exact Surveigh of the Streets Lanes and Churches contained within the Ruines of the City of London*, 1667.

**Greaves, C.,** *RN*. Admiralty hydrographer. *Chart of Brixham Roads in Tor-Bay, with the adjacent Shores and Town of Brixham Quay, showing intended breakwater* [Devon], London 1836 (C. Hullmandel's Lithography).

**Greaves, G.D.** Surveyor, worked in Natal. *Plan of part of the Division of D'Urban*, 1849 MS; Plan of country north of Pietermaritzburg, 1849 MS (with L. Cloete).

**Gredsted, F.** *Kjøbenhavn med Forstaederne*, 1875.

**Greebe, Fredrik Willem.** Publisher of Amsterdam. *Amsterdam*, 1765; J. Sels' *Kaart der Tafel Baay van Kaap de Goede Hoop*, 1784.

**Greef, Goert [Goirt] de.** Dutch cartographer and possibly surveyor, worked in Noord-Holland. *Het Gooi en Aangrenzende Gelsieden*, 1524.

**Greeff,** *Professor Doctor* R. *Ilha de São Thomé* [Gulf of Guinea], Gotha 1884 (with F.J. Araujo, used in Petermann's *Geographische Mittheilungen*).

**Greeley, Aaron.** American district surveyor. *Plan of Private Claims in Michigan Territory* (3 sheets), 1810; Military ground at Detroit, 1809.

**Greeley, Carlson [& Co.].** Atlas of Hyde Park, Illinois, 1880; Atlas of the Town of Lake, Illinois, 1883-1892; Atlas of Chicago, 1884, 1891.

**Greeley, Horace [& Co.].** 'Job Printers and Stereotypers, 29 Beekman St., New York'. Untitled map of the Chicago, Madison & Lake Superior railroad route, 1854.

**Green, —.** See **Longmans**.

**Green, Alexander Henry.** (1832-1896). Geological map of South Carr, Lincolnshire, *c.*1896.

**Green [Greene], Elizabeth.** See **Greene, Robert**.

**Green, F.A.** Route in Damaraland, 1858 MS.

**Green,** *Lieutenant* F.V. *Statistical maps of Washington and the District of Columbia* (13 maps), Washington 1880.

**Green, Francis** (1731-1791). English cartographer.

**Green, I.W.** Assistant to the Quarter Master General. *Plan of the Encampment on Finchley Common*, 1780 MS.

**Green, J.** *Astronomical Recollections* (19 plates), Philadelphia 1824.

**Green, James** (*fl.*1791-*d.c.*1829). Surveyor, produced manuscript maps of parts of England. Nottingham Canal, 1791.

**Green,** James (1781-1849). Civil engineer, active in Wales, England and Ireland. Mouths of the Taff and Ely, 1829.

**Green,** John. Pseudonym adopted by the Irish writer Bradock Mead. See **Mead,** Bradock.

**Green,** Lowthian. Tetrahedral Map of the World, 1899.

**Green,** *Captain* M. *Course of the Shut ul Arab*, 1858.

**Green,** R. See **Greene,** Robert.

**Green,** T. *Plan von München*, 1806.

**Green,** Valentine (1739-1813). Draughtsman, engraver and publisher from Salford, Lancashire. Worked in London as engraver to George III. He died in London. *Brampton Bryan Castle, Herefordshire*, 1778; *The History and Antiquities of the City and Suburbs of Worcester* (volume II), 1796 (with a map of the city and suburbs of Worcester by George Young, engraved by J. Russell).

**Green,** W. Manuscript plans of Bergen-op-Zoom, 1751.

**Green,** *Major General* William RE. Chief Engineer at Gibraltar, 1771-1782. Signed numerous surveys of Gibraltar.

**Green,** William. Chart of the Cocos Islands, 1779, published Dalrymple.

**Green,** William (1760-1823). Surveyor, artist and publisher from Ambleside. Worked on a survey of Manchester and Salford from 1787, published as *Plan of Manchester & Salford* (9 sheets), 1794. **Ref.** BAGLEY, J.J. & HODGKISS, A.G. *Lancashire: A history of the County Palatine in early maps* (1985) pp.54-55.

**Green,** William. *The Picture of England Illustrated* (2 volumes), London, J. Hatchard 1803, 1804 (printed by R. Butters).

**Green,** William. City of Melbourne, 1854.

**Green,** *Reverend* William Spotswood. New Zealand, 1884; *Selkirk Range*, 1888.

**Green & Wadham.** *City of Adelaide,* c.1860.

**Green Globe.** See **Quirini.**

**Greene,** —. *City of Jefferson and its Environs* [Indiana], Louisville, Kentucky 1868.

**Greene,** —. See **Sherrards,** Brassington & Greene.

**Greene,** father & daughter. **Ref.** TYACKE, S. *London map sellers 1660-1720* (1978) pp.117 et passim.
• **Greene** [Green], Robert (d.1688). Publisher and mapseller of London, freeman of the Merchant Taylors' Company, 'Near Ratcliff Cross', London (1673); 'at ye Rose & Crowne in ye middle of Budge Row' (1675-1688). Sold Robert Morden's *A new Map of Germany*, 1673; collaborated with Robert Morden on the publication of *A New mapp Of England Scotland and Ireland...*, 1674 (engraved by W. Binneman and F. Lamb); Wenceslaus Hollar's *A New Map of the Cities of London & Westminster...*, 1675, 1685; *A New Mapp Of The World...*, 1676; co-publication of John Adams's road map of England and Wales, 1677; *A New Map of Scotland with the Roads*, 1679; *A Mapp of Virginia, Maryland, New Jersey, New York & New England*, c.1680 and later issues, (with John Thornton, used in the Blathwayt Atlas); *The Royal Map of England...*, 1682 (engraved by F. Lamb, one copy known); A New Map of Canaan, 1684; England with Post Roads, 1686; *A Mapp of Ireland With the Roads and Baronies*, 1686; Thomas Holme's *A Map of the Province of Pennsylvania...* [cartouche title] / *A Map of the Improved part of Pennsylvania in America* [above map border], London, 1687 (with John Thornton, engraved by F. Lamb).
•• **Greene** [Green], Elizabeth (*fl.*1688-1689). Daughter of Robert Greene. Continued his business at the Rose and Crown in Budge Row, on behalf of her under-aged brother Nathan. Greene's plates passed to Philip Lea c.1689.

**Greene,** Francis V. Surveyor for the United States Northern Boundary Commission. Worked with W.J. Twining and others on a series of maps of the boundary line between the north west point of the Lake of the Woods, and the summit of the Rocky Mountains, 1872. **N.B.** Compare with Francis Vinton **Greene,** below.

**Greene,** Francis Vinton (1850-1921). Soldier, historian and engineer. Atlas

*Valentine Green (1739-1813). (By courtesy of the National Portrait Gallery, London)*

*Russian Army in Turkey*, 1877-1878, published 1879; republished as *Turquie d'Europe d'Après le Traité de Berlin, Juillet 1878*, Paris, Hachette & Cie 1914.

**Greene**, H.F. *Map of the lands of the Union Pacific Railroad Company*, 1869, 1870.

**Greene**, J.N. Civil engineer. Keweenaw Point [Michigan], 1880.

**Greenfield**, W.H. Worked for the Geological Survey Office of the Department of Mines, Queensland (from 1908). *Cyclists' road map, Brisbane and surrounding districts*, Brisbane, Surveyor General's Office 1896 (photo-lithographed by W.H. Greenfield); *Sketch map of the Etheridge Goldfield*, Brisbane, Department of Mines 1908; *Sketch Map of the Herberton & Chillagoes gold & mineral fields*, Brisbane, Department of Mines 1909 (prepared under the supervision of W.H. Greenfield); *Sketch map of east Central Queensland gold, mineral & coal fields...*, Brisbane, Department of Mines 1910; *Sketch map of the Croydon & Etheridge Goldfields*, Brisbane, Department of Mines 1911; *Sketch map of Kangaroo Hills, Star River, Charters Towers & Ravenswood gold & mineral fields*, Brisbane, Department of Mines 1912 (prepared under the supervision of W.H. Greenfield).

**Greenhill**, Henry (1646-1708). Governor of the Gold Coast; Commissioner of the Navy, 1691. Plans of forts on west coast of Africa, 1680-1682; chart of the coast of Africa, 1682.

**Greenhow**, Robert (1800-1854). Physician, scientist, linguist and historian. Translator for State Department. North West coast of America, 1840; West coast of North America, 1844 MS; *Map of the Eastern and Middle portions of North America...*, Philadelphia 1845 (drawn by George H. Ringgold, engraved by E.F. Woodward); *Map of the Western and Middle Portions of North America*, 1846.

**Greenland**, Herbert. Surveyor of Coles Chambers, King Street, Sydney. *Oxford Township French's Forest*, [n.d.].

**Greenleaf**, Jeremiah (1791-1864). *A New Universal Atlas*, Brattleboro Vermont 1840, 1842, Brattleboro and Boston 1848 (taken from David Burr's work of the same title, first published New York City, D.S. Stone *c.*1835, maps engraved by Illman & Pillbrow).

**Greenleaf**, Moses (1777-1834). Born in Newburyport, Massachusetts, began work as a storekeeper in Maine, from 1806 worked with William Dodd as a land developer. He settled in Williamsburg, Maine in 1810. From 1810-1815 worked on the compilation of a map of Maine, published as *Map of the District of Maine...*, Boston, Cummings & Hilliard 1815 (engraved by W.D. Annin); when Maine became a state in its own right he revised the map for publication as *Map of the State of Maine*, 1820; *Survey of the State of Maine*, Portland, Maine, Shirley & Hyde 1829 (map engraved by J.H. Young & F. Dankworth). **Ref.** RISTOW, W.W. *American maps and mapmakers* (Detroit 1986) pp.94-96; SMITH, E.C. *Moses Greenleaf, Maine's first map-maker* (1902).

**Greenly**, Edward (1861-1951). Geological surveyor mostly working in Northern Highlands of Scotland. Geological map of the Isle of Anglesey, Geological Survey 1919.

**Greenough**, George Bellas (1778-1855). First President Geological Society, 1811; President of the Geographical Society, 1839-1840; d. Naples. *A Geological Map of England and Wales* (6 sheets), London, Longman, Hurst, Rees, Orme & Browne 1819, 1839; Wales, 1839; *General Sketch of the Physical and Geological features of British India*, London, E. Stanford 1855 (engraved by A. Petermann).

**Greenshield**, —. *Route on the Niger*, 1885-1897 (with Hamilton).

**Greensill**, Thomas. Surveyor and architect. Plan of the Town of Australind [Western Australia], 1870.

**Greensted**, Edward (*fl.*1751-1769). Estate surveyor. *Estate plan Wateringbury*, 1769 MS.

**Greenwood**, brothers. Land surveyors from Gisburn, Yorkshire.
• **Greenwood**, Christopher (1786-1855). Surveyor, mapmaker and publisher from Yorkshire. Possibly trained locally as a land surveyor, worked later in Dewsbury, then Wakefield from *c.*1815, where he launched his survey of Yorkshire. This was followed by Lancashire. He then directed the surveys for a whole series of county maps with the aim of producing a large scale atlas of the English and Welsh counties. He established an office at N$^{o.}$50 Leicester Square, London in 1818, then N$^{o.}$70 Queen Street, Cheapside (1820), he was joined by his brother John in 1822. Some of his maps were published by (or co-published with) his father-in-law William Fowler, or from 1819 George Pringle and his son George Pringle Junior. Christopher also worked as a publisher with Thomas Sharp and William Fowler as Sharp, Greenwood & Fowler *q.v.* The incomplete county atlas set began with *Map of the County of York* (9 sheets), Hurst, Robinson & Company 1818, H. Teesdale 1828 (based

on a grid laid out by Francis & Netlam Giles, surveyed from 1815, engraved by S.J. Neele & Son); *Map of the County Palatine of Lancaster* (6 sheets), Wakefield & London, W. Fowler & C. Greenwood 1818 (engraved by S.J. Neele & Son, plates sold to Cruchley, revised, re-engraved edition published G.F. Cruchley 1836); *Map of the County Palatine of Chester* (4 sheets), Wakefield, W. Fowler & C. Greenwood 1819; *Map of the County of Middlesex* (4 sheets), London, G. Pringle & C. Greenwood 1819; *Map of the County Palatine of Durham* (4 sheets), London, G. Pringle & C. Greenwood 1820, c.1824, 1825, 1846; *Map of the County of Stafford* (4 sheets), London, W. Fowler & C. Greenwood 1820; *Map of the County of Wilts.* (4 sheets), 1820, c.1826; *Map of the County of Kent* (4 sheets), London, G. Pringle Jun$^r$ 1821, 1827, E. Ruff & Co. c.1837 (engraved by S.J. Neele & Son); preparation of other county maps continued with his brother John; later lone productions included *An Epitome of County History, Vol.I County of Kent*, 1838-1839 (published in 2 parts; there were no further volumes). **Ref.** HARLEY, J.B. *Christopher Greenwood: County map-maker and his Worcestershire map of 1822* (Worcester Historical Society, 1962).

• **Greenwood**, John (1791-1867). Surveyor and publisher, assisted his brother with the compilation of county and other maps from 1822. Returned to Yorkshire before 1838, where he continued to work as a land surveyor.

• **Greenwood & Co.** (fl.c.1821-1835). County surveyors, and later publishers, noted at 70 Queen Street, Cheapside, (1821); 174 Piccadilly (1823); Waterloo Place (1824); Regent Street, Pall Mall, (1827-1830); 21 King Street, Covent Garden (1832); Aldine Chambers, Paternoster Row (1833-1834); Burleigh Street, Strand (1834). After 1827, succeeded Greenwood, Pringle & Co. in the publication of C. & J. Greenwood's series of large scale county maps. By this time, the plates for Yorkshire were in the hands of Henry Teesdale and C. Stocking, and those for Warwickshire had been sold to Edward Ruff, other plates were sold off later. Publishers for C. & J. Greenwood's *Atlas of the Counties of England*, 1828-1834.

• **C. & J. Greenwood** (fl.c.1822-1834). Christopher and John Greenwood, worked together on the continuation of Christopher's series of large scale maps of English counties (mostly 1" to the mile), over 30 had been published by 1831, but the series remained incomplete. Other maps included *Map of London*, London, Greenwood, Pringle & Co. 1827, Greenwood & Co. 1830, Ruff & Co. 1835 and other editions to 1854 (surveyed 1824-1826, engraved by J. & J. Neele); *Map of the South East Circuit of the Principality of Wales comprising the Counties of Glamorgan, Brecon and Radnor*, London, C. Greenwood & Co. 1828; prepared a smaller scale series of county maps (3 miles to 1") published in parts under the title *Atlas of the Counties of England* (part one), C. Greenwood & Co. 1828, (part two), Greenwood & Co. 1830, (part three), Greenwood & Co. 1831, (part four) Greenwood & Co. 1834 (engraved by J. & C. Walker, J. & J. Neele, J. Dower and H. Frost); *Map of the County of Huntingdon*, London, Greenwood & Co. 1831 (the last in the incomplete series of large scale county maps).

• **Greenwood, Pringle & Co.** (fl.c.1824-1827). Publishers of N$^o.$13 Regent Street, Pall Mall, London (1824). Published the county maps of C. & J. Greenwood. George Pringle Junior left the partnership in 1827.

**Greenwood**, A. *An introductory atlas of international relations* (47 maps), London, Headley bros. 1916 (with H. Clay, included maps by H.S. Hattin).

**Greenwood**, J. (fl.c.1830-1840). Copper and wood engraver, and copper-plate printer of Hull. Engraved the map for *The Trent And Humber Picturesque Steam-Packet Companion...*, 1833; *Lincolnshire*, 1836 (taken from the map which appeared in the 1811 edition of *Cary's New English Atlas*); *Holderness* 1840.

**Greenwood, Pringle & Co.** See **Greenwood, brothers**.

**Greer**, William. *Plan Surigao Bay*, 1762, published Dalrymple 1774; *Track of the 'Nassau' through the Cocos Islands*, 1779, published Dalrymple 1792.

**Gregg**, Josiah (1806-1850). Trader and writer. *Map of the Prairies with parts of the adjoining frontier of the United States and Mexico*, c.1841; *Map of the Indian Territory Northern Texas and New Mexico*, 1844 (in Gregg's *Commerce of the Prairies...*, 1844).

**Grégoire**, Louis (1818-1897). French historian and geographer. *Planisphère Grégoire Surface* *de Sphéroid-Terrestre*, Paris 1875; *Géographie Illustrée*, before 1882; *Géographie, physique, politique et économique de la France et de ses colonies*, Paris, Garnier Frères 1883; *Géographie de la France*, Paris, Garnier Frères 1885; *Nouvel Atlas de Géographie Moderne* (43 maps), Paris, L. Grégoire 1890; *Atlas Général de Géographie Moderne*, Paris, Garnier 1891; *Atlas Universel de Géographie Physique & Politique*, Paris, Garnier Frères *c*.1892.

**Grégoire**, Pierre. New map of Spain and Portugal, 1799 (E. Mentelle, J. Stockdale, E. Chanlaire & S.J. Neale are also credited).

**Grégoire**, R.P. Plan de Lyon, Lyon 1740 (with Claude Seraucourt).

**Gregoras**, Nicephoros (*c*.1295-1360+). Byzantine scholar. Wrote a revision of part of Ptolemy's *Geographia*.

**Gregorii**, Johann Gottfried (1685-1770). German pastor, historian and geographer. *Geographia novissima* (2 volumes), Frankfurt 1708; *Curieuse Gedancken von denen... Land-Charten*, Frankfurt 1713; *Atlas Portatilis*, Nuremberg 1717; *Atlas Portatilis Germanicus*, Nuremberg 1723.

**Gregorius**, F.G. Cosmographia novissima, Paris 1816.

**Gregory**, —. Engraver of Liverpool. Dublin, 1809.

**Gregory**, brothers. English explorers from Nottinghamshire, travelled extensively in Australia both individually and together. *Journals of Australian Explorations*, Brisbane, James C. Beal 1884. **Ref.** WALDMAN, C. & WEXLER, A. *Who was who in world exploration* (1992) pp.295-296.
• **Gregory**, *Sir* Augustus Charles (1819-1905). Was taken by his family from his native Farnsfield, Nottinghamshire to Australia in 1829. Worked for the government of Western Australia as a surveyor from 1841. In 1846 he left Perth with his brothers Francis and Henry in an abortive attempt to cross Australia. He was later commissioned jointly by the Royal Geographical Society and the British government to explore and survey potential agricultural land in the Northern Territory, 1855-1857. In 1858 he began explorations of central Australia, working his way from Brisbane to Adelaide. The journals for these expeditions included many charts of coastal waters and maps of inland districts. Several accounts of expeditions were carried in the *Journal of the Royal Geographical Society*, London, from 1855. In 1859 he became Surveyor General of Queensland. *Map showing part of the route taken by North Australian Exploring Expeditions...* (2 sheets), *c*.1855; *Gregory's Journal of the North Australian exploring expedition*, 1858; his work was used as one of the sources for *Stanford's new map of Australia* (2 sheets), 1860; *Geology of Northern Australia*, 1865.
• **Gregory**, Francis Thomas (1821-1888). Explorer in Australia, brother of A.C. Gregory (above), and like him a surveyor for the government of Western Australia. *Expedition to the North-West coast of Australia* (with map), 1862.
• **Gregory**, Henry. Brother of Augustus and Francis. Also an explorer in Australia.

**Gregory**, Charles C. Civil engineer. New Brunswick, 1867.

**Gregory**, Edmund (1832-1913). Government printer and journalist in Australia. Map of Queensland, 1898; Ipswich Beds, 1899.

**Gregory**, H. American lithographer. *Map showing the Niagara and Detroit Rivers Railway...*, 1858 (with Charles Beard, lithographed by C.W. Hamilton).

**Gregory**, Henry. Publisher, 'Near the India House', London (1764). Bought William Herbert's business *c*.1775. *A New Map, or Chart in Mercators Projection of the Western or Atlantic Ocean, with part of Europe, Africa and America*, 1764; publishers and sellers of the 4th-6th editions of William Herbert's *A new directory for the East Indies* (46 charts), 1776, 1780, (56 charts) 1787 (first published by Herbert with 29 charts, 1758).

**Gregory**, I. Joint publisher of a map of Leicestershire, 1799 (surveyed by J. Whyman, engraved by J. Luffman, published W. Dawson, I. Gregory and J. Prior).

**Gregory**, Isaac. *Russian & Turkish Armies*, 1854; *War in Italy*, 1859.

**Gregory**, J.W. Compiler of *Oxford Wall Maps*, London, H. Frowde 1911.

*Sir Augustus Charles Gregory (1819-1905). (By courtesy of National Library of Australia)*

**Gregory,** *Captain* James F. Surveyor for the United States Northern Boundary Commission. Worked with others on a survey the boundary line between the north west point of the Lake of the Woods, and the summit of the Rocky Mountains (45 maps), 1872 (with A. Campbell, W.J. Twining and F.V. Greene).

**Gregory,** *Reverend* John. Opuscula, 1649 (use of globes); *The Description and Use of the Terrestrial Globe*, London 1663.

**Gregory,** John Bates. *The North Wales Coal Field*, Hope 1879 (with Jessie Price).

**Gregory,** T. *The Shropshire Gazetteer*, Wem 1824.

**Gregory**, T. Surveyor. Leasehold Estates, Wellington St., Strand [London], 1826.

**Greig**, George (1799-1863). Printer, publisher and campaigner for press freedom at Cape Town. Various plans of Cape Town and environs, used in *South African almanac and directory for the year 1830*, 1830 (and for 1831, 1832 & 1833); sketch of a small part of the Colony of the Cape of Good Hope, 1831 (engraved by Charles Cornwallis Michell).

**Greig**, N. Collector of Land Revenue (1816), and Superintendant of Quarantine (1827), Malta. *A Plan of the three Cities Senglea, Conspiqua and Vittoriosa...*, 1816 MS; *Malta* [Valletta & Floriana], 1816 MS.

**Greig**, William. Publisher of Montreal. *Plan of the City of Montreal from a Survey made by the order of the Mayor & Common Council in 1835 with the new Improvements to 1839.*

**Greipel**, Eduard von (*d*.1823). *Neueste General Karte des Österreich ob der Enns* (6 sheets), 1809.

**Greischer**, —. Publisher. Charles de Juvigny's Siege of Budapest, 1686.

**Greischer** [Gryscher], Matthias (*d.c.*1712). Engraver in Vienna. Johann Alexander Reiner's *Regni Hungariae*, Vienna 1683.

**Greive**, J.C., *junior*. *Java*, *c*.1876.

**Gremion**, —. Officer, Swiss Guards. Bohemia and Moravia, 1779.

**Grenard**, Joseph Fernand. *Asie Centrale*, 1898, 1899.

**Grenell**, *Captain* John. Served under Sir Henry Clinton. Hudson River Highlands, 1775 MS.

**Grenet**, *abbé* — (*b*.1750). French geographer, professor at Lisieux. *Atlas portatif...*, 1779-1782 (with maps by Rigobert Bonne); Italian edition by P. Santini, 1794; *Abrégé de géographie ancienne et moderne*, 1782.

**Grenfell**, *Lieutenant* A.B. Surveys of the coast of Jamaica, 1876-1879 (with others), used for Admiralty Charts 1880 and 1881.

**Grenfell**, *Reverend* George (1849-1906). Cornish missionary in Africa. Surveys of the rivers of the Cameroons, 1874; manuscript maps of the Congo, from 1880; surveys of the Congo River, 1884-1889 used in *A Map of the Congo River between Leopoldville and Stanley Falls*, Royal Geographical Society 1902.

**Grenier**, Jacques-Raymond, *vicomte* de Giron (1736-1803). French hydrographer. East coast of Madagascar, 1768; Foul Point, 1768; chart of currents in the Indian Ocean, 1770; maps for Bellin's *Hydrographie Françoise*, including: *Carte de l'archipel au nord de l'Isle de France* [Mauritius], Bellin 1776; and *Cartes du système des courants des Mers de l'Inde* (2 sheets), Bellin 1776; *A chart of Mahé and Amirantes Island...* 1776, published Laurie & Whittle 1800.

**Grenier**, L. Engraver of 'rue des Noyers Nº·33', Paris. C.V. Monin's *Amerique Septentrionale*, *c*.1830; engraver for Dépôt de la Marine, 1861, for A.-H. Dufour, 1864.

**Grenier**, Nicolas. Commissaire de la République Suisse. Environs of Geneva on Peter Martel's plan of Geneva, 1743.

**Grent**, William. English cartographer. *A New and Accurate Map of the World...*, Thomas Jenner 1625, 1632, 1641 **Ref.** SHIRLEY, R. *The mapping of the World (1983) Nº.313 pp.336-337.*

**Grenville** [Greville; Greynvile], *Sir* Richard (*c*.1542-1591). English naval Commander from Buckland Abbey, Cornwall, cousin of Sir Walter Ralegh. Commissioner of Works for Dover Harbour. In 1585 Grenville undertook an exploratory voyage to North America as part of Ralegh's plans to create a colony of Virginia in honour of the Queen. The small party left behind established a short-lived colony on Roanoke Island, North Carolina (the survivors were taken back to England by Sir Francis Drake in 1586). Dover, with proposed Pier, 1584 MS.

**Gressien**, Victor Amédée. Hydrographic engineer, sailed with Dumont d'Urville to Australia and New Zealand 1826-1829. His charts, mostly engraved by Chassant with lettering by Hacq, were published in J.S.C. Dumont d'Urville's *Voyage de la Corvette Astrolabe*, Paris, J. Tastin 1833.

*Sir Richard Grenville (c.1542-1591) by an unknown artist. (By courtesy of the National Portrait Gallery, London)*

*The Rt Hon Sir George Grey (1812-1898). (By courtesy of the National Library of Australia)*

**Gressier**, C.L. Hydrographical engineer. *Carte reduite de la baie de Todos os Santos...*, 1819-1823; *Plan de l'embouchure de la rivière de Cayenne*, 1820, and other charts of Brazil for the *Supplément* to *Neptune Françoise*, 1822-1824; Atlantique méridionale, 1834; Carte Hydrographique des parties connues de la Terre, Paris, 1835; *Carte de la partie Méridionale de la Mer du Nord*, Paris 1847.

**Gretschel**, Heinrich. Wrote theoretical works on map projections. *Lehrbuch der Karten-Projection*, Weimar 1873.

**Gretter**, P. Boller Landtafel, 1602.

**Greubel**, M. *Schulwandkarte Unterfranken*, c.1893.

**Greuter** [Greuther], Matthäus (c.1566-1638). Cartographer, astronomer, globemaker, artist, typecutter and engraver from Strasbourg. Worked in Lyon, Avignon, and Rome, where he died. *Disegno nuovo di Roma* (town plan, 8 sheets), 1618 re-issued 1626; *Frascati* (3 sheets), 1620; *Italia di Matteo Greuter* (12 sheets), 1630; 49-cm terrestrial globe, Rome 1632, republished Milan, Giovanni Battista de Rossi 1695, Rome, Calcografia della R.C.A. 1744; 49-cm celestial globe, 1636 (both globes were copied and reduced from Blaeu's 68-cm globes); *In hac coelesti Sphaera...*[26.5-cm celestial globe], Rome 1636 (copied from P. van den Keere and P. Plancius); 26.5-cm terrestrial globe, 1638 (copied from Blaeu).
**Ref.** DEKKER, E. *Globes at Greenwich* (1999) pp.344-349; Van der KROGT, P. *Globi Neerlandici* (Utrecht 1993) pp.211-213; KRETSCHMER, DÖRFLINGER & WAWRIK *Lexikon zur Geschichte der Kartographie* (1986) vol.C/1, p.277 [in German].

**Greve**, Wilhelm. Lithographer and printer of Berlin. Plans Strasbourg, 1874-1878; H. Kiepert's *Neue Karte von Bulgarien...*, Berlin, Dietrich Reimer 1877; H. Kiepert's *Carte des nouvelles frontières entre la Serbie, la Roumainie, la Bulgarie la Roumélie orientale au sud* (4 sheets), Berlin, Dietrich Reimer 1881 (engraved by Wilhelm Droysen); *Plan Bremen*, 1890; *Gross-Berlin*, 1906; *Karta des südlichen Theiles von Luzon und benachbarter Inseln auf Grundlage der Coello'schen Karta zur Reise von F. Jaugor*, [n.d.] (drawn by Richard Kiepert).

**Greville**, *Sir* R. See **Grenville**, *Sir* Richard.

**Grew**, Frederick. Engraver. *Warwickshire*, c.1870.

**Grey**, *Captain* —. Surveys on the west coast of Australia including *The Country from Gantheaume Bay to the River Arrowsmith*, used, together with those of others, by A. Arrowsmith for *Map and Chart of the West Coast of Australia from Swan River to Shark Bay...*, 1841.

**Grey**, Andre. See **Gray**, Andrew.

**Grey**, B. Draughtsman for Stephen Makreth's *A Plan of the Town of Lancaster*, 1778.

**Grey**, George (fl.1755-c.1763). Land surveyor, produced manuscript plans of various sites in England, including the river at Lancaster, 1755.

**Grey**, *Sir* George (1812-1898). British explorer and administrator, born in Lisbon, Portugal. Explored Western and north Western Australia, became Governor of South Australia, 1841-1845; Governor of New Zealand, 1845-1854, 1860-1868 (where he succeeded in improving relations between the Maoris and the English settlers); Governor of the Cape Colony, 1854-1860. *Journal of Two Expeditions of Discovery in North-West and Western Australia 1837-1839*, London 1841.

**Grey**, Herbert F. Charts of the Bosphorus & Dardanelles, 1830-1832.

**Grey**, *Major* R. *Limpopo River*, c.1891.

**Grey**, William. Survey Newcastle, 1649.

**Greynvile**, *Sir* R. See **Grenville**, *Sir* Richard.

**Grezel**, E. Letter engraver of Paris. Credited on some sheets of Victor Levasseur's *Atlas National Illustré*, Paris, A. Combette 1845 and later.

**Gridley**, Enoch G. Engraver, worked in Boston for Morse's American Gazetteer, 1797; in New York, 1803-1805; and in Philadelphia, to 1818. Engraver of *A Map of the United States of America*, for *Carey's General Atlas*, 1814; other work for Carey, 1817; Kentucky, 1818.

**Gridley**, *Lieutenant-Colonel* Richard (c.1710-1796). Surveyor, and Commander of Artillery at the Siege of Louisbourg. *A Plan of the City and Fortress of Louisbourg*, 1745, Boston, John Smibert 1746 (engraved by Peter Pelham), later London, Thomas Jefferys 1758, 1760, 1768.

**Grieben**, Karl Leopold Eberhard Theobald (1826-c.1914). Specialised in hiking maps and travel guides. Grieben's guide books (many volumes with maps and plans of Europe), London, Williams & Norgate, Berlin, A. Goldschmidt 1909-1913.

Griel, —. *Carte de la Rade Nouvelle de Cherbourg*, 1786.

Grienfeld, J.D.F. de. See **Floriantschitsch** von Grienfeld, Johann Dismas.

Grieninger, J. See **Grüninger**, family.

Grierson, family. Publishers in Dublin for 150 years, *c*.1703-1856. **Ref.** SHIRLEY, R.W. *Printed maps of the British Isles 1650-1750* (Tring and London 1988) pp.64-66; SMITH, David *Imago Mundi* 39 (1987) pp.91-92; SHIRLEY, R. *ibid.* pp.92-93.
• Grierson, George [I] (1680-1753). Born in Scotland and moved to Ireland *c*.1703. Publisher, printer and bookseller 'at the Kings Arms & 2 Bibles in Essex St., Dublin'. 'Kings Printer' (1732). Most of his output was re-issues and re-engravings of earlier maps. He was succeeded by his son George Abraham. Re-issued Sir William Petty's *Hiberniae Delineatio*, 1732; copy of Henry Pratt's *A Mapp of the Kingdom of Ireland...* (4 sheets), 1732, 1740; re-engraved Moll's *The World Describ'd...* for publication in Ireland, 1733, 1744; Thomas Salmon's *Modern History*, 1739 (pirated edition); *The Roads of the South Part of Great-Britain called England and Wales*, 1740; pirate edition of Moll's *Atlas Minor*, *c*.1745; book IV of the *English Pilot*, 1749 (engraved by James Barlow).
•• Grierson, George Abraham [II] (1728-1755). Elder son of George Grierson, succeeded his father as King's Printer on the death of his father in 1753. Two years later he was himself succeeded by his brother Hugh.
•• Grierson, Hugh Boulter Primrose (*d*.1771). Younger son of George Grierson, took over the firm on the death of his elder brother in 1755; moved to 28 Parliament Street in 1761. 'Printer to the King'. *English Pilot, Fourth Book*, 1767; compiled separate maps of Armagh, Down, Dublin, Louth, Kildare and Wicklow, 1770.
•• Grierson, Mary. Widow of Hugh Boulter greirson, managed the business on behalf of her under-age son George [III], 1771-1778.
••• Grierson, George [III] (*d*.1821). Grandson of George [I], son of Hugh Boulter whom he succeeded as King's Printer in Dublin. William Beauford's *New and Correct Irish Atlas*, *c*.1818 (most maps drawn by Beauford and engraved by Neele, the last three were drawn and engraved by J. Taylor).
•••• Grierson, George Abraham [IV]. Son of George III, succeeded his father as printer and publisher in 1821. He was joined by his brother John Grierson and Martin Keene who worked together as 'King's Printers' (*fl.c*.1822-1826). *New and Correct Irish Atlas*, *c*.1825 (with John Grierson and Martin Keene, maps by W. Beauford and J. Taylor, engraved by J. Taylor and S.J. Neele).
•••• Grierson, John. Publisher of Dublin. Worked with his brother George Abraham, and Martin Keene as 'His Majesty's Printers' in Dublin *c*.1822-1826.

Grieve, *Lieutenant* (later *Commander*) Albany Moore. Officer in the Indian Navy. Survey of India, 1845-1858; *Chart of the Gulf of Aden*, 1847 (with R. Barker and S.B. Haines), published John Walker 1852 (engraved by J. & C. Walker); *Survey of the Coast of Sindh and Cutch*, 1853 (with E. Brazier); and others.

Grieve, J. Lithographer and zincographer of 33 Nicolas Lane, London. Proposed embankments on the Thames from Fulham to Barking (7 maps and 1 plan), 1841.

Grieve, James. Australian engraver. *Proeschel's General Agricultural & Gold Fields Map of Victoria*, *c*.1858; *Map of Victoria*, 1865. **N.B.** Compare with **Slight & Grieve**.

Grieve, John. Plan for an additional pier on the Isle of Whithorn [Wigtownshire], 1792 MS.

Griffen, Joseph. *Elements of modern geography...* (with atlas), Glen's Falls N.Y. 1833 (printed by A. Smith).

Griffin, —. Publisher with John Bumpus as 'Bumpus and Griffin, 3 Skinner Street', London. *A New Map of London and the adjacent Villages...*, 1831, 1836.

Griffin, Bruce N. See **Griffing**, Bruce N.

Griffin, J.G. Surveyor of Jersey Chambers, 334 George Street, Sydney, N.S.W. *Ferrier's Estate, Rockdale*, 1892.

Griffin, James. *Rotary planisphere*, 1845.

Griffin, Peter. Publisher, map & print seller 'next y$^e$ Globe Tavern Fleet Street', London. *A Pocket Map of the Cities of London & Westminster...*, 1748 edition (engraved by Will Roades, first published R. Hulton 1731); one of the sellers of T. Kitchin and T. Jefferys's *The Small English Atlas*, 1749.

**Griffin**, Walter Burley (1876-1937). In 1911 he entered a competition for the design of a new Australian Federal Capital, later became Federal Capital Director Design and Construction. Produced several plans for the new city, including *Canberra, plan of city and environs*, Melbourne 1918 (used as the basis for several later plans).

**Griffin**, William. Printer and bookseller in Fetter Lane (1764-1767); 'Catherine St., in the Strand' (1767-1776), London. One of the sellers of Joseph Smith Speer's *The West-India Pilot...* (13 charts), 1766 edition (charts engraved by Pinald).

**Griffin**, William M. *Primary geography of the state of New Jersey*, Newark N.J., Advertiser Printing House 1884 (with C.E. Meleney).

**Griffing** [Griffin], Bruce N. (*fl.*1870s). Native of Newtown, Connecticut. Collaborated with D. Jackson Lake on some of his earlier North American county atlases. *Atlas of Pickawat County, Ohio*, 1871; *Atlas of Darke County, Ohio*, 1875, 1888; *Atlas of Brown County, Ohio*, Philadelphia, Lake, Griffing & Stevenson 1876 (with D. Jackson Lake); *Atlas of Franklin County, Kentucky*, 1882; *Map of Blackford County, Indiana*, Philadelphia, Griffing, Gordon & Co. 1886; *Atlas of Wood County, Ohio*, 1886; *An Atlas of Hancock Co. Ind.*, Philadelphia, Griffing, Gordon & Co. 1887; *Atlas of Mercer County, Ohio*, 1888; *Atlas of Fulton County, Ohio*, 1888; *Atlas of Defiance County, Ohio*, 1890. **N.B.** See also **Lake, Griffing & Stevenson**.

**Griffing, Dixon & Co.** Publishers of Philadelphia. *An Atlas of Davies County, Ind.*, 1888.

**Griffing, Gordon & Co.** Publisher of Philadelphia. B.N. Griffing's *Map of Blackford County, Indiana*, 1886; *Atlas of Hancock County, Ind.*, 1887; *Atlas of Jay County, Ind.*, 1887.

**Griffith**, father and son. Surveyors of Anne Arundel County, Maryland, USA. **Ref.** PAPENFUSE, E.C. Atlas of historical maps of Maryland, 1608-1908 (1982) pp.48-53.
• **Griffith**, Joshua (1730-1779). Deputy surveyor of Anne Arundel County, Maryland.
•• **Griffith**, Dennis (1759-1805). American surveyor from Anne Arundel County, son of Joshua, above. Active in Philadelphia. *Map of the State of Maryland...*, 1794 published Philadelphia, J. Vallance 1795, J. Melish 1813 (engraved J. Thackara and J. Vallance, included a large inset plan of Washington).

**Griffith**, E. [& Son.]. Publishers of Birkenhead. *Plan of the Borough of Birkenhead*, 1890; *New Map of Higher and Lower Bebington* [Cheshire], 1890; *New Plan of the District of Wallasey* [Cheshire], Birkenhead 1890; *Map of Hoylake and West Kirby Local Board District* [Cheshire], 1890.

**Griffith**, J.M. *Official Map of Amador County* [California], San Francisco 1866

**Griffith**, *Sir* Richard John (1784-1878). Surveyor and geologist of Dublin, worked for the Irish valuation office. His study of the geology of Ireland was on-going, and from c.1814 his constantly revised manuscript geological maps of Ireland were well known to the members of the Dublin Society and the Geological Society of Dublin. *Bog of Allen*, 1810; *Leinster Coal District*, 1814; *Roscommon*, 1817 (with William Edgeworth); *A Map of the County of Roscommon*, London, J. Cross 1825; *A Map of Ireland*, James & Luke Hansard 1830 (engraved by James Basire, published in *The Report of the State of the Poor in Ireland*); geological map of Ireland, published in *Atlas to Accompany 2d Report of the Railway Commissioners Ireland 1838*, 1838; a larger scale version of the geological map (using a base map by Lieutenant T.A. Larcom of the Irish Ordnance Survey) was published as *A General Map of Ireland to accompany the Report of the Railway Commissioners* (6 sheets at quarter-inch to the mile), Dublin, Hodges & Smith, and London, James Gardner 1839 and many later issues with major revisions in 1840, 1852, 1853 and 1855. **Ref.** DAVIES, G.L. Imago Mundi 29 (1977) pp.35-44.

**Griffith**, Samuel Young. Plan of Cheltenham, 1825.

**Griffith**, W.P. Publisher of Norfolk, Virginia. W. Mahone's *Map showing route of Norfolk & Petersburg Rail Road and its connections with Ohio & Mississippi Rivers*, 1858 (also published in Philadelphia by F. Bourquin & Co.).

**Griffith**, William, *junior*. Atlas of Gallia County, Ohio, 1874.

**Griffith & Farran**. Publisher of London. *Pictorial Geography for the instruction of children*, 1861.

**Griffiths**, Edward. South African surveyor. *Pinetown*, 1851.

**Griffiths**, Harry D. *Map of Filabusi District* [Southern Rhodesia], 1900; *Map of the Gwanda District* [Matabeli Land], 1900; *Map of the Coronation Line...* [Transvaal], London, P.H. Browne 1903.

**Griffiths**, Henry. Engraver and artist. Scotland, 1846.

**Griffiths**, Ralph. Bookseller and publisher. Published the 'Grand Magazine of Universal Intelligence', 1758-1760, containing 20 maps including: *A New and Accurate map of the English Empire in North America*, 1758; *A New Map of the Coast of Africa*, 1758; *A Plan of the City of Cologne*, 1758; *A Plan of the Harbour and Town of Louisbourg in the island of Cape Breton*, London 1758; *Map of the country round Québec with the camps of the English & French at the siege thereof*, London 1759; *Asia*, 1759; *A New and accurate Survey of the Country about the Cities of London and Westminster and the Borough of Southwark*, 1760; and others.

**Grigg**, John. Publisher of Philadelphia. *A New General Atlas... to illustrate the Universal Geography of Malte-Brun*, Phildelphia 1828, 1830, J. Laval 1832; *American Atlas*, Philadelphia, J. Grigg 1829.
• **Grigg & Elliot** [Grigg, Elliott & Co.]. Publishers of 'N⁰·14 N. 4th St.', Philadelphia. *A New General Atlas... to illustrate the Universal Geography of Malte-Brun*, 1837 edition (previously published by John Grigg); *Smiley's Atlas, For the Use of Schools and Families*, Philadelphia 1839 edition; R.M. Smith's *Modern Geography*, 1848, 1849.

**Grigham**, H.G. Surveyor associated with Geil and Harley. *Map of the counties of Clinton and Gratiot, Michigan*, Philadelphia, Samuel Geil 1864 (with D.S. Harley, J.P Harley, J.D. Nash & M.C. Wagner).

**Grigingerus**. See **Criginger**.

**Grignon**, Reynolds (d.1787). Engraved the decorative cartouche for Fry and Jefferson's *A Map of the most Inhabited part of Virginia...*, Thomas Jefferys 1753.

**Grigny**, —. Engraver for Conrad Malte-Brun, 1812.

**Grigor'yev**, Fyodor. Russian topographer. Map of the lands between the Dnepr and Dniester Rivers; assisted J.-N. Delisle with compilation of Atlas, 1735; accompanied Delisle on expedition to Obdorsk.

**Grigviger**. See **Criginger**.

**Grijalva**, Juan de (c.1489-1527). Commander of a Spanish quest for Aztec gold in Mexico. Credited with a manuscript map of Brazil.

**Grijp** [Gryp], Dirck. Engraver in Amsterdam. *Nova Virginiae Tabula*, J. Hondius 1618; engraved some plates for John Speed's *A Prospect of the Most Famous Parts of the World*, George Humble 1626; some plates for Jodocus Hondius the younger including *Nova Brabantiae Ducatus Tabula*, c.1629, re-published by Willem Blaeu, 1630 and later.

**Grilo**, Miguel. Spanish mapmaker. *Atlas geografica de España*, 1876. **N.B.** Compare with **Rubio, Grilo y Vitturi**.

**Grim**, David (1737-1826). Citizen of New York, native of Bavaria, amateur mapmaker. *A Plan of the City and Environs of New York...1742, 1743 & 1744*, 1813. **Ref.** COHEN, P.E. & AUGUSTYN, R.T. *Manhattan in maps 1527-1995* (1997) pp.62-63.

**Grim**, Franz. *Plan von Wien*, 1872.

**Grim**, M. von. See **Grimm**, Maximilian von.

**Grimaldi**, Filippo, SJ (1639-1712). Jesuit astronomer and cartographer in China. Succeeded Ferdinand Verbiest in Peking, 1688; *Constellations*, 1711.

**Grimaldi Casta**, L. *Carta politica speciale del regno d'Italia...*, Istituto Cartografico Italiano 1893 (with G.E. Fritzsche).

**Grimes**, Charles (1772-1858). Deputy Surveyor-General, New South Wales, 1791;

Surveyor-General, 1802; discovered the Yarra Yarra River, 1803. *A Topographical Plan of the Settlements in New South Wales*, Aaron Arrowsmith 1799.

Grimm, J. Siebenbürgen [Romania], 1855.

Grimm, Julius Ludwig (*fl.*1806-1834). German cartographer, worked with Karl Ritter and Heinrich Berghaus. *Palaestina*, 1830; *Atlas*, Berlin, Simon Schropp & Co. 1833, 1836; *Kleiner Schulatlas*, Berlin, Reimer 1833-1838 (with Theodor von Liechtenstern); *Karte von Hoch-Asien* (4 sheets), 1832 (with C. Ritter); *Atlas von Asien*, 1833-1854; and others.

Grimm, L. Draughtsman, worked with others on *Reymann's Special-Karte* [Schleswig-Holstein], 1821, published Glogau, C. Flemming *c.*1860.

Grimm [Grim], Maximilian von. Draughtsman and engraver of Vienna. *Grundriss der Kaiserl: Königl: Haupt und Residenz Stadt Wien*, 1799.

Grimm, Sigmund [Sigismund] (*c.*1480-1530). Printer and physician in Augsburg. Published the plan of Augsburg made by Jörg Seld, 1521.

Grimm, Simon. Engraver and publisher in Augsburg. Different views of Augsburg, 1678, 1679, 1682.

Grimmel [Grimel], Ivan. Lake Ladoga, Ingria & Carelia, 1735; Russian Karelia (7 sheets), Jacques-Nicolas Delisle 1745; maps for the Russian Academy, 1755-1770.

Grimminger, Georg Adolf (1802-1877). Lithographer in Stuttgart.

Grimoard, *General* Philippe Henri, *comte* de (1753-1815). *b.* Verdun, France. Atlas to campaigns of Turenne, 1782.

Grimstone, Edward. Translator of English edition of José de Acosta's *Historia natural de las Indias*, 1590; Pierre d'Avity's *Le monde*, 1626.

Grindlay, *Captain* Robert Melville. Surveyor and travel writer in India. *India*, James Wyld 1840.

Grineau, Charles William Bryan de (1883-1957). Artist, served in the Royal Artillery in the First World War, producing battlefield drawings for publication in *Illustrated London News*. In the Second World War he produced panoramic maps for *Illustrated London News*.
**Ref.** COOK, K.S. The Map Collector 51 (1990) p.40; The Times 21st May 1957 p.13 & 24th May 1957 p.13.

Griner, S. See **Grynaeus**, Simon.

Grinlinton, *Lieutenant* —. Plans Sevastopol, 1857, 1858.

Grinnell, Henry (1799-1874). New York merchant. Founding member and President of the American Geographical and Statistical Society. His connections with the whaling industry generated an interest in the waters of the Arctic, encouraging him to fund several exploratory voyages. *A Chart... of the American Arctic Expedition*, *c.*1851.

Gripenhielm, *baron* Carl (*c.*1655-1694). Poet and surveyor of Stockholm, Director General of Corps of Surveyors, 1683. Sueciae et Gothia, 1688 MS (scale 1:3 000 000), copied and printed in Paris 1706, later reprinted Amsterdam in 1740s; *Landt-och Siö charta öfver Siön Mählaren och dess öijar*, 1689; survey of Skärgard, 1689-1691; survey of roads from Stockholm, 1690s.

Grisco, M. See **Gruisco**, Matteo.

Grisebach, August (1814-1879). German botanist. *Die Vegetations-Gebiete der Erde*, published in *Petermanns Mitteilungen*, 1866.

Griselini, Francesco (1717-1783). Cartographer and engraver. *Città de Praga*, *c.*1760; re-drew mural maps in Sala dello Studio of the Ducal Palace, Venice 1762.

Grismand, John. One of the publishers of Michael Drayton's *Poly-Olbion*, 1622.

Grist, father and son [J. Grist & Son].
• Grist, Jonathan [John] (*fl.*1790-1839). Land surveyor of Canterbury and Old Cavendish St., London, produced manuscript maps of parts of southern and eastern England. Tillingham, 1799 MS; Foulness Island Levels, 180l; Rushfield Wood [Kent], 1804; Dunkirk, Kent, 1811; survey of the road from Sturry to Herne Bay, 1813; Great Barton etc., Canterbury 1814; estate at St Thomas' Hill [Kent], 1822; *Great Shelford and Mudgrove Estates*, 1824 (with George Grist).

•• **Grist**, George. Estate surveyor of Canterbury, son of Jonathan. *Great Shelford and Mudgrove Estates, Kent*, 1824 (with Jonathan Grist); Harbledown, 1824; (with Jonathan Grist); Winter Estate Thanet, 1824 MS.
• **J. Grist & Son**. Surveyors of Canterbury. *Hardres Court, Kent*, l837.

**Griswold**, B.J. American publisher. Bird's-eye view of Fort Wayne, Indiana, 1907.

**Grive**, *abbé* de la. See **Delagrive**, Jean.

**Griwtonn**, P.L. *Carte de L'Isle de St Domingue*, Paris 1801.

**Groc**, Alcide. *Département de la Charente Inférieure*, 1876.

**Grodecki** [Godreccius; Godretius; Grodziecki], Wacław [Wenceslaus] (*c*.1535-1591). Polish cartographer and engraver. Map of Poland, 1557, (only known example lost at Munich 1945), re-published Basle, Giovanni Oporini 1562 (used by Ortelius 1570 and others). **Ref.** KARROW, R.W. pp.280-282; MEURER, P. pp.159-160 [in German]; NIEWODNICZANSKI, T. 'The unknown specimen of the Second Edition of Wacław Grodecki's map of Poland...', *Polski Przegląd Kartograficzny* Vol.19 N°·1-2 (1987) pp.22 -28.

**Grodemetz** [Grodmetz], Jean David. Dutch map copyist and engineer. Copied Nicolas Sully's untitled survey of Zuidwest-Zeeuwsvlaanderen (1716), 1729; Baye de Gibraltar, *c*.1718.

**Grodsky**, —. Lieutenant in the Russian Army. One of the Russian representatives amongst the international team which draughted *...la frontière Bulgaro-Serb selon article 36 du traité de Berlin... 1878* (10 sheets plus title), 1879.

**Grodziecki**, W. See **Grodecki**, Wacław.

**Groeger** [Gröger], C. *Nord-friesischen Inseln*, 1888.

**Groell** [Gröll], Michael. Printer, publisher and bookseller in Warsaw. *Carte générale et nouvelle de toutte la Pologne, du Grand Duché de Lithuanie et des pais limitrofes*, 1770, 1772 (engraved by B. Folin).

**Groenewegen**, Gerrit. *Atlas van de Zeehavens der Bataafsche Republiek*, 1805.

**Groenewegen**, J. Publisher and bookseller in the Strand, London. Joint publisher, with N. Prévost, of volume V of David Mortier's *Nouveau théâtre de la Grande Bretagne*, published as *Supplement du nouveau theatre de la Grande Bretagne*, 1728. **Ref.** HODSON, D. *County atlases of the British Isles published after 1703* Vol.I (1984) pp.35, 38, 40.

**Groenou** [Groenouw], Dirck Brekens van. Dutch surveyor, worked mostly in Utrecht. Caarte van...'t Gerecht van Wijck, 1662-1667.

**Groesbeck**, Gerard van (1508-1580). Bishop of Liège, 1563; Plan of Liège used by Georg Braun and Frans Hogenberg, 1572.

**Groffier**, Valerien. Missionary maps, 1883-1886 including *Atlas des Missions catholiques* 1886.

**Gröger**, C. See **Groeger**, C.

**Grognard**, François. *Mer Méditérranée*, 1745; *Carte de l'Archipel* [Aegean], 1745 (used by J.N. Bellin in *Le Neptune François*, 1753-1754).

**Grohmann**, Paul. *Dolomit-Alpen*, 1875.

**Grol**, J. See **Groll**, J.

**Groll**, Cornelius. *Provincie Noord-Holland*, 1853.

**Groll**, Hans van. Plan of The Hague, 1603.

**Groll** [Grol], J. *Détroit de Sonde*, 1840-1841, published 1846; *Carte des Atterrages de Batavia*, 1846.

**Gröll**, M. See **Groell**, Michael.

**Groll**, Max (1876-1916). After the death of F. von Richthofen, Groll edited his maps of China and made them ready for publication. Northern China, Berlin, D. Reimer 1885; Southern China, Berlin, D. Reimer 1912; ocean depth chart, 1912.

**Grolman**, Karl Wilhelm Georg von (1777-1843). First head of the official survey of East Prussia, up to 1819.

**Grombchevsky**, Colonel —. *Sketch of the Alai Road*, 1893.

**Gronden**, A. van den. *Plan van Edam*, 1743; *Plan van Monnikendam*, 1743.

**Grondona**, Nicolas. Maps of the Provinces of Argentina, *c*.1865; *Provincia de La Rioja, Rosario de santa-Fè c.*1865; *Argentina*, 1875.

**Gronen**, E. German engraver, born Munich, worked in Winterthur. *Culmbach*, 1853; *Schweiz, c.*1877.

**Gronovius**, Jacobus (1645-1716). Dutch classicist. *Geographia antiqua*, 1697.

**Groom**, E. Forster [& Co.]. 'Map seller and stationer, 15 Charing Cross, S.W.' (London). *Bohemian Campaign*, 1866; *Rugg's Handy Reference Map of London*, 1890.

**Groom**, Samuel (*fl*.1825-1850). Surveyor, produced manuscript maps of Shropshire and parts of Wales. *Barracks Brecon*, 1825 MS.

**Groot**, Cornelius de. *Geologische kaart van Biltong*, 1887.

**Groot**, G. de [en Zoon]. Publishers of Amsterdam. Re-issued Isaak Tirion's *Atlas van Zeeland*, 1788 (with P. Schouten & R. Ottens; first published 1760).

**Groot** [Grotte], J. de. Publisher in Amsterdam. Collaborated with P. Schouten, G. Warnars, S. & G. Luchtmans of Leyden, and A. Blussé en Zoon of Dordrecht in the republication of the works of Isaak Tirion. *Nieuwe en beknopte Hand-Atlas*, 1788 (with others); *Nieuwe kleine Hand-Atlas...*, 1789; *Beknopte Atlas*, 1790; *Nieuwe en keurige Reis-Atlas door de XVII Nederlanden*, 1793, 1794 (with others, maps by Tirion).

**Grooten**, Christiaan. See **Sgrooten**, Christian.

**Grophius**, Martin Gottfried. Engraver of Augsburg. Worked for Tobias Conrad Lotter. *Atlas minor*, 1744?

**Gropp**, Captain A. *Transvaal*, 1886.

**Gros**, Professor C. Revised, with the assistance of J. Aspin, *Lavoisne's Complete Genealogical, Historical and Chronological Atlas*, London, John Barfield 1814 and later editions (first published with no maps in 1807).

**Grosdidier**, F.E. *Europe centrale*, 1870.

**Grose**, Francis (*c*.1731-1791). Antiquary and draughtsman. *Antiquities of England and Wales* (6 volumes), 1773-1787 (52 maps, altered reprints of John Seller's *Anglia contracta*, 1695, maps in the Supplement, 1787); *Antiquities of Ireland*, 1773-1811.

**Grosier**, *abbé* Jean Baptiste Gabriel Alexandre (1743-1823). *Atlas générale de la Chine*, 1777-1784.

**Grosjean**, A. *Plan de Melun*, 1889.

**Grosley**, Pierre-Jean [Jean Pierre] (1718-1785). *b*. Troyes. *Plan nouveau et correct des villes et fauxbourgs de Londres et Westminster et du Bourg de Southwark*, Lausanne 1770, 1774, Paris, 1784.

**Grosmann**, Charles W.F. Reuben H. Donnelley's Sectional *Atlas of Chicago*, 1891.

**Gross**, —. 'Strassenbaudirector' [chief road engineer]. *Plan Göppingen*, 1783; *Ost-Galizien u. Lodomerien in 14 Sectionen...*, *c*.1808.

**Gross**, A.G. von. *Historisch-Militairischer Atlas zu des Freyherrn*, 1808.

**Gross** [Grotz; Grosz], Alexander [Sandor]. Hungarian emigré, father of Phyllis Pearsall. Formed the map publishing company Geographia Ltd., which produced numerous road maps, county maps, London maps and town plans from 1911. Gross left for America *c*.1930. *The new road map of England and Wales...*, London, Geographia Ltd 1911; *Boy Scouts map of the British Isles*, 1911; *'Geographia' map of Great Britain*, 1912; *The Premier road book of Great Britain*, 1913; *Picture map of western Europe*, 1916; political and war maps for the Daily Telegraph, *c*.1916-1918; *The Daily Telegraph pocket atlas of the war*, London 1917; and many others. **Ref.** PEARSALL, P. *A-Z maps: the personal story from bedsitter to household name* (Sevenoaks, Kent, 1990).

• [**Pearsall**], Phyllis. See **Pearsall**.

**Gross**, Colonel Emanuel (1681-1742). Swiss mathematician and cartographer, worked in Prussia, Ticino, Laupen, Echallens and Modena.

**Gross**, I.M. Surveyor, worked mainly on county maps of Michigan for Samuel Geil. *Map of Kalamazoo Co., Michigan*, Geil & Harley 1861 (surveys by I. Gross, drawn by S.L. Jones, engraved by Worley & Bracher); *Map of Lapeer Co., Michigan*, Samuel Geil 1862 (with W.E. Doughty); *Map of Allegan co., Michigan*, Philadelphia, Samuel Geil 1864; and others.

**Gross**, Rudolph (*b*.1808). Work included railway maps, also: *Plan von London*, 1844; *Polytopischer Reise Atlas*, 1845-1852; Geographische Schul-Atlas, 1847; *Karte der Eisenbahnen Deutschlands*, Stuttgart, J.B. Metzler *c*.1850; *Die Wurttembergische Eisenbahn und der Bodensee* (4 maps), Stuttgart, Paul Neff *c*.1850 (lithographed by F. Malté); *Neuer Schul-Atlas*, 1862; *Neuester Atlas*, *c*.1868; *Deutsches Reich*, 1873; and others.

**Grosschmidt**, János. Maps of the salt mines of Máramaros [now Maramures, Transylvania, Romania]. Also maps of Tiszaújlak, 1768; Rónaszék, 1778-1781.

**Grosselin-Delamarche**, —. *Atlas de géographie*, 1869. **N.B.** Possibly F.A. Delamarche and A. Grosselin.

**Grossen**, Johann Gottfried. *Orbis in tabula*, *c*.1720.

**Grosset**, C.S.C. Watch Colony of Surinam, 1781.

**Grossmann**, —. *Umgegend von Ruppin*, 1860.

**Grosvenor**, H.C. *Silver Mines Arizona*, *c*.1860.

**Grosvenor**, James. Pilot. *A new chart of the mouth of the Thames*, 1786; *Chart of the sounds and channels from the Nore to Margate road*, 1786; both published in *The Channel Pilot*, Robert Sayer 1789.

**Groth**, Adolf (*b*.1676). German pastor of Livland. Surveys of the Baltic areas under the control of the Polish crown. *Ducatuum Curlandiae ey semigalliae nec non districtus Regii Piltensis*, 1730 published 1770.

**Grothaus**, Friedrich. *Plan der Stadt Barmen*, 1850.

**Grotte**, J. de. See **Groot**, J. de.

**Grouner**, Johann Samuel. Reduced edition of I.G. de Rovéréa's *Carte du Gouvernement d'Aigle...* [Switzerland], 1788 (original surveys of 1734-1744 now lost).

**Grout**, I.R. Surveyor. Clinton River, 1849.

**Groux**, Charles Jacques. His signature replaced those of others on the cartouches of the Fortin editions of some Gilles and Didier Robert de Vaugondy's maps, from *c*.1778 (the cartouches were otherwise unchanged from previous editions); *Plan de la Ville et du Nouveau Pont d'Orléans*, Paris, Imprimerie royale 1783. **Ref.** PEDLEY, M. *Bel et Utile: The work of the Robert de Vaugondy family of mapmakers* (1992) p.67 and map catalogue.

**Grove**, —. Publisher with Canaan, Swinton & Ritchie, Edinburgh. John Wood's Plan of Lanark 1825 (in Scottish Town Atlas).

**Grove**, *Captain* Carl Frederik (1758-1829). Undertook a survey of the coasts of Norway with Niels Wibe and Benoni d'Aubert [so-called *Grovske drafter*], 1791-1803, published as *Speciel Kaart over en Deel af Den Norske Kyst...* (7 sheets), various dates.

**Grove**, *Sir* George (1820-1900). Civil engineer and musicologist, best known for the *Dictionary of Music and Musicians*. Bible Atlas, 1868; co-editor of *An Atlas of Ancient Geography Biblical and Classical...* (68 maps), 1872-1874.

**Grove**, John. *Plan of Ipswich*, 1761.

**Grover**, Mrs. —. Bookseller in Pelican Court in Little Britain, London. *Carolina*, 1682.

**Grubas**, Giovanni. *Mare Mediterraneo*, 1801; *Adriatic Sea*, 1803.

**Grubb**, *Lieutenant* J.H. *Plan of Mathurin Bay*, 1818.

**Grube**, A. *Plan von der Stadt Aschersleben*, *c*.1890.

**Grube**, A.W. (1816-1884). German geographer.

**Grube**, F.W. Publisher. *Düsseldorf*, 1848.

**Grube**, Henry. Publisher, Paternoster Row, London. *The Cyclist's map of 50 miles round London...*, c.1895.

**Gruchet**, Jacques and Guillaume. Printers and booksellers of Havre de Grâce [Le Havre], France.
• **Gruchet**, Jacques. Published second edition of George Boissaye du Bocage's *Carte Ronde ou Reduite Fort*, 1679 (first published 1669); first edition of Réné Bougard's *Le Petit Flambeau de la Mer* (64 maps), 1684 (the right to print *Le Petit Flambeau* passed to Jacques Hubault).
•• **Gruchet**, Guillaume. Presumably related to Jacques. Published the third edition of *Carte Ronde...*, 1696; took over the rights to *Le Petit Flambeau* from J. Hubault, and printed new editions 1709, 1716.
•• **Gruchet**, — *veuve*. Widow of Guillaume, continued to issue Bougard's work after the death of her husband. She then published it in association with P. Faure, who took over from 1763. *Le Petit Flambeau de la Mer*, 1731, 1742 (with Pierre Faure), 1752 (with Pierre Faure).

**Gruchy**, H.G., & Co. Lithographers of Melbourne, Australia. J. Walch & Son's Map of Tasmania, 1874.

**Gruendler** [Gründler], August. Printer in Wroclaw. Martin Helwig's Silesia, 1627.

**Gruenewald** [Grunewald], C. Engraver for Wilhelm Ernst August von Schlieben, 1828-1830.

**Grueninger**. See **Grüninger**.

**Gruenthal** [Grünthal], —. (*fl.*early 18th century). Swabian geographer.

**Gruenwald**, Christoph. See **Grünewald**.

**Gruisco** [Grisco], Matteo [Mateus]. Portolan chartmaker of Majorca. Carta Marina, 1581 MS.

**Grumbkow**, — von. Mediterranean, 1799; Black Sea, 1834; *Klein Asien*, 1840; *Atlas des Preussischen Staates*, c.1866.

**Grumm**, Ladislaus. Cycling maps. *Grumm's Streckenkarten*, Vienna, Artaria 1898.

**Grumprecht**, Thaddaus E. Contributed to C.G.D. Stein's *Neuer Atlas der ganzen Welt*, 1853 edition.

**Grund**, Francis Joseph. *Handbuch und Wegweiser für Auswanderer* (with maps), Stuttgart 1846.

**Grund**, H. *Wilhelmshaven*, 1872.

**Grundemann**, *Doctor* Reinhold (1836-1924). *Missions-Weltkarte*, 1865; *Allgemeiner Missions-Atlas*, 1867-1871; *Neuer Missions-Atlas*, 1903.

**Gründler**, A. See **Gruendler**, August.

**Grundt**, Christoph Ludwig (*fl.c.*1750-1775). Surveyor.

**Grundy**, John. Father and son.
• **Grundy**, John, *the elder* (1697-1749). Father of John the younger, below. Mathematician and surveyor of Congerstone, later Spalding, Lincolnshire, produced manuscript maps of Derbyshire, Leicestershire and Lincolnshire. *Atterton Lordship*, 1729; *Plan of Spalding*, 1732; *River Witham*, 1743-1744 (with John Grundy, the younger).
•• **Grundy**, John, *the younger* (1719-1783). Surveyor and engineer of Spalding, Lincolnshire, produced manuscript maps of parts of eastern England. *River Witham*, 1743-1744 (with John Grundy, the elder), 1762; *East Fen*, 1774.

**Grüneberg**, Martin (1655-1707). Map of Friedrich-Wilhelm-Graben [Brandenburg], 1704-1707.

**Gruner**, E. *Atlas général des houillères*, c.1909-1911 (with J.G. Bousquet).

**Gruner**, Gottlieb Sigmund (1717-1778). Swiss archivist and geographer in Hesse, Thorberg, Bern, Landshut and Fraubrunnen. Created maps for his book on Swiss mountains, Bern 1760.

**Grunert**, —. Engraver for Adolf Stieler, 1869.

**Grünewald** [Gruenwald], Christoph *senior* and *junior*. German engravers. *Australien*, c.1830; *Kreis Pfalz*, 1839; *Eisenbahn-Karte v. Deutschland*, 1846.

Simon Grynaeus (1493-1541). (By courtesy of Rodney Shirley)

**Grüninger**, father and son. **Ref.** BRUMAN, H.J. 'The Schaffhausen *Carta Marina* of 1531', *Imago Mundi* 41 (1989) pp.124-132.

• **Grüninger** [Grieninger; Grueninger], Johannes [Johann Reinhard] (1480-1528). Printer, publisher and wood engraver in Strasbourg. Printed and sold *Globus mundi*, 1509 (including a world map); M. Waldseemüller's *Carta itineraria Europae*, 1520, 1527 (first published 1511); worked with Lorenz Fries on an edition of Ptolemy's *Geographia* based on the earlier maps of Martin Waldseemüller (50 woodcut maps), 1522, new edition edited by W. Pirkheimer using the same maps, 1525 (published jointly with J. Koberger of Nuremberg); Lorenz Fries's *Carta marina* (12 sheets), 1525, 1530 (based on Martin Waldseemüller's *Carta marina*, 1516, and accompanied by a commentary by Fries entitled *Uslegung der Mercarthen oder Cartha Marina*). **Ref.** KARROW, R.W. *Mapmakers of the sixteenth century and their maps* (1993) pp.193-204; *Imago Mundi* 37 (1985) pp.42-53; PASTOUREAU, M. *Les atlas français XVIe-XVII siècles* (Paris 1984) pp.375-380; JOHNSON, H.B. '*Carta Marina*': *World Geography in Strasbourg 1525* (Minneapolis 1963).

•• **Grüninger**, Christoph. Son of Johannes. Published the Latin edition of *Carta marina*, 1531.

**Grüninger** [Grueninger], Bonifacius. [Karte des] *Schwarzwaldes*, 1783, 1788.

**Grünthal**. See **Gruenthal**.

**Gruss**, Franz. Ingénieur de la Cour de Vienne. *Grund Riss der Stadt Wien* (4 sheets), 1770 (with Joseph Neussner).

**Grynaeus** [Griner; Gryner], *Professor* Simon (1493-1541). Geographer and theologian, born Veringen, worked and died in Basle. With Johann Huttich compiled *Novus Orbis Regionum*, 1532 and later (the 1532 edition published by J. Herwagon in Basle contains *Typus Cosmographicus Universalis...*, a world map attributed to Sebastian Münster and possibly Hans Holbein, the younger; the Paris 1532 edition includes Oronce Fine's *Nova Et Integra Universi Orbis Descriptio*, 1531). **Ref.** SHIRLEY, R. *The mapping of the World* (1983) N[os.]66 & 67 pp.73-75.

**Gryp**, Dirck. See **Grijp**, Dirck.

**Gryphius**, Antoine. Publisher of Lyon. J. Chaumeau's *Histoire de Berry*, 1566 (with a map by Chaumeau used later by Ortelius, Belleforest and De Jode).

**Gryscher**, M. See **Greischer**, Matthias.

**Guad**, M. See **Quad**, Matthias.

**Guadagnino**. See **Vavasore**, Giovanni Andrea di.

**Guadet**, Joseph (1795-1881). *Atlas de l'histoire de France*, 1833.

**Gualdo Priorato**, Galeazzo, *comte* de Comazzo (1606-1678). *Schau-Platz Desz Niederlandes Oder...*, Vienna 1673, republished as *Teatro del Belgio...*, 1683.

**Gualtherot**, Vivant. Printer and publisher of Paris. *Angliae descriptio*, 1545; published French and Latin editions of Gemma Frisius's version of Peter Apian's *Cosmographicus liber*, 1551, 1553.

**Gualterus**, Philippus. See **Gautier** de Châtillon.

**Guarini**, *Capitano* Giuseppe. *Pianta Topografica del Territorio di Ravenna*, Rome, Ignazio Benedetti 1770.

**Guarinoni**, Luca. Publisher in Venice. Augsburg, 1568.

**Gubernatis**, Enrico de. *Carta d'Epiro* [northwestern Greece], 1869-1875, printed 1879.

**Guchen**, Maximinus a. See **Maximinus** a Guchen.

**Gucht**, Michael van der (1660-1725). Engraver of Antwerp, worked and died in London. Title-page to Herman Moll's *System of Geography*, 1701; numerous engraved illustrations, frontispieces and titles, especially for geographical and architectural works.

**Gudeman**, E. *Gude's european war map*, New York 1914.

**Gudenov**, Pyotr Ivanovich [Peter Ivanovitsch]. N. Asia, 1668; *Siwerische Landchard*, 1690.

**Gudme**, A.C. (1779-1835). *Kiel über die Gegend von Kiel...*, Kiel 1844 (with F.H. Geerz).

**Gudmundi**, Jonas. N. Atlantic, 1570 MS.

**Güdmundsson** [Gudmundus], Jon [Jonas] (1574-1658). Icelandic chartmaker. Northern Regions, c.1650; *Delineatio Grondlandiae*, published 1706.

**Guedes**, P. *Carta hypsometrica de Portugal*, Lisbon 1906 (engraved by L. Wuhrer).

**Guedeville**, N. See **Gueudeville**, Nicolas.

**Guembel** [Gumbel; Gümble], Karl Wilhelm von. (1823-1898). German geologist. *Bayern*, 1858; *Geologische Karte von Bayern*, published 1911.

**Guenther** [Günther], E. Lithographer. *Karte vom Harz*, Berlin, W. Lubeck 1866; *Karte vom Thüringerwald*, Berlin, W. Lubeck 1866; *Karte der Insel Ruegen*, W. Lubeck 1866.

**Guenther** [Günther], Johann. Hand-Atlas, 1822; Volksatlas, 1842.

**Guenther**, O.A. *Das Schloss Fürstentein's bey Freyburg in Schlesien*, Dresden c.1820.

**Guenther** [Günther], W. *Gera und Umgegend*, 1895.

**Guérard**, family of engravers.
• **Guérard**, Nicolas, *the elder* (c.1648-1719). Artist and engraver of Paris, 'rue St. Jacques à la Perle' (1683); 'à Reine du clerge proce St. Yves' (1715). Cartouches for N. de Fer's map of the course of the Rhine (3 sheets), 1689 (with Liébaux); engraved the ornamentations and cartouches for N. de Fer's *Les Frontières de France et d'Allemagne...* (9 plates), 1689 (map and lettering engraved by Liebaux); *Alemagne*, Duval 1691; Nicolas de Fer's *Mappe-Monde*, 1694; cartouche for N. de Fer's *L'Atlas Curieux*, 1700; cartouches for Delisle's *Carte du Canada...*, 1703; N. de Fer's *Le Comté de Nice*, 1704; Delisle's *Carte de Tartarie...*, 1706; Delisle's *Carte des Comtez de Hainault de Namur et de Cambrensis...*, 1706; for père Placide de Sainte-Hélène, 1714; and others.
•• **Guérard**, Nicolas, *fils* (fl.1683-1750). Engraver. Nouvel France, 1683; Amérique septentrionale, 1700; *L'Afrique*, Paris 1705; Amédée François Frézier's *Plan Baye de Coquimbo*, 1716; *Carta de la Provincia de Quito...* [Equador], 1750 (from the work of P. Maldonado).
•• **Guérard**, Marie. Engraver, daughter of Nicolas, the elder. Signed some work.

**Guérard**, Augustin Frédéric Stanislas. Charts of Iceland, 1866-1867, including *Croquis du Mouillage de Bildal dans Arner Fiord... 1866* [Iceland], Paris, Dépôt des Cartes et Plans de la Marine 1867

**Guérard** [Guérand], Jean (d.c.1640). French pilot and hydrographer of Dieppe, 'ingenieur et geographe du roi'. Led three expeditions to north-eastern Brazil, 1596, 1603, 1612; teacher of hydrography and examiner of pilots from 1615; expedition to Africa, 1639. Chart of the coast of France (2 sheets), 1625 MS; planisphere, 1625 MS; *Description hidrographique de la France* (7 maps), 1627 MS; chart of northern Europe and the Arctic, 1628 MS; *Carte universelle hydrographique* [world map], 1634 MS; map of Northern Europe, c.1635 MS (Military Museum, Istanbul). **Ref.** MOLLAT du JOURDAN, M. et al *Les portulans: Cartes marines du XIIIe au XVIIe siècle* (1984) pp.254-259, 259 [in French], also published in English as *Sea charts of the early explorers 13th-17th century* (1984) pp.254-257, 259.

**Guérard**, Jean Eugene. *Hobart Town*, Melbourne, Hamel & Ferguson c.1870.

**Guérard**, N. See **Guérard**, family, above.

**Guerber**, —. *Carte des services à vapeur dans l'Ocean Atlantique, L'itineraire des divers services est tiré, en partie du Guide Guerber*, 1867.

**Guerich**, Doctor Georg. *Geologische Skizze von Afrika*, 1887; *Schlesien*, 1890.

**Guérin**, Christophe (1758-1830/1831). Artist and engraver of Strasbourg.

**Guérin**, Léon. Publisher of Paris. M. Vuillemin's *Atlas du Cosmos* (issued in 26 parts), 1861-1867 (engraved by S. Jacobs).

**Guérin**, Victor (1821-1890). French teacher, archaeologist, geographer and traveller, specialising in the Eastern Mediterranean and North Africa. *Ora maritima Palaestinae*, 1854; Grande carte de la Palestine, 1881; *Plan de Jérusalem*, 1884. **Ref.** BROC, N. *Dictionnaire illustré des explorateurs Français du XIXe siècle: I Afrique* (1988). p.169 [in French].

**Guerin de Lamotte**, —. 'Ingenieur Geographe'. *Carte corographique des environs de Lisbonne, dressée sous la direction de Chles Picquet...*, Paris, C. Picquet 1821 (engraved by Richard Wahl; also available in Lisbon, P. et G. Rey).

**Guéroult** [Guervaldus], Guillaume (1507-1564?). Born in Caen, later moved to Lyon where he worked with his brother-in-law the printer and publisher Balthazar Arnoullet. Author of *Premier livre des figures et pourtraitz des villes... d'Europe...* (9 maps and plans), Lyon, B. Arnoullet 1552, second and augmented edition published as *Epitome de la Corographie d'Europe...* (21 woodcut maps and plans, mostly copied by Arnoullet from Münster), Lyon, B. Arnoullet 1553, B. Bonhomme 1557. **Ref.** PASTOUREAU, M. *Les atlas français XVIe-XVIIe siècles* pp.225-227 [in French].

**Guerra**, Domenico. Italian engraver. *Pianta della Città di Napoli...*, 1815.

**Guerra**, Giuseppe (1750-c.1820). Prolific topographical engraver active in Naples; chief engraver of the topographical office in Naples from 1815 until retirement in 1817; much of his work was for Rizzi Zannoni. Engraver for Rizzi Zannoni's *Atlante Marittimo*, 1784-1792 (with Aniello Cataneo), published 1792; for Rizzi Zannoni's *Atlante Geografico* [31 maps of the Kingdom of Naples], 1788-1812, published 1808; L. Bardet di Villanova's *Pianta di Tolone e delle sue vicinanze* [Toulon and environs], 1793; Lombardy (4 sheets), 1795; Northern Italy (5 sheets), 1799; Italy (2 sheets), 1802; Petetin's map of Sicily, 1810. **Ref.** VALERIO, V. *Società uomini e istituzioni cartografiche nel Mezzogiorno d'Italia* (1993) pp.542-545.

**Guerrero**, Pedro. *Isla de Cuba*, 1720.

**Guerreros**, Juan Antonio. Rio de la Plata, Buenos Ayres 1790 MS.

**Guerry**, André Michel (1802-1866). French lawyer with an interest in social statistics, published three choropleth maps of moral statistics to accompany *Statistique comparée de l'état... des crimes... de France*, 1829 (with Adrien Balbi); other maps to accompany *Essai sur la statistique morale de la France*, 1833. **Ref.** ROBINSON, A.H. *Early thematic mapping in the history of cartography* (1982) pp.158-170.

**Guersch** [Gürsch], Carl Friedrich. Copper engraver of Berlin. Battle Plans, 1782-1800; Poland, 1791.

**Guervaldus**, G. See **Guéroult**, Guillaume.

**Guesnet**, —. *Mouillage de Scala-Nova*, 1843-1844.

**Guessefeld**. See **Güssefeld**.

**Guether** [Güther], Franz. Haardt-Gebirge, 1889; *Reliefkarte vom Schwarzwald*, Freiburg 1891-1893; *Umgebung von Karlsruhe*, Karlsruhe, J. Bielefeld 1895.

**Guetrather** [Gutrather], Odilo von (1665-1731). Benedictine monk and geographer of Salzburg. *S.R.I. Principatus et Archiepiscopatus Salisburgensis*, c.1812; *Das Hoch-Fürstl.Erzstifft Salzburg sambt angründe de Orthe*, published in his *Gebrauch der Land-Karten...*, Salzburg 1713.

**Guettard**, Jean-Étienne (1715-1786). French naturalist, geologist and doctor of medicine, b. Etampes, d. Paris. Pioneer of geological maps. *Cartes minéralogiques... des terrains qui traversent la France et l'Angleterre* (2 sheets at different scales), Philippe Buache 1746 (first attempt to give geological observations of a large area); also with Buache, geological maps of Switzerland, Egypt and the eastern Mediterranean, and north eastern North America, [n.d.]; *Atlas et description minéralogique de la France*, 1777-1780 (with A.-G. Monnet, first geological atlas of France; maps based on Cassini). **Ref.** LESSING, P. 'Early geological maps of West Virginia', *Earth sciences history: journal of the History of Earth Sciences Society* 8 (Department of Geology and Geography, Denison University, Ohio 1989) pp.14-35; ROBINSON, A.H. *Early thematic mapping in the history of cartography* (1982) pp.52-54.

**Gueudeville** [Guedeville], Nicolas (c.1654-c.1721). French geographer. Provided the information and wrote the text for Henri Abraham Châtelain's *Atlas historique* (7 volumes), Amsterdam 1705-1720; description of Europe for Pieter van der Aa's *Le Nouveau Théatre du Monde...*, Leiden 1713 (other descriptions were written by Ferrarius).

**Gueydon**, Louis Henri de. *Plan de la Ensenáda de Barragan, dans La Plata*, Paris 1830-1833.

**Guffroy**, Maurice. *Carte de la Guiane française*, Paris, Erhard frères 1901.

**Guggisberg**, *Major* Frederick Gordon RE. Director of Surveys, Gold Coast Colony. *Supplementary Sheet Tarkwa Mining Map... showing Fura-Prestea Tramway*, London, Stanford's Geographical Establishment 1905; *Map shewing the proposed Boundary between Gold Coast Colony & Ashanti*, 1906; *Wall Map of the Gold Coast Colony, Ashanti and the Northern Territories* (4 sheets), W. & A.K. Johnston Ltd 1908.

**Gugliantini**, —. *Città di Firenze*, 1826.

**Guianotus**, Francesco. Engraver. Town views from *L'Ungheria compendiata*, 1686.

**Guicciardini** [Guichiardini; Guizardino], brothers. Two of the five brothers sent from Florence by their father Iacopo to run the family mercantile business in Antwerp. Agnolo arrived in 1522, Giovan Battista and Lorenzo in 1527, later to be joined by Raffeallo and in 1539, Lodovico. It was Giovan Battista and Lodovico who became actively involved in geography and cartography.
• **Guicciardini**, Giovan Battista [Jean-Baptiste; Johann Baptista] (1508-*c*.1586). Italian merchant from Florence based in Antwerp, and later, Brussels. Brother of Lodovico. Also worked as a geographer and cartographer. World map in the form of a double-headed eagle, Antwerp 1549 (now lost), later adapted by Georg Braun 1574.
**Ref.** KARROW, R.W. *Mapmakers of the sixteenth century and their maps* (1993) pp.283-284.
• **Guicciardini**, Lodovico [Louis; Luigi] (1521-1589). Merchant and geographer from Florence, brother of Giovanni Battista with whom he worked in Antwerp. d. Antwerp. Author of *Descrittione di... tutti i Paesi Bassi...*, Antwerp, Guglielmo Silvio 1567, and editions to 1660 by various publishers in different languages (with numerous maps and town plans after Braun and Hogenberg). **Ref.** FRANGENBERG, T. 'Chorographies of Florence'in *Imago Mundi* 46 (1994) p.48, p.62 (notes 55-56), 58.

**Guidalotto**, Nicola di Mandario. Franciscan chartmaker. Sea atlas, 1646 (with 4 charts) MS.

**Guido de Pisa**. *Mappamundi* ('T-O' map), 1119.

**Guidotti**, Giovanni Lorenzo. Engraver, from Lucca, Italy, worked there and in Genoa. *Pianta de Genova*, *c*.1740.

**Guierry**, —. *Carte du Tche-Kiang*, 1874.

**Guieysse**, *Captain* —. African Islands, 1771, published Dalrymple.

**Guigoni**, Maurizio. *Atlante geografico universale*, 1875.

**Guijetus**, L. See **Guyet**, Lézin.

**Guilbaudière**, Jouhan de la. French chartmaker. Atlas of Pacific (35 charts), *c*.1696 MS.

**Guilbert**, Pierre Edouard. Served under Jules Sébastien César Dumont D'Urville on the corvette *Astrolabe*. *Carte particulière de la Baie Tasman...*, 1827 (map engraved by A. Tardieu, text by Basancon & Hacq); *Carte des Iles Loyalty*, 1827 (inset); *Carte des Iles Tonga*, 1827 (inset); *Carte d'une Partie de la Côte de New-South Wales*, published 1833 (engraved by A. Tardieu); *Carte particulière du Détroit de Cook*, published 1833 (engraved by Chassant); and other maps, all in J.S.C. Dumont d'Urville's *Voyage de la corvette l'Astrolabe executé pendant les années 1826, 1827, 1828, 1829... Atlas*, Paris, J. Tastu 1833.

**Guildford**, N.& G. Publishers of Cincinnati. Revised edition of B. Hough & A. Bourne's *Map of the State of Ohio...*, 1831 (engraved by W. Woodruff).

**Guilfoyle & Parsons**. Surveyors of Temple Court, King Street, Sydney, New South Wales. *Kebblewhite's Subdivision, Leichhardt*, 1886.

**Guillain**, *Capitaine* —. *Voyage à la Côte Orientale d'Afrique, executé pendant les années 1846, 1847 et 1848 par le brick le Ducouëdic sous le commandment de M Guillain*, Paris, Arthus Bertrand 1851 (printed by Mme. Ve. Bouchard Huzard).

**Guillaume de Conches** (*b.c*.1080). b. Conches, near Evreux, France. Composed glosses on works of Boetius and Macrobius. *De philosophica c*.1130 (2 world maps).

**Guillaume de Tripoli**. World map, 12th century.

**Guillaume-Maury**, —. *Carte Géometrique et Topographique du Département du Puy-de-Dôme*, Paris 1845 (with A. Sauly); *Carte routière du département du Puy-de-Dôme*, Clermont Ferrand 1868.

**Guillemin**, Amédée (*b*.1826). French astronomer and geographer. *Le Ciel*, before 1867; *Tableau cosmographique*, 1878.

**Guillemard**, Francis Henry Hill (*b*.1852). *The Life of Ferdinand Magellan and the first circumnavigation of the globe 1480-1521* (16 maps), London, George Philip 1890; Philippine Islands, 1894.

**Guillemaro**, Gilberto. Entrada de la Bahiá... de Panzacola [Florida], 1787.

**Guilleminot**, *Général* Armand-Charles, *comte* de (1774-1840). French diplomat. *Carte itineraire de l'Espagne et du Portugal*, 1823.

**Guillier**, A. *Carte geologique du Département de la Sarthe*, c.1874-1876.

**Guillo**, H. Prescott. Surveyor. Atlas of Brookline, Massachusetts, 1893 (with Charles L. Eliott).

**Guillon**, *Monsignor* —. 'Evêque d'Eumenie, Vicaire apostolique de Mandchourie'. *Carte de Mandchourie Méridionale (Leo Tong ou Province de Moukden)*, Paris, Adrien Launay, des Missions Etrangères 1894 (engraved by R. Hausermann).

**Guillon**, Victor. *Nouvelle Carte de France...*, Paris, A. Logerot c.1862.

**Guillot**, —. Engraver of '123, rue du Cherche-Midi', Paris. One of the engravers for L. Grégoire's *Atlas Universel de Géographie Physique & Politique*, Paris, Garnier Frères c.1892.

**Guillot**, E. Engraver of 'rue Vavin 35', Paris. One of the engravers for L. Grégoire's *Atlas Universel de Géographie Physique et Politique*, Paris, Garnier Frères c.1892; *Grand Atlas Departmental de la France et de l'Algérie*, Paris 1900; engraver of maps of the French départements (67 maps), Paris, E. Plon, Nourrit & Cie. 1914.

**Guillot**, Joseph Frédéric. *Carte routière du Département de Calvados*, 1871-1876.

**Guillotière** [Guiloterius], François [Fridericus] de la. See **La Guillotière**, François de.

**Guillyn et Cie**. Publishers of Paris. Nicolas Louis de La Caille's *Journal historique du voyage*, 1763.

**Guimpel**, Friedrich (1774-c.1830). Engraver and artist of Berlin. Worked for Daniel Friedrich Sotzmann, 1808.

**Guiter** [Guittair], Ch.A. (1756-1787). French engraver, known to have worked in Copenhagen, Denmark. Worked for C. Wessel, 1771; Laaland, 1776; *Kort over Siælland og Möen med tilgrændsende Kyster af Skaane Falster Laaland Langeland Thorsinger Fyen Samsöe og Jylland...* [Denmark], C. Wessel and H. Skanke 1777; for Société des Sciences, Copenhagen, from 1777; Fyn, 1780, 1783.

**Guiterrez**, D. See **Gutiérrez**, Diego.

**Guitet**, Mathurin (1664/1665-1745). Dutch hydrographer and merchant. Charts of the German coast, c.1708-1710; *Wad- en Buyten Kaart...*, Amsterdam 1710.

**Guittair**, Ch.A. See **Guiter**, Ch.A.

**Guittan**, Philippe. Bahia de tous les Saincts, 1647.

**Guizardino**, Johann Baptista. See **Guicciardini**, Giovanni Battista.

**Gulbrandsen**, C.C. Celestial map on the ceiling of Grand Central Terminus, New York, 1913 (with J. M. Hewlett).

**Güldenstädt** [Guldenstedt], Johann Anton (1745-1781). Naturalist and physician from Riga, Latvia. Map of the Caspian Sea, 1776; Russia and the Caucasus (volume 1), St. Petersburg 1787, (volume 2) St. Petersburg 1791; Black Sea, 1798.

**Guldenstein**, Anton Friedrich von. *Grundriss der Haupt u. Residenz-Stadt Wien mit sämmtlichen Vorstädten...*, 1832.

**Guler von Weynegg** [Weineck], Johannes Peter (1562-1637). Swiss geographer, historian and soldier. Beschreibung von Rhaetia, 1616.

**Gulick** [Gulik], A. 8th edition of *Nouvel Atlas des Enfants*, Amsterdam 1799 (with Honkoop, 10th edition published Leiden, A. & J. Honkoop 1817).

**Gullan**, Edward (1785-1851). Scottish born engraver and printer worked at Blewitt's Buildings, Fetter Lane, London. Engraved (anonymously) *Map of the Island of Capri, in the Gulf of Naples*, for issue 22 of *The Repository of Arts, Literature, Commerce, Manufactures, Fashions, and Politics*, London, R. Ackermann 1810, re-issued, with the addition of "E. Gullan, sculp.", in *Naples and the Campagna Felice...*, London, R. Ackermann 1815; *A Map of the Eastern [...Western...] Part of the Roman Empire By Thos. Kitchin Senr.* (2 maps), 1815 (a re-engraving by Gullan for E. Gibbon's *The History of the Decline and Fall of the Roman Empire*, London, T. Tegg 1815, 1819 and other editions); engraved 13 maps for W. Thorn's *Memoir of the Conquest of Java*, London, T. Egerton 1815; 9 maps for W. Thorn's *Memoir of the War in India*, London, T. Egerton 1818. **Ref.** HERBERT, F. 'Wheat revised and Thorn recognised: Maximilian of Wied's cartographer' in *Terra Incognitae* 33 (2001).

**Gullick**, William Applegate (*c.*1859-1922). Government printer, Sydney, New South Wales. *New South Wales Harbours*, 1900; *Blue Mountains from Lawson to Lithgow*, 1909.

**Gümbel** [Guembel], Carl [Karl] Wilhelm von (1823-1898). Bavarian geologist. *Geognostische Karte des Königreichs Bayern*, 1858; *Geologische Übersichtskarte von Bayern*, 1899.

**Gumilla**, José, *SJ*. Spanish missionary. Jesuit missions in the Orinoco, 1741. *Histoire Naturelle, Civile et Geographique de l'Orenoque* (with a map), Avignon, veuve de F. Girard 1758 (translated from the Spanish edition).

**Gummer**, *Staff Sergeant* P.E. RE. Military surveyor and draughtsman. Produced numerous manuscript plans of military installations on Malta, 1896-1903.

**Gumoens**, C. de. *Plan de la ville et baye de Cadiz*, 1820.

**Gumpp**, brothers. Engineers of Innsbruck, Austria.

• **Gumpp**, Johann Martin (1643-1729). Artist and architect. Coloured manuscript drawing of the Inn valley, later 17th-century.

• **Gumpp**, Johann Baptist (1651-1728). Engraver and engineer of lnnsbruck. *Mappa Der Fünff Österreichischen Herrschaften Vor dem Arlperg*, 1685.

**Gumppenberg**, Wilhelm von (1609-1675). Jesuit and historian. *Atlas Marianus* (4 volumes), Ingolstadt 1657-1659.

**Gumprecht**, Thaddäus Eduard (1801-1856). Geographer of Berlin.

**Gunby**, William. *Chart of the harbour of Wainfleet*, 1809.

**Gundling**, Jacob Paul, *Freiherr* von (1673-1731). Historian and statesman, born Hersbruck, Franconia [east of Nuremberg], died Potsdam. Historian at the College of Heralds, Berlin, President of the Royal Prussian Academy of Sciences (1718). *Land-Charte des Chvrfürstenthums Brandenburg* (2 sheets), Berlin and Amsterdam 1724 (engraved by G.P. Busch); map of the Altmark, 1724; Dukedom of Magdeburg, 1730 (engraved by G.P. Busch). **Ref.** SCHARFE, W. 'President, King's Jester and cartographer: Jakob Paul von Gundling and the first domestic map of Brandenburg 1724 (Paper presented to the 15th International Conference on the History of Cartography, Chicago 21-25th June 1993 [unpublished]).

**Gunman** [de Valle], Christopher (*d.*1685). Master in English Navy. Algiers Bay, 1664 MS; La Bouche de Valle, Guernsey, 1680.

**Gunn**, Otis B. Native of Wyandott, Kansas. Published a guide in response to the Kansas Gold Rush, entitled *New Map and Hand-book of Kansas and the Gold Mines*, Pittsburgh 1859 (included a map of routes from the Missouri River to the Kansas gold mines).

**Gunning**, P. Hydrographer, *RN*. *Table Bay, with the road of the Cape of Good Hope*, 1806 MS.

**Gunnison**, *Lieutenant* (later *Captain*) John Williams (1812-1853). Surveyor and draughtsman from Goshen, New Hampshire. Served as an artillery officer, then with the U.S. Corps of Topographical Engineers from 1838. He was killed at Sevier River. Surveys in Georgia, the

*Björn Gunnlaugsson (1788-1876). (By courtesy of the National Library of Iceland)*

pacific Northwest and Northern Lake regions, 1840-1849; explorations with H. Stansbury of Utah and the Great Salt Lake, 1849-1850. *Map of a Reconnaissance between Fort Leavenworth on the Missouri River and the Great Salt Lake in the Territory of Utah*, 1850 (with H. Stansbury and A. Carrington); *Map of the Great Salt Lake...*, 1850 (with H. Stansbury and A. Carrington); feasibility survey for a transcontinental railway along the 38th parallel, 1853. **Ref.** WALDMAN, C. & WEXLER, A. Who was who in world exploration (1992) pp.301-302.

**Gunnlaugsson**, Björn (1788-1876). Icelandic surveyor and teacher of mathematics and natural science at Bessastadir and Reykjavik. Between 1831 and 1843 he worked on surveys of the inhabited parts of Iceland on behalf of the Icelandic Literary Society. The surveys were sent to Copenhagen, where O.N. Olsen finalised the printed maps. De mensura et delineatio... Icelandiae interioris, 1834; *Uppdráttr Íslands* (4 sheets), Reykjavik and Copenhagen 1844 and later; *Uppdráttr Islands*, Copenhagen 1849.

**Gunst**, P. *Plan von Riga*, 1879.

**Gunst**, Pieter van. *Castrum Lipsiae* [Leipzig], c.1720.

**Gunten**, Nicholas. Manuscript map of Enfield Chase, Middlesex, 1658 (with Edmund Rolfe).

**Gunter**, Edmund (1581-1626). Clergyman, astronomer, mathematician and inventor. Professor of Astronomy at Gresham College, 1619. Created many measuring, navigational and surveying instruments including Gunter's Chain (22 yards, 100 link), Gunter's Line, Gunter's Quadrant and Gunter's Scale; wrote on the Cross-staffe, 1626.

**Günther**. See **Guenther**.

**Günther**, Georg. *Reichs-, Landes- und Bezirks-Strassen- dann auch Eisenbahn-Dampfschifffahrts- und Telegrafen-Karte des Erzherzogthums Oesterreich ob der Enns*, 1866.

**Gunz**, Gall Josef. Diocesan and parish maps of Vorarlberg, Austria. *Designatio Limitum Dioecesium...*, 1785.

**Guran**, Alexander (1824-1888). General and cartographer of Vienna.

**Gürich**, Georg. *Geologische Skizze von Afrika*, published in *Petermanns Mitteilungen*, 1887.

**Gurlach**, P. See **Gerlach**, P.

**Gurmontius**. See **Gourmont** family.

**Gurney**, R. Surveyor. Buckingham property, Aylesbury, 1804 MS.

**Gürsch**, C.F. See **Guersch**, C.F.

**Gur'yev**, Yemelyan. Russian topographer. Maps of various frontier military stations, 1732-1733.

**Güssefeld** [Guessefeld], Franz Ludwig (1744-1808). Cartographer from Osterburg, studied at Brandenburg, worked in Weimar and Nuremberg. d.Weimar. Worked for Homann's Heirs from 1773. *Nouvelle carte geographique du Marggraviat de Brandebourg*, 1773; maps of Austria, 1778; *Regnorum Hispaniae et Portugalliae*, 1782; *Carte über die XIII vereinigte Staaten von Nord-Amerika...*, Homann's Heirs 1784; Moldau, 1785; Russian Empire & Great Tartary, 1788; *Neuer methodischer Schul-Atlas*, Weimar 1792; *Carte itinéraire de l'Europe pour le Guide des voyageurs...*, Weimar 1793; *Charte über Königreich Schweden*, Nuremberg 1793; Weimar, 1800; West Indies, 1800; *Chart über die sæmtlichen zum Westphælischen Kreis...*, Nuremberg, Homann's Heirs 1802; for Adam Christian Gaspari, 1804-1811; *Charte vom Königreiche Sachsen...*, Weimar, Geographische Institut 1812. **Ref.** KRETSCHMER, DÖRFLINGER & WAWRIK Lexikon zur Geschichte der Kartographie (1986) vol.C/1, p.284 [in German].

**Gutbier**, Ludwig von. Lithographer and draughtsman of Glogau, for Carl Flemming. Gegend von Huhnstein und Schandau, 1857; 34th edition of C.G.D. Stein's *Neuer Atlas der ganzen Erde für die Gebildeten...*, 1877; *Neuer Atlas*, 1879 edition.

**Gutch**, George (1790-1874). Architect and estate surveyor in London. District surveyor, 1825-1874, Surveyor to the Bishop of London, 1827. *Plan of Paddington*, 1828-1852.

**Gutch & Cox**. Engravers, lithographers and

publishers of Southampton. *Map of Southampton & neighbourhood*, 1876; *Map of the New Forest*, 1884.

**Gutenhag**, *Count* of. See **Herberstein**, Sigismund.

**Guthe**, *Professor* Hermann (1825-1874). *Plan Hannover, Oldenburg &c.*, 1868; *Lehrbuch der Geographie*, 1877-1879; *Palästina*, 1890; *Bibelatlas*, Leipzig, Wagner & Debes 1911, 1926.

**Güther**, F. See **Guether**, Franz.

**Guthrie**, William (1708-1770). Scottish geographer and historian from Brechin. Studied at the University of Aberdeen, settled in London, 1730. *History of Scotland* (10 volumes), 1767; *A New Geographical, Historical & Commercial Grammar*, 1770 (maps by T. Kitchin) and many other editions up to 1843 (those from 1794 onwards carried maps by J. Russell); *The Atlas to Guthrie's System of Geography*, 1785 onwards.

**Gutiérrez** family. Spanish chart and mapmakers.
• **Gutiérrez** [Guiterrez; Gutiero], Diego, *the elder* (*c*.1485-1554). Chart and instrument maker from 1534, worked with Sebastian Cabot. Succeeded Cabot as 'Piloto major de la Casa de la Contratación', Sevilla, 1547-1554; cosmographer to the King. Portolan chart of the Atlantic with the coasts of Africa and America, 1550 MS; World, 1551 MS (based on Sebastian Cabot); Diego's charts were the basis for Hieronymous Cock's *Americae Sive Quartae Orbis Partis Nova et Exactissima Descriptio. Auctore Diego Gutiero...*, Antwerp 1562 (first printed map to name California and the largest printed map to that date). **Ref.** BURDEN, P. *The mapping of North America* (Rickmansworth 1996) pp.37-39; KARROW, R.W. *Mapmakers of the sixteenth century and their maps* (1993) pp.285-287; GOICOECHEA PORTUONDO, J.M. 'La cartografia española de América del siglo XVI', *Revista de Historia Naval* 7 (1989) pp.26-32 published by Instituto de Historia y Cultura Naval, Madrid [in Spanish]; KRETSCHMER, DÖRFLINGER & WAWRIK *Lexikon zur Geschichte der Kartographie* (1986) vol.C/I p.284 (numerous references cited) [in German]; LAMB, U. 'Science by litigation: A cosmographic feud' *Terrae incognitae* I (1969) pp.40-57.
•• **Gutiérrez**, Sancho. Son of Diego the elder, licensed as a chart and instrument maker in 1539. Cosmographer attached to Casa de la Contratación, Sevilla from 1539, succeeded his father as 'Piloto major' in 1554. **Ref.** KELSEY, Harry 'The planispheres of Sebastian Cabot and Sancho Gutiérrez' *Terrae Incognitae* Vol. 19 (1987) pp.41-58.
•• **Gutiérrez**, Diego, *the younger*. Son of Diego the elder. Spanish cartographer. Sometimes erroneously credited with the 1562 map of America (See Diego the elder).
•• **Gutiérrez**, Luis. Son of Diego the elder. Involved with his brothers in chartmaking.

**Gutierrez**, Fernando. Spanish engineer. *Plano de Madrid*, 1841-1846 (with Juan Merlo and Juan de Ribera), copied and engraved for adoption as the official town plan, 1848.

**Gutrather**, Odilo. See **Guetrather**, Odilo von.

**Gutschouen**, Gérard à [Gerard van]. Plan of Louvain in Anna Beeck's composite atlases, *c*.1700.

**GutsMuths**, J.C.F. *Vollständiges Handbuch der neuesten Erdbeschreibung*, Weimar 1819-1832 (with A.C. Gaspari, G. Hassel, J.G.F. Cannabich & F.A. Ukert).

**Gutwein**, Johann Balthasar (1702-1785). German engraver. *Ecclesiae Babenbergensi*, 1771.

**Guy**, Hezekiah Bartlett (*fl.c*.1820-1849). English estate surveyor, of Hinton St. George (1821-*c*.1824), and Crewkerne, Somerset (1824). Produced tithe and inclosure maps in Dorset, Somerset and Wiltshire. Survey at Longlond, 1824 MS.
• **Guy & Stubbs** (*fl*.1820-1843). H.B. Guy, working with Oliver Stubbs.

**Guy**, *Captain* J.M., *RN*. Hydrographer, master of HMS *Falmouth*. Sketch of Cawely & Cagayenes Is. [Philippines], 1764, published A. Dalrymple 1792.

**Guy**, M.S., *2nd Master*, *RN*. Survey of the River Derwent, Tasmania, 1861-1863.

**Guyard**, Jacques Louis. *Carte des marches, que les différentes commandes ont faits pour prendre Candie...*, 1765.

**Guyen**, —. Script engraver for Guillaume Henri Dufour, 1860s.

**Guyer, —.** French lithographer. Europe in 1815, 1830 and 1868 (3 maps on one sheet), Paris, J. Dumaine [n.d.].

**Guyet** [Gajetius; Guyeto; Guijetus; Guyetus], Lézin [Licinio; Licinus; Lécin] (1515-1580). Poet and geographer of Angers. Map of Anjou, used as a basis for *Anjou. Andegavensium ditionis vera et integra descriptio*, published by A. Ortelius, 1579, (used by Maurice Bouguereau, 1591, J. Le Clerc, 1591, Willem Blaeu, 1631 and other atlas publishers).

**Guyot,** A. *Zeekust der Banda Eilanden*, 1871.

**Guyot,** Arnaud [Arnold] (1807-1884). Swiss born geologist, taught at Princeton University, New Jersey. Mural Atlas, 1856; wall maps of ancient Greece and Italy, 1866-1867 (with H.C. Cameron); Wall Atlases, 1862-1869; Australia, New York 1866; *Central States*, for *Guyot's Geographical Series*, New York, Charles Scribner & Company 1866; *Intermediate Geography*, New York, Charles Scribner & Company 1867 and later; *Elementary Geography*, 1871; *Guyot's Grammar-School Geography*, New York, Scribner, Armstrong & Company 1874; *Guyot's New Intermediate Geography*, Scribner, Armstrong & Company 1875.

**Guyot,** Christoffe. Printer of Leiden. *Trésor de chartes*, Amsterdam, Cornelis Claesz. 1602 (French edition of *Caert-Thresoor* by B. Langenes, translated by I. de la Haye, this edition had earlier been printed at The Hague by Albert Hendricks, 1600).

**Gvosdev,** Mikhail. Russian geodesist and topographer. One of the first Russians to navigate the North American side of the straits between Asia and America, 1730-1732; discovered the Diomede Islands. Chart of Bering Strait, 1741; maps of Okhota River and Kurile Islands, 1742.

**Gwin,** Roger. Essex land surveyor. Great Henny & Bulmer, Essex, 1600 MS.

**Gwynn,** *Major*, —. *Part of Abyssinia and the Sudan*, 1901 (with Lieutenant L.C. Jackson).

**Gwynn,** George. *Plan of Ramsgate Harbour*, surveyed 1815, corrected to June 1851 and published 1851.

**Gwynn** [Gwyn; Gwynne], John (1713-1786). Architect. *Plan of London after Great Fire*, 1749 (based on Sir Christopher Wren's proposals); Proposed improvements Mansion House and London Bridge, 1766.

**Gwynn,** *Lieutenant* Walter. Road Zanesville-Florence, 1828.

**Gy,** André de. See **Chrysologue,** Noel André.

**Gyatso,** *Lama* Ugyen. Schoolteacher from Darjeeling, trained in surveying by the British. Surveys in the Himalayas from 1883. **Ref.** RAWAT, Indra Singh *Indian explorers of the 19th century* published by the Government of India.

**Gybertszoon,** E. See **Gijsbertszoon,** Evert.

**Gyer,** Edward. *Estate map Erlington, Sussex*, 1629.

**Gyger** [Geiger], Hans. Father and son of Zurich.
• **Gyger,** Hans Conrad (1599-1674). Cartographer and engineer of Zurich. *Der Uralten Loblichen Statt Zürich Grafschafften, Herrschafften, Stett und Land...als Appenzell, Apt und Statt S. Gallen*, 1620; Meijenfeldt, c.1623; worked with H.K. Huber and J. Murer on *Geometrische Grundlegung der Landgraffshafft Thurgöt* (24 sheets), 1628-1629 (completed by A. Murer in 1671); *Helvetiae, Rhaetiae & Valesiae... tabula nova et exacta*, 1637; *Helvetiae...*, 1657; many maps and plans copied by Matthäus Merian and others. **Ref.** DÜRST, A. *6. Kartographiehistorisches Colloquium Berlin 1992. Vorträge und Berichte* in: SCHARFE, W. (Ed.) Berlin 1994 pp.139-151 [in German]; KRETSCHMER, DÖRFLINGER & WAWRIK *Lexikon zur Geschichte der Kartographie* (1986) vol.C/1, pp.284-285 (many references and works cited) [in German].
•• **Gyger,** Hans Georg (1626-1687). Son of Hans Conrad Gyger (above). Painter and engraver of Zurich. *Nova descriptio ditionis Tigurinae-Neue Beschreibung der Landschaft Zürich*, 1685 (from the work of Hans Conrad Gyger).

**Gylbert,** H. See **Gilbert,** *Sir* Humphrey.

**Gyldén,** C.W. (1802-1872). *Plan of Helsingfors*, 1838; *Samling af plankarter öfver samtelige städer i Finland*, Helsingfors 1843 (includes Wiburg [Viborg], 1839; Sordavala [Sortavala], 1840; and Kexholm [Priozersk],

*John Gwynn (1713-1786). (By courtesy of the National Portrait Gallery, London)*

1843); *Suomenmaan Korko-Karta* [Finland] (6 sheets), Helsinki 1850, 1853.

**Gyōgi** [Gyoki] Bosatsu (688-749). Japanese Buddhist itinerant priest; said to have made the first map of Japan. The *gyoji* type map of Japan was named after him, and maps of this style were reproduced on plates, mirrors and artefacts up to the nineteenth century. **Ref.** UNNO, K. 'Maps of Japan used in prayer rites or as charms' in: *Imago Mundi* 46 (1994) pp.65-66.

**Gyokuran**[sai] [Sadahide], Hashimoto. See **Hashimoto** Sadahide.

**Györffy**, István (1884-1939). Hungarian ethnographer and cartographer, born Karcag, died Budapest. Bithynia, 1918; ethnographic map of Hungary, 1940 (with Pál Teleki). **Ref.** ILYÉS, Gy. *A Magyar nép tudósa* [Hungarian Scientist] Budapest (1964).

**GZ** [Gonzalez], Joseph (1696-1753). Engraver for R.P. Murillo de Velarde's *Africa arreglada a las mejores relaciones...*, for *Geographia Historica Africa...*, 1752. **Ref.** STONE, J.C. (Ed.) *Norwich's maps of Africa* (1997) N°.93 p.110. **N.B.** Compare with **Gonzalez**, Joseph.

H. C. Prussian military cartographer. *Plan der Stadt Neise, c.*1725.

H., C. See **Aurelius**, Cornelius.

H., C. de. See **Hooghe**, Cornelis de.

H., E. English Chronicle (Table Roads), 1618.

H., E.D. See **Hauber**, Eberhard David.

H., E.V.D. *Plan de la ville d'Anvers,* 1890.

H., F. *Plan de la Bataille de Laffelt, c.*1748.

H., G. *The Ichnography of Charles-Town at High Water,* B. Roberts and W.H. Toms 1739 (drawn by G.H., engraved by W.H. Toms).

H., H. *Abris der Stadt Pilsen, c.*1620.

H., J.E. Route to Gold Mines North Platte, 1839.

H., L. See **Homem**, Lopo.

H., M. *Die Stat umb Bamberg* [road map], *c.*1545.

H., padre M.A. *Italia Augustiniana, c.*1730.

H., W. *The Infallible Guide to Travellers,* 1682.

**Haack**, —. *Übersichtskarte... der Trier-Saarbrückener-Mannheimer Eisenbahn,* 1847.

**Haack**, *Professor Doctor* Hermann (1872-1966). Map and atlas editor. Worked for Justus Perthes of Gotha. Globes, Gotha, J. Perthes 1914; revised and collaborated in the publication of some of the later editions of *Stielers Hand-Atlas,* 1921-1934.

**Haacma** [Haeckma], Sjoerd Aetesz. Dutch cartographer and surveyor. Provincial surveyor of Friesland from *c.*1654-1678; collaborated with Schotanus à Sterringa on surveys of Friesland.

**Haaff**, J.M. van 't (*fl.*mid-19th century). Dutch cartographer, worked in The Hague. *Kaartjes van Europa en Nederland, c.*1840s-1860; *Kaartjes van Azie en Afrika,* [n.d.]; *Kaartjes van Amerika en Australie, c.*1840s-1860.

**Haagh**, Börje (1670s-1717). Fortification Master, City Engineer of Stockholm, 1709. Map of the Swedish Colony in North America entitled *Swenska Colonie Vthi America* [Delaware], 1696 MS (after Per Lindheström).

**Haake**, A. von. Topographer to the Post Office Department, Washington. *Post Route Map of the State of Utah...,* 1897 and later.

**Haan**, D.B. de. See **Bierens** de Haan, D.

**Haan**, Friedrich Gottlieb (1771-1827). Master of Science, teacher and craftsman from Saxony. Worked in Torgau and Dresden. Erdkugel, 1821; *Die Erde fur Freunde der Geographie* [27-cm terrestrial globe], Dresden 1822; *Die Gestirne in zwei... Planisphaeren,* 1844.

**Haan,** Gerrit de. Master cartographer for the VOC (Dutch East India Company) at Batavia [Jakarta]. Java, 1741 MS.

**Haan,** J.W.G.J. Eilerts de. Overzichtskaart van Suriname, in Verslag van de expeditie naar de Suriname-Rivier, 1910.

**Haan,** L.C.A. de. *Breda en Omstreken...* (2 sheets), Breda, J. Hermans 1853 (lithographed by A.J. van Bogaerts).

**Haan,** Laurens Feykes. *Noord Occiaen*, Amsterdam, Gerard van Keulen 1714; *...Straat Davids...*, Gerard van Keulen 1714.

**Haardt,** Vincenz von (1843-1914). Austrian geographer from Iglau, specialised in wall maps and school atlases. Worked with the publisher E. Hölzel, for whom he was scientific advisor. *Oesterreichisch-Ungarischen Monarchie*, 1879; *Geogr. Atlas für Volksschulen*, 1879; *Wandkarte der Alpen*, 1882; *Orohydrographische Wandkarte von Europa*, 1883; *Übersichts-Karte der ethnographischen Verhältnisse von Asien* (6 sheets), Vienna 1887; *Hand-Atlas von Oesterreich-Ungarn*, 1887; *Physikalisch-statistischer Schul-Atlas* (14 sheets), Hölzel 1889; *Süd-Polar-Karte* (4 sheets), 1895; Blasius Kozenn's *Geographische Schul-Atlas*, 1896 edition; *Nord-Polar-Karte* (wall map, 4 sheets), Vienna 1899.

**Haas,** George (1756-1818). Engraver. *Carte des Partages de la Pologne*, c.1800.

**Haas** [Hass], Johann Baptist. Engraver in Stauffen im Breisgau. *Birkenfeld*, 1779; *Mappa... ad Historiam Nigrae Silvae*, 1788.

**Haas** [Hass], Johann Heinrich (1758-1810). German military cartographer. *Kartenaufnahme im Großherzogtum Hessen*, 1786-1805; *Rhein, Main und Neckar* (24 sheets), 1787; *Spezialkarte von dem Odenwald, dem Bauland...*, 1808.

**Haas** [Haase; Has; Hase; Hasius; Hass], Johann Matthias (1684-1742). Historian, cartographer and theologian, worked in Augsburg and Leipzig, professor of mathematics at Wittenberg from 1719, worked for Homann's Heirs from 1730s. Specialised in historical maps and town plans, some of them published posthumously by Homann's Heirs. *Tabula expeditionis Alexandri M. Macedonis* [historical maps showing the expeditions of Alexander the Great], 1716 (engraved by G.C. Bodenehr); *Imperium Turcicum*, 1731; *Regni Sinae vel Sinae propriae Mappa et Descriptio Geographica...*, 1734; *Africa...*, Homann Heirs 1737; *Ries*, 1738; *Imperii Russici et Tartaria Universae...*, 1739 (engraved by R.A. Schneider), also sold in Utrecht, J. Broedelet 1743; *Asiae Minoris Veteris et Novae...*, 1743; *Carte de l'Asie...*, 1744; *Europa in partes suas...*, c.1743; *Hungariae... Tabula...*, 1744; Grund Staedten, 1745; *Americae Mappa generalis...*, 1746; *Europa* (4 sheets), 1746; *Atlas historicus*, 1750. **Ref.** KRETSCHMER, DÖRFLINGER & WAWRIK Lexikon zur Geschichte der Kartographie (1986) vol.C/1 pp.287-288 [in German].

**Haas,** Jonas (1720-1775). German engraver, worked in Hamburg (1744-1753) and Copenhagen (1754-1775). *Neue Charte von dem Türkischen Reich gelegen in Europa Asia und Africa*, 1747 (published in *Die Heutige Historie... des Türkischen Reichs...*, 1748); *Accurater Grundris der Kayserlichen Residentz Stadt St. Petersburg*, 1750; *Provincia Irkutensis*, 1752 (for J.G. Gmelin's work on Siberia); *Faeroischen Inseln*, Copenhagen and Leipzig, F.C. Pelt 1757 (taken from L.J. Debes's work on the Faeroes of 1673); Iceland, 1772.

**Haas,** P. One of the engravers of maps and plans for Jean Bernoulli's 3 volume work about India entitled *Historisch-geographische Beschreibung von Hindustan*, published from c.1786.

**Haas,** P. Lithographer of Washington D.C. Prepared J. Macomb's surveys of Kennebec River (5 sheets), 1826; Washington Hood's *Map exhibiting the position of several lines connected with the settlement of the Ohio boundary question*, Washington 1835; A. Lewis's *Plan of Lynn Harbour and Beach* [Massachusetts], 1837; C. Wilkes's *Chart of the Southern Coast from Tybee Bar to Hunting Id. May River*, 1838 (drawn by R.E. Johnson, J. Alden & W. May); J. Glynn's *Beaufort Harbour North Carolina...*, 1839 (drawn by H.C. Flagg); J. Glynn's *A Chart of the Entrance of Cape Fear River* (4 sheets), 1839; J. Glynn's *Cape Fear River North Caroline...* (3 sheets), 1839; J. Glynn's *Continuation of the Survey of Cape Fear River...* (2 sheets), 1839; *Map of the river*

*Sabine from its mouth on the Gulf of Mexico in the sea to Logan's Ferry...* (5 sheets), 1840 (drawn by T.J. Lee).

**Haas**, Peter F. (1754-1804). Danish engraver. Battle plans, 1782-1800.

**Haas**, Wilhelm, father & son. Typefounders and printers of Basle, creators of the 'typometry' system of map production. Printed from movable type, using cast metal for symbols and line as well as lettering. **Ref.** AKERMAN, J.R. 'A collection of Haas typographic maps' *Mapline: a quarterly newsletter* 64 (Chicago 1991-1992) pp.1- 5; HOFF-MANN-FEER, E. *Regio Basiliensis* 10, 1 (1969) pp.3-56 [in German]; HORN, W. *Petermanns Geographische Mitteilungen* 94, 2 (1948) pp.90-97 [in German].
• **Haas**, Wilhelm [I] (1741-1800). Typefounder and printer from Basle. Produced a map of the canton of Basle, Basle 1776; *Sicilien*, 1777; *Die Landschaft Basel und das Frickthal*, 1795.
•• **Haas**, Wilhelm *sohn* [II] (1766-1838). Son of Wilhelm, above, succeeded to his father's business. Letter engraver, printer, art publisher and councillor in Colmar and Basle. *Der Helvetischen Republik neue Cantone... Augstmonat*, 1798.

**Haase**, —. One of the engravers for A. Stieler's *Karte von Deutschland dem Königr. der Niederlande dem Kgr. Belgien, der Schweiz und angränzenden Ländern... in XXV Blättern*, Gotha, Justus Perthes 1836, 1859, 1862.

**Haasis & Lubrecht**. Publishers of New York. *The American Union Railroad Map of the United States, British Possessions, West Indies, Mexico and Central America*, 1870, 1872.

**Haast**, *Doctor* Julius von (1822-1887). Geologist, studied at Bonn University, travelled to, and worked in New Zealand, where some geological features carry his name. Provincial Geologist for Canterbury, New Zealand, 1861-1871. Worked with F. von Hochstetter on the geological survey of New Zealand, surveys used in Hochstetter and Petermann's *Geological and Topographical Atlas of New Zealand*, Auckland 1864; *Topographical Map of the Head Waters of the Rakaia*, 1867; *Geological sketch map of Canterbury Plains*, 1864; *Geological Sketch Map of New Zealand*, 1873 (edited by J. Hector, Director of the Geological Survey); *Provinces of Canterbury and Westland, New Zealand* 1879.

**Habash al-Hasib al-Marwazi**, Ahmad Ibn Abdallah *(fl.*9th-century). Astronomer and geographer from Persia. *Kitab al-Ajram wa al-Abad* [Book of celestial bodies and distances - known as the Mamunic tables]; *Kitab fi Marifat al-Kurah* [Book of knowledge of the globe].

**Habenicht**, Hermann (1844-1917). German cartographer, compiled maps for Adolf Stieler and Justus Perthes. One of the draughtsmen for A. Petermann's *Special-Karte von Süd-Schleswig...*, Gotha, Justus Perthes 1864 (with Debes and Welker); *Die Capstadt und Umgebung* and *Table Bay und False Bay*, in Stieler's *Hand-Atlas*, 1866; *Deutschland*, 1866; revised 20th edition of Justus Perthes' *Taschen Atlas*, 1884 and later editions; *Spezial-Karte von Afrika* (12 sheets), Gotha, Perthes 1885; wall atlas, 1888; revised editions of *Stieler's Hand-Atlas*, Gotha 1888 and later (with C. Vogel and H. Berghaus); *See-Atlas*, 1894; revised Hermann Berghaus's *Chart of the World*, 1897; *Wand-Atlas*, 1899.

**Haberer**, L. One of the cartographers for *Atlas der Urproduction Oesterreichs...* (35 sheets), Vienna, R. von Waldheim *c*.1879.

**Habermann**, Franz Xaver (1721-1796). Draughtsman, painter and engraver in Augsburg. Views of New York, Boston, Quebec, Venice and Isphahan, *c*.1780; *Augspurg, c*.1796.

**Habermel** [Habermehl], Erasmus (before 1565-1606). Astronomer and scientific instrument maker of Prague. Made armillary spheres, also created a theodolite, *c*.1600.

**Habermel**, Josua *(fl.*16th-century). Instrument maker in Germany and Bohemia. Astrolabe (with a map of Germany), 1575.

**Habersham**, Robert A. *Habersham's Sectional and County Map of Oregon...*, Portland, Oregon, J.K. Gill & Co. 1874 (engraved and printed in New York by G.W. and C.B. Colton & Co.); compiled *J.K. Gill & Co.'s map of Oregon & Washington Terr.*, 1878.

**Habert**, R. *Sicilia Insula*, [n.d.].

**Hablitscheck**, —. Engraver. *Pressburg* [Bratislava], *c*.1872.

**Hablitz,** — de. Crimée (3 sheets), St Petersburg 1790; *La Tauride, c.*1800.

**Habrecht,** Isaac. Clock and instrument makers of Strasbourg.
• **Habrecht,** Isaac [I] (1544-1620). Swiss mathematician and maker of astronomical instruments. Created the famous astronomical clock in the cathedral of Strasbourg. Small world map on a clock face, 1589.
•• **Habrecht,** Isaac [II] (1589-1633). Physician, mathematician and astronomer in Strasbourg, son of Isaac I. Untitled pair of 20-cm terrestrial and celestial globes, Strasbourg 1621 (engraved by J. van der Heyden), republished Nuremberg, J.C. Weigel *c.*1700, also republished by P. Fürst (based on the work of Plancius and Van den Keere, *c.*1612); untitled hemisphere maps in *Tractatum de planiglobium, Coeleste et Terrestre...*, Strasbourg 1628, Nuremberg, J.C. Sturm 1666. **Ref.** DEKKER, E. *Globes at Greenwich* (1999) pp.349-351; BURDEN, P. *The mapping of North America* (1996) p.273; KROGT, P. van der *Old globes in the Netherlands* (1984) p.137.
•• **Habrecht,** Abraham (*d.c.*1650). Clockmaker of Strasbourg, son of Isaac I, father of Isaac III.
••• **Habrecht,** Isaac [III] (1611-1686). Inherited his father's workshop and with it the role of 'Münster-Uhrmacher' and responsibility for the maintenance of the Strasbourg cathedral clock. Clockwork, 19-cm celestial globe, 1646. **Ref.** DEKKER, E. *Globes at Greenwich* (1999) pp.214-220.

**Hacci Ahmed** [Hadji, Ahmad] (*fl.*16th century). Geographer of Tunis. Thought to be the author of an Ottoman heart-shaped world map. Woodcut world map entitled *Cümle Cihan Nümunesi* [Depiction of the Whole World], *c.*1559 (sources thought to be cordiform maps by Johannes Werner, 1514 and Oronce Fine, 1536). **N.B.** Possibly a fictitious author created by the Giovanni Battista Ramusio circle, translated into Ottoman by M. Membré and N. Cambi, and printed by Marc'Antonio Giustiniani in Venice. **Ref.** FABRIS, A. 'Note sul mappamundo cordiforme di Haci Ahmed di Tunisi' in *Quaderni di Studi Arabi* 7 (1989) pp.3-17; *Imago Mundi* 19 (1965) pp.13-21; MENAGE, V.L. 'The map of Hajji Ahmed and its makers in *Bulletin of the School of Oriental and African Studies* 21 (1958) pp.291-314.

**Hachette,** family & company. Publishers of Paris.
• **Hachette,** Louis Christophe François (1800-1864). Established a publishing house on the basis of the stock he acquired from Jacques François Brédif in August 1826. By December he had opened his own shop at 'Rue Pierre Sarazin 12', Paris. Felix Ansart's *Atlas historique*, 1836; *Petit atlas nationale des départments de la France*, Firmin-Didot et L. Hachette [before 1838]; Henri Selves's *Géographie*, 1843; Pierre-François Cortambert's *Géographie moderne*, 1847; Achille Pierre de Meissas & A. Michelot's *Petit atlas universelle*, 1860; A. Vuillemin's *Nouveau Plan de Paris*, 1862.
•• **Hachette,** George (1828-1892). Active in the family publishing firm.
• **Librairie Hachette et Cie.** Publishers noted at 'Rue Pierre Sarazin 12' (from 1826); then a large saleroom at N°·77 (from 1858); 'Boulevard St Germain' and 'Rue Pierre Sarazin 4 & 6' (1864); 'Boulevard St Germain 79' (1869). Firm is still trading as a leading publisher of general reference works as well as atlases and travel guides. Published M.-N. Bouillet's *Dictionnaire et Atlas Universel d'Histoire et de Géographie*, 1865, 1872, 1877; E.E. Desjardin's edition of *Table de Peutinger*, 1869; Adolph Joanne's *Atlas de la Défense National*, 1870; A. Vuillemin's *Atlas topographique de la France*, 1873; F. Garnier's *Voyage d'exploration en Indochine*, 1873; L. Vivien de St. Martin's *Atlas Topographique de la France*, 1873; L. Vivien de St. Martin's *Atlas dressé pour histoire de géographie*, 1874; Elisée Reclus's *Nouvelle géographie universelle* (partwork, 19 volumes), 1875-1894; Adolph Joanne's *Guides Joanne*, 1880s and later, including *Guides Joanne. Strasbourg*, 1885 (drawn by L. Thuillier); R. Andree's *Atlas manuel de geographie moderne*, 1881, 1882; Louis Figuier's *Terre & mers: description physique du globe*, 1884; L. Vivien de St. Martin's *Australie*, 1889; F. Schrader's *Atlas de geographie historique*, 1896; *Russie Orientale et Caucasie*, 1900 (drawn by G. Bagge, engraved by E. Delaune, lettering by D. Aïtoff and engraved by Er. Dumas-Vorzet, printed by Dufrènoy).

**Haci Ahmed.** See **Hacci** Ahmed

**Haci Halifa.** See **Katib Çelebí.**

**Hack,** J.B. *Three Brothers* [South Australia], 1839; *Sources of the Para* [South Australia], 1843.

**Hack** [Hacke], William (*fl.*1682-1706). Apprenticed to Andrew Welch of the Drapers' Company or 'Thames School' of chartmakers. (1671). Hack became one of the most prolific chartmakers of the 'school', producing more than 300 manuscript charts 'At the Signe of Great Britaine and Ireland by Wapping New Stairs'; also noted at Gun Wharf (1686). A large number of his charts were bound into atlases for presentation to patrons and sponsors. Over a period of about 20 years he produced a number of copies of an atlas or rutter of Spanish charts, captured by Bartholomew Sharpe off the coast of what is now Equador in July 1681. Chart of Carolina, 1684 MS; manuscript atlas of 183 maps, 1686; *A Description of the Islands of Gallappagos*, 1687; manuscript atlas of the Pacific Coasts of South and Central America, *c.*1698 (William Clements Library); atlas of charts of North America & the Caribbean (39 manuscript charts), [n.d.] (British Library); under the name of Captain William Hacke he published *A Collection of Original Voyages*, 1699 (printed by James Knapton, illustrations by Herman Moll); atlas of the coasts and ports of the Indian Ocean, the East Indies and the China Sea, *c.*1700. **Ref.** KELLY, J. 'The pirate, the ambassador, and the map-maker' History Today 48.7 (1998) pp.48-55; The Map Collector 25 (1983) (letters) pp.45-46; SMITH, T.R. 'Manuscript and printed sea charts', in THROWER, N.J. The compleat plattmaker (1978), pp.45-100; CAMPBELL, T. 'The Drapers' Company and its school of seventeenth century chartmakers' in WALLIS & TYACKE (Eds) My head is a map (1973) pp.81-106.

**Hacke**, K. *Chemnitz-Wuerschnitzer Eisenbahn*, 1859.

**Hacke**, W. See **Hack**, William.

**Hackel**, A. Population map of Mühlviertel, Austria, 1902.

**Hackenberger**, P. Tractatus geographicarum historicarum et politicarum, *c.*17th-century MS.

**Hacker**, Anton. *Eisenbahn-und Telegraphen-Karte von Europa*, Vienna, 1857 (drawn by L. Kastner).

**Hackert**, Jacob Philipp. *Carte générale de la Parie de la Sabine*, 1780.

**Hacket**, T.R. Australian mining surveyor. *Map of the Gympie gold fields*, 1870.

**Hackett** [Hacket], Thomas (*fl.*16th-century). Bookseller at the Royal Exchange, London. Reputedly translated Peter Apian's *Cosmographicus Liber*, [n.d.], and André Thevet's *The New found World, or Antarctike*, London 1557, 1568; published Bourne's *Regiment for the Sea*, 1574.

**Hackford**, Solomon. *Chart of Boston Deeps*, *c.*1847.

**Hackhausen**, C. *Kreuznach und Umgebung*, 1875.

**Hackhausen**, J.J. *Plan der Stadt Cöln*, Coblenz 1837; *Plan der Stadt Mainz*, Coblenz, J.H. Müller 1837; *Special-Karte des Regierungs-Bezirks Coeln und dessen Angrenzungen...*, 1839.

**Hackman**, Thomas. Land surveyor of Watford, Hertfordshire. *Hartfordshire, divided into Hundreds and Parishes*, 1809.

**Hackzell**, Esaias. Swedish land surveyor. Angefåhrlig Charta öfwer Torne å och Muonio Ålfwar uti Wåsterbottns Lån, 1736 MS.

**Hacnel** de Cronenthal, —. *Guerre des Alliés*, Berlin 1821.

**Hacq**, J.M. Script engraver of 'Rue de la Harpe, No. 35', Paris. *Plan de Paris*, 1825; calligraphy for Lapie's *Carte Physique, Historique & Routière de la Grèce...*, Paris, C. Picquet 1826; A.H. Dufour's *Carte Physique, Politique & Comparée de la Turquie d'Europe...*, Paris, P.J. Lameau 1827; engraved for Louis Isidore Duperrey, 1827; Laurent Gouvion St. Cyr, 1828; lettering on P. Monnier's chart of the Island of Martinique, 1831 (chart engraved by E. Collin fils); *Carte topographique de la France* (12 sheets), Dépôt de la Guerre 1832-1847; *Cartes particulière des Côtes de France...* (6 charts), Dépôt Général de la Marine 1834, 1836; Felix Ansart, 1834; Pierre Daussy, 1838; *Plan du Golfe de la Spezia Côtes d'Italie...*, Dépôt Général de la Marine 1846 (map engraved by C.E. Collin); one of the calligraphers for *Carte de la Russie* (23 sheets), Dépôt de la Guerre 1854-1856 (printed by F. Chardon l'aîné).

• **Hacq Jacobs**. J.M. Hacq and S. Jacobs, French engravers. Duflot de Mofras's *Carte de la Côte de l'Amerique sur l'Ocean Pacifique Septentrionale....*, Paris, Arthus Bertrand 1844.

**Hacquet de la Motte,** [Georg Jakob] Belsazar (*c*.1739/1740-1815). French born physician and geographer, worked in Austria. *Oryctographia Carniolica* [Slovenia], Leipzig 1778; *Mappa Litho-Hydrographica Nationis Slavicae*, 1784.

**Hadaszczok,** Johann. *Umgebungs Karte von Friedland*, 1891.

**Hadden,** George Ernest. Surveyor with J.O. Browne. Chelmsford-Norwich Railway, 1846 MS.

**Hadden,** *Lieutenant* James Murray (*d*.1817). Officer in the Royal Artillery, sent to Quebec in 1776. His manuscript sketch maps were prepared during the American Revolution and included: *Lake Champlain* [New York/Vermont]; *Magazine at Windmill Point and South Bay*; *Ticonderoga and the Works There*; *Lake Champlain to Stillwater*; *Towards Swords Farm*; *Battle of 10th. Sept.*; all these were published in *Hadden's Journal and Orderly Books*, Albany, N.Y. 1884.

**Haddock,** Ray. One of the contributors to *Atlas of the State of Michigan...*, Detroit, R.M. & S.T. Tackabury 1873.

**Haddon,** R. Estate surveyor. Property in Princes Risborough, Buckinghamshire, 1818 MS.

**Haddon,** W.T. *Africa South East Coast. Port Natal*, 1835, published Admiralty 1838.

**Hadfield,** J. 'Crown Surveyor. George Town Demerara' from 1845. *Map of British Guiana From the latest Surveys...*, 1838, *c*.1848.

**Hadjdj** Abu al-Hasan. Portolan of the Mediterranean Sea and Africa, after 1520 [in Arabic]. **Ref.** SOUCEK, S. 'Islamic Charting in the Mediterranean' in HARLEY & WOODWARD (Eds.) The history of cartography Vol.2 Book I Cartography in the traditional Islamic and south Asian societies (1992) p.265.

**Hadji,** A. Romanian officer. *Harta principatelor unite Romane...*, Romanian Ministry of War 1863 (engraved by C. Danielis).

**Hadji,** Ahmad. See **Hacci** Ahmed.

**Hadley** brothers.
• **Hadley,** John (1682-1744). Mathematical and scientific instrument maker. Invented a double reflecting octant used for improved observations of altitude, longitude and surveying, 1730. An improved version known as 'Hadley's Quadrant' was introduced in 1734.
• **Hadley,** George (1685-1768). Scientist. Theory of trade winds, meteorological diaries, 1729-1730.

**Hadol,** —. Political cartoon map entitled *l'Europe en 1871*, Paris, Lemercier 1871.

**Haeberlin,** R. *Eisenbahnen Deutschlands*, 1859.

**Haeckma,** Sjoerd Aetesz. See **Haacma**, Sjoerd Aetesz.

**Haeduus.** See **Quintin**, Jean.

**Haefkens,** J. *Kaart van Centraal America*, Dordrecht 1832.

**Haegel,** J.J. *Wiener Welt-Industrie Ausstellung*, 1873.

**Haellström,** C.P. See **Hällström**, Carl Peter.

**Haen,** A. de. *Bergen op Zoom*, 1739.

**Haenel,** E. See **Hänel** von Cronenthal, Emil.

**Haenel,** Edward. Printer of Berlin. *Dr. Edward Stolle's Uebersichtskarte der Rübenzucker-Industrie*, Berlin, F.A. Herbig 1853.

**Haënke,** Tadao. Santa Cruz de la Sierra [Bolivia], *c*.1770 MS.

**Haensel** [Hänsel], G. *Kohlengebiete des Lugau-Oelsnitzer Beckens*, 1872.

**Haestens,** Henrijck [Hendrick]. Publisher of Leiden. Plan Rotterdam, 1599; *Vas Stupendae Magnitudinis quod ad Rhenum in Arce Palatina conspicitur... Heidelberga*, 1608.

**Haesters,** A. *Hand atlas für preußische Volksschulen...*, Essen, Baedecker 1863.

**Haeufler,** Joseph Vincenz (1810-1852). *Versuch einer Sprachenkarte der österreichischen Monarchie*, Pest [Budapest] 1846.

**Haeuw,** Ernest. *Plan de la Ville et du Port de Dunkerque... 1861*, Dunkerque 1862; *Plan de* la Ville et du Port de Dunkerque... 1878 avec indication des travaux d'aggrandissement du Port, 1878.

**Haevernick**, H. Geological Sketch Map of South East Africa, Justus Perthes 1884; Zululand, 1885.

**Haeyen**, Aelbert [Albert] (*c*.1560-1613). Dutch chartmaker and teacher of navigation in Amsterdam. *Amstelredamsche Zee-Caerten* (North Sea], 1585, 1586 and later editions (shows navigable channels in approaches to Amsterdam).

**Hafenrefferum**, Samuel. German savant and mystic. Untitled circular woodcut world map in *Monochordon symbolico-biomanticum abstrusissimam pulsuum doctrinam*, Ulm 1640.

**Haffner**, A. *Belfort et ses Environs* [N.E. France], 1871.

**Haffner**, J.Pr.W. (1836-1901). Chief of the Swedish army topographical service. *Kart over Finmarkens Amt*, Kristiania, Geografiske Opmaaling 1870.

**Hafften**, C. Bird's-eye view of Wyandotte [Kansas City], 1869.

**Hafiz-i Abru**, Shahab al-Din, Abdallah Ibn Lutfallah, al-Bihdadini (*d*.1430). Historian of Persia, whose maps appeared in an untitled and unfinished geographical work known as *Kitab Jughrafiya* [Book of Geography], 1414.

**Hafner**, Johann Christoph (1668-1754). Engraver and publisher in Augsburg. *Coblentium*, [n.d.]; *Mannheim*, *c*.1730; *Stockholmia*, *c*.1740; *Dantzig*, *c*.1740; *Ierusalem*, *c*.1740; *Berlin*, *c*.1740; *Cracau*, *c*.1740; *Augspurg*, *c*.1750.

**Hafner** [Haffner], Melchior *the elder* and *the younger* (*fl*. to *c*.1704). Engravers in Ulm and Augsburg. *Sacra Augusta Vindelicorum*, 1629; *Schwäbische Circul*, *c*.1672.

**Hagar**. See **Averill & Hagar**.

**Hagberg**, August. *Filipstads Bergslag*, 1873.

**Hagdorn**, F. *Eifel-Gebietes*, 1890.

**Hage**, Holker. Civil Engineer. *City of Harrisburg, Dauphin County, Pennsylvania*, 1862.

**Hage**, M. de la. See **Mengaud** de la Hage.

**Hageboeck**, A. Artist, publisher and printer of Davenport, Iowa. Bird's-eye views of Moline [Illinois], 1873; Minneapolis [Minnesota], 1873; St. Paul [Minnesota], 1873; Rock Island [Illinois], 1874

**Hagelgans**, Johann Georg (1687-1762). German archivist and historian. *Atlas historicus*, Frankfurt am Main *c*.1737; *Machina mundi sphaerica*, 1738.

**Hagelgans**, Johann Heinrich (1606-1647). German historian. *Beschreibung deß... Königreichs Portugal*, Nuremberg 1641.

**Hagelstam**, *Lieutenant Colonel* Otto Julius (1785-1870). Swedish military cartographer working in Finland. *Geografisk Militairisk och Statistik Karta öfver Sverige och Norrige...*, 1817-1818, published 1820, 1821 (engraved by S. Anderson).

**Hagen**, Christiaan van [der]. Engraver from Bremen, worked in Amsterdam and Leiden (1663-1695). *Lugdunum Batavorum*, 1670; City of Leiden, 1670, 1675; Caerte van de Dyckasie Slants van Voorne, 1675 (for Pieter Engelvaert).

**Hagen**, Johann Georg. *Atlas Stellarum Variabilium* (311 star maps), 1899.

**Hagenauer**, Johann Georg (1746-1835). German architect and cartographer.

**Hagenmeyer**, L. *Plan von Heilbronn*, 1877.

**Hagenow**, Karl Friedrich von (1797-1865). Geologist and cartographer. *Rügen*, 1829, 1835, 1839; *Grundriss von Greifswald*, 1842.

**Hager**, Johann Georg [*pseud*. Jakob Sincerus; Picander] (1709-1777). German philologist and geographer. *Ausführliche Geographie*, Chemnitz 1747; *Geographischer Büchersaal* (3 volumes), Chemnitz 1766-1778.

**Hagerup**, *Lt*. —. One of the surveyors for *Kart over Den Norske Kyst...* [Norway] (3 sheets), 1842-1844 (with Due and Rynning).

**Hagerups**, H. Publisher of Copenhagen. Johannes Holst's *Nyeste Geografisk Skoleatlas*, 1888.

**Haggart**, J.L. Central America, 1856.

**Häggström**, Petter. Swedish land surveyor. Käddis uti Västerbottns Läns Sodre Contract och Umeå Sochn, 1713 MS.

**Haggi**. See **Hakki**, Ibrahim.

**Haghe**, brothers. Lithographers from Belgium, both worked in London with Day & Haghe. **N.B.** See also **Day** family and companies.
• **Haghe**, Louis (1806-1885). Belgian lithographer and artist, working with William Day in London as partner in lithographers Day & Haghe 1823-1852, later became a watercolour painter. Map Isère, High Alps, 1841.
• **Haghe**, Charles (d.1888). Younger brother of Louis Haghe (above); lithographer working with Day & Haghe, and later Day & Son.

**Hagmana**, Arvid. Swedish land surveyor. Öreryd uti Jönköpings Län, 1735 MS.

**Hagnauer**, Robert. Atlas of Madison County, Illinois, 1892 (with George K. Dickson and H. Rinicker).

**Hagner**, Charles N. *Alleghany County, Maryland*, c.1820.

**Hagrewe**, J.L. Plan of Hanover, 1800.

**Hagstrom**, Carl Peter (1743-1807). Swedish surveyor and cartographer. Chartor öfwer Svea Rikes Landshöfdingedöme och Städer, 1800 (atlas with regional maps and town plans).

**Hague**, Arnold. American geologist. White Pine Mining District, 1870-1880; worked with S.G. Emmons on the geological sheets for Clarence King's *Geological and Topographical Atlas Accompanying the Report of the Geological Exploration of the Fortieth Parallel*, New York, Julius Bien, 1876 (topographical sheets prepared by Edward Freyhold); *Eureka District, Nevada* (13 sheets), 1883; *Areal Geology. Yellowstone National Park, Wyoming*, 1896 (edited by B. Willis, engraved by S.J. Kubel, in *Geological Atlas of the United States*, 1894-1945); *Geology of the Yellowstone National Park* (with an atlas of 27 sheets), Washington 1904 (some sheets were from the survey of 1896).

**Hahn**, father and son. Clock and globemakers of Onstmettingen-Kornwestheim.
• **Hahn**, Philip Matthäus (1739-1790). Worked with P.G. Schaudt and C.M. Hahn on astronomical clocks with globes.
•• **Hahn**, Christoph Matthäus (1767-1833). Worked with his father as clock and globemaker.

**Hahn**, *Doctor* Eduard. *Die Kulturformen der Erde*, 1892.

**Hahn**, Georg Gottlieb (1756-1823). Military cartographer.

**Hahn**, Gustav Leopold. Prussian military cartographer. *Carte des Environs de l'Éscaut*, 1785.

**Hahn**, J.H. von. Morava River, 1861.

**Hahn**, Johann Georg von (1811-1869). Austrian geographer and ethnographer. *Flußgebiete des Drin und des Wardar*, 1867.

**Hahn**, *Captain* Jonas. Swedish Navy. Re-issued Johann Månsson's Marine Atlas, 1748.

**Hahn**, Theodor. *Original Map of Great Namaqualand & Damaraland* [S.W. Africa] (4 sheets), Cape Town 1879.

**Hahr**, August (1802-1885). Geografisk och militar statistisk karta öfver Sverige och Norrige samt Danmark med Schleswig och Holstein, 1845; *Karta öfver Medlersta och Södra Sverige...* (4 sheets), Stockholm 1852 (engraved by L. Bernhardt); Karta öfver Sverige (10 sheets), 1853-1860; *Stockholms, Upsala, Nyköpings, Westeras, Örebo, och Carlstadts-Län*, 1856; *Statistik Karta öfver Medlersta...*, 1861; *Fysik och Politisk Karta öfver Medlersta och Södra Sverige*, 1869; *Fysik och Politisk Karta öfver Norra Sverige* (2 sheets), Stockholm 1870; *General Karta öfver jernvägskommunikationer inom Sverige, Norge och Danmark* (6 sheets), 1875; and others.

**Haidinger**, Wilhelm [William] Karl, *Ritter* von (1795-1871). Austrian geologist, one of the founders of Geographical Society of Vienna. *Süd-Amerika*, 1854; *Geognostische Übersichts Karte der Oesterreichischen Monarchie* (9 sheets), Vienna 1845-1847.

**Haig,** Captain A. *Wabbs Harbour, East Coast of Van Diemens Land,* 1860.

**Haig,** Captain George. Deputy Surveyor of the Carolinas in the 1740s. *Indian country in western Carolina,* [n.d.] copied by George Hunter, 1751 MS.

**Haigh,** —. *Map of the Circuits of Wesleyan Methodists in England & Wales,* 1824.

**Hailbrun,** F.R. de. See **Renner** de Hailbrun.

**Hailes,** J.C. Publisher, London. *Plan of country sections... City of Adelaide,* c.1845; *Plan of the two special surveys on the River Light,* 1842.

**Haillot,** W. One of the draughtsmen for Homem de Mello's *Atlas do Brazil,* Rio de Janeiro, F. Briquiet & Companhia 1909 (maps engraved by A. Simon).

**Hain,** Joseph. *Militär-Geographie,* Vienna 1848.

**Haines,** D. American engraver and script writer. Engraver of *A New Improved Map of the Seat of War* [North America], 1815; Henry Charles's *United States,* 1819; J.H. Young's *Map of the United States,* 1832 (with F. Dankworth); *Travellers Guide Through the United States...,* S.A. Mitchell 1832 and later (with J.H. Young); engraver for some of J.H. Young's 'Tourist's Pocket Maps', from 1834; one of the engravers for H.S. Tanner's *A New Universal Atlas...,* Philadelphia 1836.

**Haines,** Commander (later Captain) Stafford Bettesworth, Indian Navy. Admiralty surveyor. *The Coast from Bushire to Basadore in the Persian Gulf 1828,* London, J. Horsburgh 1831 (with G.B. Brucks); *Trigonometrical Survey of the Coast of Arabia...,* 1828 (with G.B. Brucks), published London, J. Horsburgh 1832; surveys in the Red Sea, 1830-1834 (with H.N. Pinching); *A trigonometrical survey of Socotra* [Indian Ocean], London, J. Horsburgh 1835 (with J.R. Wellsted); *Kooria Mooria Bay and Islands 1837,* London, J. Walker 1840; *Chart of the Gulf of Aden,* 1847 (with R. Barker & A.M. Grieve), published John Walker, 1852 (engraved J. & C. Walker); *A trigonometrical Survey of the Entrance to the Red Sea,* London, J. Walker 1850.

**Haines,** W.S. *Trigonometrical Survey of the Falls of Niagara; Executed for the Geological Report of the Fourth District,* c.1838 (with E.R. Blackwell).

**Haines & Son** [Haines & Co.]. Publishers of 19 Rolls Buildings, Fetter Lane, London. *A correct Plan of the City of Jerusalem...,* 1795; *A new and correct Map of the Land of Promise,* 1795; *A New and Correct Plan of the Cities of London and Westminster,* 1796; *A new & correct map of England & Wales,* 1796; T. Kitchin's *A new map of the World...,* 1797.

**Hains,** P.C. *Country between Millikens Bend Louisiana & Jackson Mississippi,* c.1863.

**Hainzel,** brothers. Astronomers in Augsburg. Astronomical observations with Tycho Brahe, 1659-1570.
• **Hainzel,** Johann Baptist (1524-1581).
• **Hainzel,** Paul (1527-1581). Councillor of Augsburg. Funded the construction of the timber frame for Tycho Brahe's celestial globe *Globus Magnus Orichalcicus,* c.1570.

**Haipolt,** John. Astronomer. Celestial globe, 1617.

**Haire,** G.W. *Plat of the Village of North Bend, Ohio,* 1868.

**Haiward** [Haiwarden]. See **Hayward,** William.

**Hajeck** [Hajek], I.F. Engraver. *Grundriss von Dresden,* 1839.

**Haji Ahmad.** See **Hacci** Ahmed.

**Haji Khalifa.** See **Katib** Çelebí.

**Hajj Abu al-Hasan.** Mid-16th century Ottoman map and chartmaker.

**Hajts,** Lajos (1866-1933). Hungarian military cartographer, commandant of the Institute for Military Cartography, 1919-1924. Worked on the introduction of photogrammetry in military mapping. **Ref.** 'Hajts Lajos' in: *Térképészeti Közlöny* [Cartographic Bulletin] (1933/1-2) [in Hungarian].

**Hakewill,** James (1778-1843). Architect and draughtsman. *Windsor and its neighbourhood,* 1813; plan of Buckingham Palace after John Nash, 1826 (engraved by J. Hawkworth).

**Hakewill**, H. *Plan of the Roman Villa discovered at Northleigh*, London, E. Lloyd c.1816.

**Hakki** [Haggi], Ibrahim. Turkish cosmographer. *Marifat-Namak* [Book of Knowledge], c.1750.

**Hakluyt**, *Reverend* Richard (c.1552-1616). English geographer, teacher, collector, editor, translator and author of travel literature. Native of Herefordshire, ordained 1578, lived in Paris between 1583 and 1588 as Chaplain to the English Ambassador to France. *Divers Voyages Touching the Discoverie of America*, 1582; *Novus Orbis* map of the New World, used in Hakluyt's translation of Peter Martyr's *De orbe novo*..., Paris 1587 (the engraving of the map is sometimes attributed to Filips Galle but it is more likely to have been engraved by Leonardo Gaultier of Paris); published accounts of many major voyages of discovery as *The Principall Navigations Voiages and Discoveries of the English Nation*..., 1589, enlarged edition 1598-1600 (title changed, with world map by Edward Wright); assisted Emery Molyneux with globe, 1592. **Ref.** BURDEN, P. *The mapping of North America* (1996) pp.77-78; WALDMAN, C. & WEXLER, A. *Who was who in world exploration* (1992) pp.303-304; DE VORSEY, L. 'Richard Hakluyt, Elizabethan voice of discovery' in *Meridian* 14 (1988) pp.5-12; QUINN, D.G. *The Hakluyt Handbook* Vol. II, Hakluyt Society (1974).

**Halácsy**, Sándor (1837-1885). Hungarian engineer and cartographer specialising in maps of Hungarian towns. Large scale map of Pest, 1873.

**Halbfass**, Wilhelm (1856-1938). Specialised in the mapping of lakes. *Tiefenkarte des Arendsees*, 1896; Eifelmaare, 1879; Hinterpommersche Seen, 1901.

**Halcon**, *Captain* José M. Spanish navy. *Ports Masingloc and Matalvi* [Luzon, Philippines], 1835, 1873; *Mindoro* [Philippines], 1834, 1898.

**Haldane**, *Captain* John. *A chart of the south east coast of Hay Nan*, 1776-1777, and *Plan of Galloon Bay*, 1776-1777, both published in *The East India Pilot*, Sayer & Bennett 1781, and in *The complete East India Pilot*..., Laurie & Whittle 1800.

**Halde**, J.B. See **Du Halde**, Jean Baptiste.

**Halder**, A. *Interlachen*, 1867, 1896.

**Halder**, Albert H. *Zoutpansberg Goldfields*, 1896.

**Halder**, Laurenz (1765-1821). Swiss engraver, born St. Gallen, worked also in Zurich. Johann Heinrich Heidegger's *Reisende durch die Schweiz*, 1787.

**Haldimand**, *Lieutenant*, later *General Sir* Frederick, Royal American Regiment. Deputy Surveyor general, later, as Commander in Chief of the Province of Quebec he authorised and endorsed many surveys of the province. Plans of islands in the St Lawrence River, c.1760s; *River Chaudière* [Quebec], 1761 MS; assisted Captain Samuel Holland with *A Plan of the Island of St. John in the Province of Nova Scotia*, 1765; Chart of the Magdalen Islands [Quebec], 1766 MS (soundings by William Hogg).

**Haldingham**. See **Richard** of Haldingham.

**Hale**, Edward E. Map of Kansas, Boston 1854

**Hale**, F. Surveyor of Chester, assistant to D.L. Miller. County atlases from 1895.

**Hale**, George. *Map of Lake County, Illinois*, St. Louis, L. Gast Bros. & Co. 1861 (with J.M. Truesdell).

**Hale**, H. Worked with A. Gallatin and K. Andree on *Ethnographische Karte von Nord-Amerika*, 1854 (in H. Lange's *Atlas von Nord-Amerika*, Braunschweig, G. Westermann 1854).

**Hale**, H. C. *Military Map Department Dakota* [Minnesota, Montana, North Dakota, and South Dakota] (31 sheets), 1895 (blueprint on cloth).

**Hale**, Nathan. Newspaper publisher of Boston. *A map of the New England states. Maine, New Hampshire, Vermont, Massachusetts, Rhode Island & Connecticut with the adjacent parts of New York and Lower Canada*, Boston 1826, 1849, 1853 (engraved by J.V.N. Throop); *Epitome of Universal Geography*, 1830.

**Hale**, Robert (b.1796, fl. to 1862). Produced manuscript maps of Essex and Suffolk. Also worked with William Cole as Cole and Hale, estate surveyors of Colchester 1813-1824, later 32 Oval Cottages, Hackney Road, London. Copford, 1826 MS; Hatfield Broad Oak, 1835 MS; Rainham Levels, 1834-1842 MSS.

**Halenbreck**, L. *Nordsee Küste von der Elbe bis zur Ems*, 1883.

**Hales**, —. British Army Officer. One of the British representatives amongst the international team which draughted *...la frontière Serbo-Turque selon article 36 du traité de Berlin... 1878* (11 sheets plus title), 1879.

**Hales**, John Groves (1785-1832). English geographer and surveyor, worked in Boston Massachusetts. *Map of Boston in the State of Massachusetts*, 1814 (engraved by T. Wightman Jr.); *A Map of Upper and Lower Canada With Part of the United States Adjoining*, 1813 (engraved by Wightman); *Map of Boston and Its Vicinity*, (engraved by E. Gillingham, published in Hales's *Survey of Boston and its Vicinity*, 1821); *County of Essex, Massachusetts*, 1825 (engraved by J. Throop); *Plan of the City of Boston and Territory Thereunto Belonging*, 1830 MS; *Northampton, Massachusetts*, 1831; Town of Roxbury, 1832. **Ref.** KRIEGER, A., COBB, D. & TURNER, A. *Mapping Boston* (1999) pp.120, 191-193, 195.

**Hales**, W.D. *Plan of Port Stanley [Canada]*, 1852.

**Halfeld** [Halfield], Henrique Guilherme [Wilhelm] Fernando. Brazilian mapmaker. *Atlas e Relatorio Concernente a Exploração do Rio de S. Francisco*, 1852-1854, published 1860; *Karte der Brasilian provinz Minas Geraes*, 1860 MS published Gotha, J. Perthes 1862.

**Halfpenny**, H.E. *Atlas of York County, New Brunswick*, 1878; *Atlas of Kennebec County, Maine*, Philadelphia, Caldwell & Halfpenny 1879; Atlas of Oxford County, Maine, 1880 (with J. Caldwell); *Atlas of Hancock County, Maine*, Ellsworth, Maine, S.F. Colby & Co. 1881; Atlas of Somerset County, Maine, 1883 (with George N. Colby).

**Halifax**, John. See **Sacrobosco**.

**Halil Bey**, Cerrahpasali [Halil Ibrahim]. Retired Brigadier General. Instructor in topography and cartography at the War Academy in Istanbul. Ottoman relief map of the Strait of Istanbul and its surroundings, 1884; relief map of Crete and Rhodes, 1885; Series of relief maps of the Balkan Seas, 1884-1886; relief map of Caucasus, 1887; relief map of Caucasus, 1895 (with H. Behçet); relief map depicting the defence establishments of Edirne and its surrounding territory, 1895; relief map of the Strait of Çanakkale and its battles, 1916. **Ref.** ÜLKEKUL, C. *Historical relief maps* (Istanbul Centre for Historical research 1998) pp.24-33, 40-43, 46-57, 68-77, 98-137.

**Hall**, —. *Watering Places of New England & Canada*, 1869.

**Hall**, Mr. & Mrs. Worked together on *Ireland: its scenery, Character &c....* (3 volumes), London, How & Parsons 1841-1843 (with a map of Ireland engraved by A. Adlard), London, Jeremiah How 1846, London, Hall, Virtue and Co. 1852 (map of Cork engraved by W. Hughes added).
• **Hall**, Anna Marie. Credited as joint author of *Ireland*.
• **Hall**, Samuel Carter. Husband of Anna, credited as joint author of *Ireland*.
• **Hall, Virtue and Co.**. Publisher of 25 Paternoster Row, London. Published the 3rd edition of Hall's *Ireland*, 1852.

**Hall**, *Reverend* Anthony (1679-1723). Fellow of Queen's College, Oxford. Thought to have written some of the earlier parts of *Magna Britannia...*, London, John Morphew and John Nutt from 1714 (with maps by Robert Morden, published as a partwork over several years). **Ref.** HODSON, D. *County atlases of the British Isles published after 1703* Vol.I (1984) pp.21-22.

**Hall**, B. Assistant surveyor in the Surveyor General's Office, Melbourne. *Part of the parish of Yering*, 1855.

**Hall**, *Commander* (later *Captain*) Basil (1788-1844). Liukiu [Ryukyu] Islands, 1818; Voyage to Corea & Loo Choo Islands, 1818 (first detailed surveys of Korean coasts).

**Hall**, C. Engraver. *A Plan of the Town and Harbour of Boston*, London, J. De Costa 1775.

**Hall,** Charles. Surveyor. Plan of the town of Niagara, 1840s.

**Hall,** Charles Francis (1821-1871). Journalist from New Hampshire. Travelled to Baffin Bay (in search of Sir John Franklin's lost expedition of 1848), and explored Frobisher Bay, 1860-1862. He died during an attempt to reach the North Pole in 1871. *Smith Sound, Kennedy & Robeson Channels*, 1871, published 1875. **Ref.** WALDMAN, C. & WEXLER, A. Who was who in exploration (1992) pp.304-305.

**Hall,** *Staff Commander* Daniel, US Navy. Portland Harbor, 1873; *The Shannon*, 1877-1879.

**Hall,** David. Printer of Philadelphia. Lewis Evans's *A General Map of the Middle British Colonies in America*, 1755.

**Hall,** E. *Velociped-turist-Karta öfver Sverige*, 1898.

**Hall,** Edward S. *Lloyd's New Military map of the Border & Southern States*, New York, H.H. Lloyd & Co 1861

**Hall,** Elias. *Mineralogical and geological map of the Coalfields of Lancashire, Yorkshire, Cheshire, and Derbyshire*, 1830.

**Hall,** F. *Map and elevation of the Shubenacadie navigation from Halifax harbour to the Basin of Mines* [Nova Scotia], c.1826 (lithographed by C. Ingrey). **N.B.** Compare with Francis **Hall** [Scotland], below, and Francis **Hall** [Canada], below.

**Hall,** Francis. *Plan of the proposed Edinburgh and Glasgow Union Canal... Surveyed... 1813 and 1814*, 1814 MS (with H. Baird), reduced edition engraved for H. Baird's *Report* by W. & D. Lizars, 1818; *Plan of the proposed Union Canal from Dumbryde to Meggetland*, 1816; *General plan of the ... Union Canal with other navigations*, 1818 (with H. Baird). **N.B.** Compare with F. **Hall** above, and Francis **Hall** [Canada], below.

**Hall,** Francis. Civil engineer in Canada. *Plan of Cobourg Harbor with Proposed Improvements*, 1835; map of the road from Queenston to Sixteen Mile Creek [Niagara], 1838; *Plan and section of the Kingston Road*, 1843 (with others). **N.B.** Compare with F. **Hall** above, and Francis **Hall** [Scotland], above.

**Hall,** George. Officer in the Royal Staff Corps. One of those involved in a military survey of northern France. *British Lines of Occupation. N. of France (after 1815)...* (10 sheets), 1818-1823 MSS.

**Hall,** Henry. Draughtsman to the Royal Engineers, Cape Town. *South Africa Compiled and Corrected...*, 1849 MS; *Outline of South Africa to 16° S. latitude reduced from H. Hall's Map*, Capetown 1857 and later versions (engraved and printed by J.A. Crew); *Map of South Eastern Africa...*, London, Edward Stanford 1859; *Map of the Eastern Frontier of the Cape Colony* (2 sheets), London, Edward Stanford 1856; *South Africa compiled from all the available official authorities...*, London, Edward Stanford c.1860 (engraved in Cape Town, printed and published in London by Stanford's); *South Africa by Henry Hall R.E.D.*, London, Edward Stanford 1871, 1875, 1878 ('Engraved at Stanford's Geographical Establishment Charing Cross', used in John Noble's *Descriptive handbook of the Cape Colony*, 1875).

**Hall,** J. Lithographer of Edinburgh. *Canada and Part of the United States*, Aberdeen, John Mathison 1834 (in *Counsel for Emigrants* and *Sequel to the Counsel For Emigrants*).

**Hall,** J. Surveyor aboard H.M.S. *Swallow*. Survey of the east coast of China, province of Shan Tung and Kyan Chau bay, c.1863 (with G. Stanley, A. Hamilton, W.N. Goalen & H.R. Harris)

**Hall,** J.B. Publisher of 66 Cornhill, Boston. *North America*, 1849.

**Hall,** J.H. American lithographer of Albany, New York. Michigan Range 37-25, 1846 (one of three maps published by Weare C. Little & Co. under the collective title *Maps of the Mineral Region of Lake Superior*).

**Hall,** *Captain* James (d.1612). Pilot from Hull. Undertook voyages on behalf of King Christian IV of Denmark to Greenland and the Arctic 1605, 1606 and 1607; and on a mercantile expedition in 1612, when Hall was killed. Reported on and charted part of the coast of Greenland. Narratives and charts were used by Samuel Purchas in *Purchas His Pilgrimes*, 1625, by Captain Luke Foxe, 1635, by Thomas Rundall, 1849, and by

*Lieutenant Colonel Lewis Alexander Hall (1794-1868). (By courtesy of the Ordnance Survey, Southampton)*

Christian Gosch, 1897. **Ref.** WALDMAN, C. & WEXLER, A. *Who was who in exploration* (1992) pp.305-306.

Hall, *Sir* James (1761-1832). Geologist from Lothian, Scotland. Undertook extensive studies of the geology of Scotland.

Hall, *Professor* James (1811-1898). Geologist and palaeontologist from Massachusetts. Appointed state geologist of New York in 1836. *Geological Map of the Middle and Western States*, c.1843; *Geology of New York*, 1843; *Map illustrating the geological features of the country west of the Mississippi River*, 1857 (drawn by Thomas Jeckyll); surveys used by W.E. Logan for the U.S. sections of his *Geological Map of Canada*, 1865.

Hall, John. *Chart of the River Humber...*, surveyed c.1821-1822, published London, J.W. Norie & Co. 1823, 1836, 1842, 1844.

Hall, *Captain* John, *RN. A chart of the south part of Sumatra*, 1794; *A new chart of the Straits of Sunda*, 1794; *The Straits of Sincapore*, 1799, all published by Laurie & Whittle 1800.

Hall, Joseph [pseud. Mercurio Britannico] (1574-1656). Clergyman and writer. Bishop of Exeter (1627-1642), Bishop of Norwich (1642-1647). *Mundus alter et idem sive Terra Australis* [included a world map with an imaginative and satirical interpretation of the southern hemisphere], London c.1605 (engraved by W. Kip), Hanover 1607, Dutch edition 1643. **Ref.** SHIRLEY, R. *The mapping of the World* (1983) Nº 251 p.267.

Hall, Joseph L. (fl.1825-1841). Contributed to the Ordnance Survey in connection with Roman sites in Roxburgh.

Hall, Leventhorpe. Government printer of Hobart, Tasmania. *West Melbourne*, 1875; *Tasmania*, 1883; *Geological features of Tasmania*, 1888.

Hall, *Lieutenant-Colonel* Lewis Alexander *RE* (1794-1868). Succeeded Thomas Colby as Superintendent of the Ordnance Survey in 1847. Hall was not a surveyor or mapmaker himself and his contribution was largely managerial. In 1854 he was succeeded as Superintendent by Major Henry Toms.

Hall, L.W. *Lake Ontario to River Mississippi*, 1800.

Hall, Mary Lucy. *Our World*, Boston, Crosby & Nichols 1864; *Our World No.II - A second series of lessons in geography*, Boston, Ginn Brothers 1872, 1875; *Our World; or, First lessons in geography for children*, Boston, S.F. Nichols 1886; *Our World reader*, Boston, Ginn & Company 1889.

Hall, Mathew, father & son. Surveyors of Essex.
• **Hall**, Mathew, *senior*. (fl.1760-d.1780). Carpenter of Maldon, mapped some woodland in Essex.
•• **Hall**, Mathew, *junior* (fl.1773-d.1800). Surveyor of Maldon, son of Mathew *senior*, produced manuscript maps of Essex. Gt. Braxted, 1777 MS; Purleigh, 1778 MS.

Hall, Mosiah. *Relief Map of Utah*, Salt Lake City, Pierce & McMillen 1895 (assisted by Charles Wright).

Hall, Ralph. Engraver of London. *Virginia* from *Historia Mundi* 1636, 1637, 1639. **Ref.** BURDEN, P. *The mapping of North America* (1996) p.308.

Hall, Richard. *A Plan of the City of Glocester*, 1782 (with Thomas Pinnell, engraved by H. Mutlow).

Hall, Richard (1806-1878). Surveyor, produced manuscript maps of sites throughout England. New College [Oxford] Estate at Adstock, Buckinghamshire.

Hall, Richard Thomas. Civil engineer, worked in South Africa. *Railway Worcester*, 1874; *South African Republic*, 1877.

Hall, Robert. Plan of Boston, Lincolnshire, surveyed 1741, published 1742 (with W. Nichols, engraved by W.H. Toms).

Hall, Rouland [Rowland]. Bible maps of Eden, Exodus, Canaan, the Holy Land and the Eastern Mediterranean, Geneva 1560.

Hall, *Captain* Sampson. *Plan of the Bay and Harbour of Rio de Janeiro...*, published Laurie & Whittle 1794; *A chart of the west coast of Sumatra*, 1794, published Laurie & Whittle 1798.

**Hall**, Samuel C. See **Hall**, Mr. & Mrs., above.

**Hall**, Samuel Read (1795-1877). *School History of the United States*, 1839 (including maps).

**Hall**, Sidney [Sydney] (*fl.*1818-1853). Engravers and publishers at 18 Strand, London from 1818; 3 Arundel Street (1830); 186 Strand (1831-1833); 14 Bury Street, Bloomsbury (1844). **Ref.** BURDEN, E. 'Gorton's Topographical Dictionary Maps by Sidney and Selina Hall' in IMCoS Journal 85 (Summer 2001). **N.B.** Compare with **Thomson** & Hall, also of 14 Bury Street, Bloomsbury (1814).

• **Hall**, Sidney [SID$^y$ HALL] (*d.*1831). Engraver for *Map of the Central States of Europe, Situate between France and Russia, describing their New Limits Conformably with the Definitive Treaty of Peace...* 1815, Charing Cross, W. Faden 1816; John Leyden's *Africa...*, 1817; A. Arrowsmith's *New Holland*, Edinburgh, A. Constable 1817; *Pacific Ocean*, Edinburgh, A. Constable & Co. 1817; *Map of the countries round the North Pole*, Arrowsmith 1818; engraver of *The Niagara Frontier*, London, Longman, Hurst, Rees, Orme & Brown 1818 (in Francis Hall's *Travels in Canada and the United States*); W.M. Leake's *Map of Egypt...*, 1818 republished by Edward Stanford 1875; *Leigh's New Map of the Environs of London*, 1819; engraver for *Leigh's New Pocket Atlas of England and Wales...*, Samuel Leigh 1820 and later; *A Guide through London...*, London, M.J. Godwin & Co. 1821, 1823; William Burchell's *A Map of the Extratropical part of Southern Africa...*, 1822; *The Pedestrian's Companion Fifteen Miles round London*, London, M.J. Godwin & Co. 1822; *Map of Van Dieman's land...*, London, G. & W.B. Whitaker 1824; *Northern part of Angusshire*, 1825, published 1832; maps for Vane's *Narrative of the Peninsular War*, 1828; *East India Islands*, London, Longman, Rees, Orme, Brown & Green 1828; *Colombia*, 1828, 1840, 1844; *Pacific Ocean*, 1828; *Plan of the Cities of London & Westminster with the Borough of Southwark*, 1828; *A New General Atlas*, 1830, 1849; *Narrative of War in Germany & France*, 1830; *An Historical Atlas*, 1830; drew and engraved the maps for John Gorton's *A Topographical Dictionary of Great Britain and Ireland* (partwork), London, Chapman & Hall 1830-1832 (the later maps were prepared by Selina [S.] Hall, the maps were re-used or re-engraved for several later atlases in the name of Sydney Hall); engraved maps for John Thomson & Co., 1831; *A New Map of North & South Wales &c.*, 1831 (in *Leigh's Guide to Wales & Monmouthshire*); *Wales* (2 sheets), 1832; *A New British Atlas*, Chapman & Hall 1833 (maps from *A Topographical Dictionary*); *Sidney Hall's British Atlas*, 1834 (maps from *A Topographical Dictionary*); and many others.

• **Hall**, Selina [S. HALL] (*fl.*1831-*d.*1853). Succeeded her husband Sidney as engraver and manager of his business. She prepared new maps and re-issued the earlier works of her husband. Completed the maps for John Gorton's *A Topographical Dictionary of Great Britain and Ireland* (partwork), London, Chapman & Hall 1830-1832, 1833; 82 maps for Murray's *Encyclopaedia of Geography*, 1834; maps for William Pinnock's *Guide to Knowledge*, *c.*1835; county map of Kent for Brady's *The Dover Road Sketch Book...*, 1837; plan of Bristol, 1840, 1850; maps for Adam & Charles Black, 1840; *Thirty-two plates of the Constellations*, 1840-1842; *A Travelling County Atlas...*, London, Chapman & Hall 1842 and many later editions (maps from *A Topographical Dictionary*); *Australia and Islands Adjacent*, London, Longman & Co. 1844; *A New County Atlas*, Chapman & Hall, 1847 (maps from *A Topographical Dictionary*); *Atlas of Australia*, 1853 (with Bartholomew and W. Hughes); *The English Counties*, Chapman & Hall 1860 (maps from *A Topographical Dictionary*); *Australasia*, 1860; and many others. **N.B.** Compare with **Thomson** & Hall, also of 14 Bury Street, Bloomsbury (1814).

**Hall**, Thomas (*fl.*1728-1739). Estate surveyor from Debenham, Suffolk. Property at Debenham, 1737; estate at Copford and Birch, Essex, 1739 MS.

**Hall**, Thomas C. *Reese River in Nevada*, 1864.

**Hall**, W. Bullia Flats, Ganges, 1848 MS.

**Hall**, William (*d.*1614). Printer of London for John Sudbury and George Humble; in partnership with John Beale. John Speed's *Theatre of the Empire of Great Britaine*, 1611-1612, also part of 1614 edition (completed after his death by Thomas Snodham).

**Hall,** *Lieutenant* William. Draughtsman of the Engineer's Department, Quebec. Produced numerous manuscript plans, and copies of earlier surveys. *Map of Part of the province of Quebec comprehending Nova Scotia, New Brunswick, the Island of Cape Breton and New England and extending westward to the River Mississippi,* Quebec 1791 MS; *Survey of the town and fortifications with part of the River St. Charles and the Plains of Abraham* [Quebec], 1791 MS; *Upper and Lower Canada....* 1796 MS; *Upper Canada,* 1799-1800; *Building lots to landing on Niagara River,* 1798; *Harbour of York,* 1800; *Plan of the City and Fortifications of Quebec,* 1800 (with John B. Duberger).

**Hall,** William [& Co.]. Publisher of New York. David H. Burr's *A New Universal Atlas,* 1835.

**Hall & Elvan.** Meridian Hill, 1867.

**Hall & Mooney.** Lithographers of Buffalo, N.Y. Joshua Lind's *Plan of the Town of Hamilton, Canada West,* 1837; *Map of Niagara Falls and Adjoining Shores,* 1839; *Niagara River and Parts Adjacent,* 1840; R. Hugunin's *Chart of Lake Erie,* 1843; C. Rankin's *Map of the Niagara, Gore and Wellington Districts...,* Toronto 1845; A. McNeilledgel's *A Nautical Chart of the North Shore of Lake Erie...,* 1848; *California & Oregon,* 1849; *Map of the County of Oxford,* 1852 (drawn by T.S. Shenston, also used in Shenston's *The Oxford Gazetteer*).

**Hall & Virtue & Co.** see **Hall,** Mr. & Mrs., above.

**Hallart,** L.N. d'. *Facies Budae Regiae Metropolis Hungariae,* Munich, M. Wening c.1688.

**Halle,** S. Engraver. Prussia, 1802.

**Halleius,** E. See **Halley,** Edmond.

**Haller,** —. *Allen County, Ohio,* 1871.

**Haller,** Hans Jakob (1549-1624). Swiss theologian, mathematician and globemaker in Alt St Johann, Zollikon and Zurich.

**Haller,** Johannes (1573-1621). Swiss historiographer and engineer in Zurich. Conducted topographical surveys which contributed to H.C. Gyger's map of the Zurich area, 1620.

**Haller,** Leopold Friedrich. Publisher in Brünn. *Maehren und Oesterreichisch-Schlesien,* 1810.

**Haller von Hallerstein,** Friedrich. German geographer and cartographer. *Neuer Hand-Atlas,* Nuremberg before 1818.

**Hallerstein,** Augustin, SJ (1703-1774). Austrian Jesuit missionary working in China c.1725. *Carte de Macao,* 1739; *Observationes... Pekini Sinarum factae,* Vienna 1768.

**Hallerworden,** Martin. Publisher. *Alt und Neue Preussen,* 1684.

**Hallett,** Holt S. *Parts of Burma Siam and the Shan States illustrating the explorations of H.S. Hallett,* c.1886 (published in the *Proceedings of the Royal Geographical Society*).

**Hallewell,** *Lieutenant-Colonel* Edmund Gilling. One of the surveyors for *Military Sketch of the S.W. part of the Crimea...* (8 sheets), 1854-1855 MSS, (1 sheet), 1856 (lithographed by J.G. & W.J. Kelly).

**Halley** [Halleius], Edmond (1656-1742). English Astronomer Royal, geophysicist and geographer. Observed the periodic comet which bears his name in 1682, and predicted its return 1758; between 1698 and 1701 commanded the *Paramore,* making observations in the north and south Atlantic; succeeded John Flamsteed as Astronomer Royal in 1720; first to use isogonic lines on map; first to map and give detailed description of the trade winds. *Catalogus Stellarum Australarum,* 1679; surveyed coasts and tides in the English Channel, 1680-1701; untitled map of the trade winds, 1686; *Draught and view of Trinidada,* 1700, used by Thornton in *The English Pilot,* 1703, and Dalrymple, 1792; *...A New and Correct Sea-Chart of the Western and Southern Ocean Shewing the Variations of the Compass,* 1701 (magnetic chart of the Atlantic); *... A New and Correct Sea-Chart of the whole World shewing the Variations of the Compass,* 1702 (magnetic

*Edmond Halley (1656-1742). (By courtesy of the Bodleian Library, Oxford)*

chart extends isogones to the Indian Ocean); *A large chart of the English Channell*, Mount & Page 1702 (map of tides in the English Channel); Astronomiae cometicae synopsis, 1705; *A Description of the Passage of the Shadow of the Moon over England in the Total Eclipse of the Sun on the 22nd Day of April, 1715 in the Morning*, 1715; *Atlas maritimus & commercialis...* (54 charts), 1728 (with N. Cutler). **Ref.** COOK, A. *Edmond Halley* (1998); ROBINSON, A.H. *Early thematic mapping in the history of cartography* (1982) pp.46-47, 69-71, 84-86 etc.; THROWER, N.J.W. *The three voyages of Edmond Halley in the 'Paramore' 1698-1701* (London 1981); THROWER, N.J.W. 'Edmond Halley and thematic geo-cartography' in THROWER N.J.W. (Ed.) *The compleat plattmaker* (1978) pp.195-228; RONAN, C. *Edmond Halley* (1970); THROWER, N.J.W. 'Edmond Halley as a thematic geo-cartographer', *Annals of the Association of American Geographers* 59 (1969) pp.652-676; ARMITAGE, A. *Edmond Halley* (1966); CHAPMAN, S. 'Edmond Halley as physical geographer and the story of his charts' in *Occasional Notes of the Royal Astronomical Society* 9, (1914).

**Hallez D'Arros**, Charles Henri Olivier. *Empire Français*, 1870.

**Halligan**, Gerald Harnett (1856-1942). Hydrographer and civil engineer. *Plan of Port Kembla*, 1895.

**Halliger**, Johann I. *Gundriss* [sic] *von Putbus*, 1834.

**Halligey**, J.T.F. Plan of Freetown, [Sierra Leone], 1881.

**Halliwell**, R. Script engraver. Engraved the title to [Charles] *Smith's New General Atlas containing distinct Maps of all the Principal Empires, ... Throughout the World*, 1808; also *Smith's Classical Atlas containing Distinct Maps of the Countries described in Ancient History...*, 1809.

**Halloran**, Alfred Laurence, *Master, RN*, HMS *Osprey. Port Moniganui, New Zealand*, 1847; Jeddo Bay, Niphon [Tokyo Bay], 1849.

**Halloran**, Henry Ferdinand (1869-1953). Australian surveyor and conveyancer. Established Henry F. Halloran & Co. (1897). *Plan of part of the Mulgoa irrigation area including the Mulgoa Irrigation Company's township*, Sydney, John Sands Lith. *c.*1890; *Port Stephens City*, 1920; Environa, *c.*1920s; *Princes Highway Estate*, 1921-*c.*1930.

**Hallowell** [Carew], Captain Sir Benjamin, RN (1760-1834). *Description of Caiffe on the coast of Syria*, 1799. **N.B.** In 1828, Hallowell adopted the name of **Carew**, to become Admiral Sir Benjamin Carew.

**Halloy**, J.B.J. d'O. d'. See **Omalius** d'Halloy.

**Halls**, Daniel *fl.*1715-d.1730). Essex estate surveyor and schoolmaster of Halstead (1715), and Colchester (1723). Survey at Coggeshall, 1720 (with W. Kendall); Great Tew, 1723; estates at Hatfield Peverel, 1723 MSS.

**Hällström** [Haellström; Hellström], Carl Peter (1774-1836). Finnish cartographer and geologist. Contributed to S.G. Hermelin's *Geographiske Chartor öfver Swerige*, 1798-1804; *Charta öfver Wasa...*, 1798; *Charta öfver Uleaborgs Höfdingedöme...*, 1798; *Charta öfver Sturfurstendömeêt Finland*, 1799; *Charta öfver Wästeras...*, Stockholm 1800; *Charta öfver Uppsala*, Copenhagen 1801; *Karta öfver Gottland...*, Stockholm, S.G. Hermelin 1805.

**Halma**, father & son.
• **Halma**, François (1653-1722). Editor, printer, publisher and bookseller of Utrecht (1674-1698), Amsterdam (1699-1710), Leeuwarden (from 1710), printer to the University of Franeker from 1701. Re-issued Mercator's edition of *Claudii Ptolemaei Tabulae geographicae...*, 1695, 1698 (with Van de Water & Strick), 1704 (with Van der Water); re-issued J. Bonfrère's *Tabula geographica Terrae Sanctae*, 1700; Nicolas Bion's *L'Usage des Globes Celestes et Terrestres...*, 1700; *Description de tout l'univers*, 1700 (pocket-sized edition of Sanson's atlas); *Geographia Sacra...*, 1704; A.Ph. De La Croix's *Algemeene Wereld-Beschryving...*, 1705; B. Schotanus à Sterringa's *Uitbeelding der Heerlykheydt Friesland*, 1718; *Tooneel der Vereenigde Nederlanden*, Hendrik Halma 1725.
•• **Halma**, Hendrik. Publisher, son of François. Edmond Halley's Isogonal world chart, 1700; François Halma's *Tooneel der Vereenigde Nederlanden*, 1725.

**Halma,** *abbé* Nicolas (1755-1828). Mathematician. Translated Ptolemy's *Geographia* into French, 1828.

**Halpen,** Patrick. Engraver of Dublin. William Richards and Bernard Scalé's *A Plan of the City and Environs of Waterford,* Dublin 1764; engraved a reduced copy of J. Rocque's plan of Dublin for Lawrence Flin 1765 (published in Walter Harris's *History of Dublin,* 1766).

**Halpin,** William C. *United Jewish Cemetery (Cincinnati, Ohio),* 1862.

**Halsall,** John. Publisher of St. Louis, Missouri. *Map of the Route to California, Compiled from Accurate Observations and Surveys by Government,* 1849 (from J.C. Frémont's surveys, engraved by J.E. Ware); *Sectional map of the territory of Kansas,* 1857.

**Halse,** —. (*fl.*19th century). English globemaker.

**Halsey,** John (*fl.*1637-c.1649). English estate surveyor, worked in Buckinghamshire, Hertfordshire and possibly Norfolk.

**Halsey,** John (*fl.*1703-d.c.1738). Surveyed estates in various parts of England. Waterden, Norfolk, 1713-1714 MS.

**Halstead,** J.T. *Carver, Carver County, Minnesota,* 1857.

**Halsted,** E.P. *Coast of Arracan, Akyab* [Burma], 1842.

**Ham,** G. Town and Harbour of Rye [Sussex, England], 1665; surveys in the parish of Boughton-under-Blean, Kent, 1665.

**Ham,** Thomas [& Co.] (1821-1870). Cartographer, engraver, lithographer and publisher of Collins Street, Melbourne (1849); 35 Swanston Street, Melbourne (1854), later Chief Engraver at Government Engraving Office, Brisbane (1868). Compiled, engraved and published *A map of Australia Felix...,* Melbourne 1847, 1849; engraved and published *Map of the purchased and measured lands... Melbourne and Geelong Districts,* Melbourne, T. Ham 1849; *Ham's squatting map of Victoria,* 1851, 1853; *Squatting districts of Australia Felix,* 1851-1861; *The squatting map of Victoria,* 1851, 1853, 1854, 1857, Melbourne, James J. Blundell & Co. 1860; *Plan of the City of Melbourne,* 1854; Queensland (4 sheets), Brisbane 1856; *Suburban and Country Lands in the parish of Alberton Gipps Land,* Melbourne, Public Lands Office 1857; lithographer for R.W. Larritt's *Plan of building allotments at Iron Bark & Long Gullies, Sandhurst,* Melbourne, Surveyor General's Office 1857; publisher for R.W. Larritt's *The township of Axedale & Suburban allotments,* Melbourne, T. Ham 1858; *The New Map of Queensland,* Brisbane, J.W. Buxton 1863; *Map of the City of Brisbane...,* 1863; *Atlas of the Colony of Queensland,* (14 maps), 1865; *East Coast of Australia Queensland Cleveland Bay,* Brisbane 1867 (chart engraved by W. Knight, hills lithographed on stone by T. Ham); *District of the Gulf of Carpentaria Northern Queensland,* Brisbane, Government Engraving Office 1868; *Map of Queensland,* Brisbane, Government Engraving and Lithographic Office 1871. **Ref.** ARDEN G. A Sketch of Port Philip.... (Thumb Creek 1891).

**Hamavi,** Shahab al-Din Yaqut. See **Yaqūt,** Shahab al-Din Abū Abdallah.

**Hamblin & Sayfang.** Printers of Queen Street, Cheapside. *Erin* [Ireland], Westminster, W. Ginger 1810 (engraved by Luffman, for E. Smedley's *Erin a Geographical and Descriptive Poem*).

**Hamdallah,** Mustawfi Qazwini, Ibn Abu-Bakr Ibn-Muhammad, Ibn Nasr (1281-1349). Persian historian and geographer from Qazwin. *Tarikh Guzidih* [Selected History], 1330; *Nuzhat al-Qulub* [Geography], 1340.

**Hamdani-al,** Abu Mohammed, al-Hasan, Ibn-Ahmad (*c.*893-c.945). Arab historian, geographer and astronomer. *Encyclopaedia al-Iklil* [The Crown]; *Sifat Jazirat al-Arab* [Geography of Arabian Peninsula].

**Hamdi,** Mej Achmet. Carte Dabbe-Obeiyad, Cairo 1875; Kordofan, 1876.

**Hamel,** Joseph. Surveyor. *City of Quebec,* 1829.

**Hamel,** P.W. Topographer and draughtsman. Expedition to South and South East Nevada, 1869.

**Hamel,** J. Publisher of 88 Collins Street West, Melbourne. Map of Australia shewing the routes of the explorers, 1863 (lithographed on stone by E. Jevezy).
• **Hamel & Ferguson.** Lithographers of 85 Queen Street, Melbourne. J.E. Guérard's *Hobart Town, c.*1870; *Plan of Grongal, Murrumbidgee River, c.*1890.

**Hamer,** —. *Atlas de l'Histoire de l'Empire Ottoman,* Paris, Bellizard, Dufour & Co. 1843 (printed by Thierry Frères). **N.B.** Compare with Joseph Hammer.

**Hamer,** Stefan. Publisher of Nuremberg. Siege of Wolfenbüttel, 1542; Plan of Wismar, 1547 (used by Braun and Hogenberg, 1572 and later); Siege of Braunschweig, 1554.

**Hamersley,** William James. Publisher. William C. Woodbridge's *School Atlas to accompany the Modern School Geography,* before 1843 (previously published by O.D. Cooke, 1821, 1823, the later, 1843 edition, was published by Belknap & Hamersley).

**Hamersvelt** [Hammersveldt], Evert [Everard] Symonsz. [Symonszoon] van (*c.*1591/1592-1653). Dutch engraver, worked in Amsterdam, 1616-1643, for Hondius, Janssonius and others, sometimes with Salomon Rogiers. His maps appear in many atlases over subsequent years. Engraved for Jean Le Clerc, 1619; for Jodocus Hondius, the younger, from *c.*1625; maps in John Speed's *A Prospect of the Most Famous Parts of the World,* 1627; for Henricus Hondius and Johannes Janssonius, 1630-1632; maps in later editions of the Mercator-Hondius Atlas, from 1633; maps in Blaeu's *Novus Atlas,* 1635; maps in Johannes Janssonius's *Novus Atlas,* 1645.

**Hamid,** Abdul *Bey.* Officer in the Turkish Army. One of the Turkish representatives amongst the international team which draughted ...*la frontière Serbo-Turque selon article 36 du traité de Berlin*... 1878 (11 sheets plus title), 1879.

**Hamilton,** —. Route on the Niger, 1885-1897 (with Greenshield).

**Hamilton,** *Lieutenant* —. Plan of Limerick, 1844.

**Hamilton,** A. Surveyor aboard H.M.S. *Swallow.* Survey of the east coast of China, province of Shan Tung and Kyan Chau bay, *c.*1863 (with G. Stanley, J. Hall, W.N. Goalen & H.R. Harris).

**Hamilton,** A. Boyd. *Maps of the District of Columbia and city of Washington* (3 maps), 1852.

**Hamilton,** *Captain* Alexander (*d.*1732). Merchant in East India Company, 1688-1723. *A New account of the East Indies* (2 volumes with 8 maps), Edinburgh, J. Mosman 1727, 2nd edition published London, Charles Hitch and Andrew Miller 1744.

**Hamilton,** Alexander Charles. *Chart of the town, harbour and approaches...* [Mahébourg, Mauritius], 1860.

**Hamilton,** Archibald, *Junior (fl.* from 1769, *d.*1792). British publisher 'near St. John's Gate', London. *An Accurate Map of the Country round Boston in New England,* 1776 (published in *Town & Country Magazine*).

**Hamilton,** C.W. Lithographer. *Map showing the Niagara and Detroit Rivers Railway...,* 1858 (by Charles Beard and H. Gregory).

**Hamilton** [formerly Buchanan], Francis (1762-1829). Doctor and cartographer, worked in India. *Map of the Dominions of the Gorkha,* 1819.

**Hamilton,** *Colonel* Frederick William. Officer of the Grenadier Guards. *Encampment of the British Forces... near Scutari May 24th 1854* [Turkey], 1854, 1856; *Sketch of the Encampment of the 1st. and 2nd. Divisions of British Troops... also French, Turkish & Egyptian Troops at Varna... 20th June 1854* [Bulgaria], 1854 (lithographed War Department 1855, sold by Williams & Norgate); *Encampment of the First Division of the British Army under the Command of the Rt. Honble. Genl. Lord Raglan G.C.B. at Aladin from the 1st to the 27th July 1854* [Bulgaria], 1854 MS, lithographed War Department 1856; *Encampment of the First Division of the British Army serving in Turkey.. at Gevrekli... 26 July to 16 August 1854* [Bulgaria], 1854; *Battle of the Alma Sep. 20th 1854,* 1854; Battle of the Tchernaya [Crimea], 1855; *Balaclava & surrounding* heights..., 1856; *Sketch of the ground South of Varna called the Heights of Galata Burun, on which was encamped part of the British Army... July and August, 1854... Corrected, and the positions of the Troops inserted...,* London, Topographical & Statistical Depot, War Department 1856.

**Hamilton**, *Captain* H.G. Sketch New England Beardy Plains, New South Wales, 1843.

**Hamilton**, *Commander* I., *RN. Sketch of Matagorda*, 1842.

**Hamilton**, James. Estate surveyor. Plan of Chelsea, 1664 (used by Thomas Faulkner in his *An Historical and Topographical Description...,* 1810).

**Hamilton**, John. Publisher. *The Edinburgh Geographical and Historical Atlas,* 1835 edition (first published by D. Lizars, 1831).

**Hamilton**, Robert. Published *Hunter's Plan of the City of Edinburgh and its Vicinity,* 1827.

**Hamilton**, S. Printer, Falcon Court, Fleet St., London. Atlas to William Guthrie's *System of Geography,* 1801 and later.

**Hamilton**, Theodore F. *The People's Pictorial Atlas,* New York, J. David Williams 1873 (with C.H. Jones); *Historical Atlas of the World...,* Chicago, H.H. Hardesty 1875 (with C.H. Jones).

**Hamilton**, W.G. *Map [of the U.S.]... for Railroad Men,* c.1870.

**Hamilton**, W.J.W. Explorer and surveyor for the New Zealand Company, attached to the survey mission of H.M.S. *Acheron.* Records and surveys from his inland expeditions of 1849-1850 were incorporated in *New Zealand Middle Island Sheet VIII Cape Campbell to Banks Peninsula,* London, Hydrographic Office 1857; explorations in the Oreti and Aparima [Jacobs] Rivers, 1850-1851, used in the *Acheron's* chart *New Zealand Middle and South Islands Sheet XI Foveaux Strait and South Id.,* London, Hydrographic Office 1857.

**Hamilton**, Walter. Officer in the Corps of Guides, Kabul. *Eye sketch of the road between Kasim Khel and Kabul via the Shutar Gardan Pass,* Simla, Quarter Master General's Dept. 1879.

**Hamilton**, *Sir* William John. Routes Asia Minor, 1836; *Phrygia*, 1840 (with V.J. Arundell).

**Hamilton, Adams & Co.** Publishers of Paternoster Row, London (1828). Amalgamated with W. Kent & Co. (1889), and Simpkin, Marshall & Co., to become Simpkin, Marshall, Hamilton, Kent & Co. *q.v..* Canaan, 1828; [Robert] *Wilkinson's General Atlas of the World,* 1831 edition; one of the publishers of John Betts's *The Family Atlas,* 1848 edition.

**Hamley**, *Major General Sir* Edward Bruce. One of the Boundary Commissioners appointed under the Treaty of Berlin 1878. *Commission Européenne de délimitation de la Bulgarie. Rectification de la Frontière Méridionale de L'ancien Sandjak de Sofie,* 1879; *...Frontière entre la Bulgarie et la Roumélie Orientale,* 1879; *Carte de la Frontière Russo-Turque entre Karaourgan et l'ancienne frontière...,* 1880.

**Hamlin**, William (1772-1869). American engraver and nautical instrument maker of Providence, Rhose Island. *A Map [of] the Town of Providence,* 1803 (from a survey by Daniel Anthony).

**Hamm**, P.E. Engraver for Henry Carey and Isaac Lea 1822-1831; one of the engravers for A. Finley's *Atlas Classica,* Philadelphia 1829.

**Hamm**, *Doctor* Wilhelm von (1820-1880). *Weinkarte von Europa,* 1869, 1873.

**Hammann**, W. *Plan von Heilbrunna* [Germany], Stuttgart c.1874.

**Hammar**, F. Transvaal, published Frederick Jeppe 1868; Zululand (6 sheets), 1906.

**Hammel**, L', —. (fl.19th century). German mathematician and cartographer.

**Hammelmann**, Hermann (1525-1595). Theologian and historian of Oldenburg. Plan of *Wesel,* and views for Braun & Hogenberg's *Civitates Orbis Terrarum,* 1574.

**Hammer**, —. Surveys of Silesia and Brandenburg, 1782-1784.

**Hammer**, A.M. *Eisenbahnkarte...Frankreich*, 1867; *Schul-Atlas*, 1868; *Königreich Bayern*, 1870; *Österreichisch-Ungarische Monarchie*, 1870; and others.

**Hammer**, Albert. *Plan von Eisleben*, Eisleben 1881.

**Hammer**, Bernhardt Frantz (fl.1746-1750). Danish geographer.

**Hammer**, Christian Friedrich (1760-1838). Military cartographer in Franconia. *Der Fränkische Kreis nebst den angrenzenden Laendern*, 1804; *Fürstenthum Würzburg*, 1805; Hohenlohe, 1806; *District von Franken um Nürnberg*, Homann Heirs 1807; *Mapa de España y Portugal*, Nuremberg 1812; Franconia, 1813; map of Württemberg, Baden and Hohenzollern, published Nuremberg, Campe 1817; *Ober-Mayn-Kreis*, 1824; *Königreich Bayern* (2 sheets), 1830; Württemberg, 1831.

**Hammer**, Christopher Alix (1720-1804). Danish cartographer and civil servant. *En Norsk Atlas*, 1773; *Christiansands Stift*, Khavn 1776; *Bergeno Stift*, Hafn 1785; *Trondheims Stift* (2 sheets), Copenhagen, Hafniae 1786.

**Hammer**, Heinrich. See **Martellus**, Henricus.

**Hammer**, Joseph. One of the sources for J. Hellert's *Plan de Constantinople...*, Paris 1836 (engraved by U. Muschani, printed by Thierry Frères, published in J.J. Hellert's *Nouvel Atlas physique, politique et historique de l'Empire Ottoman...*, Paris, Bellizard & Co. 1843).

**Hammer**, R. Grönland, Gotha, Justus Perthes 1883.

**Hammer**, W. *Königreich Polen*, 1861, 1864; *Deutschland* (9 sheets), 1881.

**Hammerbergh**, M. Dutch military engineer and cartographer, worked in Zeeland (1700), styled himself 'Ingenieur ten dienste van den Lande'. *Caerte van Vlaenderen...volghens projecht van De Heer Generael Coehoorn in 't Jaer 1699*, c.1700 MS.

**Hammerdörfer**, Karl [Carl] (1758-1794). German philosopher and historian. *Geographia historiae*, Leipzig 1785-1792.

**Hammershaimb**, Wenceslaus Franciscus de [Wenzel Franz von]. Falstria, 1676, 1682 (used by Frederick de Wit and others, Amsterdam 1680s-1690s).

**Hammersley**, *Captain* Frederick. Surveyor and draughtsman for *Military Sketch of the S.W. part of the Crimea...* (8 sheets), 1856 (with others).

**Hammersveldt**, E.S. See **Hamersveldt**, Evert Symonsz.

**Hammet**, Lacon N. *Lamelin Harbour* [Newfoundland], Admiralty Chart 1845.

**Hammett**, Charles E. *Junior*. Surveyor in Rhode Island. *Road map of the island of Rhode Island or Aquidneck*, 1849 (drawn by G.F. Turner), lithographed by Sarony & Major of New York, 1860; *Township of Newport with part of Middletown*, New York, Sarony, Major & Knapp 1860.

**Hammond**, A. Surveyor. Assisted with survey for William Gordon's map of Huntingdonshire, 1730-1731.

**Hammond**, *Captain Sir* Andrew Snape. *A Chart of Delawar Bay*, 1779, also published by Robert Sayer and John Bennett 1787.

**Hammond**, C.S. [& Co.]. Publishers of New York. *Arizona*, New York 1911; *War atlas*, 1914; *Handy almanac, encyclopedia and atlas*, 1914; *Hammond's home and office atlas of the World*, 1914; *Hammond's illustrated atlas of the World*, 1914; *Hammond's Modern atlas of the World*, 1914; *The new reference atlas of the World*, 1914; *Hammond's pictorial atlas of the World*, 1914; *Hammond's standard atlas of the World*, 1914; *The compact atlas of the World*, 1915; *Hammond's atlas of the World*, 1915; *Hammond's popular atlas of the World*, 1916; *Belgium and Franco-German frontier*, 1916; *Compact atlas of the World*, 1917; *Hammond's standard atlas of the World*, 1917; *Hammond's large scale war map of the Italian front*, 1918.

**Hammond**, *Lieutenant* George C. RN. *Bay of Bengal*, 1876; *Chittagong River*, 1877.

**Hammond**, J.T. American engraver. Drew and engraved *Traveller's Map of Michigan*,

*Illinois, Indiana & Ohio*, 1836; *Western States*, 1835 (published in a school geography atlas).

**Hammond** [Hamond], John. Former student at Clare Hall, Cambridge (c.1579). *Cantebrigia...* (9 sheets), 1592 (copper engraving by Augustine Ryther), copied by John Speed, 1611. **Ref.** LOBEL, M.D. *Imago Mundi* 22 (1968) pp.54-55.

**Hammond**, *Captain* John. *A new... chart of the North or German Sea...* (2 sheets), 1787, published Robert Sayer 1789.

**Hammond**, *Captain* R.P. *Map of the City of New York of the Pacific*, 1849.

**Hammond Publishing Co.** Publishers of Chicago. Plat book Green & Jersey Counties, Illinois, 1893.

**Hamon**, Pierre (*d.*1569). Writing master to Charles IX of France. France, 1568; World, 1568 MSS.

**Hamond**, J. See **Hammond**, John.

**Hamont**, Jean Baptiste, *sieur* Des Roches. See **Des Roches**, Jean Baptiste Hamont.

**Hampton**, William. English estate and inclosure surveyor. Plan of the manor of Allington, 1770 MS; Essex plans, 1763 MS; plans Preston Candover, 1771 MS.

**Hampton**, William (*fl.*1812-1846). Estate, bog and county surveyor in Ireland. Manuscript maps of parts of Roscommon and County Longford [Ireland], 1813 MS.

**Han maps** (206-168 BC). Anonymous topographic, military and city maps of the kingdom of Changsha Guo of the Han Dynasty (south central Hunan Province of China). **Ref.** HSU Mei-Ling *Imago Mundi* 45 (1993) pp.90-100.

**Hanbury**, William. *Parish of Kelmarsh* [Northamptonshire], 1739.

**Hanchi**, Fujii. See **Ochikochi** Doin.

**Hancko**, François. German engineer in British service. *Plan du Campament de l'Armée Alliée...* [East Flanders], *c.*1744 MS; *Plan du Campement de l'Armée Alliée devant Namur et celui de François. le 2me Aout 1746*, 1746 MS.

**Hancock**, *Captain* Benoni. Marine surveyor. Draught Cockle & St. Nicholas Gatts, 1751, published 1753.

**Hancock**, E.J. Surveyor. Thames to Collier Row Canal, 1818 MS.

**Hancock**, Henry. Survey of Los Angeles, 1858 MS (with G. Hansen).

**Hancock**, *Doctor* John. Chart of the interior of British Guiana, 1811 (surveys used in several later maps, including J. Hadfield).

**Hancock**, *Commander* John *RN*. Naval surveyor. Assisted Captain Orlebar with surveys off the north east coast of America. Surveys of the Gulf of St. Lawrence, published 1857; *Newfoundland - South Coast*, 1860-1861 (with others), published Admiralty 1862 (engraved by J. & C. Walker); *La Poile Bay...*, 1861-1862 (with others), published Admiralty 1864 (engraved by J. & C. Walker); *Catalina Harbour*, 1862 (with others) published Admiralty 1863 (engraved by J. & C. Walker); *South West Coast of Newfoundland*, 1862 (with others), published 1865 (drawn by E.J. Powell, engraved by Davies & Powell); *South East Coast of Newfoundland*, 1863 (with others), published Admiralty 1864, 1889 (drawn by E.J. Powell, engraved by Davies & Powell); *East Coast of Newfoundland...*, surveyed 1867-1871 (with others), published Admiralty 1868 (drawn by H. Stafford, engraved Davies, Bryer & Co.).

**Hancock**, Robert. Engraver. Plan of Worcester, 1764; *Plan of Derby*, 1791; T. Chantry's *Plan of the City of Bath*, 1793.

**Hancock**, William. American surveyor. *Map of St. Clair, St. Clair county, Mich.*, St. Clair, W. Hancock 1854; Wayne County, Michigan, 1854.

**Hancox**, John. Birmingham canal navigations, 1855.

**Hand**, John. Publisher and mapseller of '409 Oxford Street, near Soho Sq.', London. John Andrews's *Plans of Capital Cities of Europe*, 1771; re-issued J. De Costa's *Plan of... Boston*, 1775; John Andrews and Andrew Dury's *65 miles round London*, 1777; *30 miles round London*, 1782.

**Handcock,** Matthew. Deputy Surveyor General for Ireland, 1780s.

**Handcock,** W. Department of the Surveyor General of Lands. Estate Plan of Clanwilliams, Tipperary, 1790 MS.

**Handley,** Benjamin. Posted to Constantinople as assistant surveyor to the Turco-Persian Boundary Commission, 1852; Mohammera [Persia], 1853 MS; *Sketch shewing the Positions occupied by the French and English troops in Constantinople and its Environs*, *c*.1855.

**Handmann,** Johan Jacob (1711-1786). Swiss medal and seal engraver in Basle. Engraved for I. Bruckner's globe, 1752.

**Handtke,** Friedrich H. (1815-1879). German cartographer of Leipzig. Amongst other works he created maps which were credited by Carl Flemming to the fictitious editor Dr. Karl Sohr. *Karte von Europa* (72 sheets), 1836; *Schul-Atlas der neueren Erdbeschreibung*, Glogau, Flemming 1840 and later editions; *Vollständiger Hand-Atlas...* (80 sheets), Glogau, C. Flemming from 1844 (a supplement was added 1845 to create *Vollständiger Universal-Handatlas*, 1845-1865 and editions to 1888); *Karte von Austral-Continent...*, 1845; *Hand-Atlas des Preussischen Staats*, 1846 and later editions; contributed to *Reymann's Special-Karte* [Schleswig Holstein], 1849, 1859, 1860; Crimea (4 sheets), 1854, 1855; *Wandkarte von Palästina* (4 sheets), 1858; *Süd-Amerika*, 1872; *F. Handtke's Special-Karte der Europäischen Turkei*, Glogau, Flemming *c*.1876; *Handatlas vom Preuß. Staat*, Glogau, Flemming 1878; Africa, 1889; Switzerland, 1895; Australia, 1899. **Ref.** WOLTER, J.A., and GRIM, R.E. *Images of the World* (1997) pp.219-220.

**Handy,** —. *Copper Mines of Calareras County* [California], 1864.

**Hane,** Gerdt. Painter and draughtsman of Lüneburg. Plan of Ratzeburg, 1588, used by Braun & Hogenberg, 1590 and later.

**Hänel von Cronenthal,** Emil (1784-1843). Conducted surveys for official Prussian provincial maps including Brandenburg, 1832-1847; Westphalia, 1836-1842; Rhine province, 1843-1850; Saxony, 1851-1860.

**Haneman,** J.C. German surveyor and cartographer in Suriname and Guyana. *Kaart van de Colonie Suriname*, 1784.

**Hanemann,** August. Engraver. *Die Walachei* [Romania], Weimar, Geographical Institute 1853; for Adolf Stieler, 1857-1885.

**Hanemann,** F. (*d*.1877). Worked with E. Behm on population charts of the earth, Europe and Germany. Used in *Petermanns Mitteilungen*, 1874.

**Hangiya** Shichirōbei. Japanese publisher. Ochikochi Dōin and Hishikawa Moronobu's *Tōkaidō Bunken Ezu* [map of the Tokaido highroad], 1690.

**Hanicle,** *sieur* —. Drew Jouhan de la Guilbaudière's Mer du Sud, *c*.1696.

**Hanlon,** John. *Chart of the Road of Arran*, *c*.1787.

**Hanmer,** William (*d*.1625). Apprenticed to John Daniel of the Drapers' Company school (Thames School) of chartmakers, 1622.

**Hann,** Julius von (1839-1921). Austrian geographer and mineralogist. Worked as an editor on Heinrich Berghaus's *Physikalischer Atlas*, 1886-1892 (with O. Drude, D.G. Gerland, W. Marshall, and G. Neumayer); *Atlas der Meteorologie*, Gotha 1887.

**Hanna,** Andrew (*d*.1812). *A Plan of the City and Environs of Baltimore*, 1799 (state I, engraved Francis Shallus), 1801 (state II, published by Hanna with William Warner).

**Hanna,** *Captain* James. Fur trader, made return voyages from China to Nootka, 1785, and 1786-1787. *Chart of Part of the NW. Coast of America*, 1786, published A. Dalrymple 1789 (engraved by W. Harrison); *Plan of St. Patrick's Bay* [named by Hanna], 1786, A. Dalrymple, 1789 (engraved by W. Harrison); Plan of Sea Otter Harbour, 1790 (used in John Meares's *Voyage*).

**Hannagen,** Bartholomew. Irish estate surveyor. Estate plan of Queen's County, 1792 MS.

**Hannak,** Emanuel (1841-1899). Historische Schul-Atlas, Vienna 1886-1887 (with Friedrich Umlauft).

**Hannas**, Marx Anton (*d.*1676). Engraver and publisher in Augsburg. *Andechs, Neuberg*, [n.d.].

**Hanno of Carthage** (*fl.*before 450 BC). Undertook a voyage of exploration from Carthage through the Straits of Gibraltar and round the north west coast of Africa, probably as far as Sierra Leone.

**Hannot**, Victor. Publisher of Washington. *Street directory of Washington and Georgetown*, 1863 (with a map by W. Smith); *Map of Washington city and Georgetown*, 1866 (printed by A. Hoen & Co.).

**Hannstaedt**, F.W. *Hütten- und Gewerbekarte des Regierungsbezirks Arnsberg*, Iserlohn, Baedeker 1859.

**Hannum**, E.S. Atlas of Fairfield County, Ohio, 1866.

**Hansard**. Family of printers serving the House of Commons and parliamentary committees. Printed parliamentary reports from 1774-1889. The Hansard companies flourished for 83 years, until absorbed into the Stationery Office in 1881.
• **Hansard**, Luke (1752-1828). Native of Norwich, apprenticed to Stephen White, printer, bookseller & stationer of Norwich, 1765-1771. In 1772 he moved to London to work for Henry Hughes, printer to the House of Commons. He became manager in 1793, partner in 1794, and took over the company in 1799, becoming printer 'by order of the House of Commons', in Great Turnstile, Lincoln's Inn Fields. James Burney's *Chronological History of the Voyages and Discoveries in the South Sea or Pacific Ocean* (40 maps), 1803-1817 (maps engraved by J. Russell); Sixteen plans to illustrate Thomas Telford's report on communications between England and Ireland via south-west Scotland, 1809 (engraved by James Basire).
• **Hansard & Sons** (*fl.* from 1807). Partnership of Luke and his two younger sons James and Luke Graves. *Proposed new street from Charing Cross to Portland Place*, 1812 (engraved by J. Basire); *Marybone Park Farm*, 1812 (plan by J. Nash, engraved by J. Basire); *The proposed Regent's canal through Marylebone Park*, 1812 (engraved by J. Basire); *Plan for widening the Strand in the vicinity of Exeter Change*, 1826 (engraved by A. Arrowsmith Junior).

•• **Hansard**, Thomas Curzon (1776-1833). Son of Luke Hansard, printer of Peterborough Court, Fleet Street, London, and later Paternoster Row (from 1823). Apprenticed to Henry Hughes 1790-1797. He left the family firm in 1803 and took over the printing house of Thomas Rickaby in 1805. He went on to print the Parliamentary Debates which still cary his name. *Ostell's New General Atlas*, 1818 and later editions.
•• **Hansard**, James (1781-1849). Son of Luke, to whom he was apprenticed, becoming a partner *c.*1807.
•• **Hansard**, Luke Graves (1783-1841). Son of Luke, godson of James Basire II. Like his brother James he was apprenticed to his father and became a partner in the company *c.*1807.
••• **Hansard**, Luke Henry (*b.*1816). Son of Luke Graves Hansard. His desire to join the family firm was thwarted, and he became an attorney.
••• **Hansard**, Frederick. Printer, son of Luke Graves Hansard.
•• **James & Luke G. Hansard & Sons**. Younger sons of Luke, continued printing Parliamentary Papers, including maps. *Map shewing the Several Walks or Deliveries in the Country Districts of the Two Penny Post...*, 1830; *Map of London shewing the Boundaries of the General and Two Penny Post Deliveries...*, 1830; plans of the Ottawa River and Carillon canal (3 plans), 1830 (lithographed by J. Basire and A. Arrowsmith); Robert Kearsley Dawson's *Plans of the Cities and Boroughs of England & Wales* (2 volumes), 1832; *Plan of the Counties of Stanstead, Sherbrooke, Missiskoui, Shefford, Drummond, Megantic and part of Nicolet in the Province of Lower Canada...* [Quebec], 1837 (lithographed by S. Arrowsmith); *Map of the Gold Coast...*, 1842 (lithographed by J. Arrowsmith); *Map of the Gambia with the coast country adjoining...*, 1842 (lithographed by J. Arrowsmith); *Map of Canada shewing Generally the Several Public Works Completed or now in Progress*, 1843 (lithographed by J. Arrowsmith); *Plan of the Settlement at Port Louis, East Falkland*, 1843 (surveyed and drawn by M.R. Robinson, lithographed by J. Arrowsmith).

**Hansard**, Henry (*d.*1904). Printer, by Order of the House of Commons. Produced a large number of forestry maps, all lithographed by Standidge & Co., including maps of Hainault Forest, Essex, 1848; Alice Holt Forest, Hampshire, 1848; Parkhurst Forest,

Hampshire (7 maps), 1848; Delamere Forest, Cheshire, 1848; Crown woods and plantations at Ryton, Cumberland (2 plans), 1848; Forest of Dean, Gloucestershire, 1848; *Dean Forest Coalfield: Geological Survey*, 1848; Salcey Forest, Northamptonshire, 1848; New Forest, Hampshire (69 plans), 1848-1849; Bere Forest (6 plans), 1849; *Map of the Railways through England and Northern and Central Europe...*, 1850 (lithographed by J. Arrowsmith); *Returns-Lights and Lighthouses, British Colonies* [several sheets], 1850 (lithographed by Standidge & Co.); *Chart of the Coast of Brazil from Maranham to the River Plate*, 1850 (lithographed by J. Arrowsmith, printed for the Select committee on Africa Slave Trade); *Chart prepared with a view to show the present state of the slave trade on the west coast of Africa*, 1850 (prepared by J. Arrowsmith); *Chart of the Arctic Coast* [Canada], 1852 (lithographed by J. Arrowsmith); *East Coast of Africa from Cape Delgado to Delgado Bay; and Madagascar*, 1853 (lithographed by J. Arrowsmith); *Aboriginal Map of North America...*, 1857 (lithographed by J. Arrowsmith); *Map of railways proposed by the Bills in the session of 1859 in the Metropolis and its vicinity*, 1859; John Arrowsmith's New Zealand, 1860; plans for the new docks at Valletta, Malta, 1862; *Map of railways proposed by the Bills in the session of 1863 in the Metropolis and its vicinity*, 1863.

**Hansard**, Joseph. *The history, topography and antiquities of the County and City of Waterford...* (8 volumes), 1870.

**Hänsel**, G. See **Haensel**, G.

**Hansen**, Carlos F.V. *Plano topographico de la Gobernacionas de Formosa y del Chaco*, Buenos Aires 1889; *Mapa General de los Ferro Carrilles de la Republica Argentina...*, 1895.

**Hansen**, George. Survey of Los Angeles, 1858 MS (with H. Hancock).

**Hansen**, J.F. *Staatsbeschreibung des Herzogtums Schleswig*, 1758.

**Hansen**, Jules André Arthur (1849-1931). Parisian explorer and cartographer. Philippines, 1879-1881, 1886; topographical survey of the Grand Duchy of Luxembourg, 1883-1906; *Itinéraire de Dar es Salam aux Lacs Bangouéolo et Moéro par Victor Giraud... 1882-1884* [Tanzania], Paris 1885 (drawn by J. Hansen, with inset *Carte d'ensemble des Voyages aux Grand Lacs de l'Afrique Méridionale par V. Giraud*); maps for C. Huber's *Journal d'un Voyage en Arabie 1883-1884... Avec Atlas*, Imprimerie Nationale 1891; *Bassins du Haut-Nil et du Moyen Congo* (11 sheets), Paris 1893; *Afrique...*, 1895; *...Niger entre Manambougan et Tombockton...* (41 sheets), 1898; Djibouti, 1900; *Carte topographique du Grand-Duché de Luxembourg* (28 sheets), 1904-1907; *Lak Tanganyika*, 1908.

**Hansen**, Doctor R. *Südwest Schleswig um 1240*, 1893.

**Hansen**, V. *Physisk-Meteorologisk Atlas*, 1862.

**Hanser**, Anton. *Ober-Bayern des Königreichs Bayern*, Nuremberg c.1855.

**Hanser**, G. Specialised in travel maps of middle Europe. Schul-Atlas, Regensburg 1853; *Post und Eisenbahn Reisekarte von Mittel-Europe*, Vienna 1859.

**Hansford**, Robert. Numerous manuscript surveys for the European Danube Commission. Produced several of the manuscript maps which accompanied C.A. Hartley's report to the Commission in 1857; further surveys for the Commission, 1857-1871.

**Hanshall**, J.H. Map of Cheshire, c.1820 (used in part 16 of *The history of the County Palatine of Chester*, 1817-1823).

**Hanson**, Thomas. *A Plan of Birmingham*, 1778, 1781.

**Hanson**, R. *Map of the China Tea Districts*, London, Samuel Hanson, Son, Evison & Barter 1875?

**Hansteen**, Christopher (1784-1873). Norwegian astronomer and explorer. *Magnetischer Atlas*, Christiania [Oslo] 1819; maps of isodynamic lines (variations in gravitational fields), 1825, 1826; supervised trigonometrical and topographical survey of Norway, 1837.

**Hansteen**, *Professor Ridder* M.M. Kart over Christiania, Christiania [Oslo] 1844.

**Hanway, J.** *A survey of lands, tenements and hereditaments* [Harwich, Essex], 1709 MS.

**Hanway, J.,** *junior.* Plans of Stratford Place Property, 1732 MS; Estate of John Hope, 1732.

**Hanway, Jonas** (1712-1786). English traveller and philanthropist. *An Historical Account of British Trade Over the Caspian Sea, with a Journal of Travels, etc.* (4 volumes), 1753, German edition published Hamburg, 1754.

**Hanwell, Henry.** Northern Canada, 1778 MS; Hudson Bay, 1778 MS.

**Happel** [Happelius; Happell], **Eberhard** [Everhard] **Werner** (1647-1690). Geographer, historian and writer of Ulm. *Die Ebbe und Fluth auff einer Flachen Landt-Karten fürgestelt* [world chart], in his *Relationes curiosae*, Ulm 1675 (early attempt to show ocean currents, re-issued in his *Mundus mirabilis tripartus*, Ulm 1687, 1708); *Der Ungarische Kriegs-Roman* (6 volumes), 1685; *Thesaurus Exoticorum*, 1688 (in English and German); *A New and Exact map of Hungary*, London 1690; *...Insulschutt...*, Hamburg 1688.

**Happersberger, K.** *Plan von Mainz*, c.1877; *Provinz Rhein-Hessen*, 1900.

**Harada, T.** Geologische Karte des Comelico und des westlichen Carnia, 1883.

**Haraeus** [Hareieus; Hareio; Haren; Van der Haer; Verharius], **Franciscus** [Frans van] (c.1555-1631). Theologian and historian of Antwerp. *Terrae Globus...* [7-inch terrestrial globe], Antwerp, 1614; map for Heribertus Rosweydus's *Vitae Patrum*, Antwerp 1615; *Globus Terrestris Partimque Caelestis* [8.5-inch], c.1615; maps of the Holy Land and Mediterranean for Ortelius' *Parergon*, Antwerp, Christopher Plantin 1624; *Biblia Sacra*, 1630; other globes no longer extant.
Ref. KROGT, P. van der *Globi Neerlandici* (1993) pp.271-277, 417-420; KROGT, P.C.J. van der & SCHILDER, G. 'Het kartografische werk van de theoloog-historicus Franciscus Haraeus c.1555-1631' in *Annalen van de Koninklijke Oudheidkundige Kring van het Land van Waas* 87 (1984) pp.5-55 [in Dutch].

**Harbaugh, John.** Canal around Harper's Ferry, c.1803 (with Nicholas King).

**Harboe, F.C.L.** (1758-1811). Danish hydrographer. Aalborghus, 1791; Aastrup, 1793; Kort over Amter (4 sheets), 1793-1803; Sleswig-Holstein, 1776-1806; Atlas over Danmark, 1821 (with others).

**Harbrecht, Isaac & Josias.** See **Habrecht.**

**Harcourt,** *Colonel* — d'. *Plan of the Town of ...Gibraltar*, 1705.

**Harcourt, G.S.** Plan of Egham racecourse [England], 1836-1837 MS.

**Harcourt, Robert** (1571-1631). Explorer in Guiana with Sir Walter Ralegh. *Relation of a voyage to Guiana*, 1613.

**Hard, C. de.** Portuguese traveller. Map of Brazil for Johann Schöner, 1515.

**Hard, C.** One of those who worked for D.J. Lake on surveys for his county atlases, all engraved by Worley & Bracher, printed by Fred. Bourquin and published in Philadelphia by C.O. Titus: *Atlas of Branch Co. Michigan*, 1872; *Edwards County*, published 1872; *Atlas of Cass County Michigan*, 1872; *Atlas of St. Joseph County, Michigan*, 1872; *Atlas of Clinton County Michigan*, 1873; *Atlas of Allegan County Michigan*, 1873; *Atlas of Barry Co. Michigan*, 1873; *Atlas of Berrien Co. Michigan*, 1873; *Atlas of Van Buren County, Michigan*, 1873.

**Hard, William G.** American surveyor, worked with others for D.J. Lake on surveys for his county atlases all engraved by Worley & Bracher, printed by Fred. Bourquin and published in Philadelphia by C.O. Titus: *Atlas of Allegan County Michigan*, 1873; *Atlas of Barry Co. Michigan*, 1873; *Atlas of Berrien Co. Michigan*, 1873; *Atlas of Van Buren County, Michigan*, 1873.

**Hardacre, F.C.** Map of Lawrence County, Illinois, 1897; *Map of Crawford County, Illinois*, Vincennes, Ind. 1898 (with F.W. Lewis); *Clay County, Illinois*, Vincennes, Ind. 1901; *Historical Atlas of Jasper County, Illinois*, Vincennes, Ind. 1902; *Edgar County, Ill.*, Vincennes Ind. 1902; *Historical Atlas of Knox County, Ind.*, Vincennes 1903; Map of Adams county, Illinois, 1906; *Map of Henry County, Illinois*, 1907.

**Hardacre Map Co.** *Edgar County, Illinois*, Indianapolis 1917; *Moultrie and Shelby Counties, Illinois*, Indianapolis 1919.

**Hardcastle**, Lieutenant, —. *Battles of Mexico. Survey of the lines of operation of the U.S. Army...*, 1847; *General Worth's Operations in Texas*, 1850.

**Hardee**, T.S. *Map... of the Mississippi River*, 1874.

**Hardee**, W.J. *City of New Orleans*, 1897.

**Hardenburgh**, John L. Surveys of townships in New York State, used by S. De Witt on his *1st Sheet of De Witt's State Map of New-York*, 1793.

**Hardesty**, Hiram H. Publisher of Chicago and Toledo, produced decorative U.S. county atlases. Land ownership maps of Ohio, 1874; *Illustrated Historical Atlas of Carroll County, Ohio...*, 1874; L.Q. Hardesty's *Illustrated Atlas of Ottawa County, Ohio...*, 1874; *Sectional Map of Lake Co.*, 1875; *Historical Atlas of the World Illustrated*, 1875 (by C.H. Jones and T.F. Hamilton); *Historical Hand-Atlas Illustrated*, 1882; *Map of Colorado*, 1882; *Hardesty's Historical and Geographical Encyclopedia, Illustrated*, 1883; *Map of British Columbia*, 1884; *Map of Virginia and West Virginia*, 1884; *Utah, Colorado, Arizona, New Mexico*, 1889.

**Hardesty**, L.Q. *Illustrated Atlas of Ottawa County, Ohio...*, Chicago, H.H. Hardesty 1874.

**Hardie**, James. Editor of *The New York Magazine* from 1814. *The Philadelphia Directory and Register* (with a plan of Philadelphia) 1793 (printed by T. Dodson), 1794; augmented and corrected John Payne's *A New and Complete System of Universal Geography describing Asia, Africa, Europa and America*, New York, John Low 1798.

**Harding**, *Lieutenant* —. Survey of the harbour of St. John, New Brunswick, 1844 (with Lieutenant Kortright).

**Harding**, Carl Ludwig (1765-1834). Astronomer. *Atlas novus coelestis*, Göttingen 1822 and editions to 1856.

**Harding**, F.G. Publisher of 24 Cornhill, London, 1833. *The Arbitrator or Metropolitan Distance-Map* [London], 1833, republished by M.A. Leigh 1834 (printed by J. & C. Adlard).

**Harding**, Frederick. Survey assistant in Australia. *Allotments at North Melbourne near the town boundary at Flemington*, Melbourne, Surveyor General's Office 1855 (lithographed by J.B. Phelp).

**Harding**, *Lieutenant Colonel* George Judd RE. Conducted a military reconnaissance of Gibraltar together with a manuscript map entitled *A Military Plan of the Country in the Neighbourhood of the Fortress of Gibraltar*, 1811; later he commissioned and endorsed numerous detailed plans of sites on Gibraltar, c.1830-1850; also briefly in Malta, 1844.

**Harding**, J. *Leigh Place... Surrey*, 1724.

**Harding**, J. Desert of Atacama, Chile, 1877.

**Harding**, John. See **Hardyng**, John.

**Harding**, Joseph. Plan of Blackpool, c.1878.

**Harding**, Samuel (d.1755). Publisher and mapseller 'on the Pavement in St. Martin's Lane', London. Joint publisher with William Henry Toms of many maps, including several associated with the so-called 'War of Jenkins' Ear'. Henry Popple's *Map of the British Empire in America*, editions of 1739-c.1745 (with W.H. Toms, first published in London 1733, plates passed to S. Austen and T. Willdey c.1746); *Cartagena*, 1740; *Porto Bello*, 1740; *St. Sebastian*, 1740; Captain W. Laws's *This plan of the Harbour Town and Several Forts of Carthagena...*, S. Harding & W.H. Toms, Wills Coffee House 1741. **Ref.** BABINSKI, M. *Henry Popple's 1733 Map of the British Empire in America* (New Jersey 1998).

**Harding**, Thomas. Land surrounding the town of Kenilworth, Warwickshire, 1628 MS.

**Harding**, William Derisley. Civil engineer of Kings Lynn, Norfolk. Inclosure plan of the West Marshes at Holkham, Norfolk, 1852; *Plan of the East and West Marshes and Wells Channel* [Holkham, Norfolk], 1852.

**Harding Read**, W. British Consul, San Miguel [St Michael, Azores]. *This Chart of the Island* of St. Michael... 1806, London, William Heather, 1808.

**Hardinge,** Sir Arthur Edward. *Sketch of Roads Westwards of Constantinople...*, 1854 MS; plans of the Battle of the Alma, 1854.

**Hardinge,** W.H.F. *British Lighthouse Chart*, 1874.

**Hardisson Frères.** Publishers of Paris. *Islas Canarias - Illes Canaries* [proposed lighthouses and submarine cables], 1894; *Isla de Tenerife la primera que se publica con las carreteras - Ile de Teneriffe le première indiquant les Chemins Royaux*, 1894 (lithographed by F. Appel); *Plano de Santa Cruz de Tenerife...*, 1894 (title also in French, lithographed by F. Appel).

**Hardman,** Edward Townley (1845-1887). Government geologist of H.M. Geological Survey, Ireland, attached to the Kimberley Survey Expedition in Australia. *Geological map to accompany report on the geology of the Kimberley district*, Perth, Surveyor General's Office 1884.

**Hardmeier,** C. *Carta del Canton di Ticino*, c.1850.

**Hardy,** père et fils. Worked together as 'Ingenieurs du roi pour la construction des globes' at 'Hotel-Dieu, rue Saint-Julien-le-Pauvre', Paris. They made a variety of terrestrial and celestial globes ranging from 7 cms to 32 cms in diameter (1738-1745) including *Globe Céleste...* [32-cm], 1738 [example at Greenwich Observatory]; *Globe Terrestre...* [7-cm], 1738; armillary sphere [32-cm], 1738. **Ref.** DEKKER, E. *Globes at Greenwich* (Oxford 1999) pp.167, 352-353; PASTOUREAU, M. 'Les Hardy - père et fils - et Louis-Charles Desnos 'Faiseurs de globes' a Paris au milieu du XVIIIe siècle' in: TURNER, Roger, (Ed.) *Studies in the history of scientific instruments* (London 1989) pp.73-82; *Actes du 7eme symposium de la commission instruments scientifiques de l'Union international d'Histoire et de philosophie des sciences* (Paris 1987) [in French].
• **Hardy,** Jacques (*fl.*1738-1745). Craftsman globemaker of Paris.
•• **Hardy,** Nicolas (*d.*1744). Son and associate of Jacques. After his marriage to Marie-Charlotte Loye in 1742 he took over the management and development of the business. On his death 2 years later, the business was managed by his widow in association with his father.

•• **Hardy,** Marie-Charlotte Loye (*b.*1723). Wife of Nicolas Hardy. After his death she stayed on at rue Saint-Julien-le-Pauvre, and worked in association with her father-in-law Jacques. In 1749, she married mapmaker Louis Charles Desnos. They continued to sell Hardy globes from the same address, after *c.*1750 many were re-issued under the name of Desnos *q.v.*.

**Hardy,** *Général* —. *Reconnaissance militaire du Hundsruck... entre Rhin et Moselle* (6 sheets), Paris 1798.

**Hardy,** —. See **Graves** & Hardy.

**Hardy,** *sieur* Claude. 'Mareschal des logis du roy'. *Duché de Bretaigne*, Paris, M. Tavernier, Amsterdam, H. Hondius 1632 ('Designé par le Sieur Hardy...', printed by H. Hondius).

**Hardy,** F. *Carte... de la Forêt... de Fontainbleau*, c.1844.

**Hardy,** G. Mapseller of St. Johns, New Brunswick. James Wyld's *A Map of the provinces of New Brunswick and Nova Scotia...*, 1845.

**Hardy,** John. *New projection of the Western Hemisphere & Eastern Hemisphere*, 1776 (in *Gentleman's Magazine*).

**Hardy,** *Lieutenant* Robert William Hale, *RN*. *Sonora & Gulf of California*, 1829.

**Hardyng** [Harding], John (*fl.*1418-1457). English chronicler working in Scotland, 1418-1421. Coloured diagrammatic map of Scotland showing castles and towns, in his *Chronicle*, first version completed 1457 [British Library].

**Hare,** —. *Hare's Map of the Vicinity of Richmond and Peninsular Campaigns in Virginia*, New York, J.H. Colton 1862.

**Hare,** A.J. Publisher of Sandusky, Ohio. Bird's-eye views of Sandusky [Ohio], 1883 (lithographed by W.J. Morgan & Co.); and Lakeside [Ohio], 1884 (lithographed by Sinz & Fausel)

**Hare,** Benjamin. *Manour of Wimple, Cambridge*, 1638 MS?

*Edward H. Hargraves (1816-1891). (By courtesy of the National Library of Australia)*

**Hare**, Edward (*fl*.1762/3-.*d*.1816). Surveyor, produced manuscript maps of various sites in England. Warmington, Northamptonshire, 1755 MS; inclosure plans at Quadring, Lincolnshire, 1776.

**Hare**, Geo. H. Publisher of San José, California. Bird's-eye view of San José, 1869 (drawn and lithographed by W. Vallance Gray & C.B. Gifford).

**Hare**, L.G. County surveyor. Map of Monterey County, California, 1898.

**Hare**, *Lieutenant* William Aldworth Horne RE. *Rough sketch of ground around*

*Constantinople*, 1877; *Battle of Plevna* [Bulgaria], 1877; *Rough Sketch of Shumla* [Bulgaria], 1877; *Sketch of the coastline and road between Kum-Kaleh and Erenkui*, 1877; *Hand Sketch of proposed Line of Defence* [Hatay province, Turkey], 1878 (and 3 other similar strategic sketches of Hatay province); and others, all lithographed at the Intelligence Branch of the QMG's Department.

**Hare.** See **Harrison**, Sutton & Hare.

**Hareieus** [Hareio], Franciscus. See **Haraeus**, Franciscus.

**Haren**, Edward. *Sectional Map of... Missouri*, 1865.

**Haren**, Frans van. See **Haraeus**, Franciscus.

**Harenberg**, Johann Christoph (1696-1774). German scholar. Worked for Homann's Heirs, 1737-1750; *Imperii Turcici Europaei Terra*, 1741; *Palaestina seu Terra olim Sancta.../ La palestine ou La terre Sainte...*, Homann Heirs 1744 (engraved by J.M. Seligmann); *Carte de la Terre Sainte...*, Homann Heirs 1750.

**Harford**, *Lieutenant* William. Worked for John MacRae on surveys in the state of North Carolina, c.1830-1832, used for MacRae's *A New Map of the State of North Carolina*, 1833.

**Harge**, Jan. Wife of Pieter Straat. *Drechterlandt en de Vier Noorder Koggen* (4 sheets), Amsterdam 1735-1736; Vlieland, 1756.

**Hargrave**, I. Engraved a map of Kingston upon Hull, England, 1791 (with views drawn by I. Capes and engraved by J. Gale).

**Hargrave**, J. *Plan of Locheale...*, 1717 MS.

**Hargraves**, Edward Hammond (1816-1891). Discovered gold in New South Wales in 1851. His information was used by W.M. Brownrigg in the compilation of his map *The Gold Country*, Sydney 1851; description of the goldfields with one map published as *Australia and its goldfields*, London, H. Ingram 1855.

**Hargreave**, L. Fly River, New Guinea, 1876 MS.

**Hargreaves**, Thomas (fl.1832-1835). Surveyor of Burslem, Staffordshire, produced manuscript maps of that county. Stoke-on-Trent, 1831; *Map of the Staffordshire Potteries*, 1832 (for Dawson's *Staffordshire Potteries*, Burslem 1832).

**Hariamaya** Kyūbei. Japanese publisher. One of the publishers of *Kaihō Ōsaka Zu* [plan of Osaka], 1847.

**Harigonio**, *Fra* Bona. Worked in Venice. *Figura totius orbe...*, 1509 MS; *Nova carta marina...*, 1511. **N.B.** David Woodward has raised doubts as to the authenticity of the 1509 map and other manuscript world maps of the period. WOODWARD, D. The Map Collector 67 (1994) pp.2-10.

**Harinckhuijsen**, I. See **Haringhuijsen**, Isaac Abraham.

**Haring**, Pieter. Dutch draughtsman, assisted with Nicolas and Jacob Cruquius's *'t Hooge Heemraedschap van Delfland* (25 leaves), 1712.

**Haring**, R. *Michigan*, in C.L. Fleischman's *Der Nordamerikanische Landwirth*, New York 1848.

**Haringhuijsen** [Harinckhuijsen], Isaac [Isaacq] Abraham. Dutch surveyor and cartographer of Alkmaar, worked in Noord-Holland. *Caarte van t Eyerland, ende West Vlieland...*, 1688 MS (with A. van Twuyver and A. Bluzé).

**Hariot**, Thomas. See **Harriot**.

**Harkness**, Edson. Plot of the Town of Cornish, Ohio, 1829.

**Harkness**, Olney. *Plan and profile of the Phil. W. & Balt. R.R.*, 1860.

**Harlacher**, A.R. *Hydrographische Karte... Böhmen*, 1878.

**Harlee**, Thomas. Surveyor. Marion District, 1818; Horry District, 1820; Williamsburgh District, 1820, all for Robert Mills's *Atlas of the State of South Carolina*, 1825, 1838.

**Hårleman**, Carl *Baron* (1700-1753). Swedish surveyor. Director General military survey of Sweden, 1739.

**Harley,** D.S. & J.P. Surveyors associated with Geil and Harley. Worked on *Map of the counties of Clinton and Gratiot, Michigan*, Philadelphia, Samuel Geil 1864 (with J.D. Nash, H.G. Grigham & M.C. Wagner).
• **Harley,** J.P. Surveyor associated with Geil and Harley.
• **Harley,** David S. Surveyor associated with Geil and Harley. Partner in the firm of Geil, Harley & Siverd. Personally registered the following: *Map of the counties of Eaton and Barry, Michigan*, Philadelphia, Geil, Harley & Siverd 1860 (special survey by J.D. Nash, drawn by Geil & Jones, engraved by Worley & Bracher); *Map of Kalamazoo Co., Michigan*, Geil & Harley 1861 (surveys by I. Gross, drawn by S.L. Jones).

**Harliensis** or **Harlingensis,** P. See **Feddes,** Pieter.

**Harman,** H.J. *Map of the Routes... from Darjeeling to Shigatze*, 1882.

**Harman,** Robert. *A Plan of... Richmond* [Yorkshire] (4 sheets), 1724.

**Harmand,** *Doctor* Jules (1845-1921). French military physician, geographer and cartographer. Took part in explorations in Cambodia, Vietnam and Laos and produced several manuscript maps, 1875-1877, including *Le Sé Bang-Hieng* [Laos], 1877; *Itineraire du Mekong...*, 1877; and others. **Ref.** BROC, N. Dictionnaire illustré des explorateurs Français du XIXe siècle: Il Asie (1992) pp.234-238 [in French].

**Harmer** [Harmar], Thomas. Draughtsman and text engraver. Atlas to Captain James Cook's Voyages, 1784; *A General View of the principal Roads and Divisions of Hindoostan*, London, J. Rennell 1784; Henry Pelham's Clare (12 sheets), 1787 (with J. Cheevers); James Rennell's *A Map of Hindoostan, or the Mogul Empire* (4 sheets), London, J. Rennell, 1788 (map engraved by J. Phillips & W. Harrison); many charts and plans for Alexander Dalrymple, *c*.1803; script engraving for the Admiralty, 1814.

**Harme,** L.F. de. See **Deharme,** L.F.

**Harmin** de Mesa. See **Fabronius,** Hermann.

**Harmon,** Daniel Williams (1778-1845). Map of the interior of North America in: *A journal of voyages and travels in the interior of North America*, 1820.

**Harms,** Heinrich (1861-1933). Specialised in school maps and wall charts, including 'stumme Karten' or maps without place names. *Wandkarte von Deutschland* [wall chart], 1900.

**Harmssen,** *Captain* J.A. *Chart of... Luzon*, 1839; *Chart of the Hawaii Archipelago*, 1840.

**Harnberger,** Christoph. Ewe-Sprach-Gebiet [Gold Coast - Ghana], Gotha 1867.

**Harness,** *Lieutenant* Henry Drury, RE (1804-1883). Pioneering statistical cartographer, prepared the first known published flow maps, incorporating new techniques for the presentation of statistical information. Compiled population maps of Ireland showing the number of passengers carried by public conveyances and quantities of traffic, 1837, published in *Atlas to Accompany 2nd Report of the Commissioners appointed to consider and recommend a General System of railways for Ireland*, 1838. **Ref.** ROBINSON, A.H. Early thematic mapping in the history of cartography (1982) pp.117-119; ROBINSON, A.H. 'The 1837 Maps of Henry Drury Harness' in Geographical Journal 121 (1955) pp.440-450.

**Harney,** E.M. *Sheboygan County, Wisconsin*, 1862; *Kent County, Michigan*, 1863.

**Haro,** Gonzalo López de. See **López** de Haro, Gonzalo de.

**Harpe.** See **La Harpe.**

**Harper,** brothers. Publishers of New York.
• **Harper,** James (1795-1869). Founder of the Harper & Brothers company.
• **Harper,** John (1797-1875).
• **Harper,** Joseph (1801-1870).
• **Harper,** Fletcher.
• **Harper & Brothers.** Firm of publishers founded in 1833 by James Harper, traded from 82 Cliff St., New York, best known for *Harpers Weekly*. Other publications included Stephen Kay's *Travels and Research in Caffraria...* [S. Africa], 1834; *An improved Map of the Hudson River*, 1834; Sidney E. Morse's *A System of Geography for the use of schools...*, 1844, 1845, 1849; S.E. Morse

and S. Breese's *The Cerographic Atlas of the United States*, 1845 edition (first published by Sidney Morse, 1842); Samuel Breese's large wall map of the United States and Canada, 1847 (wax engraved by S.E. Morse); *The World*, c.1850; *Harper's Statistical Gazetteer of the World*, 1855; Benson J. Lossing's *The Pictorial Field-Book of the War of 1812*, 1869; Theodore R. Davis's bird's-eye view of Philadelphia, 1872; *Harper's School Geography*, 1876, 1877, 1878, 1880; map of the United States showing the distribution of races, 1914; *United States 1900*, New York & London 1914; *Map of the Eastern United States*, 1929 (designed by Griswold Tyng).

**Harper**, A.P. Surveyor. Map of New Zealand, 1893.

**Harper**, I. and J., brothers. New Cheltenham Guide, 1827.

**Harper**, Joshua [& Co.]. Lithographers of 18 Warwick Court, Grays Inn. James Skinner's *London with Postal Districts, Railways & Stations...*, c.1869.

**Harpff**, Philip (1611-1647). Engraver in Frankfurt. Worked on M. Merian's *Theatrum Europaeum*, including *Regenspurg*, 1644; *Lotringer Quartir*, 1644.

**Harrach**, *Oberlieutenant* Friedrich. *Plan... Würzburg*, 1845.

**Harraden**, Richard B. Artist and draughtsman. Designed the vignettes for Richard Grey Baker's *Map of the County of Cambridge...*, 1821 (vignettes engraved by D. Havell).

**Harrani-al Ahmad** Ibn Hamdan (*fl.*14th century). An Egyptian lawyer. World map *Jami al-Funun* [Comprehensive Techniques], 1330.

**Harrevelt**, E. van. Publisher and bookseller of Amsterdam. Published the later volumes (XIX onwards) of J.P.J. Du Bois's *Histoire générale des voyages*, E. van Harrevelt et Changuion 1772-1780; *Atlas portatif pour servir à l'intelligence de l'Histoire philosophique et politique des Etablissements et du Commerce des Européens dans les deux Indes*, Amsterdam, E. van Harrevelt et D.J. Changuion 1773 (part of G. Th. Raynal's *Histoire Philosophique*).
• E. van Harrevelt Erven [Heirs] (*fl.*1783).

Booksellers 'in de Kalverstraat', Amsterdam. Successors to E. van Harrevelt.

**Harrewyn** [Harrewijn; Herrewyn], family. Engravers working in Southern Netherlands and Brussels.
• **Harrewyn**, Jean-Baptiste. Engraver for A. Roggeveen, 1684; for Daniel de La Feuille, 1685; for Francisco de Aefferden's *El Atlas Abbreviado*, 1696, 1709, 1711, 1725.
• **Harrewyn**, Jacques [Jacob] (1660-1727). Copper engraver from Amsterdam, pupil of Romein de Hooghe, worked for Jacques Peeters at Antwerp, and for E.H. Fricx at Brussels, where he died. Doncker & Robijn's *Nieuwe Wereld* [polar projection on four sheets], 1687; *Plan de la Ville de Bruxelles*, Amsterdam 1697; *La Ville de Luxembourg*, Brussels, François Foppens 1697 (used, with additions, in Jean-Baptiste Chrystin's *Les Délices des Pais-Bas*, Foppens 1711, 1713); map of the Duchy of Luxembourg, 1697; engraved a series of battle maps and plans of fortifications and sieges in the Netherlands and Flanders, 1706-1709 including *Carte Particuliere des Environs de Lille, Tournay, Valenciennes, Bouchain, Douay, Arras, Bethune* (3 sheets), Brussels, E.H. Fricx 1706; *Carte Particuliere des Environs de Dunkerque, Bergues, Furnes, Gravelines, Calais et Autres*, Brussels, E.H. Fricx 1707; *Plan de la Bataille d'Oudenaerde du 11 Juillet 1708*, Brussels, E.H. Fricx 1708 (drawn by G.L. Mosburger); *Tournay*, before 1709; *Partie de l'Angleterre*, Brussels, E.H. Fricx 1709, 1744 (these represent just a few of the maps engraved by Harrewyn and later re-published together by Fricx under the title *Table des cartes des Pays Bas et des frontières de France*, 1712; also known under the title *Recueil des cartes des provinces meridionales des Pais Bas*, E.H. Fricx 1712); *Africae*, 1709 for S. Fernández de Medrano.
•• **Harrewyn**, Jacques-Gérard. Engraver of Brussels, son of Jacques.
•• **Harrewyn**, François (1700-1764). Engraver of Brussels. Also engraved for Fricx.

**Harrington**, *Professor* Mark Walrod (1848-1926). *Rainfall and snow of the United States* (with atlas), Washington, D.C., Weather Bureau 1894.

**Harriot**, T.G. Draughtsman. *Plan of the Environs of Messina* [Italy], 1808 MS.

**Harriot** [Hariot], Thomas (1560-1621). British naturalist, mathematician, and astronomer from Oxford. From 1585-1586 he was at Richard Grenville's Roanoke Island colony [Virginia], recording the area together with the artist John White. They both returned with Drake's expedition in 1586. Manuscript map of Albemarle and Pamlico Sounds [Virginia], c.1585; *A Briefe and True Report of the New Found Land of Virginia* (no maps), 1588 (used by Theodore de Bry as the first part of his *Grands Voyages*, 1590, published in English, French, German and Latin, with a map by John White); first map of the moon made by use of a telescope, 1609 MS. **Ref.** SCHWARTZ, S.I. & EHRENBERG, R.E. *The mapping of America* (New York 1980) pp.76, 77, 79, 81; SHIRLEY, John W. 'Thomas Harriot's Lunar Observations' in *Science and History: Studies in Honour of Edward Rosen* (Studia Copernicana 16, Wroclaw, Polish Academy of Sciences Press, 1978) pp.283-308.

**Harriott** [Harriett], *Lieutenant* Thomas. Officer in the Royal Staff Corps. *Plan of Cadiz*, 1810 MS (with F. Leicester); *Sketch of the Action near the Hill of Barrosa* [Spain], 1811 (used by J. Wyld in *Maps & Plans Showing the Principal Movements, Battles & Sieges.. during the War from 1808-1814...*, London 1840); one of those involved in a military survey of northern France published as *British Lines of Occupation. N. of France (after 1815)...* (10 sheets), 1818-1823.

**Harris**, A.S. Bird's-eye view of Fort Worth, Texas, 1913.

**Harris**, C. American publisher. H. Stebbins's *A Map of Worcester* [Massachusetts], 1833 (printed by Pendleton's Lithography).

**Harris**, Caleb. American surveyor. Surveys of Rhode Island, used by Harding Harris in the compilation of: *A Map of the State of Rhode-Island...*, Providence, Carter & Wilkinson 1795 (drawn by Harding Harris, engraved by Samuel Hill), and a reduced version entitled *The State of Rhode-Island...*, 1795 (engraved by J. Smither, used in Mathew Carey atlases).

**Harris**, Cyrus. *Pennsylvania*, 1796; *A Map of the State of Kentucky*, 1796; *England, Scotland, Ireland and Wales*, 1796, all published in Jedidiah Morse's *The American Universal Geography*.

**Harris**, *Captain* George, *RN*. *Survey of the Isle of Pines*, [coast of Venezuela] 1836.

**Harris**, H.A. Dumrah River, 1872; Jumboo River, 1872; Mahanuddee & Davey Rivers [India], 1872.

**Harris**, *Lieutenant* H.R., *RN*. Island of Barbados, surveyed 1861 (with J. Parsons and G. Stanley), published Admiralty 1873, 1886 (engraved by Davies & Company); one of the surveyors for *Island of Barbados...* (9 sheets), London 1875 (surveys of 1869); Chart of Carlisle Bay [Barbados], surveyed 1869 (with J. Parsons and G. Stanley), published Admiralty 1871, 1886 (engraved by E. Weller). **N.B.** Compare with H.R. **Harris**, below.

**Harris**, H.R. Surveyor aboard H.M.S. *Swallow*. Survey of the east coast of China, province of Shan Tung and Kyan Chau bay, c.1863 (with G. Stanley, J. Hall, A. Hamilton & W.N. Goalen). **N.B.** Compare with H.R. **Harris**, above.

**Harris**, Harding. Draughtsman for *A Map of the State of Rhode Island; taken mostly from surveys by Caleb Harris*, Providence, Carter & Wilkinson 1795 (engraved by Samuel Hill, a reduced version engraved by J. Smither and entitled *The State of Rhode-Island; compiled from the Surveys and Observations of Caleb Harris* was used in several of Mathew Carey's atlases); *Rhode Island and Connecticut*, Boston, Thomas and Andrews 1796 (used in Jedidiah Morse's *The American Universal Geography*). **Ref.** WHEAT, J.C. & BRUN, C.F. *Maps and charts published in America before 1800* (1969) Nos. 250, 251 and 287.

**Harris**, J. One of many sellers of G. Cole and J. Roper's *The British Atlas*, 1810; *Warwickshire; or, original delineations, topographical, historical and descriptive, of that county. The result of a personal survey. By Mr. Brewer*, London, 1818 (with the county map from *The British Atlas*).

**Harris**, J.S. Worked as a surveyor for the U.S. Coast Survey. Surveys for *Plan of Fort Jackson, showing the effect of the bombardment by the U.S. mortar flotilla and gunboats April 18th to 24th 1862*, Washington, D.C., U.S. Coast Survey c.1862 (drawn by E. Hergesheimer, lithographed by Bowen & Co.).

**Harris, John** (*fl.c.*1685-*c.*1720). English engraver at 'Bull's Head Court, Newgate Street', and 'Black and White Court in the Old Baily' (1699-1670), London. Copied Thomas Holme's map of Pennsylvania which was then published under the title *A Mapp of Y.e Improved Part of Pennsylvania in America...*, London, P. Lea 1687, London, George Willdey 1730, London, T. Jefferys Sr. *c.*1750, London, R. Marshall *c.*1770; engraved for Greenvile Collins's *Great Britain's Coasting Pilot*, 1693; *A New Map of the English Empire in America...*, London, Robert Morden 1695; advertised as one of the sellers of Joel Gascoyne's *A Map of the County of Cornwall* (14 sheets), *c.*1699; draughtsman and engraver for Christopher Browne's *Nova totius Angliae tabula*, 1700; *Buckinghamshire* for Robert Morden's *The New Description and State of England*, 1701; E. Halley's *A New and Correct Sea Chart of the Whole World...* [magnetic chart of the Atlantic] (2 sheets), 1701, Mount & Page 1758; *The City and Harbour of Cadiz... By an officer of the Fleet 1695...*, *c.*1702; *An Actuall Survey of the Parish of St. Dunstan, Stepney*, 1703; Joel Gascoyne's *An actual Survey of the Hamlet of Bethnal Green in the Parish of Stepney*, London 1703; *H. Jalliot's Map of the seat of War in Italy*, London, R. Morden & C. Browne 1704; C. Browne's *A New Mapp of Scotland, the Western, Orkney and Shetland Islands*, 1705; Henry Pratt's *Tabula Hiberniae novissima et emendatissima*, London, 1708; Benjamin Cole's *A New Map showing the Counties, Hundreds, Ecclesiastical Divisions, Towns, Villages... within twenty miles round Cambridge*, London, William Redmayne 1710; *A New and Exact Map of the Diocese of London*, 1714; David Mortier's *Nouveau théatre...*, 1715; Jonas Moore's map of the Fens for *Magna Britannia*, *c.*1715 (a partwork based on the maps of Morden); John Senex's *A New Map of the English Empire in the Ocean Of America or West Indies*, 1719; *The East Part of the River Thames*, 1720 (for R. Morden); *A Globular chart shewing the errors of plain and the deficiencyes of Mercator...*, (with John Senex and Henry Wilson), published in Nathaniel Cutler's *A general coasting pilot*, London, James & John Knapton 1728. **Ref.** CAIN, M.T. *The Map Collector* 57 (1991) p.5; RAVENHILL, W. in *Imago Mundi* 33 (1981) pp.28-31.

**Harris, John.** Mapmaker and geographer. *A View of the World in Divers Projections*, John Garrett *c.*1697. **N.B.** Possibly son of John **Harris**, above.

**Harris, Doctor John** (*c.*1667-1719). Doctor of Divinity, Fellow of the Royal Society and author. *The Description and Uses of the Celestial and Terrestrial Globes*, London, D. Midwinter and T. Leigh 1703 and later; *Navigantium atque Itinerantium Bibliotheca; or, a complete collection of voyages and travels...*, 1705 (with maps by H. Moll), 1744-1748, London, T. Osborne, H. Whitridge &c. 1764; *History of Kent*, London, T. Midwinter 1719. **Ref.** CAIN, M.T. *The Map Collector* 57 (1991) p.6.

**Harris, John.** Publisher and seller of prints and maps, N<sup>o.</sup>3 Sweetings Alley, Cornhill; N<sup>o.</sup>8 Broad St., London. *Plan of the French attacks upon the Island of Grenada, with the engagement between the English Fleet under the command of Admiral Byron and the French Fleet under Count D'Estaing...*, 1779 (engraved by J. Luffman); *...Plan of London, Westminster and the Borough of Southwark*, 1779; [John] *Andrews's new and accurate map of the country thirty miles round London*, 1782; *Action in the Bay of Boukkier* [Abukir, Egypt], 1798.

**Harris, John** (1756-1846). Publisher and bookseller. Worked in London and Bury St Edmunds, Suffolk. In 1846 he was joined by his son to form Harris & Son. Thomas Smith's *Universal Atlas*, 1802 (with John Cooke); *America*, 1822; *Africa*, for Jehosaphat Aspin, 1823, 1832.
• **Harris & Son** (*fl.*from 1846).

**Harris, John.** Master of H.M.S. *Prince Regent*. Surveys for *A Plan of Long Point Bay and Turkey-Point Harbour Lake Erie*, 1815 (drawn by A. Vidal); *Eye sketch of Big Bear River* [Sydenham River, Canada], 1815 MS; *Chart of Kingston Harbour and Entrances thereto from Lake Ontario*, 1816 MS.

**Harris, John.** *Harris's New Map of Warwickshire...*, Birmingham, Thomas Radclyffe and Son 1842 (with John James Harris), republished as *Radclyffes' Map of Warwickshire...*, Birmingham, G.F. Rose after 1851, and as *Cooke's Map of Warwickshire...*, Warwick, H.T. Cooke & Son *c.*1861.

**Harris**, John James. Worked with John Harris (above) on *Harris's New Map of Warwickshire...*, Birmingham, Thomas Radclyffe and Son 1842 and later (as above).

**Harris**, John. Maker of facsimile maps. *A Map of New-England...*, London 1865 (copy of the 'White Hills' map of 1677 from W. Hubbard's *A Narrative of the Troubles with Indians in New England*; a second, similar facsimile of the same date is a copy of the 'Wine Hills' version of the same map); *The South part of New-England as it is Planted this yeare, 1634*, London 1865 (from W. Wood's *New England Prospect*); *Typo de la carta cosmographica de G. Vopellio Medesurgensis*, London 1873 (copy of Caspar Vopell's world map).

**Harris**, Joseph (1702-1764). Assay Master of the Mint. *Stellarum fixarum Hemisphaerium Australe ... for 1690*, J. Senex 1728 (in Daniel Defoe's *Atlas Maritimus et Commercialis*); revised John Harris's 1703 work as *The Description and Use of the Globes and the Orrery*, London, Thomas Wright & Richard Cushee 1731, 1732, 1734, Thomas Wright & E. Cushee 1738, Thomas Wright & William Wyeth 1740, Thomas Wright & E. Cushee 1745, 1751, B. Cole & E. Cushee 1757, 1763, 1768, John, Francis & Charles Rivington 1783.

**Harris**, Joseph (*fl.*1800-1815). Essex estate surveyor. Wix, 1811 MS; Little Bentley and Bromley, 1815 MS.

**Harris**, Moses. *A Plan of Chebucto Harbour With the Town of Hallefax by Moses Harris Surveyor*, 1749 MS.

**Harris**, Thomas. (*fl.*1790-1802). English optician, globemaker and instrument maker of London.
• **Harris & Son** (*fl.*from 1802). Partnership of Thomas and his son William. *New terrestrial globe* [7-cm], London, *c.*1820.
• **Harris**, William.

**Harris**, Thomas. Engraver. *A Plan of the Town of Sheffield*, Sheffield, John Robinson 1797 (drawn by W. Fairbank, published in *A Directory of Sheffield*).

**Harris**, Thomas M. Officer in the Royal Staff Corps. One of those involved in a military survey of northern France. *British Lines of Occupation. N. of France (after 1815)...* (10 sheets), 1818-1823.

**Harris**, W.G. *Map of the Province of South Australia*, London, R.K. Burt 1862.

**Harris**, Walter (1686-1761). Irish historiographer. *The antient and present state of Co. Down*, Dublin 1744 (with Charles Smith).

**Harris**, Walter Burton. *Map of North-West Morocco, shewing local distribution of tribes*, London, Royal Geographical Society 1889; *Sketch Map of Tibet and western China*, 1893 (in *The Geographical Journal*).

**Harris**, Captain Sir William Cornwallis (1807-1848). Engineer, traveller and author. *Africa North East of the Cape Colony, exhibiting the relative positions of the Emigrant Farmers and Native Tribes, May 1837...*, Bombay 1838; London, John Murray 1839 (used in Harris's *The Wild Sports of Southern Africa*).

**Harrison**, family. English family of engravers and printers who settled in Philadelphia in 1794. **N.B.** The work of William Senior and William Junior is hard to distinguish and may not be correctly attributed here. **Ref.** RISTOW, W.W. *American maps and mapmakers* (Detroit 1986) passim.; HARRISON, W.I. *William Harrison, Sr. and Sons, engravers. A checklist of their works* (South Yarmouth, Mass. 1978).
• **Harrison**, William, *Sr.* (*d.*1803). Engraver at N[o.]12 Winchester St., Battle Bridge, England; Philadelphia from 1794. Four sons followed him in the same trade. Engraved for James Rennell, 1781-1788 including *A Map of Hindoostan, or the Mogul Empire* (4 sheets), London, J. Rennell, 1788 (with I. Phillips, text engraved by T. Harmer); *Plan of Monterey in California by Don Josef Tobar y Tamariz... 1786*, A. Dalrymple 1789; *Plan of Port Sn Diego on the West Coast of California*, A. Dalrymple 1789; *Plan of Port Sn. Francisco on the West Coast of California*, A. Dalrymple 1789; Charles Duncan's *Sketch of the Entrance of the Strait of Juan de Fuca*, A. Dalrymple 1790; Joseph Southern Purcell's *A Map of the States of Virginia, North Carolina, South Carolina, and Georgia...*, 1792 (published in the London edition of Jedidiah Morse's *The American Geography*); for Robert Wilkinson, 1794.
•• **Harrison**, William, *Jr.* Engraver, son of William, above. Lt. John Hunter's *Plan of*

*Port Jackson, New South Wales*, J. Stockdale 1789 (with J. Reid); for Robert Wilkinson, 1794; Samuel Lewis's *The State of New York*, Mathew Carey 1795; *A Map of the United States compiled chiefly from the State Maps*, 1795; *Map of the United States, Exhibiting the Post-Roads...*, 1796; one of the engravers for Aaron Arrowsmith and Samuel Lewis's *A New and Elegant General Atlas*, Philadelphia, John Conrad [and other Conrad companies] 1804 and later editions; Jared Mansfield's *Map of the State of Ohio*, 1806; Bartholemy Lafon's *Carte Générale du Territoire d'Orleans*, New Orleans 1806; John Strother and Jonathan Price's *Actual Survey of the State of North Carolina*, 1808 (printed by Charles P. Harrison); Plan of Philadelphia, 1811; William Watson's *A Map of the State of New Jersey*, 1812; Philip Carrigain's map of New Hampshire, 1816; Samuel Lewis's wall map of the United States, Philadelphia, Emmor Kimber 1816 (with Samuel Harrison); A. Bradley's *Map of the United States Intending Chiefly to Exhibit the Post Roads & Distances...*, Georgetown 1819; engraved A.J. Stansbury's *Map of the... routes... from Washington to Lake Ontario*, c.1827; W. Bussard's *A map of Georgetown in the District of Columbia*, 1830. **N.B.** Compare with W. **Harrison** below, and **Harrison & Reid**.

•• Harrison, Charles P. (*b*.1783). Engraver and printer. Son of William senior, above. Printer for John Strother and Jonathan Price's *Actual Survey of the State of North Carolina*, 1808 (engraved by William Harrison Junior); T.H. Poppleton's *This Plan of the City of Baltimore...*, New York 1823.

•• Harrison, Samuel (1789-1818). Third son of William Harrison senior. Entered into partnership with publisher John Melish in 1817, for whom he engraved some state maps, 1817-1818. *Map of Lewis and Clark's Track Across the Western Portion of North America from the Mississippi to the Pacific Ocean...*, 1814; engraver for D. Macpherson, Atlas of ancient geography, Philadelphia, 1806; map of Lake Ontario, 1809; for Fielding Lucas, 1812; *Map of the Seat of War*, 1814; engraved 22 of the maps for Fielding Lucas's *A New and Elegant General Atlas*, 1816; Samuel Lewis's wall map of the United States, Philadelphia, Emmor Kimber 1816 (with William Harrison); engraved and published *Map of Indiana*, Philadelphia, John Melish & Sam'l Harrison 1817 (surveys by Burr Bradley); *A Geographical Description of the World*, Philadelphia John Melish & Samuel Harrison 1818 (with a world map engraved by Samuel Harrison in 1817); Daniel Sturges's *Map of the State of Georgia*, Savannah, Eleazer Early 1818; engraver for John Melish's *United States of America...*, 1819.

•• Harrison, —. Engraver. Son of William senior, above.

Harrison, Lieutenant —. Sketch illustrating the Action fought on the 18th September 1860, by the allied armies in China, taken from the Road Survey made by Lieutenant Colonel Wolseley and Lieut. Harrison, London, War Office 1860. **N.B.** Compare with Richard **Harrison**.

Harrison, Alfred. *Map of the City of Vanceburg and Vicinity* [Kentucky], Vanceburg 1875.

Harrison, D.R. Engraver for Edmund Blunt. *The North Eastern Coast of North America from New York to Cape Canso Including Sable Island*, 1828; *Long Island Sound from New York to Montauk Point*, E. & G.W. Blunt 1830.

Harrison, E. [& Co.]. The West End Athletic Outfitters, 259 Oxford St., London. *The 'Finger Post' Bicycle Road Guide, Containing The Main Roads of England, Wales, And Part Of Scotland, Specially Compiled For The Use of Bicyclists And Tourists*, 1883 (using lithographic transfers of John Cary's county maps).

Harrison, James. Plan of Cape-Françoise [Haiti], 1793.

Harrison, John (1693-1776). Clockmaker, mathematician and instrument maker, born in Foulby, Yorkshire. Constructed the first sea-going timepiece sufficiently reliable to allow longitude to be calculated at sea. Completed 5 chronometers between 1735 and 1770. The 4th of these (the Harrison H-4 'watch') although successfully trialled at sea in 1762 and 1764 was not approved by the Board of Longitude; a copy of H-4 made by Larcum Kendall [K-1] was used on Captain Cook's second voyage, 1772-1775 and was highly praised for its accuracy. **Ref.** SOBEL, S. Longitude (1996).

• Harrison, William. Son of John, worked with him in the construction and promotion of his clocks. Undertook the sea trials of the 4th clock, 1762 and 1764.

**Harrison,** John (*fl.*1784-1815). Engraver, printer and publisher at N°·115 Newgate Street, London. Prepared a series of maps to accompany his new edition of Paul Rapin de Thoyras's *The History of England*, 1784-1789; School Atlas, 1791; America, after J.B.B. d'Anville, 1791; county maps printed on white satin for needlework in schools, 1787; county maps, 1787-1789 published together as *Maps of the English Counties* (38 maps), 1791, 1792 (maps drawn by J. Haywood, engraved by E. Sudlow and G.S. Allen), maps re-issued as *General and County Atlas*, 1815. **Ref.** CARROLL, R.A. *The printed maps of Lincolnshire 1576-1900* (1996) pp.139-141; CHUBB, T. *The printed maps in the atlases of Great Britain and Ireland* (1974) pp.223-226 (gives details of *...English Counties*). **N.B.** HODSON, D. in *Printed maps of Hertfordshire* (1974) p.68, suggests that many of the maps for the English Counties may have appeared first in The History of England which he describes as untraced.

**Harrison,** Sir John Burchmore. Government geologist. Geological map of Barbados, 1890 (with A.J. Jukes Brown, lithographed by Malby & Sons); *Map of portions of the Lower Essequibo and Cuyuni Rivers in the colony of British Guiana...*, Demarara 1905 (with C.W. Anderson, printed by Waterlow and Sons, Ltd., London); observations used in the compilation of *Map of the Northern Portion of British Guiana. Showing the geology of the courses of the principal river and auriferous areas...*, 1906.

**Harrison,** John F. American surveyor and draughtsman. *Map of the Borough of Stonington, Connecticut*, New York, 1851 (with John Bevan, printed by Sarony & Major); *Map of the City of New-York Extending Northward to Fiftieth Street* (6 sheets), New York, M. Dripps 1851.

**Harrison,** Joseph. Northern Boundary Province of Rhode Island, 1750.

**Harrison,** L. Printers, 373 Strand, London. Scripture Atlas, 1812 (with J.C. Leigh).

**Harrison,** P. Plans of Havana, Porto Bello &c., 1740; *A Plan of the Harbour of St Augustine and the Adjacent Parts in Florida*, 1742.

**Harrison,** P. Draughtsman. *A Plan & Profil of the Fortifications now erecting to defend the Town & Harbour of Newport on Rhode Island*, 1755 MS; *A Plan of the Town and Harbour of Newport on Rhode Island*, c.1755 MS.

**Harrison,** Lieutenant Richard RE. Plan of Scutari and its Environs [Turkey], 1856 (with Edward T. Brooke).

**Harrison,** Robert H. Apprenticed to Augustus Warner, 1866-1867. Maps of Tippecanoe County and Shelby & Johnson Counties, Indiana, 1866 (with Warner & Higgins, R.M. Sherman and G.P. Sanford); *Map of Warren Co., Ohio*, 1867 (with G.P. Sanford and J.S. Higgins); *Atlas of Hamilton County, Ohio*, 1869; worked on county atlases of Ohio, Wisconsin and Iowa in partnership with George Warner, 1869-1875.

• **Harrison & Warner** (*f.*1869-1875). County atlas publisher of Marshalltown, Iowa, partnership of R.H. Harrison and George E. Warner. Became Warner & Foote in 1876, and finally Charles M. Foote & Company of Minneapolis, which traded until *c*.1900. Atlas of Hamilton county, Ohio, 1869; county atlases in the states of Wisconsin, Iowa, and Missouri, 1871-1876 including *Atlas of Clinton Co. Iowa*, Clinton & Philadelphia 1874; *Atlas of Dubuque Co. Iowa*, Clinton & Philadelphia 1874; *Atlas of Muscatine County* [Iowa], Clinton 1874; *Illustrated Historical Atlas of Macon County*, Philadelphia 1875; *Atlas of Tama County, Iowa* (25 maps), 1875.

**Harrison,** Samuel. See **Harrison,** family, above.

**Harrison,** T.R. Lithographer. *Map shewing the Line of Boundary between the British and United States Territory in North America*, 1846; *Map of the Mosquito Coast*, 1847;

**Harrison,** Thomas. Crown Surveyor. Isthmus of Panama, 1857; *Map of Jamaica*, London, Edward Stanford 1873; Jamaica (2 sheets), 1873 (compiled for George Henderson & Co.); *Map of the Island of Jamaica prepared for The Jamaica Handbook under the direction of Thomas Harrison... by Colin Liddell*, 1895 (lithographed by Stanford's Geographical Establishment); and others. **Ref.** HIGMAN, B. *Jamaica surveyed: plantation maps and plans of the eighteenth and nineteenth centuries* (Kingston, 1988).

**Harrison,** W. Engraver of Washington. A.J. Stansbury's *Map of the country embracing the several routes examined with a view to a*

*John Harrison (1693-1776). A mezzotint by L. Tassaert after an oil painting by T. King.
(By courtesy of the Science and Society Picture Library, London)*

national road from Washington to Lake Ontario, Georgetown D.C. c.1827; T.W. Maurice's *Portland harbour, with the plan of a proposed Breakwater* [Lake Erie], 1829; T.W. Maurice's *Pultneyville Bay with the Plan of a Breakwater*, 1829; William Bussard's *A map of Georgetown in the District of Columbia*, Washington 1830. **N.B.** Compare with the **Harrison** family of Philadelphia, above.

**Harrison**, W.A. Acting Government Surveyor in British Guiana. *Chart of a portion of the Barima Amacura and Waini...*, 1889 (with H.I. Perkins).

**Harrison**, Walter. *A new and universal history, description and survey of the Cities of London and Westminster...*, 1776.

**Harrison**, William (1534-1593). English topographer and antiquary. *An historicall description of the Islande of Britayne*, 1577. **Ref.** SKELTON, R.A. Archaeological Journal 108 (1951) pp.109-120.

**Harrison**, William *Senior* and *Junior* (fl.1781-1830). See **Harrison**, family, above.

**Harrison & Bashworth.** Printers of Brooklyn. William Darby's *Map of the United States Including Louisiana*, 1818 (engraved by James D. Stout).

**Harrison & Co.** Printers, booksellers and publishers of Nº·18 Paternoster Row, London. W. Bardin's 9-inch *A New, Accurate, and Compleat Terrestrial Globe*, 1783; W. Bardin's 9-inch *The Celestial Globe...*, 1785 (both globes were chosen to accompany the *Geographical Magazine*).

**Harrison & Co. Limited.** Publishers of 3, Liverpool Street, London E.C. *Lincolnshire*, 1889 (taken from Cary's county map). **N.B.** Compare with E. **Harrison**, above.

**Harrison & Reid.** Engravers. Thomas George Shortland's *Track of the Alexander*, 1788-1790.

**Harrison & Sons.** Lithographers of Lancaster Court, Strand (1831), later St. Martin's Lane, London. Specialised in military and strategic maps and papers for the Foreign Office and Houses of Parliament. *Sketch Exhibiting the Claims of Boundary, on the Part of the British and American governments, under the 5th Article of the Treaty of Ghent*, 1831; *Map of Gibraltar Waters...*, after 1831; *Map to illustrate the Boundary line established by the Treaty of Washington of the 9th August 1842 Between Her Majesty's Colonies of New Brunswick and Canada and the United States of America*, 1843; *General Map of the states of Tuscany, Modena and Parma...*, c.1847; Edward Cullen's *Rio Savana or Chaparti, with the Route thence to the Atlantic*, 1851; *Map of the Island of Newfoundland shewing the limits of the various fisheries &c....*, 1857; *Sketch of Prince Edward Island...*, 1868; several maps to illustrate the U.S. and Canadian boundary survey of the Juan de Fuca Strait, 1873; *Map of the coasts visited by Sir B. Frere and places mentioned in his reports of the Special Mission to Zanzibar and Mascat for the Suppression of the Slave Trade*, 1873; *Chart of the Inshore & Deep-sea Fishing Grounds on the Atlantic Coasts of Canada and within the Gulf of St. Lawrence*, 1876; *Orkhanie Pass* [Bulgaria], after 1877; *Batoum* [Georgia], 1880; *Odessa* [Black Sea], 1880; *Map of the West India Islands*, 1881; *Map of the West Coast of Africa from Tangier to the Cape of Good Hope*, 1881; P.C. Sutherland's *Horizontal Geological section on main road from Durban to Van Reenen's Pass*, 1881; *Cyprus Famagousta Harbour*, 1881; *Map shewing the new boundary of the Kingdom of Greece 27th Nov 1881*, 1882; *Map of Servia shewing Railways, and Custom Houses*, 1883; *Map showing the Salonica Railway with proposed junctions*, 1884; *The island of Newfoundland showing the Electoral Districts...*, Foreign Office 1884 (copied from J. Arrowsmith's map); *Map of London & Suburbs*, Statistical Society 1885; *Position of contending forces on the eve of battle on Sept 19th 1885 at Slivnitza*, 1886; *Sketch map of British Guiana by Sir Robert H. Schomburgk...*, 1886 (showing British, Brazilian and Venezuelan boundary claims); *Western Asia Minor. Railways Constructed and Projected*, 1898; *Kandia, showing position of British troops* [Greece], 1899; *Projet de la carte de l'Albanie*, 1912; *Caile Ferate Romane...* [Romania], 1913; *Greece and Aegean Sea* [meteorological map], 1915; *Carte politique et ethnographique de la Pologne*, 1916; and numerous others.

**Harrison & Warner.** See **Harrison**, Robert H.

**Harrison & Watkins.** Publishers of Corio, Victoria. Fawkner's Corio, 1841.

**Harrison, Sutton & Hare.** Publishers of Philadelphia. *Atlas of Union County, Ohio*, 1877; *Atlas of Marion County, Ohio*, 1878.

**Harrower,** Henry Draper. Writer and geographer. *Handy Atlas of the World*, New York & Chicago, Ivison etc. 1884; *Captain Glazier and his lake* (with maps), New York, Chicago, Ivison, Blakeman, Taylor & Company 1886; *A pocket atlas of the world: with descriptive text, statistical tables, etc., etc.* (91 maps), New York, Ivison, Blakeman 1887 (with S. Mecutchen); *Teacher's manual to accompany the Complete school charts*, New York and Chicago, Ivison, Blakeman & Co. 1888; *The new states: a sketch of the history of the development of the States of North Dakota, South Dakota, Montana and Washington* (with map), New York, Ivison, Blakeman & Co. 1889.

**Hart,** Abraham. Publisher and partner of Edward L. Carey. **N.B.** See **Carey & Hart** under **Carey,** family and associated companies.

**Hart,** Albert Bushnell (*b*.1854). *Epoch maps illustrating American history*, 1891 and later.

**Hart,** Andrew (*d*.1621). Publisher in Edinburgh. Timothy Pont's *A New Description of Lothian and Linlithquo*, *c*.1611 or 1612 (engraved by Jodocus Hondius, used in Hondius editions of Mercator's *Atlas*, 1630 and later).

**Hart,** *Captain* Arthur FitzRoy. *Sketch of Mouth of False Emlalazi River* [Natal], 1879 (inset on map of Zululand); *Battlefield of Ulundi*, 1881.

**Hart,** Charles (*fl*.1870s). Lithographer and printer of 36 Vesey St., New York, the same address as New York publisher F.W. Beers, and engraver L.E. Neuman. He produced atlases and maps of North American counties and districts, most of them engraved by L.E. Neuman, and published in New York by F.W. Beers: *Atlas of Lapeer Co. Michigan*, 1874; *Atlas of Ionia Co. Michigan*, 1875; *Atlas of Livingston Co. Michigan*; 1875; *Map of Montcalm Co. Michigan*, 1875; *County Atlas of Shiawassee Co. Michigan*, 1875; *Topographical Map of Tuscola Co., Michigan*, 1875; *Topographical Map of Oceana Co., Michigan...*, 1876; *Topographical Map of Sanilac Co., Michigan*, 1876; *County Atlas of Muskegon County Michigan*, 1877; *Atlas of Saginaw Co. Michigan*, 1877; *Sectional Map of the State of Michigan Lower Peninsula*, Detroit, R.M. Tackabury 1877; Bird's-eye view of Anniston [Alabama], C.N. Dry 1903; T.M. Fowler's bird's-eye views of towns in North Carolina including Rocky Mount, 1907; Asheville, 1912; and High Point, 1913; and others.

**Hart,** Henry. Civil engineer, architect and surveyor, 140 Pearl St., New York. *Kalamazoo, Kalamazoo Co. Michigan*, 1853; *City of Grand Rapids, Kent Co. Michigan*, 1853; *City of Ann Arbor*, 1853; *City of Detroit*, 1835; and others.

**Hart,** J. Plan of Bradford, 1870.

**Hart,** J.J. de. *Monding der River Marowyne* [Guiana], Amsterdam 1857.

**Hart,** James. Lithographer of New York. Produced many U.S. county maps and atlases.

**Hart,** Joseph C. (*d*.1885). American cartographer. *A modern atlas of fourteen maps*, New York, R. Lockwood 1828, 1830; *A popular System of practical geography for the use of schools, and the study of maps...*, New York, Cady & Burgess 1851; *Hart's Geographical Exercises*, New York, Ivison & Phinney 1857.

**Hart,** *Captain* Reginald Clare V.C., R.E. Contributed to *Map of the country between Amanca and Cape Coast castle, then Northward to Dunkwa* [Ghana], 1873 (with others, drawn by Captains Huyshe and Buller); *Survey of the Bussum Prah* [Ghana], 1881; *Copy of an Egyptian plan found on the battlefield of Tel el Kebir*, 1882 MS.

**Hart,** Doctor W. Explorer in Sierra Leone. *Sierra-Leone und das Timméné-Land. Nach deu Forschungen der verminckschen Expedition unter E. Vohsen, Dr. W. Hart, u. E. Keller 1882*, Gotha, Justus Perthes *c*.1883.

**Hartert,** E. Route in Nigeria, 1886.

**Hartgers** [Hartgerts], Joost. Publisher and bookseller, Amsterdam 'in de Gasthuys-Steegh bezijden het Stadt-huys, in de Boeck-winckel'

(1648). *Oost en Westindische Voyagien...*, 1648 (a collection of voyages, many taken from I. Commelin); *Beschrijvinghe van Virginia, Nieuw Nederlandt, Nieuw Engelandt*, Amsterdam, 1651 (thought to have been intended for inclusion in part two of *Oost en Westindische Voyagien*). **Ref.** BURDEN, P. The mapping of North America (1996) N° 304 pp.389-390.

**Hartknoch**, Christoph (1644-1687). German historian in Kaliningrad [Königsberg]. *Alt und neues Preussen...*, 1684.

**Hartknoch**, father and son.
• **Hartknoch**, Johann Friedrich (1740-1789). Publisher, bookseller and map seller of Riga.
•• **Hartknoch**, Johann Friedrich (1768-1819). Publisher in Riga, Leipzig (after 1798) and Dresden (after 1806). Published L.A. Mellin's *Atlas von Liefland*, 1798.

**Hartl**, Heinrich (1840-1903). Austrian military cartographer, led a group of surveyors who assisted Greek military cartographers. They contributed to several maps of Greece including: B. Nicolaïdy's *Carte de la Thessalie*, Paris 1859; E. Kalergis's *Athènes et ses environs*, Paris 1863; G. Katelous's *Carte de l'ile de Crète*, Bucharest 1868; and others.

**Hartl**, Martin. One of the draughtsmen for H.J. Marx's *Generalcharte von Australien*, Vienna 1805 (with F. Swoboda); North America, 1806; *Carte de l'Empire Autrichien*, 1809-1811.

**Hartl**, Sebastian (*c.*1742-1805). Bookseller and publisher of Vienna. Johann Anton von Doetsch's *Regnum Hungariae*, 1790 edition (engraved by I.K. von Lackner).

**Hartleben**, Adolf (1835-after 1890). Publisher of Pest, Vienna and Leipzig. Published work by Gustav Freytag amongst others. *Central Afrika, nach den neuesten Forschungen bearbeitet von Dr. Joseph Chavanne*, Vienna 1882 (lithographed by G. Freytag); *Atlas von Afrika*, 1886; *A. Hartleben's Volks-atlas*, Vienna 1888 and editions to 1911; *Die Erde in Karten und Bildern, Hand-atlas*, Vienna, Pest, Leipzig 1889; *Universal-Handatlas*, Vienna, Pest & Leipzig 1891-1892; *Kleiner Hand-Atlas über alle Teile der Erde*, Vienna & Leipzig 1892 and editions to 1911; *A. Hartleben's kleiner Volks-Atlas*, 1896, 1911 (text by F. Umlauft); *A. Hartleben's Eisenbahn-Karte der Österreich-Ungarischen Monarchie*, Vienna and Liepzig, 1914.
• **Hartleben's Verlag** (*fl.* to *c.*1914). F. Umlauft's *Deutsche Runschau für Geographie und Statistik*, Vienna 1902.

**Hartley**, Sir Charles Augustus. Chief Engineer (later Consultant Engineer) for the European Danube Commission, in which capacity he prepared and commissioned numerous maps, plans and surveys of the river and its tributaries, *c.*1855-1871. *Delta of the Danube*, 1861 (lithographed by Kell Bros.).

**Hartley**, David (1732-1813). British peace commissioner. *United States boundaries sketched by B. Franklin and D. Hartley*, 1783; *Jefferson's proposed division of the Western Territory into new States*, Paris 1784; *United States east of the Mississippi River*, 1784.

**Hartley**, Henry. Assisted T. Baines with surveys for *A Map of the gold fields of South Eastern Africa...*, London, Edward Stanford, Port Elizabeth, J.W.C. Mackay 1876.

**Hartley**, Jesse. Civil engineer and dock surveyor at Liverpool. *Plan of the Liverpool Docks*, 1845 MS; The docks [Liverpool], 1848 (with John B. Hartley, lithographed by J.W. Allen).

**Hartley**, John B. Dock surveyor of Liverpool. Hull docks with proposals for new work (2 plans), 1843 (lithographed by G.P. Poore & Co.); The docks [Liverpool], 1848 (with Jesse Hartley, lithographed by J.W. Allen).

**Hartley**, William B. *Hartley's Map of Arizona from Official Documents*, New York 1863.

**Hartman**, J.W. Lithographer of San Francisco. Nathan Scholfield's *Map of Southern Oregon and Northern California*, San Francisco, Marvin & Hitchcock 1851.

**Hartmann**, A. Swiss engraver. *St. Gotthard-Strasse*, 1830.

**Hartmann**, C. *Karte vom Kreise Salzwedel*, Berlin 1878.

**Hartmann**, C.H. Publisher at Wolfenbüttel. *Australien*, 1824.

**Hartmann**, Carl Friedrich Alexander (1796-after 1831). German mining engineer. *Der treue Führer beim Schürfen...* (with atlas), Weimar, B.F. Voigt 1848; *Geographisch-statistische Beschreibung von Californien...* (2 volumes), Weimar, B.F. Voigt 1849; *Berg-und Hüttenmännischer Atlas...*, Weimar, B.F. Voigt 1860.

**Hartmann**, F. Draughtsman, used the earlier measurements of C.F. Gauss as a basis in the preparation of seven portfolios of maps of Hanover (about 180 sheets in total), 1853-1854 (with C. Tomforde).

**Hartmann**, Friedrich (*d*.1851). German military cartographer. *Post-Charte von dem Koenigreiche Hannover*, *c*.1840; *Umgebung von Lüneburg* (6 sheets), 1843; *Vogtei Auburg* (4 sheets), 1846.

**Hartmann**, Georg (1489-1564). Astronomer and instrument maker in Eggolsheim and Nuremberg. Possibly author of the 17-cm 'Ambassador's Globe' painted by Hans Holbein, before 1533 (the other candidate is Johann Schöner); 20-cm celestial globe gores, 1538; 8-cm terrestrial globe gores (copper engraving), 1547; 8-cm celestial globe gores, 1547.

**Hartmann**, *Doctor* Georg. *Karte des Cencessionsgebietes der South West Africa...*, Hamburg 1897; *Dr Georg Hartmann. Karte des nördlichen Teiles von Deutsch-Südwest-Africa... gezeichnet von Dr M. Groll* (6 sheets), London, South West Africa Co. 1904.

**Hartmann**, Georg Leonhard (1764-1828). Plan of St. Gall, Switzerland, 1809.

**Hartmann**, J.W. Road and waterworks inspector. *Übersichtskarte des Rheinstroms längs dem souverainen Fürstentums Liechtenstein*, 1842.

**Hartmann**, Jacob. *Geognostische Karte der Umgegend von Aschaffenburg* (2 sheets), *c*.1870.

**Hartmann**, Johan Georg Friedrich (1796-1834). German military cartographer.

**Hartsinck**, Johannes Jacob. *Beschrijving van Guiana of de Wildekust*, 1770.

**Hartt**, Charles Frederick (*b*.1840). American geologist. Explored the Amazon, 1865-1871.

**Hartung**, Georg. Geologist. Supplied information for J.M. Ziegler's *Physical Map of the Island of Madeira* (2 sheets), Winterthur & London 1855; *Tenerife geologisch topographische dargeestellt...*, Winterthur, J. Wurster & Co. 1867 (with Karl von Fritsch & W. Reiss); *Tenerife, nach Vorhandenen Materialien undeigenen...*, Winterthur *c*.1869 (with K. von Fritsch & W. Reiss).

**Hartwell**, *Lieutenant* S. One of the surveyors for *Map of the Siege of Vicksburg Miss.*, *c*.1863.

**Hartwig**, Eugen von. Cartographer from Glogau, now Poland. *Plan der Umgegend von Stargard* (6 sheets), Berlin, S. Schropp 1835; *Umgegend von Salzbrumn in Schlesien*, Glogau 1838 (with E. Vogel von Falkenstein); *Feurstenstein mit reinen nächsten Umgebungen...*, Berlin 1840 (with E. Vogel von Falkenstein).

**Harvey**, Arthur (1834-1905). Statistical clerk in the Finance Department, Quebec, Canada. *Map Illustrating the Course and Comparative Magnitude of the Principal Channels of the Grain Trade of the Lake Regions*, Montreal 1862 (lithographed by Duncan & Co.).

**Harvey**, Augustus F. Civil engineer and draughtsman. Nebraska gold mines, 1859 (with William E. Harvey); *Fort Kearney City* [Nebraska], St. Louis Mo. *c*.1860 (surveyed by J.H. Maxon); *Plat of Syracuse, Otoe Co.* [Nebraska], St. Louis, Mo. *c*.1860 (surveyed by W. Barnum); *Gold Region of Colorado*, 1862 (with William E. Harvey); *Nebraska as it is*, Lincoln, 'Statesman' Print Office 1869.

**Harvey**, *Captain* C.L. *Hunting Map of the Country Adjacent To Gibraltar Surveyed in 1873* (3 sheets), London, War Office 1874.

**Harvey**, *Lieutenant* Edward, RN. Plan of Vassava, 1777; *Plan of Maham. Surveyed in May, 1777 by Lieut. E. Harvey* [Mayham, Malabar Coast, Madras], London, A. Dalrymple 1781; made additions to J. Mascall's *Plan of the Harbour and Road of Suez*, London, A. Dalrymple 1782; *A Plane Chart of part of the Gulph of Persia, partly corrected by E. Harvey, 1778*, London, A. Dalrymple 1786.

*Georg Hartmann (1489-1564).*

**Harvey**, George C. Map of Wabash County, Illinois, Carmel 1888.

**Harvey**, I. *Plan of the Battle of Waterloo fought 18th June 1815*, 1834 MS.

**Harvey**, J. Publisher and bookseller, Sidmouth, Devon. Country round Beer, 1837 (in John Rattenbury's *Memoirs of a Smuggler...*).

**Harvey**, J.S., Quarter Master General's Office, Madras. Credited with the draughting of *Map of the Peninsula of India... Derived from the latest Surveys and other information collected by major F.H. Scott*, Madras 1855.

**Harvey**, Josiah (*fl.*1787-1814). In partnership with William Darton as engravers and publishers at 55 Gracechurch Street, London. **N.B.** See **Darton**, family and company.

**Harvey**, Samuel (*b.c.*1716, *fl.-c.*1785). Estate agent, auctioneer and surveyor in Essex. Land in the parish of Braxted, 1752 MS; estate survey, 1771; farm at Kelvedon, 1772; estates at Birch, 1773 MS; Bradwell Quay, 1774 MS; farm at Mayland, 1775 MS; estate at Stanway, 1785.

**Harvey**, T. *Plan of Salo Sound* [Sweden], 1811.

**Harvey**, T. (*fl.*1811-*c.*1850). Essex land surveyor of Ilford. Noak Hill, *c.*1850 MS.

**Harvey**, T. *School Atlas of Classical Geography*, Edinburgh & London, A.K. Johnston 1867 (with E. Worsley).

**Harvey**, William [*alias* 'Aleph'] (1796-1873). Author and journalist of Islington; compiler of cartoon maps and rhymes in *Geographical Fun: being humourous outlines of various countries*, London, Hodder & Stoughton 1869 (animated educational maps illustrating the political geography of Europe). **Ref.** SLOWTHER, C. *The Map Collector* 16 (1981) pp.48-50.

**Harvey**, William E. Civil engineer. Worked with Augustus F. Harvey on: Nebraska gold mines, 1859; *Gold Region of Colorado*, 1862.

**Harvie-Brown**, J.A. *Naturalists' Map of Scotland*, 1893.

**Harwar**, George. Mapmaker and mapseller recorded 'at his shop at the Long-cellar nere Hermitage Bridge' (*c*.1691). *A New & Exact Draught of the River Canada* [St. Lawrence], 1691, (the plate passed to John Thornton and was used by him in *The English Pilot. The Fourth Book*, 1689, 1698).

**Harwood**, —. *Crescent City, Iowa*, *c*.1863.

**Harwood**, C.E. *Map of Bennington County, Vermont*, 1856.

**Harwood**, T. Etcher for Samuel Lewis. Scotland, 1842.

**Harwood**, W. *A Topographical Plan of Modern Rome with the New Additions*, London, Williams & Norgate 1862.

**Harzheim**, Joseph von (1694-1763). German Jesuit and historian. *Mappa Chorographica Omnium Episcopatuum Germaniae* [map of Germany divided into Bishoprics], 1762.

**Has** [Hase], J.M. See **Haas**, Johann Matthias.

**Hase**, F. Engraver for Adolf Stieler, 1834.

**Hase**, W. Engraver for Adolf Stieler, 1857.

**Hasebroek**, J. *Paskaart... Oost Zee*, Amsterdam 1740.

**Haselberg** [Haselberg von Reichenau; Monteleporis], Johann [Johannes]. Worked on Christoph Zell's *Des Turckischen Keysers Heerzug und vornem widder die Christen...*, Nuremberg 1530.

**Haselden**, T. *Mapp of the Known World*, London, T. Page & W. Mount 1722.

**Hasell** [Hassell], *Captain* John. Mahé Island [Seychelles], 1778.

**Hasemann**, Henning. Wolfenbüttel, 1628.

**Hasenbanck**, Johann Otto (*d*.1759). Map of the mouth of the river Elbe, 1721 (with S.G. Zimmermann).

**Hasenfratz**, K. See **Dasypodius**, Conrad.

**Hashimoto** Sadahide [Gyokuransai] (1807-*c*.1879). Japanese *ukiyoe* artist and mapmaker. *Ezo Kokyo yochi zenzu* [map of Ezo], 1854 (from the surveys of Fujita Junsai); Atlas of Japan, including a map of Fuji, 1855; *Tōkaidō Gojūsan Eki Shōkei* [highroad map - part of a proposed set of bird's-eye views of the Pacific Coast], published Okadaya Kashichi (and others) 1860; *Gokaiko Yokohama oezu* [Map of the treaty port of Yokohama], 1860; worked with Tsurumine Hikoichirō to produce a provincial map of Shimotsuke (coloured, woodblock), published by Kinkōdō and Kikuya Kōzaburō 1862; plan of Sendai city, published by Matsubayashi Rōtōken and Hōshūdō 1868; *Mutsu Dewa kokugun kotei zenzu, fu Echigo hangoku* [Map of Mutsu, Dewa and a part of Echigo province], [*n*.*d*.]. **Ref.** YAMASHITA Kazumasa *Japanese maps of the Edo Period* (Japan 1998) pp.28, 79, 163, 190-191.

**Hashimoto** Sokichi [Naomasa] (1763-1836). Japanese scholar. *Oranda shin'yaku chikyu zenzu* [world map], 1791 (translated from a Dutch map).

**Hashimoto** Tokuhei. Japanese cartographer. *Fujimi On-Edo Ezu* [plan of Edo with a view of Mount Fuji], published Shimuraya Yohachi and Edoya Kichiemon, 1818.

**Hasius**. See **Haas**, Johann Matthias.

**Haskel**, Daniel (1784-1848). Established the New York School Apparatus Co., 1830. 2.5-inch, 3-inch and 5-inch diameter terrestrial globes, *c*.1834. **Ref.** WARNER, D. *Rittenhouse* Vol.2 N°2 (1987) pp.93.

**Haskell**, Brothers. Map publishers of 122 Lake Street, Chicago. *Haskell Brothers County, Railroad Station, Post Office and Index Map of Michigan*, 1880.

**Haskins**, William (1828-1896). Provincial Land Surveyor, Canada. Undertook surveys of new settlements and laid out building plots in the townships of Eden Mills, Everton, Fergus, Guelph and Hamilton, 1855-1866, including: *Tyrcathlen the Property of the*

*Revd. Arthur Palmer in the Town of Guelph Laid out into Building Lots*, 1855; *Plan of Lots in... The Town of Guelph*, 1856.

**Haskoll**, William. Printseller and publisher in the Churchyard, Winchester. F.W. Bauer's *Plan of the City of Winchester*, 1756.

**Haslett**, John James (1811-1878). Provincial Land Surveyor, Canada. Undertook numerous property and building plot surveys in the townships of Seymour, Monteagle, Belleville, Brighton, Campbellford, Cannifton, Consecon, Marmora, Rossmore, Stirling, Trenton, etc., 1844-1863, including: *Plan of the Town of Belleville County of Hastings, and District of Victoria*, 1845; *Plan of the Eastern part of a line of road and exploration extending from the Bathurst District to the Home District*, 1848; *Plan of Consecon Village*, 1853; *Plan of Islands in the River Otanabee and its lakes*, 1855; *Plan of Cape Vesey Ordnance Reserve in the Township of Marysburgh*, 1859; *Plan of the Hastings Road...*, 1863.

**Haslingen**, Heinrich Tobias *Freiherr* von. (1649-1716). Austrian General in charge of the mapping of parts of Romania and Serbia from 1682 onwards.

**Haslop** [Hasselop; Hassellup], Henry. Printer and bookseller of London, Freeman of the Stationers' Company. *Mariner's Mirrour*, 1588 (entered in the Registers of the Stationers' Company on April 3rd 1587, English version of Lucas Waghenaer's *Spieghel der Zeevaert* charts, translated and edited by Anthony Ashley, charts engraved by T. de Bry, J. Hondius, J. Rutlinger and A. Ryther, printed by John Charlewood).

**Hass**. See **Haas**.

**Hassard**, *Lieutenant* (later *Lieutenant Colonel*) Fairfax Charles RE. Plans of existing and proposed military installations at various sites in the Ionian Islands, 1844-1847; *Plan of part of Toronto Shewing the Sites of the several Barracks, Offices &c.*, 1867 (Hassard also endorsed several other military plans and surveys in Canada).

**Hassall**, Charles (1754-1814). Surveyor of Eastwood, produced manuscript maps of parts of Wales and western England. *The Road from New Port of Milford to the New Passage of the Severn and Gloucester*, J. Cary 1792 (with John Williams); Crown lands in Monmouthshire, 1803 MS.

**Hasse**, Ernst. Compiled some of the maps for R. Andree and O. Peschel's *Physikalisch-Statistischer Atlas des Deutschen Reichs* (25 maps), Bielefeld and Leipzig, Velhagen & Klasing 1878.

**Hassel**, G. *Vollständiges Handbuch der neuesten Erdbeschreibung*, Weimar 1819 and later (with A. Gaspari, J. Cannabich, J. GutsMuths and F. Ukert).

**Hassel**, James. Survey Carteret Grant, 1746.

**Hassel**, Johann Georg Heinrich (1770-1829). German statistician and geographer. One of the editors of the geographical journal *Neue Allgemeine Geographische und Statistische Ephemeriden*, published 1822-1831 (contained many maps); *Atlas der Staaten des Deutschen Bundes*, Weimar 1824-1829.

**Hassell**, *Captain* John. See **Hasell**, *Captain* John.

**Hassellup** [Hasselop], H. See **Haslop**, Henry.

**Hasselt**, W.J.C. van. Dutch surveyor. Haarlemmermeer, 1838.

**Hassendeubel**, *Captain* F. *Plan of Santa Cruz de Rosales & of the operations of the U.S. troops... during the siege and storming... 1848* [Mexico], Philadelphia, P.S. Duval's lith. 1848.

**Hassendō** Shujin. Japanese mapmaker. *Nagasaki Saiken Zu* [plan of Nagasaki], published by Hayashi Jizaemon and Takehara Kobei after 1745 (possibly a later printing of an earlier work).

**Hassenstein**, Bruno (1839-1902). Cartographer of Berlin, worked for Augustus Hermann Petermann. Population map of Siebenbürgen [Romania] for *Petermanns Mitteilungen*, 1857; Süd-Ost Australien, 1861; *Originalkarte der Nord-Abessinischen Grenzlande...*, Gotha, Justus Perthes 1864 (designed by A. Petermann); Deutsche Colonie Rio Grande do Sul [Brazil], 1867; between 1871-1879 he compiled some of the maps for the *Spruner-Menke Hand-Atlas*,

Gotha, Justus Perthes 1880 (33rd edition of Dr K. von Spruner's *Hand-Atlas*); *Die Deutschen Besitzungen in West-Afrika*, Gotha, Justus Perthes 1884; Die Deutschen Besitzungen in West-Polynesien, 1885; *Atlas von Japan*, 1885-1887; Die Adamsbrück, 1891; Buganda, 1891; and others.

**Hassert**, *Doctor* Kurt (1868-l947). Maps for Petermann's *Geographische Mittheilungen* including: *PolarKarte*, Gotha 1891; *Geologische Ubersichtskarte von Montenegro*, Gotha 1895; *Hydrographische Karte von Montenegro*, Gotha 1895; *Pflanzengeographische Karte von Montenegro*, Gotha 1895.

**Hassinger**, Hugo (1877-1952). Professor of geography at the University of Vienna. Supervised the publication of *Burgenland* (118 maps), Vienna 1941 (an atlas of the frontier land transferred from Hungary to Austria in 1921).

**Hassler**, father and son. Both worked on the United States Coast Survey.
• **Hassler**, *Professor* Ferdinand Rudolph (1770-1843). Swiss mathematician and geodesist, born in Aarau, worked also in Paris and Bern. Emigrated to the United States in 1805, and became a teacher of mathematics at West Point, he returned there as professor 1818-1832. Hassler was selected to lead the new U.S. Coast Survey in 1807. He was confirmed as Superintendent of the United States Coast and Geodetic Survey in 1816, and began with a triangulation of parts of New Jersey from 1817. He was assisted by his son (below), and by Edmund Blunt, John Abert and others. From 1818-1820 Hassler worked as Chief Astronomer to the survey of the boundary between the U.S.A. and Canada. Triangluation of Aarau, 1791-1797 (with J.G. Tralles); *Map of New York Bay and Harbor and the Environs* (6 sheets), 1843-1845 (triangulation by J. Ferguson and E. Blunt, topography by C. Renard, T.A. Jenkins and B.F. Sands, hydrography engraved by F. Dankworth and J. Knight, topography engraved by S. Siebert and A. Rolle, views engraved by O.A. Lawson), reduced one sheet version published 1845. **Ref.** ALLEN, D.Y. *Long Island maps and their makers* (1997) pp.72-82; COHEN, P.E. & AUGUSTYN, R.T. *Manhattan in maps 1527-1995* (1997) pp.122-123; HACKLER, D.L. *Mercator's World* l, 1 (1996) pp.84-88; GUTHORN, P.J. *United States coastal charts 1783-1861* (1984) pp.17-19; GUTHORN, P. *The Map Collector* 27 (1984) pp.28, 30-31.

•• **Hassler**, J.J.S. (*c*.1799-1858). Son of Ferdinand. One of many topographical assistants with the U.S. Coast Survey. Worked with his father on the triangulation of New Jersey, *c*.1817; with others on *Map of Delaware Bay and River*, published 1848 (for F.R. Hassler).

**Hasted**, Edward (1732-1812). Historian of Kent. *The History and Topographical Survey of the County of Kent* (4 volumes, folio), 1778-1799, 2nd edition (12 volumes, octavo) 1797-1801.

**Hastings**, —. Worked in partnership as a civil engineer under the name Brown & Hastings. Draughtsmen for *Map of New England exhibiting the rail roads & telegraphic lines now in operation*, Boston Almanac 1850.

**Hastings**, E.J. *The Statistical Atlas of Commercial Geography*, Edinburgh & London, W. & A.K. Johnston *c*.1887.

**Hastings**, *Commander* G.F., *RN. Nimrod Sound* [China], 1843 published Admiralty 1844.

**Hastings**, William. *Belfast*, 1864.

**Haswell**, Thomas. Chart of Salt River, Malabar coast and Oyster Rocks in Carwar Bay, 1790, published Dalrymple 1792; *Views on the Malabar Coast*, London [n.d.]; *Views in the Vicinity of Goa*, London 1799.

**Haszard**, George T. Publisher of Charlottetown. *Prince Edward Island in the Gulf of St. Lawrence compiled from the latest surveys by H.J. Cundall 1851*, 1862 (engraved by W.H. Lizars, also published in Edinburgh, Oliver & Boyd, in Liverpool, Wilmer & Smith, in London, S. Bagster & Son).

**Hata** Awagimaru [Murakami, Shimanojo] (*c*.1762-1807). Japanese explorer and cartographer. *Izu yochizu* [Map of Izu province], 1832; *Awa-no-kuni zu* [Map of Awa province], 1854; *Kamakura shogai zu* [Guide map of Kamakura], [*n.d.*].

**Hatch**, Charles. Constructed a skeleton 'nautical globe' designed to illustrate the theories of navigation, 1854 (National Maritime Museum).

*Professor Ferdinand Rudolf Hassler (1770-1843). (By courtesy of Cartographic Associates, Fulton, Maryland, USA)*

**Hatch**, Frederick H.(1864-1932). Geologist who joined the Geological Survey of Great Britain in 1886; instructor in geology at the Royal Geographical Society and worked as mining engineer in Johannesburg. *Map of the Transvaal showing the Physical Features and Political Divisions*, 1897, 1902; *A Geological Map of the Southern Transvaal*, Edward Stanford 1903.

**Hatch & Co.** Lithographers of 29 William Street, New York. *Map of the Lake Region & St. Lawrence Valley*, 1863.

**Hatchard**, father & son.
- **Hatchard**, John (1769-1849). Publisher and bookseller at 173 Piccadilly (1797); 190 Piccadilly (1801); 187 Piccadilly (1820 to date). John Hatchard was joined in business by his son in 1808. Publisher of William Green's *The Picture of England Illustrated* (2 volumes), London 1803, 1804 (maps based on J. Cary, printed by Robert Butters); *Lockie's Topography of London*, 1810 (with G. & W. Nichol and W. Miller); A. Arrowsmith's *Carte du Sud de Norwége*, 1813; J. Wyld's *Chart of the World shewing the Religion, Population and Civilization of each country*, 1815.
- • **Hatchard & Son** (*fl.* from 1808). Booksellers of Piccadilly. Sellers of *Pigot & Co.'s British Atlas*, London, Pigot & Slater 1843; *I. Slater's New British Atlas*, London & Manchester, Isaac Slater 1846, 1857.

**Hatchard**, T. *Diocese of Ruperts Land* [Labrador], London, T. Hatchard 1849.

**Hatchett**, J. One of the engravers for George Augustus Walpoole's *New British Traveller*, 1784; for William Faden, 1798; *A Chart of the Indian Ocean*, William Faden 1803.

**Hathaway**, Joshua. *Chicago*, 1892.

**Hathon**, A.E. *Map of the City of Detroit*, 1849.

**Hátsek** [Hatschek], Ignác (1828-1902). Hungarian engineer and cartographer. b. Vágújhely, d.Budapest. Made a large number of administrative and diocesan maps of Hungary from 1867 onwards; *Az Osztrák Magyar Monarchia...*, Budapest *c.*1877; *A Magyar Korona Országainak...*, Budapest 1877; county atlas of Hungary (81 maps), 1880; *Oesterreich-Ungarn*, Gotha 1884; *Ethnologische Karte Ungarischen Krona*, 1885.

**Hatter**, F.X. Engraver for J.G. Tulla's *Charte über das Grossherzogthum Baden...*, Karlsruhe, C.F. Müller *c.*1814.

**Hattin**, H.S. Maps for H. Clay and A. Greenwood's *An introductory atlas of international relations* (47 maps), London, Headley Bros. 1916.

**Hattinga**, family (*fl.c.*1740-1790). Dutch map making family of the 18th century. Compiled many maps in connection with the War of Austrian Succession, 1743-1748 and beyond including *Atlas of the Parts of Flanders and Brabant under the rule of the Republic* (8 volumes); *Atlas van Zeeland* (4 volumes), 1744-1752; *Atlas of the Frontiers in the East and North* (3 volumes). **Ref.** DE VRIES, D. 'Official cartography in The Netherlands' in Cicle de conferències sobre Història de la Cartografia - 4rt curs La cartografia dels Països Baixos (1995) pp.41-42; AARDOOM, L. Caert-Thresoor 9:4 (1990) pp.66-72 [in Dutch]; KRETSCHMER, DÖRFLINGER & WAWRIK Lexicon zur Geschitchte der Kartographie (1986) vol.C/I, p.288 (numerous references cited) [in German].
- • **Hattinga**, Willem Tiberius (1700-1764). Army physician and cartographer (initially amateur), 'burgemeester' of Hulster Ambacht, father of David Willem Carel and Anthony. Compiled an atlas of Zeeland, with regional maps and plans of fortifications, 1740s (with David).
- •• **Hattinga**, David Willem Carel [Coutry] (1730-1790). Dutch engineer and cartographer, son of Willem Tiberius, worked mostly in Vlaanderen. Zeeland, 1753 (with Anthony Hattinga); Plan der geprojecteerde nieuwe Brandwagt of Speelbattery, 1757 MS.
- •• **Hattinga**, Anthony (1731-1788). Military engineer and cartographer, son of Willem Tiberius, worked in Zeeland and Vlaanderen. Walcheren, 1750 (with D.W.C. Hattinga); Shouwen en Duiveland, 1753 (with D.W.C. Hattinga); Noordbeveland, Wolphartsdijk en Oost-Beveland, 1753 (with D.W.C. Hattinga); Tholen, 1753 (with W.T. Hattinga); Zuidbeveland, 1753 (with W.T. Hattinga), all published Amsterdam, Isaak Tirion; Zeeland (5 sheets), Amsterdam, Isaak Tirion 1753 (with D.W.C Hattinga).

**Hatton**, Frank. *Map of British North Borneo*, *c.*1880s.

**Hatton**, T. (*fl.*1750-1768). Surveyor, produced estate maps in eastern England. Lands at

Fingrinhoe Hall [Essex], c.1750 MS; *A Survey of Wait's Farm in the parish of Wickham and County of Kent...*, 1753; Mildenhall, Suffolk, 1768.

**Hauber**, Eberhard David (1695-1765). Geographer and Lutheran theologian from Hohenhaslach, Württemberg, died in Copenhagen. Worked for Homann's Heirs and the Bodenehr family. *Memmingen mit dero Gegend*, c.1720; *Atlas Würtembergicus*, 1723; author of a book on the history of cartography entitled *Versuch einer umständlichen Historie der Land-Charten*, Ulm 1724. **Ref.** KRETSCHMER, DÖRFLINGER & WAWRIK Lexicon zur Geschichte der Kartographie (1986) vol.C/1, p.288 [in German].

**Haubois**, Egberg (fl.c.1641-c.1653). Dutch cartographer, and 'stadsbouwmeester' at Groningen. Best known for large plan of Groningen entitled *Caerte van de vermaerde ende antique stad Groningen...*, 1652 (engraved by Jan Lubberts Langeweerd).

**Haubold**, G. *Schleswig, Holstein*, 1859; engraver for Heinrich C. Kiepert, 1869; *Africa*, c.1870 for Adolph Gräf.

**Haubold**, Georg (1846-1881). Map lithographer.

**Haubold**, O. Engraver for Joseph Meyer, 1867; engraved some plates in L. Ravenstein's revised edition of *Meyer's Hand-Atlas der Neuesten Erdbeschreibung*, Hildburghausen, Verlag des Bibliographischen Instituts 1872.

**Haubold**, O., *Junior*. Engraver for Heinrich C. Kiepert, 1869.

**Hauchecorne**, G. 'Agent général de chemins de fer'. Railway maps published in Brussels by Philippe Marie Guillaume Vandermaelen. *Carte des Chemins de fer d'Allemagne et des pays limitrophes*, 1862, 1864; *Chemins de fer de l'Europe* (9 sheets), 1863.

**Hauchecorne**, Wilhelm (1828-1900). Worked with E. Beyrich on the first stages of a geological map of Europe. The map was completed by F. Beyschlag and published as *Geologische Karte von Europa* (49 sheets), Berlin, D. Reimer 1881-1913, also published as *Carte géologique internationale de l'Europe*, 1894-1913.

**Haucke**, Walter. Draughtsman for W. Goering's *Topogr. Karte v. Jerusalem und Umgebung*, Gütersloh, C. Bertelsmann 1929 (for G. Dalaman's *Jerusalem v. sein Gëlande*).

**Hauducoeur**, C.P. Engineer. *A Map of the Head of Chesapeake Bay and the Susquehanna River...*, 1799 (including inset *Plan of the town of Havre de Grace*).

**Hauer**, Daniel Adam (b.1734). Engraver in Nuremberg, worked for Homann's Heirs. *Belgii Universi*, 1748; *Mare Mediterranum*, 1770; *Regni Bohemiae*, 1770; *Sevilla Regnum*, 1781; one of several engravers for Johann Wilhelm Abraham Jaeger's *Grand Atlas d'Allemagne en LXXXI feuilles*, 1789.

**Hauer**, Franz, *Ritter* von (1822-1899). Austrian geologist. *Geologische Übersichts-Karte von Siebenbürgen...*, 1861; *Geologische Ubersichtskarte der Oesterreich-Ungarischen Monarchie* (11 sheets), Vienna 1867-1871; *Geologische Karte von Oesterreich-Ungarn mit Bosnien u. Montenegro*, Vienna, A. Hölder 1884.

**Hauer**, Jacob. Lithographer and printer of Toronto. T.W. Walsh's *Map of the Talbot District*, Toronto, Scobie & Balfour 1846; Donald McDonald's *Map of the Huron District and of the Townships of Bosanquet in the Western and Williams in the London Districts*, Toronto, Scobie & Balfour 1846; Charles Rankin's *Map of the Western District Canada*, Toronto, Scobie & Balfour 1847; W. Billyard and R. Parr's *Map of the Western District in the Province of Canada*, Toronto, Scobie & Balfour 1847; S.A. Fleming's *Plan of the Town of Cobourg*, Toronto, Scobie & Balfour 1847.

**Hauer**, Johann (1586-1660). Cartographer and engraver. Silver globe, 1620 (now in Royal Library, Stockholm, facsimiles made by Adolf Erik Nordenskiöld in 1890s).

**Haug**, C. *SpecialKarte der Kreise Meseritz...*, Meseritz 1901.

**Haug**, G.F. Karte von Palästina (with inset plan of Jerusalem), Stuttgart 1825.

**Haug**, J. *Übersichts-Karte über sämmtliche Stadtund Stiftungswaldungen van Leutkirch* [Germany, forestry map], Leutkirch 1882.

**Haug**, Johann Friedrich Gottlob. *c*.30-cm terrestrial globe, Stuttgart 1797.

**Hauke**, *Lieutenant* Maurice. Polish Artillery. Plan of Mantoue [Mantua, Italy], 1800.

**Haumann**, J.E. Carte de la République des Suisses, 1777.

**Haupt**, Ernst. *Eulen-Gebirge*, Glogau 1855; *E. Haupt Karte des Riesengebirges...*, Glogau *c*.1878; *Das Reisengebirge*, Glogau, C. Flemming *c*.1890 (with F. Handtke).

**Haupt**, G. Engraver for C.L. de Launay, 1738.

**Haupt**, Gottfried Jacob. Engraver and publisher in Augsburg. *Regnum Bohemiae, c*.1730; *Danubii Fluminis, c*.1740; *Rhenus, c*.1740; *Tartariae Europae, c*.1740; *Nova et accurata Charta Archiducatus Austriaci*, Augsburg *c*.1740; *Bavariae Circulus*, after 1742.

**Hauptmann**, Ernst. *E. Hauptmann's Wegweiser durch Leipzig und Umgegend mit Berücksichtigung aller Denkwürdigkeiten des Schlachtfeldes*, Leipzig 1863.

**Hauptmann**, H.S.H. *Charte von dem Herzogthums Gotha...*, Weimar, 'Im Verlag des Geog. Instituts' 1812 (with F.W. Streit & H. von Rhein).

**Haus**, Anton (1851-1917). *Atlas... Ozeanographie und maritime Meteorologie*, Wien 1891.

**Hausch**, F. *Europäische Turkei, Griechenland*, 1876.

**Hauser**, M. *Karte von Schlesien, c*.1869.

**Hauser**, S.T. Publisher of St. Louis. Walter Washington De Lacy's *Map of the Territory of Montana* [showing gold and silver finds], 1865.

**Hausermann**, —. *Plan de Dijon*, 1875; *Ville et Port de Brest, c*.1875.

**Hausermann**, R. Draughtsman and engraver of Paris. J. Bonnat's *Carte Approximative du Volta depuis son Embouchure jusqu'à Yegiy*, 1875; *Carte ethnographique et orohydrographique de la Turquie d'Europe...*, Paris, A. Lassailly *c*.1876 (printed by Becquet); J. Bonnat's *Carte des Concessions de The African Gold Coast Company et des voies de communication...*, Paris 1879 (printed by Becquet); maps for *Atlas National*, Paris, A. Fayard, 1880, 1885, Paris, Fayard Freres 1900; *Carte des chemins de fer de la France au 31 décembre 1880*, 1881; *Egypte*, Paris, Societé Bibliographique 1882 (with A. Simon); *Possessions anglaises et françaises Golfe de Guinée*, Paris 1883; *Tunisie et Algérie Orientale*, 1883; Père Roblet's *Carte de Madagascar* (3 sheets), Paris, H. Lecene & H. Oudin *c*.1888; *France et Algérie - Tunisie*, Paris, Augustin Challamel 1891 (printed by Lemercier & Co., published in *Nouvel Atlas des Colonies Françaises...*); P. Vuillot's *Carte du Sahara et du Nord-Ouest de l'Afrique de la Méditerranée au Sénégal et au Lac Tchad*, Paris, A. Challamel 1894 (printed by Vieillemard et Fils); Monsignor Guillon's *Carte de Mandchourie Méridionale (Leo Tong ou Province de Moukden)*, Paris, Adrien Launay, des Missions Etrangères 1894; P. Vuillot's *Soudan Français et Côte Occidentale d'Afrique*, Paris 1897 (printed by Dufrenoy); *Carte des Missions Catholiques du Siam, de la Birmanie et du Laos*, Lyon, Missions Catholiques 1904; *Carte des Missions Catholique en Afrique*, Lyon, Missions Catholiques 1905; *Carte des Missions Catholique en Australie*, Lyon, Bureau des Missions Catholique 1906; *Carte de l'Inde Ecclesiastique*, Lyon, Les Missions Catholiques 1907; *Carte de l'Amérique du Sud Ecclesiastique*, Paris 1908; *Carte générale de l'Afrique equitoriale française*, Paris, A. Challamel 1910-1913; *L'Eglise catholique dans les Isles Britanniques*, Paris 1910; *L'Eglise Catholique en Chine*, Paris 1912-1913; *L'Église Catholique dans les Balkans*, Paris 1911; *Brasil Segundo*, Paris, Aillaud, Alves e Cie. 1915; M. Dubois and J.G. Kergomard's *La Guerre en Orient*, Paris, A. Challamel 1915 (printed by Monrocq); and many others.

**Hauslab**, Franz *Ritter* von (1798-1883). Austrian cartographer, associated with the 'Militär-Ingenieurakademie' in Vienna. Owner of the so-called Hauslab (or Brixener) globe [37-cm diameter, *c*.1523]. Pioneer of colour sequence contour maps which were later developed by A. Steinhauser. Hauslab's method was first tested on a map of Turkey, 1830 MS, published Lyon 1914; *General-*

*Karte des Herzogthums Steyermark* (12 sheets), 1831; school atlas, 1864-1868; *Orographische Skizze des Kaiserthumes Mexico*, 1864; *Hypsometrische Ubersichtskarte der Norischen Alpen*, Vienna, Artaria & Co. 1865; *Hypsometrische Uibersichts-Karte von Bosnien, der Hercegovina, von Serbien und Montenegro* (4 sheets), 1876. **Ref.** KRETSCHMER, I. Imago Mundi 40 (1988) pp.10-11; WALLIS & ROBINSON Cartographical innovations (1987) 3.081 p.146; ibid. 5.121 p.229; ibid. 7.031 p.290; KRETSCHMER, DÖRFLINGER & WAWRIK Lexikon zur Geschichte der Kartographie (1986) vol.C/1, p.289 [in German].

**Haussard** sisters. Engravers of the 'rue de Plâtre', Paris. In addition to engraving the bird illustrations for G. Buffon's *Histoire Naturelle*, between them they engraved at least 37 sheets of Gilles and Didier Robert de Vaugondy's *Atlas Universel*, 1757 (including many of the decorative cartouches designed by P.P. Choffard).

• **Haussard**, Elisabeth. Engraved at least 20 sheets for the Robert de Vaugondy *Atlas*, 1749-1757, including *Gouvernements Généraux du Berry, du Nivernois, et du Bourbonnois*, 1753; *Antiqua Imperia*, 1753; *Le Royaume d'Angleterre*, 1753; *Carte de la Virginie et du Maryland Dressée sur la grande carte Angloise de Mrs. Josué Fry et Pierre Jefferson*, Paris, Robert de Vaugondy 1755; *Germania Antiqua*, 1756; *Asia Minor*, 1756; for Jacques-Nicolas Bellin, 1759; and many others.

• **Haussard**, Marie Catherine. Sister of Elisabeth, also engraver of at least 7 sheets for the Robert de Vaugondy *Atlas*. *Græcia Vetus...*, 1752; *Turquie Européenne...*, 1755; *Partie de l'Amérique Septentrionale*, 1755; *Carte de la Lorraine et du barrois...*, 1756; *Partie meridionale du Cercle de Haute Saxe...*, 1756; *Carte des Grandes Routes d'Angleterre...*, 1757; *Carte du Royaume de France*, 1758; and others.

**Haussding**, Charles. *Map of Palermo, Doniphan County, Kansas*, c.1858.

**Hausse**, E. *Carte de la Haute Californie ou Nouvelle Californie*, Paris 1850.

**Hausser**, Wolfgang. Steyr, 1584.

**Haussknecht**, *Professor* C. *Routen im Orient... von H. Kiepert*, Berlin 1882.

**Haussmann**, *Baron* George Eugène (1809-1891). Administrator in the Seine 'département' of France, 1853-1870. Contributed to *Recherches statistiques sur la Ville de Paris et le département de la Seine* (6 volumes with maps), Paris, Imprimerie royale c.1853-1860 (the series was published from 1826); *Atlas Souterrain de... Paris*, 1855; *Atlas... de la ville de Paris*, 1868.

**Hausted** [Hawstead; Hostead], Daniel (*fl*.1633-1651). English surveyor, produced manuscript drainage and estate maps in the eastern Midlands. Map of lands drained by Sir Cornelius Vermuyden [sites in Lincolnshire, Yorkshire and Nottinghamshire], 1633 MS (with R.Smith and D. Pierde); estates in Brigstock and Benefield [Northamptonshire], 1634 MS (with H. Taylor).

**Hausted** [Hawsted], John (*fl*.1610-1648). Estate and inclosure surveyor of Northamptonshire. Lands at Geddington and Newton [Northamptonshire], 1610 MS (with Thomas Thorpe).

**Hautt**, father and son. Engravers in Switzerland.

• **Hautt**, David Nikolaus (1603-1677). Engraver, printer and bookseller from Strasbourg, worked in Lucerne, Vienna and Constance. *Icon. Totius Sveviae*, 1636; *Karte der Schweiz*, 1641. **Ref.** BLASER, F. Les Hautt. Histoire d'une famille d'imprimeurs, d'éditeurs et de relieurs des XVIIe et XVIIIe siècles Lucerne (1925) [in French].

•• **Hautt**, Nikolaus. Cartographer, engraver and printer of Constance, son of David. 'Terraquei orbis typus', in Daniello Bartoli's *Geographia moralibus et politicis...*, Constance 1674; *Lacus Acronianus*, 1675.

**Hauttecourt**, F. de. *Whitesand Bay* [Land's End], 1702 (with T. Tuttell).

**Haüy**, Joseph de. Austrian military cartographer of French descent. Mapped the environs of Pécs, 1686; the area of Siklós, 1687; plan of the Castle of Buda, 1687.

**Have**, J.J. ten. Dutch geographer and schoolmaster at The Hague. Compiler of geography textbooks, wall charts and atlases. *Atlas van Nederland* (12 maps), The Hague, Johannes Ykema 1891, 1905, 1909; *Volledige School-Atlas*, The Hague, J. Ykema 1897 (printed by

P.W.M. Trap), and 8 further editions to 1931; *Handelsatlas*, The Hague 1912 (lithographed by J. Smulders & Co.); *Geïllustreerde Atlas van Nederland en Oost-Indië* (22 maps), The Hague, J. Ykema c.1898, 1908; *Geïllustreerde Atlas voor de Lagere School* (42 maps), The Hague, J. Ykema c.1898, 1909, (44 maps), 1918; *Atlas van Europa* (17 maps), 1899, The Hague, J. Ykema 1912; *Blinde Atlas der Aarde* (36 maps), The Hague, J. Ykema 1908 (lithographed by J. Smulders & Co.), 1911, 1915, 1919, (37 maps), 1922, (40 maps), 1924, (42 maps), 1928, 1935.

**Have**, Nicolaas ten (d.c.1650). Teacher at Zwolle, also worked as a cartographer at Overijssel (1640-c.1650). His maps were included in Amsterdam composite atlases, eg. Covens and Mortier, 1730s and later. *Transisalania Provincia; vulgo Over-IJssel* (4 sheets), 1650 (second edition 1743), reduced and used by Joannes Janssonius, 1658 and Joan Blaeu, 1662.

**Havell**, Daniel. Engraver and aquatint artist, produced numerous plates for illustrated travel and colour-plate books. Engraved vignettes for Richard Grey Baker's *Map of the County of Cambridge, and Isle of Ely*, 1821 (the vignettes were designed by R.B. Harraden).

**Haven**. See **Delafield** and **Haven**.

**Haven**, E. de. See **De Haven**.

**Haven**, George. Nicaragua, 1856.

**Haven**, John. Publisher of 3 Broad St., New York. Oregon, Texas & California, 1846.
• **Haven & Emmerson**. Publisher of New York. *Map of the United States*, Haven & Emmerson 1846.

**Haven**, John. Publisher, 86 State St., Boston. David H. Burr's World, 1850.

**Havenga**, W.J. 'Chef over den Topographischen dienst in Nederlandsch-Indië, Batavia' [Jakarta]. *Atlas van Nederlandsch Oost-Indië*, Batavia, G. Kolff en Co. 1885; *Eiland Sumatra*, 1886; Java en Madoera, 1888.

**Haver Droeze**, F.J. *Kaart der Bataklanden en van het eiland Nijas*, Batavia 1890 (with E.W. Hedemann).

**Havergal**, *Lieutenant* Arthur *RN*. Assisted with surveys of the Dardanelles, 1872; and Gibraltar, 1875, 1885; surveys of the northeast coast of South America between the Rivers Essequibo and Moruka, 1887, published Admiralty 1888 (engraved by J. & C. Walker); *St. Lucia* (2 sheets), surveyed 1888, published Admiralty 1889 (engraved by Edward Weller).

**Haviland**, A. *The Geographical distribution of Heart Disease in England and Wales*, Edinburgh & London, W. & A.K. Johnston 1871.

**Haviland**, *Reverend* F.T. *Hauc quam videtis Terrarum Orbis Tabulam...* [facsimile of the Richard of Haldingham world map], 1869 published London, E. Stanford 1872.

**Haviland**, *Doctor* G.D. Trusan River, 1885 MS.

**Haviland**, W. One of the engravers for H.S. Tanner's partwork *A New and Elegant Universal Atlas*, 1833-1836 (published as *A New Universal Atlas*, 1836, 1839, Carey & Hart 1842, 1843).

**Havilland**, Th. de. Lieutenant 55th Regiment. Chusan, 1857 (with Lieutenant E.W. Sargent).

**Haward**, Nicholas. See **Hayward**, Nicholas.

**Hawerbault**, F. See **Horenbault**, François.

**Hawes**, *Commander* E. *Plan de Port Natal*, 1831.

**Hawes**, J.H. Draughtsman to General Land Office 1865.

**Hawes**, Lacy [L. Hawes & Co.]. Bookseller of London. Apprenticed to Charles Hitch then became his partner c.1752. Hawes bought Hitch's shares in *Chorographia Britanniae* and *Britannia* in 1765, after the latter's death. One of the sellers (with C. Hitch) of W.H. Toms's *Chorographia Britanniae*, c.1752-c.1765; one of many sellers of Gibson's edition of *Britannia*, 1772 (the last edition with maps by Morden).

**Hawich**, C. *Plan der Stadt Trier*, 1823.

*John Hawkesworth [Hawksworth] (c.1715-1773). Engraved by Thornton. (By courtesy of the National Library of Australia)*

**Hawk**, J. Siege of The Havana, 1762 MS.

**Hawkal**, Abu al-Kasim Muhammad ibn (*d.c.*977). Classical muslim geographer of the balkhi school. *Kitab surat al-ard* [Book of the picture of the earth]. Ref. TIBBETTS, G.R. 'The Balkhi school of geographers' in HARLEY & WOODWARD (Eds.) Vol.2 book I (1992) *Cartography in the traditional Islamic and south Asian societies* pp.108-136.

**Hawkes**, Thomas I & II (*fl*.1820-1850+). Surveyors of Williton, Somerset (the work of the two individuals cannot be differentiated). Made manuscript maps of parts of Somerset. *Plan of the Parish of Glastonbury in the County of Somerset*, London, Standidge & Co. 1844.

**Hawkes**, William. Publisher of N$^{o.}$59 Holborn Hill. Apprenticed to Thomas Kitchin in 1769, and later succeeded him. *Twenty Five Miles Round New York*, 1776 (engraved by John Barber).

**Hawkesworth** [Hawksworth], John (*c.*1715-1773). Doctor of law. Writer and editor of *The Adventurer*. He also compiled an account of voyages entitled *An Account of the voyages... for making discoveries in the Southern Hemisphere* (3 volumes; 17 charts), London, W. Strahan and T. Cadell 1773 (included descriptions of voyages by J. Byron, S. Wallis and P. Carteret, as well as Captain James Cook's first voyage of 1768-1771).

**Hawkin**, *Lieutenant* —. Parts of Vancouver, 1864 MS.

**Hawkins** [Hawkyns], father and son.
• **Hawkins**, *Sir* John (1532-1595). Seafarer, adventurer and slave trader from Plymouth. Became treasurer to the navy in 1573; he was knighted for his part in the defeat of the Spanish Armada. He died during a voyage with his cousin, Sir Francis Drake, to the Spanish Main. Narratives of John Hawkins' voyages were published by Richard Hakluyt in *The Principall Navigations...*, 1589; and Samuel Purchas's *Puchas His Pilgrimes...*, 1625.
•• **Hawkins**, *Sir* Richard (*c.*1562-1622). English seaman, son of Sir John; fought against the Spanish Armada, 1588; his 'Voiage' was a world-wide plundering expedition, ending in his capture in Peru. *Observations in his Voiage into the South Sea...1593*, 1622; narratives of his voyages were also published in Samuel Purchas's *Puchas His Pilgrimes...*, 1625; and John Callender's *Terra Australia Cognita*, 1766-1768.

**Hawkins**, *Corporal* — RE. One of the British draughtsmen among the international team which prepared *Croquis de la frontière Serbo-Turque selon article 36 du traité de Berlin... 1878* (part III, sheets 11-21), Southampton, Ordnance Survey Office 1879.

**Hawkins**, *Brigadier* — . Amended Le Rouge's French edition of John Mitchell's *A Map of the British and French Dominions in North America...*, 1776 edition (first published by Le Rouge in 1756).

**Hawkins**, Alfred. Sponsor of *Plan of the Military & Naval Operations... before Quebec... 1759*, London, J. Wyld '...for Alfred Hawkins Esq$^e$, Quebec' 1841 (engraved by J. Wyld).

**Hawkins**, C. Lithographer. *Ground plan of proposed improvements in the neighbourhood of Smithfield*, London, King 1851.

**Hawkins**, Charles Edward. Surveyor for the Geological Survey. Contributed to *London and its Environs* (7 sheets), Ordnance Survey 1857.

**Hawkins**, Ernest (1802-1868). *The Colonial Church Atlas*, London, Society for the Propogation of the Gospel in Foreign Parts 1842, London, Society for Promoting Christian Knowledge 1845, 1853 (maps drawn and engraved by Joshua Archer).

**Hawkins**, G. Publisher of London. *A Plan of the Action at Seatowne*, 1745.

**Hawkins**, J.E. Cartographer and lithographer 'at the Intelligence Dept. Horse Guards'. *South Africa Sheet 4* [Natal], 1879; *Map of Zululand, compiled... from surveys and... sketches made by officers during the Campaign of 1879*, 1881, 1885; *Map of the South Western Frontier of the South African Republic...*, 1884; *Sketch of route from Ambukol to Shendy* [Sudan], 1884; *Map of the Nile Provinces from the Third Cataract (Hannek) to Khartum*, Edward Stanford 1884, 1888, 1895 (for the War Office); *Map of Zululand compiled... from the Military Trigonometrical Surveys and the various*

*Topographical Sketches made by Officers during the Campaign of 1879 and from Major McKean's Boundary Survey of 1887,* 1888; *Map of Basutoland*, 1888; *Map of Matabililand and adjoining territories*, 1889; *Sketch Map to shew the various States & Tribes in S.E. Africa between Natal and the Limpopo...*, 1890; *Map of Swaziland and Adjacent Country*, 1890; *Map of South African Republic and Adjoining Territory*, 1891; *Map of Southern Zambesia* (2 sheets), 1891; *Map of part of Matabililand*, London, E. Stanford 1896.

**Hawkins**, James. Mapseller of Fenchurch St., London. Captain James Wimble's *...This Chart of his Majesties Province of North Carolina With a full & exact description of the Sea-coast, Latitudes, Capes...*, Boston, J. Wimble and London, J. Hawkins 1738 (engraved by J. Mynde, also sold in London by Mount & Page).

**Hawkins**, Richard. Survey and map of Weald Hall estate, South Weald, Essex, 1743.

**Hawkins**, Thomas B. (*fl.*1842-1848). Surveyor of Brackley, Northamptonshire. Surveyed Yates Estate, Turweston, Buckinghamshire, 1842 MS.

**Hawkins**, William (1807-1868). Provincial Land Surveyor in Canada. *Plan of the Town of Barrie In the Township of Vespra and County Simcoe*, 1833; *Plan of the town of Rippon...*, 1833; *Plan shewing the Portage Road and Indian Reserve with the Improvements thereon from Lake Simcoe to Coldwater*, 1833; *Plan of the Military Reserve at the Falls of Niagara*, 1834; *Plan of the Survey of the Northern Boundary of the Canada Company's Huron Tract in the London District*, 1834 (with R. Birdsall and S.P. Hurd); *Plan of Building Lots situate at the East End of Toronto, Township of York*, 1835; *Plan of an Exploration made Easterly of Lake Huron under the Command of Lieut Carthew R.N....* 1835, 1836; *Exploration of the Rivers Maganetawang and Pittoiwaiis*, 1837; *Military Reserve at Queenston & in the Township of Niagara*, 1838; *Map of the Proposed Road from Toronto to Saugine, Lake Huron*, 1842; *Plan of the Proposed Road from Bradford To Barrie...*, 1843; *Road from the Narrows to Coldwater...*, 1843; *West Part of Toronto*, 1849; *Queenston*, 1854; and many others including township plots. **N.B.** Also worked with Alexander Sproat as **Sproat** & Hawkins q.v.

**Hawkshaw**, *Lieutenant* John RE. Plan of the Island of Grenada, 1832 MS.

**Hawkshaw**, John. Manuscript plan of Portrush, County Antrim, 1859; Plan of Holyhead New Harbour, 1873.

**Hawksmoor**, Nicholas (1661-1736). Architect. *The Plan of the Church of St. Alban*, 1710 (engraved by J. Harris); Plan for Cambridge, c.1720.

**Hawksworth**, J. Surveyor. Plan of St. Mary Islington, 1735.

**Hawkworth**, J. Engraver. Plan of Buckingham Palace by James Hakewill after John Nash, 1826.

**Hawley**, Homer A. (*fl.*late 1850s). Worked with Silas and Frederick Beers, D. Jackson Lake and others on county surveys in New York state for John Homer French.

**Hawley**, J.S. *Map of Schuylkill County, Pennsylvania*, 1864.

**Hawstead.** See **Hausted.**

**Hawthorne**, Henry. Manuscript plans of Windsor Castle, 1576-1577; plan of Hertford Castle showing certain rooms given over to the Courts of Wards, King's Bench, Requests and Exchequer, 1582 or 1592.

**Haxo**, *General* François (1774-1838). French military engineer, chief of staff of the French army engineers in Italy. *Projets d'Anfo...* [contour survey of the shore of Lake Idro], 1801.

**Hay**, A. Engraver of [James] *Fraser's Travelling Map of Ireland*, Dublin, J. McGalshan 1852.

**Hay**, David. Published the Dublin edition of Part I of Mount & Page's *The English Pilot* (27 charts), Dublin, David Hay 1772 (the first Mount & Page edition was published 1701).

**Hay**, James. *Hay's New Plan of Musselburgh* [Scotland], 1824 (engraved by C. Thomson).

**Hay**, John. *Chart of the S.W. Part of the Island of Jersey*, Southampton 1849.

**Hay**, Sir John Drummond. British Vice Consul in Tetuan. *Sketch of Part of Morocco between Ceuta, Tetuan and Tangier*, London, Admiralty 1860 (drawn by E.J. Powell).

**Hay**, John Ogilvie. *Routes proposed for connecting China with India & Europe*, published Edward Stanford 1875.

**Hay**, Thomas. Engraver. Plan of Amsterdam for John Andrews's *Plans of Capital Cities of Europe*, 1771.

**Hayashi** Jizaemon. Japanese publisher. Hassendo Shujin's *Nagasaki Saiken Zu* [plan of Nagasaki], [after 1745] (with Takehara Kobei, possibly a later printing of an earlier work).

**Hayashi** Joho (d.1646). Route map Kawachi to Osaka, engraved 1709 (first map of Kawachi province).

**Hayashi** Kiyosuke. Japanese cartographer. Woodblock map of the Kawachi province of Japan, 1709.

**Hayashi** Shihei (1738-1793). Japanese author. *Sangoku tsuran zusetu* [Military geography of Japan and neighbouring regions], 1785 (includes maps of Korea, Ezo and Ryukyu); French edition *Aperçu général des trois royaumes* [General view of the three kingdoms] (with 5 maps), 1832.

**Hayashi-Shi** Yoshinaga. Japanese mapmaker and publisher. *Shinsen Daizōho Kyō Ō-Ezu* [city of Kyoto], 1728-1734 (large scale woodblock map of the city); *Shimpan Zōho Ōsaka no Zu* [Plan of Osaka], 1748-1751 and later (first published by Fushimiya, 1657).

**Hayden**, —. *Marion County, Indiana*, 1855.

**Hayden**, Ferdinand Vandeveer (1829-1887). American Doctor of Medicine, naturalist and geologist from Massachusetts. Appointed head of the geological survey of Nebraska, 1867. His geological expedition to the upper Yellowstone River in 1859 resulted in a stratigraphical map of Montana, Idaho and the Dakotas; *Geological report of the Yellowstone and Missouri Rivers*, 1860; with John W. Barlow and David Heap led the 1871 expedition to northwest Wyoming, 1871-1872; explorations in Colorado, 1872-1876; *Preliminary Report of the United States Geological Survey of Montana and Portions of Adjacent Territories*, 1872; *Yellowstone National Park, From Surveys Made under the Direction of F.V. Hayden*, 1872; *The Yellowstone National Park, and the Mountain Regions of Idaho, Nevada, Colorado and Utah*, 1876; *Geological and Geographical Atlas of Colorado* (20 maps and views), Julien Bien 1877 (included maps by Henry Gannett and W.H. Holmes); *Map of Yellowstone National Park, Showing Distribution of Hot Springs*, 1878, published 1883 (one of Hayden's many maps to accompany his large 1883 report on Yellowstone). **Ref.** WALDMAN, C. & WEXLER, A. *Who was who in world exploration* (1992) pp.310-311; WALSH, J. 'The exploration and mapping of Yellowstone National Park' *Meridian* 3 (1990) pp.5-21; HAINES A.L. *Yellowstone National Park: Its exploration and establishment* (Washington 1974).

**Haydon**, W. Surveyor. *Plan of San Miguel & Darien Harbour*, 1853 MS.

**Haydon**, William. Engraver of Little Mayes Buildings, London. *A Chart of the Delaware Bay and River*, 1776 (used in W. Faden's *The North American Atlas*, 1777); James Rennell's *Bengal*, Andrew Dury [n.d.] republished Laurie and Whittle 1794; *Andrews's New and Accurate Map of the Country Thirty Miles Round London*, London, John Andrews 1782 ('map drawn and engraved by J. Andrews. The Writing Engrav'd by W. Haydon').

**Haye**, G. de la. See **Delahaye**, Guillaume Nicolas.

**Haye**, I. de la. Translated *Caert-Thresoor* by Barent Langenes into French, published as *Thrésor de chartes*, Amsterdam, Cornelis Claesz. 1600 (printed by Albert Hendricks).

**Hayes**, C. Willard. *Map of the Southern Appalachians*, 1895.

**Hayes**, E.L. American surveyor, drew maps for county atlases of North America. *Illustrated Atlas of the Upper Ohio River and valley from Pittsburgh, Penn. to Cincinnati, Ohio*, 1877; *Atlas of Osceola Co. Michigan*, Philadelphia, C.O. Titus 1878; *Atlas of Mecosta County, Michigan*, Philadelphia,

C.O. Titus 1878; *Atlas of Isabella County, Michigan*, Philadelphia, C.O. Titus 1879; *Atlas of Newaygo County, Michigan*, Philadelphia, C.O. Titus 1880; *Atlas of Grand Traverse county, Mich.*, 1881; *Atlas of Leelanau county*, 1881; *Atlas of Sebastian County, Arkansas*, 1887.

**Hayes**, Isaac Israel (1832-1881). American physician and explorer from Chester County, Pennsylvania. As surgeon, he accompanied the Elisha Kent Kane expedition to Greenland, 1853-1855; led his own expedition, 1860-1861; accompanied William Bradford to the Arctic, 1869. Surveys of Smith Sound & Kennedy Channel, 1861. **Ref.** WALDMAN, C. & WEXLER, A. *Who was who in world exploration* (1992) pp.311-312.

**Hayes** [Hays], *Captain Sir* John. Surveys in Van Dieman's Land, 1798, used for *A Chart of Van Diemen's Land the South Extremity of New Holland...*, Laurie & Whittle 1800.

**Hayes**, L. des. See **Des Hayes**, Louis.

**Hayes**, M.P. *Skeleton Map Shewing the Position of the Proposed Toronto and Georgian Bay Canal with Reference to the Trade of the Great West with the Atlantic Ports...*, Toronto, Committee of the Toronto Board of Trade 1856 (printed by Thompson & Co.).

**Hayes**, S.B. American surveyor. Worked with Charles, Augustus & L.C. Warner, and others, on surveys in Indiana and Ohio, *c.*1862-1867.

**Hayes**, S.C. [& Co.]. Publishers. California, Texas, Mexico, Philadelphia 1848.

**Hayman**, Francis (1708-1779). Painter and designer from Exeter. Founding member of the Royal Academy. Designed the cartouche for Fry and Jefferson's *A Map of the most Inhabited part of Virginia...*, London, Thomas Jefferys 1753 (cartouche engraved by Reynolds Grignon).

**Hayman**, J. *Exeter*, 1805 (used in *The British Atlas*, Vernor, Hood & Sharpe, 1810.

**Hayman**, *Lieutenant* John. Officer in the 17th Regiment of Foot of the British Army from 1777. Plan of Yorktown [Virginia] and surrounding area, 1782 MS.

**Haymann**, Burkhard. *Eisenbahnkarte von Central Europa*, 1873.

**Hayn**, *Captain* — (fl.1773-1779). Austrian military cartographer, commander of the Mappierungs Corps, 1773-1779, during the mapping of the Temes Banat [Hungary].

**Hayn**, Antonio Avinea. San Quentin e Peronne, 1537.

**Hayn**, C.F. *Die Erz Dioecese Coeln* [Diocese of Cologne], Cologne, J. & W. Boisserée, 1881.

**Hayn**, H. Publisher. *Wandkarte der Kreise Ostrowo u. Adelnau* (6 sheets) 1899.

**Haynes**, John (*b.c.*1705, *fl.*-1759). 'Surveyor, Engraver, and CopperPlate Printer, near the Middle of Stonegate, York'. Prepared manuscript maps of Yorkshire and the London area. Also engraver at Michael Angelo's Head in Buckingham Court, Charing Cross, London. *A Description of the Passage of the Annular Penumbra of that great and Visible Eclipse of ye Sun Feb.18th 1737 over Great Britain & Ireland...*, 1737; *An accurate Survey of some Stupendous remains of Roman Antiquity in the Wolds in Yorkshire...*1744 (engraved by George Vertue); *A New & Exact Plan of the City of York*, 1748; *Survey of the Botanic Gardens at Chelsea*, 1753; Park Lane to Half Moon St., 1767. **Ref.** ARMITAGE, Geoff *The shadow of the Moon* (1997) pp.12, 15-19, 39-40; RAMM, H. 'An eighteenth century archaeological map from East Yorkshire' *Journal of the International Map Collectors' Society* 40 (Spring 1990) pp.20-23.

**Haynes**, John C. Associated with N.H. Crafts's *Plan of Boston*, Boston 1864 (drawn by H.M. Wightman).

**Haynes**, M.B. Atlas of Renville County, Minnesota, 1888.

**Haynes**, Tilly. New map of Boston, 1883.

**Hays**, John C. Surveyor General, California. Public Surveys California, 1855.

**Hays**, *Captain* John. See **Hayes**, *Captain Sir* John.

**Hays**, *Mr Deputy Governor* —, African Company. West African Trading Posts, London 1745.

**Hayter,** *Captain* George. Master of the *Elizabeth*. Charts of the East Indies, 1755-1778; *Chart of the passage between Po Bato and Se Beeroo...*, 1755, published Dalrymple 1792; *Chart of the West Coast of Ava* [Burma], 1757-1758, published A. Dalrymple 1784; *A chart of the China Sea*, Sayer & Bennett c.1780; *The river of Perseen*, Sayer & Bennett 1784, Laurie & Whittle 1800.

**Hayton,** —. Map for Relation of Tartar Empire, c.1520.

**Hayward,** G.J.W. Trans-Indus, 1860 MS; East Turkistan, 1870.

**Hayward,** George (*fl.*1834-1872). Lithographer of 171 Pearl Street, New York City. Maps for the annual 'Valentine's Manual' of New York City, 1841-1870 including a copy of S. Holland's *A plan of the north east environs of the city of New York...* 1757, 1859; *A Description of the Towne of mannados or New Amsterdam as it was in...* 1661, 1859 (lithographed from an anonymous manuscript of 1664); *Map of Danbury, Connecticut*, New York, E.C. Smith & E. Van Zandt c.1860 (surveyed and published by E.C. Smith & E. Van Zandt).

**Hayward,** H.F. Draughtsman. *Map of Canada West, Shewing The Post Offices & Mail Routes*, 'Prepared by Order of the Postmaster General' 1858 (lithographed by Barr & Corss); *Department of Crown Lands. ...Plan of the North Shore of Lake Superior*, Quebec, Department of Crown Lands 1863 (lithographed by W.C. Chewett & Co.).

**Hayward,** J.A. Township maps of Ohio, 1871-1872; *Map of Bellefontaine...*, Philadelphia 1871; *Map of Delaware*, Philadelphia 1871; *Map of Galion* [Ohio], Philadelphia 1871; *Map of Kokomo* [Indiana], 1872.

**Hayward,** James. *Plan of a Survey for the proposed Boston and Providence Rail-Way*, Boston, Annin & Smith, 1828.

**Hayward,** John. Engraved charts to illustrate *Voyage made in the years 1788 and 1789, from China to the north-west coast of America... By John Meares...*, London, Logographic Press 1790.

**Hayward,** John (1781-1862). American publisher and compiler of gazetteers. *Columbian Traveller, and Statistical Register* (4 maps), Boston, J. Hayward 1833; *A New England gazetteer...*, Boston, J. Hayward, also Concord N.H., I.S. Boyd & W. White 1839, Boston, J. Hayward 1841; *A Gazetteer of the United States, comprising a series of gazetteers of the several states and territories. Maine*, Portland, S.H. Colesworthy, also Boston, B.B. Mussey 1843; *A Gazetteer of Massachusetts...*, Boston, J. Hayward 1846, Boston, J.P. Jewett & Co. 1849; *A Gazetteer of Vermont...*, Boston, Tappan, Whittemore and Mason 1849; *A Gazetteer of New Hampshire...*, Boston, J.P. Jewett 1849; *United States, from the latest authorities*, Hartford, Case, Tiffany & Company 1853; *A Gazetteer of the United States of America... to which are added valuable statistical tables...*, Hartford, Case, Tiffany & Company 1853.

**Hayward** [Haward], Nicholas. Essex estate surveyor. Plan of the Estate of Searle in Dagenham, 1764 MS.

**Hayward,** S. Publisher of Bath. *Plan of the City and Borough of Bath and its Suburbs*, Bath 1852 (survey by J.H. Cotterell, engraved by Holloway & Son).

**Hayward,** *Lieutenant* Thomas, HMS *Pandora*. Track of the Ship [East Indies], 1791; Bay Selema [Ceram], 1800; Timor to Ceram, 1801.

**Hayward,** W. Chart from Kings Lynn to Wisbech, 1591 MS. **N.B.** Compare with William **Haiwarden** (above), and William **Hayward** (below).

**Hayward** [Haiward; Haiwarde; Haywarde], William (*fl.*1591-1637). East Anglian estate and fenland surveyor. Surveys at Hillington and Heacham [Norfolk], 1592; map of Fulstow and Marshchapel [Lincolnshire], 1595; *The true description of the town of Titleshall*, 1596; *A true and exact draught of the Tower Liberties...*, 1597 MS (with John Gascoyne), engraved 1742; estate survey of the Manor of Wakeringhall [Essex] (1 volume), 1598; *A Generall Plotte and description of the Fennes and other Grounds within ye Isle of Ely*, 1604 MS; surveys in Cambridgeshire, Essex, Suffolk and Norfolk, 1610-1635; survey of the southern Fens, 1635-1636. **Ref. KEAY, Anna** *The Elizabethan Tower of London - The Haiward and*

*Gascoyne plan of 1597* (London Topographical Society 2001); **N.B.** Compare with William **Haiwarden** and W. **Hayward** (above).

**Hayward and Howard**. Atlas of Brockton, Massachusetts, 1898.

**Hayward & Moore**. Publishers, Paternoster Row. John Dower's New General Atlas, 1838.

**Haywarde**, William. See **Hayward**, William.

**Haywood** [Heywood], James. Draughtsman of 'N°·3 St. Martin's Churchyard', London. *A new map of Scotland divided into counties*, 1787 (engraved by S.J. Neele), used in John Harrison's *A School Atlas*, London 1791. **N.B.** Compare with John **Haywood**, below.

**Haywood** [Heywood], John. Draughtsman of 'N°·3 St. Martin's Churchyard', London. Draughtsman for works published by John Harrison, including *Maps of the English Counties*, 1787-1791, published 1791, 1792.

**Haywood**, William. Engineer and surveyor to the City Commissioners of Sewers. *General plan of the City of London Shewing the Public Sewers...*, 1854 and editions to 1872.

**Hazama** Shigetomi (1756-1816). Japanese cartographer and astronomer.

**Hazard**, Willis P. Publisher of Philadelphia. *Hazard's rail road & military map of the southern states*, 1863 (prepared by the Committee on Inland Transportation of the Board of Trade of Philadelphia, drawn and engraved by P.S. Duval & Sons).

**Hazart**, —. Kirchen-Geschichte der gantzen Welt, 1678-1701.

**Hazelius**, Johann August (1797-1871). Swedish surveyor. Topografiska Corpsens karta öfver Sverige, 1860.

**Hazell, Watson & Viney Ltd**. Engravers and lithographers of London. Map of an allegorical path to heaven entitled *Up and Down Lines*, c.1880 (designed by W.C. Miles).

**Hazelwood**, Samuel. Surveys of estates at Brockville, Canada, 1853-1863, including *Map of the Property of James L. Schofield Esqr near the Town of Brockville*, 1860.

**Hazen**, J.C. Artist and publisher of Boston, Massachusetts. Worked with H.H. Bailey on bird's-eye views of cities in Massachusetts, including Lawrence, 1876; Lowell, 1876; and Maynard, 1879.

**Hazen**, W.B. North-West Alaska for P.H. Ray, 1854.

**Hazen**, William. Province of New Brunswick, 1791 MS.

**Hazen**, William Babcock (1830-1887). *A narrative of military service* [campaigns and battles of the Civil War], Boston, Ticknor and Company 1885.

**Hazzard**, J.H. Engraver for Thomas, Cowperthwait & Co., Philadelphia, 1850. **N.B.** Compare with J.L. **Hazzard**, below.

**Hazzard**, J.L. Cartographer and engraver. *A New Map of Central America*, Charles Desilver 1856; *A New Map of Michigan*, Philadelphia, Charles Desilver 1856 (in *A New Universal Atlas*, 1858); *Map of Lake Superior with its Rail Road & Steamboat Connection*, Philadelphia, Desilver 1857 (printed by F. Bourquin & Co.); maps for Samuel Augustus Mitchell, 1859; Maryland & Delaware, 1860; maps for the United States Coast Survey including *Preliminary Chart of Portsmouth Harbour*, 1862.

**Hazzard**, John H. *Map of Explorations and Surveys in New Mexico and Utah*, 1859.

**Hazzen**, Richard. American cartographer of Haverhill. Northern Boundary of Massachusetts, 1741 MS.

**He** Zheng (1371-1435). Chinese Admiral during the Ming Dynasty, undertook expeditions to east Africa and the Persian Gulf, 1405-1433, on the basis of which he prepared a manuscript sea chart. **Ref.** HSU, M-L *Imago Mundi* 40 (1988) pp.96-112 (includes bibliography).

**Head**, *Midshipman* H.N. Maps and charts for Lieutenant William Edward Parry's *Journal* [search for the Northwest Passage], 1824.

**Headrick**, *Reverend* James. *View of the Mineralogy, Agriculture, Manufactures, etc. of The Island of Arran* [with maps of the island], 1807; untitled map of the soils of Angus, 1813.

**Heald**, Henry. *Roads of Newcastle County* [Delaware], 1820.

**Heap**, *Captain* David Porter. Explored the Yellowstone region with Barlow and Hayden, 1871. *Map of the Country between the Yellowstone and Missouri Rivers*, 1870; *Sketch of the Yellowstone Lake and the Valley of the Upper Yellowstone River*, 1871 (with J.W. Barlow); *Montana Territory*, 1872. **Ref.** WALSH, J. 'The exploration and mapping of Yellowstone National Park' Meridian 3 (1990) pp.12-13; HAINES A.L. Yellowstone National Park: Its exploration and establishment (Washington 1974).

**Heap**, George. Surveyor. *A Map of Philadelphia and Parts Adjacent*, 1752 (with Nicholas Scull, engraved by L. Hebert); *A Prospect of the City of Philadelphia* (4-sheet panorama), 1755; *A Plan of the City and Environs of Philadelphia*, London, W. Faden 1777 (with N. Scull). **Ref.** WAINWRIGHT, N.B. 'Scull & Heap's map of Philadelphia', The Pennsylvania magazine of history and biography 81 (1957) pp.69-75.

**Heap**, Gwinn Harris. Mapmaker of Philadelphia. *Central Route to the Pacific*, 1853; *Map of the Central Route from the Valley of the Mississippi to California*, 1854.

**Heaphy**, Charles. Artist and surveyor, arrived in New Zealand with Captain Chaffers in 1839, and became draughtsman to the New Zealand Company in Auckland. *Birdseye View of Port Nicholson in New Zealand*, 1839 (with an inset chart of the harbour by E.M. Chaffers); sketches for *Chatham Islands*, 1840; *Chart of Cooks Strait and the Recent Exploratory Routes in the Northern End of the Middle Island*, 1848; sketch of gold finds on the Coromandel Peninsula, 1852; *North Island*, 1861; *Conquered Territory North Island*, 1864; *Military Settlements Waikato*, 1868.

**Heard**, Thomas (*fl.*1844). Surveyor of Ware in Hertfordshire, worked on a tithe survey in Leicestershire.

**Heard**, William (*fl.*1834-1850). Surveyor of Hitchin, Hertfordshire, worked on manuscript maps of Essex, Hertfordshire and possibly Huntingdonshire. Broomfield, 1834 MS; Netteswell [Essex], 1848 MS.

**Hearding**, W.H. Assistant surveyor in the U.S. Corps of Topographical Engineers. *Sketch of the Navigation through East Neebish Rapids River St. Mary...*[Michigan], 1853, published 1854; worked with J.H. Forster and I.L. Beghlin on surveys for *Preliminary Chart. Lower Reach of Saginaw River...Lake Huron*, 1856, published 1867 (engraved by W.H. Dougal).

**Hearn**, Walter Risley. British Vice-Consul in Oslo. *A Map Showing the Murman Coast and the Boundaries of Norway, Finland and Russia*, 1884 MS; *Map of parts of Norway, Sweden, Finland and Russia. Showing Railways Completed and Projected*, 1886 MS.

**Hearne**, J. Map of the Kingdom of Guatemala, 1824.

**Hearne**, John. *A Map of the Kingdom of Guatemala*, London, J. Hearne 1823 (engraved by J. Walker); *Plan of the bay of St. Salvador de Jiquilisco*, 1823 (engraved by J. Walker).

**Hearne**, Samuel (1745?-1792). English traveller and sailor. Explored the American Northwest while in the service of the Hudson Bay Company, 1769-1772. Made first overland journey to Arctic Ocean at Coppermine River, 1771-1772. Although captured by Jean François Galaup de La Pérouse at Fort Albany in 1772, his manuscript account was later surrendered and published as *A Journey from Prince of Wales's Fort in Hudson's Bay to the Northern Ocean. Undertaken by order of the Hudson's Bay Company, for the discovery of copper mines, a North-West passage &c., in the years 1769, 1770, 1771 & 1772*, London, A. Strahan and T. Cadell 1795 and other editions.

**Hearne**, Thomas (1678-1735). Antiquary and scholar. Second Keeper of the Bodleian Library. Edited *The Itinerary of John Leland the Antiquary...*, Oxford, T. Hearne 1710-1712 (originally compiled by Leland in 1535-1545); edited Leland's *Britannicis Collectanea* (6 volumes), Oxford 1715, London, G. & J. Richardson 1770, London, B. White 1774.

**Heart**, *Captain* Jonathan (1748-1791). First United States Regiment. *Plan of the remains of some ancient works on the Muskingum* [near the Ohio River], 1787 (published in *Columbian Magazine*).

**Heath**, Mr. —. 'Next Fountain Tavern in the Strand'. Charles Price's Chart of the Bristol

Channel, 1729. **N.B.** Compare with Thomas C. **Heath**, below.

**Heath,** *Lieutenant* G.P., *RN.* Admiralty surveyor. Tonga, 1852; *Vavu Group* [Friendly islands], Admiralty Chart 1855.

**Heath,** *Lieutenant* Joseph [I.] (*fl.*1742-1760). Draughtsman in the Drawing Room at the Tower of London. *A True and Exact Copy of the Draught of the Tower Liberties...*, 1752 (a copy of William Hayward and John Gascoyne's plan of 1597); Map of Gosport, 1758 MS; River Mersey, 1759; the Kent coast from Dover Castle to Sandgate Castle, 1760; Milford Haven, *c.*1760. **Ref.** KEAY, A. *The Elizabethan Tower of London - The Haiward and Gascoyne Plan of 1597* (London Topographical Society 2001).

**Heath,** Joseph. Surveyor. Surveys used for *A True Coppy from an Ancient Plan...* [Norridgwock Town to Cape Elizabeth, Maine], 1719, published Thomas Johnston 1753; surveys used by Thomas Johnston for *A Plan of Kennebeck & Sagadahock Rivers & Country Adjacent,* [Maine], Boston 1754.

**Heath,** *Lieutenant* (later Admiral Sir) Leopold G., *RN.* Hong Kong, 1846-1847; *Dardanells,* 1853 MS.

**Heath,** Robert. A New and Correct Draught of the Islands of Scilly, 1744, published London, R. Manby & H.S. Cox *c.*1748 (engraved by Thomas Hutchinson).

**Heath,** Thomas (1698-1773). Instrument maker and estate surveyor of London, Freeman of the Grocer's Company. Woodford Wood, Essex, 1757 MS.

**Heath,** Thomas C. (1714-1765). Globe and instrument maker 'at the Hercules, next to the Fountain Tavern in the Strand', London 1729-1735.

**Heath,** Thomas Edward. *The Twentieth Century Atlas of Popular Astronomy,* Edinburgh & London, W. & A.K. Johnston 1903; *Our Stellar Universe: a Road-Book to the Stars,* London, King, Sell & Olding 1905.

**Heath,** William. *A bird's eye view* [political cartoon of Europe], London, T. McLean 1830.

**Heathcote,** J. Norman. Surveyor. Island of St. Kilda, 1900.

**Heather,** John (*fl.*1715-1725). Estate surveyor of Essex. Manuscript surveys of estates in Fairstead and Great Leighs, 1720-1724; surveys in Stambourne, 1721; farm in Sible Hedingham, 1722; Steeple, 1724; land at Great Maplestead, 1725.

**Heather,** William (1764-1812). Engraver and chart publisher of London. He was apprenticed to the bookseller and stationer George Michell of Bond Street, becoming a freeman of the Stationers' Company in 1789. He then worked for the teacher of navigation and chart publisher John Hamilton Moore at Little Tower Hill. He opened his own business in 1793 'at the sign of The Little Midshipman, N$^{o.}$157 Leadenhall Street', later called 'Navigation Warehouse'. The building also housed John William Norie's Naval Academy. Heather was succeeded by John William Norie. S. Clements's *Chart of the Entrance to the River Thames,* 1791; charts of the coasts of Britain, 1793-1812; *A Pilot for the Atlantic Ocean,* 1795-1801; *...Chart of Plymouth Sound...,* 1798; China Seas, 1799; *A New Chart of America with the Harbors of New York, Boston &c....,* 1799; *A new map of Egypt...,* 1800; *New set of charts for harbours in the English Channel,* 1801; *Heather's complete pilot for the Northern Navigation from London to St Petersburgh,* 1801; *The New Mediterranean Pilot,* 1802 (material supplied by John Wilson); *A new chart of the world on Mercator's projection...,* 1803 (drawn by J. Norie, engraved by J. Stephenson); Andaman & Nicobar Islands, 1803; *East India Pilot* (23 charts), 1805; New North Sea Pilot, 1807; *The Marine Atlas, or Seamen's complete Pilot...* (55 charts), *c.*1808; *...chart of the Island of St. Michael... 1806* [Azores], 1808 (drawn by W.H. Read); North American Pilot, 1810; Pilot London to Spain, 1810; Pilot of the Brazils, 1811; and many others. **Ref.** FISHER, S. *The makers of the blue back charts* (2001) pp.74-83.

• **Heather & Williams** (*fl.*1796-1800). Partnership of William Heather and a Mr. Williams.

**Heatherwick,** *Reverend* A. District east of Blantyre [Nyasaland], 1877 MS.

**Heaviside,** *Captain* W.J. Thal Chotiali Route Survey, Calcutta 1880.

**Heawood,** Edward (1865-1949). Cartographic historian, worked with H.R. Mill. Buttermere, 1893; Bassenthwaite Lake, 1895; Bathymetric Survey of the English Lakes, 1895.

**Heb** [Hebb], Andrew (*fl.from* 1625-d.1648). Bookseller 'at the Bell in St. Pauls Churchyard', London, successor to Thomas Adams. Published an edition of Peter Martyr's *Decades*, *c.*1626; William Camden's *Britain*, 1637 edition (P. Holland's translation of *Britannia*, re-used the maps engraved by William Kip and William Hole for the 1607 edition, printed by Felix Kingston, Richard Young and John Legatt).

**Heber,** Johann Jakob (1666-1725). Swiss surveyor, born Basle, worked in Stein am Rhein and Lindau. Manuscript plans of Lindau, Meersburg, Salem, *c.*1700; *Vngefehrlicher Entwurff deß jetzmahligen Fürstenthumbs Liechtenstein*, 1721.

**Heberer,** F. *Umgegend von Darmstadt*, 1896.

**Heberlein,** Adolf. Geological survey of Michigan, 1870.

**Heberstein.** See **Herberstein,** Sigismund.

**Hebert,** — [I.; L.; L.J.; Lewis; W.J.]. **N.B.** Credits to the Heberts are not consistent, and the following has been compiled by inference only. Most of the earlier maps were produced at the Colonial Department and refer to L. Hebert senior as draughtsman (some were lithographed at the Quarter Master General's Office), whilst in general the later ones, credited here to L.J. Hebert, originated in the Quarter Master General's Office.
• **Hebert,** Lewis *Senior*. Geographer and draughtsman at the Colonial Department, London, 1827-1838. Drew maps for J. Pinkerton's *A Modern Atlas*, 1809-1815 (engraved by S.J. Neele); geographer for *An Actual Survey and Itinerary Of The Road From Calais To Paris...*, London, Longman, Hurst, Rees, Orme, and Brown 1814 (with G. Dupont, printed by Schulze and Dean), 2nd edition London, G. Schulze 1831 (new title page); draughtsman for *London and Liverpool Mail Road*, QMGO 1826 (surveyed by J. Easton & T. Casebourne for Thomas Telford); *Map of the Province of Antioquia in the Republic of Columbia, and its Minerals...*, London, Charles Hauswolff 1824 (from a map by Dr. J.M. de Restrepo); *West Indies*, 1824 (printed at the Lithographic Establishment, Quarter Master General's Office); *London and Liverpool Mail Road*, 1826 (compiled by Thomas Telford); *London and Morpeth Mail Road*, 1827 (surveyed by Thomas Telford); *Map of the Bights of Benin and Biafra, including the Island of Fernando-Po*, 1829; map of the Falkland Islands, *c.*1830 MS; *Mauritius or Isle de France*, 1830; *Colony of the Cape of Good Hope*, 1830 ('Printed at the Lithographic Establishment, Quarter Master General's Office, Horse Guards, London'); *Ireland*, I. McGowan 1830 (engraved by W. Milton); *Carte Physique de la partie de la Grèce continentale située entre le Aperchius l'Astropotamus et le Golph de Corinth...*, 1831; North America west of Lake Superior and the Pacific, 1831; Map showing British and American territories with land in dispute on the Maine, Quebec and New Brunswick borders, 1831; *Laurie's new map of Scotland... with all the ... canals... to the present time...*, 1833; *Map of the Island of Mauritius exhibiting the Fortifications for its defence*, 1833; New Brunswick, 1834 MS; *Map of the East and West Falkland and Adjacent Islands*, 1836 (from surveys by Captain Robert Fitzroy of HMS *Beagle* and others); *The Russian Dominions In Europe... reduced chiefly from the great Map of Russia in 107 sheets*, London, Richd. Holmes Laurie 1836 (drawn and translated by L. Hebert); *Map of the Rock and Town of Gibraltar*, 1837 MS; Trinidad, 1838 MS.
•• **Hebert,** W.J. Draughtsman for the Lithographic Establishment, Quarter Master General's Office, Horse Guards. *Chart of the River Dvina from its estuaries to the Town of Archangel*, 1831; *Chart of the Bar and Entrance of the River Dvina*, 1831.
•• **Hebert,** L.J. Draughtsman and lithographic draughtsman of London, prepared maps for the Lithographic Establishment, Quartermaster General's Office, Horse Guards, 1836-1848. *Sketch of Part of Upper And Lower Canada with references to the proposed Military Districts*, London, R. Bentley (for Henry Colburn) 1833 (drawn by L.J. Hebert, printed by B. King); Maps of Greece, 1836; *Map of the Montenegro*, 1836 (drawn in lithography by L.J. Hebert); *Sketch of Part of Mazanderan and of part of the*

*The shop of William Heather in Leadenhall Street, London, as illustrated in Charles Dickens's* Dombey and Son.

*Upper Road between Tehran and Astrabad by Major E. Darcy Todd*, 1837 ('drawn in Transfer Lithography by L.J. Hebert and Printed at the Lithographic Establishment...'); *Plan of the Positions of the Army... March 1837* [battle of Oriamendi, Spain], 1837; J.T. Crawford's *Galveston Bay*, 1837; J.T. Crawford's chart of the Brazos Santiago and the coast as far as the Rio Grande [Texas], 1837; J.T. Crawford's map of Rio Bravo del Norte [Texas], 1837; *Upper Canada...* (3 maps), 1838-1839 ('Drawn in Transfer Lithography by L.J. Hebert and Printed at the Lithographhic Establishment Quarter Master Generals Office Horse Guards'); *Sebastopol en Crimée*, c.1838; untitled map of south-western Yemen, 1839; *Disputed Territory in North America* [Quebec, New Brunswick, Maine], 1839 ('Drawn in Transfer Lithography by L.J. Hebert', printed at the Lithographic Establishment); J.L. Campbell's *Plan of Ghuznee* [Afghanistan], 1839; *Plan of Part of the Town of Ponta Delgada in the Island of St. Michael...* [Azores], 1840; *Chart of the Bay of Ponta Delgada...*, 1840; *Sketch of part of the River St. John New Brunswick...*, 1840; *Plan von Kustenschi...* [Romania], 1840; *Granville Consulate of La Manche and Isle et Villaine* [France], 1840; *Plan des Thales zwischen den Karassu-Seen und der Donau bei Boghasköi*, London, QMGO 1840 (surveyed by Vinckle); *Plan of Alexandria and environs...*, 1840 (surveyed by Lt. Nugent); lithographer of W.S. Birdwood's *Plan of Attack on the Heights and Forts near the City of Canton...*, 1841; *Map of British Guiana constructed from the Surveys and Routes of Captn. Schomburgk...*, 1842, re-issued by the War Office in April 1887; *Croquis indicatif de la frontière entre l'Algérie et le Maroc*, 1844; *Manitoba*, 1846; *Route from York Factory to Lake Winnipeg...*, 1846; *Skeleton Ethnographic Map of the Basin of the Lower Danube and Adjacent Counties*, 1848; *A Sketch by Chronometer & Compass of St. John's River by Bar. Bulow...* [Nicaragua], 1848; and others.

**Hébert**, Edmond (1812-1890). French geologist. Works on geology, physical geography, and oceanography. *Coupe de Ste. Menehould d'Ardenne*, Paris, Kaeppelin et Cie 1856.

**Hebert**, G. Geographer. Berkshire, Cornwall, Lancashire for *G. Ellis's New and Correct Atlas of England and Wales*, 1819.

**Hebert**, J.T. *Plan de la ville de Chartres*, 1866.

**Hebert** [Herbert], Lawrence. Engraver, advertised working in Philadelphia, 1748-1752. Lewis Evans' *A Map of Pensilvania, New-Jersey, New York, and the Three Delaware Counties*, 1749; Nicholas Scull and George Heap's *A Map of Philadelphia and Parts Adjacent*, 1752.

**Hebner**, Jonathan. Engraver, printer and mapseller at 15 Great Maddox St., Hanover Sq. (1821); 12 Water Street, Blackfriars (1824), London. M. Phillips' *The Grand Southern Tour of England...*, 1821, 1824.

**Hecataeus** (*fl.*500 B.C.). Historian, traveller and statesman of Miletus, Greece. Compiled an early geography in the form of a journey round the World or *Pariegesis*, c.500 BC.

**Heck**, G. *Atlas géographique*, Paris 1842; J.G. Heck's *Neuester Plan von London und Seinen Umgebungen*, Leipzig, Verlag von J.J. Weber 1851; *Plan von Berlin*, 1851; contributed to C.G.D. Stein's *Neuer Atlas der ganzen Erde*, 1852-1879; *Nord Amerika*, Leipzig 1856; *Die Schweiz* Leipzig, 1856.

**Heck**, Johann Georg (1795-1828). Cartographer. Atlas géographique, 1830.
**N.B.** Compare with G. **Heck**, above.

**Heckel**, C. *Mannheim*, c.1860.

**Heckel**, Johann Christoph. Geographer and teacher in Augsburg. *Atlas für die Jugend*, Augsburg 1776.

**Hecken** [Heck, Hecke, Heckius], Abraham van den. Engraver and cartographer. Worked at Frankenthal, 1599-1608.

**Heckenauer**, Jakob Wilhelm (d.1738). German engraver, born Augsburg, died in Wolfenbüttel. Engraver for *Augustissimi et Invictissimi Romanorum Imperatoris Caroli VI. Haereditarium Regnum Hungaria et Regiones*, Vienna 1710; *Théatre de la Guerre en Hongrie* (3 sheets), 1737; map of south-eastern Europe, 1737.

**Heckenauer**, Leonhard (c.1650-1704). Engraver in Augsburg. Engraver for Heinrich Scherer's *Atlas Novus*, c.1700; worked on

Scherer's *Geographia Hierarchica*, Munich 1703; *Geographia Politica*, Munich 1703; engraved Sherer's *Globus Terraqueus Ecclesiastico Politicus*, Munich 1703 (the frontispiece for Scherer's *Tabellae Geographicae*).

**Heckler**, J.M. (18th century). Globemaker.

**Hector**, *Sir* James (1834-1907). Doctor of medicine and geologist with John Palliser's expedition in Upper Canada, 1857-1858. He also undertook explorations in New Zealand as Otago provincial geologist from 1863. Later became Director of the Geological Survey in New Zealand. *Map and Section Shewing the structure of the Kakabeka Falls. River Kaministiquoia*, 1857 MS, published A. Arrowsmith 1858; sketch map of the route from Rainy Lake to Lake Superior, 1857 MS (with geological notes); maps and sections of lakes and rivers between Lake Superior and the south branch of the Saskatchewan, 1858 (lithographed A. Arrowsmith 1859); contributed to *Plan of Nanaimo shewing the Coal Mines...*, London, Stanford's Geographical Establishment 1863 (with C.S. Nichol); *Sketch Map of N.W. District of Otago...*, 1863; *Sketch Map of the geology of New Zealand*, 1865; *Map of District South of Taupo Lake N.Z.*, 1870; *A Map of the Colony of New Zealand*, London, Liverpool and Wellington, New Zealand 1870 (with T.A. Bowden and W. Hughes); *Manual of New Zealand Geography* (11 maps), London, G. Philip & Son 1873 (with T.A. Bowden); *Geological Sketch Map of New Zealand Constructed from Surveys and the Explorations of Dr. F. von Hochstetter, Dr. J. Haast and others...*, Wellington 1873; *Handbook of New Zealand* (2 maps), Wellington, George Didsbury 1883; and others.

**Hector**, T. Ellipto-polar map of the World, 1872.

**Heddaeus**, Adolphus. *Cities of Pittsburgh and Allegheny*, 1852.

**Hedemann**, E.W. *Kaart der Bataklanden en van het eiland Nijas*, Batavia 1890 (with F.J. Haver Droeze).

**Hedemann**, Friedrich von (1797-1866). Prussian military cartographer. *Gegend von Kiel*, 1822; *Holstein und Lauenburg*, 1827.

**Hedgrad**, S.L. L'Afrique, 1785; also geographical playing cards, *c.*1780s.

**Hédin**, —. Printer of 'rue Antoine-Dubois', Paris. One of the lithographers for E.-F. Jomard's *Monuments de la Géographie*, Kaeppelin 1842; one of the printers for F. Bazin and F. Cadet's *Atlas Spécial de la Géographie Physique, Politique et Historique de la France*, Paris, Bazin & Cadet, also J. Delalain *c.*1862.

**Hedin**, Sven Anders (1865-1952). Swedish explorer and geographer travelled extensively in Central Asia, Chinese Turkestan, Mongolia and Tibet, 1893-1909. Surveyed and mapped the area he described as 'Trans-Himalaya' to the north of the Himalayas, as well as the sources of major rivers such as the Indus and Brahmaputra. His maps and descriptions of remote and thinly settled areas were used by many subsequent travellers in the region. Produced numerous travel maps published in *Geographical Journal* (London) and *Revue de Géographie* (Paris), from 1893; *Through Asia*, 1898; *Die geographisch-wissenschaftlichen Ergebnisse meiner Reisen in Zentralasien 1894-1897*, Gotha 1900; *Results of a Journey in Central Asia 1800-1902* (8 volumes), 1904-1908; *Trans-Himalaya* (3 volumes), 1909; *Southern Tibet* (12 volumes), 1917-1922; and others. **Ref.** WIMMEL, K. 'Along the ancient silk road' in Mercator's World Vol.5 N°2 (March/April 2000) pp.36-43; WALDMAN & WEXLER Who was who in world exploration (1992) pp.314-316; EHRENSVÄRD, U. 'Sven Hedin der Kartenmacher', Meddelanden från Krigsarkivet 12 (Stockholm, 1989) pp.157-180.

**Hedraeus**, Thomas Christofferson. Swedish surveyor. Geografiska delineation... öfwer Jåmpteland, 1645.

**Heduus** [Haeduus]. See **Quintin**, Jean.

**Heeckeren**, G.P.C. van. Oostelijk [Westelijk] Deel der Kolonie Suriname, Arnhem, C.A. Thieme 1826; re-issued Brussels, Burggraaff 1826.

**Heer**, Christoph (1637-1701). Military engineer from Lauban, Silesia [now Luban, Slask, Poland], d. Dresden. After studying in Leipzig and Strasbourg in the 1650s he followed his cousins Georg and Gottfried Hoffman to Copenhagen, where he spent 2 years working for Georg, followed by 10 years with Gottfried. In 1669 he took up a post as engineer in Strasbourg. Heer made manuscript copies of Gottfried Hoffman's series of maps of Danish fortifications (now in the Royal Library), he also adopted the Hoffman technique of combining coloured manuscript maps with symbols and text applied using a hand stamp with movable type. *Fridrichsodde*, after 1653; *Nyborg* [Island of Funen, Denmark], *c.*1660; *Speculum artis muniendi lucidissimum*, Leipzig 1694; and others.

**Heer**, Daniel. German military cartographer. *Stralsund*, Johann Baptist Homann, *c.*1715; *Schlacht bey Oudenharden*, 1719.

**Heer**, *Professor* Oswald (1809-1883). Swiss botanist. *Island of Madeira*, *c.*1855.

**Heermans**, Anna A. Hieroglyphic geography of the United States, 1875.

**Hees**, A.W.M. van. Publisher of Amsterdam. *Zakatlas van Amsterdam met volledige stratenregister*, 1913.

**Hees**, Gilles van. *Kaart en Kennis Eenvoudige Aardrijkskunde voor de Volksschool*, Groningen, J.B. Wolters 1925, 1927.

**Hees**, J. *Mosel Karte. Coblenz-Trier*, Trier, H. Stephanus *c.*1896.

**Hefti**, Andreas (1862-1931). Topographer and cartographer, born Geneva, worked in Zurich and St. Gallen, and in Bern for the Swiss Topographical Bureau, 1903-1931. Produced maps of northern Zurich canton, including the so-called 'Kriegsspielkarten' (War game maps). Canton of Zurich (24 sheets), from *c.*1895. **Ref.** DÜRST, A. 'Andreas Hefti...', Cartographica Helvetica 7 (January 1993) pp.21-32.

**Hegedüs**, Johann Nepomuk (*fl.c.*1782-1788). Map of the Ödenburger Komitat [Sopron, now Hungary], 1788; also canal and drainage maps of Burgenland [E. Austria].

**Hegemann**, Paul Friedrich August. *Ostküste von Grönland*, 1874.

**Heger**, Franz Joseph (*d.*1787). German post commissioner. *Neue und vollständige Postkarte durch ganz Deutschland* (16 sheets), Nuremberg, Homann's Heirs 1764 and later (revised version of work by J.J. von Bors, title also in French), English edition on one sheet published as *A Map of the Post Roads of Germany and the Adjacent States*, William Faden 1795.

**Hegi**, *Captain* Franz (1774-1850). Painter from Lausanne, worked in Zurich and Paris as a lithographer and engraver. Aquatints for Friedrich Wilhelm Delkeskamp, 1830, and for numerous other Swiss publishers. *Schweiz* (9 sheets), 1830.

**Hegi**, Hans Kaspar (1778-1856). Swiss engraver and lithographic printer, born Fribourg, worked in Basle, Strasbourg and Zurich.

**Heid**, —. *Plan von Reutlingen*, 1890.

**Heidanus**, Carolus. See **Heydanus**, Carolus.

**Heidegger**, Johann Heinrich (1738-1823). Swiss author, bookseller and art collector in Zurich and Italy. *Reisende durch die Schweiz*, 1787 (engraved by Laurenz Halder).

**Heidemann**, F.W. *Transportkarte von Deutschland*, 1836.

**Heiden**, Carl. See **Heydanus**, Carolus.

**Heiden** [Heyden], Christian (1526-1576). Globemaker, instrument maker, scholar and teacher at Nuremberg. Globes, 1560-1570, including a pair consisting of a convex celestial globe nested inside a terrestrial globe.

**Heiden**, Gaspard. See **Heyden**, Gaspard.

**Heidmann**, Christoph. German author of Wolfenbüttel. *Palaestina sive Terra Sancta* (with 4 maps), 1625 and later; *Europa sive manuductio ad geographium*, Helmstedt 1640.

**Heijde**, Nicolaas van der. See **Heyde**, Nicolaas van der.

**Heijmenbergh**, Jan van. See **Heymenbergh**, Jan van.

**Heijder**, Otho. *Landt Carte Schonen und Blecking*, 1659.

**Heijns**, Pieter & Zacharias. See **Heyns**.

**Heiland,** Julius. *Taschen Atlas von Berlin*, 1893.

**Heiliger,** Johann Friedrich Wilhelm. 'General Inspecteur der Indirection'. *Geographische Karte der Länder zwischen der Elbe und Weser...* (6 sheets), Hanover 1812 (title also in French, with J.L. Hogrewe).

**Heilwig,** M. See **Helwig,** Martin.

**Heim,** *Professor Doctor* Albert (1849-1937). Swiss Alpine geologist of Zurich. *Geologische Karte der Schweiz*, Bern 1894 (with C. Schmidt); *Säntis-* [relief map], 1898-1903 (with C. Meili); and others.

**Heimburger,** A. Maps for Joseph Meyer, 1830-1840.

**Hein,** F. Draughtsman. Between 1871 and 1879 he compiled some of the maps for the *Spruner-Menke Hand-Atlas*, Gotha, Justus Perthes 1880 (33rd edition of Dr K. von Spruner's *Hand-Atlas*)

**Hein,** Melchior Gottfried. Publisher of Johann David Köhler's Schlesische Chronik, 1710.

**Heine,** Albert. Lithographer. *Plan vom Königlich Prinzlichen Park und Arboretum zu Muskau*, Cottbus *c.*1868; *Der obere (untere) Spreewald* (2 sheets), Cottbus *c.*1868.

**Heine,** E. *Plan der Stadt Cöthen und der nächsten Umgebung*, Koethen 1870.

**Heine,** Hermann. *Plan der Stadt Dessau und Umgegend*, Dresden 1865, 1878; *Karte der Dresdner Heide*, Dresden 1894..

**Heine,** *General* P.B. William (*b.*1827). US Army. Writer and geographer. Central America, China, Japan &c.

**Heineken,** Christian Abraham (1752-1818). Jurist of Bremen. *Karte des Gebiets der Reichs- und Hansestadt Bremen*, 1790-1793 (with J. Gildemeister).

**Heinemann,** E. von. *Braunschweig nebst Umgegend*, *c.*1835.

**Heinfogel** [Heinvogel], Conrad (1470-1530). Mathematician and astronomer. Collaborated with Albrecht Dürer and Johann Stabius on two woodcut star charts. The co-ordinates were calculated by Stabius, Heinfogel positioned the stars, and Dürer designed and drew the constellation figures. They are thought to be the first printed star maps and were used as a source by others including Gemma Frisius. *Imagines coeli Septentrionales...* [star chart], 1515; *Imagines coeli Meridionales...* [star chart], 1515 (both charts cut by A. Dürer).

**Heinrich,** P.G. *Das Gebieth der Stadt Hamburg*, 1810.

**Heinrich Petri,** Sebastian. See **Henricpetri,** family.

**Heinrichschofen,** Wilhelm (1782-1851). Publisher of Magdeburg. *Charte vom Harz*, 1833.

**Heinrigs,** Johann. *Plan von Cöln*, *c.*1867.

**Heinsius,** P. See **Heyns,** Pieter.

**Heintzelman,** *Major* S.P. Colorado River, 1851 MS; *Map of the Gadsden Purchase...*, 1858.

**Heinz,** C.H. *Taschen Eisenbahn Atlas von Mittel Europa*, *c.*1870.

**Heinzmann,** Johann Georg (1757-1802). German bookseller and publisher in Bern. *Carte des principales Routes de la Suisse...*, *c.*1795.

**Heis,** Eduard. *Atlas Coelestis Novus*, 1872; *Atlas Coelestis Eclipticus*, Cologne 1878.

**Heis,** L. *Karte des Siegthales*, 1856.

**Heiskell,** William. Printer of Winchester, Virginia. Charles Varle's *Topographical Description...* [Frederick, Berkeley, & Jefferson Counties, Virginia], 1810.

**Heiss,** J. *Carte de la France composé de 25 feuilles*, 1833 (with I.E. Hoerl, lithographed by B. Herder).

**Heitmann,** Johan Hansson (1664-1740). Hydrographer of Amsterdam. Maps of the Norwegian coast published in Holland. *Noord Zee*, 1725.

**Heitor de Coimbra.** See **Coimbra**, Heitor de.

**Heitov,** Antonio. Planta de Macao, Lisboa 1899.

**Hekel** [Hekelius], Johan Friedrich. Edited Philipp Clüver's *Introductio in universam geographiam*, 1697 edition.

**Hékimian,** G. Maps of Turkey, including: Map of the area around Trebizond, 1883 MS; Roads from Trebizond to Baiburt and Erzeroom, 1883 MS; Map of Trebizond and Sivas vilayets..., 1883 MS; *A Map Illustrative of the Ordoo-Sivas Route, 1884*, 1884.

**Heksch,** Alexander F. *Karte der Hohen Tatra*, 1880.

**Held,** —. *Plan von Heidelberg*, c.1870.

**Held,** Carl. *Mappa imperio do Brazil...*, Rio de Janeiro 1878 (with C. Brockes and A. Zittlow).

**Held,** Franz. *Das deutsche Sprachgebiet von Mähren und Schlesien* [language map of Moravia and Silesia] (4 maps on 3 sheets), Brno 1888.

**Held,** Leonz (1844-1925). Swiss surveyor and topographer. Worked for the Swiss Topographical Bureau, 1872-1920. Collaborated on *Siegfried-Atlas*, completed 1905.

**Heldensfeld,** Anton Mayer von. See **Mayer** von Heldensfeld, Anton.

**Heldring,** *Captain* Henry. 3rd Regiment, acting engineer, Pensacola, Florida. Plan of Fort George, Pensacola, 1781 MS; Harbour of Pensacola 1781; Siege of Fort George, 1781.

**Heley,** Richard. *Map of Part of the Isle of Axholme*, 1596.

**Helfrecht,** J.T.B. *Charte von dem Fichtelgebirge*, 1800.

**Heliot,** J. Rouen, 1817 (with H. Boutigny).

**Hell,** — de. *Pilote de l'isle de Corse*, 1820-1824, Paris, Dépôt général de la marine 1831; Chart of Sardinian ports used in Supplément to *Neptune François*, 1822; charts of Majorca 1823-1829.

**Hell,** Hommaire de. See **Hommaire** de Hell, Ignace Xavier Morand.

**Hell** [Höll], Miksa [Maximilian], SJ (1720-1792). Hungarian astronomer, cartographer and globemaker, from Selmecbánya [now Banska Stiavnica, Slovakia]. d. Vienna. Lunar chart, 1764; *Ungarn* (4 sheets), 1790; historical map of Hungary *Tabula Geographica Ungariae Veteris*, 1801. **Ref.** FASCHING, Antal 'A legnagyobb magyar csillagász emléke' [Memory of the greatest Hungarian astronomer], *Térképészeti Közlöny* [Cartographical Bulletin], (1932/4) [in Hungarian].

**Hellard,** Am. Th. (*b*.1846). Norwegian geologist. Structure du globe terrestre, 1878.

**Helle,** E. Engraver. Ports maritimes de la France, 1871-1898; *Direction des Cartes et Plans. Nivre*, Paris c.1880.

**Helle,** M. Publisher of Paris. *Voyages et Promenades sur la Seine*, 1887.

**Heller,** —. Engineer, drew and engraved *Karte von der Bonn-Cölner Eisenbahn und deren Umgebungen zwischen dem Vorgebirge und dem Rhein*, 1849.

**Hellermann,** —. Frankfurt an der Oder (2 sheets), 1786-1787 MS.

**Hellert,** J.J. French cartographer. From 1836 compiled a number of maps and plans, later published together as *Nouvel Atlas physique, politique et historique de l'Empire Ottoman*, Paris, Bellizart, Dufour & Co. 1843 (engraved by C. Schrieber, A. Benitz, J. Schwaerzlé and U. Muschani, printed by Thierry Frères).

**Helleweel,** Johannes [Johan]. Dutch surveyor, worked in Zeeland. *Caarte figurative... de Ambagten van Nieuwerkke, en St. Joos Land* 1699 (with Bartholomeus de Keulenaar).

**Hellfarth,** C. Lithographer of Gotha, worked with Justus Perthes. A. Petermann's *Carte ethnographique de la Turquie d'Europe*, Gotha, J. Perthes 1861; A. Petermann's *Special-Karte von Nord-Schleswig...*, Gotha, J. Perthes c.1864; *Special-Karte von Süd-Schleswig*, Gotha, J. Perthes 1864 (drawn by Debes, H. Habenicht & Welker); *Schlesien, Königreich Sachsen und nördliches Böhmen*, Gotha, J. Perthes c.1865 (from A. Stieler's map of Germany).

**Hellis,** Clément. Map of the distribution of cholera in Rouen, 1833.

**Hellman,** Gustav (1854-1939). *Klima-Atlas von Deutschland*, Berlin 1921.

**Helloco,** —, Le. Cartes des vents dans l'Océan Pacifique Méridionale, Paris, Dépôt de la marine 1764.

**Hellström,** Carl Peter. See **Hällström,** Carl Peter.

**Hellwald,** *Baron* Friedrich Anton Heller von (1842-1892). Austrian geographer and ethnologist. *Der Feldzug des Jahres 1809 in Süddeutschland*, Vienna, C. Gerold's sohn 1864; *Centralasien*, Leipzig, O. Spamer 1880; *Amerika in wort und bild*, Leipzig, H. Schmidt & C. Günther 1883-1885.

**Helm,** Charles J. Draughtsman. Revised M. Hendges's map of *Alaska*, 1906 (lithographed by Andrew Graham Co.).

**Helman,** William. *Plan of the Island and Harbour of Codgone*, 1742, published Dalrymple 1792.

**Helmcke,** G. *Karte der Prov. Sachsen*, c.1872.

**Helme,** John. Co-publisher, with M. Lownes, J. Browne & J. Busbie, of M. Drayton's *Poly-Olbion*, 1612, 1613.

**Helme,** James. Surveyor. *An Exact Plan of the Sea Coast of the Continent from Paucatuck River Eastwards to Point Judith... to Slocums Harbour* [Rhode Island], Providence 1741 MS (with William Chandler).

**Helmer,** Ágost. Hungarian geographer. Produced historical wall maps, 1890.

**Helmersen,** Gregor von (1803-1885). Russian geologist; one of the founders of the Imperial Russian Geographical Society. *Das Südliche Ural-Gebirge von Slatoust* [Ural Mountains], Berlin 1831; *Empire of Russia*, St Petersburg 1841.

**Helms,** J.C. *Plat Book of Hancock County by Townships*, Carthage 1908.

**Helmuth,** C. *Karte von Palästina*, Halle 1843; *Plan von Jerusalem*, 1843.

**Helpen,** Coenders van. See **Coenders** van Helpen.

**Helvicus,** Nicolaus. *Theatrum Historiae Universalis Catho.-Protest* (62 maps), Frankfurt am Main, Schönwetter 1644 (engraved by D. Custos).

**Helwig** [Heilwig; Helweg], Martin (1516-1574). *b.* Neisse [Nysa, Silesia], *d.* Breslau [Wroclaw]. German geographer, cartographer, teacher and diarist. Surveyed and drew *Silesia* (woodcut in four sheets), Breslau, Johann Creutzig 1561, later printings 1605, 1627, 1642, 1685, 1738 etc. to 1776 (used by Ortelius, 1570); map of ancient Italy, 1561 (taken from Ptolemy); also wrote *Erklerung der Schlesischen Mappen*, Breslau, Crispin Scharffenberg 1564. **Ref.** SMITS, J. 'For pleasure and support' in *La Cartografia dels Països Baixos* (Institute Cartogràfic de Catalunya 1994) pp.145, 147-149, 162, 253 [in English]; KARROW, R.W. *Mapmakers of the sixteenth century and their maps* (1993) pp.288-292; HEYER, A. 'Die kartographischen Darstellungen Schlesiens bis zum Jahre 1720', *Zeitschrift des Vereins für Geschichte und Altertumskunde Schlesiens* 23 (1989) pp. 177-240, 305-355 [in German]; KOCOWSKI, B. 'Calendarium Marcina Helwiga...', *Bibliotekoznawstwo* 3 (1962) pp.75-112 [in Polish].

**Hem,** Laurens van der (1621-1678). 'Advocaat, beursman' and map collector of Amsterdam. Owned a copy of the Blaeu *Atlas Maior*, which he extended to 46 volumes with maps by others, coloured to order by Dirk Jansz. van Santen, now in the National Library in Vienna (also known as the 'Atlas van der Hem' or 'Atlas of [Prince] Eugene'). **Ref.** SCHILDER, G. *The Map Collector* 25 (1984) pp.22-26.

**Hembertus,** J.H. Mapmaker of Antwerp. Siege of Ghent, 1708.

**Heming,** George H. *Seneca County, Ohio*, 1854.

**Hemingway,** G.S. Surveyor. Inclosure plan at Angmering [Sussex], 1809 MS.

**Hemingway,** William. Surveyor. Credited with a survey of Georgetown District, South Carolina, 1820, used by Robert Mills in his *Atlas of the State of South Carolina*, 1825.

**Hemming,** John. *Royal Engineers Survey operations at the Cape of Good Hope*, 1851.

**Hemming**, P. Lawson. Draughtsman in the Hydrographic Office. *Mediterranean - Ionian Sea. Santa Maura, Ithaca and Cephalonia Islands with the adjacent coast*, Admiralty 1867 (drawn from Captain Mansell's surveys of 1864-1865, engraved by J. & C. Walker).

**Hemminga**, Doco [Duco] ab (1527-1570). Geographer, theologian and mathematician of Berlicum near Leeuwarden, Friesland. Thought to have compiled a world map entitled *Tabulam geographicam totius mundi*, c.1550, mentioned in Ortelius's *Catalogus auctorum*, but now lost.

**Hemmings**, S. British cartographer. Seat of War in America, 1789.

**Hen.**, Joh. [Ioh.] Engraver. *Provinzien Nord und Sud Carolina*, Bern 1711.

**Henault**, Jean (fl.1660-1671). 'Rue St. Jacques à l'Ange Gardien', Paris. One of the sellers of A. Manesson-Mallet's *Les Travaux de Mars...*, 1671.

**Henchman**, Daniel [*pseud.* Abraham Weatherwise]. American publisher. *Father Abraham's Almanack... for 1759*, Philadelphia 1758 (with a plan of Louisbourg Harbour).

**Henckel**, *Captain* C. *Danmark, Sverige, Norge*, 1855; *Kjöbenhavn*, c.1860.

**Henckel**, *Major* George. Deputy U.S. surveyor. *New Map of the Black Hills*, Rand McNally 1877.

**Hendersen**, David. New Zealand surveyor. *Town of Dunedin*, 1861; *Proposed Site for Lunatic Asylum and Hospital Site, Otago, Geological Survey Offices* 1862; *Gold Fields Otago*, Otago 1864.

**Henderson**. See **Turner** and **Henderson**.

**Henderson**, C.W. *County of San Luis Obispo, California*, 1890.

**Henderson**, *Captain* J.S. *Sketch Map of West Cornwall showing the Relative Position of Mines*, 1907.

**Henderson**, James. *A Map of the Brazil*, 1821.

**Henderson**, *Captain* John. Map of the soil of Sutherland, 1812 (engraved by S.J. Neele); map of the soil of Caithness, 1812 (engraved by S.J. Neele); Wick & Thurso, 1812.

**Henderyk**, Catare. Captain in the French Engineers. *Plan du fort Rametkens*, c.1809.

**Hendges**, M. Draughtsman. *Mexico*, Baltimore, Friedenwald Co. 1900; *Guatemala...*, 1902; *...United States including Territories and Insular Possessions showing the extent of Public Surveys, Indian, Military and Forest reservations, Railroads, Canals, National Parks and other details...*, 1904 (engraved by R.F. Bartle & Co., lithographed by L. Restein Co.); *...Alaska...*, 1906 (revised by C.J. Helm, lithographed by Andrew B. Graham Co.).

**Hendricks** [Henrici; Henry], Albert. Printer of The Hague. J.H. van Linschoten's *Descriptio Totius Guineae Tractus, Congi, Angolae, et Monomotapae...*, 1599; Linschoten's *Navigatio ac Itinerarium...*, 1599; *Thrésor de chartes*, Amsterdam, Cornelis Claesz 1600 (French edition of the *Caert-Thresoor* by B. Langenes, translated by I. de la Haye).

**Hendricks**, Cornelius (*fl.c.*1616). Manuscript map of parts of New York, New Jersey and Pennsylvania, 1616 (presented to the States General on August 19th 1616, used as a source for later printed maps).

**Hendrikz**. [Hendrikzoon], Dircks. Dutch surveyor and cartographer, worked in Zeeland, also sheriff of Vosmeer, 1599. *Caarte ende discriptie figuratief van de zeedijcken van Groot Merschmont...*, 1579 (with Pieter de Buck, Jan Simonse and F. Horenbault).

**Hendrip**, Hans. *Travels in Greenland*, Godthaab 1875.

**Hendrixzoon**, Willem. Surveyor in Zeeland. Hoogenmoere [Zeeland], 1440 MS.

**Hendry**, James. Engineer. *Whitechapel to Thames Docks*, 1842 MS.

**Hendry**, W.A. Draughtsman. *Mackinlay's Map of the Province of Nova Scotia, including the Island of Cape Breton*, Halifax, Nova Scotia, A. & W. Mackinlay 1867 (engraved by G. Philip & Son).

**Hendschel, Ulrich** (1804-1862). German postal officer, Frankfurt am Main. *Post-und Reise-Karte von Deutschland und den Nachbar-Staaten* (4 sheets), Frankfurt 1843; *Eisenbahn-Atlas von Deutschland, Belgien un dem Elsass* (16 maps), 1844 and later; *Eisenbahn-Karte Central Europa*, 1852.

**Henecy**, —. Irish engraver. **N.B.** Compare with C. **Henecy,** below.
• **Henecy & Fitzpatrick**. Engravers for William McCrea's map of County Monaghan, Ireland (4 sheets), c.1795; McCrea's map of Donegal (4 sheets)l, 1801.

**Henecy, C.** Engraver of J. Brownrigg's *A Map of the Kingdom of Ireland*, published in W.W. Seward's *The Hibernian Gazetteer*, Dublin, A. Stuart 1789; *A Map of Antient Ireland Previous to the ... 13th... Century*, published in W.W. Seward's *The Hibernian Gazetteer*, Dublin, A. Stuart 1789.

**Heneman** [Henneman], **Johan Christoph**. (*fl.*1780-1806). Cartographer and surveyor. Plan du siege de Brunsvic, The Hague, Pierre Gosse and Daniel Pinet 1763; *Kaart van de Colonie Surinam* (140 sheets), before 1784 MS, published as *Kaart van de Colonie Surinam en deonderhoorige rivieren en Districten* (8 sheets), Amsterdam, Gerard Hulst van Keulen and Nicolas Vlier 1784 (engraved H. Klockhoff), London edition (4 sheets), William Faden 1810. **Ref.** KOK, M. 'Johan Christoph Heneman: kartograaf van Suriname en Guyana van 1780-1806' *Caert Thresoor* februari 1982, le jg [1] (1982) pp.4-12. [in Dutch].

**Henhenfeld**, Paul Pfinzing von. Das Ampt Herrspruck, 1596.

**Henion, John W.** Collaborated with Charles M. Foote in the publication of plat books of American counties, 1890-1895 including *Plat book of Dane county, Wisconsin*, Minneapolis, C.M. Foote & Co. 1890; *Plat book of Green county, Wisconsin*, Minneapolis, C.M. Foote & Co. 1891.

**Henman**, —. Salcombe to Looe, 1791 (with Smith).

**Henn, Williams & Co.** Co-publisher, with R.L. Barnes, of *Township Map of the State of Iowa*, Philadelphia 1855.

**Henneberger** [Henneberg; Hennebergen], Caspar [Gaspar; Kaspar] (1529-1600). Prussian priest and cartographer of Ehrlich near Hof. Map of Livonia, 1555 (now lost, thought to have been used as a source by Jan Portant and M. Ambrosius); *Prussiae* (9 sheets), Kaliningrad, Georg Osterberg 1576 (woodcut by C. Felbinger, used by Ortelius, 1584, 1595, by Hendrick Hondius, 1636, by the Blaeus, 1638-1640, by J. Jansson, 1652-1658).

**Henneman, J.C. van.** See **Heneman**, Johan Christoph.

**Hennepin, Louis de** SJ (1640-1701). Franciscan priest and missionary in North America, travelled with René-Robert de La Salle down the Mississippi, 1682. His maps and descriptions appeared in several editions and languages. *Carte de la Nouvelle France et de la Louisiane Nouvellement découverte*, 1683 (published in *Description de la Louisiane*, Paris, veuve Huré 1683, Paris, A. Auroy 1688); *Beschreibung der Landschafft Louisiana*, 1689; *Nouvelle découverte d'un tres grand pays situé dans l'Amérique* (3 maps), Utrecht, W. Broedelet 1697, Amsterdam, A. van Someren 1698, English edition published London 1698, used by P. van der Aa 1704, Amsterdam, A. Braakman 1704, and other editions; *Nouveau voyage d'un pais plus grand que l'Europe*, Utrecht, E. Voskuyl 1698. **Ref.** WALDMAN & WEXLER *Who was who in world exploration* (1992) pp.316-317.

**Hennequin**, *Captain* —. *Relief du Mont-Valérian et du Bois du Boulogne*, c.1872; *Carte Géologique de l'Europe*, Brussels 1875 (compiled from earlier maps by A. Dumont, H. von Dechen, Beaumont, Sydow and J.C. Houzeau).

**Hennequin, L.** Engraver 'au Dépot G.$^d$ de la Guerre. Rue St. Landry N$^{o.}$5 en la Cité', Paris. *Carte de la France divisée suivant le plan proposé a l'Assemblée Nationale*, Paris, 1789; St Cloud, 1816; *Plan de Paris divisé en XII Mairies et 48 Quartiers*, Paris, Bassel 1817 (drawn by Hennequin fils); engraved for Jean Alexandre Buchon, 1825.
• **Hennequin** *fils*. Draughtsman for *Plan de Paris divisé en XII Mairies et 48 Quartiers*, Paris, Bassel 1817 (engraved by L. Hennequin).

**Hennert,** —. *Plan des jardins et environs de Reinsberg,* 1772; Saxony and Bohemia (20 sheets), 1778.

**Hennessy,** J.M. Oro Bay [Pacific], Brisbane 1892.

**Hennet,** George (1799-1857). Surveyor of parts of England and Wales. *Map of the County Palatine of Lancashire* (4 sheets), surveyed 1828-1829, published Henry Teesdale & Co., 1830 (engraved James Bingley). **Ref.** BAYLEY, J.J. & HODGKISS, A.G. Lancashire: a history... in early maps pp.65-66.

**Hennicke,** Johann Friedrich (1764-1848). Geographer of Gotha.

**Hennigs,** C. *Eisenbahn-Karte des östlichen Europa* (4 sheets), Utrecht *c*.1875; ... *Nederland und Belgien, c*.1876.

**Henning,** J.S. *Santa Clara District, Nevada,* 1846.

**Henrichs,** Christoph. Artist, assisted with Praetorius celestial globe, *c*.1576.

**Henrici,** Alberti. See **Hendricks,** Albert.

**Henricpetri,** Adam. Bought the woodblock for Wolfgang Lazius' map of Hungary and republished it with the title *Beschreibung und Gelegenheit des Turcken Zugs in Ungern im jar 1556...,* Basle 1577 (in *General Historien...,* map first published by J. Oporin, 1556). **N.B.** Assumed to be a member of the **Henricpetri** family, below.

**Henricpetri,** family. **N.B.** See also **Petri** family. **Ref.** HIERONYMUS, F. *1488 Petri - Schwabe 1988. Eine traditionsreiche Basler Offizin im Spiegel ihrer frühen Drucke* (Basle 1997) [in German].
• **Henricpetri** [Petri], Heinrich (1508-1579). Publisher of Basle, born Heinrich **Petri,** stepson of Sebastian Münster. On being honoured by Kaiser Karl V in 1556, he adopted the name Henricpetri. Heinrich was succeeded by his son Sebastian Henricpetri.
•• **Henricpetri,** Sebastian (1546-1627). Son of Heinrich Petri, printer and publisher of Basle, worked with his brother Sixtus until *c*.1627. German edition of Olaus Magnus' *Historien...* (with a crude copy of the map of Scandinavia), 1567; took over from his father as publisher of Münster's *Cosmographia* to which he added some new woodcut maps, 1588, 1592, 1598, 1614, 1628.
•• **Henricpetri,** Sixtus (*b*.1547). Worked as a printer with his brother Sebastian until *c*.1627.

**Henricus,** —. Surveyor of Utrecht. Salland, 1312 MS.

**Henriksen,** E. Hydrographer for the Commission Européenne du Danube. Worked with E. Magnussen and C. Kuhl on *Carte du bras de Soulina,* 1897 (lithographed by J. Schenk).

**Henrion,** Albert. *Carte du Phylloxera pour l'Arrondissement de Narbonne (Aude),* Paris 1881 (drawn by M.A. Pameron).

**Henrion,** Denis (*d.c*.1640). French mathematician and cosmographer. Published a French edition of Robert Hues's treatise on globes as *Traicté des globes,* Paris 1618 (the English edition had been published in London, 1594).

**Henriot,** J.N. Steel engraver of Paris. *Nouveau plan-guide de Paris et des environs avec les fortifications,* Paris, Bouquillard 1849 (prepared by Faucheta and Fontet); *Nouveau plan complet de Paris avec ses fortifications divis en 12 Arrondissements...,* Paris, A.Bs et F. Dubreuil 1854.

**Henry** of Mainz (*d*.1153). Henry, canon of St Mary's church, Mainz is named as the author of a text bound into a 12th-century manuscript copy of the *Imago mundi,* a geographical text by Honorius. The oval mappamundi bound into the front, and previously credited to Henry, is now thought to have been made in Durham. It is held at Corpus Christi College, Cambridge, and is described as the 'Sawley map'. **Ref.** HARVEY, P.D.A. *Imago Mundi* 49 (1997) pp.33-42.

**Henry,** Prince of Portugal [Henry the Navigator; Dom Henrique] (1394-1460). Promoter of geography, exploration and discovery, especially along the coast of West Africa. **Ref.** RANDLES, W.G.L. *Imago Mundi* 45 (1993) pp.20-28; WALDMAN & WEXLER *Who was who in world exploration* (1992) pp.318-320.

**Henry,** Albert. See **Hendricks,** Albert.

**Henry**, Alexander *the elder* (1739-1824). Undertook explorations across North America and Canada. Map of Lakes and Hudson's Bay, 1775 MS; *A map of the north west parts of America*, c.1776. **Ref.** WALDMAN & WEXLER Who was who in world exploration (1992) p.317.

**Henry**, Anson G. (1804-1865). United States Surveyor General and Indian agent. Public Surveys of Washington Territory, 1863.

**Henry**, B. Engraver, worked for Alexander Dalrymple, 1770-1775. Balambangan [Borneo], 1770; Bengal, 1772.

**Henry**, D.F. One of the surveyors working in North America under the direction of J.N. Macomb. Worked with others on surveys of Eagle Harbour, Lake Superior, 1855; and Marquette Harbour, Lake Superior, 1859.

**Henry**, David (1710-1792). Editor and publisher of St. John's Gate, London. Worked for the publisher Edward Cave from c.1731, and married Cave's sister in 1736. On the death of Cave in 1754, Henry joined Edward Cave's son Richard in partnership as publisher of *Gentleman's Magazine*. When Richard Cave died in 1766, the bookseller Francis Newbery was added to the imprint. The magazine carried a number of road strip maps and canal maps between 1765-1782. Compiled *An Historical account of all the voyages round the world, performed by English navigators...*, London, F. Newbery 1774. **Ref.** CARROLL, R.A. Printed maps of Lincolnshire 1576-1900 (1996) pp.381-383; JOLLY, D.C. Maps in British periodicals Part I (1990) pp.33-87.

**Henry**, Etienne. Cartographer and publisher of Paris. *Plan de la Bataille de Jens, Gagné par la Grande Armée Française commandé par l'Emporeur Napoleon sue l'Armée Prussienne Commandé par le Roi en personne*, Paris, Bance 1807; *Plan des batailles d'Enzersdorf et de Wagram...*, 1807; *Tableau des opérations de la Grande Armée depuis le 8 août jusq'au 20 octobre 1812* (2 sheets), 1813; *Europe historique et géographique*, Paris 1838 (with A. Megret).

**Henry**, F.P. Joint publisher, with T.M. Fowler, of T.M. Fowler's bird's-eye view of Hamburg [Pennsylvania], 1889.

**Henry**, *Lieutenant* H.R., *RN*. Hydrographic surveyor aboard HMS *Rattlesnake*. Worked with T.M. Symonds on surveys for *Australia south coast. Port Phillip*, 1836, published Admiralty 1838, 1853 (engraved by J. & C. Walker).

**Henry**, Jacques. Publisher in Paris. *Nouvelle géographie de la France*, 1842; *Nouvelle géographie et statistique de la France*, 1842.

**Henry**, John. Scottish migrant who settled in Virginia in 1727. From 1766 to 1770 he prepared a map of Virginia, compiled mostly from previous maps. *A New & Accurate Map of Virginia* (4 sheets), London, Thomas Jefferys 1770. **Ref.** RISTOW, W.W. American maps and mapmakers (Detroit 1985) p.53.

**Henry**, *Professor* Joseph (c.1797-1878). Secretary of the Smithsonian Institution. From 1825 he led one of the road surveying teams which provided data for David Burr's *An Atlas of the State of New York*, 1829; *Rain-Chart of the United States...*, Washington D.C., J.F. Gedney 1870, 1872 (base chart engraved by H. Lindenkohl, data gathered from different recording stations by telegraph in accordance with a system devised by Henry); *Summer temperature chart of the United States...*, New York, J. Bien 1874 ('compiled by C.A. Schott under the direction of Prof. Joseph Henry').

**Henry**, M.S. *Map of Northampton County, Pennsylvania*, 1850.

**Henry**, *Colonel* Maurice (1763-1825). French abbot, astronomer, geodesist and engineer, worked in Mannheim, St. Petersburg, Paris, Munich and Switzerland. Published several scientific papers on the subject of geodesy. Surveyed parts of Switzerland for the Napoleonic campaigns, 1803-1813 (with J.H. Weiss).

**Hensal**, M. East Africa, Abyssinia, 1861-1864.

**Hensel**, A. Printer of Hamburg. *Karte West-Aequatorial-Afrikas zur Veranschaulichung des Deutschen Colonialbesitzes* [Cameroon etc.], Hamburg, L. Friederichsen 1884.

**Hensel**, Daniel. See **Herline** & Hensel.

**Hensel** [Henselius], Gottfried [Godofridus]. German philologist. *Synopsis universae philologiae* (with four linguistic maps of the continents engraved on copper), Nuremberg, Homann Heirs 1741; *Vorstellung des Ursprunges vom Elb-Strome*, 1741. **Ref.** WALLIS & ROBINSON *Cartographical innovations* (1987) pp.113-114; KRETSCHMER, DÖRFLINGER & WAWRIK *Lexikon zur Geschichte der Kartographie* (1986) p.763.

**Hensgen**, C. Engraver for the Geographisches Institute in Weimar. *Markgrafschaft Maehren*, 1846; *Provinz Pommern*, 1850.

**Henshall**, —. *Henshall's Illustrated topography of twenty-five miles around London*, Simpkin, Marshall & Co. 1838.

**Henshall**, J. Engraver and printer of N$^{o.}$1 Cloudesley Terrace, Islington, London. *Gravesend, enlarged from the ordnance Survey, by Robert K. Dawson*, London, Hansard & Sons 1831; Town plans for Society for the Diffusion of Useful Knowledge [SDUK] Atlas: Athens, 1832; *London*, 'Published under the Superintendence of the Society for the Diffusion of Useful Knowledge' 1835, London, Baldwin & Cradock, 1836, London, Chapman & Hall 1844 (drawn by W. B. Clarke); Milan and Turin, 1835; *Lisbon (Lisboa)*, Edward Stanford / SDUK *c*.1836 (drawn by W.B. Clarke); *Panoramic View of London*, *c*.1836 (taken from Thomas Hornor's panorama); Copenhagen, 1837; *A Map of Tunbridge Wells in the County of Kent*, 1838 (surveyed by S. Rhodes); *Birmingham*, 1839.

**Henshall**, Samuel. *Domesday; or an actual survey of South Britain*, 1799 (with a map of Kent engraved by S. Neele).

**Henshall**, W. Engraver for SDUK Atlas, 1844-1846; for George Booth's *London*, 1846, London, J. Reynolds and George Booth 1846 (with additions). **N.B.** Compare with J. **Henshall**, above.

**Hensley**, George B. *City of San Diego*, 1870?.

**Henslow**, John Stevens (1796-1861). English botanist and geologist. *Geology of the Isle of Man*, 1821.

**Hensman**, Mary. *Dante Map* [Italy], 1892.

**Henter**, —. See **Bodoki**, Mihály.

**Hentschel**, K.F.T. Western Hemisphere, 1795.

**Hentschel**, Theodore. Lithographer of 66 Swanston St., Melbourne, Victoria. *Plan of Melbourne*, *c*.1850.

**Hentzner**, Paul. Itinerario Angliae, 1598; Itinerario Germaniae, 1612, 1629.

**Henwood**, D. Engraver for Charles Smith, 1813; *Chart of the North Sea*, 1814; *Chart of the Bay of Biscay*, London, Richard Holmes Laurie 1827 (drawn by J. Outhett).

**Henze**, A. Map editor of Leipzig. Editor of a coloured lithographed map of Europe, 1900.

**Henzé**, Gustave. *La France Agricole*, Paris 1875.

**Henzler**, G. *Schul Wand Karte von Württemberg*, 1872; *Schul Wand Karte von Deutschland*, 1875.

**Hepburn**, John (*fl*.1836-1869). Surveyor. Map of the Ellon District of Aberdeenshire, Scotland, 1848.

**Hepburn**, L. Surveyor. Lands of Ballydonagh, Wicklow, 1762 MS.

**Hepburn**, Captain R. *Coast of England and Scotland*, 1812-1825.

**Herald Press**. Lithographers of Birmingham. *Map of Warwickshire*, 1890 (used in W.B. Grove and J.E. Bagnall's *The flora of Warwickshire*, London, Gurney and Jackson, Birmingham, Cornish Brothers 1891).

**Heramb**, G. Draughtsman. *Route Map of Norway*, 1895.

**Hérault**, —. French script engraver. Worked for Maurille Antoine Moithey, 1773; for Etienne André Philippe De Prétot, 1787; *Côtes d'Espagne*, Dépôt de la Marine 1793 (map engraved by Bouclet); for Edme Mentelle, 1797-1801; letter engraver for *Atlas du Voyage de La Pérouse*, 1797; text on *Gouvernement de Champagne*... (4 sheets), 1807 (map engraved by P.F. Tardieu).

**Herba**, Giovanini de l'. Itinerario delle poste... del mondo, Rome 1563.

**Herberstein** [Heberstein; Herberstain; Herberstein von Neyberg und Guetenhag], Sigismund, *Reichsfreiherr* von (1486-1566). Austrian statesman and writer from Wippach/Krain. Imperial ambassador to Moscow, 1516-1518, and 1526-1527, died in Vienna. Wrote a description of Russia entitled *Rerum Moscoviticarum Comentarii*, 1549 (with a map by Augustin Hirschvogel), Italian edition published 1550 (with a map by Gastaldi, copied from Hirschvogel), Basle edition published by Oporin, 1551, 1556, 1571 (with a woodcut map copied from Hirschvogel, an anonymous plan of Moscow was added in 1556), German editions published 1557, 1563, 1567. His description was used as a source for many other works on Russia. **Ref.** KARROW, R.W. *Mapmakers of the sixteenth century and their maps* (1993) pp.295-297; 'Zu Sigismund Herbersteins Karte von Moscovia', *Nordost-Archive* 30 (1974) pp.3-13 [in German].

**Herbert**, *Freiherr* — von. One of the surveyors for *Navigations Karte der Donau von Semlin an bis zu ihrem Ausfluss ins Schwarzen Meer...*, Vienna, Kurtzbeke 1789 (engraved by Mansfeld).

**Herbert**, —. *Karte des Hohen Himalaja*, 1832.

**Herbert**, A. *Carte du Département de l'Aisne*, c.1868.

**Herbert**, Charles E. State of Sonora, 1885.

**Herbert**, *Lieutenant* Frederick Charles *RN* (*b*.1819). *Chart of Lake Huron*, Toronto, Hugh Scobie 1850; *Lake Ontario*, Toronto, Hugh Scobie 1852 (reduced from the surveys of Captain W.F.W. Owen and A. Ford, with additions by F.C. Herbert); surveys used for *Chart of Lake Ontario*, Toronto, W.C. Chewett & Co. 1863 (surveys by Captain Owen, Captain Ford and E.M. Hodder were also used).

**Herbert**, Frederick H. Surveyor and lithographer of East Saginaw, Michigan. *Map of the City of East Saginaw, Michigan*, c.1866; *Map of Bay County, Michigan*, 1869 (compiled by B.F. Bush); *Map of East Saginaw, Michigan Showing a portion of Saginaw City and Carrolton*, 'Compiled and Published by S.R. Kirby and Frank Eastman' 1870.

**Herbert**, Humphrey. Draughtsman and surveyor. Manuscript plans of St Philip's Castle, Minorca, 1735.

**Herbert**, James. Lithographer, '3 Kennington Place, Kennington Park S.', London. *Plan of London*, London, John Hobbs & Co. 1862.

**Herbert**, *Colonel* John. *The Ichnography or Plann* [sic] *of the fortification of Charlestown* [South Carolina], 1721 MS; Province of South Carolina, 1725 MS.

**Herbert**, L. See **Hebert**, L.

**Herbert**, L. Draughtsman for *Physical & Political Map of South America*, New York, J.H. Colton 1851 (compiled by Dr. R.S. Fisher, engraved by Neale).

**Herbert**, Lewis. Draughtsman for *Whittle & Laurie's New Map of Ireland*, London, Whittle & Laurie 1824. **N.B.** Probably L. **Hebert**.

**Herbert**, Louis. Scotland, 1823. **N.B.** Probably L. **Hebert**.

**Herbert**, Thomas (1597-1642). Maps of the Indian sub-continent, 1638.

**Herbert**, William (1718-1795). Publisher, bibliographer and mapseller. Member of the Drapers Company, served with the East India Company. Traded from the 'Golden Globe under the Piazzas on London Bridge' and '27 Goulston Square near Whitechapel Bars'. Herbert sold the business to Henry Gregory c.1775. *A new and correct map of the world...*, c.1750 (drawn by W. Godson, for G. Willdey, c.1715); *A chart of the Cape of Good Hope, and parts adjacent taken geometrically in the year 1752*, published 1767, re-issued Henry Gregory 1777, fifth edition 'improved and augmented' by Samuel Dunn, published Henry Gregory 1780; re-issued some of George Willdey's maps, c.1755 (with Thomas Jefferys Sr.); republished the 8 charts of C. Southack's *The New England Coasting Pilot* as a single sheet, 1758 (with Robert Sayer); *A New Directory of the East Indies...* (28 charts), 1758, (29 charts), 1759, (38 charts), 1767, (46 charts), 1773, (56 charts),

*A trade card for William Herbert (1718-1795). (By courtesy of Map Collector Publications)*

1787 (later editions published by Henry Gregory); William Nicholson's Ceylon, 1762; worked with John Andrews and Andrew Dury on *A Topographical Map of the County of Kent* (25 sheets), London, J. Andrews, A. Dury and W. Herbert 1769, 1775, 1777, R. Sayer & J. Bennett 1779, R. Laurie & J. Whittle 1794, single sheet version engraved by T. Kitchin, 1769, R. Laurie & J. Whittle 1794. **Ref.** ALLPRESS, P. 'William Herbert - Cartographer in *IMCoS Journal* 53 (Summer 1993) pp.15-16.

**Herbertson**, Andrew John. (1865-1915). Editor for J.G. Bartholomew, Atlas of Meteorology, 1899; editor for some Oxford University Press publications, including the 'Oxford Wall Map' sets. These sets (which usually included 4 maps as listed for Asia) were compiled for Africa, North America, South America, Europe and Australia. *The World. Thermal Regions*, 1909; *The World. Mean Annual Rainfall*, 1909; *Asia. Mean Annual Rainfall* (2 sheets) [Oxford Wall Maps], London, Henry Frowde 1909 (compiled by A.J. Herbert and E.G.R. Taylor, drawn by B.V. Darbishire); *Asia* (2 sheets) [Oxford Wall Maps], London, Henry Frowde 1910 (compiled by B.B. Rogers and E.G.R. Taylor, drawn by B.V. Darbishire); *Asia. Physical Features* (2 sheets) [Oxford Wall Maps], Henry Frowde 1910 (compiled by B.B. Rogers and E.G.R. Taylor, drawn by B.V. Darbishire); *Asia. Vegetation* (3 sheets) [Oxford Wall Maps], London, H. Frowde 1910 (drawn by B.V. Darbishire); *The British Isles. Railway and Steamer Routes*, H. Frowde 1910 (drawn by Darbishire); *The World. Pressure and Winds*, 1911; *The World. Deposition and Erosion*, H. Frowde 1912 (drawn by B.V. Darbishire); and many other similar maps.

**Herbich**, Ferenc (1821-1887). Hungarian geologist, worked on the surveying and mapping of Bukowina, east of the Carpathians. Geological map of Székelyföld [Sekler land, Transylvania], 1878.

**Herbig**, F.A. Publisher of Berlin. *Dr. Edward Stolle's Uebersichtskarte der Rübenzucker-Industrie*, 1853 (engraved and lithographed by L. Kraatz, printed by Edward Haenel).

**Herbin de Halle**, Pierre Etienne (*b.*1772). *Atlas de la République française composé de 102 feuilles*, 1802 (with Pierre Grégoire Chanlaire).

**Herbitz**, abbé —. *Carte de la Moldavie* (4 sheets), Vienna 1811.

**Herbor**, P.H. *Schul Wandkarte des Kreises Saarburg* (4 sheets), *c.*1895.

**Herbord**, A. *Grundriss der Eisthäler... im Grindelwald*, *c.*1790.

**Herborn**, E. Surveyor of Bathurst, New South Wales. *Mudgee*, 1884.

**Herbort**, Albert d'. Engineer, engraver and cartographer. L. de Marne's *Plan des environs de la Ville et des forts de Bergen-op-Zoom...*, 1747, published Paris 1748, also at Augsburg, Matthäus Seutter, 1748.

**Herbst**, F. Draughtsman. *Boundary between the United States & Mexico shewing the Initial Point under the Treaty of December 30th 1853*, 1855 (surveyed by W.H. Emory, J.H. Clark, M. von Hippel and J.E. Weyss, engraved by W.H. Dougal).

**Herbst**, Johann. See **Oporin**, Johannes.

**Herd**, *Captain* James. Scottish seaman, Master of the *Rosanna* on its voyage to New Zealand, 1826-1827. *Otago or Port Oxley in New Zealand*, 1826 MS; *Wangenuiatera or Port Nicholson...*, 1826 MS; manuscript surveys of the River Shooukianga..., used as a source by Louis Isidore Duperrey in *Hydrographie: Atlas*, 1827; charts of the Thames and Jokeehangar rivers, 1827, used as insets on *A chart of part of New South Wales*, London, J.W. Norie 1838 (in *The Complete East India Pilot*); and others.

**Herdegen**, Friedrich (*d.*1837). Bavarian military cartographer. *Chur-Baiern*, 1802; *Königreich Baiern*, 1806; *Süd-Teutschland*, 1806; *Wettersteingebirge*, 1826.

**Herder**, B. [Verlag]. Lithographer and publisher of Freiburg and Karlsruhe. *Neuer allgemeiner Hand- und Schulatlas*, 1821, 1829, 1831; J.H. Weiss & J.E. Woerl's *Topographische Karte des Rheinstrom's* (19 sheets), 1828; *Atlas von Europa* (25 sheets), 1829; *Das Koenigreich Wuerttemberg, das Grossherzogthum Baden und die Fürstenthümer Hohenzollern* (12 sheets), 1831-1834; F.G.F. von Kausler's *Atlas des plus mémorables batailles, combats et sièges des temps anciens, du moyen âge et de l'âge moderne en 200 feuilles...*, Karlsruhe 1831-1837; *Carte de la France composé de 25 feuilles*, 1833 (by J. Heiss & I.E. Hoerl); J.V Kutscheit and Julius Löwenberg's *Historisch-geographischer Atlas*, 1839; lithographer and publisher of Dr J.E. Woerl's *Atlas von Central Europa...* (62 sheets), *c.*1859; V.F. Klun's *Hand- und Schulatlas über alle Theile der erde*, 1869.

**Herédia**, M.G. de. See **Erédia**, Manuel Godinho de.

**Hereford Mappa Mundi**. World map, *c.*l280, believed to be the work of Richard of Haldingham. **Ref.** HARVEY, P. *Mappa Mundi: The Hereford world map* (1996); BARBER, P. *The Map Collector* 48 (1989) pp. 2-8.

**Heremberck**, Jacques. Engraver and printer of German origin, worked in Lyon with Michel Topié. *Des Sainctes Peregrinations de Iherusalem* [the travels of B. von Breidenbach, translated into French by N. le Huen], 1488 (the engraving of the copper plates is attributed to J. Heremberck).

**Herfst**, J.J. *Officieele Kaart van den Oranje Vrijstaat...*, Bloemfontein 1891; *Atlas van den Oranje Vrijstaat...*, 1896; *Plan van de Stad Bloemfontein*, 1898; *Poat kaart van den Oranje Vrijstaat*, Pretoria, Höverker & Wormser *c.*1898-*c.*1901 (sold in Bloemfontein by W.A. Wright); and others.

**Hergesheimer**, Edwin. Cartographic draughtsman. *Reconnaissance of the Western Coast of the United States from Gray's Harbor to the Entrance of Admiralty Inlet*, 1853 (engraved by G. McCoy, E.F. Woodward & W. Smith); *Preliminary Chart of North Edisto River* [South Carolina], U.S. Coast Survey 1853 (with J.R.P. Mechlin); *Map of Virginia Showing the distribution of its Slave Population...*, 1861; *Plan of Fort Jackson, showing the effect of the bombardment by the U.S. mortar flotilla and gunboats April 18th to 24th 1862*, Washington, D.C., U.S. Coast Survey *c.*1862 (surveyed by J.S. Harris, lithographed by Bowen & Co.); *Map showing the operations of the national forces... capture of Atlanta Georgia Sept. 1 1864*, Washington, Coast Survey Office 1864 (lithographed by C.G. Krebs); *Part of Arlington, Virginia*, 1864 (with R.E. McMath); *Upper Geyser Basin, Fire Hole River, Wyoming Territory*, Geological Survey of the Territories 1871; Geological Survey of Yellowstone Lake, 1871; *Parts of Idaho, Wyoming & Montana*, 1871; *Blackwell's, Ward's, and Randell's Islands and adjacent shores of the East and Harlem Rivers*, U.S. Coast Survey 1885.

**Hergt**, C. *Palästina*, Weimar 1869.

**Héricourt**, R. de. Voyage dans le pays d'Adel, 1840.

**Hérigone**, Pierre (*fl.*17th century). French mathematician and geographer.

**Heriot**, George (1759-1839). *Travels through the Canadas containing a description of the picturesque scenery*, London, Philips 1807 (with a map and views by George Heriot).

**Hérissant**, family.
• **Hérissant**, Jean Thomas (*d.*1772). Publisher and bookseller of Paris, 'rue St Jacques à St Paul & St Hilaire'. *Atlas moderne ou collection de cartes* (37 maps), 1762 (with J. Lattré).
• **Hérissant**, veuve. Widow of Jean Thomas Hérissant. Issued *Atlas céleste de Flamstéed*, 1776 and later; L.C. Desnos's *Atlas national et général de la France*, 1792.

**Hérisset** [Hérriset], Antoine (1685-1769). Engraver of Paris. *Carte des Environs de Mardick et de Dunkerque*, 1752; *Ville et Port de Toulon*, 1752; *Plan de La Rochelle.. en 1573*, 1755; *Plan de Versailles* for Louis Charles Desnos 1767; engraver for Louis Charles Desnos's *Atlas générale méthodique*, 1768.

**Hérisson**, Charles Claude François (1762-1840). Geographer. *L'Europe* (4 sheets), 1808; *La Mappe-monde*, *c.*1817; *France*, 1818; *Océanie*, 1837 (corrected by A.R. Frémin, 1854).

**Hérisson** [Hérrison], Eustache (*b.*1759). French hydrographical engineer and geographer; pupil of Rigobert Bonne. *Plan de la Ville de Genève*, 1777 (engraved by C.B. Glot); *Les Gouvernements de Normandie et du Maine Perche*, Paris 1783 (engraved by Perrier and André); *Gouvernements d'Anjou, de Saumurois et de Touraine*, Paris, R. Bonne 1784 (engraved by Perrier and André); *Gouvernements de Berri, de Nivernois et de Bourbonnois*, Paris, R. Bonne 1784 (engraved by Perrier and André), these last three maps were republished in *Atlas de France...*, Paris 1790; *Gouvernement de l'Orléanois*, Paris 1784 (engraved by Perrier and André); maps for l'abbé Grenet, 1785; *Carte générale de la France...*, 1791; *Atlas de poche de géographie universelle*, 1799 *Amérique*, Paris, Basset le jeune 1806, revised 1807, *c.*1819, 1823; *Carte Générale de l'Europe...*, Paris, Basset 1806; *Atlas du Dictionnaire de Géographie Universelle* (51 maps), Desray 1806, 1809; *Atlas portatif...*, Paris, Desray 1806, 1807, 1811; *L'Ireland*, Desray 1810 (engraved by C.B. Glot; *Nouvel atlas portatif*, 1811; *Carte d'Allemagne Divisée d'après le traité de Paix...1809*, Paris, Basset 1811, 1812; *Carte de l'Empire français et de l'Allemagne...*, 1811; *Carte générale d'Italie dressée par Hérisson*, 1812, revised edition published by Basset the younger 1821; *Carte Routière du Royaume de France et Des Pays Limitrophes...*, Paris, Basset 1815; *Atlas de la Géographie Universelle*, 1816; *Carte routière du royaume de France...*, 1816 and later editions; *World Atlas*, 1818; *Carte de l'Afrique divisèe...*, 1821. **Ref.** STONE, J. (ed.) *Norwich's maps of Africa* 2nd edition (1997) N°·125 p.144.

**Herkenrath**, A. German historian. Revised edition of J.E. Braselmann's *Bibel Atlas*, 1881 (first published 1868).

**Herklots**, G. *Illustrations of the Roads throughout Bengal* (112 sheets), Calcutta 1828.

**Herkner**, Jósef. (1802-1864). Polish geographer. *Atlas*, Warsaw 1850, 1853, 1856, 1862, 1867; *Mapp Królestwa Poskiego*, Warsaw 1855.

**Herkt**, Otto (1885-after 1924). *Australien*, 1884; *Afrika*, *c.*1886; *Samoa Inseln*, 1889.

**Herline & Hensel**. Engravers and printers of Philadelphia. J.T. Palmatary's *Chicago* (2 sheets), Chicago, Braunhold & Sonne 1857; *A geological and topographical map of Pennsylvania and New Jersey*, Philadelphia, Charles Desilver 1857 (engraved by W. Williams, printed by Herline & Hensel); *Map of Calhoun county, Michigan...*, Philadelphia, Geil, Harley & Siverd 1858 (surveyed by Bechler & Wenig, 'Engraved on Stone by Herline & Hensel'); *Map of the counties of Genesee & Shiawassee, Michigan*, Philadelphia, Geil & Jones 1859 (surveys by J.W. Stout, J.D. Nash & C. Wilson); *Topographical Map of the Counties of Ingham & Livingston, Michigan*, Philadelphia, Geil, Harley & Siverd 1859 (surveys by I.C. Freed, J.D. Nash, A. Jackson & C. Wilson); *Birds eye view of Egg Harbour City*, Egg Harbour City, F. Scheu 1865.

**Herman**, —. 'Rue S.e Catherine au grand S.t Louis', Orleans. Sold C. Inselin's *Plan de la Ville d'Orleans...*, *c.*1680-1706.

**Herman**, Enrique. Pilot Manila to Acapulco, 1730 MS.

**Herman**, Leonard David. Silesian cartographer, c.1812.

**Herman**, Lichenstein. The first to print Ptolemy's *Geographia* (no maps), Vicenza 1475.

**Hermann**, A. See **Herrman**, Augustine.

**Hermann**, Benedikt Franz Johann (1755-1818). Mining map of the Urals, from 1797.

**Hermann**, H.S. Printer of Berlin. *Übersicht der Administrativ-Eintheilung und der Ortsbevölkerung der Neuen Nördlichen Provinzen des Griechischen Königreiches*, Berlin, Dietrich Reimer 1883 (compiled by H. Kiepert, drawn by W. Droysen); Arktisches Amerika, Berlin 1883.

**Hermann**, M. *Eisenbahn-Karte von Mittel-Europa*, Glogau, Flemming 1865; *Reise Karte von Mittel Europa*, 1867.

**Hermann**, W. *Vulkankarte der Erde*, 1856.

**Hermannides**, Rutger. Professor of history at University of Harderwijk. *Britannia Magna et Hibernia nova Descriptio*, Amsterdam, Aegidius Janssonius Valckenier 1661 (with 31 town plans mainly after John Speed), Dutch edition published Middelburg 1666.

**Hermans**, H. *Atlas der Algemeene en Vaderlandsche Geschiedenis*, 1880 and later editions.

**Hermans**, J. Publisher of Breda. L.C.A. de Haan's *Breda en Omstreken*, 1853.

**Hermelin**, Samuel Gustav, *Friherr* (1744-1820). Publisher of Stockholm. *Geographiske Chartor öfver Swerige* (18 maps), 1797-1805 (compiled from the work of A. Swab, 1795, C. Wallman, 1796, C.M. Robsahm, 1796-1797, C.P. Hällström, 1798-1804 and C.G. Forsell, engraved by F. Akrel, G. Broling and S.J. Neele); *Charta öfver Storfurstendömet Finland*, Stockholm 1799.

**Hermite**, L'. See **L'Hermite**.

**Hermitte**, — d'. Baie de Bombétok, 1732; Baie d'Antongil [Madagascar], 1733.

**Hermon**, Royal Wilkinson (1830-1907). Public Land Surveyor and draughtsman in Canada. Survey in the township of Mono Centre, 1860; *New Map of the County of Huron Canada West*, R.W. Hermon, R. Martin & L. Bolton 1862 (lithographed by W.C. Chewett & Co.); land in Caledon East, 1863; estates in Listowel, 1863-1864; Township of Lount [Ottawa District], Toronto 1878.

**Hermoso**, J. Spanish letter engraver. I. Fernández Flórez's *Plano de la Ria de Muros y de Noya*, 1838 (drawn by J. Espejo, map engraved by C. Noguera), amended edition published 1868.

**Hermundt**, Jacob (*fl.*1697-1701). Engraver in Vienna. *Archiducatus Austriae inferioris* (8 sheets), 1697 (with Jacob Hoffmann).

**Hernandez**, —. *Plan de la Baie de Mexillones, Bolivie*, 1862.

**Hernandez**, E. One of the engravers for General Carlos Ibáñez e Ibáñez de Ibero's *Plano parcelario de Madrid por el Instituto Geográfico y Estadístico*... (3 sheets), 1897 (with F. Noriega).

**Hernandez**, J. *Map of Adams, Middlesex County, New Jersey*, 1870.

**Herndon**, *Lieutenant* W.L., *USN*. *Exploration of the Valley of the Amazon* (3 volumes with 8 maps), Washington 1854 (with L. Gibbon).

**Hero** [Heron] of Alexander (*fl.*AD 62). Mathematician and accomplished inventor. Wrote on measurement and on the *dioptra* (geared plane-table).

**Herodotus** (c.485-c.425 BC). Geographer and historian from Halicarnassus, Greece, lived in southern Italy from 444 B.C. He strongly criticised the geographers of his time and was a firm advocate of an empirical approach to cartography based on exploration.
**Ref.** HARLEY & WOODWARD (Eds.) The history of cartography Vol.I (1987) pp.135-137.

**Heroldt**, Adam. Instrument maker. Globe, Rome 1649.

**Herouville**, L. d'. Carte céleste, 1743.

**Herport**, *Colonel* Johan Anton (1702-1757). Swiss military geographer and engineer. Born Bern. Worked in Benr, Sicily, Württemberg and Morges.

**Herrade**, *Abbess of Landsberg*. Zone map published in *Garden of Delights*, c.1180.

**Herrera**, A. Alcedo y. See **Alcedo** y Henera, Antonio.

**Herrera**, Francisco Xavier de. Engraver of Manila. Plano de Manila, 1819; Y. de Aragon y Abollado's *Plano Chorografico de la Provincia de Pangasinan... En la Ysla de Luzon...* [Philippines], Deposito Topografico de Manila 1821.

**Herrera**, *Don* Juan de. Chief engineer at Cartagena. *Plano dela Bahia de Cartagena...*, 1741; source for Harbour of Cartagena & Zisipata Bay, published by T. Jefferys 1762.

**Herrera y Tordesillas**, Antonio de (1559-1625). Historiographer of the Indies under Philip II of Spain. *Descripcion de las Indias Occidentales* (14 maps), Madrid 1601, Amsterdam 1622, used in Herrera's *Historia General...*, Nicolas Rodriguez 1726. **Ref.** BURDEN, P. *The mapping of North America* (1996) pp.241-248.

**Herrewyn**. See **Harrewyn** family.

**Herrich**, A. *Wandkarte des Weltverkehrs* (4 sheets), Glogau 1850, 1894; *Schweiz*, 1895; *Nordpol*, 1896; *Dislokationkarte des Deutschen Heeres und seiner Grenznachbarn* [Central Europe], Glogau, C. Flemming c.1900 (compiled by A. Herrich, edited by H. Müller).

**Herrick**, J.K. Associated with Allan Bell & Co.'s *A New General Atlas*, 1837.

**Hérrison**, E. See **Hérisson**, Eustache.

**Herrle**, Gustav (1843-1902). American hydrographer. *Gnomonic chart constructed for use in great circle sailing and for the approximate solution of problems in nautical astronomy*, Washington D.C. 1878.

**Herrlein**, Edward. Lithographer. [J.T.] Lawson's *map from actual survey of the gold, silver & quicksilver regions of Upper California*, 1849; engraver for H.F. Walling's *Map of New London County, Connecticut*, Philadelphia, William E. Baker 1854 (printed by Wagner & McGuigan).

**Herrliberger**, David (1697-1777). Swiss engraver and publisher in Zurich and Maur. *Topographie der Eydgnossenschaft* [Switzerland], 1754-1758. **Ref.** SPIESS-SCHAAD, H. *Turicum* I (1977) pp.40-44 [in German].

**Herrman**, Albert (1886-1945). *Die Alten Seidenstraßen zwischen China und Syrien*, Berlin 1910; *Historical and Commercial Atlas of China*, Cambridge [Massachusetts] 1935.

**Herrman** [Hermann], Augustine (1621-1686). Bohemian trader, diplomat and cartographer. Herrman arrived in N. America sometime before 1633, moving to New Amsterdam in the 1640s. He became a denizen of Maryland in 1661, where, in 1662, he was granted land by Lord Baltimore in exchange for the work he undertook mapping Virginia and Maryland. Manuscript surveys of Virginia and Maryland, 1659-1670, published as *Virginia and Maryland As it is Planted and Inhabited this present year 1670* (4 sheets), London, A. Herrman and T. Withinbrook 1673 (engraved by William Faithorne, sold by John Seller). **Ref.** PAPENFUSE, E.C. *Atlas of historical maps of Maryland, 1608-1908* (1982) pp.11-15, 18-19; RISTOW, W.W. 'Augustin Herrman's map of Virginia and Maryland', in RISTOW, W.W. *A la Carte* (1972) pp.96-101.

**Herrman**, O. Lithographer. Wilhelm Ernst August von Schlieben's Atlas, 1825-1830.

**Herschel**, family.
• **Herschel**, *Sir* [Frederick; Friedrich] William [Wilhelm] (1738-1822). Born in Hanover, worked as an astronomer in England, where he built a large reflecting telescope. World (2 sheets), MS; map of Mars, 1777; discovered Uranus, 1781.
•• **Herschel**, *Sir* John Frederick William, *Baronet* (1792-1871). English astronomer, son of Sir [Frederick] William Herschel (above). Charted the stars of the southern hemisphere, 1834-1838; Cape Observations, 1847; Outlines of Astronomy, 1849; articles in *Encyclopaedia Britannica*.
• **Herschel**, Caroline Lucretia (1750-1848). Born in Hanover, worked in England as an astronomer, and assisted her brother Sir William Herschel. She discovered comets and star clusters in her own right.

*Sir William Herschel (1738-1822). (By courtesy of the William Herschel Museum, Bath, England)*

**Herschenz**, T.G. *Plan der Stadt Halle*, 1866.

**Herteln**, Zacharias. Publisher. With Thomas von Wiering, Adam Olearius' Voyages, Hamburg 1696.

**Herterich**, C.H. Engraver for Heinrich Carl Wilhelm Berghaus. *Persische Golf*, 1832; *Syrien*, 1835.

**Hertslet**, Lewis. *Map of Russia in Europe shewing the accession of territory By Conquest, Treaties, &c. from 1763 to 1836*, [n.d.] MS.

**Hertz**, Wilhelm (1822-1901). Publisher in Berlin. *Nordafrikanisches Gestadeland* (2 sheets), *c.*1850.

**Hertzberg**, Heinrich (1859-1931). Engraved history wall charts for J. Perthes, Gotha, early 20th-century (with H. Haack).

**Hertzel**, C.L. Halle, 1791.

**Hervagius**, Johannes. See **Herwagen**, Johann.

**Hervé**, —. Lithographer. *Plan Général du Port et de la Ville de Charbourg...*, Paris, E. Savary et Cie, and Cherbourg, F. Simon mid-19th century (with L. Le Breton).

**Hervet**, —. *Plan de Versailles...*, 1768, published Faden 1778.

**Hervey**, Frederic. *Hervey's New System of Geography*, London 1785.

**Herwagen** [Hervagius], Johann [Johannes]. Printer and editor of Basle. Published Johann Huttich and Simon Grynaeus's *Novus orbis regionum*, 1532.

**Herwarth**, Johann Eberhard Ernst (1753-1838). German military cartographer.

**Herz**, Charles Christophe (*d*.1879). French geographer.

**Herz**, Johann. Jurist in Vienna. *Finanz-Karte des Königreiches Böhmen* (2 sheets), *c*.1846; *Statistisch-topographische Finanz-Karte des Königreiches Boehmen*, *c*.1850.

**Herz**, Johann Daniel *the elder* (1693-1754). German engraver and publisher of Augsburg. *Mappa geographica Regni Bohemiae*, 1717 (with Michael Kauffer); engraved the decorations and explanations for J.C. Müller's *Mapa geograph. totius-Regni Bohemiae...* (25 sheets), 1720 (with M. Kauffer); *Entwurf der Haupstadt im Palestina und Residenz der Jüdischen Könige*, Augsburg *c*.1720; Jerusalem (copper engraving), Augsburg 1735 (with printed explanation); view of Jerusalem, *c*.1750.

**Herz**, Norbert (1858-1927). Theoretical works on map projections, including *Lehrbuch der Landkartenprojektionen*, Leipzig 1885.

**Herz**, W. German lithographer. *Umgegend von Tunis*, 1832; *Kreis Marburg*, 1839; *Provinz Posen*, 1839.

**Herzberg**, Heinrich. Maps for HC. Kiepert's *Neuer Handatlas*, before 1875.

**Herzfeld**, Gottfried. *Eisenbahn Atlas Deutschlands*, 1880.

**Hesiod** (*fl*.8th century BC). Supported concept of circular world with outer ocean.

**Heskett**, James. Publisher of N⁰·13 Sweetings Alley, Royal Exchange, London. Re-issued [John] *Andrews's New and Accurate Map of the Country Thirty Miles Round London...*, 1806 and later states to 1820 (engraved by J. Andrews and W. Haydon, first published by John Andrews, 1782).

**Heslop**, —. *Map of Ontario and Steuben Counties* [Albany], 1798.

**Hess**, —. *Topographie*, 1785.

**Hess**, *Lieutenant* Emanuel, Royal American Regiment. Fort Littleton at Port Royal, South Carolina 1758 MS (possibly drawn during the revolutionary war).

**Hess**, F. American surveyor. *Map of Champaign Co. O.* [Ohio] (2 sheets), S.H. Matthews 1858; *Map of Dodge County, Wisconsin*, 1860 (with Burhans & Scott); *Map of Dodge County* (2 sheets), Chicago 1869; *Map of Oakland County, Michigan*, S.H. Burhans 1872.

**Hess**, H. *Reise-Karte für das Salzkammergut*, *c*.1889.

**Hess**, *Feld-Marschall Freiherr* Heinrich von (1788-1870). Military cartographer.

**Hess**, Ludwig. Engraver. Spain and Portugal for Adam Christian Gaspari, 1809; *Nord-America entw. u. gez. von C.G. Reichard 1818*, 1823 (in *Stieler's Hand-Atlas*).

**Hess**, Oskar (1863-1921). Engraver for Justus Perthes of Gotha.

**Hesse**, L.A.C. *Atlas Minimus Universalis*, 1806.

**Hessel**, Gerritsz. See **Gerritsz.**, Hessel.

**Hesselbach**, W. Topographical engineers office [U.S.]. *Map of northeastern Virginia and vicinity of Washington*, Washington 1862 (with J. Young, used in C.D. Cowles's *Atlas*).

**Hesselgren**, Abraham (1671-1751). Swedish surveyor. Forestry divisions maps, Uppland 1732.

**Hesseln**, Robert de. 'Censeur royale et géographe de la ville de Paris' (1784). *Nouvelle topographie de la France* (1 sheet), 1780 (engraved by G.-N. Delahaye); *Plan figuré des cinq premières Divisions de la Nouvelle Topographie du Royaume de France* (1 sheet), 1780; *Nouvelle Topographie. Premier degré de détail...* (9 sheets), 1784 (drawn by J.A. Dulaure, engraved by G.-N. Delahaye); *Région Centre* [France] (3 maps), Paris 1784 (engraved by G.-N. Delahaye, published in *Nouvelle Topographie... La France en ses 81 régions*); *Premiére Carte de la Nouvelle Topographie Contenant La France Divisée en IX Regions* (1 sheet), Paris 1786; *Nouvelle topographe contenant la France divisée en 9 regions* (9 sheets), c.1786.

**Hessels,** Gerrit (*b*.1609). Son of Hessel Gerritsz. Like his father, he worked for the VOC (Dutch East India Company), c.1632-1637. He is reported to have worked in Batavia as a mapmaker, clerk and copyist, c.1632-1637. Chart of the island of Ceram Laut [Moluccas], 1633. **N.B.** See also Hessel **Gerritsz.**

**Hesselus Gerardus.** See **Gerritsz.**, Hessel.

**Hessler,** Carl. *Karte der Umgegend von Cassel*, Leipzig, G. Lang c.1897.

**Hessler,** *Senior* and *Junior*. Possibly father and son.
• **Hessler,** J.G. *Senior. Grundriss von Friedrishstadt-Dresden*, 1837 (revised by Oscar Hessler 1848); *Grundriss Haupt-u. Residenz-Stadt Dresden*, 1837 (revised by Oscar Hessler 1849); *Grundriss von Neustadt Dresden mit Autonstadt und den Scheunen-Höfen*, 1837 (revised by Oscar 1852).
•• **Hessler,** Oscar *Junior*. Revised the works of J.G. Hessler. *Grundriss von Friedrishstadt-Dresden*, 1848 (from J.G. Hessler, 1837); *Grundriss Haupt-u. Residenz-Stadt Dresden*, 1849 (from J.G. Hessler, 1837); *Grundriss von Neustadt Dresden mit Autonstadt und den Scheunen-Höfen*, 1852 (from J.G. Hessler, 1837); *Plan der Königl. Haupt-und Residenzstadt Dresden*, Dresden, R. Kuntze c.1865.

**Heteren,** J. van. *Oostzaandam* (2 sheets), 1794 (with J. Oostwoud).

**Hett,** Richard. Bookseller of London. Held a part share in the 1730 edition of Camden's *Britannia*, from 1730-1748 (E. Gibson edition with maps by R. Morden). **N.B.** Hett's name never appeared on the title page of this work.

**Hettema,** H. Jr. Compiled school atlases, all published at Zwolle by W.E.J. Tjeenk Willink. *Historische Schoolatlas*, 1896 and editions to 1939 (from 1917 it was entitled *Groote Historische Schoolatlas*); *Schoolatlas...*, 1896 and editions to 1939; *Schoolatlas der Vaderlandsche Geschiedenis...*, 1897; *H.B.S. Uitgaaf van den Historischen Schoolatlas...*, 1917 and later (reduced edition of *Groote Historische Schoolatlas*); *Kleine Schoolatlas*, 1906 and later editions; and others. **Ref.** KOEMAN, C. Atlantes Neerlandici Vol.VI (1985) pp.173-180.

**Hettner,** *Doctor* Alfred (1859-1941). German geographer and journal editor. *Peru und Bolivien*, 1890; *Kordillere von Bogata*, 1892; text for Otto Spamer's *Grosser Hand Atlas*, Leipzig 1896, 1900.

**Hettwer,** E. *Salzburg*, c.1877.

**Heubeldinck,** Marten. *Oost-Indische ende West-Indische Voyagien*, 1617-1619.

**Heuglin,** Theodor von (1824-1876). German geographer and explorer. *Reisen in Nord Ost Afrika*, 1857; *Ost-Afrika*, 1864; *Nil-Quellgebiet*, 1865; *Karte von Æthiopien*, 1868; *Nilgebiete*, 1869; *Suakin und Berber*, 1869; *Ost-Spitsbergen*, 1871.

**Heumann,** —. One of the surveyors for *Karte von dem Herzogthum Oldenburg...*, 1782-1799, engraved 1804 (drawn by C.F. Mentz, engraved by G.H. Tischbein).

**Heumann,** Georg Daniel (1691-1759). Draughtsman, engraver and publisher of Nuremberg. J. Hagelgans's *Atlas Historicus*, c.1737; *Francofurti*, 1738; *Göttingen*, 1745.

**Heunisch,** Adam Ignaz V. (1786-1863). Cartographer of Karlsruhe. *Lieues de France*, 1817; *Baden*, 1819; *Nassau*, 1822; *Neuer Hand-Atlas über alle Theile der Erde*, Karlsruhe & Baden, Marx 1827 with editions to 1836; *Die deutschen Bundesstaaten*, Karlsruhe & Baden 1828; *Allgemeiner Schul-Atlas...*, Karlsruhe, Marx 1830 and later; *Taschen-Atlas über alle Theile der Erde...*, Karlsruhe & Baden 1830 with editions to 1843.

**Heurdt,** A. van. *Comitatus Meursiensis*, de Wit c.1690.

**Heuschling,** Xavier. *Atlas du Royaume de Belgique*, 1844 (with Philippe Vandermaelen).

Heusel, F. *Plan von Alzey*, 1894.

Heusinger, Johann Heinrich Gottlieb (1766-1837). *Handatlas über alle bekannte*, 1809, 1810.

Heusser, Enrico. *Pianta di Messina*, c.1860.

Heuvel, B. van den. Dutch surveyor and cartographer, worked in Zuid-Holland. *Nieuwe caerte vande verdroncken waert van Zuyt-Hollandt*, 1686 (with Mattheus van Nispen and Abel de Vries).

Heuvelink, H.I. *Plattegrond der Stad Arnhem*, 1853.

Heuzé, Gustave (1816-1907). *La France Agricole*, 1875.

Hevelius [Hewelcke], Johannes [Jan] (1611-1687). German astronomer of Danzig. *Tabula Selenographica, Sive Lunae Descriptio*, 1647; Prodromus cometicus, 1665; Cometographia, 1668; *Firmamentum Sobiescianum sive Uranographia*, Danzig 1687, 1690.

Hevenesi, Gábor [Gabriel], SJ (1656-1717). Hungarian cartographer, Vice-chancellor of Vienna University. *Parvus Atlas Hungariae* (40 sheets), Vienna 1689 (with F.A. Colloredo); manuscript county atlas of Hungary, [*n.d.*] (University Library of Budapest). **Ref.** FALLENBÜCHL, Zoltán *Atlas Parvus Hungariae, Térképtudományi tanulmányok* [Cartographic Studies] (1959) [in Hungarian].

Hevia, Deogracias. *Provincia de Salamanca*, 1860.

Hewelcke, Jan. See **Hevelius**, Johannes.

Hewes, Fletcher Willis. Statistical cartographer. *Statistical Atlas of the United States* [10th census], New York, Charles Scribner's 1883 (with Henry Gannett); *Citizen's Atlas of American Politics 1789-1883*, 1883, 1888, 1889, 1892. **Ref.** SCHWARTZ, S.I. & EHRENBERG, R.E. *The mapping of America* (1980) p.310.

Hewett, —. Engraver. *Canada*, published in *Counsel for Emigrants*, Aberdeen, John Mathison 1834.

Hewett, *Lieutenant* (later *Captain*) William, RN. Admiralty Charts. Undertook numerous surveys of the the North Sea coasts, 1822-1841, including: Collifirth Voe, 1822; Inner Gabbard, 1824; River Humber to Lowestoft including the Wash, 1825; Lynn & Boston Deeps, 1828; Cromer, 1828; *Trusthorpe to Flamboroug Head*, 1830 (engraved by J. & C. Walker); Flamborough Head to Spurn Point, 1830; Tynemouth to Dogger Bank, 1831; Blakeney to Lowestoft, 1832; Lowestoft Inner Shoal, 1835; soundings in North Sea, 1839; survey of the Galloper Bank [Goodwin Sands], 1841 (with Captain Bullock & George Thomas), 1842; surveys used for a map of the the east coast of England and Scotland from the Wash to the Moray Firth, 1854 (engraved by J. & C. Walker); and many others. **Ref.** ROBINSON, A.H.W. *Marine cartography in Britain* (1962) pp.138-139, 193-194.

Hewett, *Sir* William. *Plan of Gariah Harbour on the Malabar Coast*, 1756.

Hewitt, E.A. *Map of the Counties of Bourbon, Fayette... Kentucky*, 1861 (with G.W. Hewitt).

Hewitt, G.W. *Map of the Counties of Bourbon, Fayette... Kentucky*, 1861 (with E.A. Hewitt).

Hewitt, J. Lithographer. *An Outline Atlas of the World*, Bristol 1850.

Hewitt, John H. *Sectional map of Cook County, Ill.*, 1868.

Hewitt, N.R. Engraver of London, noted at Queen St., Bloomsbury (1812); 10 Broad Street, Bloomsbury (1814); Grafton St. East Tottenham Court Road (1817); 1 Buckingham Place, Fitzroy Sq. (1819-1821). Mathews & Leigh's *Scripture Atlas*, 1812; *Plan of the parishes or Division of St. Giles in the Fields and St. George Bloomsbury* 1815; J. Aspin's *A Chart of New South Wales*, 1816, 1821, 1830; *Scotland*, 1815, (and other maps used in John Thomson's *A New General Atlas*, 1817 and later); engraved some additional maps for John Thomson's *The Traveller's Guide through Scotland, and its islands*, 1818 edition, 1829 (first published 1814); map of Ireland in James Wyld's *A General Atlas*, London, Baldwin & Cradock and Edinburgh, John Thomson & Co. 1819; *Kincardine Shire* and *Roxburghshire*, c.1822 (both used in John Thomson's *The Atlas of Scotland*, 1832); engraved maps to accompany the 1822 edition of Abbé Gaultier's *A Complete Course of Geography*..., John Harris & Son 1822, 1825 (maps by J. Aspin); ...*This Plan of the United Parishes of St. Giles in the Fields and St. George*, 1824, 1828.

**Hewitt**, Robert (*b*.1665, *fl*.-1716). Estate surveyor in various parts of England. Beaumont, Essex 1688 MS.

**Hewlett**, E.G.W. *Lancashire and Cheshire reduced from the Ordnance Survey*, London, E. Stanford 1904 (with C.E. Kelsey).

**Hewlett**, J. Monroe. Designer, with C.C. Gulbrandsen, of the celestial map on the ceiling of Grand Central Terminal, New York, 1913.

**Hewson**, Thomas R. Townships of Hodgins & Anderson [Canada], Toronto 1878.

**Hexamer**, Ernest [and Son] (*fl*.1857-1915). Publishing company of Philadelphia, founded by Ernest Hexamer and William Locher in 1857. Ernest Hexamer & Son specialised in insurance maps and atlases of Philadelphia until it was taken over by Sanborn Map Company in 1915.

**Hexham**, Henry (*c*.1585-*c*.1650). Soldier and scholar. 'Quartermaister to the Regiment of Colonell Goring'. Translator of an English edition of the Mercator-Hondius atlas published under the title *Atlas Or A Geographicke description of the Regions, Countries and Kingdomes of the World*..., Amsterdam, H. Hondius & J. Janssonius 1636.

**Hexham** [Hexam], John (*fl*.1587-1596). East Anglian surveyor. Castle Rising Chase, 1588 MS; map of the fenland between Peterborough and Wisbech, *c*.1589 MS.

**Heybech** [Heybeck], Nicolaus von. 'Magister' of Klosterneuburg. Credited by some with authorship of the 'Trier-Koblenzer Fragment' [map of the Rhineland], 1437 (previously attributed to Cusanus); possibly made a celestial globe which passed to Cusanus in 1444.

**Heyberger**, Josef. *Plan der... Stadt München*, 1866; *Karte von Königreich Bayern*, *c*.1867; *Pianta di Roma*, 1867.

**Heybrock**, J.M. *Nautischer Hand-Atlas*, 1857.

**Heydanus** [Heidanus; Heiden; Van der Heyden], Carolus [Carel; Carl]. Cartographer, possibly also a map collector. Credited by Ortelius with a map of Germany published in Antwerp, Hieronymus Cock *c*.1565 (no longer extant). **Ref.** KARROW, R.W. *Mapmakers of the sixteenth century and their maps* (1993) p.293.

**Heyde** [Heijde], Nicolaas van der. Dutch surveyor and cartographer, worked north Holland and Amsterdam as 'generale opsiener vande Schutsluysen en Fortificatie'. Issued reduced version of Dirck Abbestee's polder map *Karte van t'Koegras*, 1676.

**Heyden**, Carl van der. See **Heydanus**, Carolus.

**Heyden**, Christian (1526-1576). See **Heiden**, Christian.

**Heyden**, Christian Frederick [Friedrich] van der. *Französische Küsten*, *c*.1758; *Brittischer Übermacht zur See*, Augsburg, Tobias Conrad Lotter *c*.1759.

**Heyden**, F.V. *Geological and Geographical Atlas of Colorado*, 1877.

**Heyden** [Heiden; à Myrica], Gaspard [Jaspar] van der (*c*.1496-after 1549). Engraver, goldsmith and instrument maker of Louvain. Viewed as a key figure in the early development of globes in the Netherlands. Engraved *c*.25/30-cm terrestrial and celestial globes for Franciscus Monachus, Antwerp *c*.1527 (no copies known); *c*.25/30-cm terrestrial globe for Gemma Frisius, *c*.1529 (no copies known); pair of 37-cm globes with Gemma Frisius and Gerard Mercator, Louvain *c*.1536-1537. **Ref.** DEKKER, E. *Globes at Greenwich* (1999) pp.340-342; KROGT, P. Van der *Globi Neerlandici* (1993) *passim*.; SMET, A. de. 'L'orfèvre et graveur Gaspard van der Heyden et la construction des globes à Louvain dans le premier tier du XVIe siècle' in: *Der Globusfreund* 13 (1964) pp.32-48 [in French].

**Heyden**, Jakob van der (1573-1645). Engraver and publisher in Strasbourg. Wolfgang Lazius's *Austriae chorographia*, Strasbourg 1620 (drawn by Lazius in 1563, publication arranged by M. Bernegger); engraved and published Isaac Habrecht's untitled terrestrial and celestial globes, Strasbourg 1621; re-engraved Caspar Vopel's map of the Rhine under the title *Rhenus Bicornis, hoc est, Totius Rheni tractus* *delineatio...*, 1621, 1630 (drawn by Vopel in 1555); *Bavariae olim Vindeliciae*, 1622; Danube, 1627; *Lotharingiae Ducatus*, 1630; Magdeburg, 1631; *Rheni duplici origine... delineatio nova*, 1636.

**Heyden**, Pieter van der (*c*.1530-1572). Engraver of Antwerp. Credited with the engraving of maps for the Polyglot Bible edited by Arias Montanus, and published in Antwerp by Christopher Plantin, from 1569.

**Heyden**, T. van der. Dutch engraver and draughtsman. *Ein gedeelte van de Heerlyckheyt van Maerseveen*, *c*.1690s (with Philibert Bouttats).

**Heydon**, John. *John Heydon's Map of Plymouth, Devonport, Stoke... and the Neighbourhood*, 1870.

**Heydt**, Johann Wolfgang (*b*.1702). German draughtsman and engraver born in Indonesia. *Allerneuster Geographisch und Topographische Schau-Platz von Africa und Oost-Indien...* (115 maps, plans and views), 1744. **Ref.** LANDWEHR, J. *VOC A bibliography of publications relating to the Dutch East India Company 1602-1800* (Utrecht 1991) N°469 pp.264-470.

**Heylin** [Heylyn], Peter [Petrus] (1599/1600-1662). English cosmographer, geographer and Chaplain to Charles I. *Microcosmus, or a Little Description of the Great World*, 1621, expanded to become *Cosmographie in foure bookes...*, Henry Seile 1652 (maps engraved by W. Trevethen), and editions to 1682. **N.B.** Philip Burden notes that after the death of Henry Seile in 1662 the plates passed to Philip Chetwind, whose name appears from 1666. Henry's widow Anne Seile, re-published using a different set of plates engraved by Robert Vaughan, 1663 and editions to 1677. **Ref.** BURDEN, P. *The mapping of North America* pp.394-395; *ibid.* p.379.

**Heymann**, Carl. *Karte von Marokke...*, Berlin 1887; *Kaiser Wilhelms Land und Bismarck Archipel*, Berlin 1893.

**Heymann**, Eduard. *Karte der Umgebung von Troppau*, 1873.

**Heymann**, Ignazio (1765-1815). Austrian geographer and Head Postmaster in Trieste. Italie, 1798; *Italia*, Trieste 1799; *Deutschlands Postkarte...*, Trieste 1800; *Vue de la Ville et du Port-france de Trieste*, Trieste 1802; Italie (4 sheets), 1806; *Carte des Postes d'Allemagne* (4 sheets), 1808; *Map of Germany and adjoining countries*, London, J. Stockdale 1809.

**Heymenbergh** [Heijmenbergh], Jan [Johan] van. Dutch surveyor and cartographer, worked in Noord-Holland, 1662-1670, town surveyor of Alkmaar. Grandson of Gerrit Dirckz. Langendijck. *Karte van des Graeffelijckhts end d'Oochduijne... van Huijsduijne*, 1662 MS (with Dirck Abbestee).

**Heyne**, Benjamin. *Map of the Circars*, 1814; *Map of Mysore*, 1814.

**Heyne**, Charles. *Map of part of Virginia, Maryland and Delaware*, New York, E. & G.W. Blunt 1861.

**Heyns**, father and son. **Ref.** KOEMAN, C. *Atlantes Neerlandici* (1967-1971) Vol.II pp.131-134, Vol.III pp.37, 40-41, 71-75.
• **Heyns** [Heijns; Heinsius], Pieter (1537-1598). Poet and teacher of Antwerp, friend of Ortelius, father of Zacharias. Translated Ortelius's *Theatrum* into Dutch, published as *Theatre oft Toonneel des Aerdtbodems...* (53 maps), Antwerp 1571, (69 maps), 1573; composed the Dutch language rhyming text for Philippe Galle's miniature edition of Ortelius, entitled *Spieghel der Werelt* (72 maps), Antwerp 1577, (83 maps), Antwerp 1583 (maps engraved by Philippe Galle); French edition (not in rhyme) published as *Le Miroir du Monde* (72 maps), Antwerp, Christopher Plantin 1579, (83 maps), Antwerp, Christopher Plantin 1583 (used the same maps by Philippe Galle, not to be confused with Z. Heyns's miniature atlas of the same title, below, which carried woodcut maps and was published in Amsterdam, 1598).
•• **Heyns** [Heijns], Zacharias (1566-1638). Bookseller and publisher from Antwerp, settled in Amsterdam *c*.1595 'in de Hooft Deughden' or 'enseigne des Trois Vertus'. Son of Pieter Heyns (above). Produced his own version of *Le Miroir du Monde; Ou Epitome Du Theatre D'Abraham Ortelius* (80 woodcut maps), Amsterdam 1598, re-issued at Arnhem, J. Janssonius 1615; *Den Nederlandtschen Landtspiegel* (36 maps), 1599 (reduced, Dutch language edition of *Le Miroir du Monde*).

**Heyse**, D. Lithographer ['graveur-lithograaf'] and cartographer for the Dutch Topographical

*Peter Heylin (1599/1600-1662), frontispiece to* Historical Tracts, *1681. (By courtesy of the National Library of Australia)*

Service at The Hague. A. Baedeker's *School-Atlas* (22 maps), 1840; *Kaart van de Province Noord-Holland*, Amsterdam 1855; maps for J. Pijnappel's *Atlas van de Nederlandsche Bezittingen in Oost-Indië*, The Hague, K. Fuhri 1855 and Amsterdam, P.N. van Kampen 1865; worked on P.M. van Carnbée's *Algemeene Atlas van Nederlandsch Indië*, from 1863; lithographic drawing for G. Mees's *Historischer Atlas van Noord-Nederland...* (16 maps), Rotterdam, Verbruggen en van Duym 1865, reduced edition (12 maps), Leiden, Sijthoff 1881.

**Heyse**, P.F. (1816-1879). Map lithographer of The Hague.

**Heysham**, W. *Map of Calcutta showing the latest improvements as existing in 1856*, London and Calcutta, R.C. Le Page & Co. 1859.

**Heyteman**, *Captain* Johann Hansson (1664-1740). Norwegian mapmaker of Christiania [Oslo]. Zoen-water [Oslofjord], Amsterdam, Gerard van Keulen 1735; Noordzee, Amsterdam, Gerard van Keulen 1740.

**Heywood**, family. Manchester family whose members rose from humble origins to become well-known publishers. Using the new process of lithography they were able to mass produce inexpensive educational material and make it available for the first time to the whole population. Most of the cartographic output was derived from the work of others. Heywood publications were also available through London partners Simpkin, Marshall & C$^{o.}$. **Ref.** SMITH, D. 'John Heywood and others' in *IMCoS Journal* 69 (1997) pp.23-30.

• **Heywood**, Abel. Following his dismissal from a job as a warehouse boy, Abel took over the agency for the *Poor Man's Guardian* and later the Chartist paper *Northern Star*. Rose to become Commissioner of Police for Manchester, and Mayor in 1862. *Plan of London*, 1862.

•• **Heywood**, Abel II. Son of Abel Heywood. Joined his father in business in 1864.

•• **Heywood**, Abel and Son (*fl.* from 1864). Publishers and stationers of '56 & 58 Oldham Street, Manchester'. The partnership was formed when Abel II joined his father in business. *Abel Heywood & Son's Series of Penny Guide Books*, from 1868, with editions to 1905 carrying the title *Abel Heywood & Son's Guide Books* and an increased price of 2$^d$ (also published in London by Simpkin, Marshall, Hamilton, Kent & C$^{o.}$); *Abel Heywood & Son's Handy Plan of London*, c.1870 (also published in London by Simpkin, Marshall & C$^{o.}$).

• **Heywood**, John (1804-1864). Brother of Abel I. Began work as a weaver; in 1839 joined Abel in the paper, printing and publishing trade. In 1846 set up on his own as a stationer and newspaper seller at 170 Deansgate, Manchester, becoming the largest newspaper distributor and bookseller outside London. By 1859 the company had taken more space on the opposite side of Deansgate, and by 1866 had built their own 6-storey premises. '143, Deansgate & 3, Brazennose S$^{t.}$ Manchester', (1865); '141 & 143 Deansgate & Excelsior Works, Manchester', (1875). John senior was also Chairman of the Poor Law Guardians of Chorlton (Manchester). Acquired and republished H. Teesdale's *The Travelling Atlas of England & Wales*, c.1858, and many later editions (originally published as *New Travelling Atlas*, H. Teesdale 1830).

•• **Heywood**, John II (1832-1888). Son of John Heywood above; joined his father and took over at his death in 1864. Noted at '143 Deansgate, & 3 Brazenose S$^t$. Manchester' (1864); '141 & 143 Deansgate' (1868, 1880); '141 & 145 Deansgate & Excelsior Works, Manchester' (1875); 'Excelsior Buildings, Ridgefield, Manchester and 18 Paternoster Square, London' (1882). Continued publication of *The Travelling Atlas*, also: *The Tourist's Atlas of England and Wales*, c.1864 (maps from *The Travelling Atlas*; also published in London by Simpkin, Marshall & C$^{o.}$); *John Heywood's National Atlas*, c.1867 (maps engraved by John Bartholomew & Co.); *John Heywood's County Atlas of England and Wales*, 1868, c.1882 (maps from *The Travelling Atlas*); *J. Heywood's British Empire Atlas*, 1878; *John Heywood's Atlas and Geography of the British Empire*, c.1879 (compiled by Thomas Higman); *Heywood's Map-Drawing Made Easy*, c.1879 (one of several 'outline' and 'memory' maps and atlases); *John Heywood's County Atlas of Wales*, c.1879 (unlike the others, an original work); *New Maps of the Counties of England and Wales from the Ordnance Survey*, c.1879; *Pictorial Map of Manchester and Salford*, 1886; *J. Heywood's Railway Map of England and Wales*, 1892.

**Heywood**, C. Chart Merjee River, 1803.

**Heywood, J.** See **Haywood, J.**

**Heywood,** *Captain* Peter (1773-1831). Officer in the Bombay Marine, served aboard HMS *Bounty*, HMS *Dedaigneuse* and HMS *Leven* (1822). Charts of Ceylon, India, East Indies and China, 1798-1806; *Plan of Venloos Bay* [Sri Lanka], London 1805; chart of the River Plate, published Faden 1817; 'Plan of False Bay' [Cape of Good Hope], 1822 (surveyed by W.F.W. Owen, engraved by J. Walker), published in Horsburgh's *Book of Directions*, 1832.

**Hezeta,** Bruno de. Spanish mariner, left Spain in 1774. Set out from Mexico in 1775 in command of the *Santiago* as part of a three-ship exploration fleet along the northwest coast of America. *Carta reducida de las Costas y Mares Septentrionales de California...*, 1775 MS; *Plano del Puerto de la Trinidad...*, 1775; *Plano de la Bahia de la Asumpcion o entrada de Ezeta*, 1775.

**Hhaggy** Ahmed of Tunis. See **Hacci** Ahmed.

**Hibbart,** Thomas. English surveyor in Derbyshire. Worked with Samuel Barton on several surveys of settlements in Derbyshire (4 maps), 1640 MSS.

**Hibbart,** W. Engraver. *Plan of the City of Bath*, Bath, W. Frederick 1780; Five miles round Bath, Bath, W. Frederick 1787. **N.B.** Compare with W. **Hibbert**, below.

**Hibbert,** Lieutenant —. Province of La Rioja, 1839.

**Hibbert,** *Doctor* Samuel. *Geological map of the Shetland Isles*, 1820 (engraved by Lizars), enlarged edition published in Hibbert's *A Description of the Shetland Isles*, 1822 (engraved by C. Thompson). **Ref.** BOUD, R.C. 'Samuel Hibbert and the early geological mapping of the Shetland Islands' in The Cartographic Journal Vol.14 (1977) pp.81-88.

**Hibbert,** W. Five miles round Bath, Bath 1773. **N.B.** Compare with W. **Hibbart**, above.

**Hibsch,** J.E. *Geologische Karte des Mittelgebirges*, 1895.

**Hickenlooper,** Andrew (1837-1904). *Map of the city of Cincinnati, as redistricted, for use of members of Common Council and city officers*, Cincinnati, 1872.

**Hickeringill,** *Captain* Edmund. Map of Jamaica, in *Jamaica Viewed*, John Williams 1661.

**Hickey,** Benjamin. Bookseller, Nicholas Street, Bristol. John Rocque's *A Plan of the City of Bristol* (4 sheets), 1743, 1750.

**Hicklin,** John. Excursions in North Wales, 1847.

**Hickling,** C. Publisher of Boston, Massachusetts. William Jenks' *Explanatory Bible Atlas and Scripture Gazetteer, geographical, topographical and historical*, 1847.

**Hickmann,** *Professor* Anton Leo (1834-1906). Bohemian [Czech] statistician and geographer. *Industrial-Atlas des Königreiches Böhmen* (12 sheets), Prague, H. Mercy 1862-1864; *Geographisch-statistischer Universal-Taschen-Atlas* [pocket atlas], Vienna, Freytag & Berndt 1894; *Geographischer-Statistischer Taschen-Atlas von Österreich-Ungarn*, Vienna 1895, 1897; *Universal Taschen Atlas*, 1899; *Prof. A.L. Hickmann's geographisch-statistischer universal-taschenatlas*, Vienna and Leipzig, G. Freytag & Berndt, 1915.

**Hicks,** F. Tapestry weaver. Five tapestry maps, *c*.1570, now in York Museum and Bodleian Museum, Oxford.

**Hicks,** *Captain* F.R. Officer in the Uganda Rifles. Sketch map of the area to the east of Lake Albert..., 1899 MS.

**Hicks,** *Captain* John. Bay of Bengal, 1794.

**Hicks,** Nathan. *Sterling, Whiteside Co., Illinois*, 1865.

**Hicks,** R.J. Ngamiland, 1892 MS.

**Hicks-Judd Co..** Publisher of San Francisco. J.A. Finch's *Alaska and Pacific coast map*, 1898.

**Hielm,** B. *Kart over Christiania*, 1848.

**Hiemcke,** *1ste Luitenant Ingenieur* A.H. Platte grond van de stad Paramaribo

[Surinam], Amsterdam 1804 (engraved by J.C. Visser); Plan der Situatie van het Oostelijke [Westelijke] gedeelte der Kolonie Surinamen, Amsterdam 1819 (engraved by J.F. Lange), also in Surinaamsche Almanak..., 1820, Amsterdam and Paramaribo, E. Beyer en C.G. Sulpke, 1820; Platte grond van de stad Paramaribo, Paramaribo 1850 (lithographed by Petit).

**Hieronimi**, Ferenc Ottó (1803-1850). Hungarian military engineer and cartographer, born Györ, died Buda. In charge of the mapping of rivers in Hungary, taking over the survey of the Danube from Pál Vásárhelyi.

**Hieronymus**. See **Jerome**, *Saint*.

**Hietzinger**, Karl Bernhard, *Freiherr* von (1786-1835). Austrian military cartographer.

**Higby**, William R. *The World on a globular projection drawn by William R. Higby at the school of I.H. Johnson, Bridgeport, Connecticut 1839*, MS; *Map of Connecticut*, Bridgeport, Connecticut 1842 MS.

**Higden** [Hygden; Hugeden], Ranulph [Ranulf; Ranulphus] (*fl.*1299-1364). English monk and chronicler of St. Werburgh's Abbey, Chester. The name Higden is a generic term for a series of ovoid mappamundi which accompanied manuscript copies of his *Polychronicon* (many copies survive, *c.*21 of which contain a map, the quality and style of which varies). **Ref.** BARBER, P. *Imago Mundi* 47 (1995) pp.13-17; WOODWARD, D. in: HARLEY and WOODWARD (Eds.) *The history of cartography* Vol.I (1987) pp.312-313.

**Higeosky**, —. Russian Army Officer. One of the Russian representatives amongst the international team which draughted *...la frontière Serbo-Turque selon article 36 du traité de Berlin... 1878* (11 sheets plus title), 1879.

**Higgie**, George. *Higgie's map of the Isle of Bute*, 1886.

**Higgins**, brothers. Natives of Stepney, near Newtown, Connecticut. Two of the young surveyors who assisted D. Jackson Lake on his surveys of Ohio and Indiana between 1863-1873. Belmont County, Ohio, 1868 (with D. Jackson Lake); Butler County, Ohio, 1868 (with D. Jackson Lake); also maps of Montgomery, Hardin and Marion Counties, Ohio, 1869.

• **Higgins**, Jerome Silliman. Brother of R. Thornton Higgins. Surveyor for county maps with D. Jackson Lake from 1865, and one of the founding partners of Chicago atlas publishers Warner & Higgins *q.v.* (later Warner, Higgins & Beers (with Augustus Warner, John Hobart Beers and William Hermon Beers). Map of Boone and Clinton counties, Indiana, Philadelphia, Cowles & Titus 1865 (with Augustus Warner); Warren County, Ohio, 1867 (with G.P. Sanford and R.H. Harrison); *Atlas of Coles County, Illinois*, 1869 (with Augustus Warner); *Atlas of Kendall County & State of Illinois*, Chicago, Warner & Higgins 1870, 1871; *Subdivisions of the public lands, described and illustrated...*, St. Louis, Higgins & Co. 1894.

•• **Higgins, Belden & Company** (*fl.*1874-1877). Partnership of J.S. Higgins and Howard R. Belden, formed in 1874 and based at Lakeside Building, Chicago. It was possibly absorbed into J.H. Beers & Co.. Produced ten or more illustrated atlases of counties in Indiana, Michigan and Wisconsin, from 1874 including *An Illustrated Historical Atlas of Elkhart County, Indiana*, Chicago 1874; *Illustrated Historical Atlas of Henry County, Indiana*, 1875; St. Joseph County, 1875; Kent County, Michigan, 1876; and many other similar titles.

• **Higgins**, R. Thornton. Surveyor from Connecticut, collaborated with his brother J. Silliman Higgins, Charles and Augustus Warner, and D.J. Lake on surveys of counties in Ohio and Indiana, from 1868.

• **Higgins, Bro. & Co.** *Map of Moniteau County* (4 sheets), Chicago, 1878.

**Higgins**, J. Printer, St. Michael's Alley, Cornhill, London. Moorgate to London Bridge, *c.*1831.

**Higgins**, J.L. Carlton House Terrace, 1827 (drawn in lithography by J.L. Higgins from drawings by John Nash, lithographed by Engelmann & Co.).

**Higgins**, S.W. Draughtsman of Detroit, United States Deputy Surveyor. County maps of Michigan & Wisconsin, 1832-1850; *Map of Michigan, Ohio, Indiana, Illinois, Missouri, Wisconsin & Iowa...*, Buffalo, O.G. Steele 1846; Official Map of San Francisco,

1849; *Map of Calhoun County* [Michigan], c.1850; *Map of Jackson County* [Michigan], c.1850; *Map of Washtenaw County*, c.1850 (all with C. Douglass and Bela Hubbard, engraved by W.J. Stone).

**Higgins**, W.H. *Chart of the Ancient Constellations and all the Chief Stars*, London 1883.

**Higgins**, W.M. *Atlas of the Earth*, 1836.

**Higgins & Ryan**. Publishers of Indianapolis. *New Topographical Atlas and Gazetteer of Indiana... Together with a Railroad Map of Ohio, Indiana, and Illinois*, 1870.

**Higgins, Belden & Co.** See **Higgins** brothers (above).

**Higginson**, —. Mining Districts of Nevada, 1865.

**Higginson**, H. *Map of the settled districts of South Australia*, London, E. Stanford 1857 (with J.W. Painter).

**Higginson**, J.H. *Higginson's Map of New York and Vicinity*, New York 1860.

**Highman**, Frank. *Plan of the City of Salisbury* [Wiltshire], 1884.

**Highmore**, John. Travels through the Principal Cities of Europe, 1782.

**Higinbotham & Robinson**. Lithographers and publishers of 99 Pitt Street, Sydney, New South Wales. *Paddington - Parish of Alexandria*, c.1880; *Yachting and excursion map of Port Phillip and the Surrounding country*, 1886; *West Botany, parish of St George*, c.1890; Map of Port Jackson, 1893 (with a sheet of signals and maritime flags).

**Higman**, Thomas. Compiler of John Heywood's *Atlas and Geography of the British Empire*, Manchester, J. Heywood c.1879.

**Hikawaya** Zensaku. Japanese publisher. Suigashi's *Dainihon Dōchū Kōtei Saiken Ki* [travel map of Japan] (colour printed woodblock), 1837 edition (with Akitaya Ryōsuke, from an original by Kikuoka Ryōsen first published 1722).

**Hilacomilus** [Hylacomilus], pseud. See **Waldseemüller**, Martin.

**Hilal**, Muhammad. See **Ibn Halal**, Muhammad.

**Hilarides**, Johannes [Jan Jacobs Jaapix] (1649-1725). Painter, collector, linguist, writer, engraver and publisher. Designed title page for B. Schotanus à Sterringa's *Friesland*, 1717.

**Hilbert**, J. Engraver and printer of Hull. John Scott's chart and sailing directions for the River Humber, 1734.

**Hilbrants**, G. See **Hopper**, Joachim.

**Hildebrand**, Emil (1848-1919). *Atlas till allmänna och svenska historien* [historical atlas of Germany and Sweden], Stockholm, P.A. Norstedt & söner förlag 1883, 1895 (with N. Selander).

**Hildebrandsson**, Hugo Hildebrand (b.1838). Swedish geographer and meteorologist.

**Hildner**, —. *Plan de la Ville de Berlin* (4 sheets), before 1749.

**Hildt**, George H. *Map of the United States West of the Mississippi*, St. Louis, Missouri, Leopold Gast & Co. 1859 (with D. McGowan).

**Hilgard**, *Professor* Eugene Woldemar (1833-1916). Geologist and agriculturalist. State geologist in Mississippi. Wrote numerous works on agriculture, viticulture and cotton production in the United States, especially Mississippi, Louisiana and California. Many of the reports contained maps. *Geological Map of Mississippi*, New York, J.H. Colton 1855; *Reclamation of the alluvial basin of the Mississippi River* (with chart), Washington 1878; *Geography of Mississippi*, Cincinnati, Van Antwerp, Bragg & Co. 1880; Agricultural map of the Coalville Region, Washington, 1883; Yakima Region, 1883.

**Hilgard**, Julius Erasmus (1825-1891). American hydrographer, later Superintendent of United States Coast and Geodetic survey. Triangulation for *Cat and Ship Island Harbors*, 1850 (with F.H. Gerdes, topography by W.E. Greenwell, engraved by Sherman & Smith); *Preliminary Chart of Key West Harbor...*, 1851; verified *Reconnaissance of*

*Tampa Bay, Florida...*, 1855; triangulation for *Grand Island Pass, Mississippi*, 1855 (topography by W.E. Greenwell, drawing by S.B. Linton, lithographed by L.S. Rosenthal); triangulation for *St Louis Bay and Shieldsboro Harbor, Mississippi*, 1857 (with S.A. Gilbert, topography by W.E. Greenwell, drawn by S.B. Linton, lithographed by C.B. Graham); *Patuxent River*, 1859 (with others); *Magothy River to Potomac River*, 1862 (with many others); *Alaska and Adjoining Territory*, 1884.

**Hilhouse** [Hillhouse], William [Wilham]. Land surveyor. *Map of British Guiana...*, 1827 published 1828, 1834, London 1836; Massarooney River, 1834.

**Hilkens** [Hielkes], Anna. *...Oost Zee...*, 17th-century, published J. van Keulen *c.*1744.

**Hill**, family. Surveyors of Canterbury, Kent.
• Hill, Thomas (*fl.*1674-*c.*1703). Surveyor of Canterbury. Produced manuscript maps in Kent, London and Sussex, brother of Francis, father of Jared. *Newington and Cheriton* [Kent], 1683; *The Manors of Shortfare and Horton Court in Monks Horton*, 1687.
•• Hill, Jared (*b.*1687, *fl.*-1745). Surveyor of Canterbury, son of Thomas, nephew of Francis. Produced manuscript maps of sites throughout England. *A Mappe and description... of the lands of Bishop Eubrooke in ye Parish of Cheriton...* [Kent], 1713; Rye Harbour, 1717 MS; the manor of Bolingbroke, showing land in Anderby, East Kirkby, Stickford, Stickney and West Keal [Lincolnshire] (2 plans), 1719 MSS; *A Map of lands belonging to Woodchurch Place* [Kent], 1729; estates at Chingford, Essex, 1735 MS; surveys at Leyton, 1739 MS; 94 acres at Walthamstow, 1739 MS.
• Hill, Francis (*fl.*1698-*d.*1711). Surveyor of Canterbury, brother of Thomas, above. Produced manuscript maps of Kent and Sussex. Rye map of Salts & Bench, 1700 (with Samuel Newman); map of the town and harbour of Rye [Sussex], 1702 MS (with S. Newman); *A Map of the Parsonage Land of Saltwood* [Kent], 1707; *A Map of lands belonging to Woodchurch Place...*, 1729; Palm Tree Farm, Parish of Lining, Kent, 1730 MS.

**Hill**, Albert J. *Map of New Westminster District* [British Columbia], 1898.

**Hill**, B.F. *Geological Map of... West Texas*, 1904.

**Hill**, C.H. *Plan of Islington Parish*, London, T. Starling 1822 (reduced from R. Dent's survey of 1805-1806).

**Hill**, Clement. Surveyor General of Prince Georges County [Maryland], 1720-1730.

**Hill**, G.D. Surveyor General. Dakota Territory, 1861.

**Hill**, G.F. Map of the island of Trinidad, 1836 MS.

**Hill**, Henry. *Kuril Islands* [Kamchatka-Hokkaido], 1855.

**Hill**, J. Engraver. T. Girtin's *Barnard Castle in the County of Durham*, London, R. Ackermann 1800.

**Hill**, J. *Map of Vermilion Lake, St. Louis Co., Minnesota*, 1866.

**Hill**, J.H. Publisher of Burlington, Vermont. Vermont, 1840.

**Hill**, James S., *RN*. Served with Captain Owen Stanley aboard HMS *Britomart*, which arrived in New Zealand in 1840. *Plan of the harbour of Akaroa*, 1840 (with O. Stanley).

**Hill**, Johann Jakob (1730-1801). Military cartographer of Darmstadt, specialised in forest maps.

**Hill**, John (18th century). English astronomer. *Urania*, 1754.

**Hill**, John. See **Hills**, John.

**Hill**, Joseph. Plan of Salford, 1740 MS.

**Hill**, Luke M. *Plan of the City of Georgetown and Environs* [Guyana], 1893.

**Hill**, M.B. Plan of Bristol, *c.*1775.

**Hill**, Nathaniel (1708-1768). 'Globe Maker, & Engraver at the Globe & Sun in Chancery Lane, Fleet Street, London'. Began as an apprentice to Richard Cushee at the Globe and Sun in 1730, married Elizabeth (probably Cushee's daughter, E. Cushee), and later traded at the Globe and Sun as engraver, instrument maker, estate surveyor and globemaker. Survey of Barking Marshes, Essex, 1742 MS; engraved 3 Welsh language maps of the Holy Land for Richard Morris's *Y Bibl Cyssegrlan*, Cambridge, John Bentham 1746 (for SPCK): William Vincent's *Scarborough*, 1747; engraved Lewis Morris' *Plans of Harbours, Bars, Bays and Roads in St. George's CHannel*, 1748; engraver of John Warburton's ... *Hertford Shire*... London, c.1749; Warburton's ... *Middlesex*..., 1749; John Wing's *Survey of the North Level... of the Fenns*, 1749; *A New Terrestrial Globe* [one of a 3-inch pair], 1754 (republished with the imprint changed to J. Newton, 1783); also thought to have made 3-inch, 9-inch, 12-inch and 15-inch globes.

• **Hill**, Elizabeth. Probably née Cushee. Continued the business at the Globe and Sun until it passed to Nathaniel's apprentice Thomas Bateman c.1772.

**Hill**, Peter [& Co.] (*fl.c.1800-c.1820*). Publisher and bookseller of Edinburgh. *New General Atlas*, 1814; *Travelling Map of Scotland*, 1820.

**Hill**, Robert T. *Reconnaissance Map of the Big Bend Country, Texas*, 1899 MS; *Topographic Atlas of the United States: Physical Geography of the Texas Region*, Washington D.C., U.S. Geological Survey 1900.

**Hill**, *Sir* Rowland (1795-1879). Originator of pre-paid postage and the 'Penny Black' postage stamp. *Map of the roads, near to the spot where Mary Ashford was Murdered* [Erdington, Birmingham], 1817 (with George Moorcroft); survey of Birmingham, 1819. **Ref.** CAMPBELL, T. *The Map Collector 50 (1990), p.31.*

**Hill**, S. & Son. Surveyors of Croft, Lincolnshire. Survey of the estates of Earl Fortescue in Threckingham etc. [Lincolnshire], 1848.

**Hill**, Samuel. Engraver of Nº·2 Cornhill, Boston, Massachusetts (1794). Engraved for Jedidiah Morse's *The American Universal Geography* and *The American Gazetteer*, 1789-1796; engraver of *A New Map of New Hampshire*, 1791 (for Jeremy Belknap's *History of New Hampshire*); *Plan of the City of Washington in the Territory of Columbia*, 1792 (from Andrew Ellicott's manuscript plan, a rival plate of this map was engraved by J. Thackara & J. Vallance); *Plan & Elevation of the Tontine Crescent... Boston*, 1794; for Mathew Carey's *The General Atlas*, 1795-1796; maps for Carey's *American Atlas*, 1795; for John Malham's *The Naval Gazetteer*, Boston, W. Spottswood and J. Nancrede 1795, 1804; *An Accurate Map of Europe*, 1795; Caleb Harris' *A Map of the State of Rhode Island*, 1795 (drawn by Harding Harris); Osgood Carleton's *A Plan of Boston*, 1796 used in John West's *Boston Directory*, 1803; Harding Harris' *Rhode-Island and Connecticut*, Boston 1796; *Atlantic Hemisphere...*, 1797; *A Correct Chart of the West India Islands*, 1797; *Map of the District of Maine Massachusetts...* (4 sheets), 1801, 1802 (drawn by G. Graham, engraved by J. Callender & S. Hill), Boston, B. & J. Loring 1802; O. Carleton's *Map of Massachusetts Proper...* (4 sheets), Boston 1801 (drawn by G. Graham, engraved by J. Callender & S. Hill), Boston, B. & J. Loring 1802, revised edition published Albany, Amos Lay 1822. **Ref.** WHEAT & BRUN *Maps and charts published in America before 1800 (1978) many entries and descriptions.*

**Hill**, Samuel W. American geologist, worked for Charles T. Jackson on geological maps of Michigan including: *Plan of the Underground Work of the Lac Labelle Mines*, 1847 (lithographed by Ackermann); *Underground Works of the Northwest Mines...*, 1847 (lithographed by Ackermann); *Topographical and Underground Plan of the Cliff Mine Situated on Keweenaw Point...*, 1847; *Geological Map of the Isle Royale Lake Superior Michigan*, 1847 (with J.W. Foster, J.D. Whitney and W. Schlatter); Geological maps of Lake Superior, 1855; *Geological Map of the Trap Range of Keweenaw Point*, Philadelphia, R.L. Barnes 1863 (with William H. Stevens & C.P. Williams); *Map of Isle Royale, Lake Superior, Michigan, shewing some of its geological and vein phenomena*, New York, John J. Bloomfield 1871; and others.

**Hill**, Thomas. See **Hill**, family, above.

**Hill**, William. 'Totius terrarum orbis tabula nova', in *Dionysii Orbis descriptio*, London 1679.

*Sir Rowland Hill (1795-1879). Frontispiece to H.W. Hill's* Rowland Hill and the Fight for Penny Post. *(By courtesy of the National Library of Australia)*

**Hillebrands,** A.J. *Atlas van de Vereenigde Staten van Noord Amerika*, Groningen, J. Oomkens 1849.

**Hiller,** Karl (1869-1943). Cartographer for Justus Perthes.

**Hillestrom,** —. *Costa Oriental d'Africa. Provincia de Moçambique*, Lisbon 1890.

**Hillestrom,** Christopher. Trollhättan, Sweden, 1765 MS.

**Hillhouse,** William. See **Hilhouse,** William.

**Hilliar,** H. *Borough of Liverpool*, 1851.

**Hilliard, Gray & Co.** Publishers of Boston, Massachusetts, probably successors to Cummings & Hilliard *q.v.*. [Joseph Emerson] *Worcester's Modern Atlas*, 1821 (another work by J.E. Worcester was published by Cummings & Hilliard of Boston).

**Hillock,** —. Contributor to Thomas Jefferys's *American Atlas*, 1775.

**Hills,** Commander E.H. *Trondhjem Bay*, 1873.

**Hills,** Graham H. *Chart of the River Mersey* (9 sheets), 1864; *Liverpool Bay*, 1877-1880.

**Hills,** Henry. 'Printer to His Majesty for his Household and Chapel'. *The exact description of the City of Buda, with several Encampments, Approaches and batteries of the Imperial Bavarian and Brandenburgh forces...* (2 sheets), London, 1686.

**Hills** [Hill], John. Mapseller 'in Exchange Alley in Cornhill', London. John Seller's *A Mapp of New England*, 1676; John Oliver's *A Mapp of the Cityes of London & Westminster & Burrough of Southwark*, c.1680.

**Hills,** Lieutenant John (fl.1777-1817). Assistant engineer, surveyor and draughtsman, trained in the Drawing Room of the Tower of London, served with the Engineers until 1784. After stays in New Jersey and New York City, he settled in Philadelphia in 1786. Many manuscript Revolutionary War plans were drawn or copied by Hills, 1777-1782. *A Collection of Plans in the Province of New Jersey*, c.1776-1781; *A drawn plan of the Peninsula of Chesopeak Bay* (3 sheets), 1781; plans for William Faden, 1784-1785; C. Varle's *...This Plan of the City and its Environs* [Philadelphia], c.1794, (engraved R. Scot of Philadelphia); *This Plan of the City of Philadelphia and Its Environs...*, 1797 (engraved by John Cooke of London, England); *A map exhibiting the different stage routs, between the cities of New York, Baltimore, and parts adjacent...*, Philadelphia, E. Savage 1800; *A Plan of the City of Philadelphia and Environs Surveyed... 1801, 2, 3, 4, 5, 6, & 7* (9 plates), 1808 (engraved by William Kneass); North America, 1811 MS; and many other manuscript surveys. **Ref.** GUTHORN, P. *John Hills, Assistant Engineer* (Portolan Press 1976); GUTHORN, P. *British maps of the American Revolution* (1972) pp.24-27.

**Hillyer,** H.L. *Camden Co., Georgia*, c.1868.

**Hilpert & Chandler.** Printers and engravers of Chicago. B.A.M. Froiseth's *Salt Lake City, prepared expressly for Crofutt's Salt Lake City Directory 1885*, 1885.

**Hilscher,** A. *Kreis Landsberg*, 1892; *Kreis Saatzig*, c.1895.

**Hilten,** Jan van. Printer of Amsterdam. *Luxenbourg*, 1646.

**Hiltensperger,** Johann Jost (1711-1792). Woodcutter, engraver and printer of Zug, Switzerland.

**Hilton,** J. Engraver. *A Plan of the Antient City of Canterbury*, 1752 (surveyed by William and Henry Doidge); Edward Jacob's *A Plan of the Town of Faversham* [Kent], c.1774.

**Hilton,** William (d.1675). Sailor from Charlestown, Massachusetts. His explorations around Cape Fear were recorded on a manuscript map by Nicholas Shapley entitled *Discovery Made by Wm. Hilton of Charles Towne*, 1662 (now lost, a copy is in the British Library); and in his *A Relation of a Discovery lately made on the Coast of Florida*, 1664 (description only).

**Himburg,** Christian Friedrich (d.1801). Publisher of Berlin. *Kenntniss des gestirnten Himmels*, 1777; *Halbkugeln der Erde*, 1786.

**Hime,** *Colonel* Albert Henry. Royal engineers. *Natal Harbour Works,* 1876 MS; Boundary Survey... [Orange Free State and Natal] (9 sheets), 1884 (with N.W.E. Fannin, surveys by Orpen and G. Tatham).

**Himmelreich,** —. *Rudolstadt und Umgegend,* c.1894.

**Himmerich,** *Colonel* Johann. Military cartographer of Hamburg. *Elbe-strom,* Amsterdam, Covens & Mortier, *c.*1730.

**Hincke,** *Captain* P.A.W. von. *Plan der Stadt... Magdeburg,* 1809.

**Hind,** *Professor* Henry Youle (1823-1908). Geologist from Nottingham, studied at Cambridge, also in Germany and France. Moved to North America in 1846 and settled near Toronto where he taught while continuing his own studies. Served as geologist with the 1857 Canadian Red River exploring expedition. *...Canoe Route between Fort William and Lake Superior and Fort Garry and Red River...,* 1858 (published in Hind's *Narrative...* of the expedition); *Map of the Country between Red River & Lake Winnipeg on the East and the Elbow of the South Saskatchewan on the West,* J. Arrowsmith 1858; *Map of the Country from Lake Superior to the Pacific Ocean...,* John Arrowsmith 1860; geological survey of New Brunswick, 1864; *...Estuary of the St Lawrence,* 1877; and other maps and river profiles. **Ref.** WALDMAN & WEXLER Who was who in world exploration (1992) p.323.

**Hind,** John Russell. English astronomer. *School Atlas of Astronomy,* 1855, 1860 (maps engraved by A.K. Johnson).

**Hinder,** Thomas. Chart coast of North America, 1732 MS.

**Hindermann,** Emanuel. Lithographer and publisher of Basle. Map of Malta, [n.d.] (engraved J. Locherer).

**Hinderstein,** G.F.D. van. See **Derfelden** van Hinderstein, Gijsbert Franco von.

**Hingenau,** Otto von (1818-1872). *Geologische Übersichts-Karte von Mähren u. österr. Schlesien,* Vienna 1852.

**Hingeston,** Mileson. Bookseller of London. One of the many sellers of E. Gibson's edition of William Camden's *Britannia,* 1772 (last edition with Robert Morden's maps).

**Hinks,** Arthur Robert. Secretary to the Royal Geographical Society, 1915-1945. Working drawing of Mount Everest, 1933-1945 (with H.F. Milne, taken from a 1933 flight over Everest and a photogrammetric survey of 1935).

**Hinman,** Russell. *Map of Washington city and environs,* Cincinnati, Van Antwerp, Bragg & Co. 1881 (published in *The Eclectic atlas and hand-book of the United States).*

**Hinman.** See **Mitchell** & Hinman.

**Hinman & Dutton.** Publishers of 6 North Fifth Street, Philadelphia. Sellers of Mitchell & Young's *A Map of the World on Mercator's Projection...,* 1837 and later (engraved by J.H. Young, F. Dankworth, E. Yeager and J. Knight); J.H. Young's Wisconsin, 1838.

**Hinrichs,** Johann Conrad (1765-1813). Publisher and seller of books and maps at Leipzig. *Bataille de Marengo,* Rienicke et Hinrichs 1801; *Neue Bellona...,* 1802; Carte d'Allemagne, 1803; Hamburg, 1810; *Hydrographische Carte von Europa,* 1811; *L'Empire Français* (48 sheets), 1812; *Neue politische-Militärische und Post-Karte vom ganzen Französischen Reiche welche England, den Rheinbound, die Schweiz, das Konigreich Italien und Spanien...*(4 sheets), 1812 (title also in French); C.G.D. Stein's *Neuer Atlas der ganzen Erde...,* 1819 and later; K. Vogel's *Schulatlas,* 1837.

**Hinrichs,** L.E. *Hinrichs' Groote teekenatlas,* Gorinchem, J. Noorduyn & Zonen 1901 and later editions.

**Hinrichs,** Oscar. Publisher of Chicago. *Guide Map of Central Park* [New York City], 1875; *Map of the United States of Mexico,* 1893.

**Hinton,** *Captain* —. Texan Navy. Surveys used for *San Luis Harbour* [Texas], Admiralty 1844, also published as *Port de Saint-Louis... 1843,* Paris, Dépôt-Général de la Marine 1856.

**Hinton,** C. Publisher of 1 Ivy Lane, Paternoster Row, London. Described as the publisher of many of the maps which appeared in *The Panorama: Or, Traveller's Instructive Guide: Through England and Wales*, London, 1820 (on the title page J. Wallis is described as printer, and W.H. Reid as publisher); an 1822 edition of R. Miller's set of geographical cards bears the imprint of C. Hinton and J. Wallis.

**Hinton,** I.T. [J.T.]. (*fl.c.*1828-1831). Publisher in London. J. Roger's *Panorama of London*, 1830.
• **Hinton & Simpkin & Marshall.** *Map of the states of Virginia and Maryland*, London, I.T. Hinton & Simpkin & Marshall 1831 (engraved and printed by Fenner, Sears & Co., also other maps and views used in different editions of J.H. Hinton's *History*); J.H. Hinton's *The History and Topography of the United States*, 1832. **N.B.** See also **Simpkin & Marshall**; compare with J.H. **Hinton**.

**Hinton,** John (*d.*1781). Bookseller and publisher 'at the Kings Arms, S*t.* Pauls Church Yard' (1745-1752), Kings Arms, Newgate Street (1752-1765), Kings Arms, 34 Paternoster Row (1766-1781). Published volumes 1-39 of the *Universal Magazine...*, 1747-1766 (which contained maps by Bowen and Kitchin); commissioned a series of large scale county maps from Emanuel Bowen and Thomas Kitchin, 1749 (plates sold to John Tinney *c.*1752), these were later to be published as *The Large English Atlas*, 1760 and later.
**Ref.** HODSON, D. *County atlases of the British Isles published after 1703*, Vol.II (1989) pp.97-102; *ibid.* Vol.III (1997) pp.53-61.

**Hinton,** *Reverend* John Howard (1791-1873). Theologian and historian. *An Atlas of the United States of North America*, London, Simpkin & Marshall, Philadelphia, T. Wardle 1832; *The History and Topography of the United States*, London, Hinton & Simpkin & Marshall 1832, Boston, Samuel Walker 1834, London, J. Dowding 1842 (many maps engraved by James Archer, and Fenner, Sears & Co.). **N.B.** Compare with I.T. **Hinton** (above).

**Hipparchus** of Nicea (*c.*190-post 126 BC). Greek astronomer and geographer of Rhodes. Criticised Eratosthenes's map of the known world by use of Pythagoras' theorem. Proposed division of equator into 360°, worked on the principles of latitude and longitude and established several forms of map projection. Reputedly made a celestial globe. **Ref.** WALLIS & ROBINSON *Cartographical innovations* (1987) pp.176-177 and other entries; DICKS, D.R. *The geographical fragments of Hipparchus* (London 1960).

**Hippel,** Mauricio von. One of the surveyors for *Boundary between the United States & Mexico...*, 1855 (drawn by F.Herbst); *Plano Topographico de la Ciudad de Merida*, 1864-1865.

**Hipschmann** [Hibschmann; Hübschmann], Johann Philipp Jakob (*d.*1655). Engraver for Johann Hoffmann in Nuremberg. *Britannische Inseln* (2 sheets), *c.*1650.

**Hipschmann** [Hibschmann; Hübschmann], Sigismund Gabriel (*b.*1639). Engraver of Nuremberg, worked for Johann Hoffmann. Untitled world map in Georg Christoff von Neitzschitz's *Welt-Beschauung*, 1666; *l'Alsace* (2 sheets), 1674; *Mappe-monde Geo-Hydrographique* (after Nicolas Sanson), Nuremberg, Johann Hoffmann 1675; *Circulus Suevicus* (2 sheets), Nuremberg, J. Hoffmann 1676; *Brandenburg* (2 sheets), Nuremberg, J. Hoffman *c.*1677; *Aigentlicher Grundtriess Der Inssel Schiedt in Ungern den Herumbgrentzenten Haubt Vestung und Örter, gengen der Türcken*, Nuremberg 1680; *Totius Fluminus Rheni*, *c.*1680; also engraved for Johann Christoff Baer, 1681.

**Hiranoya** Mohei. Japanese publisher. Joint publisher (with Echigoya Jihei) of Shikata Shunsui's map of the Nishikawa Canal, 1863.

**Hirawi,** Ali, Ibn-Abu-Bakr, Ibn Ali. 12th-century Persian geographer and traveller.

**Hire,** P. de la. See **La Hire**, Philippe de.

**Hire,** George. *Boonting Islands* [Moluccas], 1831.

**Hirsch,** *Professor Doctor* Adolph (1830-1901). German born astronomer and geodesist, worked in Neuchâtel.

**Hirschfeld,** C.L. Publisher of Leipzig. *Plan von Dresden*, *c.*1872.

**Hirschfeld,** Gustav (1847-1895). Geographer

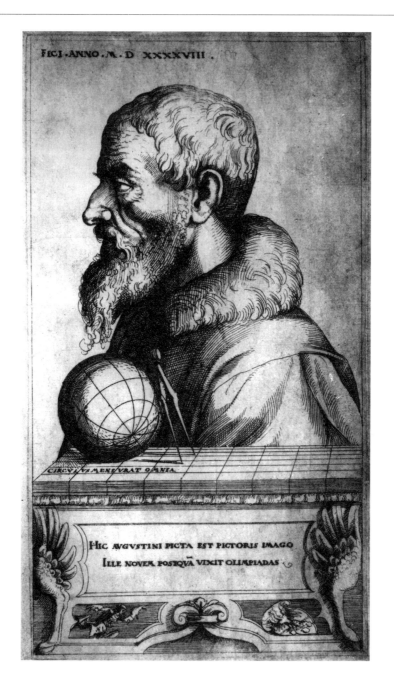

*Augustin Hirschvogel (1503-1553). (By courtesy of Austrian National Library)*

and archaeologist, professor at the University of Königsberg, [Kaliningrad], Prussia.

**Hirschgarter**, Mathhias. See **Hirtzgarter**, Matthias.

**Hirschhorn**, —. *Hirschhorn's Business Map of London and Suburbs*, Letts, Son & Co., 1880.

**Hirschko**, Carlos. Mamore ò Madera, 1782.

**Hirschmann**, L. *Wandkarte von Bayern* (4 sheets), Regensburg *c.*1868 (with G. Zahn); *Atlas für Volkes Schulen*, *c.*1872; *Wandkarte von Europa*, *c.*1875 (with G. Zahn).

**Hirschvogel** [Hirschfogel; Hirsvogel; Hirssfogel], Augustin (1503-1553). Painter, engraver and cartographer from Nuremberg, son of glass-painter Viet Hirschvogel. Worked in Nuremberg, Ljubljana (after 1536) and Vienna (from 1544), where he died. Map of the Turkish borders, 1539 MS (lost); map of upper Austria, 1542 (later published by Gerard de Jode as *Beschreibung des Erczherzogtumb Oesterreich ober Enns durch Augustin Hirsfuogel*, 1583); Carinthia, 1544 (later used by De Jode and Ortelius); maps and illustrations for Sigismund von Herberstein's description of Russia, 1546, 1547, 1549; plans of Vienna including a circular plan with fisheye view of the fortifications and surrounding countryside, 1546-1549; Saxony, 1550; *Hanc Viennae...* (6 sheets), 1552 (prepared from the earlier plans); surveys of Hungary 1552-1553 resulting in a manuscript map, published posthumously as *Nova et hactenus non visa regnoru[m] atque provintiarum... descriptio* (12 sheets), Nuremberg, Hans Weigel 1565 (copied by Ortelius, 1570 and de Jode, 1578); and others. **Ref.** FISCHER, K. *Cartographica Helvetica* 20 (1999) pp.3-12 [in German]; KARROW, R.W. *Mapmakers of the sixteenth century and their maps* (1993) pp.294-301; KRETSCHMER, DÖRFLINGER & WAWRIK *Lexicon zur Geschichte der Kartographie* (1986) vol.C/1, p.301 (numerous references cited) [in German]; SCHWARZ, K. *Augustin Hirschvogel. Ein deutscher Meister der Renaissance* (Berlin 1917) [in German].

**Hirt**, C.Z. *Schulwandkarte des Kaiser Wilhelms Kanals* (4 sheets), *c.*1895.

**Hirth**, Georg (1841-1916). German cartographer, working with August Petermann at Justus Perthes's 'Geographische Anstalt' in Gotha.

**Hirtzgarter** [Hirschgarter], Matthias (1574-1653). Swiss theologian, mathematician and astronomer in Zollikon. *Raetiae veteris*, 1616; *Detectio dioptrica corporum planetarum verorum* (includes detailed map of the moon), 1643.

**His**, J. de. Engraver of *Description de la partie des Indes Orientales qui est sous la domination du Grand Mogul*, 1663 (in Melchisedec Thévenot's *Relations de divers Voyages Curieux*, 1666).

**Hiscocks**, F.E. & Co. Australian publisher. *New Victorian Counties Atlas*, Melbourne 1874.

**Hishikawa** Moronobu (1618-1694). Japanese *ukiyoe* artist and mapmaker. Draughted *Zushu Atami ezu* [Pictorial map of Atami], 1681; *Tokaido Bungen Ezu* [Itinerary map of Tokaido Highway], published by Hangiya Shichirōbei 1690 (with Ochikochi Dōin, from surveys by Hōjō Ujinaga).

**Hisinger**, W. Geological map of Southern Sweden, 1830.

**Hislop**, *Commandant* —. Guyana, 1802.

**Hitch**, Charles (*d.*1764). Publisher and bookseller 'at the Red-Lion, in Pater-noster-row'. Apprentice and son-in-law of Arthur Bettesworth, in partnership with Bettesworth 1733-1739, then with his own former apprentice Lacy Hawes 1752-1764 (to whom his share of *Britannia* was sold in 1765). Shareholder in Thomas Badeslade and William Henry Toms's *Chorographia Britanniae*, 1745 edition (with W.H. Toms), *c.*1746 edition (with W.H. Toms & John Clark), *c.*1749 edition (with W.H. Toms & William Johnston), *c.*1752 and later (with L. Hawes); one of the many sellers of E. Gibson's version of Camden's *Britannia*, 1753 edition (with maps by Robert Morden).

**Hitchcock**, father and son.
• **Hitchcock**, Edward (1793-1864). State geologist of Massachusetts. *Geological Map of Massachusetts*, 1832 (printed by Pendleton's Lithography and used in Hitchcock's *Geology of Massachusetts*, 1833, 1835); Survey of the State [Massachusetts], 1837 and 1841.
•• **Hitchcock**, Charles Henry (1836-1919). Son of Edward (above), expert on the geology

of the eastern United States, professor of geology at Dartmouth College. *Geological Map of the United States*, 1874 (with William P. Blake); contributed to Francis A. Walker's *Statistical Atlas of the United States*, 1874; worked with H.F. Walling on his *Atlas of the State of New Hampshire*, New York, Comstock & Cline 1877; also associated with the Vermont Geographical Survey.

Hitchcock, A. *Borough of Shrewsbury*, 1832.

Hitchcock, De Witt C. Draughtsman and engraver. *Map of railways in New England and part of New York... for the Pathfinder Railway Guide*, Boston, Snow & Wilder 1847; *Map of Honduras and San Salvador, Central America. Showing the line of the proposed Honduras Interoceanic Railway By E.G. Squier*, New York 1854 (lithographed by Sarony & Co.).

Hitchcock. See **Marvin** & Hitchcock.

Hixson, William W. [& Co.] [Hixson Map and Litho. Co.] (*fl.*1896-1930s). Initially worked for George Ogle, then began producing his own numerous cheap and simple county and township maps of the mid-west, which were made up into county plat books. From his base in Rockford, Illinois, Hixson published plat books and county atlases in many editions, covering the whole of Illinois and maps of other states. *Map of Waushara County, Wisconsin*, 1896; Racine and Kenosha Counties, Wisconsin, 1899; *Map of Boone County, Illinois*, Rockford, W.W. Hixson & Co. 1899; *Map of Champaign County, Illinois*, Hixson Map & Litho. Co. 1902; *Map of Bureau County, Illinois*, Hixson Map Co. 1902; *Map of Christian County, Illinois*, Taylorville, Hixson Map Co. 1902; *Map of Carroll County, Illinois*, Hixson Map & Litho. Co. 1903; *Map of Alexander and Pulaski Counties, Illinois*, Hixson Map Co. 1904; *Plat Book of Champaign County, Illinois*, Rockford 1919; *Plat Book of Edwards County, Illinois*, 1919; *Plat Book of Bureau County, Illinois*, W.W. Hixson & Co. 1923; and many others. **Ref.** CONZEN, M.P. *Imago Mundi* 36 (1984) pp.26-28.

Hjertstedt, Friedrich. Swedish surveyor. Charta öfwer gräntse Tullarme, 1744.

Hjort, C.A. Swedish surveyor. *Charta öfver Wästerås Höfdingedöme*, Stockholm 1800; Västmanland, 1800.

Hjort, Daniel. Swedish surveyor. Worked with Andreas Bureus on a wall map of Scandinavia entitled *Orbis Arctoi nova et accurata delineatio* (6 sheets), 1626.

Hjorth, F. *Kort over Alands Havet*, 1861; *Kort over den Finske Bugt*, 1861.

Hoar, J. *Nashua and Nashville Villages* [New Hampshire], 1842.

Hoare, Edward. Engraver of London, in partnership with James Reeves.
• Hoare & Reeves. Engravers, publishers and mapsellers of London. '13 Little Queen Street, Lincoln's Inn Fields & 45 Kirby Street, Hatton Garden' (1822-1823); '45 Kirby Street & 90 Hatton Garden' (1823); 90 Hatton Garden (1823-1825); 14 Warwick Court, Holborn (*c.*1825-1826); 15 Warwick Court, Holborn (from 1830). G. Thompson's *A New Map of London and its Environs*, 1822, 1823, 1824, 1826, 1827; *London with 320 References: A List of Hackney Coach and Watermen's Fares*, 1823, 1824, 1826; William Ebden's *Twelve Miles round London*, 1823, *c.*1835; [Orlando] *Hodgson's Guide through the British Metropolis*, Hodgson & Co. 1824 edition; *Hodgson's New Map of the County of Kent*, Hodgson & Company 1824 and later issues; T.L. Murray's *An Atlas of the English Counties...*, 1830, 1831, 1832 (maps copied from Ebden's). **Ref.** SMITH, D. *Imago Mundi* 43 (1991) pp.52-53.

Hoare, W.S. *Italy*, 1863; *South Africa*, 1863; *France*, 1865.

Hoart, C.T. Surveyor and draughtsman. Worked mostly with Werner on 'astronomical and topographical observations' of islands in the Indian Ocean. Hoart draughted the resulting plans. *Plan of St. John de Nova Islands made from Astronomic and Topographic Observations by Messrs Werner & C.T. Hoart* [Madagascar], 1829 MS; *Plan of Agalega Island...* [Mauritius], 1829; *Plan of the North Entrance of Peros Banhos Islands...*, 1829 MS; *Plan of Saloman Islands*, 1829 MS; *Plan of the Six Islands...*, 1829 MS; *Plan of Cotivie Island...* [Seychelles], 1829 MS; *Plan of Providence...* [Seychelles], 1829 MS; *Plan of the Eagle Islands...*, 1829

MS; *Plan of Diego Garcia...*, 1829 MS; *Chart of the Chagos Archipelago* [Indian Ocean], 1829 (not with Werner).

**Hobbs**, E.D. City surveyor. *A plan of the city of Louisville and its environs in 1831* [Kentucky], Louisville 1831.

**Hobbs**, Lieutenant Henry. *Plan of Abercromby Heights and environs of Port of Spain* [Trinidad], 1803 MS (and other surveys and manuscript plans in the West Indies).

**Hobbs**, John Stratton (*fl*.1825-*d*.1865). Joined John Norie's hydrographic department in the 1820s, and worked as chart compiler, subsequently for J.W. Norie & Wilson. Following the retirement of Norie *c*.1839, Hobbs became supervisor of the hydrographic department for Charles Wilson. *North Sea*, 1845; *A Chart of the East Coast of England, from Dungeness to Newcastle...* (4 sheets), 1849; *A Chart of the Indian and part of the Pacific Oceans...*, 1850; *English Channel*, 1851; *The Cattegat, the Sound, and the Great and Little Belts...*, 1852; *Straits of Malacca*, 1852; *North Sea*, 1854, 1857; *A chart of the coast of Guayana, &c....*, 1854; *South Atlantic Ocean*, 1856, 1874; *A general chart... England &c. southward and round the World*, 1859; *Atlantic*, 1860; *Coast of the Cape Colony*, 1863, 1874, 1885; and many others.

**Hobbs**, John (*fl*.1687-1699). Estate surveyor of Essex, Middlesex and Surrey. *Mucking & East Tilbury*, 1687 MS.

**Hobbs**, John [& Co.]. Publishers of '20 Little Russel Street, Bloomsbury W.C.' *Plan of London*, 1862 (lithographed by James Herbert).

**Hobbs**, S.B. Surveyor. *Seychelles: Island of Mahé*, 1900.

**Hobday**, Captain J.R., RN. *Mandalé* [Burma], 1886; *Andaman Islands*, 1888; *Upper Irrawaddy*, 1892.

**Hobler**, George Alexander (1864-1935). Railway engineer and administrator, Commonwealth Railways. *Inspection tour North Western, Eastern and Kimberley Divisions of West Australia*, 1920; *Plan shewing proposed resumption along Darwin railway*, 1929.

**Hobley**, Charles William. Assistant Deputy Commissioner, East Africa Protectorate. Explorations and surveys in East Africa. *Sketch map showing boundary between Uganda and East Africa Protectorates*, 1902 MS; *East Africa Protectorate. Sketch shewing proposed Masai Reserves and connecting road*, 1904; *East Africa Portectorate. Sketch of the Rift valley showing proposed temporary land settlement*, 1904.

**Hobson**, Captain Samuel. *A Ground Plot of Londonderry*, sold by Richard Chiswell *c*.1690.

**Hobson**, William. *Charlestown, Doniphan County, Kansas*, *c*.1857.

**Hobson**, William Colling (*d.c*.1878). From *c*.1832 worked on the preparation of a map of Ireland which was later prepared by J. & C. Walker and published as *To Her most gracious Majesty Queen Victoria, This Map of Ireland...* (4 sheets), Liverpool, E. Holt 1838 (also sold in London by J. & C. Walker); *...map of the County Palatine of Durham...*, 1840 (engraved by J. & C. Walker); *Yorkshire* (4 sheets), 1843 (engraved by J. & C. Walker); *Hobson's Fox-Hunting Atlas* (42 maps), London, J. & C. Walker 1849 and later (maps are lithographic transfers from those of J. & C. Walker's *British Atlas*).

**Hochdanz**, Emil. *Türkischer Kriegsschauplatz*, 1877; *Dislokationskarte der Russischen Armee...*, Stuttgart *c*.1880.

**Hochholzer**, Hugo. *Map of the... Comstock Lode*, 1865.

**Hochreiter**, E. *Ethnographical map of Bohemia*, 1883.

**Hochstetter**, Anthony. Surveyor and publisher. *Plan of the City of Norwich*, 1789.

**Hochstetter**, Doctor Ferdinand von (1829-1884). Austrian geographer and mineralogist. Served aboard *Novara* on an Austrian world voyage of scientific exploration. The expedition arrived at Auckland, New Zealand in 1858, and Hochstetter stayed on to undertake numerous explorations and surveys. *Geologisch-Topographischer Atlas von Neu-Seeland* (6 sheets), 1863 (maps compiled by

*Robert Hoddle (1794-1881). (By courtesy of the National Library of Australia)*

A.H. Petermann from the surveys, sketches and observations of Hochstetter, J. von Haast and others), also published as *Geological and Topographical Atlas of New Zealand*, Auckland 1864.

**Hock**, F. *Karte vom Herzogthum Sachsen*, Coburg 1844.

**Hocquart**, — *le jeune*. *Nouvelle carte de France routière et postale*, Audin 1827. **N.B.** Compare with E. **Hocquart**, below.

**Hocquart**, A. Engraver of relief for J. Andriveau, 1829.

**Hocquart**, Edouard *(b.1789)*. 'Marchand d'estampes' and engraver of Paris. *Plan de la Ville de Paris*, Paris, J. Moronval 1827 (engraved by Charmont); C.V. Monin's *Carte*

*William Hodges (1744-1797). Taken from J.C. Beaglehole ed.,* Journals of Captain J. Cook, *Cambridge Press for the Hakluyt Society 1961 (By courtesy of the National Library of Australia)*

*générale des postes et des routes de la France,* 1828; A. H. Dufour's *Atlas départmental de la France,* 1835; C.V. Monin's *Atlas universel de géographie ancienne et moderne,* 1837; Thuillier's *Nouvelle carte des postes de la France,* 1840.

**Hoctomanno,** *Conte.* See **Freducci,** Hoctomanno.

**Hodder,** Edward Mulberry *MD* (1810-1878). Commodore of the Royal Canadian Yacht Club. *The Harbours and Ports of Lake Ontario, in a Series of Charts Accompanied by a Description of Each by Edward M. Hodder...,* Toronto, Maclear & Co. 1857 (plan compiled from surveys by Captain Owen and Lieutenant Herbert); surveys of the 'harbours and ports of the lake' used on *Chart of Lake Ontario...,* Toronto, W.C. Chewett & Co. 1863.

**Hoddle,** Robert (1794-1881). Native of London, arrived in New South Wales in 1823 and was appointed assistant surveyor in Sydney. From 1851-1857 he served as Victoria's first Surveyor General. He was also responsible for laying out the plan of Melbourne in 1837. *Province of Victoria*, 1853. **Ref.** SCURFIELD, G. *The Hoddle Years* (1995).

**Hodge**, Robert. See **Darton** & Hodge.

**Hodge, Allen and Campbell**. Publishers of New York. *The New-York Directory...*, 1789, (with a plan of New York by James McComb Jr., engraved by Cornelius Tiebout). **Ref.** COHEN, P.E. & AUGUSTYN, R.T. *Manhattan in maps 1527-1995* (1997) pp.90-91.

**Hodgen**, Robert S. *Map of St Clair County, Illinois*, Buffalo, Joseph W. Holmes 1863; worked with J.W. Holmes (as Holmes & Hodgen) on *Map of the County of Macon, Illinois*, Philadelphia 1865; *Coles Co., Illinois*, 1869.

**Hodges**, Sir James. Bookseller and publisher 'at the Looking Glass over against St. Magnus Church, London Bridge' (1742). *A New History of Jamaica*, 1740 (includes map); Joint publisher (with T. Cox) of the second edition of J. Cowley's *A New and Easy Introduction to the Study of Geography*, 1742; one of the sellers of T. Badeslade and W.H. Toms's *Chorographia Britanniae...*, 1742 edition; one of the publishers of Thomas Hutchinson's *Geographia Magnae Britanniae*, 1748, 1756; one of the publishers of Thomas Kitchin's *Geographiae Scotiae*, 1749, 1750, 1756; one of the publishers of William Camden's *Britannia*, 1753 (with Robert Morden's maps).

**Hodges**, N. Sikh Territory, 1846 (with Captain C. Wade).

**Hodges**, William (1744-1797). Landscape and portrait artist aboard the *Resolution* on Captain Cook's second voyage, mostly working on coastal views. Chart of South Georgia, 1775 MS. **Ref.** DAVID, A. *The charts and coastal views of Captain Cook's voyages...1772-1775* (The Hakluyt Society 1992) pp.lix-lxi.

**Hodges & McArthur**. Dublin sellers of *Ewing's New General Atlas*, 1825 edition (first published Edinburgh, Oliver & Boyd 1817).

**Hodges & Smith**. Publishers of Dublin. Richard Griffith's *A General Map of Ireland to accompany the Report of the Railway Commissioners* (6 sheets), 1839 and later (also sold in London by James Gardner), a U.S. edition was published Philadelphia, Durkan, Beehan & Maher 1860; Robert Kane's *The Industrial Resources of Ireland* (4 maps), 2nd edition 1845 (the first edition had no maps); *Map of Ireland, Shewing the New Fishery Districts*, 1849.
• **Hodges, Smith & Co**. *Midland Great Western Railway, Ireland*, 1850; *Map of Meath and ward Hunting District*, 1862.
• **Hodges, Figgis & Co**. *Map of Proposed Irish Tramways and Light Railways*, 1884; G.A. Dagg's *Devia Hibernia*, 1893; *Lansdowne maps of the Down Survey*, 1920.

**Hodgkin**, R.S. *St. Clair County, Illinois*, 1863.

**Hodgkinson**, J. See **Hodskinson**, Joseph.

**Hodgson**, family and company. Printers and publishers of London from *c*.1808. **N.B.** The degree of involvement of each individual is uncertain, as are the relationships between them. **Ref.** SMITH, D. *Imago Mundi* 43 (1991) pp.48-58.
• **Hodgson**, William. Printer at 20 Strand (1808); 25 Fleet Street (1820-1821); 10 Newgate street (1822); 48 Lothbury (1825).
• **Hodgson**, Bernard. Printer at 25 Fleet Street (1820-1823); 43 Holywell Street (1821); Church Cottage, Thames Ditton (1825).
• **Hodgson and Company** (*fl.c*.1822-1830). A loose association of members of the Hodgson family and others, with a variety of addresses. Hodgson and Company were known at 25 Fleet Street (1822); 10 Newgate Street (1822-1824). William and Bernard may have been in partnership at Fleet Street from 1820, and at Holywell Street. Orlando Hodgson was associated with the company in 1825. Published early editions of William Ebden's county maps e.g. *Hodgson's New Map of the County of Dorsetshire*, London, Hodgson and Company, 1824 (engraved by Hoare & Reeves, for editions published by William Cole 'late Hodgson & Company', the titles were amended to begin *Ebden's New Map of the County of...*, 1825); *The Pocket Tourist & English Atlas* (43 maps), 1820; *Hodgson's Guide through the British Metropolis...*, 1823 (engraved by J. Reeves); *Hodgson's New Map Fifteen Miles Round London*, 1823 (engraved

by J. Reeves); J. Wallis & W.H. Reid's *The Panorama of England and Wales*, 1825 edition (first published as *The Panorama Or, Traveller's Instructive Guide...*, W.H. Reid 1820).

• **Hodgson**, Orlando. Printer, printseller and publisher of London. Thought to have been part of Hodgson and Company in 1825. Published in his own right at 21 Maiden Lane (1825-1828); 10 Cloth Fair (1832-1835); 111 Fleet Street (1836-1844). He was in partnership with G. Biggs, 1842-1843. *The Pocket Tourist & English Atlas*, 1827 (one of the re-issues of Joseph Allen's *A Geographical Game..*, 1811); re-issue of John Shury's *Plan of London from Actual Survey*, 1838 (first published 1832); *Leigh's New Pocket Road-Book* bound with *Leigh's New Atlas of England & Wales*, 1842, 1843 (with G. Biggs of 421 Strand); re-issue of M.A. Leigh's *New Plan of London*, 1843 (first published c.1827). **Ref.** BERESINER, Y. The Map Collector 30 (1985), pp.40-41.

**Hodgson**, C.J. Surveys for *Plan of the Battle of Sabraon fought on the 10. February 1846* [Lahore, Punjab], London 1846 (with others).

**Hodgson**, Caspar W. *Map of the Philippine Islands*, London, George G. Harrap & Co. 1908 (engraved by A. Briesmaster).

**Hodgson**, J.A. Hurriana District [India], 1810-1811; Bettiah Frontier, 1815.

**Hodgson**, K.M. Topographical survey of the sources of the Marañon and Mantaro Rivers, Central Peru, 1929.

**Hodgson**, *Lieutenant* Robert. Moskito Shore [Nicaragua], 1760 MS.

**Hodgson**, Thomas. Survey of Oxfordshire, 1766 published (4 sheets), A. Dury 1767 (engraved by T. Jefferys).

**Hodgson**, Thomas (*fl*.1821-1839). Westmorland surveyor. Westmoreland (4 sheets), surveyed 1823-1825, published 1828 (engraved W.R. Gardner).

**Hodgson**, Thomas. Bought the copyright to H.G. Collins's *Illustrated Atlas of London...*, and republished it with the title *London at a glance: An illustrated atlas of London...*, 1859, 1860 (first published 1854).

**Hodgson**, William. See **Hodgson**, family (above).

**Hodgson**, William. Many surveys in the English Lake District. *Estates in the Lake District in the County of Cumberland for Sale*, Carlisle, C. Thurnham & Sons, 1865 (lithographed by C. Thurnham); *Plan of Scarness and Braidness Estates* [Bassenthwaite, Cumberland], 1865; *Particulars and Conditions of Sale of Extensive and Valuable Estates... situate, lying and being in the County of Cumberland* (7 plans), Carlisle, C. Thurnham & Sons, 1866; *Plan of Highside Syke and Spouthouse*, 1866; and others.

**Hodler**, Emil. *Interlaken und Umgebung*, c.1878.

**Hodskinson** [Hodgkinson], Joseph I (1735-1812). Engineer, surveyor and engraver of Arundel Street, Strand, London. Worked in different capacities on maps of areas throughout England. Engraver for Thomas Jefferys's Bedfordshire (8 sheets), c.1765 (surveyed by J. Ainslie & T. Donald); engraver for *County of Cumberland Surveyed* (6 sheets), T. Jefferys 1774 and later (surveyed by T. Donald & J. Ainslie 1770-1771), reduced version (1 sheet) published J. Hodskinson & T. Donald 1783; Yorkshire (20 sheets), T. Jefferys 1771-1772 (surveyed by J. Ainslie, T. Donald & J. Hodskinson 1767-1770); survey of East and West Molesey, Thames Ditton and Walton-on-Thames, 1781; Wells, 1782; surveys for *The County of Suffolk* (6 sheets), London W. Faden 1783, 1820 (engraved by W. Faden), reduced edition (1 sheet), W. Faden 1787.

• **Hodskinson**, Joseph II (*d*.1800). Worked for his uncle, Joseph I.

**Hoeckner** [Höckner], Carl. Engraved for Joseph Meyer's *Universal-Atlas*, from 1830 published from 1833; for C.F. Wieland, 1842-1846.

**Hoedl**, Leopold Joseph. Publisher of Vienna. *Einleitung in die alte und neuere Geographia*, 1734.

**Hoeff**, D. van der. *Strait of Sikakap, Sumatra*, 1860.

**Hoefnagel** [Hofnagel; Houfnagel; Huefnagel; Hufnagel], father and sons.
• **Hoefnagel,** Joris [Georg; Joeris] (1542-1600). Painter, poet, miniaturist and topographical draughtsman from Antwerp. Travelled widely and produced nearly 100 views for Braun and Hogenberg's *Civitates orbis terrarum,* 1572-1618 (including *Bourges,* 1575); Cadiz and environs, used by Ortelius 1584. **Ref.** NUTI, L. 'The mapped views by Georg Hoefnagel: the merchant's eye, the humanist's eye' *Word & Image: a journal of verbal /visual enquiry* 4 (April-June 1988), pp.545-570, published Taylor & Francis, London; POINTER, S. *The Map Collector* 4 (1978) pp.7-14; POPHAM, A.E. 'George Hoefnagel and the *Civitates Orbis Terrarum*', in *Maso Finiguerra* Vol. I (1936), pp.183-201.
•• **Hoefnagel,** Jakob (1575-1630). Revised many of his father Joris's plans for Braun and Hogenberg. Vienna (6 sheets), 1609 (only one copy known).
•• **Hoefnagel,** Johann. Engraver, brother of Jakob Hoefnagel.

**Hoeg,** Anders (1727-1796). Danish hydrographer and teacher of navigation. *Soekaart over Nord-Söen,* 1769.

**Höegh,** —. Surveys for *Kaart ove Dannevirkestillingen mellem Slien og Reidedalen* [Schleswig-Holstein], 1863 (surveyed in 1861, drawn by Weyen).

**Hoeius.** See **Hoeye.**

**Hoehne,** Hermann. *Karte von Teplitz-Schönau,* 1873.

**Hoelzel,** Eduard. See **Hölzel,** Eduard.

**Hoen,** family and company.
• **Hoen** [Hohn], August (1817-1886). Lithographer and map printer from Höhn, Germany. Emigrated to the U.S.A. with 8 younger brothers and sisters, and his lithographer cousin Edward Weber in 1835. They settled in Baltimore. August worked with Edward Weber, trading as E. Weber & Company. Hoen took control of the company on the death of Weber in 1848, and changed the name to A. Hoen & Company in 1853. In 1849 he married Caroline, widow of Edward Weber. **N.B.** Walter Ristow, in *American maps and mapmakers (1986) p.300,* describes Weber as the uncle of August and Ernest.
•• **Hoen,** Albert Berthold. Son of August, inherited the company of A. Hoen & Co. on the death of his father in 1886.
• **Hoen,** Ernest. Younger brother of August. Involved with A. Hoen & Company.
• **A. Hoen & Company** (*fl.*1853-1981). Engravers, lithographic printers and publishers of Second Street, Baltimore. By 1882 they had built a new building on Lexington, Holliday and North Streets. When these premises were destroyed by fire in 1901 the company moved to Chester, Chase and Biddle Streets. Printers of maps and charts for the Federal Government. The company was absorbed by John Lucas Printing Company in 1981. ...*a Portion of Oregon Territory,* 1852, 1872; A. Paul's *Revised Location of the Boundary Avenues authorized by the ordinance of the Mayor & City Council of Baltimore,* 1853; *Sketch of the Public Surveys in Michigan,* 1855; *Monk's New Map of Central America, Ucatan & Florida...,* 1857; *Moule's New American Map...,* Jacob Monk 1857; Martenet's *Howard County, Maryland,* 1860; V.P. Corbett's *Map of the seat of war...* 1861 [Washington], 1861; A. Faul's *Swann Lake and Aqueduct of the Baltimore City Water Works,* 1862; Victor Hannot's *Map of Washington City and Georgetown,* 1866; Davoust Kern's map of York, Pennsylvania, 1879; R.D. Irving's *The Copper-Bearing Rocks of Lake Superior,* U.S. Geological Survey 1885 (adopted a new technique for geological illustration); *Berkely Springs* [West Virginia], John Moray 1889; *Atlas...Venezuela and British Guiana* (76 maps), 1897; *Alaskan Boundary Tribunal. Atlas...*(25 maps), 1903; A.K. Lobeck's *Physiographic Diagram of the United States,* A. Nystrom and Company 1921; printed the map supplements for *National Geographic,* 1929-1975; *Soil Map of the United States* (12 sheets), 1935.

**Hoen,** Dirck Jansz. [Janszoon]. Dutch chartmaker. Zuyder Zee, 1560.

**Hoepli,** company. Publishers of Milan. Produced school atlases and wall charts. G. Roncagli's *Atlante Mondiale Hoepli di geografia moderna fisica e politica* (80 maps), Milan 1894.

**Hoerl,** I.E. *Carte de la France composé de 25 feuilles,* 1833 (with J. Heiss, lithographed by B. Herder).

**Hoerold,** G. *Karte von den Bergwerken... in Ober Schleisien,* 1874.

*Joris Hoefnagel (1542-1600), engraved by Henricus Hondius. (By courtesy of Rodney Shirley).*

**Hoese,** J.B. de la (*fl.*1880s). Engraver for the 'Institute Cartographique Militaire', Brussels.

**Hoest,** Georg. Morocco, 1781.

**Hoet,** Gerard (1648-1733). Dutch engraver and draughtsman. Designed embellishments for Bernard Du Roy's *Nieuwe caerte vande Provincie van Utrecht,* Amsterdam, Nicolaas Visscher 1696; frontispiece for François Halma's *Geographia Sacra,* Amsterdam 1704.

**Hoeye,** father and son. Engravers and publishers of Amsterdam.
• **Hoeye** [Hoeius; Hoijaeus], François [Franciscus; Francoys] van den [de la] (*c.*1590-1636). Copper engraver and publisher of Amsterdam. Father of Rombout, to whom his plates passed. Map of Holland, 1620; engraved Wassenaer's world map (6 sheets), *c.*1629 (the only known copy is dated 1661); re-issued a world map by J. Hondius, 1630 (first published by Hondius *c.*1625); world map on Mercator's projection after Willem Jansz. Blaeu (2 sheets), *c.*1630; Germany, 1632. **Ref.** SCHILDER, G. *Three world maps by François van den Hoeye of 1661, Willem Jansz. Blaeu of 1607, Claes Jansz. Visscher of 1650* (Amsterdam 1981) (with facsimile plates); SCHILDER, G. *Imago Mundi* 31 (1979) p.46.
•• **Hoeye,** Rombout van den (1622-1671). Printer, publisher, engraver and mapseller of Amsterdam 'inde Kalverstrate inde dri Roosen Hoeie'. Son of, and successor to, François. Some of the plates passed to H. Allard sometime before 1666. *Leo Belgicus,* 1636; panorama of London, *c.*1640.

**Hofacker,** A. *Karte von Düsseldorf und Umgebung,* 1874.

**Hofel,** *Professor* B. Map of Vienna, Sollinger 1850.

**Höfel,** Blasius (1792-1868). Engraver of Vienna, noted for the use of acid etched printing plates.

**Hofer** [Hoffer], Andreas. Engraver of Nuremberg. Engraved *Atlas Homannianus,* 1762 (one of the title pages for *Atlas Germaniae,* drawn by J.J. Preisler).

**Hofer,** J.J. *Übersichtsplan der Stadt Zürich,* 1879.

**Hofer & Co.** Printer of Zurich. *Schiffahrtstraßen von Central-Europa,* 1918; *Mapa topográfico de la propiedad y fábrica La Farga,* 1920.

**Hoff,** Heinrich. Publisher of Mannheim. *Dr. Karl Glaser's Atlas über alle Theile der Erde,* 1841-1842, 1846; *Dr. Carl Glaser's Topisch-Physikalischer Atlas,* 1844, 1846; *Dr. Carl Glaser's Schul-Atlas,* 1846; *Kleiner Atlas der neuesten Erdbeschreibung* (26 maps), 1848.

**Hoff,** Heinrich Ernst von (1782-1851). *Charte von Koenigreich Würtemberg* (6 sheets), 1812?, 1819; *Gegend von Tübingen,* 1822.

**Hoff,** Karl Ernst Adolf von (1771-1837). Geographer and geologist of Gotha. *Thüringer Wald,* 1807; *Deutschland...,* Gotha, J. Perthes 1838.

**Hoffensberg,** F. *Nyeste Skole-Atlas,* Copenhagen, C.W. Stincks Forlag *c.*1854.

**Hoffensberg,** J. *Haand og Skole Atlas, c.*1876 (with C.C. Brix).

**Hoffer,** Andreas. See **Hofer,** Andreas.

**Hoffgaard,** Jens [Jans]. Danish mapmaker, produced charts and maps of Iceland and its coasts. *Island med sine fortooninger havner fiorder &c.,* 1723 MS.

**Hoffman.** See also **Hoffmann, Hofman** or **Hofmann.**

**Hoffman** brothers. Military engineers active in Denmark. **Ref.** DAHL, B.W. 'The military engineer Gottfried Hoffman and his works in Denmark 1648-1687' in *Fortress* 3 (1995) pp.3-12.
• **Hoffman,** Georg (*d.*1666). Native of Silesia, student in Leipzig and Strasbourg. In royal service in Denmark from 1643, appointed military engineer of Copenhagen in 1647. Designed the towns and fortifications of Frederiksodde and Sofieodde.
• **Hoffman,** Gottfried (*c.*1631-1687). Military engineer and cartographer from Lauban in Silesia [now Luban, Slask, Poland]. Studied at Leipzig and Strasbourg, became clerk of works for the Copenhagen fortifications in 1648 under the guidance of his elder brother Georg. City engineer of Copenhagen (from 1658), first head of the newly formed Danish

Royal Engineers (1684), commander of the Citadel of Copenhagen (from 1685). One of the first in Denmark to use trigonometry in his surveys. Produced many battlefield maps, as well as plans and maps of Danish fortifications, many of these were later copied by his cousin Christoph Heer. Hoffman and Heer used a distinctive style, with text and symbols added to the coloured manuscripts using a stamp with movable type. Gottfried produced more than 250 maps and plans, which include a survey of the island of Bornholm [Baltic], 1653; plan of Helsingborg, *c.*1653; map of the battle with the Swedish army, 1658; maps of the provinces and waters of Denmark, from 1660; maps of the battles between the Danish and Swedish armies in Scania during the war of 1675-1679, from *c.*1679.

**Hoffman**, C.A. *Res-Karta öfver Sverige*, Stockholm, P.A. Huldberg 1856.

**Hoffman**, C.L. ...*Karte des Herzogthums Oldenburg*, 1852.

**Hoffman**, Carl (1866-1900). Cartographer specialising in maps of the Alps. *Stubaier Gruppe*, [n.d.].

**Hoffman**, Charles F. Surveys for the California Geological Survey. Worked with J.T. Gardner on surveys of the Sierra Nevada, 1863-1867 (used by the Geological Survey of California for their map of 1868); Map of San Francisco Bay, 1867; State of California, 1873, revised 1877.

**Hoffman**, E. See **Hoffmann**, Ernst von.

**Hoffman**, Friedrich. See **Hoffmann**, Friedrich.

**Hoffman**, J.D. Topographer in North America. *From Fort Smith to the Rio Grande*, 1859; *From the Rio Grande to the pacific ocean*, 1859; recorded the topography for Henry Gannett's *Maryland - District of Columbia - Virginia, Washington sheet*, 1886 (with D.J. Howell); topography for *California Honey Lake Sheet*, U.S. Geological Survey 1891 (with Lieutenant G.M. Wheeler's survey of the 1870s).

**Hoffman**, J.F.C. See **Hoffmann**, Johann Friedrich Carl.

**Hoffman**, Johann. See **Hoffmann**, Johann.

**Hoffman**, Karl Friedrich Vollrath. See **Hoffmann**, Karl Friedrich Vollrath von.

**Hoffman, Pease & Tolley**. Lithographers of Albany, New York. E. Jacob's *Map of the City of Albany...*, Albany, Sprague & Co. and New York, M. Dripps 1857.

**Hoffmann**, *Lieutenant* —. Prussia, 1831.

**Hoffmann**, August. *Plan der Stadt Breslau*, 1868.

**Hoffmann**, C. *Karte der centralen Ortler Gruppe*, 1875; *Spanien und Portugal*, 1876.

**Hoffmann**, Carl. Prussian military cartographer. *Umgegend Breslaus* (4 sheets), 1820.

**Hoffmann**, Carl (1802-1883). Publisher of Stuttgart. *Reise- Post-und Zoll-Karte von Deutschland*, 1834; *Atlas des gestirnten Himmels*, 1839.

**Hoffmann** [Hoffman], Ernst von (1801-1871). German explorer and mineralogist. Geological maps of Ural, 1831.

**Hoffmann** [Hofman], Ferenc (1828-1900). Hungarian surveyor and cartographer, director of the Triangulation Office, 1874-1897. Worked on the triangulation survey of Buda. Received the Order of the Knight's Cross for his work on the cadastral survey of Hungary.
**Ref.** *Kataszteri Közlöny* [Cadastral Bulletin], (1900/10) [in Hungarian].

**Hoffmann**, *Professor* Friedrich (1797-1836). German geographer and geologist. Geologische Karte, Prague, Joseph Jüttner 1820; *Geognostische Charte vom nordwestlichen Deutschland* (24 sheets), Berlin 1829; Carta geologica delle Alpi Apiano, 1833; *Geognostische Karte von Sicilien...*, 1839 (drawn by C. Zirbeck, engraved by F.W. Kliewer); *Liparische Eilande*, 1844.

**Hoffmann**, George Heinrich. German cartographer. *Palaestina*, 1833.

**Hoffmann**, Heinrich Albert (1818-1880). Publisher of Berlin. Published Julius Löwenberg's *Historisch-geographischer Bilder-Atlas für die Jugend*, 1844.

**Hoffmann**, J.J. *Kreis Mosbach* 1896; *Kreis Offenburg*, 1897.

**Hoffmann,** Jacob. Engraver and publisher of Vienna. *Archiducatus Austriae Inferioris* (8 sheets), 1697 (with Jacob Hermundt).

**Hoffmann,** Johann (1629-1698). Map publisher of Nuremberg. *Nova et Exacta Totius Regni Hungariae Delineatio,* 1664 (engraved by W. Pfann); *Ungarn, Siebenbürgen, Wallachey, Moldau und Angrentzende Türckische Länder...* (with 16 inset plans and views), 1675 (engraved by J. Azelt); *Mappe-monde Geo-Hydrographique...,* 1675 (after Jaillot, engraved by Sigismund Hipschmann); *Circulus Suevicus* (2 sheets), 1676 (engraved by S.G. Hipschmann); *Districtus Norinbergensis,* 1677; *Brandenburg* (2 sheets), *c.*1677; *Geographisches carten-spiel von Europa,* 1678 (playing-card maps of Europe); *Geographisches carten-spiel von Asia, Africa, und America, c.*1678 (another set of playing-card maps); *Grundriss der kayserlichen residenz-Stadt Wien, mit der türkischen Belagerung...* (map of the Turkish siege of Vienna), *c.*1683; *Rheinstrohm,* 1689; German edition of Olfert Dapper's Africa, 1689; 4th edition (in German) of P. Duval's *Geographiae Universalis...,* 1694; and many others. **Ref.** SZANTAI, L.. *Atlas Hungaricus* (1996) pp.229-233 [annotations in Hungarian, short notes in English].

**Hoffmann** [Hoffman], Johann Friedrich Carl (1733-1793). Bavarian military cartographer. *Fürstenthum Bayreuth,* 1780 MS; *Militärische Karte des Fürstentums Bayreuth...,* 1799 (with J.C. Stierlein).

**Hoffmann,** Johann Georg. Administrative maps [Waldbücher] for the forests of Franconia, 1722-1755.

**Hoffmann,** Johann Wilhelm. Publisher of Weimar. Otto von Kotzebue's *Entdeckungs-Reise in die Süd-See,* 1821; *Navigators-Inseln,* 1830.

**Hoffmann,** John Isidore. *Plan of the Witwatersrand,* 1904.

**Hoffmann** [Hoffman], Karl Friedrich Vollrath von (1796-1842). German geographer and private lecturer in geography at the University of Munich. Also worked in Stuttgart as Superintendent of the Geographical Institute for Cotta Verlag. Co-editor, with H. Berghaus, of the geographical periodical *Hertha,* 1825-1827. *Atlas über alle Theile der Erde für Schulen,* 1835; *Himmels-Atlas,* 1835-1837; *Karte vom Königreiche Würtemberg und dem Grossherzogthume Baden,* Stuttgart, J. Scheible's Buchhandlung 1836 (lithographically engraved on stone by W. Pobuda and J. Rees, lithographic printers Pobuda, Rees & Comp.); Orbis terrarum antiqua, 1841.

**Hoffmann,** Otto. Bavarian railways map, 1890.

**Hoffmann,** Wolfgang. Printer of Frankfurt am Main. Beschreibung von Schweden, F. Hulsius 1632; M. Merian's *Topographia Alsatiae,* 1644.

**Hoffmann & Campe.** Publishing house in Hamburg. *Juetland,* 1848; *Schleswig* (4 sheets), 1848; *Holstein & Lauenburg,* 1849.

**Hoffmeister,** F.L. *Plan von Basel,* 1800; German town plans, *c.* 1820.

**Hoffmeister,** George B. Chart of Barcelona, Admiralty 1839.

**Hoffmeyer,** Fritz. *Schleswig Holstein,* 1867; *Marburgs Umgegend,* 1894.

**Hoffmeyer,** Nils Henrik Cordulus (1836-1884). Danish meteorologist. Meteorological maps.

**Hofman,** E. *Karte des noerdlichen Ural,* 1850.

**Hofman,** F. See **Hoffmann,** Ference.

**Hofman,** Hans de. Danish publisher. Issued editions of Erik Pontoppidan's *Den Danske Atlas,* from 1763.

**Hofman,** —. Draughtsman. *Sketch of the Island of Capri,* 1810.

**Hofman,** —. *Plan der Stadt Worms,* 1891.

**Hofman,** Elias (d.1592). Untitled map of environs of Frankfurt am Main (2 sheets), Cologne 1583 (engraved by HW [Hans Weyrich]).

**Hofmann,** F. Anton. Engraver. Battle plans, 1836-1850.

**Hofmann**, Franz. Printer of Prague. *Prag* (2 sheets), *c.*1815.

**Hofmann**, X.A. von. *Paris und Umgegend*, *c.*1870.

**Hofmeister**, Johannes (1721-1800). Bookbinder, bookseller and publisher of Zürich 'an der Rosengass / Rue de la Rose'. Reprinted J. Murer's *...Statt Zürych...* (6 woodblocks), 4th edition 1766, 1790 (first published 1576). **Ref.** DÜRST, A. 'Die Planvedute der Stadt Zürich von Jos Murer, 1576' in *Cartographica Helvetica* 15 (1997) pp.23-37 [in German with summaries in English and French].

**Hofnerus**, Erasmus Sabinus. See **Fabronius**, Hermann.

**Hofnagel**, Georg. See **Hoefnagel**, Joris.

**Hofrichter**, R. *Kreis Leobschütz*, 1892.

**Hogan**, Charles [& Co.]. *Hogan's Commercial Map of London and the Home Counties*, 1904, 1908.

**Hogan**, John Sheridan (1815-1859). *Canada: An Essay*, Montreal, B. Dawson, also London, Sampson, Low 1855 (contained T.C. Keefer's *Map of the Province of Canada, and the Lower Colonies*).

**Hogan**, *Captain* M. Straits of Boeton, published Laurie and Whittle 1796; Chart of Straits westward of New Guinea, 1796, published Laurie and Whittle 1798.

**Hogan**, William. *Map of Tasmania in 1859*, [n.d.], 2nd edition 1875 (engraved by W. & A.K. Johnston).

**Hogane**, James T. *Map of the city of Davenport and its suburbs, Scott County, Iowa*, 1857 (with Lambach).

**Hogard**, Henri. *Carte des Vosges*, 1845.

**Hogben**, father and son. Estate surveyors of Kent.
• **Hogben**, Thomas (1703-1774). Surveyor and schoolmaster of Smarden. Undertook numerous surveys of estates, mostly in Kent. *Plan of Goldwell Farm in the Parish of Aldington...*, 1738; land at Old Romney and Midley, 1749; *A Mapp and Ad measurement of a Farm at Chartway Street...*, 1750; High Halden and Bethersden, 1753; *Lydd: a map of Jack's Court*, 1753; marshland at Newchurch, 1754; 121 acres in Ash-next-Wingham, Kent, 1757; Ruckinge, Orlestone and Warehorne..., 1757; *A Map and Measure of York Farm in Lower Gillingham... in the County of Kent*, 1763; estate at Westwell and Challock, 1766; Wickhambreaux, 1766; and others.
• **Hogben**, Thomas & Henry. Father and son working together. Ivychurch and St. Martins, New Romney..., 1767; Saltwood, Stanford and Postling, Kent, 1768.
•• **Hogben**, Henry (*fl.*1759-*d.*1822). Surveyor, worked initially with his father Thomas. Produced manuscript maps of Essex and Kent. Estate at Boxley, 1771 MS; *Romney Marsh divided into its several Waterings...*, 1775; East Tilbury *c.*1775-1796 MS; lands at Challock and Westwell, 1777 MS; estate at Bearsted [Kent], 1778 MS; Crown lands at Dover, 1784; *Survey of Woodlands, the property of the Dean and Chapter of Rochester Cathedral* [Chatham and Gillingham, Kent], 1787; *A Map of the Manors, Messuages, Lands and premises in the several parishes of ... Hartlip...*, 1788; *A map of Emetts Farm, Kingsmill Land and Pierce Land at Smarden*, 1790; *The Isle of Thanet*, Margate, J. Warren 1802 (engraved by J. Ellis); and others.

**Högborn**, A.G. Geologiska Karta öfver Jamtlands Län, 1894.

**Hogeboom** [Hoogeboom], Andries. Dutch engraver in Amsterdam. *Exactissima Helvetiae, Rhaetiae Valesiae...*, N. Visscher [n.d.] (used in the Janssonius townbooks, 1682 and by A. Braakman in *Atlas Minor*, 1706); N. Visscher's *S. Imperium Romano-Germanicum oder Teutschland...*, *c.*1684; N. Visscher's *Groningae Et Omlandiae...*, *c.*1684.

**Hogenberg**, family. **Ref.** MEURER, P. 'The Cologne map publisher Peter Overadt' in *Imago Mundi* 53 (2001); MEURER, P. *Corpus der älteren Germania-Karten* (Alphen aan den Rijn 2000); STEMPEL, W. 'Franz Hogenberg (1538-1590 und die Stadt Wesel: Mit einem Beitrag zur Biographie' in PRIEUR, J. *Karten und Gärten am Niederrhein* (Studien und Quellen zur Geschichte von Wesel 18, 1995) pp.37-50; HELLWIG, F. (ed.) *Franz Hogenberg-Abraham Hogenberg, Geschichtsblätter* (Nördlingen 1983).
• **Hogenberg**, Hans (*d.*1544). Map publisher in Mechelen, 1520.

•• **Hogenberg**, Remy [Remegius; Remigius] (1536-1589). Engraver and publisher from Mechelen, possibly the son of Hans (above) and brother of Frans (below), came to England as a refugee. G. Mascop's map of the bishopric of Münster, 1568; engraved a copy of Mercator's *Flandriae recens exactaq. descriptio*, c.1570 (the original dated from c.1540); engraved 9 county maps for Christopher Saxton's *Atlas of England and Wales*, 1575-1578, published 1577-1579; John Hooker's *Isca Damnoniorum... Vulgo Excester* [Exeter], c.1587.

•• **Hogenberg** [Hoogenberg; Hoghenberghe], Frans [Franciscus] (c.1538-1590). Flemish artist, copper engraver and publisher from Mechelen, possibly son of Hans, above. Worked in Mechelen, London (1568-1569), and Cologne from 1564, where he died. Engraved maps for Ortelius' *Theatrum orbis terrarum*, 1570; T. Stella's *Mansfeldiae, Saxoniae totius, nobilissimae, nova et exacta chorographica descriptio*, 1570; joint publisher with Georg Braun of *Civitates orbis terrarum* (first 4 volumes), 1572-1588; *Itinerarium Orbis Christiani*, 1579/1580; *Jerusalem*, 1584; *S. Augustin* (taken from B. Boazio) and *Americae et proxima regionum* (both in *Relation oder beschreibu[n]g der Rheiss und Schiffahrt auss Engellandt*, Cologne 1589); engraved a single-sheet copy of C. Vopel's map of the Rhine, Cologne, Peter Haack 1590; and others. **Ref.** BURDEN, P. *The mapping of North America* (1996) pp.88-90; GOSS, J. *Braun & Hogenberg's The city maps of Europe* (London 1991).

••• **Hogenberg**, Johannes (fl.1594-1614). Engraver of Cologne, son of Frans Hogenberg by his first wife. He was bought out of the family business c.1594 by Agnes Lomar, his stepmother. He worked for other Cologne publishers, and with Matthias Quad. *Germania*, Johann Bussemacher 1595; engraved historical maps for S. Broelmann's *Epideigma*, 1608.

•• **[Lomar]**, Agnes. Second wife of Frans Hogenberg. Continued the business after his death. Published the 5th volume of *Civitates orbis terrarum*, 1598.

••• **Hogenberg**, Abraham (c.1585-c.1653). Painter, engraver, and publisher of Cologne, son of Frans by his second wife Agnes Lomar. He took over the Hogenberg firm c.1610. Engraved a map of the environs of Cologne, c.1615; Volume 6 of *Civitates orbis terrarum*, 1617; had a share in the printing of Johannes Gigas's *Prodromus geographicus* [atlas of the Bishopric of Cologne], 1620; maps for Caspar Enns's *Fama Austriaca*, 1627; Johann Noppius's *Aacher Chronik*, 1632.

**Hogg**, Alexander (fl.1778-1805). Publisher and bookseller of 'Kings Arms, N°·16 Paternoster-Row', London. John Hamilton Moore's *A New and Complete collection of Voyages and Travels...*, 1778; George H. Millar's *The New and Universal System of Geography*, 1782; G.W. Anderson's *A New Authentic and complete collection of Voyages round the World* [published in 80 weekly parts], 1784-1785; *A New & Correct Plan of the Cities of London and Westminster with the Borough of Southwark*, 1784 (engraved for W. Thornton's *New, Complete, and Universal History... of London* and re-used in Richard Skinner's *New and Complete History... of London*, A. Hogg 1796); *A New General Chart of the World*, 1784; George Augustus Walpoole's *The New British Traveller*, 1784 (maps engraved by T. Conder); *Historical Descriptions of New and Elegant Picturesque Views of the Antiquities of England and Wales* (partwork), c.1787-1789 and similarly titled publications (re-using Thomas Kitchin's maps from the *London Magazine*); *The New and Complete English Traveller*, 1794; *The Antiquities of England and Wales*, 1795 (re-used the Kitchin plates). **N.B.** Donald Hodson, in *County Atlases of the British Isles published after 1703* entry N°·269 Vol.III (1997) pp.105-120, 165-170 suggests that George Augustus Walpoole, Henry Boswell and other 'contributors' to *The New British Traveller*, *Historical Descriptions* and other Hogg publications were inventions of Alexander Hogg.

• **Hogg & Co.** Publishers of 16 Paternoster Row, 1805-1818.

**Hogg**, James (fl.to c.1874). Printer and publisher of 4 Nicolson Street (1856), Edinburgh. *Peninsula of Mt. Sinai*, 1849; *Business Man's Note-Book* (with maps), c.1856.

**Hogg**, Thomas (fl.1807-1814). Land surveyor of 34 Castle Street, Holborn. Made manuscript maps of Cambridgeshire, Essex and Middlesex. Tolleshunt Major, c.1807 MS.

**Hogg**, William, RN. Master HMS *Canceaux*. Plans of islands in the St. Lawrence River. Soundings for a chart of the Magdalen Islands, 1766 (surveyed by F. Haldimand, endorsed by Haldimand and S. Holland).

**Hoggar**, Robert Syer. Engineer. *Plan of the City of Oxford*, 1850, later used to show the extent of cholera in the city in 1854.

**Höggmayr**, *Frater Magister* Angelus (1680-1739). Bavarian Augustinian friar and historian. *Germania Augustiniana*, c.1730; *Italia Augustiniana*, c.1730; *Gallia Augustiniana*, c.1730; *Hispania Augustiniana*, c.1730.

**Hogius**. See **Hooghe**.

**Hogrewe**, Johann Ludwig (d.1814). Hanoverian military engineer and cartographer. *Topographische Landesaufnahme des Kurfürstenthums Hannover*, 1764-1786 (with G.J. du Plat); Map of the canal between Manchester and Runcorn, 1777; *Geographische Karte der Länder zwischen der Elbe und Weser...* (6 sheets), 1812 (title also in French, with J.F.W. Heiliger).

**Hohagen**, F. Plan of Cusco, Paris 1861.

**Hohe**, C. *Bad Neuenahr und Umgebungen*, 1860.

**Hoheneck**, G.A.E. von. See **Enenckel** von Hoheneck, Georg Acacius.

**Hohenegger**, Ludwig. Geological maps of Germany, 1861-1866. *Geognostische Karte des Nord-Karpathenien Schlesien...*, Gotha, J. Perthes 1861; *Geognostische Karte des ehemaligen Gebietes von Krakau*, Vienna 1866.

**Hohenkerk**, L.S. Draughtsman. Chart of the mouth of the Waini River, c.1897; *Map of British Guiana*, Georgetown, C.K. Jardine 1910.

**Höhm**, *Oberleutnant* —. Draughtsman. Maps for *Karte und Plane zu den Grundsätzen der Strategie erläutert durch die Darstellung des Feldzuges von 1796 in Deutschland*, [n.d.].

**Hohn**, August. See **Hoen**, August.

**Hohn**, Karl Friedrich. Bavarian historian. *Atlas von Bayern*, 1840, 1841.

**Höhnel**, Ludwig von (1857-1942). Austrian hydrographer. Works on African lakes for Petermann's *Geographische Mittheilungen*, 1889, 1893 (volumes 35 & 39). **Ref.** KRETSCHMER, I. 'Die kartographischen Ergebnisse der Teleki-Höhnel Entdeckungsreise 1887-1888', *Mitteilungen der Österreichischen Geographischen Gesellschaft* 130 (Wien, 1988) pp.39-67 [in German].

**Hohoff**, T. *Deutsches Reich*, 1878.

**Hoijaeus**, F. See **Hoeye**, François.

**Hoijer**, Frederick Herman. Norvegiae, Sueciae et Daniae, Stockholm 1685 MS.

**Hoinckhusen**, Bertram Christian von (1651-1722). German surveyor and cartographer. Made the first comprehensive survey of Mecklenburg, c.1700 MS (now at 'Mecklenburgisches Landeshauptarchiv', Schwerin).

**Hoirne**, Jan van. See **Hoorn**, Johannes à.

**Hois**, Jacob. Chart seller of Copenhagen. J.T. Reinke & J.A. Lang's *Zee Kaart van't Helgoland...*, c.1787.

**Hoit**, David. Deerfield & Springfield, Massachusetts, 1794.

**Hojeda**, Alonso de. See **Ojeda**, Alonso de.

**Höjer**, M. *Karta öfver Värmlands Län*, 1878.

**Hōjō** Ujinaga (1609-1670). Japanese soldier and scholar, official of the Japanese Shogunate. Conducted a survey of the coastal roads of Japan, 1651 (surveys used by Ochikochi Dōin and Hishikawa Moronobu for their map of the Tokaido highroad, published by Hangiya Shichirōbei 1690); directed the official surveys of Edo soon after the big fire of 1657, published by Ochikochi Dōin as *Kambun Gomaizu* [The Five Kambun Era Plans of Edo], c.1670-1673; *Shoho Nihon zu* [Map of Japan in the Shoho period], c.1656. **Ref.** YAMASHITA Kazumasa *Japanese maps of the Edo Period* (Japan 1998) pp.27 and 188-189; ARIMA, Seisuke *Hojo Ujinaga to sono heigaku* [Hojo Ujinaga and his military science], 1936 [in Japanese].

**Hokusai**, Katsushika. See **Katsushika Hokusai**.

**Hol**, Leinhart. See **Holle**, Lienhart.

**Holbein**, family.
• **Holbein**, Hans *the elder* (1465-1524). Artist.

•• **Holbein**, Ambrosius. Elder brother of Hans (below). Possibly the author of a map of the island of Utopia, 1518 (drawn to illustrate the third edition of Thomas More's *Utopia*).

•• **Holbein**, Hans, *the younger* (1497-1543). Artist, painter and engraver from Augsburg. Son of Hans the elder. Best known in England for his portraits, which include Henry VIII and his wives. Decorations for world map in Johann Huttich and Simon Grynaeus's *Novus orbis regionum*, Basle 1532; painted a terrestrial globe as part of the setting for 'The Ambassadors', 1533; woodcut border designs for versos of maps in the Ptolemy of 1535 (etc.). **Ref.** INGRAM, E.M. *The Map Collector* 64 (1993) pp.26-31.

**Holbrook family**, [Holbrook & Co.]. American manufacturers and distributors of globes and school apparatus. **Ref.** WARNER, D. *Rittenhouse* Vol.2 No.3 (1987) pp.94-98 [with references].
• **Holbrook**, Josiah (1788-1854). American orrery and globemaker. School orrery, 1830; 5-inch terrestrial globe, *c.*1840.
•• **Holbrook**, Alfred. Son of Josiah. Joint founder of Holbrook & Co.
•• **Holbrook**, Dwight (*c.*1817-1890). Son of Josiah. Joint founder of Holbrook & Co.
•• **Holbrook & Co.** (*fl.*1840s). Alfred and Dwight set up the company of Holbrook & Co., Berea, Ohio in the 1840s. They manufactured a 5-inch wooden globe, *c.*1840s; ceiling mounted school orrery, 1846. **Ref.** WARNER, D.J. 'Holbrook's hemisphere globes' *Rittenhouse: Journal of the American Scientific Instrument Enterprise* 1 (Hastings-on-Hudson, NY, 1986) pp.3-6.
•• **Holbrook Apparatus Mfg. Co.** Set up in Connecticut by Dwight Holbrook in 1854. 3-inch and 5-inch globes produced in large quantities for schools; *Eight Inch Terrestrial Globe...*, 1857.
•• **Holbrook School Apparatus Co.** (*fl.*1855-*c.*1860). Globemakers of Hartford Connecticut. Set up by Dwight Holbrook in 1855 to act as distributors for Holbrook globes and apparatus.
••• **Holbrook**, Charles W. Son of Dwight, took over the family business in the 1870s. 5-inch, 8-inch and 12-inch Terrestrial globes. *Chas. Holbrook's 12-in Globe*, *c.*1882; *C.W. Holbrook's New 5-Inch Terrestrial globe...*, George Gardner *c.*1900.

**Holbrook**, George. *A Survey of the Coast of Newfoundland from Bonaventure Head to Rocky Bay by George Holbrook & William Bullock R.N....*, T. Hurd 1822 (engraved by J. Walker).

**Holbrook**, H. *Town of Dillon, Phelps Co., Missouri*, St. Louis 1860.

**Holbrook**, J. Railway Survey Essex, 1835 MS.

**Holbrook & Son**. *Holbrook's map of Portsmouth, Portsea, Landport, Southsea & suburbs...*, 1895.

**Holbrooke**, W.H. Lithographer and publisher of 2 Crow Street, Dublin. *Holbrooke's Railway and Parliamentary Map of Ireland*, 1846 (with trading statistics for 1837 and occupation statistics for 1841, also published in London by J. McCormack).

**Holcroft**, J. Draughtsman. *Map of Part of the State of Michigan Showing the Pine Lands of the St. Mary's Ship Canal Company*, 1858.

**Holden**, G.B. *Monroe County, Wisconsin*, 1858.

**Holden**, M. *Small Celestial Atlas*, 1818.

**Holdich**, *Colonel Sir* Thomas Hungerford (1843-1929). Geographer and military cartographer. *Abyssinia*, 1869; maps of India and Afghanistan, 1879-1898.

**Holding**, *Reverend* J. Province of Tanimbé [Madagascar], 1870 MS.

**Holditch**, George. Noted at Lynn Regis [King's Lynn]. *Chart of the Eastern Coasts of England*, 1810.

**Holdredge**, Sterling M. Guide Book Pacific, 1865-1866.

**Holdsworth**, —. See **Robinson**, Son & Holdsworth.

**Holdsworth**, *Lieutenant* John Kelly RA. *Chart of the Black Sea and surrounding countries, Shewing the Telegraphic Lines...*, 1856.

**Holdsworth**, Samuel. Bookseller of London. James Fraser's *Guide through Ireland*, 1838 (also sold in Dublin, William Curry Jr. and Edinburgh, Fraser & Co.).

**Hole**, William (*fl.*1600-1646). Engraver of maps, portraits, music, title pages and illustrations. General map of England and Wales, and English and Welsh county maps (after

Christopher Saxton) for George Bishop and John Norton's edition of William Camden's *Britannia*, 1607 (with William Kip); *Virginia Discovered and Discribed by Captayn John Smith...*, 1612; maps and title page in Michael Drayton's *Poly-Olbion*, 1612; maps of Goshen and Arabia in Sir Walter Ralegh's *History of the World*, 1614.

**Holgarth**, Ludwig, *Graf* von (*fl.*1804-1810). Cartographer of Vienna.

**Holgate**, George. Town surveyor in South Africa. *Pietermaritzburg*, 1873.

**Holinshed** [Hollingshead], Ralph [Raphael] (*d.*1580). *Chronicles of England, Scotland and Ireland*, 1577.

**Holkema & Warendorf**, Van. Publishers of Amsterdam. *Atlas van Nederland voor Wielrijders*, 1898.

**Hollack**, Emil. *Übersichtskarte von Ostpreussen*, 1908.

**Holland**, Frederick W. Surveyor of 177 Pitt St., Sydney, New South Wales (1858-1859). *Throsby Estate, situated at Newtown...*, 1857.

**Holland**, Frederick Whitmore (1837-1880). Canadian geographer.

**Holland**, *Lieutenant* Henry (*d.*1798). Surveyor in Canada. One of the surveyors for *Plan of the Settlement at the Bay of Quinte*, 1784; surveys at Adolphustown, 1784 (with others); surveys at Fredericksburgh, 1784 (with others).

**Holland**, John. Surveyor. River Colne & Tributary, 1842 MS.

**Holland**, *Lieutenant* John Frederick (1760-1845). Draughtsman and map copyist, worked for Samuel Holland. *Plan des environs de Cataraqui, recu du Major Ross le 12 Sepr*, 1783; *Plan of old Fort Frontenac and Town Plot of Kingston*, 1784 (surveyed, drawn and copied by J.F. Holland, endorsed by F. Haldimand and S. Holland); *Plan of the Ottawa or Grand River. Copied from an original...*, *c.*1790 (endorsed by S. Holland); *A Map of the Province of Quebec...*, 1791 (prepared for Samuel Holland); copied Augustus Jones's *Plan of Eleven townships fronting on Lake Ontario*, *c.*1791 (endorsed by S. Holland); and others.

**Holland**, *Captain* Nathaniel. Coastal surveys of North America from 1774, used for many published charts including: *A New Chart of the Coast of North America from Port Royal Entrance to Matanza Inlet, Exhibiting the Coast of Georgia &c.*, London, Laurie & Whittle 1794; charts in Jefferys' *North American Pilot*, Laurie & Whittle 1795; *A New Chart of the Coast of North America from Currituck Inlet to Savannah River*, London, Laurie & Whittle *c.*1800; *A New Chart of the Coast of North America from New York to Cape Hatteras...*, London, Laurie & Whittle 1809. **N.B.** Compare with Samuel **Holland**.

**Holland**, Philemon (1552-1637). Translated William Camden's *Britannia* into English, published as *Britain, Or A Chorographical Description Of The Most flourishing Kingdoms...*, G. Bishop & J. Norton 1610, A. Heb 1637; associated with the Latin edition of Speed's *Theatre of the Empire of Great Britain*, 1616.

**Holland**, *Captain* (later *Major*) Samuel (1728-1801). Dutch artillery officer and military engineer, moved to England in 1754 and joined the Royal American Regiment. He practised surveying in Quebec, New France, New England and New York, rising to become Surveyor General for the Northern District of North America in 1764. As well as undertaking many surveys and maps himself, his signature endorses the work of other draughtsmen and cartographers in North America. Holland was succeeded as Surveyor General of Canada by his nephew Joseph F. Bouchette. *Plan of Louisbourg, the Harbour...*, 1758; *A Plan of Québec and environs... during the siege... in 1759* MS published *c.*1780 (with J.F.W. Des Barres and Hugh Debbieg); *A Sketch of St. John's Harbour and Part of the River*, *c.*1760 MS; *Explanation of a Fort projected at Déchambault*, 1761; *A Plan of the Island of St. John in the Province of Nova Scotia*, 1765 (with F. Haldimand, Lt. Robinson and T. Wright, surveys used by Jefferys, 1776, by Le Rouge 1778 and by Faden 1778); *The Provinces of New York and New Jersey with Part of Pensilvania and the Governments of*

*Trois Rivieres and Montreal*, London, Sayer and Jefferys 1768, 1775, 1776 (engraved by Thomas Jefferys, and included by him in his *The American Atlas*, and Sayer & Bennett's *The American Military Pocket Atlas*, also derivatives); untitled manuscript plan of New York, c.1776; *A Topographical Map of the Province of New Hampshire*, London, William Faden 1784 (with the assistance of others); *A Map of Part of Canada for the use of His Majesty's Secretary of State* [Quebec, Ontario, New York, Pennsylvania, Ohio], 1790; *Plan of the Tongue of Land laying between a part of Ottawa River and part of River St. Lawrence...*, 1795; *A New Map of the Province of Lower Canada...*, London, William Faden 1802, 1813 (engraved by Allan & Wilson), J. Wyld 1829, 1838, 1840, 1843. **Ref.** SEBERT, L.M. 'The first maps of the eastern townships', Association of Canadian Map Libraries and Archives Bulletin 77 (Ottawa, 1990) pp.1-5; BOSSE, D. 'Samuel Holland's maps of New Hampshire in the William L. Clements Library' Mapline 56 (December 1989) pp. 1-3; RISTOW, W.W. American maps and mapmakers (Detroit 1986) pp.25, 28, 52, 57, 59; GUTHORN P. British maps of the American Revolution (1972), pp.27-29; CHIPMAN, W. 'The life and times of Major Samuel Holland, Surveyor General, 1764-1801' Ontario History Vol.21 (1924) pp.11-90.

**Hollander**, M. Carte géologique de la Corse, 1877.

**Hollar**, Wenceslaus [Wenzel; Vaclav] (1607-1677). Prolific Bohemian artist, etcher and engraver from Prague, one of few to use etching for map production. He was taught engraving by Matthäus Merian in Frankfurt. From 1636, apart from a spell in Antwerp (1645-1650), he worked mostly in London until his death. Hollar produced illustrations, views and frontispieces in addition to maps. He was appointed Iconographer to the King in 1660. Plans of Düren, 1634; *Oxforde*, London c.1643; plan of Hull, c.1643; untitled allegorical map of Great Britain and Ireland, 1643; *The Kingdome of England & Principality of Wales Exactly Described* [the so-called 'Quartermaster's Map'] (6 sheets), London, Thomas Jenner 1644, and later issues; *A New & Exact Mappe of England...*, London, Francis Eglesfield 1644 and editions to 1688; *A New & Exact Map of Ireland...*, London, Peter Stent 1653, J. Overton 1669; *Westminster and London*, 1655; *Chorographica Terrae Sanctae Descriptio*, 1657; *A New Map of the Kingdom of Hungaria...*, London, Peter Stent 1664, John Overton 1673, 1683; *A New Map of Barkshire with all the Hundreds...*, London, J. Overton 1666; *A New and Exact Map of America and Ilands...* (copperplate engraving), London, Thomas Jenner 1666; *A Map of Both Cities, London and Westminster, before the Fire*, John Overton c.1666; *A Map or Grovndplot of the City of London and the Suburbes thereof...*, London, John Overton 1666 (a second very similar plate exists, carrying the same title); *A Generall Map of the whole City of London with Westminster & all the Suburbs...*, 1666; John Leake's *An Exact Surveigh of the Streets Lanes and Churches contained within the Ruines of the City of London...*, 1667; *A New Mapp of the Kingdome of England and Principalitie of Wales*, c.1670; plans of the fortifications of Tangier, 1669; engraved some of the maps for Richard Blome's *Britannia*, and *Speed's Maps Epitomised*, 1673; *London. A New Map of the Citties of London Westminster and ye Borough of Southwarke with their Suburbs...*, Robert Greene and Robert Morden 1675 and other editions; *London*, London, James Clark c.1675; and others. **Ref.** JACKSON, P. 'Some notes on Hollar's Prospect of London and Westminster taken from Lambeth' in London Topographical Record 26 (1990), pp.134-137; BANNISTER, D. The Map Collector 21 (1982) pp. 22-24; PENNINGTON, R. A descriptive catalogue of the etched work of Wenceslaus Hollar 1607-1677 (Cambridge 1982).

**Holle**, Gottfried Friedrich Ludewig [Louis] (1817-1895). Publisher of Wolfenbüttel. *Harz*, 1849; *Niederlande* (8 sheets), c.1850; *Hand-Atlas von Nord-Amerika* (40 sheets), 1853. **N.B.** Compare with L. **Holle**, below.

**Holle**, K.F. Taalkaart van Sumatra, Batavia 1887.

**Holle**, L. Publisher and lithographer of Wolfenbüttel. *Compendiöser Hand- und Wand-Atlas der neuesten Erdkunde*, 1843; *Vollständiger Schul-Atlas der neuesten Erdkunde* (29 maps), 1847, (27 maps) 1851, 1855; *Kleiner Schulatlas der neuesten Erdkunde* (8 maps), 1847 and later editions; *Nieuwe Generale Karte van het Koningrijk der Nederlanden* (8 sheets), Utrecht 1850; *Amerika*, 1851; *Wand-Karte des Kaiserthumes Oesterreich* (6 sheets), c.1868;

*Wenceslaus Hollar (1607-1677), a self portrait made in 1647. (By courtesy of Rodney Shirley)*

*Wandkarte des Preussischen Staates...* (7 sheets), *c.*1868; *Schul-Wand Karte von Europa*, *c.*1868; *Schulwand Karte von Palestina* (4 sheets), 1868; *Grosse Wand Karte der Planigloben...*, *c.*1868. **N.B.** Compare with Gottfried **Holle**, above.

**Holle** [Hol; Holl; Holm], Lienhart [Leonardum]. Printer of Ulm. Published the Ulm edition of Ptolemy's *Cosmographia* (32 woodcut maps), 1482 (the type and woodblocks passed to Johann Reger *c.*1484, who republished the work in an expanded edition, 1486). **Ref.** CAMPBELL, T. *The earliest printed maps 1472-1500* (1987) pp.135-138.

**Holliday,** *Captain* —. Draught of the Bristol Channel, completed before 1728 (kept in the Merchants' Hall, Bristol).

**Holliday,** Thomas (*fl.*1818-1855). Surveyor, produced manuscript maps of Yorkshire. *Complete Treatise on Practical Land Surveying*, 1838.

**Holliday, Wise & Co.** Publisher of New York. *Plano de Manila*, 1899.

**Hollingsworth,** S. *North America including the British Colonies and the Territories of the United States*, Edinburgh 1787.

**Hollingworth,** H.G. Survey Po-Yang Lake, 1868 MS.

**Hollingworth,** J. (*fl.*1757-*d.*1774). Surveyed estates in eastern English Home Counties. Chishall, 1769 MS; Little Waltham, 1776 MS; Pebmarsh, 1807 MS (drawn by nathaniel Kinderley).

**Hollingworth,** T. Publisher in Lynn. *A Map of the Great Level of the Fens...*, London & Lynn 1751.

**Holloway,** *Captain* later *Major General Sir* Charles *RE*. Rivers Thames & Medway, *c.*1795 MS; *Plan of the Castle of Kelletbahar* [Dardanelles], 1799 (and other manuscript plans of fortifications in the Dardanelles); signed many plans of military sites on Gibraltar, 1810-1817.

**Holloway,** H.R. *Isle of Wight*, *c.*1840.

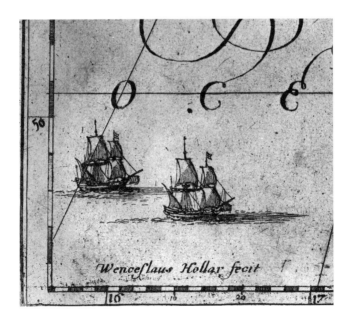

*Signature of Wenceslaus Hollar (1607-1677), from* A New Mapp of the Kingdome of England and Principalitie of Wales, *c.1670 (By courtesy of Map Collector Publications).*

**Holloway, Harry D.** (*fl.*1838-1869). Surveyor of Ringwood, Hampshire, prepared manuscript maps of Dorset, Hampshire and Wiltshire. Inclosure plans in the New Forest, 1851; Ashley Lodge and attached Crown Lands, 1853; Broomy Lodge and attached Crown Lands, 1853, 1854; Sloden inclosure in Ashley and Broomy Walks etc. [New Forest], 1855.

**Holloway, J.** *Plan of the City of Bath*, 1848.

**Holloway, T.** Hampshire surveyor. Rhinefield Sandys inclosure [New Forest], 1809; Burley Rails [New Forest], 1810; New Burley inclosure, 1810.

**Holloway, T & H.** Surveyors of Christchurch, Hampshire. Burley, 1838 MS. **N.B.** Compare with H.R. **Holloway**, H.D. **Holloway** and T. **Holloway**.

**Holloway, Thomas.** *Atlas, c.*1898.

**Holloway & Son,** Lithographers of Bath. *Plan of the City and Borough of Bath and its Suburbs*, Bath, S. Hayward 1852 (survey by J.H. Cotterell); *District 5 Miles round Bridgwater*, 1853. **N.B.** Compare with J. **Holloway**, above.

**Hollstein, —.** Draughtsman. Plan of the town of Karlobag, Croatia, *c.*1714 MS.

**Holm, F.** Engraver. *Karta öfver Gotheborgs och Bohus Län*, 1859.

**Holm, G.F.** *Skizze Kaart over Kong Christian den IX-Land* [Greenland], 1884-1885; Greenland, 1888.

**Holm, Leonard.** See **Holle**, Lienhart

**Holm, Saemundur Magnusson** (1749-1821). Icelandic cartographer. *En Deel af Vestre Skaptafells Syssel*, 1770.

**Holm** [Campanius Holm], family [The Campanius family of Stockholm].
• **Holm** [Campanius], **Johannes** (1601-1683). Swedish missionary and chaplain to the Swedish settlement in the Delaware valley, 1642-1648. Translated Luther's Small Catechism into the Delaware Indian language, Stockholm, Thomas Campanius Holm 1696 (included *Nova Suecia* - a map of New Sweden by Pehr Lindheström).
••• **Holm, Thomas Campanius.** Copperplate engraver and book publisher in Stockholm. Grandson of Johannes Campanius Holm, above. Published his grandfather's catechism in 1696; used his grandfather's notes to compile a history and description of New Sweden, or Pennsylvania entitled *Kort Beskrifning Om Provincien Nya Swerige uti America...* (9 plates including 5 maps engraved by Holm, at least one based on Lindheström), Stockholm 1702. **N.B.** Not to be confused with Thomas Holme q.v. (1624-1695), who is also associated with early maps of Pennsylvania. **Ref.** ERLING, P.A. *The Map Collector* 42 (1988) p.39.

**Holman, A.J.** [& Co.; Holman Company]. Publishers of Philadelphia. *New Biblical Atlas with index* (15 maps), 1898; *Indexed atlas to the Holy Bible* (15 maps and plans), 1904.

**Holmberg, Henrik Johann** (1818-1864). *Ethnographische Skizzen Über die Völker des russischen Amerika*, Helsinfors, H.C. Friis 1855.

**Holmberg, Johann Christoph.** Publisher of Stockholm. Carl Gustaf Tornquist's charts of naval actions in the West Indies and United States - *Grefve grasses sio batailler*, 1787.

**Holme, John** (1785/6-1813). Land surveyor of Colchester, Essex. West Bergholt, 1810 MS.

**Holme, Randle F.** Hamilton River, Labrador, 1887 MS; Peninsula of Labrador, 1888.

**Holme** [Holmes], **Thomas** (1624-1695). Captain in Cromwell's army from Lancashire. Official in the Down Survey of Ireland, later emigrated to the United States and was appointed Surveyor General for Pennsylvania in 1682. Laid out a plan for the future city of Philadelphia with William Penn, 1682; *A Portraiture of the City of Philadelphia...*, London, J. Thornton & Andrew Sowle 1683; *A Map of the Province of Pennsylvania...* [cartouche] / *A Map of the Improved part of Pennsilvania in America* [title above map border], London, Robert Greene & John Thornton 1687, London *c.*1705 (this map was copied by John Harris and published under the title *A Mapp of Y.e Improved Part of Pennsylvania in America...*, London, P. Lea 1687 and later). **Ref.** CORCORAN, I. *Thomas Holme 1624-1695. Surveyor General of Pennsylvania* (1992).

**Holmes, —.** Engraver. Lewis Krazeisen's *Plan of London, Westminster and Borough of Southwark*, 1793.

**Holmes, J.B.** *Map of... Harlem, New York*, 1874.

**Holmes, J.L.** Printers of 315 Pitt Street, Sydney. *Chiswick Park, Auburn*, 1884; *Emerson's Junction Estate, Kogarah*, 1886.

**Holmes, John** (fl.1696-1721). Surveyor of Eltham. Surveyed estates in various parts of England. Layer Marney [Essex], 1698 MS; Hadleigh, 1709 MS.

**Holmes, John** (1703-1759). Surveyor of Eltham. Eltham parish, with the King's lands in Lee, Mottingham and Bexley..., 1749. **N.B.** Possibly son of John Holmes, above. Compare also with William **Holmes**, below, also of Eltham.

**Holmes, Joseph W.** American mapmaker. *Map of Hancock County, Illinois*, Buffalo 1859 (with Charles R. Arnold); *Map of Pike County, Illinois*, 1860 (with C.R. Arnold); *Map of Madison County, Illinois*, Buffalo 1861 (with Arnold); published Hodgen's *Map of St Clair County, Illinois*, 1863.
• **Holmes & Arnold.** Publishers of Buffalo, New York. J.W. Holmes and C.R. Arnold's Map of Madison County, Illinois, 1861.
• **Holmes & Hodgen.** Partnership of J.W. Holmes and R.S. Hodgen. *Map of the County of Macon, Illinois*, Philadelphia 1865

**Holmes, P.** *Map of Henry County, Illinois*, Chicago 1860.

**Holmes, Philip.** Engraver. John Overton's *A new and most Exact map of Africa...*, 1666.

**Holmes, S.F.** *Sheffield proposed Improvements*, 1873.

**Holmes, Thomas** (fl.1700-1706). Estate surveyor of eastern England. Quendon 1702 MS.

**Holmes, Thomas** (fl.c.1775-1800). Estate surveyor of Essex & Hertfordshire. Paglesham, 1775 MS.

**Holmes, W.H.** Engraver for Samuel Augustus Mitchell's Mexico, 1859; M.F. Maury's wall map of the United States, Washington D.C. 1860; atlas to Chamber's Encyclopaedia, Philadelphia 1869.

**Holmes, William.** Survey of Eltham Court Farm, Woolwich, 1828 MS.

**Holmes, William Henry** (1846-1933). Landscape artist, became Director of the National Gallery of Art at the Smithsonian Institution in 1906. Relief map of Yellowstone National Park, 1876; produced topographical views, land classification and geological maps for F.V. Hayden's *Geological and Geographical Atlas of Colorado*, J. Bien 1877; *Panorama from Point Sublime* [Grand Canyon], 1882.

**Holst,** *Captain* Jacob. Danish sea captain. Two charts of parts of the coast of Madagascar, 1738, used by Dalrymple 1792.

**Holst, Johannes.** *Nyeste Skoleatlas*, Copenhagen 1874; *Nyeste Geografisk Skoleatlas*, Copenhagen, H. Hagerups 1888.

**Holsteen, S.** *Sectional map of southern Minnesota*, 1855.

**Holstein, Lukas** (1596-1661). Geographer from Hamburg, studied in Leiden. Travelled with his friend Philipp Clüver. Restored some of the maps in the gallery of maps in the Vatican. Prepared unpublished amendments to Dudley's *Dell'arcano del mare*.

**Holster Atlas.** Colloquial name for Sayer and Bennett's *The American Military Pocket Atlas*, 1776.

**Holt, Adam** (b.c.1690/1, fl.-1743). Essex estate surveyor of Wanstead. Survey of the manor of Aldersbrooke, Wanstead, 1723; estates at Epping, 1743 MS.

**Holt, Edward.** Publisher of Hanover Street, Liverpool. J. & C. Walker's *To Her most gracious Majesty Queen Victoria, This Map of Ireland... is... Most Respectfully Dedicated...*, 1838 (also sold in London by J. & C. Walker).

**Holt, George L.** Stationer and bookseller of Cheyenne, Wyoming. *Holt's New Map of Wyoming*, 1884 (engraved and printed by G.W. & C.B. Colton).

**Holt, Thomas** (fl.1822-1836). Irish surveyor. *Plan of the City and Suburbs of Cork*, Dublin 1834; *Plan of the Liberties or County of the City of Cork*, Cork 1837.

**Holt,** Warren (*fl.*1862-1875). Publisher and mapseller of San Francisco, 305 Montgomery Street; later 411 Kearney St. (1869); 607 Clay Street (1873). Agent for J.H. Colton (1861). [Minard H.] *Farley's Map of the Newly discovered Tramontane Silver Mines in Southern California and Western New Mexico...*, 1861; L. Ransom's *New Map of the State of California...*, 1862 (lithographed by Louis Nagel); *De Groot's Map of Nevada Territory...*, 1863; L. Ransom and A.J. Doolittle's *A New Map of the State of California and Nevada Territory*, 1863, 1868; A.W. Keddie's *Map of Owen's River Mining Country*, 1864; *Map of the State of Nevada*, 1866 (lithographed by Grafton T. Brown); *Map of the States of California and Nevada*, 1869, reduced edition 1873, 1875 (compiled by C.D. Gibbes, J.H. Von Schmidt and A.W. Keddie, lithographed by S.B. Linton); also agent for J.H. Colton. **Ref.** RISTOW, W.W. American maps and mapmakers (Detroit 1986) p.461.

**Holthausen,** Friedrich Wilhelm. Dr. Carl Glaser's *Topisch-physikalischer Atlas*, Mannheim, Hoff 1844, Stuttgart, Krais und Hoffmann 1855.

**Holtrop,** Willem. Dutch cartographer and publisher of Amsterdam. In partnership with the widow of Steven van Esveldt from 1778 until he took over the business in 1784. Zakatlas, 1763-1789; composite pocket atlas, *c.*1780 (maps by G. & A. van Huissteen, T. Crajenschot and S. van Esveldt); *Zak-Atlas, of Leidsman des Reizigers / Atlas Portatif, ou Guide des Voyaguers* (68 maps), *c.*1794 (composite atlas with maps by Holtrop and others); *Atlas du Rhin, de Bonn jusqu'à Bâle*, 1798; pocket atlas of the Netherlands, Luxembourg and Belgium, *c.*1835. **N.B.** See also **Esveldt**-Holtrop.

**Höltzl,** Abraham. Austrian engraver. *Ducatus Würtemberg*, 1650.

**Holtzman,** R.O. *Map of the city of Washington and environs*, Washington, A.G. Gedney 1885.

**Holtzwurm** [Holzworm; Holzwurm; Holzwurmb], brothers.
• **Holtzwurm,** Israel (*d.*1617). Austrian engineer and surveyor. topographical map of Carinthia, 1611, 1650; topographical map of Upper Austria, from 1616 (after his death it was completed and published by his brother Abraham).
• **Holtzwurm,** Abraham. Completed the survey of Upper Austria which had been begun by his brother in 1616. Published as *Archiducatus Austriae superioris, regio supra Anisum cognominatus* (4 sheets), Regensburg 1628 (reduced version used by Joan Blaeu, 1662).

**Holwell,** John (1649-*c.*1686). Surveyor of London. *A Sure Guide to the Practical Surveyor*, 1678.

**Holywood,** John. See **Sacrobosco**, Johannes de.

**Hölzel,** Eduard (1817-1885). Publisher and lithographer from Prague, set up a publishing house in Olomouc in 1844, moving to Vienna in 1861 where he worked with Blasius Kozenn and other geographers. They specialised in school atlases, wall charts and other geographical publications. Eduard died in Salzburg, but the company went on to publish atlases in many languages well into the 20th century. B. Kozenn's *Geographischer Schul-Atlas für Gymnasien, Real-und Handelsschulen*, 1861 (later published with the title *Geographischer Atlas für Mittelschulen*, also re-edited by V. v Haardt and published under the same title with more than 50 maps, 1881, and many other editions in Czech, German and other languages of the Austrian Monarchy); Hungarian language history atlas by G. Jausz (32 sheets), 1873; Löytved's *Plan of Beyrouth...*, Vienna 1876; *Marche progressive de la Russie sur le Danube* (7 maps), 1878; *Mittelheilungen der K.K. geograph. Gesellschaft*, 1878; *Physikalischer-statistischer Hand-Atlas von Oesterreich-Ungarn* (25 sheets), 1882-1887; V. v Haardt *Physikalisch-statistischer Schul-Atlas* (14 sheets), 1889; history atlas by F.W. Schubert & W. Schmidt (36 sheets), 1889; J. Chavanne and F. Ritter von Le Monnier's *Physikalisch-statistischer Hand-Atlas von Oesterreich-Ungarn* (25 sheets with 39 maps), 1882-1887; Haardt's *Uebersichts-Karte der ethnographischen Verhältnisse von Asien*, 1887; F. Ritter von Le Monnier's *Sprachenkarte von Österreich-Ungarn* (4 sheets), 1888; A Penck and E. Richter's *Atlas der österreichischen Alpenseen* (15 sheets), 1895-1896; V. von Haardt's *Süd-Polar-Karte* (4 sheets), 1895; F. Noë's *Geologische Übersichtskarte der Alpen*, 1890; and many others. **Ref.** KRETSCHMER,

DÖRFLINGER & WAWRIK *Lexicon zur Geschichte der Kartographie* (1986) vol.C/1, pp.310-311 (numerous references cited) [in German].

**Hölzel**, Hugo Joseph (1852-1895). Cartographer and publisher of Vienna.

**Holzhalb**, Johann Rudolph (1723-1806). Engraver of Zürich. *Glarner-Land*, 1774; *Isles de Spizbergue*, 1779; for Hans Heinrich Körner 1785.

**Holzhauer**, C. *Plan von Brandenburg*, 1880.

**Holzinger**, Georg (*fl*.1810-1820s). Publisher of Vienna. Issued 2nd edition of Jacob Auspitz's *Beer Haluchot* ['Explanation of the Plates'], 1818 (with 5 maps engraved by Gottfried Prixner, thought to be the first Hebrew atlas).

**Holzmüller**, Heinrich. Views of Nördlingen and Amsterdam (reduced from the work of C. Anthonisz., used in Sebastian Münster's *Cosmographia*, 1550).

**Holzschuher**, Georg (*d*.1526). Town councillor of Nuremberg, one of those involved in the commissioning of Martin Behaim's globe of 1492. He compiled an account of it in 1494.

**Holzwurm**. See **Holtzwurm**.

**Homann** and heirs. Family and firm of mapmakers and publishers in Nuremberg. They worked with several of the major geographers and scholars of the day and produced numerous atlases and collections of maps, as well as publishing the work of others. **Ref.** FELDMANN, H.-U. 'Cartography in Switzerland 15th-18th-century' in *La cartografia dels països de parla Alemanya* (Barcelona 1997) pp.58-60, 203-204; HEINZ, M. *Imago Mundi* 49 (1997) pp.104-115; HEINZ, M. *Imago Mundi* 45 (1993) pp.45-58; KRETSCHMER, DÖRFLINGER AND WAWRIK *Lexicon der Geschichte der Kartographie* (1986) vol.C/1 pp.315-318 (numerous references cited); SANDLER, C. *Johann Baptista Homann, die Homännischen Erben...* reprint edition (Amsterdam 1979).

• **Homann**, Johann Baptist [Baptista] (1664-1724). Self-taught copper engraver from Oberkammlach near Mindelheim, Schwabia. Engraved for David Funck, Jacob von Sandrart and others from the early 1690s, then founded his own cartographic and publishing establishment in 1702. By 1707 he was compiling his own atlases. He became geographer to the Holy Roman Emperor in 1715, and was a member of the Prussian Royal Academy of Sciences. He died in Nuremberg. *Neuer Atlas ueber die gantze Welt...* (33 maps), 1707, 1710; *Potentissimae Helvetiorum Reipublicae Cantones Tredecim...*, 1714; J.P. Nell's *Neuvermehrte Post-Charte durch gantz Teutschland*, c.1714; *Großer Atlas über die ganze Welt* (126 maps), 1716; J.C. Eisenschmidt's *Tabula novissima totius Germaniae...* (4 sheets), 1716 and later; *Atlas methodicus...*, c.1719 (with J. Huebner); *Atlas Novus Terrarum Orbis Imperia* (2 volumes), c.1720; J.G. Doppelmayr's *Basis Geographicae Recentioris Astronomica...*, c.1724; and many others.

•• **Homann**, Johann Christoph (1703-1730). Publisher, engraver and mapseller of Nuremberg. Son of, and successor to, Johann Baptist. He ran the business from 1724 until his own death. *Tabula geographica Europae Austriacae generalis...*, c.1725; *Statuum Maroccanorum*, 1728; *Thuringiae tabula*, 1729; and others.

••• **Homann Heirs** [Heredes Homanniani; Héritiers de Homann; Homännische Erben] (*fl*.from 1730). On the death of Johann Christoph Homann the company became known as Homann Heirs. Initially it was run by the geographer Johann Michael Franz, and the engraver Johann Georg Ebersberger (son-in-law of J.B. Homann). Collaborators included Johann Matthias Haas, Tobias Mayer and Georg Moritz Lowitz. In time, shares passed to Jacob Heinrich Franz, and Barbara Ebersberger and her husband G.P. Monath. Franz Ludwig Güsseseld joined the company in 1773. By the early 19th century control was with Georg Christoph Franz and Friedrich Albrecht Monath. In 1813, by which time the firm had issued more than 900 maps, Homann Heirs passed to Georg Christoph Franz Fembo whose son Christoph Melchior Fembo closed the company c.1852. *Grosser Atlas*, 1731; J.M. Haas's *Africa secundum...*, 1737; J.G. Doppelmayr's *Atlas coelestis...* (30 maps), 1742; *Homannisch-Hasischer Gesellschaftatlas* (18 maps), 1747; *Atlas historicus...*, 1747; T. Mayer's *Germaniae mappa critica...*, 1750; *Atlas Silesiae...* (20 maps), 1750; *Atlas Scholasticus Maior* (36 maps), 1753; *Atlas Germaniae Specialis* (125 maps), 1753; *Atlas Compendiarius...*, 1753; *Atlas geographicus maior...* (2 volumes, 275 maps), 1753/1759;

*Johann Baptist Homann (1664-1724). Portrait by J. Kenckel, engraved by W. Winter (By courtesy of Rodney Shirley)*

*Atlas scholasticus* (18 maps), 1754; *Natur- und Kunstatlas* (61 maps), 1760; *Compendiöser Atlas von Deutschland* (30 maps), 1762; *Atlas Novus Republicae Helveticae* (20 maps), 1769 (with maps by G. Walser); F.L. Güssefeld's *Electorat de Brandenbourg...*, 1773; *Atlas regni Bohemiae* (15 maps), 1776 (based on the atlas of Johann Christoph Müller); *Atlas der sämmtlichen Niederlande* (17 maps), 1788; *Atlas von der Oesterreichischen Monarchie* (18 maps), 1803; *Atlas von Spanien*, c.1805; D.F. Sotzmann's *Der Schlesische Atlas...*, 1813; and many others. **N.B.** See also **Ebersperger** family, **Fembo** family, and **Franz** family etc..

**Homann**, E. Publisher of Kiel. *Der Nord-Ostsee-Kanal Unter Benutzung der Generalstabskarte...*, 1886 (drawn by H.B. Jahn, lithographed by Dorn & Merfield).

**Homby**, *Captain* —, *RN*. Chart Coast Guard Service, 1844.

**Home**, David Milne (1805-1890). Scottish advocate, amateur geologist and cartographer, helped with the promotion of the Ordnance Survey of Scotland. Maps of Scottish districts. **Ref.** BOUD, R.C. 'David Milne Home...', The Cartographic Journal 29 (June 1992), pp.3-11.

**Home**, H. *The Rhenish Atlas*, 1845.

**Home**, J. *History of the Rebellion* [Scotland] (with 4 maps including 3 battle plans), 1802.

**Home**, *Commander Sir* James Everard *Bt*. Corrected George Peacock's surveys of 1832, published by the Admiralty as *San Juan de Nicaragua*, 1833 (engraved by J. & C. Walker); *River Para* [Brazil], 1838.

**Home** [Hume], John (1733-c.1809). Estate surveyor in the northeast of England. Many estate plans in Scotland including the lands of Tilliesnaught, property of Thomas Forbes in the parish of Birse, Aberdeenshire; *Plan of all the Lands and Baronys of Troup*, 1767 MS.

**Home**, *Lieutenant Colonel* Robert *RE*. *Sketch Map to illustrate Probable Concentration and Lines of Operation in the event of An Occupation of the Principalities by Russia* [Romania], QMG's Department 1876; directed the preparation of *Survey of Defensive Position near Bulair shewing the Lines constructed by the Anglo-French Army in 1855...* [Turkey], 1876 (drawn by G.A. Cockburn & H. Chermside); *Plan of Gallipoli and environs...*, 1877.

**Home**, *Colonel* R. Trans-Caucasia, 1877.

**Homem**, family (*fl*.16th-century). Portuguese chartmakers. **Ref.** KRETSCHMER, DÖRFLINGER & WAWRIK Lexicon zur Geschichte der Kartographie (1986) Vol.C/I pp.318-319 [in German]; CORTESAO & TEIXEIRA da MOTA Portugalliae monumenta cartographica (Lisbon 1960) passim.
• **Homem**, Lopo (*d.c.*1572). 'Master of Sea Charts' in Lisbon from 1517; appointed official cartographer to the King of Portugal, 1529. World atlas, 1519; Europe and the Mediterranean, *c*.1550; Atlantic, *c*.1550; planisphere, 1554 MS.
•• **Homem**, Diogo [Diego] (*fl*.1530-*c*.1576). Son of Lopo Homem. Having fled Portugal, he worked in London from *c*.1547, later also worked in Venice. 24 manuscript works extant including portolan chart *Carto do Mediterraneo*, between 1557 and 1576; world atlas, 1558; different maps and atlases of Europe and the Mediterranean, 1559, *c*.1561, 1563, *c*.1566, *c*.1571, 1572, 1574, 1576; world atlas, 1561; world atlas, 1568; eastern mediterranean, 1569; La carta del navigare, Venice, Paolo Forlani 1569 and other editions (first sea chart engraved on copper); and others.
•• **Homem**, André. Son of Lopo and brother of Diogo, worked in Antwerp. *Universa ac Navigabilis Totius Terrarum Orbis Descriptio...* (planisphere on 10 sheets), Antwerp 1559 MS (now in Bibliothèque Nationale, Paris).

**Homem**, Lourenço. *Carta... das principaes estradas de Portugal*, 1808.

**Homem de Mello**. Brazilian cartographers, possibly father and son.
• **Homem de Mello**, *Barão* Ignacio Francisco M. *Altitude Comparada dos Pontos Culminantes do Systhema Orographico Brasiliero*, Rio de Janeiro *c*.1875; *Carta Physica do Brasil*, Rio de Janeiro 1876; *Atlas do Imperio do Brazil* (33 maps), Rio de Janeiro, Angelo Agostini e Paulo Robin 1882 (with F.A. Pimentas Bueno, engraved by C. Lomellino de Carvalho, printed by Lithographia Paulo Robin et Companhia).
•• **Homem de Mello**, *Doctor* Francisco. Civil

engineer and draughtsman at the polytechnic school of Rio de Janeiro. Worked with Barão Homem de Mello on *Atlas do Brazil* (176 page octavo), 1908, ('first printing', 33 maps and 66 pages of text), Rio de Janeiro, F. Briguiet & Companhia 1909 (maps drawn by F. Homem de Mello, W. Haillot, Marcellino R. Da Silva, and Olavo Freire, all engraved by A. Simon of Paris).

**Homeria**, Diego. Spanish hydrographer. Sea chart, 1673 MS.

**Hommaire de Hell**, Ignace Xavier Morand (1812-1848). French geological engineer and explorer, worked on the construction of the Lyon-Marseilles railway. Travelled in south Russia, the Crimea and the Caspian Sea, and later the Black Sea, Kurdistan and Persia. *Carte géologique de la Russie méridionale*, Paris 1844; 'Atlas' (2 volumes) to *Les Steppes de la Mer Caspienne, le Caucase, la Crimée et La Russie méridionale*, 1844; Geological Section Crimea, 1855; *Voyage en Turquie et en Perse... 1846, 1847 et 1848...* (with maps and sketches), Pierre Bertrand 1859 (maps engraved by F. Delamare, lithographed by Lemercier). **Ref.** BROC, N. *Dictionnaire illustré des explorateurs Français: II Asie* (1992) pp.242-244 [in French].

**Hommeyer**, Heinrich Gottlob. *Zeichnung der Schweiz nach einer geometrischen Construction...*, Berlin 1804.

**Homolka**, Josef [József] (1840-1907). Hungarian cartographer of Moravian descent. From 1868 worked on the Hungarian State Survey, and in 1873 became director of the Cartographic Department of the Hungarian State Printing Office. Revised the school atlas of Ferenc Ribáry, 1873, 1881; *A Magyar Szent Korona Országai* [The Lands of the Holy Crown of Hungary], 1881; *Nagy Magyar Atlasz* [Great Hungarian Atlas], 1901.

**Homoródi**, Lajos (1911-1982). Hungarian engineer and cartographer. Modernised state surveys and photogrammetry. Chairman Hungarian Cartographic Society 1960-1980. **Ref.** KARSAY, F. *Magyar geodéziai és kartográfiai irodalom* (Budapest 1983) [in Hungarian].

**Honaert**, J. von. Weener mit Weener Meer, 1673.

**Hondius** [Hondt; d'Hondt], family of engravers, globemakers and publishers. **Ref.** Van der KROGT, P. 'Commercial cartography in the Netherlands...' passim in *La Cartografia dels Països Baixos* (Barcelona 1994); Van der KROGT, P. *Globi neerlandici* (1993) passim.; HOOKER, B. 'New light on Jodocus Hondius' great world Mercator map of 1598' *Geographical Journal 159* (March 1993), pp.45-50; SCHILDER, G. *The Map Collector 59* (1992) pp.44-47; BARBER, P. *The Map Collector 52* (1990), pp.8-13; KRETSCHMER, DÖRFLINGER & WAWRIK *Lexikon zur Geschichte der Kartographie* (1986) vol.C/1 pp.319-320 (numerous references cited) [in German]; SCHILDER, G. *The Map Collector 32* (1985) pp.40-43; KOEMAN, C. *Atlantes Neerlandici* (1967-1971) Vol.II pp.136-146; KOEMAN, C. *Imago Mundi 19* (1965) pp.108-110.

• **Hondius**, Jodocus [Josse de; Joost] *the elder* (1563-1612). Flemish engraver and instrument maker from Wakken, Belgium. He began learning the craft of engraving at the age of 8, settled in London 1583 and married Colette van den Keere (sister of Pieter) in 1587. In 1593 he moved to the Netherlands, set up an engraving workshop in the Kalverstraat, Amsterdam, and steadily moved into publishing and bookselling. Jodocus died in Amsterdam. The business was continued by his widow Colette (from 1612), later by his sons Henricus and Jodocus the younger, and by his son-in-law Johannes Janssonius II. Engraved some plates for *Mariner's Mirrour*, London 1588 (English translation of Lucas Jansz. Waghenaer's *Speculum nauticum*); thought to have engraved H. Broughton's *A mapp of the north part of the equinoctial...* [North Polar projection], c.1588; *Typus orbis terrarum*, London 1589; *Americae novissima descriptio*, 1589; *Typus Angliae*, London 1590; *Angliae et Hiberniae...*, 1592; engraved E. Molyneux's pair of 24-inch globes, London c.1592; *Nova totius Europae descriptio...*, 1595; pair of 3.5-inch globes, Amsterdam 1597; *Typus Totius Orbis Terrarum...* [the World on Mercator's projection, the so-called 'Christian Knight' map], c.1598; B. Langenes' *Caert-Thresoor*, Middelburg 1598 (with Pieter van den Keere), republished Amsterdam, Cornelis Claesz 1599; worked with P. Plancius on a 14-inch celestial globe, Amsterdam c.1598; pair of 14-inch globes, Amsterdam 1600; pair of 8.5-inch globes, Amsterdam 1601; acquired Gerard Mercator's atlas plates in 1604, added to them, and re-issued *Atlas* [known as 'Mercator-Hondius' atlas], 1606 and many subsequent editions in different languages to

*Jodocus Hondius the elder (1563-1612), right, with Gerard Mercator (1512-1594), a double portrait engraved for inclusion in the Mercator/Hondius series of atlases from 1619. (By courtesy of the Universiteitbibliotheek, Amsterdam)*

1666 (new text for the maps was written by Petrus Montanus, husband of his sister Jacomina); *Atlas Minor* (Latin text, 152 maps), 1607 (reduced version of *Atlas*, with Johannes Janssonius and Cornelis Claesz.), eight further editions with French, Latin or German text and varying numbers of maps were published up to 1621; *Nova et exacta totius orbis terrarum descriptio* (12 sheets), 1608; began work on a pair of 21-inch globes, 1611 (unfinished at his death); plates for John Speed's *Theatre of the Empire of Great Britaine*, 1612.

• **Hondius**, Colette van den Keere (*d.*1629). Sister of Pieter van den Keere, widow of Jodocus Hondius. Continued the business after his death in 1612, while their sons were still minors.

•• **Hondius**, Elisabeth (*b.c.*1588). Daughter of Jodocus the Elder and Colette van den Keere. Married Johannes Janssonius in 1612, both assisted her mother Colette, and Johannes was later to work with Elisabeth's brothers Jodocus and Henricus.

•• **Hondius**, Jodocus, *the younger* (1594 or 1595-1629). Son of Jodocus the elder, continued father's business (initially with brother Henricus) at 'Calverstraat in den Wackeren Hondt by het Stadhuys' [sub signo Canis Vigilantis in Platea Vitulina, prope Senatoriam Domum], Amsterdam; after *c.*1621 had own business. Married Anna Staffmaecker, 1621. Shortly after his death 36 map plates were sold to Willem Jansz. Blaeu (34 of which formed the nucleus of the Blaeu atlas series). Completed his father's pair of 21-inch globes, 1613 (with A. Veen); *Novissima ac exactissima totius orbis terrarum descriptio magna cura & industria... delineata*, c.1618 (engraved by Jodocus the elder c.1611); engraved and issued Adriaen Veen's Scandinavia, 1613; re-issued 9-inch, 14-inch and 21-inch globes, 1618; *P. Bertii Tabularum Geographicarum*, 1616 and later (Peter Bertius's version of B. Langenes's *Caert Thresoor* with newly engraved maps); pair of 17-inch globes, 1623 (with Johannes Janssonius, celestial globe designed by A. Metius, plates engraved by Abraham Goos, a cousin of Jodocus); series of continent wall-maps, 1623; *Magnae Britanniae et Hiberniae Tabula*, Amsterdam c.1625. **Ref.** STEVENSON, E.L. & FISHER, J. *The map of the World by Jodocus Hondius 1611* (New York 1907) with facsimile.

•• **Hondius**, Henricus [Hendrick] (1597-1651). A younger son of Jodocus Hondius, the elder. Assisted his mother and his brother (1619-1629). Started on his own in 1621 on the Dam 'sub signo Atlantis'. On the death of his brother (1629) he returned to parental home on the 'dam in den Wackeren Hondt'. In the 1640s Henricus seems to have left Janssonius to develop the Atlas. P. Merula's *Cosmographiae generalis...*, 1621 edition; republished the Mercator-Hondius *Atlas*, 1623 and editions to 1633; worked in partnership with his brother-in-law Johannes Janssonius on further editions of *Atlas* 1633 and editions to 1637, after 1638 the title changed to *Atlas Novus*, 1638 and other editions to 1644 (also with French text); re-issued Willem Jansz. Blaeu's *Nova Universi Terrarum Orbis Mappa...*, Amsterdam 1624 (second state of the map issued in 1605). **N.B.** Not known to have connections with Henricus **Hondius**, publishers of Delft (see below).

**Hondius**, Henricus (*fl.*1624-1651). Engraver and printer of Delft, at 'den Rinck bij de Kolff' (1624-1644). Probably not related to Henricus Hondius of Amsterdam (see above). Issued reprints of Floris Balthasar and Balthasar Florisz. van Berckenrode's *Ware afbeeldinghe vant Dijckgraefschap van Delfland* (10 sheets), 1624, 1630, 1644 and 1648 (first published 1611).

**Hondius**, Wilhelmus [Willem] (*c.*1597-1660). *b.* The Hague, *d.* Danzig. Engraver for Guillaume Levasseur de Beauplan in Danzig, 1648-1650. Siege of Smolensk (16 sheets), 1634; *L'Isle Anthony Vaz* [Brazil], Nicolaes Visscher 1640; engraved Marcin German's maps of the Polish salt mines *Miasto Wieliczka*, 1645 (surveyed by German, drawn and engraved by Hondius); for Daniel Zwicker 1650.

**Hondt**. See **Hondius** family, above.

**Hondt**, Pieter de (1696-1764). Publisher and bookseller of The Hague. Republished some of the work of Janssonius including *Accuratissima Orbis Delineatio...*, 1740; Georg Horn's *Description exacte de l'Univers...*, 1741 (first published by Janssonius, translations published as *Atlas of the Antique World*, 1741 and *A compleat body of ancient geography*, 1741; *Atlas van het aloude Holland...*, 1745 (copied from M. Alting); *Atlas de la Hollande...*, 1745 (from M. Alting); Jacques Nicolas Bellin's World, 1750.

**Hone**, William. British publisher. Plan of St. Peter's Field, Manchester [scene of the Peterloo Massacre], 1819.

**Honervogt**, Jacques [Jacob] (*c.*1590-*c.*1663). Native of Cologne, worked as an engraver and publisher in Paris from 1608 'Chez Iac Honervogt rue St. Iacques a lanseigne de la Ville de Cologne'. After 1654 he worked with Gérard Jollain, who took over after his death. *Nova totius terrarum orbis geographica...*, 1625 (after Jodocus Hondius), later state published Jollain 1655; *Noua America Descriptio*, Paris 1640; *Novissima Descriptio Angliae Scotia et Hiberniae*, Paris c.1625-1650, later state published Jollain, c.1650. **Ref.** LOEB-LAROQUE, L. in: KÖHL, P.H. & MEURER, P.H. (Eds.) *Florilegium Cartographicum* Leipzig, Bad Neustadt a.d.S. (1993) pp.85-89 [in German].

**Honey**, E.R. *New Haven in 1641*, 1880.

**Hongxian**, Luo (1504-1564). *Guang Yu Tu* [enlarged terrestrial atlas], *c.*1555.

**Honius**, Henricus. Engraver of Haarlem, worked in Italy for Claudio Duchetti. Africa, 1579.

**Honkoop**, A. & J. Publishers, printers and booksellers of Leiden. J.S. Stavorinus's *Reize van Zeeland over de Kaap de Goede Hoop... In De Jaaren MDCCLXVIII-MDCCLXXI*,

*Johannes Honter (1498-1549). (By courtesy of Laszló Gróf)*

1793, 1797-1798; *Nouvel Atlas des Enfants*, 10th edition 1817 (8th edition was published Amsterdam, Gulick & Honkoop 1799).

**Honorat**, Barthélemi. *Sainte Bible*, 1585 (with a map of the Mediterranean, Caspian Sea and Persian Gulf by A. Du Pinet).

**Honoré**, François l' [et Cie.]. Publisher of Amsterdam. *Voyages du Baron de la Hontan dans l'Amerique Septentrionale* (2 volumes), 1705, (3 volumes), 1728; involved in the publication of the first part of Henri Abraham Châtelain's *Atlas historique...*, 1705-1735. **N.B.** See also **Châtelain** family.

**Honorius**, Julius (*fl*.5th century). Cosmographer. Wrote *Cosmographia* (a rather inaccurate world geography).

**Honorius** Augustodunensis [of Autun]. *Imago Mundi*, c.1129.

**Honstein**, —. *Duché de Savoye*, 1751.

**Honter** [Honterus], Johannes [János; Jan Coronensis] (1498-1549). Hungarian scholar, cartographer, scientist, cosmographer and engraver, son of Georg Gras, born and died Brassó [now Brasov, Transylvania, Romania]. From 1530 to 1533 he learned woodcutting and printing in Basle, where he published his first works, he later published from his own press in Kronstadt [Brassó]. Best known for *Rudimentorum cosmographiae...* (text with two maps) 1530, and many later editions; cut the woodblocks for a pair of celestial maps *Imagines constellationum...*, 1532, used in Ptolemy's *Almagest*, 1541; *Chorographia Transsilvaniae-Sybenbürgen* (2 sheets, woodcut), printed Basle 1532 (known from one surviving example, now in Széchényi Library, Budapest), copied by Münster, 1544 and others; *Rudimenta Cosmographica* (14 woodcut maps), 1542 (heavily revised edition of the 1530 work), published in many later editions some using their own woodblocks *e.g.* Christoffel Froschauer, Zurich 1546 and later. **Ref.** KARROW, R.W. *Mapmakers of the six-*

teenth century and their maps (1993); pp.302-315 includes many references; MOSKOPP, J. 'Eine Karte aus dem *Atlas Minor* des Johannes Honter: Syria', *International Yearbook of Cartography* Vol. 28 (1988) pp.83-93 [in German]; SZATHMÁRY, T. *Descriptio Hungariae* (Italy 1987) [in Hungarian]; KRETSCHMER, DÖRFLINGER & WAWRIK *Lexikon zur Geschichte der Kartographie* (1986) vol.C/1, pp.320-321 [in German]; HRENKÓ, P. 'Honter 450 éves térképe Erdélyről' [The 450-year-old map of Transylvania by Honter], *Geodézia és Kartográfia* [Geodesy and Cartography] 1983/2 [in Hungarian]; ENGELMANN, Gerhard *Johannes Honter als Geograph* (Cologne 1982) [in German]; KISH, G. *Imago Mundi* 19 (1965), pp.13-21.

**Honya** Kihei. Publisher of Takasaki, Japan. Map of the volcanic eruption of Mt. Asama (woodblock), 1883.

**Hon'ya** Matasuke. Japanese publisher. Colour map of the shrines and temples of the Shimosa Province (woodblock), *c*.1849.

**Hood**, Edwin C. County plat books in Wisconsin, Minnesota and Iowa, 1887-1899 including: *Plat book of Richland county, Wis.*, Minneapolis, C.M. Foote pub. co. 1895; *Plat book of Vernon county, Wisconsin* Minneapolis, C.M. Foote pub. co. 1896; *Plat book of Jefferson county, Wisconsin*, Minneapolis, C.M. Foote pub. co. 1899.

**Hood**, M.L. Lithographer of Hobart. *Walch's Plan of the City of Hobart Town*, Hobart Town, J. Walch & Sons *c*.1870 (drawn by Albert Reid).

**Hood**, R.V. Lithographer and publisher of Liverpool St. *Plan of the City of Hobart Town*, 1854; *Plan of the unsettled part of Tasmania*, 1865; also Admiralty charts.

**Hood**, Robert. Rondout and Oswego Railroad, Kingston, New York, 1871.

**Hood**, Thomas (*fl*.1577-1604). An early 'Thames school' chartmaker with a shop in the Minories, London. He was also a lecturer in mathematics and navigation and argued for the use of flat charts for navigation at sea in preference to globes. *The use of both the globes, Celestiall and Terrestriall*, London, Thomas Dawson 1592; *The Marriners guide, set forth in the form of a dialogue...*, 1592; chart of the North Atlantic and West Indies, 1592 MS; chart of north-east Atlantic, 1592; chart of the Bay of Biscay to the Channel, 1596 MS; charts of English coastal jurisdiction, 1604. **Ref.** SMITH, T.R. 'Manuscript and printed sea charts' in THROWER, N.J. *The compleat plattmaker* (1978).

**Hood**, Thomas (d.1811). Publisher in partnership with Thomas Vernor from 1794. Worked with him and Sharpe on the publication of G. Cole & J. Roper's *The British Atlas* (partwork), from 1804. **N.B.** See **Vernor**, Hood & Sharpe.

**Hood**, *Lieutenant* Washington (1808-1840). U.S. War Department, Bureau of Topographical Engineers. *Map exhibiting the position of several lines connected with the settlement of the Ohio boundary question*, Washington D.C. 1835 (lithographed by P. Haas); *Ohio Boundary* (3 maps), 1835 (directed by A. Talcot, surveyed by W. Hood, R.E. Lee and W. Smith, drawn by W. Hood and J.R. Irwin); *A Map of the Extremity of Cape Cod including the Townships of Provincetown and Truro...*, Washington 1836 (engraved by W.J. Stone); *Core Sound, North Carolina*, 1837 (surveyed by J. Kearney, T.J. Lee, L. Sitgreaves and A.M. Mitchell, drawn by W. Hood, engraved by W.J. Stone); *Map of the seat of War in Florida...*, Washington D.C. 1838 (engraved by W.J. Stone); Lake Nicaragua, 1838; *Map of the United States Territory of Oregon, west of the Rocky Mountains...*, Washington 1838; Map of the Yellowstone, 1839.

**Hoof**, *Kaptein-Ingenieur* Hendrik van. Topographische kaart van de linie van den IJssel van Arnhem tot de Zuiderzee (31 sheets), 1783 MS (with J.H. Hottinger, M.A. Snoeck, H.J. van der Wijck, J.A. van Kesteren).

**Hooftman**, Gilles Egidius (1521-1581). Antwerp merchant, shipowner & scholar. Said to have suggested the atlas format to Abraham Ortelius.

**Hoog**, — de. Engraver. *La Ville de Candie...*, Paris, Antoine de Fer1669. **N.B. Compare with Romein de Hooghe**

**Hooge.** See **Hooghe**.

**Hoogeboom**, A. See **Hogeboom**, Andries.

*Romein de Hooghe (1645-1708). Copperplate by J. Houbraken after H. Bos. (By courtesy of Rodney Shirley)*

**Hooghe** [Hogius; Hooge], Cornelis de (1540-1583). Dutch engraver, illegitimate son of Charles V; beheaded at The Hague. Map of Norfolk for Christopher Saxton, 1574; *Hollandia...*, 1565 (based on a 1537 map by Jacob van Deventer, engraved for Ludovico Guicciardini's *Descrittione di tutti i Paesi Bassi...*, 1567 and later).

**Hooghe** [Hogius; Hooge], Romein [Romanns; Romain; Romanus; Romeyn; Romyn] de (1645-1708). Dutch engraver, designer and painter. Ennobled by John III Sobieski of Poland in 1675; worked at Dordrecht in 1693. English Ports, 1667; engraved world map for Frederick de Wit, 1668; Flanders, 1670; *Luxemburgum...* [siege plan], Amsterdam, Nicolaes Visscher II 1684, 1685; view and panorama in Blaeu townbooks, 1682; *Atlas maritime... Cartes Marines A L'Usage des Armées Du Roy de la Grande Bretagne...*, P. Mortier 1693, Dutch edition with the sub-title *Zee Atlas*, 1694; frontispiece to de N. de Fer's *Atlas Royal*, 1695; also signed frontispieces to many of the volumes of 'Le Petit Beaulieu'.

**Hoogkamer**, Willem Hendrik. Engraver of Groningen. George Thompson's *Plan van de Kaap Stad...*, 1828.

**Hoogstraten**, M.J. van. *Kaart der Havenwerken te Vlissingen*, 1875.

**Hoogwerf**, J. *Bij de kaart Eenvoudige Aardrijkskunde voor Het Christelijk Onderwijs* (3 volumes - I. Nederland, II. Europa, III. Wereldeelen), Groningen, P. Noordhoff 1928 (atlases of blind maps).

**Hooiberg**, T. *Kaart van... der Nederlanden*, 1845.

**Hook**, Andrew. Issued third state of James Turner's *Chart of the Coasts of Nova-Scotia and Parts Adjacent*, Philadelphia 1760.

**Hooke**, Robert (1635-1703). Philosopher, astronomer and pioneer of experimental science, secretary to the Royal Society. Surveyor of the City of London from 1666. Prepared a plan for the redevelopment of London after the Great Fire, 1666; associated with the preparation of Moses Pitt's ill-fated *English Atlas*, 1680. **Ref.** ROSTENBERG, L. *The Map Collector* 18 (1982) pp.20-25; ROSTENBERG, L. *The Map Collector* 12 (1980) pp.2-8.

**Hooker**, John. *Isca Damnoniorum... Vulgo Excester* [Exeter], *c.*1587 (engraved by Remy Hogenberg, later copied for Braun and Hogenberg's *Civitates orbis terrarum*, vol. VI 1617). **Ref.** RAVENHILL, W. & ROWE, M. 'A decorated screen map of Exeter based on John Hooker's map of 1587' in: GRAY, ROWE & ERSKINE (eds.) *Tudor and Stuart Devon* (1992) pp.1-12.

**Hooker**, *Sir* Joseph Dalton (1817-1911). British botanist, travelled extensively in New Zealand, Tasmania, Morocco, the Antarctic and the Rockies in search of specimens. Between 1847 and 1851 he was in the Himalayas of Nepal and eastern Bengal. *Independent Sikkim*, 1861.

**Hooker**, William (*fl.*1804-1846). Instrument maker, engraver, printer and map publisher from Philadelphia. Worked in New York from 1822 as 'Instrument Maker & Chart Seller to the U.S. Navy and Agent for the Nautical Store, 202 Water Street' [E.M. Blunt] (1822-1830); later '...Removed to... N$^{o.}$312 Pearl Street'. In 1816 he married Eliza, the daughter of E.M. Blunt, becoming sales agent for Blunt in New York, 1822-1830. He engraved for the Blunt company until 1846, also for Humphrey Phelps, Peabody & Company, and A.W. Wilgus. One of the engravers for Aaron Arrowsmith and Samuel Lewis' *New and Elegant General Atlas*, Philadelphia, John Conrad 1804 and later editions; for Pinkerton, 1804; worked with Gideon Fairman on N. Bowditch's *Chart of the Harbours of Salem...*, 1806; engraved some of the maps for Lawrence Furlong's *American Coast Pilot*, Blunt 1809; for Edmund M. Blunt *Plan of the City of New York*, 1817, 1830; engraved maps for W. Darby's *A Tour from the city of New York to Detroit...*, New York, Kirk & Mercein 1819; *Hooker's New Pocket Plan of the City of New York*, 1824 and editions to 1840s; revised T.H. Poppleton's *Plan of the City of New York*, Prior & Dunning 1826, Prior & Brown 1829 (first published Prior & Dunning 1817); *New Pocket Plan of the City of New York*, 1833.

**Hoole**, John. London bookseller and publisher. Collaborated with Henry Overton 'at the White Horse without Newgate', London *c.*1724-1734. Joint seller (with H. Overton) of I.V. Kircher's *The Travellers Guide or Ogilby's Roads Epitomized...*, *c.*1725; Dublin, 1730; *A Pocket Map of London*,

*Westminster and Southwark...*, H. Overton and J. Hoole 1731 (engraved by R.W. Seale); *Nine New & Accurate Maps of the Southern Counties of England...* (2 copperplates to make 1 sheet), H. Overton and J. Hoole *c.*1732 (engraved by R.W. Seale); *The Roads of England according to Ogilby's Survey*, H. Overton and J. Hoole *c.*1732 edition (engraved by S. Nicholls, first published by Henry Overton *c.*1712). **Ref. HODSON, D.** County atlases of the British Isles published after 1703 Vol.I (1984) pp.182-183.

**Hooper, Charles** [& Co.]. Publisher of 'White Hart Court and Alderman's Walk, London E.C.'. Co-publisher for many of the Kelly's county maps including *Hertfordshire*, *c.*1894 (lithographic transfer from F. Bryer's of 1874, also used for Kelly's Directories); *Kelly's Map of the Suburbs of London*, 1894.

**Hooper, H.S.H.** Assistant to the Master of HMS *Queen*. *Plan of the Island of Gozo* [Malta], 1851 MS.

**Hooper, Samuel** (*fl.*1756-1793). Publisher, bookseller and stationer at New Church, Strand; 25 Ludgate Hill (1770-1793); 212 High Holborn, London. He was succeeded by his widow Mary (below). Published John Abraham Collet's *A Compleat Map of North Carolina...*, 1770 (engraved by J. Bayly); sold Joseph Smith Speer's *The West India Pilot* (26 charts), 1771, (28 charts), *c.*1785 (first published 1766, charts in this edition engraved by J. Bayly and Prinald); *Plan of Bombay*, 1772; *A Plan of Madura* [Madras], 1772; *Plan of Surat Castle* [Bombay], 1772; John Byres's *Plan of the Island of St. Vincent*, 1776 (engraved by J. Bayly); John Byres's *Plan of the Island of Tobago*, 1776 (engraved by J. Bayly); *Attack on Gibraltar*, 1782; *A Plan for a Navigable Canal from London to Norwich & Lynn*, 1785; Francis Grose's *The Antiquities of England and Wales* (partwork) 1773-1787 and later; *An Index Map to the Antiquities of Scotland*, 1791; *A New Map of the County of Hertford...* (part I only), 1793 (with W. Flexney, part II was published the following year by Mary Hooper).

• **Hooper & Wigstead** (*fl.*1784-1797). Publishers of 'N⁰·212 High Holborn, facing Southampton Street, Bloomsbury Square' (1797). The imprint appears on some editions of *The Antiquities...*, 1784, 1785, 1797; W. Darrell's *The History of Dover Castle* (with plan and views), 1797.

• **Hooper, Mary.** Widow of Samuel Hooper; continued his business after 1793 from 'N⁰·212 High Holborn', London; *A New Map of the County of Hertford...*, 1794 (engraved by F. Vivares).

**Hoorenhout, Jacques.** Dutch surveyor. Historical map of Zeeland, 1540.

**Hoorn, Abraham van.** Printer and publisher of Rotterdam 'op de Visschersdijk achter de Beurs' (1684). Re-issued Jan Jansz. Stampioen's *Het Hooge Heemraedtschap van Schielandt*, 1684 (first published 1660).

**Hoorn** [Hoirne; Horen], **Johannes à** [Jan de Beeldsnyder van]. Wood engraver and mapseller from Hoorn, worked in Antwerp. *Caerte van de Oosterscher Zee* [Germania Inferioris tabula], Antwerp 1526 (thought to be the oldest known surviving north European sea chart, three fragments only survive). **Ref. KARROW, R.W.** Mapmakers of the sixteenth century and their maps (1993) p.316; **HOFF, B.** van't 'Jan van Hoirne's map of The Netherlands and the 'Oosterscher Zee' printed in Antwerp in 1526' in Imago Mundi 11 (1954) p.136. **N.B.** Thought to be the father of Joost Jansz. **Beeldsnyder** van Bilhamer.

**Hoorn, Melchisedech van.** Dutch surveyor and engraver, 'pren'esnijer ende scilder... burger van Antwerpen'. Plan of Utrecht, 1569 (thought to be the oldest printed plan of Utrecht, used by Braun and Hogenberg).

**Hoornhoven, C.A. van.** See **Hornhovius, Cornelis Anthonisz.**

**Hoover, H.S.** *Atlas of Bremer County, Iowa*, Waverley, Iowa, H.S. Hoover & W.P. Reeves 1875.

**Hooyer, G.B.** *Atlas behoorende bij Hooyer Krijgsgeschiedenis van Nederlandsch Indië*, 's-Gravenhage, Gebroeders van Cleef, and Batavia [Jakarta], G. Kolff & Co. 1895 (with 59 maps and plans).

**Hope, George. T.** Secretary for the Jefferson Insurance Company of New York City, later President of the Continental Fire Insurance Company. Began to work on a large-scale map of New York City, the first designed specifically to allow fire risk to be assessed more easily. He was too busy to complete the work himself and appointed an engineer,

*George T. Hope. (By courtesy of the Library of Congress, Washington DC)*

William Perris, to continue with the surveys and the draughting of the map of the city, while Hope chaired a committee to direct the project. The committee specified the scale to be adopted, and stipulated the details that were to be shown such as construction materials. They also devised a set of symbols and colours which would represent the information on the map. These requirements set the standards for fire insurance plans for many years to come. The first volume of maps was published as *Maps of the City of New York surveyed under the directions of insurance companies of said city*, New York, Perris Company 1853 (surveyed and drawn by William Perris.

**Hope**, William C. *Accurate map of Barbados*, 1834.

**Hopfgarten**, — von. Prussian military cartographer and editor. *Atlas von dem zu Süd-Preussen gehörigen Posener Kammer-Departement*, Berlin 1799 (with Sotzmann).

**Hopkins**, brothers and company. Surveyors and publishers of Philadelphia. **Ref.** RISTOW,

W.W. *American maps and mapmakers* (Detroit 1986) pp.259-260, 400 & 443.

• **Hopkins**, Griffith Morgan *Jr* (*fl*.1859-1900). North American civil engineer and surveyor, founded a publishing company *c*.1865. Associated with numerous city and county atlases in North America, both as surveyor and publisher. Began producing county wall maps for R.P. Smith from *c*.1857, published under different names from the same address in Minor Street, Philadelphia. *A Topographical Map of Lincoln Co. Maine*, Philadelphia, Lee & Marsh 1857; *Map of Cuyahoga County Ohio...*, Philadelphia, S.H. Matthews 1858; *Map of Adams County, Pennsylvania*, Philadelphia 1858; surveys used for *Clark & Tackaburys' New Topographical Map of the State of Connecticut*, Philadelphia 1859; [Richard] *Clark's Map of Litchfield County, Connecticut...*, Philadelphia 1859; maps of counties in Pennsylvania and New Jersey, 1860 and 1861; Warren County, New Jersey, 1860 (with Henry W. Hopkins); Armstrong County, Pennsylvania, 1861 (with H.W. Hopkins); map of Perry, Juniata and Mifflin counties, Pennsylvania, 1863; Fayette County, Pennsylvania, 1865; *Atlas of the county of Suffolk, Massachusetts*, Philadelphia, G.M. Hopkins & Co. 1873-1875 (with H.W. Hopkins); *City atlas of Providence, R.I.* (3 volumes), Philadelphia, G.M. Hopkins 1875; *Atlas of Baltimore county, Maryland*, 1877; *Atlas of Danbury, Conn.*, 1880; *Atlas of Bryn Mawr and vicinity...*, 1881; *Atlas of the city of Philadelphia*, from 1885; *Atlas of city of New Haven, Connecticut...*, 1888; *Atlas of the city of Norfolk, Va. and vicinity*, 1889; and many others.

• **Hopkins**, Henry W. (*b*.1838). Younger brother and work associate of Griffith Morgan Hopkins. Took over his brother's company in 1900 and ran it until his retirement in 1907, the company was then run by G.B.C. Thomas. Warren County, New Jersey, 1860 (with Griffith M. Hopkins); Armstrong County, Pennsylvania, 1861 (with G.M. Hopkins); *Atlas of Delaware County, Pennsylvania*, 1870; *Atlas of Germantown, Pennsylvania*, 1871; *Atlas of the county of Suffolk, Massachusetts*, Philadelphia, G.M. Hopkins & Co. 1873-1875 (with G.M. Hopkins).

• **G.M. Hopkins Company** (*fl*.1865-1953). Publishing company of 320 Walnut Street, Philadelphia. It was founded by Griffith and Henry Hopkins in 1865, and specialised in maps and atlases of counties in Maryland, Pennsylvania and Virginia. They later produced urban real estate atlases for U.S. cities, especially in New England, New York, Pennsylvania and New Jersey. The Hopkins company was bought by the Franklin Survey Company of Philadelphia in 1943. *Combined Atlas of the State of New Jersey*, 1873 (published in two versions); *Atlas of Baltimore County, Maryland and Environs* (2 volumes), Philadelphia, G.M. Hopkins 1876, 1877; *Atlas of fifteen miles around Baltimore*, Philadelphia 1878; *Atlas of fifteen miles around Washington*, 1878, 1879; *Atlas of the cities of Pittsburgh & Allgheny*, 1882; *A complete set of surveys and plats of the properties in the city of Washington*, 1887; *Map of the District of Columbia*, 1887, 1891; *Real estate Plat of Washington, D.C.*, 1892; *Real estate plat book of Washington* (3 volumes), 1893-1894 (supplement published 1896); *The vicinity of Washington, D.C.*, 1894; *Atlas of Hudson county, New Jersey*, 1908, 1919; and many others.

**Hopkins**, B.B. Publisher, 66 South Fourth Street, Philadelphia (1815). Publisher of C.W. Bazeley's *New Juvenile Atlas*, 1815.

**Hopkins**, *Sergeant* E. R.E. Surveyor and draughtsman. Survey of Sambro Island, 1881; Macnabs Island [Halifax], 1883 MS; survey of the town of Halifax with harbour and surrounding country (24 sheets), 1885-1886 MS (also 2 other surveys of the same date, each on 8 sheets); *Co. Halifax, Nova Scotia... 1886*, Southampton, Ordnance Survey Office 1889; Nova Scotia County, Halifax 1889; *Index Sheet to Six Maps of Vancouver Island, British Columbia*, *c*.1900.

**Hopkins**, Edward. Map of the township of Great Sandal, Yorkshire, 1607 MS.

**Hopkins**, Roger [and Sons]. Civil engineers of Plymouth. The Island of Portland (92 plans), 1832 MSS.

**Hopkinson**, John. Synopsis Paradisi, Leiden 1598 (with map).

**Hopkinson**, John. Geologist and climatologist, prepared papers for the *Transactions of the Hertfordshire Natural History Society and Field Club* which incorporated the following maps: *Map of Hertfordshire showing the*

*River Basins and the Position of the Rainfall Stations*, 1880 (later adapted for publication in A.R. Pryor's *A Flora of Hertfordshire*, 1887); outline map of Hertfordshire, 1881; *Map of Hertfordshire showing its Climatological Stations*, 1889 (published in *Transactions..*, 1890).

**Hoppach**, C. *Plan de la ville d'Ostende avec ses environs et le campement des trouppes des alliés... le 17 juin 1706* in Eugène Henri Fricx's *Table des cartes des Pays Bas et des frontières de France...*, Brussels 1712.

**Hoppe**, —. Engraver for Johann Georg Friedrich Hartmann, 1824.

**Hopper**, H. Carmer. *Map Showing the route of the Salt Lake, Sevier Valley, and Pioche Railroad*, 1873.

**Hopper** [Hopperus], Joachim (1523-1576). b. Sneek, Friesland, d. Madrid. Jurist and member of the Great Council of Mechelen (1554); member of Privy Council in Brussels (1561); Keeper of the Great Seal of Philip II (1566). Thought to have used manuscript maps of ancient Friesland made by J. van Deventer as the basis for his *Antiquae Frisiae...*, used by Ortelius as an inset on *Frisia Occidentalis*, 1579.

**Hoppner**, *Captain* H.P., *RN*. Hydrographer. Northwest Passage, 1824-1825.

**Hopson**, —. *Handy Map of the Crouch End Cricket Fields*, c.1897.

**Hopton**, Arthur (1589-1614). English topographer. *Speculum Topographicum or the Topographical Glasse...* (with 3 woodcut maps), London, Simon Waterson 1611.

**Hopton**, Richard. *Plans of Leamington Spa*, 1834.

**Horatius**, Andreas Antonius. Drew maps and plans for João José de Santa Teresa's *Istoria delle guerre del regno del Brasile*, Rome, F. Corbelletti Heirs 1698, Antonio de Rossi 1700.

**Horatius**, Juriaen Janssen. Rivier van Siera Lione, Sierra Leone, 1666.

**Hore**, Edward C. Maps of Tanganyika, 1879-1889.

**Horen**, Georg Christian von. Draughtsman and engraver of Bern. *Carte des Communications Télégraphiques du Régime Européen dressée d'après des documents officiels...*, Berne 1880 (printed by F. Lips).

**Horen**, J. van. See **Hoorn**, Johannes à.

**Horenbault** [Horenbout; Hawerbault], François [Francies]. Dutch cartographer and surveyor, worked in Zeeland, styled 'caertemaecker en ingenieur'. *Caarte ende discriptie figuratief van de zeedijcken van Groot Merschmont...*, 1579 (with Pieter de Buck, Jan Simonse, and Dircks Hendrikz.).

**Horenbault**, Jacques. Flandre maritime, c.1620.

**Horman**, Robert. Mouth of the River Thames, 1580 MS.

**Horn** [Honorius Regius; Hornius], *Doctor* Georg [Georgius] (1620-1670). Professor of history and geography at Leiden. Wrote an introduction for the second edition of *Accuratissima Orbis Antiqui Delineatio* (53 maps), Amsterdam, J. Janssonius 1653, 1654, 1660, J. Janssonius van Waesbergen 1677, 1684, The Hague, P. de Hondt 1740, English edition entitled *A full and exact Description of the Earth, or Ancient Geography...*, Amsterdam, J. Janssonius van Waesbergen 1700, London, Timothy Child 1700, The Hague, P. de Hondt 1741, French edition entitled *Description exacte de l'Univers, ou l'ancienne Géographie...*, The Hague, P. de Hondt 1741; text also used in the 6th volume of *Novus Atlas*, J. Janssonius 1658.

**Horn**, Hosea B. *Horn's overland guide from the U.S. Indian Sub-agency, Council Bluffs, on the Missouri River, to the city of Sacramento, in California*, New York, J.H. Colton 1852.

**Horn**, J. à. See **Hoorn**, Johannes à.

**Hornbeck**, H.B. Map of St. Thomas [Virgin Islands], Copenhagen 1835-1839, 1846.

**Hornberger**, C. *Die Goldküste...*, 1873.

**Hornbogen**, P. Associated with J.I. Kettler's *Generalkarte... Ostafrika* (12 sheets), c.1892 (capacity unknown).

**Horne,** J. Fiji Inseln, 1877-1878.

**Horne,** James. *Plan... of proposed canal between Loch Long at Arrochar and Loch Lomond at Tarbet,* 1821; *Plan showing the ground permanently occupied by the Edinburgh and Glasgow Union Canal through Hopetoun Estate...* (3 plans), 1822 MSS; *Plan of the ground permanently occupied by the Union canal through Riccarton...,* 1833 (copied from a plan by Thomas Grainger, 1824); *Plan of the road... from Borrowstoness to ... Champony,* 1837 (lithographed by Leith & Smith).

**Horne,** John (1848-1928). Scottish geologist, President of Royal Society of Edinburgh (1915-1919). Joined the Geological Survey in 1867, for which he worked until his retirement. Collaborated with Benjamin Peach. Together they produced many famous works reporting their surveys. Surveys of the southern uplands and north-west Highlands, from 1867 (with B.N. Peach); survey of Dalradian [north-east Scotland], 1877-1885. **Ref.** SUTTON, J. 'John Horne' *Dictionary of national biography: Missing persons* (1993) pp.330-331.

**Horne,** Robert (fl.1666-1691). Bookseller of London 'by the Royal-Exchange'. *Carolina Described,* London, Robert Horne 1666 (thought to be the first printed map of Carolina, published in *A Brief Description of the Province of Carolina...*). **Ref.** BURDEN, P. *The mapping of North America* (1996) N°.392 pp.497-499; CUMMING, W.P. *The southeast in early maps* (1958) 60 pp.147-148.

**Horne,** T.H. *The Lakes of Lancashire, Westmorland and Cumberland,* London 1816 (engraved by H. Mutlow).

**Horne,** Thomas. Co-publisher with others of Edward Hatton's *A New Map of the Cityes of London, Westminster and the Burrough of Southwark...,* 1707.

**Horne, Thornthwaite & Wood.** Publishers of London. T.K. Mellor's *The Handy Map of the Moon, c.*1886.

**Horneck,** Kane W. *Plan of English Harbour with its environs including Falmouth Harbour and Monks Hill* [Antigua], 1752; plans of fortifications on Antigua, 1752 (with D.E. Baker); Hugh Fort on St Mary's Isle [Scilly Isles], 1744.

**Horner,** *Le père* — (d.1880). French missionary from Alsace, began work on Réunion, was later sent to East Africa, and founded hospitals, missions and orphanages in Zanzibar. Map for *Voyage à la côte orientale d'Afrique...,* 1873.

**Horner,** *Professor* Johann Caspar (1774-1834). Swiss astronomer, physicist, instrument maker and councillor in Zurich and Hamburg.

**Horner,** Robert E. Produced revised editions of Thomas Gordon's *A Map of the State of New Jersey,* 1849, 1850, 1853, 1854 (first published 1828).

**Horner,** William B. *Railway & route map to the Gold Regions of Kansas & Nebraska,* 1859 (lithography by J. Gemmell); *The gold regions of Kansas and Nebraska: being a complete history of the first year's mining operations...* (with 2 maps), Chicago, W.H. Tobey & Co. 1859.

**Hornhovius** [Van Hornhoff; Van Hoornhoven], Cornelis Anthonisz. Dutch pastor and administrator at St. Laurensabdij at Oostbroek, de Bilt 1599. *Florentissimi Trajectini principatus typus...* (2 sheets), J. Hondius 1599.

**Hornius,** G. See **Horn,** Georg.

**Hornor,** Thomas (1785-1844). Surveyor, landscape gardener, artist and inventor from Kingston-upon-Hull, trained as a land surveyor and engineer by his brother-in-law William Johnson. Based at 3 Church Court, Inner Temple (1812), 2 Robert Street, Adelphi (1822), 10 Royal Terrace (1825). He died in New York. *Plan of the Parish of Clerkenwell,* surveyed 1807-1808, published 1813; *Plan of the town and parish of Kingston upon Thames,* 1813; surveys in South Wales, 1814-1820; *Plan of the Neath and Swansea Junction Canal,* 1818; sketched a panorama of London from the top of St Paul's Cathedral, 1821 (painted by E.T. Parris for display to the public in the 'Colosseum' in Regents Park from 1829). **Ref.** HYDE, R. *The Map Collector* 39 (1987) pp.37-39; HYDE, R. *Imago Mundi* 29 (1977) pp.24-31.

**Hornung,** D. Biblische Geschichtskarte (8 sheets), Leipzig 1854.

**Horowitz**, Pinchas. Jewish cartographer, worked in Jerusalem. Eretz Israel, 1840; Eretz Israel and Syria [Hebraische Palaestine und Syria Handkarte], Jerusalem 1908.

**Horrebow**, Niels (1712-1760). Danish author. *Nachrichten von Island...*, Copenhagen 1753, english edition published as *The Natural History of Iceland* (with map of Iceland), 1758.

**Horrel**, C.C. Surveyor in Victoria, Australia. *East Melbourne*, 1858.

**Horsburgh**, *Captain* James, *RN* (1762-1836). Publisher and 'Hydrographer to the Honble. East India Company'. Strait of Macasser, 1800; *Atlas of the East Indies and China Sea*, 1806-1821 (with Penelope Steel); East India Pilot, 1817 (with Penelope Steel); *Chart of the Straits of Gaspar, Straits of Banca and adjacent areas of the China and Java Seas*, 1819 (engraved by J. Bateman); re-issued M. Mackenzie's *Treatise on Maritime Surveying*, 1819; worked on *Indian Atlas*, published from 1827-1833, continued by John Walker 1854; *Passages through the Barrier Reefs...*, 1830-1832 (used by Norie in *The Complete East India Pilot*, 1838); *Book of Directions*, 1832.

**Horsey**, Samuel. *New Map of the Isle of Wight*, *c.*1840.

**Horseley**, John (1685-1732). Began working on *Northumberland* (2 sheets), completed after his death by Robert Cay, 1753.

**Horsely**, *Captain* —. Based at Liverpool. Lagos & its channels, 1789 (for Robert Norris).

**Horsfield**, Robert. Bookseller and publisher of London. Abbé Lenglet-Dufresnoy's *Geographia Antiqua et Nova*, 1768 edition; one of many publishers of William Camden's *Britannia*, 1772 (last edition with Robert Morden's maps).

**Horsfield**, Thomas. Island of Banka, 1822; Island of Java, 1852.

**Horsley**, Benedict (*fl.*1648-1702). Surveyor of York. Plan of York, 1694, published Pierce Tempest 1697.

**Horsley**, John.(*fl.*1728-1734). Map of Roman Britain, 1732, copied as *Britanniae antiquae tabula geographica*, 1775, and used in *A complete body of Ancient Geography*, Laurie & Whittle 1795 (second English translation of d'Anville's *Géographie ancienne et abrégée*), also copied in *The British Antiquary*, 1794. **Ref.** MACDONALD, Sir George, 'John Horsley, Scholar and Gentleman', Archaeologia Aeliana 4th series, X (Newcastle-upon-Tyne 1933), pp.1-57.

**Horsley**, *Lieutenant-Colonel* William Henry. Chief Engineer, Department of Public Works, Madras. Map of Madras Presidency, 1861.

**Horst**, J.T. Charte von Grün und Holtzteichen von Accumer bisz Benser-Syhl, 1736; Charte Insel Norderney, 1743.

**Horstmann**, F.H.W. *Situations-Plan des Thier-Garten bei Berlin*, 1833.

**Horstmann**, Nicholas. Rio Essequibo to Rio Negro, 1743.

**Horta**, Juan de. *Mapa del Reyno de Navarra*, 1772 (used by T. López).

**Horton**, family. Practised together as estate surveyors producing manuscript surveys of parts of Essex and Kent.
• **Horton**, Jonathan (*fl.*1806-1823).
• **Horton**, Simon (*fl.*1816-1823). Surveyor of Buckland, near Dover, Kent. Plan of Great Hougham Court, 1816; Lexden Manor, 1819 MS.

**Horton**, *Lieutenant* James. Officer in the Royal Staff Corps. *Reconnaissance of the Country lying between the Guadajira and Albuhera Rivers, the Guadiana and the Sierra de Montsalud* [Badajoz, Spain], 1812 MS (with others); one of those involved in a military survey of northern France entitled *British Lines of Occupation. N. of France (after 1815)...* (10 sheets), 1818-1823 MSS.

**Horton**, James. Ascension Harbour... 1833 [Caroline Islands], Admiralty 1840.

**Horton**, *Surgeon Major* J.A.B. Maps of Dassay and Ashante Gold Coast, 1882.

**Horton & Leonard**. Printers. *John Wentworth's subdivision* [Chicago], 1868; *Provident Homestead Association lots for sale* [Chicago], 1870.

**Horváth** [Pálóczi-Horváth], Ádám (1760-1820). Hungarian surveyor, cartographer and poet. Survey and mapping of the Estate of Pomáz, 1783; maps of the River Dráva, 1795 (Történeti Múzeum, Budapest). **Ref.** HRENKÓ, Pál 'Pálóczi-Horváth Ádám földmérö', Geodézia és Kartográfia [Geodesy and Cartography], (1980/4) [in Hungarian].

**Horwood**, Richard (c.1758-1803). Surveyor and publisher. *Plan of the Cities of London and Westminster & the Borough of Southwark...* (large-scale insurance plan prepared for the Phoenix Company in 32 sheets), 1792-1799 also published William Faden 1807, 1813, 1819 and possibly James Wyld; plan of Liverpool (6 sheets), 1803. **Ref.** LAXTON, P. introduction to The A to Z of Regency London [reduced facsimile volume] (1985); LAXTON, P. 'Richard Horwood's plan of London: a guide to editions and variants, 1792-1819' London Topographical Record 26 (1990) pp.214-263.

**Hose**, C. *Sketch Map of Baram District, Sarawak, Borneo... 1884-1892*, London 1893.

**Hoser**, Joseph Karl Eduard (1770-1848). Physician and cartographer of Prague. Personal physician to Archduke Karl. *Karte des Riesen-Gebirgs* [Polish-Czech border], Vienna 1806.

**Hosford**, Oramel. Contributed to *Atlas of the state of Michigan*, 1873.

**Hōshūdō** —. Japanese publisher. Hashimoto Gyokuran's plan of Sendai city, 1868 (with Matsubayashi Rōtōken).

**Hosken**, *Lieutenant* Henry, *RN*. *Amsterdam Island*, 1874; *Goro Is.* [Fiji], 1874.

**Hosking** [Hoskins], William (1800-1861). Railway surveyor of Essex and London. Harwich railway, 1840-1844 MS.

**Hoskold**, Henry Davis. *Memoria general y especial sobre las minas, metalurgia, leyes de minas, recursos, ventajas etc., de la explotación de minas en la Republica Argentina* (with maps and plans), Buenos Aires, Impr. del *Courrier de la Plata* 1889; *Mapa topografico de la Republica de Argentina* (10 sheets), London & Liverpool, George Philip & Son 1895.

**Hoskyn**, *Captain* Richard *RN*. Master of HMS *Porcupine*, later as Superintendent of Charts at the Hydrographic Office, he directed the draughting of charts, c.1869-1873. *Deep Sea Soundings in the North Atlantic from Ireland to Newfoundland*, 1862 (with J. Dayman, surveys prior to the laying of telegraphic cables);

**Hoskyn**, *Commander* Richard F., *RN*. Commander HMS *Myrmidon*. Directed surveys of the Red Sea, 1871-1872. *Ireland. East Coast* (charts), 1857 published Admiralty 1860; Banks Strait, 1888; corrected Frederick John Owen's Tasmania, 1893, all for the Admiralty.

**Hosmer**, John. Surveyor in Gloucestershire. Outline map of the Forest of Dean, 1833; many plans and surveys for the Dean Forest Commissioners, 1834; Encroachments in Ruardean Walk [Forest of Dean], 1834 (lithographed by C. Ingrey); the iron and coal districts of the Forst of Dean, 1841; Ruardean Walk: the Delves and Harry Hill inclosures, 1844 MSS; and other Dean Forest inclosures.

**Hosokawa** Minamoto no Yasuyoshi. Japanese mapmaker. *Kan'ei Kaisei Sakai ō-Ezu* [plan of Sakai city], published by Suharaya Mohei (and others) 1851.

**Hospin** [Hospein], Michael (1565-after1610). German court painter and mapmaker. *Hohenloe und Berlichingen*, 1589 MS; *Kirchberg an der Jagst*, 1607 MS; map of the Bishopric of Würzburg, 1616.

**Hossard**, Paul Michael (1797-1862). French engineer and geographer. Collaborated with others on the Pyrenees section of Carte d'État Major, 1852-1857.

**Hosse**, —. Engraver. *Karte der Hohenzollernschen Lande* (9 sheets), Berlin, Topographical Section, Royal Prussian General Staff 1863 (with Ruzek and A. Meyer).

**Hossinger**, T. *Isochronenkarte des Weltreiseverkehrs von Berlin aus 1812 und 1912* [traffic map centred on Berlin], [n.d.].

**Host**, S. de L'. See **Sorriot** de L'Host.

**Hotan** (1654-1738). Japanese Buddhist priest and scholar. *Nansenbushu bankoku shoka no zu* [world map], woodcut version published by Bundaiken Uhei, 1710. **NB.** Rokashi, the name of the author on the map above is presumed to be Hotan.

**Hotchkiss**, *Major* Jedediah (1828-1899). Natural scientist, surveyor, cartographer and teacher. Schoolmaster in Virginia, 1847-1861; from 1861 during the American Civil War served as a topographical engineer in the Army of Northern Virginia. From 1868 to 1870 he acted as topographical engineer for the Board of Survey of Virginia, preparing topographical, geological and general maps of the counties. Many maps, letters and papers are now held by the Library of Congress. Tygart's Valley, 1860 MS; Shenandoah Valley, 1862; *The Battlefields of Virginia. Chancellorsville*, 1867 (with William Allan); *Preliminary map of Lunenburg county, Va.*, Staunton Va. 1871; *Map of Virginia*, Staunton Va. 1874; geological map of Virginia, 1876; *Map of Rockingham county, Virginia...*, Staunton Va. 1877; maps and text for *Historical Atlas of Augusta County, Virginia*, 1885; 123 maps and sketches in *Atlas to accompany the Offical Records of the Union and Confederate Armies*, 1891-1895. **Ref.** ROPER, P. *Jedediah Hotchkiss: Rebel mapmaker and Virginia businessman* (Shippensburg, PA, 1992); ROPER, P. *The Map Collector* 49 (1989) pp.2-8; BOSSE, D. 'Jedediah Hotchkiss: Stonewall Jackson's topographer' in: *Mapline* No.22 (The Newberry Library, Chicago, June 1981).

**Hotchkiss**, William Otis (*b*.1878). Geologist and geographer in Wisconsin. *Wisconsin* (2 sheets), 1911; *Limestone road materials of Wisconsin*... (with maps), Madison, Wisc., The State 1914 (with E. Steidtmann); *Mineral land classification, showing indications of iron formations...*, Madison, Wisc., The State 1915 (assisted by O.W. Wheelwright and E.F. Bean); *A brief outline of the geology, physical geography, geography, and industries of Wisconsin*, Madison, Wisc., The State 1925 (with E.F. Bean).

**Hotes**, Heinrich. *Plan... Oldenburg*, 1867.

**Hotot**, Gilles. 'Imprimeur Ordinaire du Roy et de la Ville', Orléans, France. *Plan et profil au naturel de la ville d'Orléans*, *c*.1648.

**Hotta** Nisuke (1744-1826). Japanese astronomer, surveyed the coastal navigation route from Edo to Hokkaido. Portolan, 1799 MS; terrestrial and celestial globes, 1808.

**Hotta** Satonobu. Japanese publisher. Woodblock map of Mt. Daisen, Hōki Province, [late Edo].

**Hottinger**, J.H. Swiss-born engineer and cartographer, compiled maps of Drenthe, Groningen and Overijssel, *c*.1783 (with others).

**Hottinger**, *Professor* Johann Heinrich (1681-1750). Swiss orientalist, theologian and engineer in Holland, Marburg and Heidelburg.

**Houard**, J. See **Howard**, John.

**Houblou**, L. Lithographer. Route du Simplon, 1820.

**Houbraken**, J. Contributed to I. Tirion's *Atlas van Zeeland*, 1760.

**Houdaen**, Vincent. Publisher of Kalkar, Cleves. P. Laicksteen's *Antiquae...* and *Novae urbis Hierosolymitanae topographica delineatio* [old and new Jerusalem], 1570 (used in Braun and Hogenberg's *Civitates orbis terrarum*, 1574, and by De Jode *c*.1580); C. Sgrooten's *Peregrinatio filiorü Dei* [travels of the Apostles] (copper engraving on 10 sheets), 1572.

**Houdaille**, *Colonel*, Charles. French geographer on the Ivory Coast. *Carte de la mission Houdaille*, 1900. **Ref.** BROC, N. *Dictionnaire illustré des explorateurs Français de la XIXe siècle: I Afrique* (1988) pp.174-175 [in French].

**Houdan** [D'Houdan], François d' (1747-1828). Engraver. Partner in, and engraver of, Dumez and Chanlaire's *Atlas nationale de la France*, 1790 and many later editions; Jean Denis Barbié du Bocage's *L'Ecosse avec ses Isles*, 1800; *Environs de Versailles* (12 sheets), *c*.1808; also engraved for 'Dépôt de la Marine' and 'Dépôt de la Guerre'; maps for J. Peuchet and P.G. Chanlaire's *Description topographique et statistique de la France*, Paris, P.G. Chanlaire and Veuve Dumez 1810.

**Houel**, Jean (1735-1813). *Voyage Pittoresque des Isles de Sicile, de Malte & de Lipari*, 1782.

**Hough**, Benjamin. American surveyor, worked with A. Bourne. Original Surveys of the Townships of Michigan, 1815; *Map of the State of Ohio from Actual Survey*, Chillicothe, B. Hough & A. Bourne, and Philadelphia, J. Melish 1815 (engraved by H.S. Tanner, based on J.F. Mansfield's map of 1806).

**Hough**, Charles C. Publisher. John Atkinson's *Forest of Dean*, 1845; John Atkinson's *The Forest with High Meadow and Great Doward Woods*, 1845 (engraved by J. & C. Walker).

**Hough**, Franklin B. (1822-1885). Physician, natural scientist, and statistician from Martinsburg, New York. Superintendent of the New York Census, 1855. Worked with John Homer French. *A history of St. Lawrence and Franklin Counties, New York*, Albany, Little & Co. 1853; edited R.P. Smith's *Gazetteer of the State of New York*, 1872 edition.

**Houghton**, Douglass. State geologist of Michigan, 1837-1845. Directed at least four county surveys within the state undertaken by S.W. Higgins, C. Douglass and Bela Hubbard: *Map of Calhoun County*, 1844; *Map of Jackson County*, 1844; *Map of Lenawee County*, 1844; *Map of Washtenaw County*, 1844; also associated with *Geological Map of Townships in the Northern Peninsula of Michigan subdivided by D. Houghton*, 1845 (engraved by E. Weber & Co.); *Geological Map of Township Lines in the Northern Peninsula of Michigan... under... Houghton's contract for surveys with reference to Mines and Minerals*, 1845 (surveyed by W.A. Burt, lithographed by E. Weber & Co.)

**Houghton**, Dugdale. *London & Birmingham Canal*, c.1835.

**Houghton**, George Lampton. American geographer. Published 8-inch, 12-inch and 18-inch globes, 1902.

**Houghton**, J. *China Sea*, 1821.

**Houghton**, Jacob Jr. *The mineral region of Lake Superior*, Buffalo, O.G. Steele 1846.

**Houghton**, M. *Plan of the Channel between Sumatra and Lucepara*, 1818 MS.

**Houghton**, W. Lithographer of 162 New Bond Street, London. *Map of British Honduras*, c.1865.

**Houiste**, François (*b.*1794). French engraver. *Carte de France physique et administrative*, 1838.

**Houlanger**, —. *Plan des Forteresse de Vallete* [Malta], 1645.

**Houlen**, P. van. Hambantotte Salt Pans, 1813 MS.

**Houlston**, R.F. *Houlston's New Pocket Map of the Beaufort Hunt*, Chippenham 1904.

**Houlston & Sons**. Publishers of Reading, Berkshire. Wiltshire, 1869; Reading, c.1880.

**Houlston & Sons**. Publishers of '65 Paternoster Row E.C.' (1870); '7, Paternoster Buildings, E.C.' (c.1877), London. *Environs of Hull*, 1870; *Houlston & Sons New series of District Handy Maps from the Ordnance Survey*, 1870; *Houlston & Sons Handy Map of London*, c.1877; *Environs of Ripon* [Yorkshire], 1877 (drawn by John Bartholomew); *Huntingdon and neighbourhood*, 1877 (drawn by John Bartholomew); *Map of Weston-super-Mare*, 1880; *Plan of the Town of Weston-super-Mare*, c.1892.

**Houlston & Wright**. Publishers of London. *Surrey*, 1868 (drawn by J. Bartholomew); *Environs of Cambridge*, 1870.

**Hourlier**, Geneviève. Widow of Antoine de Fer. Continued her husband's business from 1673, and on her retirement in 1687 passed the stock to her son Nicholas. **N.B.** See **Fer**, de, family.

**Hourst**, Lieutenant de vaisseau Émile Auguste Léon (1864-1940). French explorer on the Niger and the Yang-Tse rivers. Maps for *Sur le Niger et au pays des Touareg*, 1898; *Instructions nautiques. Chine: Haut Yang-Tse-Kiang et affluents*, 1904; *Seconde mission Hourst: dans les rapides du Fleuve-Bleu. Voyage de la première cononnière française sur le Haut Yang-Tse-Kiang*, 1904. **Ref.** BROC, N. *Dictionnaire illustré des explorateurs Français de la XIXe siècle: I Afrique* (1988) pp.176-177; *ibid.* Vol.II Asie (1992) pp.244-245 [in French].

**House, W.M.** *North Dakota and Richland County Chart* (17 maps), Chicago, Rand McNally & Co. 1897.

**House & Brown.** Publishers of Hartford, Connecticut. *The United States*, 1847; Mexico, Texas, etc. 1847, 1849.

**Housman**, John. Pocket Plan of Manchester & Salford, 1800.

**Houston**, John. American surveyor. *Map of the Town of Torrington, Litchfield County, Connecticut*, Philadelphia, Richard Clark 1852.

**Hout**, H.I. van den. Contributed to D. Buddingh's *Zak-Atlas van het Koningrijk der Nederlanden*, 1842.

**Houte**, Drs P. van. *Aardrijkskundige Repetitie- en Werkboek* (blind maps in 2 volumes), Rotterdam, Nijgh & Van Ditmar 1929, 1930 (with C. Wiskerke).

**Houten**, G. van. Designed engraved title page to Jacob Robijn's *Zee atlas*, 1683.

**Houten**, Van. Cocoa makers. *Petit Atlas.... Offert par les fabricants du Cacao van Houten* (20 map cards with data on the reverse), 1910.

**Houtman**, brothers. Dutch navigators and contemporaries of the Blaeu family. Their observations contributed greatly to the celestial works of the period. **Ref.** KROGT, P. van der *Globi Neerlandici* (1993) many references; WALDMAN & WEXLER *Who was who in world exploration* (1992) pp.327-328.
• **Houtman**, Cornelis de [Cornelius] (*c*.1565-1599). Navigator and trader from Gouda. Led expeditions to the East Indies in 1595-1597, and 1598-1599. He was killed in Sumatra. Chart of Mossel Bay and Flesh Bay [Cape of Good Hope], published in Lodewijcksz' *Premier livre de l'histoire de la navigation auz Indes orientales*, 1598; Descriptio Hydrog. [Dutch route to E. Indies], 1597 published Claesz. 1598.
• **Houtman**, Frederik de (*c*.1571-1627). Dutch navigator, brother of Cornelis above, with whom he sailed on the East Indies expeditions 1595-1599. After imprisonment in Sumatra 1599-1601 he returned to Holland, was later to serve as an administrator in South East Asia (1605-1623). Houtman provided a great deal of information for celestial charts and globes of the period.

**Houtrijve**, J. van Jr. Publisher of Dordrecht. *Atlas van Noord Nederland* (10 maps), 1839; *Atlas van het Koninkrijk der Nederlanden* (13 maps), 1840.

**Houve**, Paul de la. Publisher and engraver of Paris. Copied J.H. van Linschoten's *Itinerario*, 1600 (first published 1594); made and published a copy of the Plancius world map *Orbis terrarum...*, *c*.1600 (Plancius edition published 1596); *Nova et exactissima insulae Siciliae descriptio*, 1600; V. Paletino's *Nova descriptio Hispaniae* (4 sheets), 1601 (first published by H. Cock, 1553); C. Sgrooten's *Nova celebrerrimi Ducatus Gueldriae* (6 sheets), 1601 (first published H. Cock, *c*.1563); Plan de la Ville de Florence, Paris 1601; Plan de Messine, Paris 1601; P. Plancius's Italiae, Illirici... descriptio, 1620.

**Houwens**, Jan. Dutch engraver, printer and publisher, worked in Rotterdam. Published the revised re-issue of Pieter van den Keere's 1611 world map *Nova Totius Terrarum Orbis Tabula*, *c*.1689; re-issued Jacob Quack's *Afbeeldinge van de Maes van de Stadt Rotterdam tot in Zee...* (first published 1665), 1698. **Ref.** SHIRLEY, R. *The mapping of the World* (1983) n°412 pp.433-435.

**Houze**, Antoine Philippe. French publisher. Atlas historique de France, 1840; Atlas historico de España, 1841; *Atlas universel historique et géographique*, Paris 1848-1849, Paris, Lebigre-Duquesne 1859.

**Houzeau**, Jean Charles. (1820-1888). Astronomer and geographer. Director at the Observatory at Brussels. *Uranométrie Générale* (with 5 star maps), 1878.

**Houzé de l'Avlnoit**, Auguste J.G. *Côte Occidentale d'Islande*, 1859; *Carte Géologique de l'Europe*, 1875.

**Hovart**, J.A. Reis Oost-Indië, 1789-1792 MS.

**Hove**, F.H. van. Engraver. *Bathe*, 1713.

**Hovell**, William Hilton (1786-1875). English sea captain, settled near Sydney in 1820s. Undertook many explorations including an expedition to southeastern Australia with

Hamilton Hume, 1824. *Mr Hume's sketch of a tour performed by W.H. Hovell and himself from Lake George to Port Phillip*, 1825. **Ref.** WALDMAN & WEXLER Who was who in world exploration (1992) p.328.

**Hoven**, F.H. van. Engraver. Frontispiece for A. Cellarius's *Atlas Coelestis*, 1660.

**Hoven**, Georg Christian von (1841-1924). German cartographer of Württemberg. Worked also in Stuttgart and Paris, and for the Swiss Topographical Bureau in Bern, 1896-1921.

**Hövinghoff**, J. (18th century). Danish engraver and globemaker.

**Hovius**, L.H. *fils*. Printer of 'Place Cathedrale', Saint-Malo. R. Bougard's *Le Petit Flambeau de la mer*, Saint-Malo 1785 (first published Le Havre 1684).
• **Hovius**, L. Re-published *Le Petit Flambeau...*, Saint-Malo 1817.

**How**, Jeremiah. Publisher and partner in How & Parsons.
• **How & Parsons**. Publishers of 132 Fleet Street, London. Mr. & Mrs. Samuel Carter Hall's *Ireland: its scenery, Character &c....* (3 volumes), 1841-1843 (with a map of Ireland engraved by A. Adlard).
• **How**, Jeremiah. Publisher of 209 Piccadilly, London. Published a 'New' edition of *Ireland*, 1846.

**How**, John. London printer. John Thornton's *English Pilot... Third Book*, 1703, 1711.

**Howard**, *Commander* Frederick. Admiralty surveyor. Undertook many surveys of the coast of South Australia, 1863-1873 published as Admiralty Charts.

**Howard**, Henry (*fl*.1837-*c*.1850). Surveyor of Winchester, Hampshire, produced manuscript maps of various parts of England. Stowe Estate, 1843 MS; Hamlets of Great & Little Pollicott, 1848; Chandos Estate, 1848.

**Howard**, Horton. Publisher of Columbus, Ohio. Alfred Kelley's *Topographical Map of the State of Ohio*, 1828 (engraved by William Woodruff).

**Howard** [Houard], John. Surveyor and publisher of Whitehaven. *A Plan of the Town & Harbour of Whitehaven in the County of Cumberland*, 1791 (engraved by J. Russell).

**Howard**, Nicholas. Essex estate surveyor. 40 acres at Dagenham, 1764 MS.

**Howard**. See **Hayward** & Howard.

**Howden**, J.A. *Howden & Odbert's Atlas of Warren County Pennsylvania*, Washington, Pa., J.A. Howden & A. Odbert 1878.

**Howe**, H. *Rhenish Atlas*, 1845.

**Howe**, Henry (1816-1893). *Historical Collection of the State of New York*, New York, S. Tuttle 1841, 1842, 1846 (with J.W. Barber); *Historical Collection of the State of New Jersey*, 1844, 1852, 1862 (with J.W. Barker); *Historical Collections of Virginia*, Charleston, S.C. Babcock & Co. 1845, 1846, 1849; *Historical Collections of Ohio*, Cincinnati 1847, 1852, 1869 and later; *Historical Collections of the Great West*, Cincinnati, H. Howe 1857, 1873; *Our Whole Country* (with maps), Cincinnati, H. Howe 1861 (with J.W. Barker).

**Howe**, Henry G. *Oil District of Pennsylvania*, 1866.

**Howe**, Samuel Gridley (1801-1876). Atlas of the United States printed for the use of the blind, Boston, N.E. Institution for the Education of the Blind 1837.

**Howe**, Thomas. Strait of Singapore, 1759; chart of the south coast of Madeira, 1762 (used by Dalrymple 1792).

**Howe & Spalding**. Publishers of New Haven, Connecticut. *Carta esferica que comprende los rios de la Plata, Parana, Uruguay y Grande y los terrenos adyacentes a ellos*, New Haven 1819 (drawn by M.L. Picor, engraved by A. Doolittle & N. & S.S. Jocelyn); Jedidiah Morse and Sidney Morse's *A New Universal Atlas of the World*, 1822.

**Howell**, Alfredo. *Plano... da Enseada do Quicembo*, 1888.

**Howell**, *Major* C.W. Produced civil war maps including: *Map of the battlefield of Spottsylvania C.H....*, Washington 1865; *Map of the battle fields of North Anna... 1864*,

1865 (lithographed by J. Bien); *Map of the battle fields of the Tolopotomoy, and Bethesda Church... 1864*, 1865 (lithographed by J. Bien); *Map of the battle fields of the Wilderness... 1864*, 1865 (lithographed by J. Bien); *Map of the country in the vicinity of Todds Tavern...1864*, 1865.

**Howell**, D.J. American civil engineer and landscape architect. One of the topographers for *Maryland - District of Columbia - Virginia*, U.S. Geological Survey 1886 (with J.D. Hoffmann); *Map of the property of Rosslynn Development Company, Alexandria Co. Virginia*, Washington D.C., D.J. Howell 1890 (lithographed by Bell Litho. Co.); *Map showing route of District of Columbia Suburban Railway: Sept.1892*, Washington D.C. 1892 (lithographed by Bell Litho. Co.); *Index map to Washington County plats, District of Columbia*, Washington, 1892 (lithographed by Bell Litho. Co.); *Massachusetts Avenue Heights, Washington D.C.*, Washington, Thos. J. Fisher & Co. 1911; *Oak View D.C.* Washington D.C. Thos. J. Fisher & Co. 1912.

**Howell**, George W. Joint compiler (with C.C. Vermeule) of *A Topographical Map of a Part of Northern New Jersey...*, Geological Survey of New Jersey 1882 (lithographed by J. Bien).

**Howell**, Henry H. *Geological Survey of Great Britain* (10 sheets), 1816-1864.

**Howell**, Reading (d.1827). Surveyor in Pennsylvania, produced the earliest maps of the whole state. In 1788 he was exploring and surveying the Delaware and Lehigh rivers, by 1790 he was receiving state funds to enable him to undertake a survey of Pennsylvania. *A Map of the State of Pennsylvania* (wall map, 4 plates), London, James Phillips 1792, Philadelphia, Emmor Kimber 1806, 1816, 1817, reduced edition 1811 (source for several maps of Pennsylvania, including one for *Carey's American Atlas*, 1795); *To the Legislature and the Governor... A Map of Pennsylvania, & Parts connected...*, c.1792, 1796 (engraved by J. Trenchard) and later editions; map of the United States, 1794 (no example known).
**Ref.** RISTOW, W.W. *American maps and mapmakers* (Detroit 1986) pp.108-110; see WHEAT & BRUN *Maps and charts published in America before 1800* (1978) for descriptions.

**Howitt**, Alfred William (1830-1908). Australian explorer and naturalist. *Howitt's track of second journey north of Cooper's Creek*, 1862.

**Howitt**, Samuel. Engraver of Buckingham Place. New South Wales, 1816; Chart of New South Wales & Van Diemen's Land. 1828.

**Howland**, C.W. *Atlas of Abington & Rockland, Mass.*, New York, Comstock & Cline 1874 (with W.A. Sherman).

**Howland**, H.G. *Atlas of Hardin County, Ohio*, 1879.

**Howlands**, W. American engraver. Western States, 1853.

**Howlett**, Bartholomew (1767-1827). Engraver from Louth, Lincolnshire. Views and map of Lincolnshire in William Miller's *A Selection of Views in the County of Lincoln...*, 1805.

**Howlett**, Samuel Burt (1794-1874). Surveyor, produced maps of sites in England, Ireland and Wales. Became chief draughtsman at the Ordnance Office. Drew, copied and endorsed numerous maps and plans. Stow Maries, 1818 MS; proposals for defence works in the dockyard at Milford Haven, 1829; routes of proposed railways in Ireland, 1836; Hurst Peninsula, Hampshire, 1842; land on Plumstead Marshes, 1855.

**Howlett & Brimmer**. Publishers of London. *The French Capital... During the Imperial reign of Buonaparte*, c.1810.

**Hoxton**, Walter. Ship's master and cartographer. *Mapp of the Bay of Chesepeack, with the Rivers Potomack, Potapsco North East and part of Chester*, published London 1735 (advised navigators of the north-east current now known as the Gulf Stream), republished W. Mount & T. Page c.1740.

**Hoyau**, Germain (c.1525-1583). Editor, engraver and printer of 'rue Montorgueil, au Chef Sainct Denys', Paris. *Icy est le vray pourtraict naturel de la ville, cité, université & Fauborgz de Paris...* (woodcut), 1553 (engraved by Olivier Truschet), facsimile published as: *Plan de Paris sous le règne de Henri II*, 1877.

**Hoyle**, John. Engraver. *New Mapp of the City of Norwich*, 1728.

**Hsi-ju.** See **Ricci**, Celestino.

**Hsu** Lun. Defence atlas of China, 1538.

**Huang** Peng-nien (1823-1891). Atlas of Hopei, 1884.

**Huang** Shang. Chinese astronomer. P'ing-chiang t'u [City map of Suchow], *c*.1193; Ti Li Thu [General map of China], *c*.1193 (thought to be the earliest printed map of China, later to be engraved on stone by Wang Chih-Yuan in 1247).

**Huang** Ts'ien-jen. World map with China at its centre, 1767.

**Huart**, Johannes Marinus (1809-1855). Engraver and lithographer of Bergen op Zoom, Netherlands. Worked for Philipp Franz von Siebold, 1840.

**Huart**, M.L. *Kaart van de Hoofdplaats Buitenzorg en Omstreken*, 1880.

**Hubault**, Gustave (*b*.1825). Atlas de géographie, 1873.

**Hubault**, Jacques. Bookseller and printer of Havre de Grâce [Le Havre], France. Produced the second and subsequent two editions of Réné Bougard's *Le Petit Flambeau de la Mer* (64 maps), 1690, (67 maps), 1691, 1694 (first published Jacques Gruchet 1684, publication rights had passed to Guillaume Gruchet by 1709).

**Hubbard**, Bela. American geologist and surveyor. Worked with S.W. Higgins and C.C. Douglass on surveys of Michigan counties including *Map of Calhoun County*; *Map of Washtenaw County*; *Map of Jackson County*; *Map of Lenawee County*, all engraved by W. Stone and published Washington city, 1844; *Geological map of a district of the Ontonagon...*, 1846 (with Higgins).

**Hubbard**, John. *Rudiments of Geography*, 1807.

**Hubbard**, Josiah (*fl*. early-17th century). Map maker of Hull. Credited with maps of Greenland, based on the surveys of James Hall.

**Hubbard**, William (1621-1704). Scholar and cleric from Essex, England, settled in New England *c*.1635, studied and taught at Harvard University. Best known for *A Narrative of the Troubles with the Indians in New-England*, Boston, John Foster 1677 (containing John Foster's woodcut map of New England entitled *A Map of New-England, being the first that was ever here cut...*, known as the 'White Hills' map; the London edition of *A Narrative...* included a similar map printed from a different block and known as the 'White Hills' map due to differences in spelling). **Ref.** WHEAT & BRUN *Maps and charts published in America before 1800* (1978) n°.144 p.29; ADAMS, R.G. *William Hubbard's 'Narrative', 1677: A Biographical Study* Bibliographical Society of America Papers 33 (1939) 25-39.

**Hubbard**, William (*fl*.1822-1861). Surveyor of Dartford, Kent. Produced manuscript maps of the eastern Home Counties. *The estate of Black Fenn Farm...*, 1822; Kevington, St. Mary Cray, Kent, 1825; property at Luddesdown, Kent, 1833; Dartford & Crayford Ship Canal, 1835; *Elmers End Farms situate in the parish of Beckenham*, 1836.

**Hübbe** [Huebbe], Heinrich (1803-1871). Cartographer. Prepared maps for Adolf Stieler, 1824-1828; *Jreland* [sic], 1824 (used in Stieler's *Hand-Atlas*, Gotha, Justus Perthes 1827, with title changed to *Ireland*); Nord-Americanische Freistaaten, 1824; Stieler's *Hand-Atlas*, 1834, 1840, 1843, 1848; *Grundriss von Hamburg's Häfen...*, Hamburg, 1839.

**Hübbe**, Joaquin. *Mapa de la peninsula de Yucatan*, [n.d.] (with Andres Aznar Perez), a revised edition by C. Hermann Berendt was published Paris, Régnier 1878.

**Hübbe**, S.G.S. See **Huebbe**, S.G.S.

**Hubbill**, *Captain* Nathaniel. *A rough draught of New Haven fort...*, 1781.

**Hüber**, Blasius. See **Hueber**, Blasius.

**Huber**, C. Alpine panoramas, 1869-1872.

**Huber**, Charles. *Journal d'un Voyage en Arabie 1883-1884... Avec Atlas*, Imprimerie Nationale 1891 (maps by J.A.A. Hansen).

**Hüber, D.** Austrian engraver. *Postkarte der Österreichischen Monarchie*, 1824; *Pontus Euximus*, 1832; *Krakowa*, 1833; *Wien's Umgebungen*, 1840.

**Huber, D.** Engraver. *Topographische Karte von Wiens Umgebungen auf 4 Meilen im Umkreise*, Vienna, Artaria & Co. *c.*1866 (drawn by F. Orlitsch). **N.B.** Compare with D. Hüber, above.

**Huber,** *Professor* **Daniel** (1768-1829). Swiss mathematician, astronomer, scholar, surveyor and librarian of Basle. Surveyor to the canton of Basle, 1813-1824 (surveys later used and published by F. Baader).

**Huber, E.** Publisher of Philadelphia. Issued Samuel Lewis' United States, 1816.

**Huber, Emil.** Selkirk Range, British Columbia, 1891.

**Huber, Gaspar.** Engraver of Kilkenny. Siege of Kilkenny, 1645.

**Huber, Hans Kaspar** (1566-1629). Collaborated with Johannes Murer and H. Gyger on *Geometrische Grundlegung der Landgraffsafft* [sic] *Thurgöv* (24 sheets), 1628-1629 (completed by A. Murer in 1671).

**Huber, J.** *Die Sardinische Monarchie*, 1855; *Polen mit den angrenzenden Ländern*, Nuremberg 1858; *Schul- und Reise Karte Deutschland*, 1867.

**Huber, Johann Heinrich** (1677-1712). Engraver of Zurich. Worked also in Vienna and Leipzig. *Toggenburg*, 1710; thought to have collaborated with Emanuel Schalch on Johann Jakob Scheuchzer's *Nova Helvetiae tabula Geographica* (16 sheets), 1712.

**Huber, Josef Daniel** (1730-1788). Austrian military cartographer. *Scenographie... Wienn* (24 sheets), 1769-1772, engraved 1777.

**Huber, Konrad.** Captain A.A. Humphreys's *Military Map of the Peninsula of Florida south of Tampa Bay*, 1856 (engraved on stone by K. Huber).

**Huber, R.** Commander of Artillery. *Empire Ottoman, division administrative* (4 sheets), F. Loeffler 1899; *Empire Ottoman, carte statistique des cultes chrétiens* (4 sheets), Baader & Gross *c.*1920.

**Huber, Thomas.** *Profil durch Deutschland*, 1887.

**Huber & Cie.** Publisher of Bern. *Plan der Stadt Bern / Plan de la Ville de Berne*, 1868 (drawn by G. Anselmier, engraved by R. Leuzinger); *Quartier und Strassen Plan der Stadt Bern*, 1884.

**Huber & Cie.** Publisher of Frauenfeld. *Karte der Walliser Mundart*, Frauenfeld 1913.

**Huberinus, Mauritz.** Globemaker of Nuremberg. *Globum Coelestis et Terrestris fabrica et usus*, 1615.

**Hubert, Henry** (1879-1941). French geologist, sent to survey Dahomey, French West Africa, later sent to Sudan. *La carte géologique du Dahomey*, 1908; *Le relief de la boucle du Niger*, 1911; *Mission scientifique au Soudan*, 1916; geological map of West Africa (5 sheets), 1917-1926. **Ref.** BROC, N. Dictionnaire illustré des explorateurs Français de la XIXe siècle: I Afrique (1988) pp.177-178 [in French].

**Hubert, R.** Engraver of Caen. Maps for Samuel Bochard's *Geographiae sacrae*, 1646.

**Huberti, Adrian** (*fl.*1605). Engraver of town views including Ville d'Amiens, Arras, Cales, Groningen &c., 1605.

**Hubertis, L.A.** See **Uberti,** Lus Antonio degli.

**Huberts,** *Doctor* **Wilhelmus Jacobus Arnoldus** (*b.*1829). *Nieuwe Geographische Atlas der geheele aarde*, Groningen & Arnhem, P. Noordhoff and J. Voltelen *c.*1870, Arnhem, J. Voltelen *c.*1877; *Historisch-Geographische Atlas der Algemeene en Vaderlandsche Geschiedenis*, Zwolle, W.E.J. Tjeenk Willink 1870, 1873, 1880, 1886; *Atlas der Oude Geschiedenis*, Zwolle, W.E.J. Tjeenk Willink 1885 (with Dr. E. Mehler, lithographed by W.B. Kuypers); *Atlas voor de Vaderlandsche Geschiedenis*, Zwolle, W.E.J. Tjeenk Willink 1888. **Ref.** For details of atlases see KOEMAN, C. Atlantes neerlandici (1967-1971) Vol.II pp.152-153; ibid. Vol.VI, pp.15, 182-183.

**Hubertz, J.R.** Medical map of Denmark showing the number of deranged persons per

1000 population, 1853 (published in *Annales Médico-psychologique*, 1854).

**Hubinger**, Fr. Alan (*fl.*19th century). Globemaker. 130-cm manuscript terrestrial globe, 1824

**Hübl**, Artúr von (1852-1932). Cartographer and engineer, born Nagyvárad [now Oradea, Transylvania, Romania]; commander of the Military Cartographic Institute at Vienna from 1916; set up and organised the Cartographic Institute in Rio de Janeiro (1921-1924). Early user of photogrammetric surveying techniques.

**Hubley**, *Lieutenant-Colonel* Adam (1744-1793). Pennsylvania Regiment. Manuscript sketch maps of army encampments, 1779-*c.*1781. **Ref.** GUTHORN P. *American maps and mapmakers of the revolution* (1966) p.23.

**Hübner**, Adolf. *Natal & Orange Fluss Orange-Freistaat*, Gotha 1871; *Süd-Afrika*, 1872; *Transvaal*, 1877.

**Hübner**, F. German engraver. Maps for C. Sohr's *Vollständiger Hand-Atlas*, Leipzig 1848.

**Hübner**, Johann. See **Huebner**, Johann.

**Hubner**, Martin. Professor of History at the University of Copenhagen. *An accurate Map of the Kingdom of Norway*, 1755 (in Erik Pontoppidan's *The Natural History of Norway*).

**Hübschmann**, —. Publisher in Munich. Lamont's *Magnetische Karten von Deutschland und Bayern*, 1854.

**Hübschmann** [Hueschmann], Donat (1540-1583). German engraver. J. Sambucus's *Vngariae Tanst. descriptio* (2 sheets), 1566.

**Hübschmann**, Gustav. *Mittel Europa*, 1869.

**Huby**, L. French engraver. *Carte des Environs de Saint-Omer*, St. Omer, Baclé 1815.

**Huc**, *abbé* Evariste-Regis (1813-1860). French missionary and explorer from Caylus, near Toulouse. Visited China, Mongolia and Tibet, 1839-1846. *Souvenirs d'un voyage dans la Tartarie et le Thibet*, Paris 1851. **Ref.** BROC, N. *Dictionnaire illustré des explorateurs Français de la XIXe siècle: II Asie* (1992) pp.247-248 [in French]; WALDMAN & WEXLER *Who was who in world exploration* (1992) pp.329-330.

**Huchet de Cintré**, H.M.F. *Plan du Port de Teavarua... 1859* [Society Islands], Paris 1862.

**Huddart**, *Captain* Joseph (1741-1816). East India Company Captain, hydrographer and civil engineer, Elder Brother of Trinity House, Fellow of the Royal Society. Several charts published by Sayer and Bennett, and Laurie & Whittle. The rivers Kennebeck and Sheepscut [Maine], *c.*1777, published 1783; *A Plan of Cape Bona Esperance*, 1778, Sayer & Bennett 1784, Laurie & Whittle 1794; *A chart of the west coast of Sumatra from old Bencoolen to Buffaloe Point...*, 1778; *A new hydrographical survey of the north and St George's Channels*, 1779; *A survey of the Tigris from Canton to the Island of Lankeet*, 1786; *A sketch of the Strait of Gaspar*, 1788; *A new hydrographical survey of the north coast of Ireland and the west coast of Scotland from Tory Island to Cape Wrath*, 1790; *The Coasting Pilot, for Great-Britain and Ireland...*, London 1794; **Ref.** HUDDART, W. *Unpathed waters: account of the life and times of Joseph Huddart FRS, at one time a Captain in the Service of the Honourable East India Company...* (Quiller Press, London 1989).

**Huddleston**, John. *Parish of Aldingham, Lancashire*, 1848.

**Hudjakov**, E. Russian publisher. J. Trescott's *Mappa Gubernii Irkutensis*, 1776; J. Islenieff's *Mappa Fluvii Irtisz...*, 1777; J. Islenieff's *Tabula Exhibens Cursum Fluvii Irtisch*, 1780.

**Hudson**, George (*fl.*1847-*c.*1850). Surveyor, of Woolwich. Dagnall Hill Farm, & Pedley Property, 19th century MS.

**Hudson**, Henry (*c.*1550-1611). English navigator in Dutch employ. Made four voyages seeking a short route to China by way of the Arctic Ocean. On his last voyage in 1610 entered the bay which bears his name. An account of his expedition, with charts, was published by Hessel Gerritsz. **Ref.** POWYS, L. *Henry Hudson* (1927); ASHER, G.M. *Henry Hudson the navigator* (1860).

*Henry Hudson (c.1550-1611) on his last voyage. From a painting by Hon J. Collier, engraved by W. Greatbach. (By courtesy of Valerie Scott)*

**Hudson**, John. *Geographia veteris scriptores Graeci minores*, 1698.

**Hudson**, John. Inclosure plan at Helpringham, Lincolnshire, 1774; inclosure award at Nettleham, Lincolnshire, 1777 (with William Thistlewood); Kyme Eau [Lincolnshire], 1792.

**Hudson**, John. *A complete guide to the Lakes* [English Lake District], 1843.

**Hudson & Goodwin**. American publishers of Hartford, Connecticut. Moses Warren and George Gillet's *Connecticut*, 1812.

**Hue**, Gustave. *Atlas de géographie militaire*, Paris 1879.

**Huebbe**, H. see **Hübbe**, Heinrich.

**Huebbe** [Hübbe], S.G.S. Australian Stock Routes, 1896.

**Hueber** [Hüber], Blasius (1735-1814). Tyrolean farmer and surveyor, worked as assistant to Peter Anich on surveys of the northern Tyrol. On the death of Anich he completed the surveys for the *Atlas Tyrolensis*. With A. Kirchebner as assistant he then undertook several surveys in western Austria. *Atlas Tyrolensis* (21 sheets), Vienna 1774 (begun by Peter Anich, surveys completed by Hueber 1766-1769), reduced edition (9 sheets), Paris 1808; surveys of the county of Nellenburg, 1780-1785 (with A. Kirchebner); *Provincia Landvogtiae Superioris et Inferioris Sueviae* (2 sheets), Vienna 1782 (with Kirchebner, surveyed 1775-1780); *Provincia Arlbergica* (2 sheets), Vienna 1783 (with Kirchebner, surveyed 1771-1774); *Karte über die Kais. Koenig. Vorderoesterreichische Grafschaft Nieder- und Oberhohenberg*, 1788 (with Kirchebner, surveyed 1786); *Karte über die Kais. Koenig. Vorderoesterreichische Reichsgefürstete Markgrafschaft Burgau*, 1793 (with Kirchebner, surveyed 1788-1793); *Neueste General-Karte von Tirol*, Vienna, Artaria & Co. 1806. **Ref.** FISCHER, H. 'Vermessungen und Kartierungen in Titol und in Vordösterreich, 1760 bis 1793' in *Cartographica Helvetica* 19 (1999) pp.37-45 [in German with summaries in French and English].

**Huebinger**, Melchior. *Map of the city of Lyons, Clinton Co., Iowa*, Davenport, Iowa, Huebinger c.1892; *Atlas of the State of Iowa*, Davenport, Iowa, Iowa Publishing Co. 1904; *Atlas of the town city of Peoria and environs...*, Davenport, Iowa, Iowa Publishing Co. 1909; *Map of the city of Burlington, Iowa*, Des Moines, Iowa Publishing Co. 1910; *Huebinger's map and guide for river to River Road*, Des Moines, Iowa, Iowa Publishing Co. c.1910; *Huebinger's map and guide for Omaha-Denver transcontinental route...*, Des Moines, Iowa Publishing Co. c.1911; *Huebinger's automobiles and good road atlas of Iowa*, Des Moines, Iowa Publishing Co. c.1912; *Huebinger's map and guide for Des Moines, Ft. Dodge, Spirit lake and Sious Falls highway*, Des Moines, The Iowa Publishing Co. c.1912; *Tri-county Farmers Institute atlas covering parts of Scott, Muscatine and Cedar Counties, Iowa*, Davenport, Iowa, Huebinger Publishing Co. c.1921.
• **Huebinger & Company**. *Plat book of Clinton County, Iowa*, Davenport Iowa, M. Huebinger & Co. 1894.
• **Huebinger Brothers**. Publishers of Davenport, Iowa. *Atlas of Johnson County Iowa...*, c.1900.
• **Huebinger Surveying and Map Publishing Company**. *Standard historical atlas of Jasper County, Iowa*, 1901.
• **Huebinger Publishing Co.** Publishers of Davenport, Iowa. M. Huebinger's *Tri-county Farmers Institute atlas covering parts of Scott, Muscatine and Cedar Counties, Iowa*, c.1921.

**Huebner**, Johann. Father and son.
• **Huebner** [Hübner], Johann (1668-1731). Historian and geographer of Hamburg. Worked with Johann Baptist Homann, and made contributions to several Homann atlases. *Kleiner Atlas Scholasticus*, c.1715; *Methodischer Atlas*, 1719; *Museum geographicum*, 1726; *Atlas Minor*, Homann c.1730; worked with the Ottens family, 1735; *Iohann Hübners bequemer Schul-Atlas*, Homann 1754.
•• **Huebner**, Johann. Son of Johann, above. Annotations to *Museum geographicum*, 1726.

**Huefer** [Hüfer], H. Novaya Zemlya, 1874.

**Huerne de Pommeuse**, —. Lac de Nicaragua, 1833.

**Hues,** Robert (1553-1632). English mathematician. Published a theoretical work on the use of globes entitled *Tractatus de Globis et eorum Usu...*, London, Thomas Dawson 1594 and later, Dutch edition translated by J. Hondius, published Amsterdam, Cornelis Claesz. 1597 (issued as a user guide for Hondius globes), French edition, Paris, D. Henrion 1618, English edition, London, J. Chilmead 1638 (based on the revised Dutch version published by J.I. Pontanus). **Ref.** Van der KROGT, P. *Globi Neerlandici* (1993) pp.219-229 *et passim*.

**Huet,** A. *Amsterdam*, 1874.

**Huet,** Luis. *Plano de la Villa de Pansacola*, 1781.

**Huet,** Pieter Daniel. *Carte de la situation du Paradis Terrestre...*, Amsterdam, P. Mortier 1705 (in *Atlas antiquus..*, Amsterdam 1705, re-issued Amsterdam, Covens & Mortier c.1730, map also used in N. Sanson's *Atlas Nouveau*, Amsterdam, Covens & Mortier c.1727 and P. de Hondt's *Description exacte de l'Univers ou l'ancienne Geographie*, The Hague 1741).

**Hüfer,** H. See **Huefer,** H.

**Hufnagel.** See **Hoefnagel,** family.

**Hufty,** S. American engraver of Philadelphia. For Fielding Lucas of Baltimore, 1823; for Matthew Carey, 1823.

**Hugeden,** R. See **Higden,** Ranulph.

**Hugel,** *Baron* Charles (1796-1870). Map Punjaub, 1846 MS, published John Arrowsmith 1847.

**Hughes,** —. See **MacAuley** & **Hughes.**

**Hughes,** Andrew. 'A draught of South Carolina and Georgia', *c.*1745 (in Thornton & Fisher's *The English Pilot. The Fourth Book*, London, Mount & Page 1764 and later).

**Hughes,** Edward. *The Hand Atlas for Bible Readers*, London, Varty & Owen *c.*1848; *An Introductory Atlas of Modern Geography for the use of Schools and Families*, London 1851; *An Atlas of the Bible Lands*, 1852; *A School Atlas of Physical, Political and Commercial Geography*, London 1853 (maps compiled and engraved by E. Weller); *A Tabular View of European Geography*, London, T. Varty 1854.

**Hughes,** *Colonel* G.W. Engineer. Survey of the Panama Railway showing settlements and soundings, Hydrographic Office 1849, (also published by Heinrich C. Kiepert 1856 and F. Ferrer, 1859).

**Hughes,** *Reverend* Griffith. Rector St. Lucy's, Barbados. *A Map of the Island of Barbados*, in Hughes's *Natural History of Barbados*, 1750.

**Hughes,** Hugh. Welsh artist. *Dame Venodotia, Alias Modryn Gwen* [cartoon map of North Wales], Carnarvon, H. Hughes *c.*1835 ('designed, drawn, engraved and published by H. Hughes, Printed by A. Miller, Liverpool'), Caernarvon, H. Humphreys *c.*1840 (drawn on stone by J.J. Dodd), Carnarvon, H. Humphreys *c.*1847 (lithographed by Maclure, Macdonald & Macgregor of Liverpool).

**Hughes,** Hugh. Engraver and publisher of 15 St Martins-le-Grand, London. *A New Map of London, Westminster, Southwark and their Suburbs for Coghlan's Picture of London*, 1847, 1848.

**Hughes,** J.H. Private surveyor in Tasmania. Hobart, 1837; Monmouth and Bucks Counties, Tasmania, 1837.

**Hughes,** James. *Town of Crawford, Orange County, New York*, 1863.

**Hughes,** John. *Baltic Sea*, 1854.

**Hughes,** John T. *Doniphan's expedition: conquest of N[ew] M[exico]* (5 maps), Cincinnati, J.A. & U.P. James 1848.

**Hughes,** Luigi (*b.*1836). Italian geographer. *Nuovo atlante geografico*, Rome 1889.

**Hughes,** M. & J. Surveyors. *Map from Palisades to Paterson* [New Jersey], 1867.

**Hughes,** Mathew. *City of Orange, New Jersey*, 1872; *Essex County, New Jersey*, 1874.

**Hughes**, Michael. *Plumstead Township, Pennsylvania*, 1859; *Farm Map of the Town of Wallkill, Orange City*, Philadelphia 1862; *Town of Newburgh, Orange County, New York*, 1864.

**Hughes**, *Lieutenant Colonel* Philip RE. *Plan of part of the Niagara Frontier*, 1814; *Plan of the Country round Fort Erie Shewing the Entrenchments &c thrown up by the Enemy in August 1814*, 1814.

**Hughes**, Price (d.1715). Welsh adventurer, coloniser and planter, travelled amongst the Indians. He was murdered in 1715. *A Map of the Country adjacent to the River Misisipi copy'd ... from the original Draught of Mr Hughes*, 1720 (copied by Alexander Spotswood, the lost original had been drawn by Hughes in 1713). **Ref.** CUMMING, W.P. The southeast in early maps (edition revised by L. De Vorsey, 1998) pp.22-23.

**Hughes**, R.B. Draughtsman for *Map of the Rivers Parana and Paraguay* (4 sheets), Liverpool, W. Forshaw Lith. 1842.

**Hughes**, Samuel (fl.1837-1846). Civil engineer, produced railway and tithe maps of various parts of England. *Essex Railway Survey*, 1835; Robert Tyas' *Geology of England & Wales*, 1841.

**Hughes**, T. Publisher of Stationers Court, London. B. Lambert's *Plan of the Cities of London and Westminster with the Borough of Southwark...*, 1806 (engraved by Neele, used in Lambert's *The history and survey of London and its environs*).

**Hughes**, Thomas. *Union Township, New Jersey*, 1860.

**Hughes**, W. Engraver of London. J.J. Forrester's *The Portuguese Douro and the Adjacent Country...*, 1848, 1852.

**Hughes**, W. Engraver of London. George Virtue's *The Environs of London*, c.1847, 1851; *Eslick's Patent Dissected Map of England & Wales* (jigsaw puzzle), mid-19th century; map of Cork for S.C. Hall's *Ireland* (3rd edition), London, Hall, Virtue & Co. c.1852. **N.B.** Compare with W. **Hughes**, above, and William **Hughes** (below).

**Hughes**, William (1818-1876). Geographer, teacher, cartographer and publisher of London. Noted at 9 Wharton Street, Pentonville, (1838); King's Head Court, Shoe Lane, (1839-1840); 3 King's Head Court, Gough Square, (1841-1843); Aldine Chambers, 13 Paternoster Row (1851). Teacher of geography and map drawing at St. John's College, Battersea (from 1840), cataloguer of geography books in the British Museum, 1841-1843, lecturer and later professor of Geography (from 1864) at King's College, London and other London colleges. Fellow of the Royal Geographical Society. Wrote several theoretical works on geography, as well as national and county geographies, he also traded as a map publisher. *The Illuminated Atlas of Scriptural Geography*, Charles Knight 1840; maps for Adam and Charles Black and for Edward Stanford, 1840-1853; map of Ascot race course for *The Sporting Review*, 1841 *Atlas of Constructive Geography*, 1841; map of Ireland for W.H. Bartlett's *The Scenery and Antiquities of Ireland*, George Virtue 1843; *Map of British and Roman Yorkshire*, 1847; *Australasia*, 1847; *Map of Jerusalem and the adjacent country*, 1848; revised *Moule's English Counties*, c.1848; *The emigrant's map of the world...*, 1850; maps for George Philip's *Imperial General Atlas*, 1853 (with John Bartholomew and Augustus Petermann); *General Atlas*, 1855; completed S. Maunder's *Treasury of Geography*, 1856; edited *Philip's Family atlas of physical, general and classical geography*, 1858; *The National Gazetteer of Great Britain And Ireland* (partwork), 1863-1868 (W. Hughes is credited as engraver), republished as *A New County Atlas Of Great Britain And Ireland...* (68 maps), London, Virtue c.1873 (maps used again in A.H. Keane's *A New Parliamentary and County Atlas of Great Britain And Ireland...*, London, J.S. Virtue & Co. Limited c.1886); *The new comprehensive atlas of modern geography*, 1870; *Philips' Industrial Map of the British Islands*, George Philip 1873; *Season-chart of the world*, 1875; and many others. **Ref.** CARROLL, R.A. The printed maps of Lincolnshire 1576-1900 (Lincoln 1996) pp.318-319; VAUGHAN, J.E. 'William Hughes 1818-1876' in FREEMAN, T.W. (Ed.) Geographers: biobibliographical studies Vol.9 (1985) pp.47-53.

**Hughson**, David (*pseud.*). The name under which Edward Pugh compiled his works about London. **N.B.** See **Pugh**, Edward.

**Hugo d'Aceso**, F. *Ville du Havre*, 1887; *Ville de Marseilles*, 1888; *Ville de Nantes*, 1888.

**Huguenin**, W.U. Military surveys of the provinces of Groningen, Friesland, Drenthe en Overijssel, 1820-1824.

**Huguetan**, brothers (*fl.*1687-1703). Worked in partnership as publishers and booksellers of Amsterdam. **Ref.** KOEMAN, C. *Atlantes Neerlandici* (1967-1971) Vol.II pp.117, 154; *ibid.* Vol.III pp.5-8; *ibid.* Vol.IV pp.423.
• **Huguetan**, Marc (1655-1702). Bookseller and publisher from Lyon, settled in Amsterdam in 1686. Associated with Pieter Mortier's counterfeit edition of Sanson's *L'Atlas Nouveau*, Amsterdam 1692, 1696 (copied from the French edition, some were published under the name Georges Gallet); also Jaillot's *Neptune François*, 1693 (with Pieter Mortier).
• **Huguetan**, Jean Henri.
• **Huguetan**, Pierre.
• [Pierre Mortier & Company]. Partnership of Pieter Mortier and Marc Huguetan. **N.B.** See **Mortier**, Pieter.

**Hugunin**, Robert. Draughtsman in the Clerk's Office of the District Court of the United States for the Northern District of New York. *Chart of Lake Erie*, 1843 (lithographed by Hall & Mooney).

**Huibnette**, L.F. Draughtsman. *Plan de la Ville de Metz, avec ses Projets*, 1804 MS.

**Huilier**, L'. See **L'Huilier**.

**Huilier**, P. Engraver in Paris for P. Duval's *Carte de l'Italie...*, 1663; P. Duval's *L'Amerique autrement le Nouveau Monde et Indes Occidentales*, 1664.

**Huissteen**, G. & A. van. Maps used in a composite atlas compiled by Willem Holtrop, 1780.

**Hulbert**, Ed.J. Topographical and mining engineer. *Geological and topographical map of the mineral district of Lake Superior, Michigan*, E. Warner & E.J. Hulbert 1855, 1864 (with J.C. Booth, engraved by H. Colton).

**Huldberg**, P.A. Publisher of Stockholm. C.A. Hoffman's *Res-Karta öfver Sverige*, 1856.

**Hulet**, John. Estate surveyor of Buckinghamshire, Kent and Essex. Parish maps of Essex, 1654 MS.

**Hulett**, James. Engraver. *Plan of Prague*, 1742; *City of Jerusalem*, 1744.

**Hulett**, W.E. *Every Stranger His Own Guide to Niagara Falls*, Buffalo 1844 (with maps of Niagara Strait and Niagara Falls).

**Hulett**, Willem. Engraver for Engelbert Kaempfer's *Histoire Naturelle, Civile, et Ecclesiastique de l'Empire du Japon*, 1729.

**Hulki**, Dabistanli Zekeriya. Relief map of the Ottoman Rumelia Belfur and its surroundings, 1894; relief map of Thessallia, 1898. **Ref.** Ülkekul, C. *Historical relief maps* (Istanbul Centre for Historical Research 1998) pp.60-61, 65-67.

**Hull**, Abijah. Surveyor. Plan of Detroit, 1807.

**Hull**, Edward (*b.*1829). Geological Survey Ireland, 1867; *Geological Map of Ireland*, 1878.

**Hull**, George L. *Morristown, Morris County, New Jersey*, 1874.

**Hull**, *Commander* Thomas A., *RN*. *Arctic Sea. Barrow Point and Point Moore*, surveyed 1852-1853, published as Admiralty Chart 1853 (engraved by J. & C. Walker); one of the surveyors for the Admiralty chart of the Ionian Sea Channels of Corfu, 1863-1864, published 1865; Atlantic Ocean Pilot, London, Admiralty 1868; Current Charts of the Pacific, 1872; and others.

**Hull**, W.H. Samangia or Keyser's Bay, 1818 MS; *Chart of Caloombyan Harbour*, 1818 (with W.H. Johnston), published J. Horsburgh 1819; Samangia Bay, 1819.

**Hullmandel**, Charles (1789-1850). Lithographic printer of many maps and views. *A Skeleton Map shewing the different divisions of the army...* 1817, 1818, 1819, & 1820 [India], *c.*1821; City of Bunarus, 1822; Vicinity St. Leonards, 1829; *Geography and Geology of Lake Huron*, 1824; *Plan of the Principal Settlements in Upper Canada*, 1823; also numerous illustrations and plates for travel books including a map of Natal for Captain Allen Gardiner's *Narrative of a*

*Journey to the Zoolu Country in South Africa*, 1836.

**Hullmandel & Walton.** Lithographers, London. California, 1849; South Essex Estuary Plans, 1852.

**Hülmer,** Johann. Member of Homann Erben firm [Homann's Heirs].

**Hulot,** *baron* Etienne. (*b*.1857). Secretary of the French Geographical Society. *Voyage de l'Atlantique au Pacifique*, 1888.

**Huls,** Joa. Plan of Jerusalem, *c*.1480.

**Hulsbergh,** Henry. Engraver. *A New Mapp of the Bay and Towne of Cadiz with all it's Fortifications... 15th. of August 1702...*, London, David Mortier, *c*.1703; Nicolas de Fer's *Théâtre de la guerre en Savoie*, 1703; *A Map of the World Corrected from the Observations communicated to the Royal Societys of London and Paris. By John Senex*, 1711; maps for Richard Blome's *England Exactly Described*, Thomas Taylor 1715 (later edition of *Speed's Maps Epitomis'd* of 1681); *Plan of Preston*, 1715; Thomas Taylor's *The Principality of Wales*, 1718; *Plan of Claremont*, 1720; *South America Corrected from the Observations communicated to the Royal Society's of London & Paris. By John Senex*, [n.d.]; *Asia Corrected from the Observations communicated to the Royal Society at London and the Royal Academy at Paris By John Senex FRS*, [n.d.] (from *c*.1740 the Senex maps were re-issued in *The English Atlas* by Mary Senex).

**Hulsius,** Friedrich, *junior* (*c*.1580-*c*.1660). Engraver and publisher in Frankfurt. Maps for Johann Ludwig Gottfried including *Inventarium Sveciae*, Frankfurt am Main 1632; *Alsatia Inferior* and *Wirtenberg Ducatus*, in Gottfried's *Archontologia Cosmica*, 1638 and later.

**Hulsius** [Hulst], Levinus [Lieven] (*c*.1546-1605). Publisher, geographer and instrument maker of Frankfurt am Main. Published a series of accounts of voyages under the overall title *Sammlung von sechs und zwanzig Schiffahrten in verscheidene Fremde Länder...* (26 parts, 18 of which were published posthumously), Frankfurt am Main or Nuremberg 1590-1650 [various editions]; *Erste Schiffart...*, 1598, 1650; *Totius orbis terrae*, 1598; translated Ortelius's Epitome into German as *Ausszug aus des A. Ortelii Theatro*, 1604.
• **Hulsius,** Levinus [Heirs]. After his death, publication of the collection of voyages begun by Levinus Hulsius was continued by his widow and family until *c*.1650.

**Hulst van Keulen.** See **Keulen**.

**Hultström,** F.A. Swedish surveyor. Ribbenås, 1855-1859 MS.

**Hulton,** J.G. Saudi Arabia & Yemen, 1836.

**Hulton,** Robert. Print and mapseller 'at the Corner of Pall Mall, St. James's', London. *A Pocket Map of the Cities of London & Westminster and the suburbs thereof...*, 1731 edition (engraved by Will Roades), later republished by P. Griffin.

**Humann,** Carl. *A Special Atlas, without a general title, of Asia Minor with statistical tables*, Vienna 1872; *Neuw Karte von Bulgarien*, 1877; *Syrien*, 1890.

**Humann,** Rudolph. French naval officer, surveyed the Mekong river basin. Map for *Exploration chez les Moï*, 1888-1889. **Ref.** BROC, N. *Dictionnaire illustré des explorateurs Français de la XIXe siècle: II Asie* (1992) pp.249-250 [in French].

**Humbert,** Abraham (1689-1761). Military cartographer.

**Humbert,** C.I. von. Prussian military cartographer. *Belagerung von Maynz*, 1793; *Plan von der Insel Potsdam*, Berlin 1800; *Plan der Stadt Hersfeld*, 1830; *Kurfürstenthum Hessen* (4 sheets), *c*.1830; and others.

**Humbert,** Pierre. Amsterdam bookseller and publisher. Dutch edition of Amédée François Frézier's voyage to the South Sea, Amsterdam 1717.

**Humbert,** Pierre, *junior*. City Surveyor. Plan of Boston, 1895.

**Humble,** father and son. Publishers and booksellers of London.
• **Humble,** George (*d*.1640). Publisher, book and print seller of The White Horse, Pope's Head Alley (1610-1627), and Pope's Head

*Alexander von Humboldt (1769-1859). Portrait by Charles Louis Bazin. (By courtesy of the National Library of Australia)*

Palace (1627-1640). Father of William Humble (below). Nephew and partner of John Sudbury, with whom he published John Speed's *Theatre of the Empire of Great Britaine*, 1612, 1614-1616, published further editions alone until 1631; pocket version of Speed's *Theatre* published as *England Wales and Ireland* (57 maps, mostly by Pieter van den Keere), 1619; *A Mapp of the Sommer Ilands once called the Bermudas...*, 1626 (engraved by Abraham Goos, used in *A Prospect...*, 1627); *England Wales Scotland and Ireland described...* (63 maps), 1627 and later (expanded edition of *England Wales and Ireland*); *A Prospect of the most famous parts of the World*, 1627 and later (Speed's companion volume to the *Theatre*, printed by John Dawson).

•• **Humble**, William (1612-1686). Son of George Humble, above. Publisher and bookseller of Pope's Head Palace. Published further editions of John Speed's works, from 1646 until 1654; plates sold to William Garrett c.1658, from whom they probably passed immediately to Roger Rea, Sr. & Jr.

**Humble**, Philip. Barcelona, c.1690 MS; Plan of the fortifications at Campredon [north Spain], c.1690 MS.

**Humboldt**, Alexander von [baron Friedrich Heinrich Alexander von] (1769-1859). Prussian natural scientist, astronomer, explorer and geographer. Invented system of isothermal lines for representing temperature distribution, 1817. Travelled throughout Europe 1796-1798, and the Americas 1799-1804 (with Aimé Bonpland). His massive contributions to science included the affects of altitude, geomagnetism, geology, meteorology and astronomical observations. Explored and charted the upper Orinoco and the Rio Negro, 1800; General map of Spanish America, 1803, published as *Carte Générale du Royaume de la Nouvelle Espagne*, 1811; *A Map of New Spain*, 1804, published 1811; *Voyage aux régions équinoxiales du Nouveau Continent* (35 volumes), Paris 1805-1834; *Atlas Géographique et Physique du Royaume de la Nouvelle Espagne...*, Paris, F. Schoell 1811; *Atlas géographique et physique des régions équinoxiales du Nouveau Continent*, Paris 1814-1834; *Carte des lignes Isothermes...*, 1817; *Atlas zu Humboldt's Kosmos*, Stuttgart, Bromme 1851-1853, French edition published as *Atlas du Cosmos*, Paris 1867. **Ref.** WALDMAN & WEXLER *Who was who in world exploration* (1992) pp.331-332; THROWER, N.J.W. *The Map Collector* 53 (1990) pp.30-35; KRETSCHMER, DÖRFLINGER & WAWRIK *Lexicon zur Geschichte der Kartographie* (1986) vol.C/1 pp.321-322 (many references cited) [in German]; *Cartographic Journal* 4 (1967) pp.119-123.

**Hume**, *Reverend* Abraham. *Liverpool ecclesiastical and social*, 1858; *Dr. Hume's Religious Map of England*, 1860.

**Hume**, Alexander. Two plans of Nagasaki, 1762 published Dalrymple 1792 (from an earlier Dutch manuscript).

**Hume**, Hamilton (1797-1873). Australian explorer, undertook several expeditions from 1814 onwards including a crossing of the southeast corner of Australia with William Hilton Hovell, 1824. *Mr Hume's sketch of a tour performed by W.H. Hovell and himself from Lake George to Port Phillip*, 1825. **Ref.** WALDMAN & WEXLER *Who was who in world exploration* (1992) p.333.

**Hume**, Robert. *Plan of Part of the British Settlement of Honduras between Rio Hondo and Sibun*, Belize 1858 (surveys used by Alfred Usher for his *Map of British Honduras*, 1888).

**Humelius**, Johannes (1518-1562). German mathematician, teacher and geodesist. Coloured manuscript maps of the forests of Saxony. In 1560 he was commissioned by Elector Augustus of Saxony to undertake a survey of the Saxon lands. The map was completed by his pupil, B. Scultetus in 1568.

**Humfrey**, John. Supposed author of the earliest separate map of Battle of Bunker Hill: *Sketch of the Action between the British Forces and American Provincials on the heights of the Peninsula of Charlestown* [Massachusetts], 1775 (engraved by William Faden and Thomas Jeffreys).

**Humfrey**, *Major* John Hambly. Quartermaster General's Department. *The Country between St. Sebastian & the French Frontier* [Pyrenees], James Wyld c.1823.

**Hummel**, Bernard Friedrich. *Handbuch der alten Erdbeschreibung*, Nuremberg 1785-1793.

*Hamilton Hume (1797-1873). (By courtesy of the National Library of Australia)*

**Humphrey**, Hannah. Publisher of 18 Old Bond Street, London; later 37 New Bond Street. James Gillray's *A new Map of England & France. The French Invasion or...* [comic map of England], 1793.

**Humphreys**, *Captain* Andrew Atkinson (1810-1883). Topographer and hydrographer, Chief of Engineers U.S. Army. Directed many military and geological surveys. *Map of the routes examined and surveyed for the Winchester and Potomac Rail Road, State of Virginia*, 1831, 1832; Surveys of Mississippi Delta, 1850-1851; *Military map of the Peninsula of Florida south of Tampa Bay*, 1856 (engraved on stone by Konrad Huber); Military Department of Oregon, 1858; *Report on the physics and hydraulics of the*

*Mississippi River...*, 1861 (with others); *Army of the Potomac*, 1869; *Geological Atlas*, 1876-1881; *From Gettysburg to the Rapidan. The Army of the Potomac, July 1863 to April 1864*, New York, C. Scribner's Sons 1883.

**Humphreys**, Clement. County surveyor. *Map of the Northern Portion of San Francisco County*, San Francisco, B.F. Butler's Lithography 1852.

**Humphreys**, Daniel. Publisher of Phildelphia. *A Plan of the Attack of Fort Sulivan the Key of Charles Town in South Carolina*, 1776.

**Humphreys**, Hugh. Publisher of Castle Sqaure, Caernarvon. Published *Dame Venodotia, alias Modryb Gwen; A Map of North Wales*, c.1840, c.1847 (devised by artist Hugh Hughes and drawn on stone by J.J. Dodd, first published by H. Hughs, c.1835).

**Humphreys**, William. Surveyor. Caecil [Cecil County, Maryland?] County from survey of the roads in 1792.

**Humphrys**, William (1794-1868). Engraver from Dublin, worked in Philadelphia. Engraved the illustration of Columbus which appeared on the title page of H.S. Tanner's *A New American Atlas...*, 1823 and editions to 1859.

**Humrich**, —. *Umgebung von Coblenz*, 1883.

**Hunaeus**, *Professor Doctor* Georg Christian Conrad (1802-1882). German geologist and mathematician. *Provinz Hannover*, 1864.

**Hunckel**, G. Lithographer of Bremen. *Hansestadt Bremen*, 1837; *Karte... der Weser*, 1846; *South Australia*, c.1850.

**Hundeshagen**, Bernhard (1784-1858). German architect, librarian and cartographer. *Stadt und Festung Maynz mit ihren Umgebungen*, 1815; *Bonn*, c.1840.

**Hüner**, —. One of the surveyors for *Karte von dem Herzogthum Oldenburg...*, 1782-1799, engraved 1804 (drawn by C.F. Mentz, engraved by G.H. Tischbein).

**Hünerwadel** [Huenerwadel], Gottlieb Heinrich (1769-1842). Military cartographer.

**Hunfalvy**, János (1820-1888). Hungarian geographer. President of the Geographical Society, Budapest. *Magyar kezi atlasz*, 1865.

**Hunger**, Johann Michael. Swiss cartographer from Rapperswil. Map of the region of Glarus, 1682.

**Hunnius**, Ado. Cartographer. *Map of Kansas*, 1870; *Missouri & New Mexico*, 1873.

**Hunsen**, J. *Empire de Djambi* [Sumatra], 1878.

**Hunt**, C.C. Explorer. *Western Australia*, Perth 1865.

**Hunt**, Charles. Printer of 36 Vesey Street, New York City; an address shared by F.W. Beers, his publishers, and the lithographer L.E. Neuman. Printers of F.W. Beers's *State Atlas of New Jersey...*, 1872. **N.B.** Compare with Charles **Hart**.

**Hunt**, F.W. George Woolworth Colton's *Historical Atlas*, 1860.

**Hunt**, Freeman [& Co.]. Publisher of New York. *The American Popular Atlas*, New York 1835; T.G. Bradford's *Atlas Designed to Illustrate the Abridgement of Universal Geography, Modern & Ancient...*, 1835 (also published in Boston by W.D. Ticknor, and in Philadelphia by Desilver, Thomas & Company).

**Hunt**, H.A. Commonwealth meteorologist. *Average rainfall map and isohyets of New South Wales*, Melbourne 1910 (drawn by W.B. Hicks).

**Hunt**, James (fl.c.1633-1674). East Anglian estate surveyor. Mannor of Welles [Norfolk], 1668 MS.

**Hunt**, John. *Draught of St George's Fort* [Phippsburg, Virginia], 1607.

**Hunt**, John P. *City of Philadelphia*, 1875.

**Hunt**, *Captain* Phineas. *Plan of the island of Carnicobar*, 1769, published Dalrymple 1792; *A new chart of the Andaman and Nicobar Islands*, 1799, published in Laurie & Whittle's *The Complete East India Pilot...*, 1800.

**Hunt,** Richard S. *Map of Texas* (large folding map), New York, Joseph Hutchins Colton 1839 (with Jesse F. Randel, published with a guide to Texas).

**Hunt,** Robert. *Memoir of the Geological Survey of Great Britain*, 1855-1882.

**Hunt,** Thomas Carew. *Chart of the Bay of Ponta Delagada Island of St. Michael* [Azores], 1840; *Plan of part of the Town of Ponta Delgada showing the position of the forts 1839* [Azores], 1840.

**Hunt,** Thomas Sterry (1826-1892). American geologist and chemist. *Sketch of the Geology of Canada*, 1856.

**Hunt,** Uriah. Philadelphia publisher. *Cabinet Atlas*, 1830.
• **Hunt,** Uriah & Son. Publishers of Thomas H. Burrowes's *State-Book of Pennsylvania*, 1846, 1847.

**Hunt & Eaton.** Publisher of New York. *The Columbian Atlas of the World We Live In*, 1893; Map of Texas, Oklahoma and Indian territory, c.1895.

**Hunt & Stevens.** *Hunt & Stevens' Map of the City of Sydney*, 1868.

**Hunte,** John. Pilot of Plymouth. South West Ireland, published Hessel Gerritsz. 1612.

**Hunter,** A. Engraver. Newcastle-Maryport Canal, 1795.

**Hunter,** B.J. Surveyor in New York state. Map of Montgomery County, New York, 1853; map of Oswego County, New York, 1854 (both with S. Geil).

**Hunter,** Benjamin. A Chart from the Island of Tenedos, 1816 MS.

**Hunter,** C.M. *Atlas of boroughs and towns in Philadelphia, Bucks and Montgomery Counties on line of North Pennsylvania R.R.*, Philadelphia, J.D. Scott 1886; *Atlas of the city of Williamsport, Penna.*, Philadelphia 1888.

**Hunter,** E. *Joliet, Illinois*, 1859.

**Hunter,** George. Surveyor General of the Carolinas. Map of the Cherokee settlements of the southern Appalachians, 1730 MS; Indian country in western Carolina, 1751 MS (copied from the work of George Haig).

**Hunter,** H.L. Manuscript maps and views of Corsica, 1794.

**Hunter,** *Reverend* Henry (d.1802). *The History of London and its Environs* (2 volumes with maps and plans of the city, the docks, the river, canals and the Home Counties), J. Stockdale 1811 (first issued as a partwork 1796-1799).

**Hunter,** James. *Plan of the Isle aux Noix at the North End of Lake Champlain*, 1780.

**Hunter,** John. Draughtsman and engraver. *A Plan of the City of Chester* (copper engraving), 1782, also in 2 sheet version, 1789.

**Hunter,** *Lieutenant* (later *Captain*) John, *RN* (1738-1821?). Scottish naval officer, Master of the *Eagle*, rose to become Vice Admiral in 1810. Made various voyages to Australia and was appointed Governor of New South Wales, (1795-1800); *A Sketch of the Navigation from Swan Pt. to the River Elk at the head of Chesapeak bay*, 1777; *Plan of the River Delaware and Philadelphia*, 1777 (later used in *Atlantic Neptune*); *New York, East River, part of Hudson River...*, 1777; New York Harbour, 1779 (later used by Des Barres in *Atlantic Neptune*); *Sketch of Sydney Cove*, 1788 published 1789; *Chart of the Coast between Botany Bay and Broken Bay...*, 1788; *Plan of Port Jackson New South Wales*, 1788, J. Stockdale 1789 (engraved by J. Reid & W. Harrison); Stewarts Island, 1791; Plan du Comté de Cumberland [NSW], 1802, published in *Neptune François*.

**Hunter,** Joseph. Stikine River [British Columbia / Alaska], 1877.

**Hunter,** M. *Plan and view of Asab* [Red Sea], 1783; *Plan of the road of Jasques on the coast of Persia*, 1783; *Plan & view of Gogo in the Gulph of Cambay*, 1784; *Plan and view of Moha* [Red Sea], 1784.

**Hunter,** *Captain* R.L. *Wabaay Harbour [Ceram]* [Moluccas], 1840.

**Hunter**, Thomas. Lithographer and printer of 716 Filbert St. Philadelphia. *Combination Atlas Map of St. Clair County Michigan*, 1876; atlas of Santa Clara County, California, 1876 (compiled by T.H. Thompson); atlas of Sonoma County, California, 1877 (compiled by T.H. Thompson); atlas of Alameda County, California, 1878 (compiled by T.H. Thompson); atlas of Solano County, California, 1878 (compiled by T.H. Thompson); *Plan of the City of Philadelphia*, 1881 (a reprint of a directory map of 1797).

**Hunter**, William. Publisher. 23 Hanover Street, Edinburgh. Plan of Edinburgh, 1836.

**Hunter**, *Sir* William Wilson (1840-1900). *Atlas of India*, 1894.

**Hunter & Beaumont**. Publishers of Frankfort, Kentucky. *Ohio Navigator*, 1798.

**Huntington** family. Engravers and publishers of Hartford, Connecticut.
• **Huntington**, Nathaniel Gilbert (1785-1848). *A System of Modern Geography: for schools, academies and families designed to answer the two-fold purpose of a correct guide to the student*, New York 1833, Hartford, E. Huntington & Co. 1833, Hartford, R. White 1835, Raleigh N.C., Turner & Hughes *c*.1836, Hartford, R. White and Hutchinson & Dwier 1836; *Huntington's School Atlas... to... Accompany the System of Modern Geography by Nathaniel G. Huntington A.M.*, New York 1833, also published in Hartford, Connecticut, Reed and Barber 1833, 1838.
• **Huntington**, Eleazor (1789-1852). Engraver and publisher of Hartford, Connecticut. Engraver (with A. Willard) of *United States*, for Frederick Butler's *A Modern Atlas*, Wethersfield, Connecticut, Deming & Francis 1825; *Map of the United States*, Hartford, 1826, 1831; publisher of N.G. Huntington's *A System of Modern Geography...*, Hartford 1833; publisher of A. Daggett's *Map of Connecticut*, 1836, 1837 editions (first published by Daggett *c*.1827).
• **Huntington**, F.J. [& Co.]. Publisher of Hartford (1832) and 174 Pearl Street, New York (1833-1838). S. Griswold Goodrich's *Atlas, Designed to illustrate the Malte-Brun School Geography*, Hartford 1832, New York 1836; *Huntington's School Atlas... to... Accompany the System of Modern Geography by Nathaniel G. Huntington A.M.*, New York 1833; engraver for E.H. Burritt's *Atlas, Designed to Illustrate the Geography of the Heavens*, 1835; *Middle with part of the Southern & Western States*, New York 1838 (for the ninth edition of an atlas by Nathaniel Huntington).

**Huntington & Savage**. Publishers of 216 Pearl Street, New York; later became George Savage. S.G. Goodrich's *A National Geography for Schools*, 1845, 1846, 1848; S.G. Goodrich's *A Comprehensive Geography and History*, Huntington & Savage, Mason & Law, 1850, later published by J.H. Colton 1855.

**Huntley**, William (*fl*.1830-*c*.1844). Estate and tithe surveyor of Essex and Kent. Lexdon Heath [Colchester], 1830 MS.

**Hunton**, J. Surveyor. *Williamsport, Shawnee County, Kansas*, 1857.

**Huntte**, John (*fl*.1572-*c*.1575). Essex estate surveyor. Sturmer Mere [Essex and Suffolk], *c*.1575 MS.

**Huot**, Jean Jacques Nicolas. Revised Conrad Malte-Brun's *Géographie universelle*, 1830, 1837; *Géographie du Moyen Age*, 1846 (with Malte-Brun); *Carte géologique de la Crimée... pour accompagner le voyage dans la Russie Méridionale par Mr Anatole de Demidoff*, Paris, Ernest Bourdin 1853 (engraved by Pierre Tardieu, printed by Chardon ainé et fils); *Karte der Krim* [Crimea], 1855.

**Hupel**, August Wilhelm (1737-1819). Clergyman and historian of Livonia. *Topographischen Nachrichten von Lief-und Ehstland*, Riga 1774.

**Hurd**, D.H. [& Co.]. *Town and City Atlas of the State of New Hampshire*, 1892; *Town and City Atlas of the State of Connecticut*, Boston 1893.

**Hurd**, Davis. Surveyor and engineer. *Map Exhibiting the Farmington, & Hampshire & Hampden Canals... Connecticut River to Canada*, New Haven, N. & S.S. Jocelyn 1828.

**Hurd**, Samuel Proudfoot (1793-1853). As Surveyor General he commissioned surveys and endorsed numerous maps and plans of Canada. *Plan of the Survey of the Northern Boundary of the Canada Company's Huron Tract in the London District*, 1834 (with R. Birdsall and W. Hawkins)

**Hurd,** Captain Thomas, RN (c.1757-1823). Successor to Alexander Dalrymple as Hydrographer to the Board of the Admiralty, 1808-1823, in charge of the Royal Navy's surveying service. Hurd also stimulated the chart publishing industry by extending the availability of Hydrographic Office charts to the mercantile marine. *A Plan of the Sea Coast to the Westward of Penobscot Bay in North America...*, 1775; *Coast of North America from Penobscot to St. Johns*, 1775; first exact survey of Bermuda, 1783-1797, chart published as *The Bermuda Islands*, 1827; survey of Brest and the Brittany coast, 1804; chart of Falmouth, 1806; *Charts of the English Channel* (volume I of the 'Channel Atlas'), London, Ballintyne & Byworth 1811; *Charts of the coasts of France, Spain & Portugal* (volume II of the 'Channel Atlas'), London, Ballintyne & Byworth 1811; Australia & Tasmania, 1814; published some charts by Captain W.H. Smyth, 1822 (engraved by J. Walker, republished in *The Hydrography of Sicily, Malta and the Adjacent Islands*, The Admiralty 1823); *Chart of the Island of Ceylon...*, 1822. **Ref.** FISHER, S. 'Captain Thomas Hurd's survey of the Bay of Brest during the blockade in the Napoleonic Wars', Mariner's Mirror 79 (1993) pp.293-304.

**Huré,** —, veuve de Sébastian. Publisher of 'Rue St Jacques à l'image St. Jérome, près St. Séverin', Paris. Louis de Hennepin's *Description de la Louisiane*, 1683.

**Hureau de Senarmont,** Henri. One of the compilers of *Carte géologique des environs de Paris*, 1865.

**Hurez,** J.F.J. Publisher of Cambrai. Plans of an attack on Antwerp, 1817.

**Hurford,** Mrs John. Noted at Altrincham, Cheshire. A compendious chart of Ancient History and Geography, 1830.

**Hurlbert,** J. Beaufort. *Physical Atlas with coloured maps of the Dominion of Canada*, Ottawa 1880; Manual of Biblical Geography, 1884.

**Hurlbert,** Jesse Lyman. *A Bible Atlas. A Manual of Biblical geography and history*, Chicago, Rand McNally 1928. **N.B.** Compare with J. Beaufort **Hurlbert**, above.

**Hurley,** R.C. *Tourist's Guide to Hong Kong*, 1896.

**Hurn,** Joris van. Plan of The Hague, 1616 (with Cornelis Bos, used by Braun and Hogenberg in *Civitates orbis terrarum*).

**Hurrell.** See **Sorrel** & Hurrell.

**Hurst.** See **Longmans**.

**Hurst,** G. *Town and Royal Harbour of Ramsgate*, 1822 (with R. Collard).

**Hurst,** John. Publisher of Wakefield. Christopher Greenwood's *Map of the County of York*, 1817 (with Robinson, Son & Holdsworth, etc.).

**Hurst,** W. Lithographer of 17 Fulham Place, Paddington and 48 Bedford Row, London. Map of parts of Durham and Yorkshire, showing the Stockton and Darlington, Wear Valley and Redcar, and proposed Middlesbrough and Guisborough, railways, 1851 (compiled by R.J. Semple).

**Hurter,** father and son. Cartographers of Memmingen, Germany.
• **Hurter,** Christoph (1576/1577-1635). Architect of the city of Memmingen. *Herae amnis* (3 sheets), 1619; *Alemaniae sive Sveviae superioris Chorographia* (4 sheets), 1625 (used by Willem Blaeu, 1634 and later).
•• **Hurter,** Johann Christoph (1613-after 1680). Notary and cartographer of Memmingen, son of Christoph. *Geographica provinciarum Sveviae descriptio. Schwaben in XXVIII. übereintreffenden Tabellen vorgestellt* (28 sheets), Augsburg 1679; *Memmingen* [plan], 1680 MS.

**Hurtig,** A. Plan of Prague (9 sheets), 1884.

**Hurus,** Pedro. Spanish translator of Saragossa, Spain. Bernhard von Breydenbach's *Peregrinatio in Terram Sanctam*, 1498.

**Hus,** —. Engraver. Plan of Hamburg, 1651.

**Husen,** Franciscus. Engraver. *Siciliae Regnum*

in the Mercator *Atlas portatif*, Amsterdam, Henri du Sauzet 1734.

**Huske**, John (*c*.1721-1773). *The Present State of North America*, London 1755 (with a map based on John Mitchell's, engraved by Thomas Kitchin).

**Husman**, Johann. Swedish engraver. *Samsoe*, 1675; *Scania*, 1677.

**Hussey**, T.J. *A Planisphere of the Stars*, *c*.1830 (with F. Dawson).

**Hussey**, William (*fl*.1805-1855). Manuscript maps of Gloucestershire, Buckinghamshire and Surrey. Land near Horsenden, 1806 MS; Tilford House & Farms, 1855 MS.

**Husson**, I.M. Bookseller of The Hague. *Nouveau Plan de la Ville de Luxembourg Avec Tous les Ouvrages que le General de Bauffe y a fait...*, 1743 (engraved by De St. Hilaire Mallet).

**Husson**, Pierre [Pieter] (1678-1733). Publisher and bookseller at the corner of the 'Kapel brugh', The Hague. Took over from Anna Beeck as publisher of battle plans when she left The Hague in 1717. *Le Theatre de la Guerre dans tout le monde*, 1706; *Les XVII Provinces*, 1706; *Les tablettes guerrières*, 1709 (for Daniel de La Feuille); *Variae tabulae geographicae...*, *c*.1710 (maps by many different makers); *Plan de la ville et citadelle de Dunkerque*, Schenk *c*.1713.

**Hustler**, John. *Plan of the Canal from Leeds to Liverpool*, 1788.

**Huszár**, Mátyás (1779-1843). Hungarian surveyor, cartographer and astronomer. Director of the survey of the lowland region of central Hungary from 1818; Director General of the Danube Survey Office, 1822-1829; survey and mapping of the area of the Körös rivers, from 1830. *Declinatio Teritorii Possesionis Oroszfalu*, 1805. **Ref.** KÁROLYI, Zs. 'Huszár Mátyás szerepe' [The role of Mátyás Huszár], in *Technikatörténeti Szemle* [Technological History Review] (1975) [in Hungarian].

**Hutawa**, brothers. Draughtsmen, lithographers and publishers of St. Louis, Missouri. Draughtsmen for *Plan of the City of St. Louis*, St. Louis, Charles Friederich & Company 1838, 1842, J. Hutawa 1851, N.L. Wayman 1853; draughtsmen for *Map of that part of the State of Missouri... called Platte Country*, St. Louis, E. Hutawa 1842 . **Ref.** RISTOW, W.W. *American maps and mapmakers* (Detroit 1986) pp.257, 451-452, 462.

• **Hutawa**, Edward. Lithographer and publisher of 'N<sup>o.</sup>7 S. Third Street between Market & Walnut', St. Louis. Publisher of E. & J. Hutawa's *Map of that part of the State of Missouri... called Platte Country*, 1842; *Oregon Territory*, 1843; *Sectional Map Of the State of Missouri...*, E. Hutawa 1844 (compiled by E. Hutawa, engraved by J. Hutawa); *Atlas of the County of St. Louis, Missouri*, 1847 (showing original land ownership patterns); *Mexico and California*, 1848.

• **Hutawa**, Julius. 'Lithographer, Map Publishing Office. N. Second Street 45, St. Louis, Missouri', established *c*.1835. Worked with Edward (above). J.C. Frémont & J.N. Nicollet's *Map of the City of St. Louis*,1846; *Map of Mexico, New Mexico, California & Oregon*, 1847, 1863 (in *Missouri Republican*); *Map and profile Sections Showing Railroads of the United States*, 1849 (drawn and lithographed by J. Hutawa); *Map of Chicago and Vicinity Compiled by Rees & Rucker, Land Agents*, 1849 (drawn by William Clogher); publisher of *Plan of the City of St. Louis*, 1851 edition (first published 1838); lithographer for I. Williamowicz's *Map of the Cairo and Fulton Railroad*, *c*.1853; compiled *Map of the United States Showing the Principal Steamboat Routes and Projected Railroads Connecting With St. Louis*, 1854 (in the *Missouri Republican*); C.T. Uhlmann's *Atlas of the County of St. Louis, Missouri*, 1862 (showing current land ownership patterns); *Vicinity of Mexico*, 1863.

**Hutcheon**, T.S. Civil engineer. *Plan of the Burgh of Elgin*, 1855.

**Hutchings**, James Mason (1820-1902). *Hutchings' tourist guide to the Yo Semite Valley and the big tree groves for the Spring and Summer of 1877*, San Francisco, A. Roman & Co. 1877.

**Hutchings**, W.F. Surveyor. *A Map of the County Palatine of Chester... 1828-1829*, London, H. Teesdale & Co. 1830 (with W. Swire, engraved by J. Dower); *A Map of the County of Staffordshire... 1831-1832*, London, H. Teesdale 1832 (with J. Phillips, engraved by J. Dower).

**Hutchings & Rosenfield.** Publishers of San Francisco. Overland mail routes, 1859; for De Groot, 1860; *Correct Map of the Gold Diggings on Fraser's and Thompson's Rivers*, c.1870.

**Hutchins**, John. *The History and Antiquities of the County of Dorset* (folding map, 59 plans & views), 1774, (184 engraved maps in 4 volumes), 1796-1815.

**Hutchins**, John Nathan (c.1700-1782). Almanack compiler in the 1750s-1770s. Almanack for 1759 (with a plan of Louisbourg, 1758).

**Hutchins**, *Captain* Thomas (1730-1789). Assistant engineer 60th Foot Regiment, served as geographer to the southern continental army during the Revolutionary War. Accompanied, and mapped, Colonel Henry Bouquet's march into the area west of the Ohio River in 1764. As Geographer of the United States (from 1785) he was in charge of surveying public lands and was director of the General Land Office. *A topographical plan of that part of the Indian-Country through which the army under the command of Colonel Bouquet marched in the year 1764*, 1765 (published in William Smith's *An Historical Account of the Expedition under the Command of Henry Bouquet Against the Ohio Indians*, Philadelphia 1765); *Plan of the Battle near Bushy Run*, 1765; survey for *A Plan of the Rapids in the River Ohio...*, 1766 MS (with Harry Gordon) published 1778; *A Topographical Description of Virginia, Pennsylvania, Maryland, and North Carolina*, London, J. Almon 1778 (with accompanying map, *A New Map of the Western Parts of Virginia, Maryland and North Carolina*, (engraved by J. Cheevers); *A Plan of Lakes Ponchartrain and Maurepas and the River Ibberville and also of the River Mississippi from its mouth to the River Yazou*, 1779; chart of the Coast of West Florida, 1781; surveys in Georgia, 1782; *Plat of the seven ranges of townships*, surveyed 1785-1786. **Ref.** RISTOW, W.W. *American maps and mapmakers* (Detroit 1986) pp.38-39, 73, 75; See WHEAT & BRUN *Maps and charts published in America before 1800* (1978) for descriptions; GUTHORN, P. *British maps of the American Revolution* (1972) pp.30-31.

**Hutchinson**, Ebenezer. Engraver of Hartford, Connecticut. Engraved two editions of J. Whitelaw's *A Correct Map of the State of Vermont...*, 1821, 1824 (first edition engraved by A. Doolittle, 1796).

**Hutchinson**, Thomas. Engraver and draughtsman. *Great Britain and Ireland with ye Judges Circuits*, 1747, 1789, 1805; engraver for Robert Heath's map of the Islands of Scilly, c.1748; *Geographia Magnae Britanniae...*, 1748, 1756 and later (sold by S. Birt, T. Osborne, D. Browne, J. Hodges, I. Osborne, A. Miller, J. Robinson). **Ref.** HODSON, D. *County atlases of the British Isles published after 1703* Vol.II (1989) Nos.205-206 pp.46-52.

**Hutchinson**, *Lieutenant* W.C. India, District Hoshiyarpoor, 1853 (2 miles to inch); and others.

**Hutchison**, *Lieutenant* (later *Commander*) John, RN. Surveyor. Port Jackson, 1857, 1859; North Australia, 1864; Port Adelaide, 1869; Ports on Gulf St Vincent, 1870; and others.

**Huthwaite**, Samuel. Estate surveyor. Annsby, Lincolnshire. 1748.

**Hutson**, George, *junior* (fl.1771-d.1794). Surveyor, mapped estates in various parts of England.

**Hutter**, Franz Xaver. Engraver of Augsburg. *Polen*, 1796; *L'Europe* (6 sheets), 1805; *Provinz Neuburg*, 1808; *Baden*, 1812.

**Hutter**, Josef (1790-1866). Engraver and publisher of Augsburg. *Lauingen*, 1822; *Augsburg*, c.1826.

**Hutter**, Otto. *Umgebung von Kempten*, 1867.

**Huttich**, Johann. Alsatian scholar. Translated Ptolemy's *Geographia* from the Greek, Strasbourg 1525; worked with Simon Grynaeus on *Novus Orbis Regionum*, 1532 and later editions, (the Basle, J. Herwagon 1532 edition contains *Typus Cosmographicus Universalis*, the world map attributed to Sebastian Münster and Hans Holbein the younger, the Paris edition of 1532 contains Oronce Fine's *Nova, Et Interga Universi Orbis Descriptio*, of 1531); material on Alsace used by Sebastian Münster in *Cosmographia*, 1544 etc..

**Hutton**, Charles. Publisher. *Plan of Newcastle* (2 sheets), 1770, 1772.

**Hutton**, *Doctor* Charles (d.1823). Professor of mathematics at Royal Military Academy, Woolwich. First to use contours in Britain for his map of Schiehallion, Scotland, 1777 (with Nevil Maskelyne).

**Hutton**, Christopher Clayton. Intelligence officer attached to MI9, the escape branch of British Military Intelligence set up in 1939. Liaised with John Bartholomew on the production of light and easily hidden escape maps, printed on silk, for use by the British armed forces. **Ref.** BOND, B. *The Map Collector* 22 (1983) pp.10-13; HUTTON, C.C. *Official secret* (London 1960).

**Hutton**, E.W. New Zealand geologist. Geological map of the Province of Otago, 1875.

**Hutton**, Isaac. Engraver of Albany, New York. Simeon De Witt's *A Plan of the City of Albany... MDCCXCIV*, 1794.

**Hutton**, James (1726-1797). Scottish geologist. *Theory of the Earth*, 1785. **Ref.** CRAIG, McINTYRE and WATERSTON *James Hutton's theory of the Earth: The last drawings* (Edinburgh 1978).

**Hutton**, N.H. *...El Paso & Ft. Yuma Wagon Road*, 1857-1858.

**Hutton**, William Rich. American surveyor. Assisted E.O.C. Ord with surveys for *Plan De La Ciudad de Los Angeles*, 1849 MS.

**Hutton & Corrie**. County of Dumfries Ordnance Survey, 1859.

**Huttula**, F. Engraver of *Caucasus cum adjacentibus regionibus*, 1833 (in C.G. Reichard's *Orbis terrarum*).

**Huusmann**, Johan (d.1711). Engraver of Copenhagen.

**Huvé**, Louis. Said to have worked with Claude Prévost on a plan of Chartres, engraved on wood by R. Rancurel as *Portraict ou plan de la ville de Chartres*, Paris, M. Sonnius 1575 (published in Belleforest's *La Cosmographie universelle*).

**Huvenne**, J. One of the draughtsmen and surveyors for Vandermaelen's *Carte Topographique de la Belgique... en 250 Feuilles*, c.1854 or 1857; *Carte topographique de Bruxelles*, 1858.

**Huyberts** [Huijberts], K. Engraver. Worked on Sanson's map of the River Meuse, used in a pirated edition of Jaillot's *Atlas Nouveau*, Amsterdam 1696 and de Fer's *Atlas Royal*, 1699.

**Huybrechts**, & C$^{ie}$. *Plan de la ville d'Anvers et ses environs*, Antwerp, R. Huybrechts & C$^{ie}$ 1899.

**Huyett**. See **Parker** & **Huyett**.

**Huygens**, Christiaan (1629-1695). Dutch mathematician, astronomer and physicist, Fellow of the Royal Society, member of the Académie Royale des Sciences. Worked with others at the Académie on the establishment of a standard meridian of longitude.

**Huygens**, *Captain* Henry, RE. Manuscript atlas for Clive of India, 1765-1766.

**Huyn de Réville**, *Captain* —. Plan topographique de l'Albanie, [n.d.] MS.

**Huys**, François. Engraver of *Nova Totius Terrarum Orbis Geographica Ac Hydrographica Tabula*, Amsterdam, Jodocus Hondius c.1625, François van den Hoeye 1630, C. de Jonge 1664, Van Keulen 1680s.

**Huys**, Pieter (c.1519-1581). Artist and engraver of Antwerp, for Christopher Plantin, 1568.

**Huyser**, C.J. de. *Kaart... van Noord Amerika*, 1772; plans of Arabia, in Carsten Niebuhr's *Description de l'Arabie*, 1774.

**Huyshe**, *Captain* George Lightfoot. Rifle Brigade. *Sketch of Road from Prince Arthur's Landing, Thunder Bay, L. Superior to Lake Shebandowan as traversed by the Red River Expeditionary Force* [Canada], 1870; *Map of the country between Amanca and Cape Coast Castle...* [Ghana], 1873 (with Captain Buller and R.C. Hart); Map in two parts showing route of British troops from Cape Coast Castle as far as Faisoowah, 1873, 1874.

**Huyssen van Kattendyke**, W.J.C. West Coast Kiusiu [Japan], 1860.

**Hyakuga** (18th century). Japanese geographer. Map of Yamashiro Province [Kyoto], 1778.

**Hyde**, Albert A. *Driving Chart of Hartford and Vicinity, 15 Miles Around*, 1884.

**Hyde**, E. Belcher. *Miniature Atlas of the Borough of Manhattan in One Volume*, 1912.

**Hyde**, G.W. *Sectional Map of Oregon and Washington*, New York 1856.

**Hyde**, J. Mapseller of London, 'under ye North Piazza of ye Royal Exchange'. One of the sellers of Charles Price's *A Correct Map shewing all the Towns, Villages, Roads, the Seats of Ye Nobility and Gentry with whatever else is remarkable within 30 Miles of London*, 1712.

**Hyde**, W. & Co. American publishers in Boston. Barnum Field's American School Geography, 1832.

**Hyde**, William. See **Hyde**, Lord & Duren.

**Hyde & Co.** Publishers of Brooklyn Borough, New York. *Map of the Borough of Brooklyn, City of New York*, 1901.

**Hyde, Lord & Duren**. Publishers of Portland, Maine. *Atlas to Warren's System of Geography...*, 1843.

**Hyett**, William (*fl*.1800-1821). Worked for the Ordnance Survey under William Mudge, 1815. Instructor in surveying, Royal Military College, Woolwich, 1824.

**Hygden**, Ranulf. See **Higden**, Ranulph.

**Hyginus**, Gromaticus (1st century BC). Roman land surveyor or *agrimensor*. Wrote a treatise on land surveying, included in the so-called *Corpus agrimensorum*. **Ref.** DILKE, O.A.W. in HARLEY & WOODWARD (Eds.) *The history of cartography* Vol.I (1987) pp.216-218; DILKE, O.A.W. *Greek and Roman maps* (Thames & Hudson 1985) pp.93-100.

**Hyginus**, Gaius Julius (*fl*.1st century AD). Librarian to Augustus. Wrote a Latin astronomical poem *Poeticon astronomicon* (used as a source for many later works on the constellations).

**Hygman**, Nicolas. Printer of Paris. *Le Grant Voyage de Iherusalem*, Paris, Françoys Regnault 1517 (with Oronce Fine's woodcut map of Jerusalem).

**Hylacomylus**, M. See **Waldseemüller**, Martin.

**Hyllested**, Charles, *junior*. *Cotton growing regions of the United States*, 1875.

**Hylton**, Edmond Scott. Engineer, surveyor and draughtsman. *A Plan of the Town and Harbour of Saint John's In Newfoundland with Fort William and parts adjacent*, 1750 MS (with J. Bramham); *A Plan of Aqua Fort Harbour scituate near Ferryland in Newfoundland*, 1752 MS; *A Plan of Cape Broil, Capeling Bay & Ferryland Harbour...*, 1752 MS.

**Hynes**, James E. Publisher in Toronto. *Conlin, Bonney & Co.'s New Map of Upper Canada or Canada West*, 1861.

**Hynmers**, Richard. Translated Willem Jansz. Blaeu's *De Groote Zee-spiegel* into English, published as *Sea Mirrour* (111 charts from the Dutch edition), from 1625.

**Hyōhyō** Sanjin. Japanese mapmaker. Map of Gokinai Shōran, Dokumujishoro 1841.

**Hyōshiya** Ishirōbei. Japanese publisher. *Eiri Edo Ō-Ezu* [Plan of Edo], 1684.

**Hyozo**, Kobo. See **Kabo** Hyozo.

**Hyrne**, Edward. *A New and Exact Plan of Cape Fear River from the Bar to Brunswick... 1749*, MSS, London, T. Jefferys 1753.

**Hyslop**, J.M. Vestiges of Assyria, 1855 (with James Felix Jones).

**I.** Often indicates letter **J** on early printed maps.

**I.** *Carte topographique d'Allemagne*, c.1780.

**I., A.F.** Constantinople, 1570; Jerusalem, 1559.

**I., B.** See **Jenichen**, Balthasar.

**I. E.S.** See **Stein**, I.E.

**I., G.H.** *Plan de la Ville... de Besançon*, 1788.

**I Hsing** (672-717). Chinese astronomer and cartographer.

**Iacubiska**, —. See **Jacubicska**, Stephan.

**Iago**, *Reverend* William. *Ecclesiastical Map of Cornwall*, 1877.

**Iakinth**, *Archimandrite* Nikita Yakovlevich Bichurin. Manuscript maps of Tartary, 1823-1825.

**Ibáñez e Ibáñez de Ibero**, *General* Carlos (1825-1891). Founder and Director General of the Geographical and Statistical Institute, Madrid. Directed the preparation of *Plano Parcelario de Madrid por el Instituto Geográfico y Estadístico...* (3 sheets), 1879 (engraved by F. Noriega and E. Hernandez); *Mapa Topográfico de España...* (12 sheets), 1875-1880 (engraved by P. Peñas).

**Iasolino**, Giulio. See **Jasolinus**, Julius.

**Iberville**, Pierre Le Moyne, *sieur* d' (1661-1706). French naval officer. Made several voyages of exploration to the Gulf of Mexico and Lower Mississippi between 1698 and 1702, and founded the first French settlements in Louisiana. Information collected by Iberville was used by Guillaume Delisle to locate the Mississippi Delta correctly on his maps, in manuscript at first on *Carte des environs du Missisipi*, 1701 and *Carte du Canada et du Mississipi*, 1702, then on *Carte du Mexique et de la Floride...*, 1703, and later printed maps. **Ref.** WALDMAN & WEXLER *Who was who in world exploration* (1992) pp.335-336; SCHWARTZ & TALIAFERRO *The Map Collector* 26 (1984) pp.2-6; McWILLIAMS, R.G. 'Iberville's Gulf Journals' (1981).

**Ibn Abd al-Hakam** [Abu al-Qasim Abd al-Rahman ibn Abdallah ibn Abd al-Hakam] (*d*.871). Wrote on the concept of a world landmass in shape of a bird.

**Ibn al-Arabi**, Muhyi al-Din Muhammad Ibn Ali (1165-1240). Arab mystic philosopher of Andalusia. *al-Futuhat al-Makkiyah* [Meccan openings], [n.d.]; *Insha al-Dawa'is* [Production of Spheres], [n.d.].

**Ibn al-Nadim**, Abu al-Faraj, Muhammad Ibn-Ishaq (*d.c*.995). Wrote a summary of Muslim culture *Kitab Fihrist* [Index of Arabic Books], 988 (translated into English by Columbia University Press, 1970).

**Ibn al-Wardi**, Abu Hafs Zayn ai-Din (1290-1349). Historian who lived in Aleppo.

*Mukhtasar fi Akhbar al-Bashar* [a chronicle]. **N.B.** Should not be confused with Ibn al-Wardi, Siraj al-Din (d.1457), see below.

**Ibn al-Wardi**, Siraj al-Din, Abu Hafs Umar (*d.*1457). Geographer. *Kharidat al-Ajaib wa Faridat al-gharaib* [The pearl of wonders and the precious gem of marvels], [n.d.].

**Ibn Batuta**, Abu Abdallah, Muhammad Tanji (1303-1377). Arab explorer and geographer from Tangier. Travelled extensively across the Middle East, Asia, southern Europe and northern Africa covering some 75,000 miles over some 28 years. *Tuhfat al-Nuzzar wa Gharaib al-Amsar* [Gift of the Observers and the oddities of the Regions], [n.d.]. **Ref.** WALDMAN & WEXLER *Who was who in world exploration* (1992) pp.336-337.

**Ibn Faqih**, Abu Bakr, Shahab al-Din Ahmad Ibn-Muhammad (9th-10th century). Geographer from Hamadan, Persia. *Kitab al-Buldan* [The book of the cities/regions], 902.

**Ibn Hawkal.** See **Ibn Hawqal**.

**Ibn Halal** [Hilal], Muhammad (*fl.*13th century). Arab astronomer. Celestial globe, 1275.

**Ibn Hawqal** [Ibn Hawkal], Abu al-Qasim Muhammad (10th century). Traveller and geographer of the Balkhi school of Islamic geography. Travelled extensively throughout the Islamic world. Produced an important work on Islamic geography *Surat al-Ard* [The configuration of the World], 988. **Ref.** HARLEY & WOODWARD (Eds.) *The history of cartography Vol.2 Book I Cartography in the traditional Islamic and South Asian societies* (1992) *passim.*.

**Ibn Husayn**, Siyd Ali (*d.*1562). Turkish astronomer, hydrographer and navigator. *Mohit* [Guide to the Indian Ocean], 1554.

**Ibn Khaldun**, Abu Zayd, Abd al-Rahman, Muhammad (1331-1405). Arab historian born in Tunis, died in Cairo. Travelled the Islamic Mediterranean and wrote a history which included a chapter on geography with references to Ptolemy and al-Idrisi entitled *Kitab al-'Abar* (with a world map), translated by F. Rosenthal as *The Muqaddimah: An Introduction to History*, 1958. **Ref.** HARLEY & WOODWARD (Eds.) *The history of cartography* Vol.2 Book I *Cartography in the traditional Islamic and South Asian societies* (1992) *passim.* pp.170-171.

**Ibn Khurdadbih**, Abu al-Qasim, Muhammad (*d.*912). Early Persian geographer. *Kitab al-Masalik wa Mamalik* [Book of routes and countries], 846 (based on Ptolemy's *Geographia*).

**Ibn Majid** [Ahmad ibn Majid] (*fl.*1460-1500). Produced portolan charts.

**Ibn Rustah**, Abu Ali Ahmad, Ibn-Umar (*fl.* early 10th century). *Kitab al-Alaq al-Nafisa* [On mathematics, geography and astronomy].

**Ibn Said al-Maghribi**, Abu al-Hasan Ali (1213-1286). Arab geographer, born in Spain, died in Tunis. His geographical works were used by later authors.

**Ibn Yunus**, Abu al-Hasan Ali Ibn-Abd al-Rahman (10th century). Astronomer. Produced a world map on silk, *c.*1008; *al-Zij al-kabir al-Hakimi* [Hakimite tables], [n.d.].

**Ibn Zarqalah** [Zarqello-al; Azarquiol] (*d.*1100). Astronomer of Toledo who designed a universal astrolabe.

**Ibrahim Ibn-Said** al-Sahli al-Wazzan (*fl.*11th century). From Valencia, Spain. With the help of his son Muhammad he built the earliest preserved Islamic celestial globe, 1080.

**Ibrahim Muteferrika** [Ibrahim Efendi Mutafarrikah] (1674-1745). Ottoman diplomat of Hungarian origin. Founder of the first Ottoman printing house in Istanbul. Map of Marmara Sea, 1719-1720; *Bahriye-i Bahr-i Siyah* [Map of the Black Sea], 1724-1725; *Memalik-i Iran* [Countries of Iran], 1729-1730; Katib Çelebi's *Tuhfetü 'l-kibar fi esfarü 'l-bihar* [Tribute to the Great Men of Sea] (5 maps and drawings), 1729; *Iklim-i Misr* [Country of Egypt], 1730; *Tarih-i Hind-i Garbi...* [History of Western Indies] (4 maps and diagrams), 1730; *Jahan-Numa* [panorama of the World] (52 maps and drawings), 1732.

**Ichikawa** Yoshikazu. Japanese *ukiyoe* artist and mapmaker. *Gokaikō Yokohama no Zu* [plan of Yokohama], published by Azumaya Shinkichi 1860-1861.

**Ide**, A.W. *Ide's New Map of Montana*, Helena 1890.

**Ide**, L.N.N. Publisher of '138½ Washington Street, Boston', Massachusetts. Joint publisher (with Matthew Dripps of New York) of I. Slatter and B. Callan's *Map of the City of Boston, Massts.* [wall map], 1852 (lithographed by F. Mayer).

**Ides**, Evert [Everard; Everardus] Isbrand [Ysbrants] (*c.*1660-1705). Diplomat, travelled to Siberia and China. His map *Multum emendavit* was used in *Nova tabula Imperii Russici*, 1704, English edition of the book containing this map, 1706.

**Idiaquez**, Eduardo. *Plano de un proyecto de camino de La Paz al Rio Caca*, 1886; *Zongo y Challana; memoria sobre el camino para ligar La Paz con el puerto Ballivian en el rio Kaka*, La Paz, Le Razon 1886 (with Carlos Bravo); *Mapa elementario de Bolivia*, La Paz 1894; *Provincia de Caupolican*, 1899; *Mapa de la República de Bolivia*, 1900.

al-**Idrisi**, Abu 'Abdallah, Muhammad Ibn Muhammad [al-Sharif] (1100-*c.*1150-1165). Arab geographer and cartographer, born in Ceuta, Morocco, educated in Cordoba and travelled the Iberian Peninsula, France, England, North Africa up to Asia-Minor. In 1138, he went to Palermo, Sicily at the request of Roger II, Norman king of Sicily, where he lived most of his lifetime. Towards the end of his life he returned to North Africa and probably died at Ceuta. His major geographical work entitled *Nuzhat al-Mushtaq fi Ikhtiraq al-Afaq* [The Pleasure of the Anxious traveller in exploring the World], 1154 (known as *Tabula Rogeriana* or *Kitabi-i-Rujir* [the Book of Roger], described a large planisphere on silver plate); *Rawdat al-Faraj wa Nuzhat-al muhaj* [Gardens of pleasure and recreation of the souls], [n.d.] (the existence of this work has been questioned). **Ref.** MAQBUL, A.S. 'Cartography of al-Sharif al-Idrisi' in HARLEY & WOODWARD (Eds.) *The history of cartography* Vol.2 Book I *Cartography in the traditional Islamic and South Asian societies* (1992) pp.156-174; AARNIO, P. 'Idrisin maailmankartta', in *Marhaba: Vuosikirja Suomalais-arabialainen yhdistys ry* [Yearbook of the Finnish-Arabic Society], (Helsinki 1982) pp.10-16 [summary in English]; AMUNDSEN, S. *The Geographical Magazine* 51 (1979) pp.353- 359.

**Ienefer**, *Captain.* See **Jenifer**, John.

**Ieremin**, —. See **Yeremin**, —.

**Ienichen**, Balthasar. See **Jenichen**, Balthasar.

**Iglehart**, N.P. [& Co.]. *New map of Chicago comprising the whole city*, 1858.

**Iglesias**, Miguel. *Carta Hidrografica del Valle de Mexico*, 1862.

**Ignaz**, George, *Freiherr* von Metzburg. See **Metzburg**, Georg Ignaz.

**Ignat'yev**, Stepan. Russian topographer. District maps in the province of Moscow, 1728-1729.

**Ihering**, *Doctor* Heinrich von. Maps of Brazil, 1887.

**Ihn**, Godfrid. Falsterbo Stad, 1751 MS.

**Ikeda** Toritei [Tori] (1788-1857). Japanese *ukiyoe* artist and mapmaker. *Shinano no kuni oezu* [Map of Sinano province], 1835; *Tango no kuni oezu* [Map of Tango province], [n.d.]; *Tempō Kaisei Shūchō Kyō Ezu* [Plan of Kyoto], published Takehara Kōbei 1841.

**Ikkandō** —. Japanese publisher. Sea chart of the routes connecting Osaka and the eight provinces (woodblock), 1863.

**Ikku** —. Mapmaker of Osaka. *Dainihon no zu* [Character map of Japan], *c.*1823].

**Ilacomilus**, M. See **Waldseemüller**, Martin.

**Ilens**, *Doctor* —. Plan de la guerre de Flandres, Strasbourg 1750.

**Iles**, John Alexander Burke. Island of Nevis, 1871.

**Iliff**, John W. Publishers of Chicago. Imperial Atlas, 1892.

**Iliffe & Son**. Map printers of Coventry and London. *Warwickshire*, 1893 (taken from an original of 1862, by John Bartholomew & Son); *The Way-About Series of Gazetteer Guides No.6. The Way About Sussex...*, 1896; *...The Way About Hertfordshire...*, 1896, 1898; and others.
• **Iliffe, Sons, & Sturmey, Ltd.** Publishers of '3, St. Bride's Street, E.C.', London. Revised edition of *Warwickshire*, 1899 (published in *...The Way About Warwickshire...*); *... The*

*Way About Hertfordshire...*, 1899; *Illustrated Guide to Lincolnshire*, 1900 (with a map by Gall & Inglis).

**Il'in**, A.A. See **Ilyin**, Alexander Afinogenovich.

**Ilive**, father and son. Printers of London.
• **Ilive**, T. Reeve Williams, John Thornton and Robert Morden's *A New Map of Virginia, Maryland, Pensilvania, New Jersey, Part of New York and Carolina*, London, 1698.
•• **Ilive**, Jacob. Son of T. Ilive. *A Plan of the Ward of Aldgate*, 1739 (engraved by R.W. Seale); *A Plan of the Ward of Aldersgate*, 1740 (engraved by R.W. Seale).

**Illes**, Stephan. Born in Bratislava, worked in Jerusalem. *Jerusalem aus der Vogelschau* [Bird's-eye view], Vienna 1873.

**Illis**, Thomas. *A description of the Land belonging unto Christ Church Canterbury called Walworth manour in the Parish of Newinton*, 1681, published by the London Topographical Society, 1932.

**Illman** family.
• **Illman**, Thomas. English engraver working in London *c*.1824; moved to Hudson Street, New York, where he founded the engraving firm of Illman & Pillbrow. *Kentucky and Tennessee*, 1834; *Louisiana*, New York 1834; engraved for David H. Burr, 1835 (maps published in *A New Universal Atlas*).
• **Illman & Pillbrow**. Partnership of Thomas Illman and Edward Pillbrow, working as engravers, printers and publishers at Boston, Massachusetts. *City of New-York*, New York, David H. Burr 1832; Oregon Territory, 1833; completed D.H. Burr's unfinished world altas which was published as *A New Universal Atlas*, New York City, W. Hall & Co. 1835 (maps re-used by Jeremiah Greenleaf for his atlas of the same title, 1840); Samuel Walker's North America, 1846.
• **Illman & Sons**. Publishers. Lithographic work for Captain S. Eastman's Indian tribes of the United States of America, 1853.
•• **Illman**, G. Son of Thomas.
•• **Illman**, H. Son of Thomas.
•• **Illman Bros**. (*fl*.1860-1861). Founded an engraving business in Philadelphia.

**Ilyin** [Iljin; Illyn], Alexander Afinogenovich (1832-1889). Russian cartographer and publisher 'on the corner of the Ekaterinovsky Prospect and masterskii Boulevard n°·43', St Petersburg. Published the journals *Universal Traveller*, from 1867, and *Nature and People*, from 1879. Also published many maps, atlases, wall maps and text books. After his death, the company was run by his two sons until 1917. Map of Finland, *c*.1865; map of eastern Europe, 1868 (by Nestor Terebenev); *Province de St Petersburg*, *c*.1870; Russian Empire, 1871; produced many railway, post and travel maps of Russia, 1871-1887; Kiev oblast', 1875; *Podrobnyj atlas vsech Castej sveta* [world atlas], 1876-1884; eastern Europe with Finland and Poland, 1884; plan of St. Petersburg, 1884 (drawn by N. Voznesensky); Asiatic Russia, 1880; School Atlas, 1888; road map Vistula, 1889.

**Imai** Hachikuro (1790-1862). Japanese surveyor, worked for the feudal lord of Matsumae (now Hokkaido). Set of maps of Hokkaido, 1841.

**Imamura** Yoshikage. Japanese cartographer. Route map to Kompira Shrine from land and sea (woodblock, colour map), published Awaya Bunzō 1778.

**Imbault**, A. *Plan de la Ville du Mans*, 1862 (with Desgranges).

**Imbert**, Anthony. French naval officer, marine painter and lithographic artist who settled in New York in 1824. Prepared illustrations, including maps, for Cadwallader D. Colden's *Memoir... of the completion of the New York Canals*, New York 1825 (claimed as the earliest successful use of lithography in North America).

**Imbert**, *Captain* Duca A. *Adriatisches Meer* (4 sheets), 1867-1875; *Brindisi Harbour*, 1875 [Admiralty].

**Imbert**, J. Leopold. *Carte des possessions Angloises Dans l'Amérique Septentrionale*, Paris 1777.

**Imbert des Mottellettes**, Charles. *Atlas syncronistique... Histoire moderne de l'Europe*, 1834.

**Imfeld**, [Franz] Xaver (1853-1909). Swiss engineer and cartographer, worked in Zurich, and for the Swiss Topographical Bureau in Bern. *Alpen Panorama*, 1878; *Mont Blanc*, 1896; *Relief-Karte der Centralschweiz*, 1898. **Ref.** IMHOF, E. *Die Alpen* 57, 3 (1981) pp.128-130 [in German].

**Imlay**, Gilbert. *A Topographical Description of the Western Territory of North America* (2 volumes, no maps), London 1792 and other editions including: New York, Samuel Campbell 1793 (the New York edition contains *A Plan of the Rapids of the Ohio*, after T. Hutchins, in volume 1, *A Map of the State of Kentucky...*, after J. Filson, in volume 2, and a map of the South East United States), second English edition (3 maps - Kentucky, after Filson, Rapids of the Ohio, after T. Hutchins, and a Map of the Western Part of the United States), Debrett 1793, third English edition (4 maps including a map of Tennessee), 1797 (maps engraved by Thomas Conder).

**Imle**, *Ober-Lieutenant* —. *Schul-Atlas*, 1862; *Neuester Atlas... der Erde*, 1868.

**Immanuel**, Friedrich (1857-after 1935). German military historian. Alpine maps, also Northern India, 1892-1894; *Insel Sachalin*, 1894.

**Imme**, *Lieutenant* F.M. Draughtsman in the Royal Prussian Invalid Corps. *Carta Generale della Sicilia*, 1800 (taken from S. von Schmettau's map of 1719-1721).

**Imray**, family and companies. **N.B.** See also **Blachford** & Imray. **Ref.** FISHER, S. *The makers of the blue back charts* (2001); FISHER, S. *The Map Collector* 31 (1985) pp.18-23; ROBINSON, A.H.W. *Marine cartography in Britain* (Leicester 1962) pp.114-126; WILSON, E. *The story of the 'Blueback Chart' published by Imray, Laurie, Norie & Wilson*, London 1937.

• **Imray**, James, *senior* (1803-1870). City of London stationer, initially based in Cheapside as a bookbinder and bookseller, and later in Budge Row. Joined chart publisher Michael Blachford in partnership as Blachford & Imray at 116 Minories, London, 1836-1846. Imray sold his stationery business and bought Blachford's share in the company in 1846, becoming sole proprietor of the chart publishing business as 'James Imray at 116 Minories' (1845-1850); 'N⁰·102 Minories, London' (1850-1854). Published loose charts, pilots, and sailing directions. They also sold shipping guides, manuals and nautical instruments. James Imray himself managed the chart publishing side of the business, which supplied high quality charts for all maritime routes including those to the East Indies and the Antipodes. James Gordon's *Lunar and Time Tables*, 1849; Captain Henry Wolsey Bayfield's *Gulf and River of St. Lawrence*, 1850, 1853 (Bayfield's original charts date from 1828); *Chart of the South Pacific Ocean...*, 1851; *General chart of the South Atlantic*, 1852; *A new chart of the Bay of Biscay*, 1852; *Chart of the Grecian Archipelago and Ionian Islands*, 1852 (from the surveys of W.H. Smyth, R. Copeland and T. Graves); *Chart of the west coast of South America*, 1853; *Chart of the English Channel*, 1854; *A chart of the east coast of North America*, 1855; *Eastern Passages to China and Japan*, 1866; and numerous others.

•• **Imray**, James & Son (*fl.*1854-1899). James Imray & Son, 102 Minories (1854-1870); also N⁰·89 Minories and N⁰·1 Postern Row (from 1860). Partnership of James Imray and James Frederick Imray. James Imray's interests in the company passed to his son on his death in 1870. When James Frederick died in 1891 the business was left in trust to his eight children and was managed by his cousin Alfred Imray, and Henry Jenkins. James Imray & Son Ltd was bought from the trustees by William and Charles Wilson, and James Cutbill and Herbert Parbury Imray.

•• **Imray**, James Frederick, *junior* (1829-1891). Fellow of the Royal Geographical Society, chart publisher and author of sailing directions and coastal directories (with W.H. Rosser). Worked with his father as a partner in James Imray & Son, managing the side of the business which handled books and nautical instruments, he took over management of the business in 1870. *The coast of the Cape Colony* (2 sheets), London 1854, 1869; *The Lights and Tides of the World*, 1866 and editions to 1957 (with W.H. Rosser); *South coast of America showing the navigation round Cape Horn*, 1871; *Western Pacific*, 1872; *North Pacific*, 1872; *South Pacific*, 1879; *South and east coasts of Australia*, 1879.

••• **Imray**, James Cutbill. Elder son of James Frederick Imray. Worked in the family business. Joint editor (with W.R. Kettle) of *The Pilot's Guide for the River Thames*, 1905.

••• **Imray**, Herbert Parbury. Son of James Frederick Imray. Involved in the family business.

•• **Imray**, Alfred. Cousin of James Frederick Imray, after whose death in 1891 he was

appointed joint manager (with Henry Jenkins) of James Imray & Son.
••• **James Imray & Son, Ltd. and Norie & Wilson** (fl.1899-1903). Chart and nautical book publishers of 156 Minories.
••• **Imray, Laurie, Norie & Wilson Ltd.** (fl. from 1904 to date). Publishers of nautical charts, originally at 156 Minories. Incorporated the Imray, Norie, Wilson, Laurie and Kettle family businesses, directed by Charles and William Wilson, James and Herbert Imray, and Daniel and William Kettle. Their work was ultimately superseded by that of the Admiralty, and the company largely turned to specialist charts for fishing and yachting.

**Inagaki** Koro. Japanese cartographer. *Sekai bankoku chikyu zu* [World map], 1708.

**Inberg**, Isak Johan (1835-1893). Finnish cartographer. Finland (4 sheets), 1876.

**Incelin**, C. See **Inselin**, Charles.

**Inciarte**, *Doctor Ingenieur* Felipe d'. Map of the coastal area of Venezuela and Guyana from the Orinoco to the Essequibo estuary, 1779, copied by the Ordnance Survey 1898.

**Indar** [Yndar], Ramon. Spanish military cartographer. *Nueva descripción geográfica del principado de Cataluña*, Barcelona 1824 (engraved by Estruc); *Mapa de Cataluña...*, 1859, 1860s.

**Indeisseff** [Indeisev], —. *Plan de Moscou*, 1852.

**Indicopleustes**, Cosmas. See **Cosmas** Indicopleustes.

**Indicott**, John. *A plan of Kennebeck River and the forts thereon, built by the forces raised for the defence of the eastern frontiers of the province of Massachusetts Bay...*, Boston 1754.

**Inga**, Athanasius. Credited with *West-Indische Spieghel*, Amsterdam, B. Janszoon & J.P. Wachter 1624 (maps engraved by A. Goos). **N.B.** According to P. Burden in The Mapping of North America (1996), the existence of Inga, supposedly a Peruvian from Cusco, is unlikely.

**Ingalls**, *Captain* Rufus, Quartermaster, United States Army. Salt Lake City to San Francisco, 1855; Routes to Fort Vancouver, 1858 MS.

**Inganni**, G. *Città di Piacenza*, 1855.

**Ingber**, O. *Umgebung von Frauensee*, 1899.

**Inger**, W.D.S. Printer, Frankfurt am Main. Pays Bas, 1784.

**Inghirami**, Giovanni (1779-1851). Astronomer. *Carta geometrica della Toscana*, 1830.

**Inglefield**, *Admiral* Edward Augustus *RN* (1820-1894). Commander of H.M.S. *Phoenix* on an expedition in search of Sir John Franklin's lost Arctic expedition, 1852-1854. *Chart Shewing the North West Passage...*, Hydrographic Office 1853 (drawn by W.H. Fawekner, 2 versions, one lithographed by J. Walker, the other by Day & Son); *Upernivik Harbour*, 1855; *Smith Sound... 1852*, 1875.

**Inglis**, family. **N.B.** See **Gall** [Inglis], family.
• **Inglis**, Robert (d.1887). Printer in partnership with his father-in-law James Gall from 1848.
•• **Inglis**, James Gall (d.1939). Son of Robert, also involved in Gall & Inglis.
••• **Inglis**, Robert Morton Gall. Son of James Gall Inglis. Worked with Gall & Inglis.
•• **Inglis**, Harry Robert Gall (d.1939). Son of Robert, also involved in Gall & Inglis.

**Inglish**, Robert (d.1695). *View of Chelsey Colledge by Kip*, 1694; Plan Chelsea Hospital, c.1695.

**Ingraham**, *Captain* Joseph. Master of the American brigantine *Hope* on a voyage to the Pacific and China, 1790-1792. North-West Coast of America, 1790-1792 MS; manuscript charts and views of the Islands of Hawaii, 1791; *Hope's Track from the Sandwich Islands towards China*, 1791 MS; Queen Charlotte Island, 1791; Vancouver Island, 1792.

**Ingraham**, Robert G. *Map of New Bedford & Fairhaven* [Massachusetts], New Bedford, Taber & Co. 1857.

**Ingram**, Herbert (1811-1860). Printer, bookseller and publisher from Boston [Lincolnshire]. Began in business in Nottingham with his brother-in-law Nathaniel

Cooke in 1834, he later worked from premises in Crane Court off Fleet Street, London. With engravers James and Henry Vizetelli he founded the *Illustrated London News* in 1842. The paper issued numerous engraved views, panoramas and maps of all parts of the World (including many drawn and engraved by John Dower). E.H. Hargraves's *Australia and its Goldfields*, 1855.

• **Ingram, Cooke & Co.** Publishers of London. Partnership of Herbert Ingram and Nathaniel Cooke. *The illustrated hand-book to London and its environs*, 1853 (includes maps by John James Dower).

**Ingram**, I. Engraver. *Planisphere... Celeste*, 1752. **N.B.** Compare with J. Ingram, below.

**Ingram**, John. Engraver. Nicolas Louis de La Caille's *Carte du Cap de Bonne Esperance et de ses Environs 1752*, 1755 (published in La Caille's *Diverses observations, astronomiques et physiques faites au Cap de Bonne Esperance...*); *Carte minéralogique de l'Election d'Estampes*, 1757.

**Ingram**, *Captain* Thomas Lewis. Customs officer in Ghana. *Plan of the island of Saint Mary in the River Gambia*, 1859 MS.

**Ingrey**, Charles. Publisher and lithographer of 310 Strand (1820); 131 Fleet Street (1839), London. Also worked with G.E. Madeley as Ingrey & Madeley. Lithographer for *Plan of the Principal Settlements of Upper Canada*, 1820 (drawn by G.E. Madeley); F. Hall's *Map and elevation of the Shubenacadie navigation from Halifax Harbour to the Basin of Mines* [Nova Scotia], c.1826; plan of the Chester and Birkenhead railway, 1830 (for George Stephenson, engineer); *Township of Goderich* [Canada], c.1830; *Township of Guelph* [Canada], c.1830; *Plan of Heligoland from a Map found on the Island...*, c.1830; plan of Stable Court between St.Martin's Lane and Adelaide Street [London], c.1830; *Labyrinthus Londinensis or the Equestrian Perplexed. A Puzzle*, 1830-1837 (with F. Waller, a geographical game based on the streets of London); maps of the Forest of Dean, 1834; *Plan of the exterior of the Forest showing parts proposed for cultivation* [Forest of Dean], 1835; A.F. Gardiner's *Counry of Natal*, 1836; five maps of townships in the Huron Tract of Canada, c.1839.

• **Ingrey & Madeley** (*fl.*1825-1828). Lithographers, of 310 Strand, London. Partnership of Charles Ingrey and G.E. Madeley. Ingrey and Madeley are also found trading independently from addresses in the Strand. *De Beauvoir Town* [Hackney], 1825; *Proposed improvements at Charing Cross and St Martins Lane*, c.1826; *Map of the Townships in the Province of Upper Canada*, 1827; *Map of part of the Province of Upper Canada*, c.1828. **N.B.** See also G.E. **Madeley**.

**Inks**, William C. *Plan of the Town of Franklin, Missouri*, 1857.

**Innes**, C.E. Civil engineer at Grand Rapids, Michigan. *Map Muskegon River*, c.1840 ('Surveyed, Drawn & Compiled Under the Direction of Wm. P. Innes by C.E. Innes & Tinkham', lithographed by E. Mendel).

**Innes**, F.W. Surveyor, Survey Office, Tasmania. Hobart, 1865.

**Innes**, James. Publisher of London. *Ancient cities of London and Westminster in the early part of the Reign of Queen Elizabeth*, 1849 (with N. Taperell, after George Vertue's *Civitas Londinium... MDLX*, 1737).

**Innes**, John C. City engineer. *Map of the City of Kingston / County of Frontenac / Canada West*, New York, Snyder Black & Sturn 1865.

**Innes**, Robert L. Civil engineer in Canada. *Plan of the County of Hastings*, Belleville, Mackenzie Bowell 1860 (lithographed by W.A. Little, published in *Directory for the County of Hastings*); *Plan of Consecon Village*, 1866 (copied from a plan by J.A. Haslett, 1853).

• **Innes & Macleod.** Civil engineers, Robert L. Innes and Henry Augustine Macleod. Township surveys in Hastings County, including Belleville, 1861; Madoc, 1864; Belleville, 1865 and 1866.

• **Innes & Simpson.** Civil engineers, Robert L. Innes and George Albert Simpson. Surveys in Belleville, Hastings County, 1867.

**Innes**, Robert S. *Map of Kalamazoo Co., Mich.*, 1870.

**Innes**, *Lieutenant* Thomas. Plan for Dock at Yarmouth Haven, 1759 MS; Caledonian Harbour, San-Blas-Kays, c.1760 MS.

**Innes,** William P. *Map Muskegon River*, c.1840 ('Surveyed, Drawn & Compiled Under the Direction of Wm. P. Innes by C.E. Innes & Tinkham', lithographed by E. Mendel).

**Innevelt,** Matheus Cornelis [van]. *Het Eiland van Dordrecht...*, 1641 MS.

**Innys,** William & John. Publishers and booksellers of London.
• **Innys,** William (d.1756). Bookseller and publisher of 'Princes Arms, St. Pauls Churchyard', London. Older brother of John (below). Pieter Kolb's *The present state of the Cape of Good Hope*, 1731, 1738; Emanuel Bowen's *A Complete System of Geography*, 1744, 1747; Bowen's *A Complete Atlas*, 1752 (used maps from *A Complete System...*).
• **Innys,** John (1695-1778). Younger brother of William (above). London bookseller and publisher, initially apprenticed to his brother; set up on his own in some other trade c.1730. Brought together a collection of maps entitled *A General System of Cosmography...*, c.1749 (now at Holkham Hall in Norfolk). **Ref.** GROVE & WALLIS The Map Collector 56 (1991) pp.12-21; LOWENTHAL & WALLIS The Map Collector 63 (1993) pp.10-11.

**Inō** Tadataka [Chukei] (1745-1818). Japanese astronomer, cartographer and official surveyor of the Shogunate; spent 16 years undertaking a survey of Japan, later completed and published by the Shogunate. *Dainihon Enkai Yochi Zenzu* [Set of maps of the coast of Japan], 1821 (first Japanese-produced map of Japan to have meridians and parallels after original surveys); *Nihonkoku chiri sokuryo-zu*, c.1821/1824; *Nishi Nihon* [Map of western Japan] (3 sheets), 1865. **Ref.** YAMASHITA Kazumasa *Japanese maps of the Edo Period* (Japan 1998) pp.27 and 56; WALLIS & ROBINSON *Cartographical innovations* (1987) entry N°4.072 p.173; ibid. entry N°4.122 p.183; SAITO T., SATO S. & MOROHASHI T. 'Meiji shoki sokuryo shi shiron: Ino Chukei kara kindai sokuryo no kakuritsu made', *Chizu* 17:2 (1979) pp.25-33; 16:2 (1978) pp.34-40; 16:1 (1978) pp.34-40; 15:3 (1977) pp.1-13; HOYANAGI Mutsumi *Ino Tadataka no Kagakuteko Gyoseki* (Tokyo 1974) [in Japanese].

**Inokhodtsev,** Piotr Borisovich (1742-1806). Russian astronomer, member of the St. Petersburg Academy of Sciences from 1779, and the Russian Academy from 1785. Geographical study of the Steppes in the basins of the Volga, Don and Shat, 1769-1775 (with Lovits).

**Inoue** Shi. Japanese publisher. Joint publisher (with Yoshidaya Binzaburō) of Miki Kōsai's map of Echigo Province, 1868-1869.

**Inoue** Sōbei. Produced manuscript maps of Japanese villages (15 sheets), 1836 (with Shōya Sadasaburō).

**Inselin** [Incelin; Inslin], Charles (fl.1680-1715). French engraver 'de Charleville' (c.1680); 'rue S.t Jacques' (1713). Engraved numerous maps which appeared in the publications of Nicolas de Fer 1693-1709 and later, also engraved for Duval, 1703-1704. *Plan de la Ville d'Orleans...*, Orleans, Herman c.1680-1706 and later; *Carte nouvelle curieuse du Royaume d'Espagne*, 1690; *Campredon ville forte de Catalogne...*, *Les environs de Palamos et de Girone...*, and *Barcelonne ville et port fameux d'Espagne...*, all published Nicolas de Fer 1692; engraver for François Froger, 1699; de Fer's *Atlas Curieux*, 1700-1705; engraved for père Placide de Sainte Hélène, 1703; N. de Fer's *Les Jonctions des deux grandes Rivieres de Loire et de Seine par le Nouveau Canal d'Orleans et celuy de Briare...*, 1705 and other editions; for Guillaume Delisle, 1707; *La ville d'Orleans avec ses Environs...*, Paris, Inselin 1713 (reduced edition of a map published c.1680), de Beaurain 1720; *Carte des Royaumes D'Angleterre D'Ecosse Et D'Irlande*, Jaillot 1715; *Carte Generale de L'Afrique Contenant les Principaux Etats...*, published Paris, Crépy 1735; and many others.

**Inskeep,** —. See **Bradford** & Inskeep.

**Inskip,** George H. Corrected *Vancouver Island... from a Russian chart*, 1856.

**Inskip,** R.M., *RN*. Admiralty Hydrographer. *Puget Sound*, 1846-1849; Port Simpson, 1856.

**Inslin,** C. See **Inselin,** Charles.

**Insprugger,** Sebastian *SJ*. Austrian Jesuit and historian. *Austria mappis geographicis distincta* (2 volumes), Vienna 1727-1728.

**Insulander** och Gillberg. Geometrisk karta öfwer Gästrickland, 1789.

**Insulanus.** See **Delisle,** Guillaume.

**Inverarity,** *Captain* David, *RN.* Hydrographer. *South Coast of China,* surveyed 1793, published 1801; Ceylon, 1800; *Chart of Mazambique harbour,* 1802, London, A. Dalrymple 1806; *Chart of the NW Coast of Madagascar,* 1802 published London, A. Dalrymple 1806; *Passandava... with the Bays of Marbacool and Chimpaykee on the N.W. Coast of Madagascar,* 1803, London, A. Dalrymple 1806; *Majambo Bay on the N.W. Coast of Madagascar,* 1803, London, A. Dalrymple 1806; *Delagoa Bay,* 1806; *Chart of Goa and Murmagoa Roads,* 1812, London, J. Horsburgh 1816; and others.

**Iona,** Benjamin Ben. World, 1160-1173.

**Iredell,** Abraham (1751-1806). Deputy Surveyor General, West District, Upper Canada. *The above is a plan of the Town of Chatham...,* 1795; *The above is a Plan of a Tract of Land called the Huron Reserve,* Detroit 1796; *The above is a plan of a Meridian Line, in the Township of Malden...* (2 sheets), 1797 MS; *Plan shewing the Site of the Military Post of Amherstburg,* 1797; and other township surveys.

**Ireland,** T. Lithographer of Montreal. Missouri to Walla Walla, Oregon Territory, 1851; P. Fleming's *Map shewing the Line of The Montreal, Ottawa, and Kingston Trunk Line Railway Company,* 1851.

**Iriarte,** H. [y Cia.]. Lithographic printers of Mexico City. Baja California, 1858.

**Irmédi-Molnár,** László (1895-1971). Hungarian professor and cartographic researcher, worked for the state cartographic institute and edited *Térképészeti Közlöny* [Cartographic Bulletin]. Organised the faculty of cartography at Eötvös Lóránd University, Budapest, of which he took the chair in 1951.

**Irmer,** Anton. *Plan der Stadt Zerbst,* 1880.

**Irminger,** *Captain* later *Admiral* C.L.C. (1802-1888). Danish admiral and hydrographer. Works on ocean currents, 1854-1855; including *Carte des Temperatures et des Courants... entre les Shetland et le Groenland,* 1855.

**Irrgang,** Oswald (1880-1902). Lithographer of Leipzig.

**Iruhae,** Carol. Orbis terrae antiquae, 1827.

**Irvine,** Alexander. Survey of the boundaries of Virginia & North Carolina, 1728 (with William Mayo).

**Irvine,** H. American surveyor. Connecticut town plans, both published by Richard Clark of Philadelphia. *Map of the Town of Guilford...,* 1852; *Map of the Town of Waterbury...,* 1852.

**Irving,** Benjamin Atkinson. *Atlas of Modern Geography,* London, Simpkin & Marshall 1862.

**Irving,** Roland Duer (1847-1888). *Atlas of the Geological Survey of Wisconsin* (75 sheets), Madison 1877-1802 (with T.C. Chamberlain and M. Strong); *The Copper-Bearing Rocks of Lake Superior,* 1885 (for the U.S. Geological Survey).

**Irving,** Washington (1783-1859). American writer, author of works about the life and voyages of Christopher Columbus (with maps), 1831, 1849 and other editions; *The Rocky Mountains...,* 1837 (with two maps by B.L.E. Bonneville).

**Irwin,** D.T. Plan of Esquimalt and Victoria Harbours, showing defences... [British Columbia], 1879.

**Irwin,** George. *Reference Map, Street List... Dublin,* 1853.

**Irwin,** *Lieutenant* James R. Draughtsman for sheet II of *Ohio Boundary,* 1835 (surveyed by W. Hood and R.E. Lee).

**Irwin,** R. and Samuel. *Forest County, Pennsylvania,* 1868.

**Irwin,** R.H. *Geological Map of Coolgardie* (4 sheets), 1899.

**Irwin,** S.M. *Venango County, Pennsylvania,* 1860.

**Irwin,** William. Chart of the River Kenmare, 1749-1751, used on *A New Map of the Kingdom of Ireland,* 1763 (in Bowen and Kitchin's *Large English Atlas*).

**Isaacs**, Abraham. Draughtsman of New York. *A Plan of the Harbour and City of Louisbourg* [Nova Scotia], *c.*1748.

**Isakov**, Leonty Borisovich (*d.*1745). Russian surveyor. Surveys for the Bakhmut and Tora salt factories, 1740-1742; boundary surveys, 1742; roads from Moscow to St. Petersburg, 1744-1745 (with Safanov).

**Isbister**, Alexander Kennedy. Expedition to Peel River [Canada], 1839-1841.

**Ischirkoff**, A. Language map of Bulgaria, 1915.

**Iselburg** [Isselburg, Yselburg], Peter (1568 or 1580-*c.*1630). Engraver and publisher of Nuremberg. *Maintz und Oppenheim*, 1620; *Heydelberg*, 1622.

**Isenburg**, —. *Plan von Düppel...1864*, 1878.

**Isengrin** [Isingrin], Michael (1500-1557). Swiss printer of Basle. Sebastian Münster's edition of Solinus's *Polyhistor* (with 20 woodcut maps), 1538, 1543 (with Heinrich Petri); published Sebastian Münster's pirated editions of Aegidius Tschudi's description and map of the Swiss Alps (wall map on 9 woodblocks), 1538 (German and Latin editions, map republished from the same blocks as *Nova Rhaetiae atq[ue] totius Helvetiae...*, 1560, with additions by Konrad Wolfhart).

**Isenmenger** [Eisenmenger; Siderocrates], Samuel (1534-1585). German historian and geographer. *Libellus geographicus*, Tübingen 1562.

**Isennachensis**, Jodocus. See **Trutvetter**, Jesse.

**Ise-No-Kimi** (*fl.*683 AD). First known Japanese surveyor. Provincial surveys (now lost).

**Iseya** Chūbei. Japanese publisher. Joint publisher (with Yoshidaya Bunzaburō) of *Bunken On-Edo Ō-Ezu* [plan of Edo], 1848.

**Ishiguro**, family. Japanese surveyors and cartographers of the late Edo Period. Five generations of the family worked as surveyors.
Ref. YAMASHITA Kazumasa *Japanese maps of the Edo Period* (Japan 1998) pp.28 and 225.
• **Ishiguro** Nobuyoshi (1760-1836). Arithmetician and land surveyor in Etchū Province. Also surveyed and produced maps of neighbouring provinces.
••• **Ishiguro** Nobuyuki. Grandson of Nobuyoshi. Manuscript route map of Tonami County, Etchū Province, 1839-1841.

**Ishikawa** Tomonobu [Ryūsen] (*fl.*1684-1715). Japanese *ukiyoe* artist and cartographer, worked in Edo [Tokyo]. Made woodcut and coloured maps of Japan to a style which persisted for many years. *Honchō Zukan Kōmoku* [Map of Japan], 1687; *Bankoku sokai zenzu* [European type world map], 1688; *Edo zukan komoku* [Plan of Edo with text], 1689, 1693, 1695; map of Japan, 1689 (later used by E. Kaempfer and J.C. Scheuchzer); *Hōei Tōto Zukan* [plan of Edo], 1705, 1706 (published by Sagamiya Tahei); *Dai Nihonkoku Ō-Ezu* [map of Japan], Yamaguchiya Gonbei 1728. Ref. YAMASHITA Kazumasa *Japanese maps of the Edo Period* (Japan 1998) pp.27, 48-49 and 98-99; WALLIS & ROBINSON *Cartographical innovations* (1987) entry N°6.081 p.250.

**Ishikawaya** Wasuke. Japanese publisher. Tomonari Shōkyoku's tourist plan of Osaka, [late Edo].

**Ishimura** Tei Ichi. *Dai Nihon seidzu* [Map of Japan], 1876.

**Ishizuka** Saiko (*d.*1780). Japanese scholar, with interests in Chinese and geography. *Enkyu bankoku chikai zenzu* [World map in stereographic projection], [n.d.].

**Isidore** of Seville [Isidorus Hispalensis] (*c.*560-636). Archbishop of Seville (602-636). *Etymologiae* or *Origines*, an encyclopedia with section on geography. T-O map, first printed edition Augsburg, Günther Zainer 1472 (regarded as the first western printed map). Ref. KRETSCHMER, DÖRFLINGER & WAWRIK *Lexicon zur Geschichte der Kartographie* (1986) Vol.C/1 pp.330-331 [in German]; STEVENS W.M. 'The figure of earth in Isidore's *De natura rerum*' in *Isis: Official Journal of the History of Science Society* 71 (1980), pp.268-277; CAMPBELL, T. *Earliest printed maps 1472-1500* (1997) pp.108-111.

**Isingrin**, Michael. See **Isengrin**, Michael.

**Isle**, De l' family. See **Delisle**.

**Islenyev** [Islenev; Islenieff; Isslenieff], Ivan [Johann] Ivanovich (1738-1784). Russian officer, astronomer and cartographer. Russia (31 sheets), 1753-1786; *Mappa Flumii Irtisz*, 1777; *Tobol'sk*, 1780; *Mappa Generalis Gubernii Asowiensis*, 1782.

**Isler**, John B. *Map of Caroline County, Maryland*, 1875.

**Ismael**, G. Engraver. *Cram's Rail Road, County and Township Map, of Ohio, Michigan, Indiana, and Kentucky*, 1870.

**Israel**, A.B. Treatise on the Use of Globes, St. Louis 1875.

**Israel**, *Captain* S. Damaraland & Namaqualand, 1885-1886 MS.

**Issel**, Arturo (*b*.1842). Italian naturalist and geographer. Schizzo geologico nel Finalere, 1885.

**Isselburg**, Peter. See **Iselburg**, Peter.

**Issleib**, —. Engraver of *Russische Ost-See-Provinzen Livland, Esthlund umd Kurland* in L. Ravenstein's revised edition of [Joseph] *Meyer's Hand-Atlas der Neuesten Erdbeschreibung*, Hildburghausen, Verlag des Bibliographischen Instituts 1872. **N.B.** Compare with the geographer Wilhelm **Issleib**, below.

**Issleib** [Ißleib], Wilhelm. German geographer and cartographer. Maps for Joseph Meyer's *Grosser Hand-Atlas*, 1850; *Volks Atlas über alle Theile der Erde*, Gera, Issleib & Rietschel 1868 (with E. Amthor); *Atlas... Staaten Deutschlands*, 1869; *Provinz Brandenburg*, *c*.1870; *Eisenbahnkarte von Central-Europa*, *c*.1870; *Atlas von Oesterreich-Ungarn*, 1871; *Toutes les Parties de la Terre*, 1873; *Repetitions-Atlas...*, 1875; and others.

**Isslenieff**. See **Islenyev**, Ivan.

**Istakhri**, Abu Ishaq, Ibrahim Ibn Muhammad al-Farsi [Karkhi] (*d*.957). Prominant Persian geographer and cartographer, represented the Balkhi school of Islamic geography. *al-Masalik wa al-Mamalik* [Routes and Countries], [n.d.]. **Ref.** ALA'I, Cyrus The Map Collector 60 (1992) pp.2-8.

**Istumi** [Yatarō] (1805-1882). Japanese mathematician and surveyor. Coastal survey around Edo [Tokyo] Bay, 1839.

**Isupov**, Mickael (*b*.1698). Trained at the Moscow Mathematical and Navigation School. Worked with Vassily Leushinsky on the mapping of Mazhaisk, Vereya, Borovsk and Vyazma, 1720.

**Itamiya** Zenbei. Japanese publisher. Joint publisher (with Kawachiya Tasuke) of colour printed woodblock plan: *Sesshū Ōsaka Zenzu* [plan of Osaka], 1844-1848.

**I'Toya** Zuiemon. World portolan, *c*.1590 MS (now lost, but later copied by Takami Senseki).

**Ittar**, Sebastiano. Malta, 1791.

**Iturriaga**, José (1699-1766). Maps of Venezuela, 1753-1762 MS.

**Ivachinzov**, Nicolas (1819-1871). Russian hydrographer and traveller. Caspian Sea, 1866-1869.

**Ivanov**, —. One of the calligraphy engravers for Pyadyshev's *Atlas géographique de l'Empire de Russie, du Royaume de Pologne et du Grand Duché de Finlande...*(83 sheets), 1820-1827, 1829, 1834.

**Ivanov**, —. Buchara and Afghanistan, 1884.

**Ivanov**, Ivan Nikiforovich (1784-1847). Russian hydrographer. Surveys of the White, Barents and Kara seas, 1821-1829.

**Ivashintsov**, Nikolai Alexandrovich (1819-1871). Russian hydrographer. Surveys of the Baltic Sea, 1848; commanded surveys of the Caspian Sea, 1853-1868.

**Ive**. see **Ivy**, Paul.

**Ivens**, Roberto (1850-1898). Spanish explorer in Africa. *Curso do Rio Zaire*, 1885; *Portugesisches Expedition durch Süd-Afrika*, 1887.

**Iverque**, *Baron* d'. See **Grante**, James A.

**Ives**, Edward. Maps of a voyage from England to India, and Persia to England, 1773.

**Ives**, Frederick E. Printer of Philadelphia. Perfected half-tone screen, 1881.

**Ives,** Joseph Christmas (1828-1868). Military cartographer and topographical engineer from New York City, served with the U.S. Army Corps of Topographical Engineers from 1853. Surveys for a proposed Pacific railroad, 1853; Florida, 1856; maps of the Colorado River in his *Report Upon the Colorado River of the West...1857 and 1858* (with maps and a navigational guide to the Colorado River), Washington 1861 (maps lithographed by Julius Bien). **Ref.** WALDMAN & WEXLER *Who was who in world exploration* (1992) pp.341-342.

**Ives,** William. *...Plats of Isle Royale, Michigan,* 1848

**Ivey.** See **Ivy,** Paul.

**Ivison,** —. Partner in a number of publishing companies of New York and Chicago. Addresses included 111 Lake Street, New York. Amongst other work, the Ivison companies re-published later editions of the works of the Coltons. *J.H. Colton's American School Quarto Geography with an atlas... drawn by G. Woolworth Colton,* 1865 edition.
• **Ivison, Phinney & Co.** (1857-c.1863). *Colton & Fitch's Modern School Geography,* Ivison & Phinney 1857, 1859, Ivison, Phinney & Co. 1860, 1863; Joseph C. Hart's *Geographical Exercises,* 1857; *Colton & Fitch's Introductory School Geography,* Ivison & Phinney 1859 edition, Ivison, Phinney & Co. 1863 edition.
• **Ivison, Phinney, Blakeman & Co.** (1867, 1868, 1871, 1887). *Colton & Fitch's Modern School Geography,* Ivison, Phinney, Blakeman & Co. 1867, 1868, 1871; *The Source of the Mississippi,* New York & Chicago, Ivison, Phinney, Blakeman & Co. 1887.
• **Ivison, Blakeman, Taylor & Co.** (1875-1887). William Swinton's *A Complete Course of Geography,* Ivison, Blakeman, Taylor & Co. 1875; William Swinton's *A Grammar School Geography,* New York & Chicago, Ivison, Blakeman, Taylor & Co. 1880; H.D. Harrison and W. Swinton's *A descriptive atlas of the United States,* New York & Chicago, 1884; Henry D. Harrower's *Handy Atlas of the World,* New York & Chicago, 1884; Harrower's *Captain Glazier and his lake* (with maps), 1886; and others.
• **Ivison, Blakeman & Co.** (1887-1888). Publisher of New York. H.D. Harrower & S. Mecutchen's *A pocket atlas of the world: with descriptive text, statistical tables, etc., etc.* (91 maps), 1887; H.D. Harrower's *Teacher's manual to accompany the Complete school charts,* 1888; H.D. Harrower's *The new states: a sketch of the history of the development of the States of North Dakota, South Dakota, Montana and Washington* (with map), New York, Ivison, Blakeman & Co. 1889.

**Ivory,** T. Engraver. *Scotland,* 1803 (in W. Chalmers's *The Gazetteer of Scotland,* Dundee 1803, 1806).

**Ivoy,** Colonel — d'. *Plan de la bataille ganiée sur la Plesne de Hochstette* [Blenheim] (2 sheets), The Hague 1704 (drawn by Jan van Cal); *Plan de la glorieux Battaille donné le 13 aout 1704 prés de Hochstett,* The Hague, Anna Beek 1704 (drawn by J. van Vianen); *Plan of the Lines of Brabant forced July 18, 1705 by the Army of ye Allies,* published London 1789; Tirlemont, 1705.

**Ivry,** Content d'. 'Architect du Roi'. *Plan... de l'Eglise de la Magdelaine* [Paris], 1761.

**Ivy** [Ive; Ivey; Ivye], Paul (*fl.c.*1570-d.1604). Surveyor and engineer specialising in fortifications and harbours. Coasts of Western Europe (attributed), *c.*1590; Younghall [Ireland], [n.d.] MS; Cork, *c.*1600 MS; Coasts of England, 1600 MS; coasts of France, England, Ireland and Low Countries, *c.*1600 MS; Kinsale Harbour [Ireland], 1601 MS; Cork Harbour, 1602.

**Izard,** John Grafton. *Plan of the Parishes of St Giles in the Fields and St George, Bloomsbury,* 1890.

**Izmailov** [Ismaylov], Gerasim Grigorovich (fl.1775-1786). Russian navigator based at Unalaska, 1770s, conducted explorations of the coasts of Kamchatka and Alaska. Met Captain James Cook, 1778, and provided him with charts; navigated Bering Strait, 1783-1786. Kamchatka, 1775; Kodiak and Afognak Islands (inset on Wilbrecht's map of Alaska), 1787. **Ref.** DAVID, A. *The Map Collector* 52 (1990) p.3.

**Izōsai** Ikkei. Japanese cartographer. *Kairiku Dōchū Zue* [bird's-eye view map of Yamagata, Niigata and Fukushima, Japan], published by Jinryūdō [late Edo].

**Izumiya** Hanbei. Japanese publisher. Doi Empei's map of the eight provinces of the Kantō area, 1849.

**Izumiya** Konjiō. Japanese publisher. Map of river conservation works in the Kantō area (woodblock), *c.*1743.

**Izumiya** Shōjirō. Japanese publisher of plans and tourist guides. *Kaisei On-Edo Ezu* [plan of Edo], 1826.

**Izumiya** Tōschichi. Japanese publisher. Suzuki Naokichi's plan of the shrines and temples of Edo, *c.*1784.

**Izumoji** Izuminojō. Japanese publisher. Coloured, woodblock plan of Honjo and Fukagawa Districts, Edo, 1761.

**Izumoji** Izuminojō. Japanese publisher. One of the publishers of Yana Nizaemon's *Zōho Shokoku Dōchū Tabisuzume* [travel map of Japan], 1830.

**Izumoji** Yūzaburō. Japanese publisher of the Edo Period. Joint publisher (with Yamamoto Heikichi) of a plan of the fire services in Edo.

*The interior of Nicolas De Fer's shop, at the sign of "La Sphere Royale" in Paris, 1705. (By courtesy of Jonathan Potter)*

**J.**, H. See **Joyce**, Henry.

**J.**, J. *Carte Générale du Danemark*, 1866.

**J.**, J.H. See **Huebner**, Johann.

**J.**, G.H. *Plan de la ville et citadelle de Besançon*, 1786, 1788, Paris, Jean c.1790.

**Jablonowski**, Aleksander (1829-1913). *Atlas Historyczny...* [Historical Atlas of Poland and the Ukraine], Cracow 1889-1904.

**Jablonowski**, *Prince* Joseph Alexander (1711-1777). Patron of Polish cartography.

**Jabotynsky**, Z. Hebrew cartographer in London. Eretz Israel [Land of Israel], London, Hebrew Publishing Company, 1925 (with S. Perlman).

**Jachnick**, *Lieutenant* —. Military surveyor. *Plan von der Feste... Koenigstein*, 1792; *Plan der Belagerung Maynz*, 1794.

**Jacinot**, Dominique. L'usage de l'astrolabe avec un traité de la sphère, Paris, Barbie 1545.

**Jack**, A. Lithographer of Strasbourg. Belagerung bei Kehl, 1870; Belagerung bei Strassburg, 1870.

**Jäck** [Jaeck], Carl (1763-1808). Draughtsman, copper engraver and publisher of Berlin, worked with D.F. Sotzmann and S. Schropp. *Karte vom Fürstenthum Halberstadt, den Grafschaften Wernigerode u. Hohenstein und der Abtey Quedlinburg bearbeitet...*, Berlin, S. Schropp 1788 (drawn by Rosenberg), revised by C.L. von Oesfeld, 1794; *Deutschland* (16 sheets), 1792; *Ost-Preussen* (25 sheets), 1796-1810; Schlacht bey Pirmasens, 1797; *General Karte von dem Königl: Preussischen Staaten*, Berlin 1799; *Insel Potsdam* (4 sheets), 1800; *Topographische Karte... von Westphalen...* (22 sheets), 1805-1815; *Neueste Postkarte von Deutschland*, c.1806 and later (title also in French); *Karte von Ost-Preussen nebst Preussisch Litthauen und West-Preussen nebst den Netzdistrict*, c.1807; *Champ de Bataille d'Eylan* [1807], Berlin 1810; and others.

**Jack**, *Major* Evan Maclean, RE (1873-1951). Served in the Royal Engineers from 1893 in Gibraltar and St. Helena until 1903 when he joined the Ordnance Survey. He was twice sent to work on boundary commissions in Central Africa. General Staff Officer in the Geographical Section of the War Office from 1913, and officer in charge of the survey section of the British Expeditionary Force, 1914-1918. During the war this section developed into a series of Field Survey Companies comprising 5000 officers and men by 1918. They were responsible for spotting, ranging, surveying, draughting and printing numerous trench maps in the field. He served as Director General of the Ordnance Survey, 1922-1930. **Ref.** CHASSEAUD, P. The Map Collector 51 (1990) pp.24-32.

**Jack**, Robert Logan (1845-1921). Government geologist and explorer in Australia. *Report on the geology and mineral* *resources of the district between Charters Towers goldfields and the Coast* (92 maps), Brisbane, Government printer 1879; *Geological Map of Queensland* (2 sheets), Brisbane 1886, 1892; *Geological sketch maps to accompany the second report on the tin mines near Cooktown*, Brisbane 1890 (engraved by W. Knight); *Map of Chillagoe... Mining District*, 1898; Geological map of Charters Towers, 1898, 1899; *Map showing the Palmer and Normanby Rivers...*, 1898; Part of Gympie Goldfield, 1899; Queensland (6 sheets), 1899.

**Jack**, Thomas C. Plan of Edinburgh, 1836.

**Jackson**, *Captain* —. Plan of Lake George, 1756 (used as an inset on William Brasier's *A Survey of Lake Champlain... 1762*, Sayer & Bennett 1776).

**Jackson**, —. Engraver of N⁰·19 Abbey St., Dublin. Late edition of John Rocque's Dublin, 1780. **N.B.** Compare with Zachariah **Jackson**.

**Jackson**, *Lieutenant* —. Survey from Midnapur to Nagpur, 1818; *Plan of the town of Singapore*, 1828.

**Jackson**, —. *New Map of Oil Creek Township... Pennsylvania, c.1870.*

**Jackson**, *Lieutenant* —, *RN*. Honolulu Harbour, 1881.

**Jackson**, —. *Map of the Badsworth Hunt*, c.1892.

**Jackson**, *Lieutenant Commander* A.L. Survey of the west coast of Gozo Island [Malta], 1917.

**Jackson**, Charles T. Geologist in the United States, prepared a report for Congress on the mineral reserves of Michigan. *Geological Map of Keweenaw Point, Lake Superior, c.*1847; *Geological Map of Isle Royale Lake Superior* [Michigan], New York 1849 (used by John Farmer as an inset on his map of Michigan and Wisconsin, 1853); together with maps by others, the above were published in Jackson's *Report on the geographical and mineralogical survey of the mineral lands of the United States in the State of Michigan*, Washington 1850.

**Jackson**, Clements Frederick Vivian. *Geological sketch map of Leonora, Mt. Maragaret* (with accompanying plans and sections), Perth, Geological Survey of Western Australia 1904.

**Jackson**, Frederick. *Nottingham, c.1892.*

**Jackson**, Frederick George (1860-1938). Leader of the Jackson-Harmsworth Polar Expedition, 1895-1897. Arctic Regions, Philadelphia 1897; *Franz Josef Land*, Royal Geographical Society 1897.

**Jackson**, George (*fl.*1761-*c.*1789). Surveyor of Richmond, Yorkshire, produced manuscript maps in Yorkshire. *A Plan of the... Mannors of Mortham, Rookby...*, 1765; *His Majesty's lead mines* [Grinton, Yorkshire], 1768; *The manor or lordship of Hooton*, 1782; *Crown estate at Sutton Grange and Greenthwaite Grange*, 1784.

**Jackson**, Henry (*fl.c.*1872). Chief surveyor of Wellington province, New Zealand.

**Jackson**, *Captain* H.M. *A Particular Plan of Lake George surveyed in 1756*, London, Robt. Sayer & Jno. Bennett 1776 (used in William Faden's *American Atlas*).

**Jackson**, *Captain* [later *Lieutenant-Colonel*] Hugh Milbourne. Surveyor general of the Transvaal. Sketch Plan Prahue, 1881; *The Settlement of Sierra Leone in January 1884*, London, War Office 1884; Anglo-Siamese Boundary Commission [Malayan-Thai boundary], 1890; *Map of Transvaal Colony*, 1902.

**Jackson**, J.W. *Plan of Madras and its suburbs*, 1870.

**Jackson**, John. Philippine Islands, 1814.

**Jackson**, John. *Mazari Bay* [north coast Africa], 1847.

**Jackson**, Joseph (*fl.*1812-*c.*1852). Surveyor, produced manuscript maps of parts of eastern England. *The Open Fields in the Parish of Sutton*, 1840; *Wistow, Huntingdonshire*, 1852.

**Jackson**, *Lieutenant* Lambert Cameron. *Part of Abyssinia and the Sudan*, 1901 (with Major Gwynn).

**Jackson**, *Lieutenant Colonel* Louis Charles. Manuscript surveys of the Faro River, Nigeria, 1903; Sketch map of Yola showing the divisions into Hausa and Fula districts [Nigeria], 1904.

**Jackson**, Luke. Engraver. *A map of that part of Pensylvania now the principle seat of war in America...*, London 1777 (from a survey by N. Scull).

**Jackson**, *Major* M. Plan of the Siege of Rangoon, 1852.

**Jackson**, P.J. *New postal address map of Carlisle*, 1880.

**Jackson**, Peter. In 1833 he joined the firm of Fisher, Son & Co., to which he succeeded c.1843. Republished J. & C. Walker's map of Lancashire, 1843 (first published Fisher, Son & Co. 1831).

**Jackson**, Robert. Publisher of Dublin. John Walker's *Elements of Geography*, 1788 (also published in London by James Phillips).

**Jackson**, Samuel. Surveyor in New South Wales. *Plan of the central & western divisions of Luddenham...*, Sydney, Allan & Wigley Lith. 1859; *Plan of the eastern division of the Luddenham Estate...*, Sydney, Allan & Wigley 1859.

**Jackson**, W. Publisher of Oxford. Isaac Taylor's *Oxford*, 1751 (engraved by G. Anderton); River Witham, 1751.

**Jackson**, W. Engraver. *Field of Waterloo* (2 sheets), London 1815 (surveyed by S. Wharton).

**Jackson**, W.H. Pony Express Route, 1860-1861 (with H.R. Driggs).

**Jackson**, William [& Co.]. Publisher. W. Ebden's *The Pedestrian's Companion, Fifteen miles round London*, 1822 (with M.J. Godwin, engraved by Sidney Hall); *A guide through London and the surrounding villages*, 1826.

**Jackson**, *Captain* William A. Mining engineer. *Map of the Mining District of Virginia*, Fredericksburg, W.A. Jackson, and Philadelphia, H.S. Tanner 1836; *A Map of the Mining District of California*, New York, T.A. Mudge 1850 (published with a twelve page guide for gold hunters), revised edition published New York, Lambert & Lane 1851 (with an enlarged guide).

**Jackson**, Zachariah. Publisher of Dublin. John Payne's *Universal Geography*, 1792 (maps carry the note 'engraved for Jacksons Edition of Payne's *New System of Universal Geography*').

**Jackson & Cowan**. Publishers of Glasgow. John Lothian's *County Atlas of Scotland*, from 1827 (title page is dated 1826, also published in London by J. Duncan).

**Jacme**, Mestre. See **Cresques**, Jafuda.

**Jacob**, A.A. Plan of Nurbudda Mineral Districts, Surat [India], 1854.

**Jacob**, Carl. *Physikalische Schulkarte von Plauen*, c.1896.

**Jacob**, E. American surveyor. *Map of the City of Albany*, Albany, Sprague & Co., and New York, M. Dripps 1857 (lithographed by Hoffman, Pease & Tolley); *Map of the City of Poughkeepsie* [New York], 1857.

**Jacob**, Edward. *An Accurate Plan of the Town of Faversham, Kent*, 1770; *A Plan of the Town of Faversham*, 1774 (engraved by J. Hilton, in *History of the Town and Port of Faversham*).

**Jacob**, F.J. Blue Rapids,... Kansas, 1870.

**Jacob**, H. *Jacob's 1878 official map of Prince Edward county, Virginia*, 1879.

**Jacob**, J. Publisher of Vienna. Nova Silesiae Theatrum, 1750.

**Jacob**, *Captain* J. *Plan of the Battle of Meanee* [India], 1843.

**Jacob**, Léon (1858-1935). Explorer and railway engineer. Colonie du Gabon et du Congo (3 sheets), 1892.

**Jacobi**, C. Cartographer. Bible Atlas, 1891.

**Jacobi**, Ludwig Hermann Wilhelm. *...Karte des Regierungs Bezirks Arnsberg*, 1858.

**Jacobi**, W.J. *Map of Liberia*, Washington 1916.

**Jacobs**, J.S. *Theatrum Bellorum a Cruce Signatis*, 1842.

**Jacobs**, Rudolf. *Karte der Oestlichen Erdhälfte*, 1838; Erdkarte (8 sheets), 1848.

**Jacobs**, S. Engraver at 1 Rue de Condé, Paris. Map of Van Diemen's Land, 1847; maps for Carl Müller's *Geographi Graeci Minores*, 1855; maps for John Murray, 1859-1866 (drawn by Carl Müller); India, 1861; engraver for Vuillemin's *Atlas du Cosmos* (issued in 26 parts), Paris, L. Guérin 1861-1867. **Ref.** TALBERT, R.J.A. *Imago Mundi* 45 (1994), pp.128-150.

**Jacobs & Barthelemier**. Engravers. Europe Railroads, 1861.

**Jacobs & Ramboz**. Steel engravers of Paris. *Geological Map of Canada*, Paris, Gustave Bossange 1869 (compiled and drawn by Robert Barlow).

**Jacobsen**, F. Surveyor. Grand Plan Crysolite mine Iviktout [Greenland], Godthab, 1860.

**Jacobsen**, Françoijs. Chart of Tasman's Voyages, c.1666.

**Jacobsen**, Jan Cornelis. Dutch cartographic draughtsman, worked in Zeeland, 1629. *Caerte van de Hooge Heerlijcheyt van Oosterlant*, 1629, 1660 (with Cornelis Stavenisse).

**Jacobsz.** [Lootsman; Lootsman brothers; Theunitsz.] family (*fl.*1643-1717). Marine chartmakers, booksellers, publishers and printers of Amsterdam. Used the name Lootsman to distinguish themselves from others with the same name. **Ref.** KOEMAN, C. *Atlantes Neerlandici* (1967-1971) Vol.IV pp.223-265.

• **Jacobsz.** [Lootsman], Anthonie [Theunis] (1606/1607-1650). Founded a printing and publishing house 'op het Water in de Lootsman tussen de Oude en Nieuwe Brugh', Amsterdam. Compiled and published a mariner's guide entitled *De Lichtende Columne ofte Zeespiegel*, 1644, 1649, 1650, English edition, 1649 (the plates were sold to Pieter Goos in 1650 and used by him in his own *Zeespiegel*); *'t Nieuw Groot Straets-Boeck*, 1648 and many later editions (later French and English editions credit Jacob and Casparus, or Casparus, as publishers); *Nieuwe Lees-Caert*, 1648.

• **Jacobsz.** [Robijns; Weduwe van Theunis Jacobsz.], Lijntje (1602-1689). *Née* Robijns, married Anthonie Jacobsz. in 1631. After his death in 1650 she managed the business until his sons were old enough to take over. Completed and published Anthonie's reduced version of the Zeespiegel as *'T nieuwe en Vergroote Zee-Boeck...* (87 charts), 1652; *Nieuwe Lees-Caert*, 1653 edition.

•• [**Theunitsz.**] [Lootsman], Jacob (*d.*1679). Son of Anthonie [Theunis] Jacobsz. Produced a mariner's guide with new maps (to replace the plates that had passed to Pieter Goos), entitled *Nieuw en Groote Lootsmans Zee-Spiegel*, 1654 and many later editions, French editions from 1666 (with Casparus), English editions from 1668 (with Casparus); published *Nieuwe Water Wereldt Ofte Zee Atlas*, 1666, 1676 (with Casparus), 1678, English edition published jointly with Casparus 1671, 1677.

•• [**Theunitsz.**] [Lootsman], Casparus (1635-1711). Although his correct family name was Theunitsz. (son of Anthonie [Theunis] Jacobsz.), Casparus professionally styled himself Lootsman. Caspar collaborated with Hendrik Goos and Hendrik Doncker in publishing the text of *Zee-Spiegel*, c.1680, each binding up his own charts and title page. Joint publisher, with Jacob, then sole publisher, of the *Nieuw... Zee Spiegel*, 1654 and later, French editions, 1666 and later, English editions, 1668 and later; from 1671 he was joint publisher of Jacob's *Nieuwe Water Wereldt*, and continued to publish it after his brother's death, with editions in Dutch, 1681, French, 1681 and English, 1681, 1684, 1685, 1688, 1689, 1694.

••• [**Conynenberg**], Jacob. Son of Jannetje, sister of Jacob and Casparus Theunitsz., assisted Caspar in the continuation of the family business after Jacob Theunitsz.'s death in 1679.

**Jacobsz.**, Pieter. Dutch monk, surveyor and cartographer at Thabor, Ost-Friesland, worked also in Noord-Holland. Plan of the environs of Alkmaar, 1532 (with Maerten Cornelisz. and Symon Meeuwiz).

**Jacobszoon**, Anthony. See **Jacobsz.**, family.

**Jacobszoon**, Jan. *De Caerte váder Zee*, 1541.

**Jacobszoon,** Laurens. Publisher of Amsterdam. Joint publisher (with G. Rooman of Haarlem) of a Bible (4 copperplate maps by Peter Plancius), 1590, 1592.

**Jacobus** Angelus, of Scarparia. See **Angelo,** Jacopo d'.

**Jacobus,** Belga. See **Bos,** Jacob.

**Jacobus** de Mailo (fl.16th century). Hydrographer.

**Jacobus,** Russins. Chartmaker of Messina. Mediterranean, 1564, 1588 MSS.

**Jacomo,** Joannis. Italian chartmaker. Portolan chart, c.1500 MS (National Maritime Museum, Greenwich).

**Jacot-Guillarmod,** Charles (1867-1925). Worked on geological maps, and on revisions to the so-called *Siegfried-Atlas* [Swiss Alps].

**Jacotin,** *Colonel* Pierre (1765-1828). French cartographer and chief of topographical engineers. Went as part of Napoléon Bonaparte's expedition to Egypt, where from a base in Cairo he organised a series of surveys, 1799-1801. Survey of Palestine, 1799; *Basse Egypte*, 1800; *Carte générale de l'Egypte* (47 sheets, part of *Description de l'Egypte*), Paris 1807 and later; *Egypte* (3 sheets), 1818; Palestine, 1818; compiled *Atlas de l'Egypte*, engraved 1808, published 1828. **Ref.** GODLEWSKA, A. 'The Napoleonic survey of Egypt' *Cartographica* 25:1&2 (1988); GODLEWSKA, A. *The Map Collector* 24 (1983) pp.2-8.

**Jacoubet,** T. *Atlas générale... de Paris*, 1836.

**Jacovlev** [Yakovlev; Jakowiew], —. *Plan von Samarkand...*, Gotha 1865 (in Petermann's *Geographische Mittheilingen*).

**Jacquemart,** *Lieutenant* Alfred. Topographical surveyor, Sénégal and Niger regions, 1879 and later; *Atlas colonial: éd. populaire et classique*, Paris, C. Bayle 1890 (with H. Mager).

**Jacquemin** [Jacquemart], C. *Plan... de la ville de Lyon*, 1747.

**Jacquemont,** Victor (1801-1832). French natural scientist and explorer from Paris. Travelled in North America and Hawaii, 1826-1827; India, 1828-1832. He died in Bombay. *Voyage dans l'Inde pendant les années 1828 à 1832* (6 volumes), 1835-1841. **Ref.** BROC, N. *Dictionnaire illustré des explorateurs Français di XIXe siècle: II Asie* (1992) pp.254-258.

**Jacques,** A.T. County of Dudley, New South Wales, 1870.

**Jacques,** João Candido. *Carta Geographica do... Rio Grande do Sul* [Brazil], 1893.

**Jacquot,** E. *Carte agronomique de... Toul*, 1860; *Carte géologique du département du Gers*, 1869.

**Jacubasch,** *Doctor* —. *Karte von St Andreasberg*, 1890.

**Jacubicska** [Iacubiska], Stephan (1742-1806). Austrian military. cartographer. Plan de Vienne (4 sheets), [n.d.].

**Jadis,** Léon. *Les pays Cathare...* [Languedoc], 1890.

**Jadrow** [Jadrov], Jan [Ivan Alekseevic]. *Mappa Królestwa Polskiego...*, Warsaw 1863, 1868.

**Jaeck,** Carl. See **Jäck,** Carl.

**Jaeck,** W. Engraver, 1860.

**Jaeger** [Jäger], father and son. Papermakers, publishers and booksellers of Frankfurt am Main.
• **Jaeger,** Johann [Jean] Wilhelm [Guillaume] Abraham (1718-1790). Artillery officer, engineer, cartographer, publisher, bookseller, clockmaker and instrument maker from Nuremberg. Moved to Frankfurt am Main in 1745, and in 1748 he married into a family who owned a powdermill, where he began to make paper in 1777. He bought a bookshop in Frankfurt in 1762, and soon after he began work on his *Grand Atlas d'Allemagne* which retained its importance into the nineteenth century. He died in Frankfurt and was succeeded by his son. *Les Etats de la Saxe* (12 sheets), 1779; *Grand Atlas d'Allemagne* (81 sheets), from 1768; *Le Théâtre de la guerre Russie-Turquie* (6 sheets), 1770. **Ref.** KRETSCHMER, DÖRFLINGER & WAWRIK *Lexicon zur Geschichte der Kartographie* (1986) Vol.C/1 pp.353 [in German]; GROSSE-STOLTENBERG, R. 'Grand Atlas

*Victor Jacquemont (1801-1832). (Taken from N. Broc, Asie)*

d'Allemagne edited by Johann Wilhelm Abraham Jaeger, Frankfurt am Main 1789' in *Imago Mundi* 28 (1976) pp.94-104.

•• **Jaeger**, Johann Christian. Publisher of Frankfurt am Main, son of Johann Wilhelm. Worked with the family business and continued publication of his father's *Grand Atlas d'Allemagne*, c.1782-1803; *Pays-Bas* (6 sheets), 1784; *Neuer Kriegs-Schauplatz*...[Austria, Russia, Turkey] (2 sheets), 1788.

••• **Jaeger'sche Papiergroßhandlung** (fl.1803-1922). After 1803 the company remained in the hands of the family for a while.

**Jaeger**, Carl Friedrick Julius (1815-1857). Lithographer. *Kaart van Java*, [n.d.].

**Jaeger** [Jäger], E. Publisher of Stuttgart. *Schul-Atlas*, 1862.

**Jaeger**, F.W. Map publisher of Hamburg. *Grosser Schul-Atlas*, Hamburg 1838; *Kleiner Schul-Atlas*, 1837.

**Jaeger** [Jäger], Franz Anton (1765-1835). Priest and historian. *Das hohe Rhoengebirg in Franken*, 1802.

**Jaeger**, Godfrey [& Company]. Publishers. *Historical Hand-Atlas... of the United States and the Provinces of Canada... and the History of Ottawa County, Ohio*, 1881 (maps by Rand McNally from *New Railroad* and *County Map of the United States and Canada*, 1876).

**Jaeger**, J. *Zakatlas der Nederlanden*, 1843; *Zakatlas van Europa*, 1843, *c.*1860; *Goedkope Schoolatlas der Nederlanden*, 1848; *Uitgebreide Hand-Atlas der geheele wereld...*, 1849; *Atlas van het Koningrijk der Nederlanden en zijne Bezittingen*, 1850, 1856, 1877, 1878; *Zak-Atlas der geheel aarde*, 1853; *Zak-Atlas van het Koningrijk der Nederlanden en het Groot-Hertogdom Luxemburg*, 1853; *Groote Schoolatlas*, 1854; *Goedkope Schoolatlas*, 1858.

**Jaeger**, Johann Wilhelm. See **Jaeger**, father and son.

**Jaegersberg**, Johann Christopher Jaeger von (*d.*1750). Danish surveyor. *Charte over Eilandet St. Croix udi America*, *c.*1750 MS (with Johann Cronenberg). **Ref.** HOPKINS, D. *Imago Mundi* 41 (1989) pp.44-58.

**Jaegerschmid**, Alexander. Engraver of Paris. *Quartier de Ste Catherine... Cuba*, 1834.

**Jaettnig** [Jättnig], Carl [*elder* and *younger*]. Engravers of Berlin.
• **Jaettnig**, Carl, the *elder* (*fl.c.*1795-1835). Copper engraver and publisher of Berlin. *Pommern*, 1789; *Atlas Posener Kammer-Departement*, 1798; *Magdeburg*, 1799; *Berlin*, 1800; *Suedpreussen* (13 sheets), 1802-1803; *Ostfries-und Harrlingerlande* (2 sheets), 1804; [*Die*] *Preussischen Staate* (30 sheets), 1820; one of the engravers for A. Papen's *Topographischer Atlas des Königreichs Hannover und Herzogthums Braunschweig...* (80 sheets), Hanover, 1832-1847.
•• **Jaettnig**, Carl, the *younger* (*fl.c.*1824-1850). Engraver. *Grundriss von Berlin*, *c.*1850.

**Jaettnig** [Jättnig], Ferdinand [*elder* and *younger*]. Engravers of Berlin. *Berlin*, 1812; *Deutschland*, 1816; [*Das*] *Abendlaendisch-Roemische Reich*, 1823.

**Jaettnig** [Jättnig], Wilhelm (*fl.c.*1830-1840). Engraver of Berlin. *Gegend um Potsdam*, 1833; for Heinrich Berghaus, 1835-1850; *Asien* (4 sheets), 1837; *Wege Karte der Insel Rügen*, *c.*1850; for Justus Perthes, 1852.

**Jaffe**, *Rabbi* Mordechai ben Abraham (*c.*1535-1612). Talmudist, astronomer and geographer. Wrote *Levushim* (10 volumes), 1590, 1604. Map *Zurat Hagvulot* [Shape of the Borders], 1603.

**Jaffray**, *Reverend* John. Sea coast about Peterhead, 1739.

**Jagellonicus Globe**. Anonymous globe, *c.*1510 (now in Crakow University Library).

**Jagen**, family of 18th-century Dutch engravers in Amsterdam.
• **Jagen**, Isak van (*fl.*1682-1727). Dutch engraver, etcher, draughtsman and artist. Engraved Nicolas and Jacob Cruquius's *'t Hooge Heemraedschap van Delflant*, (25 sheets) 1712 (with L. van Anse, T. Doesborgh, Jan and Jacob Deur, Gijsbert Schouten and Pieter Ruyter).
•• **Jagen**, Cornelis van (*fl.*1706-1744). Engraver of Amsterdam. Son of Isak, father of Jan.
••• **Jagen**, Jan van (*c.*1710-*c.*1796). Dutch designer, mapmaker and engraver, son of Cornelis; worked in Amsterdam, 1732-1796. Engraved Melchior Bolstra's *Kaart van de Beneeden rivier de Maas en de Merwede van de Noord Zee tot Hardinksveld* (7 sheets), 1738-1745 (with David Koster); worked for the Ottens firm *c.*1741; W.A. Bachiene's *Atlas tot opheldering der hedendaagsche historie*, 1785 (after Emanuel Bowen).

**Jäger**. See also **Jaeger**.

**Jager**, *Doctor* G. Weltkarte, Gotha 1865.

**Jäger**, Johann Wilhelm Abraham. See **Jaeger**, father & son.

**Jager**, P.J. *Hémisphères Célestes*, *c.*1867; *Hémisphères Terrestres*, *c.*1867.

**Jager**, Robert. *A true and lively description of His Majestie's Roade, the Downes, and his Highness' towne and Port of Sandwich...*, *c.*1624 MS, copy, 1629; *A New Haven for Sandwich*, 1634.

**Jahiz**, Amr Ibn-Bahr, Basri (777-869). Arab author from Basra [Iraq].

**Jago**, Richard Howlett. Royal Military Surveyor. Survey of the coast from Blyth to

Hartley [Northumberland], 1797; roads, towns and martello towers in the counties of Surrey, Kent and Sussex, 1801; Sheerness, 1801; *Plan of St John's*, 1802; map of Fort de France [Martinique], 1802; Dublin, 1803; plan for a defensive tower at Dover, 1804; Stepney and London City, c.1805 (copied from William Hayward and John Gascoyne).

**Jagor**, Fedor (1817-1900). Luzon, Berlin 1872; Philippine Islands, 1875.

**Jahn**, G.A. Publisher of Halle, Germany. Edition of C.L. Harding's *Atlas novus coelestis*, 1856.

**Jahn**, H.B. Draughtsman for *Der Nord-Ostsee-Kanal*, Kiel, E. Homann 1886 (lithographed by Dorn & Merfield); *Karte von Kiel*, 1890.

**Jahn**, Jaroslav Jiljí (1865-1934). *Geologischtektonische Übersichtskarte von Mähren und Schlesien*, Vienna 1911.

**Jahncke**, —. *Reise-Atlas des Norddeutschen Bundes*, 1869.

**Jahns**, M. *Atlas zur Geschichte des Kriegswesens von der Urzeit bis zum Ende des XVI* [battle atlas], 1879.

**Jaillot family. Ref.** DAVIS-ALLEN & REINHARTZ *Imago Mundi* 42 (1990) pp.94-98; SHIRLEY, R. *Printed maps of the British Isles 1650-1750* (1988) pp.73-76; KRETSCHMER, DÖRFLINGER & WAWRIK *Lexikon zur Geschichte der kartographie* (1986) pp.353-354 [in German] (cites references); PASTOUREAU M. *Les Atlas Français* (1984) pp.229-292 [in French]; PASTOUREAU M. *Imago Mundi* 32 (1980) pp.65-72 [in French].

• **Jaillot**, Alexis-Hubert (c.1632-1712). Sculptor, print and map seller, publisher, and geographer from Saint-Claude, Franche-Comté. Trained as a sculptor, and set out for Paris in 1657 with his brother Simon (d.1681), a carver of ivory. In 1664 he married Jeanne (d.1675), daughter of the map publisher Nicolas Berey, by whom he had seven children. When in 1668, Alexis and Jeanne purchased the geographical stock-in-trade of Berey Sr. and Berey Jr. from the estate of the latter, Jaillot took over his father-in-law's shop '…joignant les grands Augustins aux deux globes' and continued to publish his works. Much of Jaillot's output was based on the maps of Nicolas Sanson, in the form of editions prepared for him by Adrien and Guillaume Sanson. In 1676 he married Charlotte Orbanne, and further enlarged his family. Alexis died in Paris, and his works were republished and amended by his son, grandson and granddaughters, and later by J.C. Dezauche (in *Atlas Géographique et Universel*, 1789) and the Basset family (c.1792-1800). *Carte de Franche-Comté et du Comté de Montbeliard*, 1669; re-issued Willem Jansz. Blaeu's wall map of Africa, 1669; from c.1670, Alexis arranged for Adrien and Guillaume Sanson to rework the maps of Nicolas Sanson on 2 sheets, these formed the basis of *Atlas Nouveau*, 1681, 1684, 1689 (editions of 1692, 1696 and 1698 were pirated by P. Mortier in Amsterdam); *Carte Particulière des Postes de France*, 1689 (engraved by Cordier); *Atlas françois...* (1 or 2 volumes), 1695 and later (many of the maps, which varied in number from copy to copy of the *Atlas*, were reduced from the Sanson maps of *Atlas Nouveau*); *Nova Transilvaniae principatus Tabula...* (ethnic map), 1696; Palestine, 1697; *La France Divisée par provinces...*, 1701; *Carte particulière des environs de Paris*, 1706 (first published by Maquart, 1685); Liste générale des Postes de France, 1708.

•• **Jaillot**, Bernard Jean Hyacinthe (1673-1739). Son of Alexis-Hubert Jaillot by his first wife Jeanne Berey. On the death of his father he took over the business at the above address, selling some of the stock back to his stepmother, Charlotte Orbanne to cover the rent at 'les grands Augustins aux Deux Globes'. Plans de Paris (4 sheets), 1713; Postes de France, 1713; *Germaniae l'Emoire d'Allemagne...*, c.1715; *Le Royaume de France*, 1717; republished *Atlas françois*, 1719, 1721 and later; *Nouvelle Carte des Postes de France*, 1726 and later; France écclésiastique (4 sheets), 1731; *Plan de Luxembourg*, 1735; Diocèse de Bayeaux, 1736.

••• **Jaillot**, Bernard-Antoine (d.1749). Géographe du roi (1728). Son of Bernard Jean Hyacinthe Jaillot. In 1731 he bought the geographical stock of A.H. Jaillot from his step-grandmother Charlotte Orbanne. Worked with his brother-in-law Jean Baptiste Renou de Chauvigné-Jaillot. Map of Switzerland (4 sheets), 1717; republished *Atlas françois*, 1734 edition; *Plan de Luxembourg*, 1741 (engraved by Coquart); Routes of the Pretender, 1747; *Carte des*

*Alexis-Hubert Jaillot (c.1632-1712). (By courtesy of Rodney W. Shirley)*

*Postes de France...*, 1748, 1765 (drawn and engraved by P. Aveline).

••• **Jaillot,** *Demoiselles* [Antoinette Charlotte, Jeanne Nicole, Marie Marguerite and Françoise] (*fl.*1749-*c.*1763). Daughters of Bernard Jean Hyacinthe and Marie Marguerite De La Salle, as sisters and successors to Bernard-Antoine, they ran the business from 1749. Françoise married J.B. Renou de Chauvigné in 1755.

••• **[Renou de Chauvigné-Jaillot],** Jean Baptiste Michel (1710-1780). Géographe du roi (1757). As husband of Françoise Jaillot he was active in the family business from *c.*1757 at 'le quai et à côté des Grands Augustins'. In 1781 the maps, plans, plates and proofs were sold. Some plates were melted down, but parts of the stock were bought by Jean Claude Dezauche and Louis Charles Desnos. Prague, 1757; *Carte des Postes de France,* 1771 (revised by Desnos, 1782); *Recherches Critiques Historiques et Topographiques sur la Ville de Paris* (5 volumes), 1775.

**Jaime,** Jean Gilbert Nicomède (1858-*c.*1894). French naval officer. Surveys in Tonkin, 1884-1885; Niger, 1889. **Ref.** BROC, N. Dictionnaire illustré des explorateurs Français di XIXe siècle: I Afrique (1988) pp.180-181 [in French].

**Jain World Map.** Map of the World according to the tenets of Jainism: *Manuslyaloka* [The World of man], 15th century. **Ref.** WHITFIELD, P. The image of the World pp.30-31.

**Jaïn,** François Gédéon. *L'Empire Français, c.*1805.

**Jaisse,** L. de la. See **Lemau** de la Jaisse, Pierre.

**Jailly,** A. *Les Grandes Routes Commerciale du Monde,* 1868.

**Jainin,** —. Engraver. *Carte du Département de l'Eure,* 1808.

**Jakeman & Carver.** Map of Herefordshire, 1890; *Plan of the City of Hereford...,* 1890.

**Jakubicska,** *Captain* Stephan (1742-1806). Maps and plans of the environs of Vienna, including *Grundriss der Haupt und Residenzstadt,* Vienna, Artaria 1789.

**Jamar,** A. Publisher of Brussels. Atlas générale de géographie moderne, Tarlier 1854.

**Jamard,** T. *Routes de la Comète qui a été observée pendant les années 1531, 1607, 1682 et qui doit reparoitre en 1757 ou 1758,* 1757; *Route qui doit tenir pendant le mois de Mai 1759 le Comète...,* 1759.

**James,** *Lieutenant* Edward Renouard *RE.* Manuscript plans of military sites on Malta including *Plan of the shore on either side of French Creek,* 1854 MS; *Carte Générale du territoire Cedé par l'Empire de Russie à l'Empire de Turquie d'après le Traité de Paris... 1856...,* 1857 MS; *Carte topographique et spéciale de la frontière tracée entre les empires de Russie et de Turquie en Bessarabie d'après le Traité de Paris...* (33 sheets), 1857 MS.

**James,** Edwin (1797-1861). Botanist and geologist with Major S.H. Long's expedition across America. *Account of an Expedition from Pittsburgh to the Rocky Mountains: performed in the years 1819 and '20* (2 volumes with atlas), Philadelphia, H.C. Carey and I. Lea 1822-1823.

**James,** Ernest A.H. *Railway Map of part of South America, c.*1890.

**James,** Fred. One of the editors of *Imperial Royal Canadian World Atlas: An atlas for Canadians,* Chicago, Geographical Publishing Co. 1935 (with Lloyd Edwin Smith and Frederick K. Branom).

**James,** H.E.M. *Map of Manchuria,* London 1887 (in *Proceedings of the Royal Geographical Society*).

**James,** H.F. Lithographic press, Ridgefield, Manchester. Plans of Market Street, Manchester, 1821, 1822.

**James,** *Colonel Sir* Henry (1803-1877). Worked for the Ordnance Survey in Ireland under T. Colby before 1850. Director General of the Ordnance Survey in Great Britain and Ireland, 1854-1875. Introduced the technique of photozincography for the production of Ordnance Survey maps, *c.*1860. **Ref.** OWEN & PILBEAM Ordnance Survey. Map makers to Britain since 1791 (1992) pp.53-66; OLIVER, R. The Map Collector 54 (1991) pp.36-37.

**James,** *Captain* Horton, *RN.* Admiralty surveyor. *Ascension Harbour* [Caroline Islands], 1840.

**James, J.A. & U.P.** American publisher based in Cincinnati. *Texas*, 1836; John T. Hughes's *Doniphan's expedition: conquest of N[ew] M[exico]* (5 maps), 1848; *James' railroad and route book for the western and southern states*, Cincinnati 1854 (compiled by J. Griswold).

**James, J.O.N.** Surveyor. *Map of the China Coast*, Calcutta 1860; *North-East Frontier of Bengal* (6 sheets), 1865; *India*, Calcutta, Surveyor General's Office 1870.

**James, Joseph.** *Survey of the Boundary between British and Jamoo Territories...* [India], 1851 (with others).

**James, S.A.** *Keokuk County, Iowa*, 1861.

**James, T.R.** Compiled *Victoria. Skeleton Map of Telegraph Circuits*, Melbourne 1871 (drawn by J.W. Payter).

**James,** *Captain* Thomas (c.1593-1635). English navigator, sailed to Hudson Bay in 1631 aboard *Henrietta Maria*. James Bay, where he sheltered that winter, is named after him. *The Platt of Sayling for the discoverye of a Passage into the South Sea*. 1631. 1632., 1633 (in *The Strange and Dangerous Voyage of Captain Thomas James*). **Ref.** BURDEN, P. *The mapping of North America* (1996) p.298.

**James,** *Lieutenant Colonel* Thomas. Commandant of Artillery in North America. Inset on Faden's *A Plan of the attack of Fort Sulivan*, 1776.

**James, U.P.** See **James, J.A. & U.P.**

**James, W.D.** Surveyor. *Province of Ogadyn* [Abyssinia and Somaliland], 1855; *Routes in Somaliland* (13 sheets), 1885 MS.

**James, William** (d.1827). *Map of maj. gen. Ross' route with the British column, from Benedict, on the Patuxent river, to the city of Washington, 1814*, London 1818; *Map of the Straits of Niagara*, 1818 (engraved by H. Mutlow); *Part of Lake Ontario and of the River St. Lawrence*, 1818 (engraved by J. Walker, all the above published in W. James's *A Full and Correct Account of the Military Occurences of the Late War*, London, W. James 1818).

**Jameson, T.** *A Geographical Chart of Europe* (4 sheets), 1793.

**Jameson,** *Doctor* W. *River Napo* [South America], 1861 MS.

**Jamet, —.** *Topographical Map of St. Lucia*, 1847 (with Detaille).

**Jamieson, Alexander.** Astronomer. *A Celestial Atlas*, London 1822; *An Atlas of... Maps of the Heavens*, 1824.

**Jamieson, G.** RN. Assisted W.E. Archdeacon with the surveys for *North Sea Heligoland...*, 1887 published 1888 (drawn by Archdeacon, engraved by Davies & Co.).

**Jamieson, John.** *Roman Military Way in Muir of Lour*, 1785.

**Jamieson, J.B.** *City of Helena... Arkansas*, 1859.

**Jamnitzer, Wenzel** (1508-1586). Goldsmith of Nuremberg. Made a 28-cm brass terrestrial globe, 1566.

**Jan** of Stobnicza. See **Stobnicza**, Johannes de.

**Janellus, Johannes.** See **Gianelli**, Giovanni.

**Janesson, Nicolas** (d.1617). Cartographer and engraver of Amsterdam.

**Janicke, A.** Engraver. John Fiala's *General map of the United States & their territory between the Mississippi & the Pacific Ocean... Showing routes... Post Routes... gold, silver and copper regions in Kansas, Nebraska and Arizona...*, 1859 (engraved on stone by A. Janicke, lithographed by A. McLean).
• **Janicke & Co.** Lithographers, 3rd St., St. Louis. *Pike's Peak Gold Region*, 1859.

**Jankó** [Jenei]**, Sándor** (1866-1923). Hungarian forestry engineer and cartographer, born Vasvár, died Vönöck. Pioneer of photogrammetric cartography, published a work on the subject in 1917. **Ref.** BENDEFFY, L. 'Emlékezés Jankó Sándorra' [In Memoriam], in *Geodezia es Kartografia* [Geodesy and Cartography], (1973/4) [in Hungarian].

**Jankowsky, J.** *Russo-Turkish War Map*, 1877.

**Jann**, Franz Xaver. Priest and teacher of Augsburg. Revised edition of A. Desing's *Universal-Historie auf der Land-Karten*, 1781.

**Janney**, J.D. and E. Publishers. Lucas County, Ohio, 1861.

**Janocki**, J.D. Catalogued Polish cartographic collection in Zaluski Library, 1775 (published later by Eduard Rastawiecki as *Mappografia*, Warsaw 1846).

**Janot**, *Doctor* E. *Mapa Scienna Europy*, c.1874 (translated from B. Kozenn's map of 1869); translated a world map by B. Kozenn into Polish (4 sheets), c.1875; *Austryacko-Wegierska Monarchia*, Wiedeh c.1876.

**Janota**, Eugeniusz Arnold (1823-1878). Cartographer of Cracow. **N.B.** Compare with E. **Janot**, above.

**Jansen** [Jenson], H.J. and Perronneau (fl.1791-1804). Publishers and booksellers of Paris, 'Place du Museum' (1797); 'Rue des Saints-Pères N°·1195' (June 1797); 'Rue des Maçons-Sorbonne N°·406', 'Rue des Postes N°·6, près celle de la Vieille-Estrapade'. F. Levaillant's *Carte de la Partie Méridionale de l'Afrique*, c.1792; *Pays entre Mer Noire et Caspienne*, 1795; *Voyage Par Le Cap De Bonne-Espérance A Batavia, A Bantam Et Au Bengale, En 1768, 69, 70 Et 71, Par J.S. Stavorinus...* (2 volumes), Paris 1798 (translated from Dutch by H.J. Jansen).

**Jansen**, J. *Charte von Friedrichstadt*, 1851.

**Jansen**, J.F. *Kaart van Europa*, 1874; *Wandatlas van Nederland*, 1876.

**Jansen**, V. One of the printers for F. Bazin and F. Cadet's *Atlas Spécial de la Géographie Physique, Politique et Historique de la France* (32 plates), Paris, F. Bazin, F. Cadet and J. Delalain c.1862.

**Jansma**, T. *Schematische Teekenatlas van Nederland*, 1907; *Schematische Teekenatlas van Europa en de Wereld-deelen*, 1909.

**Janson**, —. Printer in Paris, for Ramón Escudero, 1861; *Atlas Géographique de la République du Pérou par Mariano Felipe Paz Soldan*, Paris, Auguste Durand 1865 (P. Arsène Mouqueron's French edition of Soldan's atlas of 1861).

**Jansonius**, Guilielmus. See **Blaeu**, Willem Jansz.

**Janssen**, Claes. Dutch surveyor and cartographer, worked in Zeeland, lived at St. Joosland. *Caerte metinge en afteykeninge... van West Craeyert...*, 1676.

**Janssen**, Jules (1824-1907). French physician, astronomer and geologist. Travelled in South America with the Grandidier brothers (1857-1858), and in the Far East and the Pacific. *Mission au Japon*, 1876; *Observations magnétiques dans la presqu'île de Malacca*, 1876; *Rapport à l'Académie sur la mission en Océanie*, 1883. **Ref.** BROC, N. *Dictionnaire illustré des explorateurs Français du XIXe siècle: II Asie* (1992) pp.259-261 [in French].

**Janssen**, Léon. Editor of a new edition of Manuel Godinho de Erédia's *Malacca*, Brussels 1881 (first published 1613).

**Janssen**, P.J.C. (b.1824). French astronomer and geographer.

**Janssonius** [Jansson; Jansz.; Janszoon] family. Influential in map publishing in the Netherlands throughout the seventeenth century, initially in Arnhem, then in Amsterdam. **Ref.** KROGT, P. van der *Koeman's Atlantes Neerlandici* Vol.I (1997) passim.; KROGT, P. van der *Globi Neerlandici* (1993) passim.; KROGT, P. van der 'Commercial cartography in the Netherlands' in *4rt curs La cartografia dels Països Baixos* pp.101-108 et passim.; KRETSCHMER, DÖRFLINGER & WAWRIK *Lexikon zur Geschichte der Kartographie* (1986) Vol.C/1 pp.356-357 [in German]; KOEMAN, C. *Atlantes Neerlandici* (1967-1971) Vol.II pp.158-204; ibid. Vol.IV pp. 266-275.

• **Janssonius** [Jansz. I; Jan Jansz. of Arnhem], Johannes (fl.c.1597-1629). Printer, publisher and bookseller of Arnhem. Father of Johannes Janssonius. Associated with Cornelis Claesz. in the publication of Pieter Bertius's *Tabularum geographicarum Contractarum...*, 1600-1603; collaborated with the Hondius family in editions of the Mercator-Hondius *Atlas minor*, 1607-1621 (sold also by Cornelis Claesz. 1607, 1610, 1613, 1614); second edition of Zacharias Heyn's *Le Miroir du Monde...*, 1615; Jan François Le Petit's *Nederlandsche Republycke...*, 1615 (with town plans after Guicciardini); Cornelis Wytfliet's *Descriptionis Ptolemaicae augmentum*, 1615; Magini's version of Ptolemy's *Geographia*, 1617 edition.

•• **Janssonius** [Jansz. II; Jansz. of Amsterdam], Johannes [Jan; Joannes] (1588-1664). Born in Arnhem, the son of Johannes Janssonius. In 1612 he married Elisabeth Hondius, daughter of Jodocus Hondius the elder and Colette van den Keere. In the same year he founded his own business in Amsterdam as a publisher of atlases, maps and globes 'Opt Water in De Pascaerte' ['Dwelling upon the waterside by the Old Bridge at the sign of the sea mappe']. 1612 was also the year that Jodocus Hondius the elder died, and Johannes Janssonius assisted his widowed mother-in-law Colette with the continued publication of the Mercator-Hondius *Atlas*. When Colette and her son Jodocus the younger died in 1629, Johannes Janssonius and another son of Colette's, Henricus Hondius, worked together on a complete revision of the Mercator-Hondius *Atlas*. This became known as *Atlas Novus* from 1638 and was published in several languages over many years. In the early 1640s Henricus Hondius turned his interests elsewhere, but Janssonius continued expanding the *Atlas*, culminating in the 11 volume *Atlas Maior* with editions in Latin, French, Dutch, German and English. Maps of France and Italy, 1616; re-issued Pieter van den Keere's 4-inch terestrial globe, 1620 (first published 1613); 6-inch terrestrial globe by Abraham Goos, 1621; with Jodocus Hondius Jr published a pair of 17-inch globes, 1623, 1636, 1648 (terrestrial globe by Abraham Goos, celestial globe by Adriaan Metius); republished the 10-inch terrestrial globe by Petrus Plancius, 1627, 1645 editions (engraved by Pieter ven den Keere and Abraham Goos, first issued 1614); *Atlas Minor*, 1628 and later (engraved by Abraham Goos and Pieter van den Keere, editions in Latin, French, Dutch and German); *Theatrum Universae Galliae*, 1631; *Theatrum Imperii Germanici*, 1632; *Le Nouveau Phalot de la Mer* [sea atlas], 1635, 1637; *Appendix Nova Atlanti*, 1647; *Petri Bertii Beschreibung der gantzen Welt*, 1650 (German text edition of *Tabularum geographicarum*); *Accuratissima Orbis Antiqui Delineatio*, 1652, new edition with text by Georg Horn published 1653, 1654, 1660; *Lichtende Columne ofte zee-spiegel*, 1652; *Descriptio Maris Mediterranei*, 1654; *Lighting Colomne or Sea Mirrour*, 1654; *Theatrum urbium* [Janssonius Townbooks] (8 volumes), 1657 (re-issue and expansion of Braun and Hogenberg's *Civitates* series, he had bought the plates in 1653); *Atlas Maior* (11 volumes), 1658 and later; *Atlas Coelestis seu Harmonia Macrocosmica*, 1660, 1661.

••• **Janssonius Heirs** [Haeredes Joannis Janssonii; Janssonius van Waesbergen; Janssonius haeredes] (1664-1674). After the death of Johannes Janssonius the estate was split three ways between his two surviving daughters, Elisabeth and Maria, and the children of their dead brother Jodocus (see below). Major publications such as the *Atlas Maior* and *Atlas Minor* were not to be divided, and remained in print under the joint ownership of the Janssonius Heirs. The business was managed by Johannes Janssonius van Waesbergen, who later acquired much of the plate stock when the assets were dispersed at auction in 1674. Republications included *Atlas contractus*, 1666; *Atlas Coelestis seu Harmonia Macrocosmica*, 1666.

••• **Janssonius**, Jodocus [Joost] (1613/1614-1655). Son of Johannes Janssonius II. Married Francina von Offenberg in 1642. Jodocus died before his father, so it was his widow and children who inherited his third of the estate and the family business.

••• **Janssonius**, Francina (d.1665). Widow of, and part successor to Jodocus Janssonius.

••• **Janssonius**, Jodocus *weduwe en erven* [veuve, et héritiers]. Francina, widow of Jodocus, and daughters Sara & Susanna, who were subsequently partners in Janssonius Heirs. *Histoire de la vie de Frédéric Henry*, 1656 (editions in Dutch and Latin).

•••• **Janssonius**, Sara (d.1670). One of the Janssonius Heirs, through her father Jodocus. She married Elizius Weyerstraten who was briefly in business association with her uncle Johannes Janssonius van Waesbergen.

•••• **Janssonius** (1645-1669), Elisabeth. One of the Janssonius Heirs through her father Jodocus. Married to Jurriaan van Kempt.

•••• **Janssonius**, Susanna. One of the Janssonius Heirs through her father Jodocus. In 1667 she married her cousin Gillis Janssonius van Waesbergen, who was active in the Waesbergen family business.

•••• **Janssonius**, Johannes (d. after 1663). Son of and part successor to Jodocus Janssonius. Not known to be actively involved in the business.

••• **Janssonius**, Maria (*b*.1620). As daughter of Elisabeth Hondius and Johannes Janssonius she held a one third share in Janssonius Heirs, but was not active in the business. She married Johannes van Almeloveen.

••• **Janssonius** [Janssonius van Waesbergen], Elisabeth. Daughter of Elisabeth Hondius and Johannes Janssonius. In 1642 she married Johannes Janssonius van Waesbergen, who became active in Janssonius Heirs through Elisabeth's one third share.

••• **Janssonius van Waesbergen**, Johannes I (*c*.1616/1617-1681). Bookseller and publisher from Rotterdam. Born Johannes van Waesbergen he adopted the name Janssonius in 1642 on his marriage to Elisabeth, the daughter of Joannes Janssonius II. Father-in-law to Willem Goeree. Johannes van Waesbergen was a bookseller and publisher in his own right, but also managed the firm of Janssonius Heirs, he later bought many of the Janssonius plates at auction in 1674. Waesbergen was succeeded in the business by his two sons. *Atlas Minor*, 1673 (Latin edition), 1676 (Dutch edition); *Atlas sive Cosmographicae Meditationes*, 1673, 1676 (from Jan Cloppenburgh's plates, first published in 1630); Georg Horn's historical atlas *Accuratissima Orbis Antiqui Delineatio*, 1677, 1684; intended joint publisher, with Steven Swart, of Moses Pitt's *The English Atlas*, Oxford 1680 (curtailed by Pitt's bankruptcy).

•••• **Janssonius van Waesbergen**, Johannes II (*d*.before 1725). Son of Elisabeth and Johannes Janssonius van Waesbergen. On the death of his father, he and his brother Gillis took charge of the Waesbergen business on behalf of their mother. In principle, Waesbergen's materials were auctioned in 1694, but the imprint of Johannes and Gillis appears on the title pages of later works including: Georg Horn's *Description of the Earth or Ancient Geography*, 1700 (English edition of Horn's historical atlas); *La Guerre d'Italie*, 1701, 1702.

•••• **Janssonius van Waesbergen**, Gillis (*d*.1708). Son of Elizabeth and Johannes Janssonius van Waesbergen, married his cousin Susanna, the orphaned daughter of Jodocus Janssonius in 1667. In 1682, Gillis left the business in the hands of his brother, and set up on his own in Danzig.

**Janssoon van Delff**, Mathijs. See **Been**, Matthijs Janszoon de.

**Jansz.**, Evert van Schaick. *Chaerte ende beschrijvinghe vande Middelweert...*, 1571 (with C. van Berck).

**Jansz.**, Harmen and Marten (*fl*. early-17th century). Dutch chartmakers. Possibly worked with Augustijn Robaert, their neighbour in 'den Witten Os', Edam, north Holland. Manuscript charts, 1604-1630; map of the Atlantic, 1604; World (6 sheets), *c*.1606; World, 1610; Europe, *c*.1618.

**Jansz.**, Jan. See **Janssonius**, Johannes.

**Jansz.**, Willem. See **Blaeu**, Willem Jansz.

**Janszoon**, C. Published a Bible, Delft 1587 (with maps copied from French Bibles).

**Janszoon**, G. See **Blaeu**, Willem Jansz.

**Janszoon**, J. See **Janssonius** family.

**Janszoon**, L. Schenk. See **Schenk**, Leonard Jansz.

**Januensis**, Nicolo de Canerio. See **Caverio**, Nicolo di.

**Janvier**, Antide (1751-1835). Parisian clockmaker. Titles attributed to A. Janvier could be the work of J.D. Janvier (below).

**Janvier**, *sieur* Jean Denis [Robert]. French geographer and cartographer of 'Rue St. Jacques et l'Enseigne de la Place des Victoires', Paris (an address also used by J.B. Nolin). *Carte du gouvernement militaire de L'Isle de France*, Paris, J.B. Nolin 1746 (engraved by G.D. Chambon); *Les Royaumes de Suede, de Danemarck at de Norwège...*, 1749; maps for Jean de Beaurain's *Atlas de géographie ancienne et moderne*, 1751; France, 1751 (with S.G. Longchamps); *L'Afrique divisée en tous ses Etats* (4 sheets), Paris, 'chès les Sieurs Longchamps et Janvier' 1754; *L'Amérique divisée en tous ses pays et Etats* (4 sheets), Paris, Longchamps et Janvier 1754 (with S.G. Longchamps); *Isles Britanniques*, Bordeaux 1759; *Le royaume de Hongrie*, 1759; *...d'Espagne et de Portugal*, 1760; *Les Couronnes du Nord comprenant les Royaumes de Suède, Danemarck et Norwège*, Paris, Jean Lattré 1760 (engraved by P.P. Choffard); *L'Europe divisée en ses principaux Etats suivants les nouvelles*

*Observations astronomiques*, Paris, Lattré 1760; *L'Amérique en ses principaux Etats*, Paris, Lattré 1760; *Partie méridionale des Pays-Bas*, Paris, Lattré 1760, 1784; Mappemonde (4 sheets), Lattré, 1761; *L'Afrique*, 1762; *L'Amerique meridionale*, 1762; *L'Amerique septentrionale...*, 1762; *Les Royaumes de Pologne et de Prusse avec le Duché de Curlande...*, Paris, Lattré 1774 (drawn and engraved by F.N. Martinet); *Mappemonde*, 1775; maps for Santini's *Atlas Universel...*, 1776; William Faden's *The General Atlas*, 1778; *L'Asie divisée en ses principaux Etats*, Paris, Lattré c.1780; *Plan de Paris*, Jean Lattré, 1783-1790.

**Janvier**, Robert. See **Janvier**, Jean Denis.

**Janvrin**, Daniel. Guernsey, 1810 (engraved by Gale & Butler); *Jersey, c.*1814.

**Jappé**, J.H. *Kaart van de Provincie Groningen* (4 sheets), 1835.

**Jaques**, Herbert. *Map of Mount Desert Island, Maine*, 1896 and later editions (with Edward L. Rand and Waldron Bates).

**Jaques**, John & Son. Publishers of 102 Hatton Garden, London. Educational card game based on county maps: *Skits - a Game of the Shires, c.*1890.

**Jaquier**, —. Draughtsman for Samuel Engel's *Mémoires et observations géographiques...*, Lausanne 1764-1765 (maps of American Northwest).

**Jaquier**, Francesco. French clergyman and professor of mathematics in Rome *c.*1770. Wrote text *De origine progressu geographiae* which was appended to some editions of *Geographia antiqua*, Rome 1774 edition and later (the historical atlas of Christophorus Cellarius, first published 1686).

**Jaquy**, François. Published a map of the Garden of Eden in a French Bible, Geneva, F. Jaquy, A.Davodeau & J. Bourgeois 1560, Geneva, F. Jaquy 1562.

**Jaraczewski**, J. (1798-1867). Polish military topographer and surveyor.

**Jardin**, J.R. de. Mappa de la provincia de Goyaz [Brasil], 1875; Rio Araguaya, 1879.

**Jardine**, W. Johnston. Plan of Edinburgh, 1850; *A View of the Coast of Japan and Islands*, 1859 ('drawn on stone by C.D.M. Strickland).

**Jardot**, A. *Carte...des Chemins de fer de l'Europe*, 1842.

**Jarman**, George. Engraver. *An exact delineation of the Cities of London and Westminster and the Suburbs thereof*, London, A.E. Evans & Sons 1857 (copied from Faithorne and Newcourt's plan of 1658); for James Reynolds, 1862; for Edward Stanford, 1863.

**Jarman**, Richard. Draughtsman and engraver. [Henry G.] *Collins' Illustrated Atlas of London*, 1854; *Reynolds' Map of Modern London Divided into Quarter Mile Sections*, 1857 and editions to *c.*1868; *Collins' Standard Map of London*, Stanford *c.*1859; *New map of Portsmouth, Portsea, Landport & Southsea...*, 1864.

**Jarmslinshi**, W. *L'Empire Ottoman en Europe Vilayet Salonique Sandjak Drama*, Cavalla 1877 MS (later lithographed by Harrison & Sons).

**Jarolimek**, Ludwig. One of the cartographers who contributed to *Atlas der Urproduction Oesterreichs...*, published Vienna, R. von Waldheim after 1876.

**Jarrad**, *Lieutenant* F.W. *RN*. Admiralty surveyor. Surveys around the Ionian Islands, 1863-1864; *Approaches to Rangoon River*, 1877; *Bay of Bengal*, 1879.

**Jarrett**, John. Estate surveyor. Abney Estate, 1767 MS.

**Jarrett & Co. Lith.** Lithographer of Sydney. *Rebellion in the Soudan*, Sydney, Campbell Brothers 1880s; Ashfield, 1880; Waverley, *c.*1880; Lakes Land, Picton, 1885.

**Jarrold**, Samuel. 19th-century Quaker printer and publisher of Norwich. Produced temperance maps of Norwich and Great Yarmouth.
• **Jarrold & Sons**. To date still in business in Norwich publishing topographical and guide books. *Plan of Norwich*, 1887.

**Jartoux**, Pierre, *SJ* (1699-1720). French astronomer and missionary in China. Made contributions to *Allgemeinen Reichskarte unter der gegenwärtigen Dynastie* (woodcut, 28 sheets), 1717, (copper etching), 1719, (woodcut, 32 sheets), 1721 (with J.B. Régis, E.X. Friedel, J. Bouvet and de Mailla).

**Jarves**, James Jackson. Newspaper editor and American diplomat based in Honolulu, later Florence. *A Correct Map of the Bay of San Francisco and the Gold Region*, Boston, James Munroe & Company 1849 (lithographed by J.H. Bufford & Company).

**Jarvis**, J.F. *Jarvis' map of Washington*, Washington 1897.

**Jarwood**, *Captain* —. Plan of Delagoa Bay (inset on John William Norie's Cape of Good Hope, 1831).

**Jasinski**, Yakub. Polish surveyor. Military maps, *c*.1791.

**Jasolinus** [Iasolino], Julius [Giulio] (*c*.1537/1538-1622). Physician, active in Naples from 1563. Wrote *De rimedi naturali che sono nell'Isola di Pithecusa. Hoggi detta Ischia*, Naples 1588, with map *Ischia aenaria hodie Ischia* (used by Ortelius from 1590, also by Joan Blaeu, 1647, 1662).

**Jaspers**, Jan Baptist (1620-1691). Engraver associated with Moses Pitt's *English Atlas* project, 1680-1683.

**Jastrzebowsky**, Albert. *Carte climatologique de Varsovie*, 1846.

**Jättnig**. See **Jaettnig**.

**Jaubert**, Amédée de (1779-1847). Historical geographer. *Voyage en Arménie et en Perse fait dans les années 1805 et 1806*, 1821; revised Georg von Meyendorff's *Carte du Khanat de Boukhara*, 1843. **Ref.** BROC, N. Dictionnaire illustré des explorateurs Français du XIXe siècle: Il Asie (1992) pp.261-262 [in French].

**Jaugeon**, N. *Sciences Du Jeu Du Monde Ou La Carte Generale Contenante Les Mondes Coeleste Terrestre Et Civile...* (world map in six sheets), Paris 1688. **Ref.** SHIRLEY, R. The mapping of the World (1983) N°.538 pp.535, 540-541.

**Jauncey**, F., *RN*. Admiralty surveyor. Port Ta-Outze, 1832; Cum-Sing-Mun Harbour, Admiralty Chart 1840 (with J. Rees).

**Jausz**, György (1842-1888). Hungarian cartographer. Compiled a Hungarian language historical atlas (32 sheets), Vienna, Hölzel 1873-1878.

**Jay**, G.M. Le. See **Le Jay**, Guido Michael.

**Jayhani**, Ahmad Ibn-Muhammad, Abu Abdallah (*fl*.10th century). Persian statesman and scientist. *Ashkal al-'Alam* [The shapes of the World] (with maps).

**Jean**, family. Geographers, publishers and 'Marchands d'Estampes' of Paris. Noted at various numbers in 'Rue St. Jean de Beauvais' including 'N°.4' (1790-1793); 'N°.32' (1798-1804); 'N°.10' (1805-1812); 'N°.8' (1808). Most works are credited simply 'Jean', including: *Introduction à la géographie...*, after 1791; *Carte de France*, Mondhare et Jean 1793; J.B. Poirson's *Carte itinéraire de la France...*, 1798; Sanson's *Carte d'Espagne et de Portugal...*, 1800; *Carte des environs de Paris...*, 1801; *Plan Routier de la Ville et Faubourg de Paris en 12 Municipalités...*, 1803; *Plan Routier de la Ville et Faubourgs de Paris...*, 1804, 1811.

• **Jean**, Pierre (d.1805?). Worked with L.J. Mondhare at 'rue S.t Jean de Beauvais N°.4' (1791-1793), then at addresses above. Republished B.J.H. Jaillot's *Plan de Luxembourg*, L.-J. Mondhare et P. Jean 1790 and later (first published 1735); *Plan de la ville et citadelle de Besançon*, 1790 edition; *Plan de la Ville et Faubourg de Paris...*, L.-J. Mondhare et P. Jean 1791, 1792.

•• **Jean**, — (d.1839). J.B. Poirson's *Plan de la Bataille d'Austerlitz...*, 1806; one of several sellers of Thomas Lopez's *Carte des royaumes d'Espagne et de Portugal où l'on a marqué les Routes de Postes...*, 1808; J.B. Poirson's *Carte de la France...*, 1811; Bonissel's *Carte de l'empire français... avec les royaumes d'Espagne, de Portugal, d'Italie et de Naples, la Confédération du Rhin, l'Yllirie et la Dalmatie, c*.1811; *Carte du Théatre de la Guerre entre la France et la Russie...*, 1812 (re-used a map edited by Crépy in 1769); re-issue of Brion de la Tour's *l'Afrique*, 1814; *l'Amérique*, 1818 (late re-issue of Jean Baptiste Nolin's map); *Afrique*, 1829; *Océanie*, 1837; and others.

**Jeanes**, B. *Geographical Synopsis of Europe*, 1840.

**Jeannel**, R. *Afrique orientale*, 1914 (with C. Alluaud).

**Jeanneret**, —. See **Pierre** & Jeanneret.

**Jeanneret**, Georges. Text and engravings for *Un Séjour a l'Île de Saint-Pierre* (20 wood engravings), Paris, Neuchatel & Geneva, 1878 (drawings by G. Guillaume).

**Jeans**, J.W. *Handbook for Finding the Stars*, London 1848.

**Jeanviliers**, —. See **Jenvilliers**, —.

**Jeaurat**, —. French engineer, worked with others on sheet 110 (Verdun) of the Cassini *Carte de France*, 1754-1758.

**Jebb**, Richard. *A Plan of the Ellesmere Canal...*, 1796 (with A. Davies, engraved by Neele).

**Jeckl**, Jakob A. Cartographer of Vienna. *Post-Karte des Herzogthums Kaernten*, 1786; *Geistliche Karte über die dermalige Diözeseintheilung im Lande Inner-Oesterreich*, Vienna 1787.

**Jeckyll**, Thomas. Civil engineer and draughtsman. *Map illustrating the general geological features of the country west of the Mississippi River*, 1857; *Map of the United States and their Territories between the Mississippi and the Pacific Ocean and part of Mexico...*, Washington D.C., S. Siebert 1857 (both maps published in *Report of the United States and Mexican Boundary Survey*, Washington, Cornelius Wendell 1857).

**Jeekel**, *Lieutenant* C.A. Officer in the Royal Dutch Navy. *Chart of the Boundary Line between the English and Netherlands Possessions on the West Coast of Africa*, 1868 MS (with F.M. Skues); *Map of the former Dutch possessions on the Gold Coast*, 1873.

**Jeffers**, *Lieutenant* W.N. *Honduras Eisenbahn Projet*, 1853; *Port Caballos*, 1853; *San Salvador*, 1858; *Mexico*, 1863.

**Jefferson**, father and son. **Ref.** HALLISEY, J. 'Thomas Jefferson, cartographer: A father's legacy to his renaissance son' in: *Mercator's world* Vol.1 No.3 (1996) pp.22-27; VERNER, C. *Imago Mundi* 21 (1967) pp.70-94; VERNER, C. *The Fry and Jefferson Map* (Princeton 1950).

• **Jefferson**, Peter (*d.*1757). Surveyor and planter in Virginia. Father of Thomas Jefferson (below). With Joshua Fry surveyed the boundary between Virginia and North Carolina, 1749-1751; with Fry compiled *A Map of the Inhabited part of Virginia...*, 1751, published London, T. Jefferys *c.*1753, 1754 (*most* is added to the title), 1755 and later.

•• **Jefferson**, Thomas (1743-1826). United States President; son of Peter Jefferson (above). *A Map of the country between Albemarle Sound and Lake Erie*, 1786 (in *Notes on the State of Virginia*, 1787); plan for the City of Washington, 1791 MS.

**Jefferson**, John. *Jefferson's New and Improved Map of the Isle of Man*, 1840, 1848; *Commons Dreim Gill and Cronk na Mala in the Parish of Lezayre* [Isle of Man], 1857 MS.

**Jefferson**, T.H. Geographer and publisher, New York. *Map of the Emigrant Road from Independence Mo. to St. Francisco California* (4 sheets), 1849.

**Jeffery**, *Commander* James, RN. Admiralty Charts, 1861-1869; including *Arasaig Harbour*, 1861; *East Coast of Australia... Moreton Bay to Sandy Cape*, 1863; *Loch Moidart*, 1865.

**Jefferys** [Jeffreys], John, *junior* (fl.1720-1750). Land surveyor and engraver, 'Teacher abroad of Writing, Arithmetic, Algebra, Geometry and Geography etc.' of 'Chapel Street Broad-Way Westminster' (not known to be related to Thomas Jefferys). *A New Map of all the Rivers of England and Wales...*, 1744; *A New Map (Laid down from Surveys by ye Wheel) Of all the Great or Post Roads and principal Cross Roads throughout England and Wales...*, 1750, 1754; *A Journey through Europe*, 1759 (reputedly the first known map game).

**Jefferys**, Thomas (*c.*1710-1771). Engraver, geographer, and publisher of London. Noted 'in Red Lyon Street near St. John's Gate' (1732-1750); 'at y$^e$ Corner of S$^{t.}$ Martins Lane near Charing Cross' [N$^o$.487 Strand, or 5 Charing Cross] (from 1750). Began his

*An advertisement for the Thomas Jefferys map and print shop 'at ye Corner of St. Martins Lane near Charing Cross' in London. In the foreground are examples of Jefferys' stock which included books and atlases by Halley, Senex and Moll as well as loose sheets, maps, globes and instruments. (By courtesy of Jonathan Potter)*

working life as a copper engraver and turned increasingly to geography and map publication. He was appointed 'Geographer to the Prince of Wales' 1746, and later 'Geographer to the King'. His earlier work, much of it concerned with North America and the West Indies, was compiled from existing sources, but in the early 1760s he began commissioning original surveys for a new series of English county maps. The expense of this probably contributed to his bankruptcy in 1766. Jefferys partly recovered from his financial problems with the financial support of Robert Sayer (who became a partner in some of his works and who was later to take control of *The Small English Atlas*, *The West India Atlas* and *The American Atlas* etc.). In 1769 or before, Jefferys was also joined in partnership by William Faden, who succeeded him at the Strand premises, in part of the business and as 'Geographer to the King'. *New and Exact Plan of the City's of London and Westminster... Borough of Southwark...*, 1735; maps for Edward Cave's *Gentleman's Magazine*, 1746-1757; S. Parsons and J. Bowles's map of Staffordshire, 1747; *A Plan of all the Houses destroyed and damaged by the Great Fire which began in Exchange Alley, Cornhill...*, 1748; collaborated with Thomas Kitchin on *The Small English Atlas* (first published in weekly parts), 1748-1749, 1751, Sayer & Bennett, J. Bowles and C. Bowles *c*.1775, Robert Sayer 1787, Bowles & Carver 1795 (an extended edition was published concurrently as *An English Atlas...*, Sayer & Bennett 1776, Robert Sayer 1787, Laurie & Whittle 1794, 1796, R.H. Laurie 1822, 1824); engraved maps for Thomas Salmon's *A New Geographical and Historical Grammar...*, 1749; acquired the plates for, and re-issued, G. Willdey's edition of Christopher Saxton's *The Shires of England and Wales*, *c*.1749 (with one new map); with Robert Sayer and John and Carrington Bowles he re-issued J. Rocque's *A New Plan of the City and Liberty of Westminster... the City of London and Borough of Southwark*, 1749; engraver and publisher for Joshua Fry and Peter Jefferson's *A Map of the most Inhabited part of Virginia...*, published *c*.1753 and later amended editions; *A Map of the Most Inhabited Parts of New England...*, 1755; *A New Map of Nova Scotia, and Cape Britain...*, 1755; William Gerard De Brahm's *Map of South Carolina and a Part of Georgia*, 1757; *A New and Accurate Map of the Kingdom of Ireland...*, 1759 (also published by Faulkner & Smith in Dublin); *The natural and Civil History of the French dominions in North and South America*, 1760; *A Description of the Maritime parts of France* (86 plates), 1761; *A Description of the Spanish Islands and Settlements on the Coast of the West Indies*, 1762; began a series of large scale county maps from new surveys beginning with Benjamin Donn's *A Map of the County of Devon...* (12 sheets), London, 1765; *The County of Bedford...* (8 sheets), *c*.1765 (new survey by J. Ainslie & T. Donald commissioned by Jefferys), republished W. Faden 1804; *A New Plan of the City of London*, 1766; Oxfordshire (4 sheets), A. Dury 1767 (surveyed by Thomas Hodgson in 1766); *A General Topography of North America and the West Indies*, 1768 (with Robert Sayer); Huntingdonshire (6 sheets), 1768, R. Sayer & T. Jefferys 1768, W. Faden 1804; A. Armstrong's map of County Durham (4 sheets), 1768; *The County of Buckingham Surveyed...* (4 sheets), A. Dury 1770, R. Sayer 1788, R.H. Laurie *c*.1818 (surveyed by J. Ainslie & T. Donald, 1766-1768); *A Plan of the River Thames and of the intended canal from Monkey Island to Isleworth*, 1770; Westmorland (4 sheets), 1770, W. Faden 1800 (surveyed by J. Ainslie & T. Donald); *The County of York* (20 sheets), 1771-1772 (published posthumously, surveyed by J. Ainslie, T. Donald & J. Hodskinson, 1767-1770, plates acquired by R. Sayer); maps published posthumously in Sayer & Bennett's *The West India Atlas*, 1775; and others. **Ref.** DICKINSON, G. *The Map Collector* 50 (1990) pp.32-34; HODSON, D. *County atlases of the British Isles published after 1703* Vol.II (1989) pp.57-80, ibid. Vol.I (1984) pp.146-149; PEDLEY, M. *The Map Collector* 37 (1986) pp.20-22; KRETSCHMER, DÖRFLINGER & WAWRIK *Lexicon zur Geschichte der Kartographie* (1986) Vol.C/1, p.366 [in German]; VERNER, C. *Imago Mundi* 21 (1967) pp.77-83; HARLEY, J.B. *Imago Mundi* 20 (1966), pp.27-48.

• **Jefferys**, Thomas, *junior* (*fl*.1771-*c*.1783). Son of Thomas, above. Worked for a while in partnership with William Faden as Jefferys & Faden, successors to part of the business of Thomas Jefferys senior (above). William Faden took control some time in 1783.

• **Jefferys & Faden** (*fl*.from 1773). Map publishers of London (see above). **N.B.** See **Faden, William**.

**Jeffreys, Charles.** Chart of Princess Charlotte Bay [Australia], c.1816.

**Jeffreys, J.** See **Jefferys, John** *junior*.

**Jeffs, W.** 'Librairie Etrangère de la Famille Royale', 15 Burlington Arcade, London. *Carte de Londres*, 1862; *Carte indispensable pour les étrangers*, 1862.

**Jehenne,** *Captain* Louis Auguste. French naval officer. St. Margaret Bay & New Férolle Cove [Newfoundland], 1847 (used on Admiralty Chart 299, 1866); Croc Harbour [Newfoundland], 1847 (used on Admiralty Chart 279, 1873).

**Jehotte, L.** Engraver. *Carte géologique... du Département de l'Orte*, 1801.

**Jehuda** ben Zara. Jewish portolan maker from Catalonia or Majorca, fled to Italy c.1492. Portolan chart of Europe, North Africa and the Mediterranean, c.1497 MS [Vatican Library, Rome].

**Jelle, I.** *Cleve und Umgegend*, Düsseldorf c.1848.

**Jelly, J.C.,** *RN*. Port Denison, Queensland, 1860 MS.

**Jelowiecki, Edward.** Polish military surveyor, 1820-1831.

**Jemme** van Dokkum. See **Frisius, Gemma.**

**Jemmett, William** (*fl.*1765-1787). Estate surveyor of Ashford, Kent. *A Map and Admeasurement of Perey Field Farm in the Parish of Peasmarsh...*, 1765 MS; Estate at Roxwell, Essex, 1773 MS.

**Jenderich, Christian.** *Wahre eigentliche Abbildung des Fürstl. Württemberg Hauses Sibyllen-Orth...*, c.1670.

**Jendrich, —.** Draughtsman. *Stadt Berne gegen Mittag. La Ville de Berne...*, 1757.

**Jenei, S.** See **Jankó, Sándor.**

**Jeney, Lajos Mihály.** Hungarian military cartographer, directed the survey and mapping of Erdély [Transylvania] from 1763. Also made surveys in Croatia, maps of which are now housed in the Imperial War Museum at Vienna. *Neue Situations-Charte des Großfürstentums Siebenbürgen nebst angrenzen-den Theilen der Moldau und Walacher* [Transylvania] (4 sheets), 1775 MS (reduced from earlier maps by D.T. de Fabris); Mining map of Transylvania, 1785.
**Ref.** BENDEFFY, L. 'Két ritka erdélyi térkép' [Two rare maps of Transylvania], *Geodezia és Kartográfia* [Geodesy and Cartography], (1973/6) [in Hungarian].

**Jengstrema, L.** Publisher of St. Petersburg. World, 1847.

**Jenichen** [B.I.; Ienichen], Balthasar (*d.*1599). Goldsmith, copper engraver and publisher of Nuremberg. Thought to have engraved Johannes Criginger's *Chorographia nova Misniae et Thuringiae*, 1568; *Nova Totius Palestina*, 1570; Cyprus, 1571; Sebastian Rotenhan's *Chorographia nova Franciae...*, 1571; *Thunis inn Africa* 1573; *Venetia*, 1575; Roma, c.1590.

**Jenifer** [Ienefer], *Captain* John. Darien, Isle of Pines, 1686 MS; *A Draught of the Golden... Islands...*, 1700, 1720.

**Jenig, Wolfgang Paul** (*d.*1805). Publisher of Nuremberg. Obtained and republished the copper-plate gores for J.G. Doppelmayr and J.G. Puschner's pair of 32-cm globe, 1791 (first published 1728, the terrestrial globe had been revised by Jenig).

**Jenings, William.** Moedin & Bengal Artillery. Plan of Madura, Southern India, 1755 MS.

**Jenkins, H.L.** Assam and Burma, 1827-1869.

**Jenkins, Henry Davenport** (*b.c.*1844). Stepgrandson to James Imray, worked for the Imray company from 1863 as author, editor and compiler of sailing directions (with James Frederick Imray). Joint manager of the Imray company from 1891, after the death of James Frederick Imray. *Atlantic Ocean Pilot*, 1884; *North Pacific Pilot*, 1885; *Indian Ocean Pilot*, 1886; *Pilots Guide for the English Channel*, Jenkins 1903; and others.

**Jenkins, John** (*fl.*1798-1834). Estate surveyor. Estate in St. Alban's Street belonging to St. George's, Windsor, 1825; Rectory of Wraysbury [Buckinghamshire], 1829 MS; Glebe Land Chapter Windsor, 1834.

**Jenkins, S.** Lizard to Land's End, 1810.

**Jenkins, T.A.** Topography for F.R. Hassler's *Map of New York Bay and Harbour*, 1844-1845 (with others).

**Jenkinson,** Anthony (*c*.1525-1611). Navigator, astronomer, merchant and cartographer of the Mercer's Company of London, chief factor of the London-based Muscovy Company from 1556. Travelled in Ottoman Empire, 1553, Russia, 1557-1559 and Persia via Russia, 1561-1564. Compiled *Nova absolutaque Russiae, Moscoviae, et Tartariae descriptio* (4 sheets), 1562 (engraved on copper by Nicholas Reynolds, used by Ortelius, 1570 and later, the de Jodes, 1578 and later, and others, one known extant copy at Wroclaw). **Ref.** RETISH, A.B. 'A foreign perception of Russia' insert in *The Portolan* (Summer 1995); BARON, S.H. 'The lost Jenkinson map of Russia' *Terrae Incognitae* 25 (1993) pp.53-66; KARROW, R.W. *Mapmakers of the sixteenth century and their maps* (1993) pp.317-320; WALDMAN & WEXLER *Who was who in world exploration* (1992) pp.347-348; MEURER, P. *Fontes cartographici Orteliani...* (1991) pp.175-176 [in German]; *The Map Collector* 48 (1989) p.38; BARON, S.H. 'William Borough and the Jenkinson map of Russia, 1562' in *Cartographica* 26:2 (Summer 1989) pp.72-85.

**Jenkinson,** Henry Irwin. *Jenkinson's Smaller Practical Guide to North Wales*, London, Edward Stanford 1878.

**Jenks,** Washington. *Weber County* [Utah] (4 sheets), 1888.

**Jenks,** *Reverend* William (1778-1866). *Explanatory Bible Atlas and Scripture Gazetteer, geographical, topographical and historical*, Boston, C. Hickling 1847.

**Jenkyns,** Francis. *A Plan for St. Ives Bay*, 1800.

**Jenner,** F. von [von Aubonne]. *Carte des Cantons Bern*, 1820.

**Jenner,** Thomas (*fl*.1618-*d*.1673). Bookseller, engraver and publisher of London, Freeman of the Grocers' Company (1619). At the 'Whit beare in Cornwale' [Cornhill] (from 1618); the 'White Beare by ye Exchange' (*c*.1622-1624); 'South Entrance of the Royal Exchange' (1643, 1644, 1666); 'at the Royal Exchange' (1672). The business and shop were acquired on his death by John Garrett, who continued to issue his maps. W. Grent's *A New and Accurate Map of the World...*, 1625; amended and republished *A Direction for the English Traviller*, 1643 (with maps engraved by Jacob van Langren, first published by Mathew Simmons, 1635); *The Kingdome of England & Principality of Wales Exactly Described* [the so-called 'Quartermaster's Map'] (6 sheets), 1644 and later (etched by Wenceslaus Hollar); *A Booke of the Names of all the Hundreds contained in the Shires of the Kingdom of England*, *c*.1644 (used reworked maps from *A Direction...*); *Description and Plat of the Sea-Coasts of England*, 1653; *A Book of the Names of all Parishes, Market Towns, Villages, Hamlets, and smallest Places, in England and Wales*, 1657, 1662, 1668 (used reworked maps from *A Direction...*); State 1 of *A new and exact map of America...*, 1666 (engraved by W. Hollar) used later in *A Map of the Whole World*, 1668. **Ref.** CARROLL, R.A. *The printed maps of Lincolnshire 1576-1900* (1996) N° 11 pp.33-36; TYACKE, S. *London map sellers 1660-1720* (1978) p.118; WORMS, L. in *The Map Collector* 34 (1986) pp.5-7; ROSTENBERG, L. *English publishers in the graphic arts 1699-1700* (New York 1963) pp.26-35.

**Jenney,** *Captain* —. *Battlefield of Chattanooga*, 1864.

**Jennings,** Cyrus. *Kentucky City*, 1855.

**Jennings,** D. *Use of Globes*, 1747.

**Jennings,** J., *junior*. *Plan of the Township of Ellington* [Huntingdon], 1774.

**Jennings,** John. Collaborated with John Leake in his *An Exact Surveigh of the Streets Lanes and Churches contained within the Ruines of the City of London...*, 1667 (engraved by Wenceslaus Hollar).

**Jennings,** Joseph C. *City of Dubuque, Iowa*, 1852.

**Jennings,** *Captain* R.H.W. *Baluchistan, Eastern Persia &c*. (6 sheets), Dehra Dun 1886.

**Jennings,** Robert. Publisher of London. Rest Fenner's *Pocket Atlas, Classical*, 1828; *Pocket Atlas of Modern Geography*, 1830; R. Fenner's *Pocket Atlas of Modern & Antient Geography*, London, *c*.1836.

**Jennings, W.H.** *Map of the Hocking Valley and Straitsville Coalfields* [Ohio], Cincinnati c.1874.

**Jennings, William**, father and son. Estate surveyors of Dorset.
• **Jennings, William** (fl.c.1768-d.1799). Produced manuscript maps of south western England. Hillfield & Scutts Farm [Dorset], 1792 MS.
•• **Jennings, William**, *junior* (fl.1792-d.1854). Surveyor of Evershot, Dorset, son of William (above).

**Jennings & Chaplin**. Publishers and booksellers of 62 Cheapside, London. One of the sellers of Thomas Moule's partwork *The English Counties Delineated*, from c.1831 (first published George Virtue, 1830).

**Jenny, Ludwig V.** (1719-1797). Austrian military cartographer.

**Jenotte**, —. Engraver for Claude-Joseph Drioux, 1865.

**Jenour, A.C.**, *RN*. British naval surveyor. Surveys used for *Table Bay breakwater and docks* [Cape of Good Hope], Admiralty Chart 1870, 1898 (with W.E. Archdeacon, drawn by A.J. Boyle, engraved by Davies, Bryer & Co.).

**Jensen, Harald.** *Frederiksborg...*, Copenhagen c.1876.

**Jensen, Hans Nicolai Andreas** (1802-1850). Danish theologian and historian. Germany, 1847; Sleswig, 1847.

**Jensen, J.L.** *Skole-Atlas...*, Copenhagen 1867 (with C.C. Brix).

**Jensen, Nicolas**. Printer of Venice. Credited with De situ orbis terrarum, 1473 (based on the work of Gaius Julius Solinus).

**Jenson, H.J. and Perronneau**. See **Jansen, H.J. and Perronneau**.

**Jentzsch, Alfred Karl** (1850-1925). German geologist. Geological maps of Prussia. *Elbing*, 1879; *Höhenschichten Karte Ost- und West-Preussens*, 1891; *Norddeutsche Tertiar*, 1913.

**Jenvilliers** [Jeanvilliers] —. French engraver. J.B. Nolin's *Direction du Mans divisée...*, 1710, 1720; J.B. Nolin's *Isle et Royaume de Sardaigne...*, 1717; *Galliae Christianae Aquitanica...*, 1718; Fr. L. de La Salle's *Diocese de Sées*, B. Jaillot 1718; *Gouvernement general du Lyonnais...*, Jaillot 1721 (both published in editions of the Jaillot *Atlas françois*); script on André Desquinemare's *Plan générale de la Forest de Fontainebleau...* (2 sheets), 1727 (map engraved by Q. Fonbonne); J.B. Nolin's *Galliae Chritianae Lugdunensis...*, 1728.

**Jep, I.** German engraver. Göttingen, c.1610.

**Jeppe, C.F.W.** (1833-1898). Surveyor of 'the Mining Department', Transvaal. *Map of the Transvaal* (4 sheets), 1868, 1877, 1889; Kaap Goldfields, Edward Weller 1888; Witwatersrand, 1888; *Die neue Grenze... der Südafrikanischen Republiek*, 1892; *Map of the Southern Goldfields*, 1894; *Jeppe's Map of the Transvaal or S.A. Republic and Surrounding Territories...* (6 sheets), Edward Stanford 1899 (with F. Jeppe, lithographed in Winterthur by Wurster, Randegger & Cie).

**Jeppe, Frederick**. Editor in the Surveyor General's Department, Transvaal. Fellow of the Royal Geographical Society. *Map of the South African Republic...*, London, S.W. Silver 1877, 1880 (engraved by E. Weller); *Jeppe's Map of the Transvaal or S.A. Republic and Surrounding Territories...* (6 sheets), Edward Stanford 1899 (with C.F.W. Jeppe, lithographed in Winterthur by Wurster, Randegger & Cie).

**Jerez, F.M.Y.** See **Moreno** y Jerez, Federico.

**Jermyn, G.A.** *Plan of the Great Western Railway*, 1838.

**Jerome, Saint** [Hieronymus] (c.348-420). Early Christian scholar, writer and translator of Bible into Latin. Although sometimes credited with maps of Palestine and Asia, those drawn to illustrate his writings, known as 'Jerome maps', date from the 12th century.

**Jérome, J.** Engraver of '38 Rue Monge', Paris. *Environs de Hanoi*, 1885; one of the engravers for L. Grégoire's *Atlas Universel de Géographie Physique & Politique*, c.1892; *Carte générale du Tonkin*, 1898.

**Jerrard**, Paul. Lithographer, 206 Fleet St., London. Gold regions of California, 1849.

**Jersey**, T. Chart of the River Thames, 1750 MS.

**Jervis**, father and son.
• **Jervis**, *Captain* (later *Lieutenant-Colonel*) Thomas Best (1797-1857). Geodesist, Director of the Topographical & Statistical Depôt, War Department. Prepared and supervised the preparation of many military and strategic maps and plans, particularly in association with the Crimean War. Operations in Konkan Province, 1824; *Chinese Plan of the city of Peking* (4 sheets), 1843 MS; *Present Seat of War* (21 sheets) [Crimean War], 1854; *Map of the present Seat of War* (6 sheets), 1854; *Map of Circassia and the Russian Territories North of the Kuban...*, 1855 (drawn by W.P. Jervis); *Map of Europe shewing the Actual Boundaries of the various Empires and States as settled by Treaty*, London, Messrs Williams & Norgate 1856; *Map of the Principal Military Communications of the Kaucasus & Contiguous Frontier provinces...*, London, Williams & Norgate 1856-1857 (taken by Jervis from a map prepared by the Imperial Army of the Caucasus, 1847); *Khiva and the Aral Sea*, 1857.
•• **Jervis**, *Chevalier* G. Son of Thomas Best Jervis. New Cycloidal Projection, 1895.

**Jervis**, F.P. Berea District, Basutoland, 1881; Leribe District, 1881.

**Jervis**, James T. Engineer, of 9 Victoria Street, Westminster. Torrington & Okehampton Railway, 1897.

**Jervis**, William Paget. Draughtsman, probably related to Thomas Best Jervis, above, whose memoirs he edited in 1898. *Map of Circassia and the Russian Territories North of the Kuban... astronomical observations by ... T.B. Jervis*, 1855; Geological map of the Crimea, 1855; Ethnological map of the Crimea, 1857; geological notes on Italian maps, 1860-1889.

**Jervois**, *Lieutenant*, J.G. RE. Military surveys and plans at St. John's, Newfoundland, 1851.

**Jervois**, *Captain* (later *Lieutenant-Colonel*) *Sir* William Francis Drummond, *RE* (1821-1897). As Deputy Director of Works (Fortifications) he endorsed many maps and plans. Keiskan Hoek, 1850 MS; *Military Sketch of part of British Kaffraria* [South Africa] (3 sheets), London, J. Arrowsmith 1850; surveys at the Cape of Good Hope, 1851; manuscript plan of Portsmouth, 1857; defence surveys at Point Levis, Canada, 1865; fortification surveys in Quebec (24 sheets), 1865-1866 (superintended by Jervois, zincographed at the Ordnance Survey); *Memorandum with reference to the improvements of the Defences of Malta and Gibraltar* (with maps), 1868; superintended fortification surveys in Canada, including Montreal (83 manuscripts), 1868-1869.

**Jesse**, *Captain* William. *Plan of the Harbours... of Sevastapol*, 1839.

**Jesse**, *Lieutenant* William Howard RE. *Ground Plan of the theatre Gibraltar*, 1844 MS.

**Jessop**, family. Civil engineers. **Ref.** BENDALL, S. *Dictionary of land surveyors and local mapmakers of Great Britain and Ireland 1530-1850* (2nd edition 1997) pp.276-277.
• **Jessop** [Jessope], William I (1744-1814). Civil engineer from Devon. Produced engineering and inland navigation plans, mostly in the midland and northern counties. *Plan of the River Trent*, 1782; *A Plan of the Ashby-de-la-Zouche Canal*, 1792; *River Foss with the line of the intended navigation and drainage*, 1792 MS; plan River Avon & Frome, 1792 (with W. White), William Faden 1793; proposed harbour improvements at Bristol, 1803 (with W. White, engraved by William Faden).
•• **Jessop**, Josias (1781-1826). Engineer trained by his father William I, above. Produced canal, railway and tramway maps.
•• **Jessop**, William II (1783-1852). Civil engineer, son of William I, with whom he worked.

**Jeuney**, Amos. Plan of Sparta, Ohio, 1815 MS.

**Jevezy**, E. Lithographer of Melbourne. Colony of Victoria, 1863; *Map of Australia shewing routes of the explorers*, Melbourne, J. Hamel 1863 (lithographed on stone by E. Jevezy).

**Jewell**, —. Publisher. *Map of the City of New Orleans*, 1870.

**Jewell**, R.J. *A Map of the Gold Fields of South-Eastern Africa*, 1876.

**Jewett,** C.F. *Atlas of Columbia County, Pennsylvania,* 1876.

**Jewett,** E.R. Publishers of Buffalo, New York. J.H. Fairbanks's *A map of the vicinity of Niagara Falls,* 1857.

**Jewett,** J.P. [& Co.]. Publishers of Boston. *A Gazetteer of Massachusetts,* 1849 edition (first published by John Hayward 1846); J. Hayward's *A Gazetteer of New Hampshire,* 1849; E.B. Whitman and A.D. Searl's map of eastern Kansas, 1856.

**Jewett,** Joseph. Publisher of Baltimore. *The Universal School Atlas,* 1832.

**Jewett,** W.P. *Sectional Map of the State of Minnesota,* 1878.

**Jezl,** Jacob. *Victoria soden 2 September Ao 1686...* [battle at Budapest], *c.*1687.

**Jia** Dan (730-805). Chinese cartographer of the Tang dynasty. Map of China on silk, 801.

**Jikoan,** Mabuchi. See **Mabuchi** Jikoan.

**Jilek,** August von (1818-1898). Oceanographer.

**Jimbo** Kotora (1867-1924). Japanese geologist and mineralogist. *Hokkaido chishitsu-kosan zu* [Geological map of Hokkaido with the location of useful minerals], 1896; *Hokkaido zenzu* [Map of Hokkaido], 1896.

**Jinryūdō** —. Japanese publisher. Izōsai Ikkei's bird's-eye view map of Yamagata, Niigata and Fukushima, Japan, [Late Edo].

**Jireček,** Josef (1825-1888). Czech historian and map editor. *Královstwí Ceské...,* W. Praze 1850; *Druhé vydáni,* Ve Vidni 1875.

**Jisbury,** J. Maps of Burma, 1863-1871.

**Jiseiken,** Okada. See **Okada** Jiseiken.

**Joalland,** Jules (1870-1940). French military officer and explorer in Africa. *De Zinder au Tchad et la conquète du Kanem,* 1901. **Ref.** BROC, N. *Dictionnaire illustré des explorateurs Français du XIXe siècle: I Afrique* (1988) pp.181-182 [in French].

**Joami** Kise (*d.*1618). Japanese artist. Map of Japan engraved on silver, *c.*1600.

**Joanne,** father and son. French geographers and writers, produced departmental maps and guides for travellers into the 20th century.
• **Joanne,** Adolphe Laurent (1823-1881). *Atlas historique et statistique des chemins de fer Français,* Paris 1859; *Atlas de la France* (95 maps), Paris, Hachette 1870 and later (maps engraved by Erhard); *Atlas de la Défense Nationale,* Paris, Librairie Hachette et Cie. 1870; and many others.
•• **Joanne,** Paul Bénigne (*b.*1847). Son of Adolphe Laurent Joanne, continued his father's maps and guides. *Guides Joannes. Strasbourg,* Paris, Hachette 1883, 1885 (drawn by L. Thuillier); *Nord,* 1891; *Guides Joanne. Environs de Plombières,* Hachette 1894 (drawn by L. Thuillier); *Pas de Calais,* 1894; *Alsace-Lorraine. Carte départmentale,* 1919; and many others.

**Joannes** de Stobnicza. See **Stobnicza,** Johannes de.

**Joannis,** Georgius. See **Giovanni,** Giorgio.

**João,** Bartolomeu. Engineer. Map of Madeira, *c.*1654 MS.

**Joaõ,** Pessoa. *Afbeelding der Stadt en fortressen van Parayba,* Amsterdam, Claes Jansz. Visscher *c.*1635.

**Jobard,** Ambroise. Lithographer and surveyor, worked in Amsterdam and Brussels, 1827-1830.

**Jobbins,** John Richard. Engraver, lithographer and publisher of 3 Warwick Court, Holborn, London. *The Environs of London to the Extent of 30 Miles from St Pauls, with all the Railways to 1842,* 1842, 1844; *A New Map of Great Britain,* 1843; *Plan of the Town of Brighton,* 1843; W.J. Curtis's *Trinidad Railway,* 1848, 1871.

**Jobin,** André. Surveyor *Carte de l'Île de Montreal...,* 1834; *Map of the City of Montreal,* 1834.

**Jobin,** Bernhard. Printer and publisher of Strasbourg. T. Schöpf's *Inclitae Bernatum Urbis* (9 sheets), 1578 (engraved by M. Krumm and J. Martin).

**Jobit,** *Lieutenant* E. French explorer in Congo region. *Le cours inférieure de la*

*Likouala-aux-herbes*, 1900; *La mission Gendron au Congo français*, 1901. **Ref.** BROC, N. *Dictionnaire illustré des explorateurs Français du XIXe siècle: I Afrique* (1988) p.182 [in French].

**Jobson**, Francis (*fl.*1579-1602). Land surveyor in Ireland from *c.*1579. Limerick, 1587; Provincial maps of Munster, 1589 MS; Cork, *c.*1589; Map of Ulster, *c.*1590 MS; treatise on the fortification of Ulster, 1598; map of Connaught, 1602.

**Jobson**, Richard. English traveller. *The Golden Trade or a discovery of the River Gambra* [Gambia], 1623 (used by Samuel Purchas in *Purchas His Pilgrimes*, 1625).

**Jocelyn**, brothers. Engravers and publishers of New Haven, Connecticut. Worked together as N. & S.S. Jocelyn in the 1820s. M.L. Picor's *Carta esferica que comprende los rios de la Plata, Parana, Uruguay y Grande y los terrenos adyacentes a ellos*, New Haven, Howe & Spalding 1819 (with A. Doolittle); engraved maps for Sidney Edwards Morse's *An Atlas of the United States...*, 1823; *A Plan of Stratford, Con.*, 1824; *The World*, 1825; Davis Hurd's *Map Exhibiting the Farmington, & Hampshire & Hampden Canals...*, New Haven, N. & S.S. Jocelyn 1828; engraved D.W. Buckingham's *Map of the City of New Haven*, 1830; and others.
• **Jocelyn**, Nathaniel (1796-1881). Draughtsman and engraver specialising in maps and banknotes. In the 1820s worked with his brother as engravers N. & S.S. Jocelyn. Journeyed to Europe with Samuel F.B. Morse, 1829-1830.
• **Jocelyn**, Simeon Smith (1799-1879). Engraver of maps and portraits. Simeon gave up engraving in 1858.
•• **Jocelyn, Darling & Co.** Publisher of D.W. Buckingham's *Map of the City of New Haven*, 1830 (engraved by N. & S.S. Jocelyn).
••• **Jocelyn, Draper, Welsh & Company.** Became the American Bank Note Company in 1854.

**Jocelyn**, A.H. Publisher in New York. Fermin Ferrer's *Government map of Nicaragua...*, 1856.

**Jode**, de, father and son. **Ref.** KRETSCHMER, DÖRFLINGER & WAWRIK *Lexicon zur Geschichte der Kartographie* (1986) Vol.C/I pp.366-367 [in German]; SHIRLEY, R. *The mapping of the World* (1983) entry N°85 pp.96-97; *ibid.* N°100 pp.113-114; *ibid.* N°124 pp.146-147; *ibid.* N°165 pp.184, 186; *ibid.* N°184 pp.202-203; KOEMAN,C. *Atlantes Neerlandici* (1967-1971) Vol.II pp.205-212; ORTROY, F. van *L'Oeuvre cartographique de Gérard et de Corneille de Jode* (Amsterdam 1963) [in French].
• **Jode** [Judaeis; Judaeus; Iuddeis], Gerard de (1509-1591). Engraver, printer, printseller, publisher and cartographer from Nijmegen, became active in Antwerp *c.*1550. Father of Cornelis de Jode. Re-engraved Giacomo Gastaldi's *Universalis exactissima atquae non recens modo...*, 1555 (first compiled 1546); B. Musinus's map of Europe (6 sheets), 1560; *Germania* (4 copper plates), 1562; F. Alvares Seco's *Portugalliae...*, 1565; 17 Provinces, 1566; *Hungariae Typus*, 1567; wall map of Africa, 1569; *Angliae Scotiae et Hibernie Nova Descriptio*, 1570; *Speculum Geographicum...*, 1570; *Nova totius terrarum orbis descriptio...*, 1571 (after Ortelius, 1564, engraved by Johann and Lucas Doetichum); wall map of America, 1576 and later; wall map of Asia (9 copper plates), 1577; *Speculum Orbis Terrarum*, 1578 (several maps engraved by Johan and Lucas van Doetichum, some maps dating from 1569), re-issued by Cornelis 1593; wall map of Europe (6 copper plates), 1584 (engraved by Lucas and Johann van Doetichum). **N.B.** Koeman describes the 1555 world map as an Italian print with a de Jode credit pasted on. Shirley on the other hand, describes it as a re-engraving.
•• **Jode** [Judaeis], Cornelis de (1568-1600). Son of Gerard de Jode. Engraver, publisher and scholar of Antwerp. Following de Jode's death, plates passed into the possession of Jan Baptist Vrients, then in 1612 to Jan Moretus. *Gallia occidentalis*, 1592; *Speculum Orbis Terrae*, 1593 (a re-issue of his father's *Speculum* of 1578); *Africae Nova Delineatio* (9 sheets), 1596.

**Jode**, Arnold de. Engraver. Plans, 1713.

**Jodot**, Marc. *Carte Industrielle du Département du Nord*, 1830-1834.

**Joeck**, —. Copper engraver of Berlin. One of the engravers for *Topographisch oeconomisch und militærische Charte des Herzogthums Mecklenberg Schwerin und des Fürstenthums Ratzeburg...* (16 sheets), 1788.

**Joel**, A. *Plan of Wolsingham Park Estate in the Parish of Wolsingham and County of Durham 1863*, 1864.

Johann von Armsheim                                    John of Holywood

*Gerard de Jode (1509-1591). Portrait by Hendrik Goltzius. (By courtesy of the Plantin/Moretus Museum, Antwerp)*

**Johann** von Armsheim. See **Schnitzer**, Johannes.

**Johannes** de Sacrobosco. See **Sacrobosco**, Johannes.

**Johannes** de Villadestes. See **Viladestes**, Johanes de.

**Johannes** Utinensis [Giovanni da Udine] (*d*.1366). Mapmaker from Udine, Italy. Compiled a 'T-O' world map, *c*.1350.

**Johannes** von Gmunden (*c*.1384-1442). Mathematician and astronomer of the Vienna school. Thought to have made a plan of Vienna, upon which the so-called *Albertinischer Plan* was based. **Ref.** KRETSCHMER, DÖRFLINGER & WAWRIK *Lexicon zur Geschichte der Kartographie* (1986) p.10.

**Johannes**, Magnus. See **Magnus**, brothers.

**Johannesen**, *Captain* E.H. Karisches Meer, Gotha, Justus Perthes 1870; Novaya Semlya, 1871; West-Sibirisches Eismeer, 1879.

**John** of Ephesus (*fl*.6th century). Nubia.

**John** of Holywood [Halifax]. See **Sacrobosco**, Johannes.

**John**, Jacob. American geographer. *The United States. Geographical amusement*, 1806.

**John**, Jesse. *Map of San Francisco*, 1857.

**Johnes** [Johns], *Lieutenant* E. Owen, *RN*. Assisted with Admiralty charts of African coasts, 1822-1826. *Survey of Port el Roque, Terra Firma*, 1827-1828.

**Johns**, D.J. *Plan of the Isthmus of Tehuantepec* [Mexico], James Wyld 1851.

**Johns**, William, *Master RN*. Admiralty charts of Australia, 1823-1831. Rows Channel, 1827 (published on a chart of Moreton Bay, 1848).

**Johnson**, —. See **Franks** & Johnson.

**Johnson**, —. See **Gaston** & Johnson.

**Johnson**, *Captain* —. Survey of Bhopal, 1821.

**Johnson**, Alvin Jewett and associated companies. **Ref.** LOURIE, I.S. 'The atlases of A.J. Johnson' in *The Portolan* 49 (Winter 2000-2001) pp.7-17; **N.B.** Compare with D. Griffing **Johnson**, below.
• **Johnson**, Alvin Jewett (1827-1884). Publisher at 111 Broadway, New York (1858); 172 William Street, New York (1859); 133 Nassau Street, New York (1860). Produced atlases, maps and encyclopaedias. Originally a book canvasser in the mid-western United States, went to New York in 1856 and started his own publishing company, A.J. Johnson, in 1858, later incarnations included; Johnson and Browning, 1859-1862; Johnson and Ward, 1862-1866; A.J. Johnson, 1866-1879; Alvin J. Johnson and Son, 1879-1880; Alvin J. Johnson and Co., 1881-1887. In its early years the firm was closely related to the firm J.H. Colton. Johnson is best known for his *...Family Atlas...*, which was sold by subscription through door-to-door book canvassers. Originally using the maps of J.H. Colton, it was updated almost every year, the maps being replaced by Johnson's own. Different editions of the atlas can be used to track the development of railways throughout the world and the development of the states of the USA. By 1866 all Colton's maps had been replaced, and Johnson was running the company himself. *Johnson's New Copper Plate Map of Michigan and the Great Lakes*, New York, A.J. Johnson, Chicago, Rufus Blanchard 1858 (printed by D.& J. McLellan); *Johnson's New Illustrated and Embellished County Map of the Republics of North America* (6 sheets), 1859 and editions to 1861, later updated and printed by J.H. Colton; *Johnson's New Illustrated (Steel Plate) Family Atlas... Under the supervision of J.H. Colton and Alvin Jewett Johnson*, 1860, Johnson & Browning 1861, 1862, 1863, Johnson & Ward 1864, A.J. Johnson 1866 with (almost) annual updates to 1887; *Johnson's map of the vicinity of Richmond and peninsular Campaign in Virginia*, New York 1862; *Johnson's New Universal Cyclopedia of Useful Knowledge*, 1879-1887; and many others.
•• **Johnson & Browning** (*fl.*1859-1862). Publisher of Richmond and New York, A.J. Johnson and Ross C. Browning. *Colton's General Atlas*, 1859 edition; re-published, simultaneously with J.H. Colton, *Colton's Connecticut with portions of New York and Rhode Island*, 1859 (previously published 1854, 1855, 1857); *Johnson's New Illustrated (Steel Plate) Family Atlas*, 1862.
•• **Johnson & Ward** (*fl.*1862-1866). 'Successors to Johnson & Browning'. When Browning withdrew from the business he was replaced by Benjamin P. Ward as financial backer. Johnson bought out Ward's share in 1866. *Johnson's New Illustrated (Steel Plate) Family Atlas*, 1862, 1863, 1864 editions.
•• **A.J. Johnson** (*fl.*1866-1879). A.J. Johnson in sole charge of the company publishing his *Family Atlas*.
•• **Alvin J. Johnson & Son** (*fl.*1879-1880). Alvin's son joined the business in 1879. He managed the company after the death of his father in 1884.
•• **Alvin J. Johnson & Co.** (*fl.*1881-1887). Publishers of Johnson's *Family Atlas*. From 1884 run by Alvin's son.

**Johnson**, D. Griffing. Geographer, publisher and engraver of New York and Washington. 80 Nassau Street, New York (1847); Trinity Buildings, 111 Broadway, New York. *Johnson's Illustrated & Embellished Steel Plate Map of the World on Mercator's Projection*, New York, D. Griffing Johnson 1847 (compiled, drawn & engraved by D.G. Johnson), republished New York, J.H. Colton 1849 (title changed to begin *Colton's Illustrated...*); *Map and Chart of the New World*, 1853; *Railroad map of Europe*, 1854;

*New Illustrated and Embellished County Map of the Republics of North America*, New York and Washington, D.G. & A.J. Johnson 1857; *Railroad... map of Illinois*, 1858; *Railroad and Township Map of the Middle States*, 1860; *County and Railroad Map of the United States* 1863. **N.B.** The editors have been unable to clarify the relationship between D.G. **Johnson** and A.J. **Johnson** and associated companies.

**Johnson**, E. *A Plan of the Level of Ancholme as surveyed in the years 1767 and 1768* [Lincolnshire], London 1791 (with J. Dickinson, 'reduced by E. Johnson and I. Dalton'). **N.B.** Compare with Edward **Johnson**, below.

**Johnson**, E. *Berwick to North Sunderland*, William Heather 1812. **N.B.** Compare with Edward John **Johnson**, below.

**Johnson**, E. *Plan of the Town of Hertford*, Cambridge 1830 (engraved by Storer).

**Johnson**, E.J. *Chart of the coasts of Northumberland and Durham*, 1819 (engraved by J. Eastgate). **N.B.** Compare with Edward John **Johnson**, below.

**Johnson**, E.V. *Prairie Region*, 1879; *Map of the Dominion of Canada 1882* (6 sheets), Montreal 1882 (drawn by A.M. Edmonds, with a table of Canadian railways, lithographed by The Burland Lithographic Co.); *Map Shewing the Railways of Canada* (3 sheets), Montreal, The Sabiston Litho. & Publishing Company 1891 (drawn by A.M. Edmonds).

**Johnson**, Edward. Inclosure plan of Tetney, Lincolnshire, 1778 MS; Goxhill Manor, 1785 MS; Barrow-upon-Humber, 1785 MS.

**Johnson**, *Commander* Edward John, *RN*. Admiralty charts Farne Islands to Berwick, 1831; East Coast of England, 1831-1866.

**Johnson**, Edwin F. Civil Engineer, chief engineer of the Northern Pacific Rail Road. *Map of the proposed Northern Route for a Railroad to the Pacific*, Hartford, Connecticut, E.C. Kellogg & Co. 1853.

**Johnson**, Francis W. Plot of Harrisburg, Texas, 1826.

**Johnson**, George, of Portaferry. See **Johnston**, George.

**Johnson**, George, *RN*. *Apia Bay in the Island of Upolou* [Samoa], 1843; *Port Underwood in Cloudy Bay* [New Zealand], Admiralty Chart 1840.

**Johnson**, *Colonel* Guy (*c*.1740-1788). Arrived in North America from Ireland *c*.1756, became superintendent of Indian affairs in 1774. Map of Colonel Bouquet's Expedition against the Ohio Indians, 1764 MS; *Map of the Country from the 40th. Deg of Latitude to Lake Ontario comprehending New York, Jersey & Pennsylvania*, 1768; *Map of the Northern Colonies with the Boundary Line established between them and the Indians at the treaty...1768*, 1769 MS; map of the Six Nations, 1771 MS (an engraved copy was published in E.B. O'Callaghan's *A Documentary History of the State of New York*, 1849-1851).

**Johnson**, Henry. *The World on Mercator's Projection* (9 sheets), 1889.

**Johnson**, Isaac (1754-1835). Surveyor and antiquary of Woodbridge, Suffolk. Surveys of Essex and Suffolk estates, from 1790; including Gestingthorpe, 1804. **Ref.** BLATCHLY, J. & EDEN, P. *Isaac Johnson of Woodbridge 1754-1835: that ingenious artist* (Ipswich 1979).

**Johnson**, J. Publisher of 'St. Paul's Church Yard', London. *A Map of Surinam*, 1791 and *A Map of Guiana &c.*, 1793 (both engraved by T. Conder, both in J.G. Stedman's *Surinam*).

**Johnson**, J. *A New and Improved Map of the Island of Antigua...*, Smith Elder & Co. 1829 (engraved J. & J. Neele, another issue in *An Historical Account of Antigua*, 1830).

**Johnson**, Jacob. Publisher of 147 Market Street, Philadelphia. Joseph T. Scott's *A Geographical Dictionary of the United States...*, 1805; *The United States Geographical Amusement*, 1806; [Richard] Brookes' *General Gazetteer Improved*, 1806 (1812 edition published by Johnson & Warner); and others.

**Johnson**, John. Anglicised form of **Janssonius**, Johannes *q.v.*

**Johnson**, John. East Anglian land surveyor. *Descriptio Omium metarum et bundarumpro*

*Hundreda de Wisebech* [Wisbech, Cambridgeshire] 1597.

**Johnson**, *Lieutenant* John, RE. Manuscript military plans of Fort George, Guernsey, 1792 MS.

**Johnson**, John Hugh. Cartographer at 1 Campbell Street, Liverpool, Fellow of the Royal Geographical Society from 1854. Worked for Philip, Fullarton and Stanford. *Oceania or Islands in the Pacific Ocean... Polynesia, Malaysia and Australasia*, Edinburgh, London & Dublin, A. Fullarton & Co. 1870; *Philips' Map of Ireland and its Railways*, London & Liverpool, G. Philip & Son 1870.

**Johnson**, John Morris [& Sons]. Lithographers of London. 'J.M. Johnson & Sons, Steam & Chrome Printers, 3 Castle St. Holborn, & 56 Hatton Gardens, E.C.' London... for the *Post Office Directory*, 1868 and editions to 1877 (from an original by B.R. Davies, 1859); *Post Office Map of Warwickshire*, Kelly & Co. 1868 (from an original by Benjamin Rees Davies, 1845); maps for *The Post Office directory for the six home counties...*, London, Kelly & Co. 1870 (from B.R. Davies originals); *Kelly's Guide to London*, 1871; *Post Office map of Hertfordshire*, London, Kelly & Co. 1871 (based on a map by Benjamin Rees Davies, 1845); F. Bryer's *Post Office Map of Hertfordshire*, 1874 (eventually replaced the Davies map in the Kelly *Directory*); and others.

**Johnson**, Joseph. London bookseller. One of the many sellers of E. Gibson's edition of Camden's *Britannia*, 1772 (with maps by Morden). **N.B.** Compare with Joseph **Johnson**, publisher, below.

**Johnson**, Joseph. Publisher of 72 St. Paul's Churchyard, London. William Paterson's *Narrative of Four Journeys into the country of the Hottentots and Caffraria...*, 1789 (with a map of southern Africa engraved by T. Conder); John Aikin's *England Delineated...* (43 maps), 1790, 1795, 1800, 1803; *United States of America*, 1791.

**Johnson**, *Midshipman* later *Lieutenant* Robert E. (*fl.*1838-1842). Served aboard the U.S. Brig *Porpoise* with Charles Wilkes and others on shore mapping for the U.S. Coast Survey. Acted as one of the surveyors for several charts and as one of the draughtsmen for *Chart of the Southern Coast from Tybee Bar to Hunting Id. May River*, 1838 (with J. Alden and W. May, lithographed by P. Haas). **Ref.** GUTHORN, P. *United States coastal charts 1783-1861* (1984) pp.47, 128, 208.

**Johnson**, Rowland (*fl.*1559-*d.*1587). Surveyor of the King's [Queen's] Works. Manuscript plan of Berwick, *c.*1575; Norham, 1576; Portsmouth Harbour, *c.*1584 MSS.

**Johnson**, Samuel L. *Handy Map of Eastbourne*, 1894.

**Johnson**, T. One of the engravers for David Mortier's *Nouveau théâtre de la Grande Bretagne*, 1715-1728.

**Johnson**, T. Draughtsman for Henry Gregory's *Trincomalay*, 1787; *Subec*, 1787.

**Johnson**, Thomas. See **Johnston**, Thomas.

**Johnson**, Thomas. Publisher of Manchester. *Johnson's Atlas of England: with all the Railways* (42 maps), 1847; *A New General Atlas*, *c.*1850.

**Johnson**, *Staff Commander* Valentine F. *RN*. One of the surveyors for the Admiralty chart of *Mediterranean Sicily*, 1864-1870.

**Johnson**, W. [& Son]. *Canals of Yorkshire, Lancashire, Derbyshire & Cheshire*, 1825.

**Johnson**, *Reverend* W.P. *Yao country* [Africa], 1822.

**Johnson**, W.W. *West coast*, 1864 MS.

**Johnson**, William. Anglicised form of **Blaeu**, Willem Jansz. *q.v.*

**Johnson**, William father and son. Surveyors of Manchester. *Map of the Parish of Manchester in the County of Lancaster... 1818-1819* (4 sheets), Manchester, William Johnson 1820; *Manor of Turton* [Lancashire], 1830; *The estate of the late James Greene esq....*, 1832 (with J. Ashworth and R. Dalglish).

• **Johnson**, William. Surveyor and publisher of Manchester. *Plans of Manchester*, 1818-

*Archibald Johnston (d.1925). (By courtesy of the Royal Geographical Society, London).*

1819, published 1870-1871.

•• **Johnson**, William. Son of William, above. Worked with his father on estate and railway surveys. *G. Bradshaw's Map of canals, Navigable Rivers, Railways, etc. in the Southern Counties of England* (3 sheets), Manchester & London 1830.

**Johnson**, William (fl.1806-1840). Scottish surveyor, map editor and draughtsman. Prepared maps and town plans, some of which were used in John Thomson's *Atlas of Scotland*, 1820-1832. *Linlithgowshire*, 1820 (engraved by Hewitt, with an inset of the town of Linlithgow, for Thomson's *Atlas*); *Peebles-Shire*, 1821 (engraved by W. Dassauville, with an inset of the town of Peebles, for Thomson's *Atlas*); *Plan of the Island of Lewis*, 1821 (lithographed by Forrester & Ruthven); *Kirkcudbright Shire* (2 sheets), 1821 (engraved by John Moffat); *Haddington*, 1822 (inset on *Haddington-Shire* in Thomson's *Atlas*); *Lanarkshire* (2 sheets), 1822 (engraved by Sidney Hall); ...

Western Isles (3 sheets), J. Thomson 1822-1823 (engraved by W. Dassauville) *Skye Island &c. Part of Inverness Shire*, 1824 (engraved by S.I. Neele & Son); contributed to J. Thomson's map of Perthshire, 1827 (engraved by S. Hall); ...*Ayrshire* (4 sheets), J. Thomson 1828 (engraved by Thomas Clerk); Inverness-shire (2 sheets), c.1830 (engraved by J. & G. Menzies); *Plan of the Lordship of Lochaber...*, 1831 (lithographed by A. Forrester); *Map of Part of Inverness-Shire*, 1831 (lithographed by Ballantine); *Gellatly's new map of the country 12 miles round Edinburgh*, 1834, 1836, 1840 (engraved by John Gellatly); and others.

**Johnson**, William P. *Map of Morgan County, Ohio*, McConnelsville 1854.

**Johnson & Browning**. See **Johnson**, Alvin Jewett.

**Johnson & Ward**. Publishers. See **Johnson**, Alvin Jewett.

**Johnston** [Johnson] family and companies.
Ref. SMITH, D. *Journal of the International Map Collectors' Society* 82 (Autumn 2000) pp.9-19.
• **Johnston**, Sir William (1802-1888). Founder of the Johnston publishing house. Lord Provost of Edinburgh, 1844-1851. Apprenticed to the James Kirkwood, Scottish engraver, watchmaker, globemaker and publisher, moving on to open a workshop at 6 Hill Square, Edinburgh in 1825, where he was soon joined by his brother Alexander Keith.
• **Johnston**, Alexander Keith I (1804-1871). Geographer, Fellow of the Royal Geographical Society from 1842. Younger brother of Sir William, joined the firm in 1826, having been apprenticed to J. Kirkwood and Sons 1820. Produced many maps in his own name, most of them engraved and published by W. & A.K. Johnston. Styled himself *Geographer at Edinburgh in Ordinary to the Queen*. First maps appeared in *A Traveller's Guide Book*, 1830; *The Physical Atlas*, 1848 (after Heinrich Berghaus, first British atlas to give a synoptic view of physical geography); Palaeontological Map of the British Isles, 1850 (with E. Forbes); *Johnston's Geological & Physical Globe...*, 1851; *The Atlas of Scotland*, 1855; *Physical Chart of the Pacific Ocean...*, 1855; *Atlas of the United States of North America...*, 1857 (with Professor Henry Darwin Rogers); commissioned by Stanford to produce a series of wall maps of the continents, c.1857 e.g. *Stanford's Library Map of Europe*, 1858, 1866, 1882 (engraved by W. & A.K. Johnston); *Stanford's Library Map of Australasia*, 1859; *The Royal Atlas of Modern Geography*, London and Edinburgh, William Blackwood and Sons 1861 (compiled by A. Keith Johnston, engraved and printed by W. & A.K. Johnston 1861; *Keith Johnston's General Atlas*, c.1862; *Johnston's military sketch map of central Europe showing the fortifications, battlefields, railway and river communication &c.*, c.1866; later editions of half-crown, shilling, and sixpenny Atlases, 1869; *New Cabinet Atlas*, 1873; *People's Pictorial Atlas*, 1873; and many others.
• **W. & A.K. Johnston** (fl.1826-1901). Founded by Sir William Johnston. Engraver and printer, later publisher of 6 Hill Square, Edinburgh (from 1825); 160 High Street, Edinburgh (from 1826); 107 George Street (from 1835); 4 St. Andrew Square (from 1837); Edina Works, Easter Road (from 1879). Also London offices at 74 Strand (1869); Paternoster Row (1873); 6 Paternoster Buildings (1878); 5 White Hart Street, Warwick Lane (1884). Appointed engravers to King William IV, 1834, engravers to Queen Victoria from 1837. Acquired the business of W.H. Lizars, c.1862. Appointed Edinburgh agents for the Ordnance Survey in 1866. Extensive and varied cartographic output, especially thematic maps and globes. Many maps, atlases and globes were produced in collaboration with other publishers, including A.H. Andrews, Rand McNally, Weber Costello and A.J. Nystrom in the United States. *A Traveller's Guide*, 1830; Atlas to Sir Archibald Alison's *History of Europe*, 1833-1842; plan of Edinburgh, 1834; *Johnston's Road Map of Scotland*, 1841; *National Atlas*, 1843; *Atlas of European Battles 1789-1815*, 1848; globe, 1851; *The Atlas of Physical Geography* (18 maps), 1852; *Atlas of Classical Geography*, 1853; *The Physical Atlas of Natural Phenomena*, 1856 (a thematic world atlas); *Johnston's Map of the United States*, 1861; *Royal Atlas*, 1861; *Elementary Atlas*, 1862; *General Atlas of Modern Geography*, 1862; *Map of New Zealand*, 1864; *Handy Royal Atlas of Modern Geography*, 1868; *Half-crown Atlas of Physical Geography*, 1870; *A School Atlas of General and Descriptive*

*Geography*, 1873; *Keen's New Map of the Watering Places of Kent*, 1876; *Diocesan Map of England*, 1878; *Half-crown Historical Atlas*, 1880; George Phillips Bevan's *Statistical Atlas of England, Scotland and Ireland*, 1882; *General Map of Australasia*, 1890; *Atlas of Commercial Geography*, c.1892; *The Cosmographic Atlas*, 1893; *Atlas of India*, 1894; *The World-Wide Atlas of Modern Geography*, 1900.

•• **Johnston**, Alexander Keith [II] (1846-1879). Son of Alexander Keith I. Trained as a draughtsman with the family company. Worked with Stanford's before returning to run the geographical department of Johnston's London office in 1869. Fellow of the Royal Geographical Society from 1869. Geographer to the Paraguay Survey, 1873-1875.

• **Johnston**, Thomas Brumby (1814-1897). Younger brother of Sir William and Alexander Keith Johnston. Geographer and partner in firm of W. & A.K. Johnston from 1852; brought the company to London in 1869. Fellow of the Royal Geographical Society from 1871, appointed Geographer to the Queen for Scotland in 1877. *Geological Map of Scotland by Archibald Geikie*, Edinburgh & London, W. & A.K. Johnston 1876.

•• **Johnston**, Archibald (d.1925). Publisher of 18 Paternoster Row, London; eldest son of Thomas Brumby Johnston. Fellow of the Royal Geographical Society from 1875. Worked initially with Longman, 1865-1867, then managed the London office of Johnston's from 1869. Between 1876-1882 he was partner in W. & A.K. Johnston, by 1892 he had his own publishing business based at Amen Corner.

•• **Johnston**, Thomas Ruddiman. Publisher of Edinburgh, 3rd son of Thomas Brumby Johnston, Fellow of the Royal Geographical Society from 1881. Partner in W. & A.K. Johnston, 1876-1878. Set up his own business at Murrayfield in 1881. *Oceania* (shows the track of the Challenger Expedition of 1874), 1880.

••• **Ruddiman Johnston & Co.** (fl.1881-1889). Independent business interests of Thomas Ruddiman Johnston at Waverley Works, Murrayfield. Sold out to the Educational Supply Association in 1889, remaining its manager until 1895. *The Metropolitan Railway Company's New Pocket Map of London*, 1889.

•• **Johnston**, James Wilson (d.1906). Publisher of Edinburgh, 4th son of Thomas Brumby Johnston. Fellow of the Royal Geographical Society from 1887. Partner in W. & A.K. Johnston from 1876, director of W. & A.K. Johnston Ltd. from 1901.

•• **Johnston**, George Harvey (d.1921). Geographer of Edinburgh, 6th son of Thomas Brumby Johnston. Partner in the family firm from 1886, Fellow of the Royal Geographical Society from 1897, appointed Her Majesty's Geographer for Scotland in 1897.

••• **W. & A.K. Johnston Ltd.** (fl.from 1901). Limited company from 1901, managed by James Wilson Johnston and George Harvey Johnston. 18-inch terrestrial globe, 1907 (also published in Chicago by Weber Costello); and others

**Johnston**, Andrew. Draughtsman and engraver, worked in Round Court, London. *The Plan of Edinburgh*, 1700 (after Frederick de Wit); Samuel Parker's *Plan of the City's of London, Westminster and Borough of Southwark...*, 1720 (published in John Senex's *A New General Atlas...*, 1721); *A New Map of the North Part of Scotland* [also ... *South Part*], 1722 (in 1722 Edmund Gibson edition of William Camden's *Britannia* with Robert Morden maps).

**Johnston**, *Lieutenant Colonel* Duncan Alexander, *RE*. Succeeded Colonel John Farquharson as Director General of the Ordnance Survey, 1899-1905.

**Johnston** [Johnson], George. Pilot of Portaferry. *Strangford River* [Ireland], Dublin 1755 (engraved by I. Ridge); *Ramsgate to Rye*, 1757.

**Johnston**, *Sir* Harry Hamilton (1858-1927). British diplomat and colonial officer based in Africa. Vice-Consul for the Cameroons Proctectorate. *Sketch Map of Mount Kilimanjaro*, Royal Geographical Society 1885 (lithographed by E. Weller); *How Africa should be divided*, 1886 MS; *Sketch Map of the Cameroons Region illustrating its chief physical features*, 1886 (lithographed by Harrison & Sons); *Map of the Rio del Rey and District lying between Old Calabar and the Cameroons Mountains...*, 1887 (lithographed by Harrisons & Sons); *Map of the Niger Delta*, Royal Geographical Society 1888; *Map of River Shire showing extent of*

*British Proctectorate...* [Malawi], 1890 MS; *Plan of the township of Tshiromo surveyed by H H Johnston...*, 1894 MS (drawn by T.H. Lloyd).

**Johnston**, *Captain* J.E. Officer in the Topographical Engineers, one of the surveyors for the U.S. Coast Survey. Contributed to *Map of Delaware Bay and River...*, 1848 (with others); triangulation surveys for *Mouth of Chester River* [Chesapeake Bay], published 1849 (with J. Ferguson, topography by H.L. Whiting, R.D. Cutts and J.C. Neilson).

**Johnston** [Johnstone], *Master* (later *Captain*) James, *RN*. Charts of the West Coast of America including: Plan of Port Brooks, 1787, published A. Dalrymple 1789 (engraved by W. Harrison); Plan of Calamity Harbour, 1787, published A. Dalrymple 1789 (engraved by W. Harrison); Plan of Rose's Harbour, 1787, published A. Dalrymple 1789 (engraved by W. Harrison); Plan of Snug Corner Cove in Prince William Sound [Alaska], 1787, published A. Dalrymple 1789 (engraved by W. Harrison).

**Johnston**, James. Survey of the River Tees from Stockton to the sea, Tees Navigation Co. 1848 (lithographed by J. Jordison).

**Johnston**, John. Chief draughtsman, Department of the Interior, Canada. *Johnston's New Topographical Map of the Whole Dominion of Canada...* (7 sheets), Burland Desbarats Lithographic Company 1874; *Map Shewing the Townships Surveyed in the Province of Manitoba and North-West Territory in the Dominion of Canada...*, 1874 (engraved by H.H. Lloyd & Co.); *Map of Part of the North West Territory including the Province of Manitoba exhibiting the several tracts of country ceded by the Indian Treaties...*, 1875; *Map of the Country to be traversed by the Canadian Pacific Railway...*, 1876 ('Photo-Lith. by the Burland-Desbarats Lithographic Company'); *Dominion of Canada. General map of part of the North West Territory...*, 1880, 1882, 1883; *General Map of part of the North West Territory and of Manitoba*, Montreal, Burland Lithographic Company 1881 (with Lindsay Russell); *Map of Part of the Province of Manitoba Shewing Dominion Lands...*, Montreal, Burland Lithographic Co. 1881; *Map illustrating the relation of the proposed Red River Valley Railway to the Manitoba Railway System and to the Red River*, Ottawa 1887; Map showing Yukon and the coast from Yakutat Bay to Fort Simpson, 1887 published 1888; Map of the Porcupine River above Fort Yukon and the Mackenzie River estuary...*, 1888; *General Map of the Northwestern Part of the Dominion of Canada...*, Jacob Smith 1898 (published posthumously, engraved by F. Drebes, lithographed by The Mortimer Co.); and others.

**Johnston**, Robert. Plan of the lands of Cullen [Scotland], 1797.

**Johnston** [Johnson], Thomas (1708-1767). Heraldic painter, publisher and engraver of maps, book-plates and sheet music, at Brattle Street, Boston, Massachusetts. William Burgis's ...*Plan of Boston in New England*, c.1728; credited as engraver on *A New Plan of Ye Great Town of Boston in New England in America...*, Boston, W. Price 1732 (a later state of John Bonner's plan of 1722); *A Chart of Canada River...*, 1746; map of the area from Norridgwock Town to Cape Elizabeth and Falmouth Neck to Pemaquid Point [Maine], 1753 (from surveys by Joseph Heath, Phineas Jones and John North); *A Plan of Kennebeck, & Sagadahock Rivers, & Country Adjacent...*, Boston, T. Johnston 1754 (surveyed by J. Heath in 1719, engraved by T. Johnston), British edition published as *A Plan of Kennebek & Sagadohok Rivers with the adjacent Coasts... By Thomas Johnston 1754*, London, A. Miller 1755 (engraved by T. Kitchin); Samuel Blodget's *A Prospective Plan of the Battle fought near Lake George... 8th of September 1755*, 1755; ...*Plan of Hudsons Rivr:...*, 1756 (surveyed by T. Clement); *Quebec*, 1759; ...*A Plan of Part of Lake Champlain*, 1762; plan of the town of Pownall, 1763. **Ref.** HITCHINGS, S. 'Thomas Johnston', in *Boston prints and printmakers 1670-1775: a Conference held by the Colonial Society of Massachusetts* (University of Virginia Press, 1973) pp.83-131; WROTH, L.C. 'The Thomas Johnston maps of the Kennebec Purchase', in: *In tribute to Fred Anthoensen, master printer* (Portland, Me. 1952).

**Johnston**, Thomas. Plan of the fields of Pinkie and Inveresk, 1778; plan of roads from Linton to Queensferry, 1788.

**Johnston**, Thomas Brumby. See **Johnston** family.

**Johnston**, W.H. Chart of Caloombyan Harbour, 1818, J. Horsburgh 1819 (with W.H. Hull).

**Johnston**, William. Succeeded Anne Clark (widow of John Clark), as a bookseller 'at the Golden Ball in St. Paul's Churchyard', and as partner in Thomas Badeslade and William Henry Toms's *Chorographia Britanniae*, c.1748 (first published 1742); Thomas Hutchinson's *Geographia Magnae Britanniae*, 1756 edition.

**Johnston**, William. *Geo-Hydrographic Survey of the Isle of Madeira*, 1788, published William Faden 1791.

**Johnston**, William. Engraver of New-Bern, North Carolina. Engraved J. Price's *A Description of Occacock Inlet...* [North Carolina], New Bern, F.X. Martin 1795; J. Price and J. Strother's *...the Sea Coast and Inland Navigation from Cape Henry to Cape Roman...*, New Bern 1798.

**Johnston**, *Lieutenant Colonel* William. *St. George's and the environs to Richmond Heights...* [Grenada], 1804 MS.

**Johnston**, *Lieutenant* William Henry RE. *The camp of Gibraltar*, 1901-1904 (with E.M. Paul).

**Johnstone**, A. *Naturalists' Map of Scotland*, c.1893.

**Johnstone**, James. See **Johnston**, James.

**Johnstone**, Quintin. Provincial Land Surveyor, Canada. *Plan of Ayton*, 1855; surveys at Harrisburg, 1855; plot divisions at Dungannon, 1867; survey at Limerick, 1857; land survey at Wollaston, 1857; land at Herrick, 1867; township plots at Cainsville, 1867 (with Lewis Burwell); Township of Plummer, Ontario, 1878.

**Joho**, Hayashi. See **Hayashi** Joho.

**Joliet**, Louis. See **Jolliet**, Louis.

**Jolís**, José [Giuseppe] SJ. *Saggio sulla storia naturale della Provincia del Gran Chaco* [Central South America] (with maps), Faenza 1789.

**Jolivet** [Jolivetus], *père* Jean [Joannes]. Priest, cartographer, and geographer to François I, and Henri II. Map of the Holy Land, 1544 (whereabouts unknown); untitled map of Berry [central France] (copper engraving on 6 sheets), Bourges 1545 (one copy only extant); *La carte generalle du pays de Normãdie* (2 vellum sheets), 1545 MS (Bibliothèque Nationale); *Description de la haulte et basse Picardye* (woodcut), Paris, Olivier Truschet c.1559; *Nouvelle description des Gaules, avec les confins Dalemaigne et Italye* (woodcut on 4 sheets), Olivier Truschet 1560 (several issues recorded, used by Ortelius, the de Jodes and Bouguereau). **Ref.** KARROW, R.W. Mapmakers of the sixteenth century and their maps (1993) pp.321-323; MEURER, P. Fontes cartographici Orteliani... (1991) pp.176-178 (numerous references cited) [in German].

**Jolivet**, Maurice Louis. French architect. Survey of West Wycombe Park [Buckinghamshire], 1752; *Pianta della Città di Napoli*, c.1755.

**Jolivetus**, Joannes. See **Jolivet**, Jean.

**Jollage**, Paul. *A Plan of Pontefract*, 1742.

**Jollain**, family and companies. **N.B.** The editors have been unable to clarify the relationships between the Jollains with any certainty. Also, as many of the works are simply credited to 'Jollain', correct attribution is difficult. Most of the following are as credited by the Bibliothèque Nationale, but attributions vary according to source.
• **Jollain**, Gérard [I] (*d*.1683). Publisher and printseller of 'rue S. Jacques à la ville de Colloigne [Cologne]', Paris (1653); 'Rue St. Jacques à l'Enfant Jesus', Paris. Worked with Jacques Honervogt, and took over from him c.1663. Copied many maps and town views (some after Braun and Hogenberg). P. Mariette's *Plan de la Ville cité université fauxbourgs de Paris...*, 1653; *Trésor des cartes géographiques...*, c.1659, 1667 (also published by N. Picart, 1651, first published by J. Boisseau 1659); *L'Amérique Francoise*, c.1665; *Artesiae Comitatis. L'Artois*, 1666; *Carte de l'Isle de France, Normandie, Vexin et Hurpois*, c.1666; *Noua America Descriptio*, Paris, 1666 (first published Honervogt 1640); *Comitatus Flandriae nova*

*tabula*, 1667; *Carte des environs de Sas et Pais Gantois*, 1668; *Nouvelle et Exacte Description du Royaume de Pologne et des Estats qui dependent*, 1669; *Planisphère ou Carte Generale du Monde...*, c.1670; P. Duval's *Nouvelle et Exacte description du Royaume de France*, c.1670; V. Le Febure's *Le Lion Belgique des Pais Bas...*, Paris, 1672 (first published Lagnet 1668); *Nowel Amsterdam en l'Amerique*, 1672; P. Duval's *Nouvelle Description de l'Egypte*, 1672; *Description particulière du diocèse de Bayeux*, 1675 (engraved by R. Michault); P. Duval's *La France Avec ses Acquisitions Enrichis des Armes de ses Provinces*, Paris, G. Jollain c.1680; plan of Luxembourg, Paris c.1680. **Ref.** BURDEN, P. *The mapping of North America* (1996) pp.331, 337, 418, 442, 533.

•• **Jollain**, Gérard [II] (1638-1722). Printseller 'à l'Enfant Jésus, rue St. Jacques vis à vis de la rue de la Parcheminerie. Son of Gérard I, above. *Nouvelle, générale et Très exacte description du Duché d'Anjou*, 1669; *Strasbourg, Ville Capitale d'Alsace...*, c.1670; *La Touraine. Turonensis ducatus*, Paris, G. Jollain 1687 [?]; *Nouvelle Carte de Savoye, Piémont et Monteferrat...*, Paris, G. Jollain 1691.

•• **Jollain**, François [Francis], *l'ainé* (1641-1704). Mapseller and publisher, 'rue St. Jacques à la Ville de Cologne', Paris. Later issues of J. Boisseau's world map (8 sheets), 1650s and/or later; Republished Nicolas Berey's *Nova Totius Terrarum Orbis Geographica Ac Hydrographica...*, Jollain 1653, F. Jollain 1669 (first published Berey, 1640); P. Duval's *Royaume de la Chine et ses Provinces*, 1672; P. Duval's *Nouvel description de l'Egipte*, 1672; Jean Petite's *Description particulière du diocèse de Bayeux*, 1675; Jean Courtalin's *Siam ou India Capitale du Royaume de Siam*, F. Jollain 1686; *Nouvelle Carte D'Angleterre avec les Royaumes D'Ecosse et D'Irlande*, F. Jollain 1689. **Ref.** SHIRLEY, R. *Printed maps of the British Isles 1650-1750* (1988) p.78; SHIRLEY, R. *The mapping of the World* (1983) N°362 pp.385-387.

•• **Jollain**, Jacques (b.1649). Engraver and 'marchand d'estampes, rue St. Jacques, Paroisse Saint-Benoit l'Etoile', Paris. Son of Gérard I. *Carte générale de la Haute et Basse Alsace*, 1675; *Carte d'Ukraine...*, 1686; *Le Duché de Savoye*, 1686; *La Picardie. Picardie Vera et Inferior*, 1689. **N.B.** Some sources assert that Jacques Jollain is a grandson of Gérard I, giving the dates 1665-1710.

•• **Jollain**, François Gérard (1660-c.1735). 'Marchand en taille douce, à l'Enfant Jésus à la Ville de Cologne', Paris. Son of Gérard I. L. de Pontigny's *Carte des Isles Britanniques*, Paris, F.G. Jollain 1689; *Carte de la guerre en Italie...*, 1701.

••• **Jollain**, Jean François (d.1719). Son of François Gérard Jollain.

**Jollain**, Claude. *Trésor des cartes géographiques*, 1667. **N.B.** Probably Gérard **Jollain** I.

**Jolliet** [Joliet], Louis, SJ (c.1645-1700). French-Canadian explorer, fur trader and cartographer from Beaupré, Quebec. Studied navigation, hydrography and music at the Jesuit college in Quebec from 1656. By 1668 he was a fur trader on the Great Lakes. He undertook several expeditions including a descent of the Mississippi to the mouth of the Arkansas River in 1673. He explored the Gulfof St Lawrence and the coast of Labrador. Royal Hydrographer of New France from 1697. *Nouuelle Decouuerte de Plusieurs Nations Dans la Nouuelle France En l'année 1673 et 1674*, 1674 (copied from an original manuscript by Jolliet). **Ref.** WALDMAN & WEXLER *Who was who in world exploration* (1992) pp.351-352; SCHWARTZ & EHRENBERG *The mapping of America* (1980) pp.126, 128.

**Jollois**, J.P.J. *Domremy*, 1821; *Carte générale de Soulosse*, 1843.

**Joly**, Joseph Romain (1715-1805). French cartographer and publisher. *Atlas de l'ancienne géographie*, 1801.

**Jomard**, Edmé-François (1777-1862). Geographer and cartographer in Paris. As a graduate of the Ecole Polytechnique he was selected to serve as a geographer and mapmaker on Napoleon's expedition to Egypt in 1798. Founder member of the Paris Geographical Society, 1821; founded and organised the Département des Cartes et Plans at the Bibliothèque Royale in 1828, which later became the basis of the Bibliothèque Nationale collection. For most of his adult life Jomard was a key figure in the promotion of geographical research and exploration, particularly in Africa. Contributed to *Description de l'Egypte* (20 volumes), 1807-1822, in particular *Carte topographique de l'Egypte* (47 sheets), 1818; *Géographie de la France...*, Paris 1832 (with V. Parisot); *Petite géogra-*

phie de la France et des Colonies, 1833; Les Monuments de la géographie... (21 facsimile maps on 81 sheets, also facsimile globes including ones by M. Behaim and J. Schöner), Kaeppelin 1842-1862 (most maps lithographed by E. Rembiélinski, or Hédin); Coup d'Oeil sur l'Î de Formose, 1856. **Ref.** GODLEWSKA, A. 'Jomard: The geographic imagination and the first great facsimile atlases' in: WINEARLS, J. (Ed.) Editing early and historical atlases (Toronto 1995) pp.109-135; HERBERT, F. The Map Collector 41 (Winter 1987) pp.22-23; KRETSCHMER, DÖRFLINGER & WAWRIK Lexicon zur Geschichte der Kartographie (1986) vol.C/1, p.368 [in German]; PELLETIER, M. 'Jomard et le Département des Cartes et Plans: organisation et développement d'une collection' Bulletin de la Bibliothèque Nationale 4e ann. 1, (mars 1979), pp.18-27 [in French].

**Jombert**, family. Publishers and booksellers of Paris.
• **Jombert**, Jean [I].
•• **Jombert**, Claude (1678-1733). Son of Jean. Traded at 'quai des Grands-Augustins, l'image Notre Dame, rue St. Jacques'.
••• **Jombert**, Charles Antoine (1712-1784). Publisher, printer, book and map seller of Paris. 'l'Image Notre Dame rue St. Jacques' (1735-1741); 'Quai des Augustins' (1742-1751); later 'rue Dauphin'. Son of Claude.
•••• **Jombert**, Claude Antoine fils, l'aîné (fl.1769-d.1788). Publisher, book and map seller of Paris. Eldest son of Charles Antoine.
••••**Jombert**, Louis Alexandre le cadet. Publisher, book and map seller of 'rue Dauphin', Paris, 2nd son of Charles Antoine, brother-in-law of Pierre Didot with whom he worked.
•• **Jombert**, Michel (c.1686-1758). 2nd son of Jean II.
•• **Jombert**, Jean [II] (1697-1762). 3rd son of Jean I.

**Jombert**, A. Plan des Sièges de la dernière Guerre de Flandres, Paris 1751. **N.B.** Compare with **Jombert** family, above.

**Jombert**, Jean. Bookseller 'pres les Grands Augustins à l'image Nostre Dame', Paris. Recorded as one of the sellers of Le Neptune François, c.1773. **N.B.** Compare with **Jombert** family, above.

**Jomini**, baron Antoine Henri de (1779-1869). Swiss military historian, worked in Paris and St. Petersburg. Atlas... des guerres de la Révolution, 1811-1816, 1840; Atlas portatif, 1840; and others.

**Jones**, family. Globe and instrument makers and sellers of London.
• **Jones**, John. Founder of John Jones & Sons, father of William and Samuel.
• **John Jones & Sons**. Makers and sellers of optical and mathematical instruments 'at the sign of Archimedes, 30 Lower Holborn', London. Succeeded by W. & S. Jones.
•• **Jones**, William (1763-1831). Son of John and brother of Samuel. Worked with his father as John Jones and Sons, later in partnership with his brother as W. & S. Jones. Designed the celestial globes which were made by the Bardin company and sold by W. & S. Jones.
•• **Jones**, Samuel (1769-1859). Brother of William. Partner in W. & S. Jones.
• **W. & S. Jones** (fl.1782-1859). Globe and instrument makers and sellers of 30 Holborn, London. Sold Bardin globes, later collaborated with the Bardin family on the series of 'New British Globes'. A Correct Globe with the new Discoveries [7-cm], London, W. & S. Jones c.1780; 18-inch globe pair, 1798 (celestial globe designed by William Jones, terrestrial globe based on a map by Arrowsmith); sellers of The New Twelve Inch British Celestial [Terrestrial] Globe (pair), 1800; makers of an orrery (with a 1.5-inch terrestrial globe), c.1800.

**Jones**. See **Sherwood**, Neely & Jones.

**Jones**, Lieutenant —., RN. Plan of Puerto Cavallo, 1741, published Robert Sayer & John Bennett 1779.

**Jones**, Lieutenant —. Ordnance Surveyor, working in Ireland. Waterford, 1841.

**Jones**, Abner Dumont (1807-1872). Illinois and the West, with a township map..., Boston, Weeks, Jordan and Comany, also Philadelphia, W. Marshall & Company 1838.

**Jones**, Augustus (c.1763-1836). Provincial Land Surveyor in Canada. Manuscript surveys including survey of Stamford, 1778 (with philip Frey); Bertie Township, 1788; Survey of the River Trent [Lake Ontario], 1791; Plans of Eleven Townships fronting onto Lake Ontario, c.1791 (copied by J.F. Holland); The Six Nations Indian Lands

*Southeasterly part Gr. River Niagara District*, 1791; *Jones Lines from the River Etobicoke (Partial) meant to enclose the Toronto Purchase*, 1797; *A Survey of the Grand River*, 1807; and many others.

**Jones**, Benjamin. Engraver of Philadelphia. Map of Rhode Island, 1778 (published in John Marshall's Life of Washington, 1807); Charles Varle's *Map of Frederick, Berkeley, & Jefferson Counties in the State of Virginia*, 1809.

**Jones**, Benjamin Richards (*fl.*1804-1810). Land surveyor and teacher, born Massachusetts, moved to Dennysville, Maine c.1803. Reputedly compiled the first American chart of the eastern Maine border area. *A Map & Chart of the Bays Harbours, Post Roads, and Settlements in Passamaquoddy & Machias...*, Boston c.1810, 1818 (engraved by Thomas Wightman), revised 1824.

**Jones**, C. Craven Property Bayswater, 1779.

**Jones**, Charles H. Compiled and edited *The People's Pictorial Atlas*, New York, J. David Williams 1873 (with Theodore F. Hamilton); *Historical Atlas of the World...*, Chicago, H.H. Hardesty 1875 (with Theodore F. Hamilton).

**Jones**, E. Worked with W. Guthrie on revisions to R. Brooks's *The General Gazetteer...*, 4th edition, Dublin, J. Williams 1776; 8th edition, Dublin, P. Wogan 1808.

**Jones**, E. Engraver of West Square. *A General Sketch of the Roman Stations and Roads in Monmouthshire and Wales*, 1801 (with C. Smith); *A Two Sheet Map of the Principality of Wales divided into Counties*, 1804 (with C. Smith); one of the engravers for *Laurie & Whittle's New Traveller's Companion*, 1806-1810; engraver for Abraham Rees's *Cyclopaedia*, 1806; *Asia*, London, C. Smith 1808; José de Espinosa y Tello's *Carta General para las Navegationes a la India Orientale por el Mar del Sur y el Grande oceano que sepera el Asia de la America* (6 sheets), London 1812; James Playfair's New General Atlas, 1814. **N.B.** Compare with **Jones, Smith & Co.**

**Jones**, E. *Islands of Japan*, 1813.

**Jones**, E. Lithographer of 128 Fulton St., New York. Seat of War in Mexico, for J. Goldsborough Bruff, 1847 (with G.W. Newman).

**Jones**, E.L. Mapa de Patagonia, Buenos Aires, 1858 MS.

**Jones**, Edward. Printer and publisher in the Savoy, London. Published maps of Ireland by Captain Thomas Phillips including *A Map or Draught of Athlone and the grounds thereabouts*, 1691 (also sold by Robert Morden and Philip Lea); *Map or Draught of Galloway*, 1691; *Limericke*, London, R. Morden & P. Lea 1691; printer of *A Draught of the Irish Army at the Battle of Aghrim in Ireland*, 1691; *A Plan of the March of the Armies, and of the Ground on which the Battle was fought in Hungary the 19th of August 1691*, London, Randal Taylor 1691.
**N.B.** Titles are as they appear in announcements as quoted in TYACKE, S. *London map sellers 1660-1720* (1978) entry N$^{os.}$202, 203, 206, 207, 214 pp.51-53

**Jones**, Edward. Engraver. *Map of south Italy and Adjacent coasts* (4 sheets), London, A. Arrowsmith 1807; *Map Exhibiting the great post roads, physical and political divisions of Europe...*, London, A. Arrowsmith 1810, 1820; *A New Map of Mexico and Adjacent Provinces...*, London, A. Arrowsmith 1810; *Outlines of the physical and political divisions of South America delineated by A. Arrowsmith...* (6 sheets), 1811 and other editions, also used in George A. Thompson's *The geographical and historical dictionary of America and the West Indies*, 1816 (translated from A. Alcedo y Henera's work of 1786); *Hydrographical chart of the World...*, A. Arrowsmith 1811; *Ireland*, London, A. Arrowsmith 1811, 1821, 1832, 1843, 1846; *...Map of the Physical Divisions of Germany, Exhibiting The Post Roads, Canals &c...* (6 sheets), London, A. Arrowsmith 1812. **N.B.** Compare with E. **Jones** [of West Square] and **Jones, Smith & Co.**

**Jones**, Ephraim. American surveyor, whose surveys of the Kennebeck and Sagadahock Rivers were used by Thomas Johnston as one of the sources for his map of 1754.

**Jones**, Evan. *A New and Universal Geographical Grammar...* (2 volumes), London, G. Robinson & T. Evans 1772.

**Jones, F.** Engraver of *Western Hemisphere*, Edinburgh, Macredie Skelty & Co. 1821 (drawn by N. Coltman for Dr. Playfair's atlas). **N.B.** Compare with E. **Jones** of West Square.

**Jones,** *Commander* Felix, *RN*. See **Jones,** James Felix.

**Jones, G.W.** Dyail. Army engineer in North America. Manuscript maps and plans of Fort George Majabigwaduce, 1779-1780; *Sketch of the neck and Harbour of Majabigwaduce*, 1779 [Clements Library].

**Jones, George.** Publisher of Ave Maria Lane, London. *London*, 1814 (engraved by Thomson), 1815 (in *Encyclopaedia Londinensis*); Monmouth, 1817.

**Jones, H.A.** *Map of Springfield, Massachusetts*, New York 1851 (with M. Smith); draughtsman for *Map of That Part of the City and Country of New-York North of 50th St.*, New York, Matthew Dripps 1851. **Ref.** COHEN, P.E. & AUGUSTYN, R.T. *Manhattan in maps 1527-1995* (1997) pp.126-127.

**Jones,** *Major General Sir* Harry David *RE*. Plans of the defences of Sevastapol and reports on the Crimean War, 1855; *General Map of the Aland Isles* (5 sheets), 1856.

**Jones, Henry.** Surveyor. Allotments Footscray, Victoria, 1858.

**Jones, Hugh** (*fl.*1773-1774). Land surveyor of Billericay, Essex. Estate at Steeple, 1773; farm at Buttsbury [Essex], 1774.

**Jones, J.** Engraver in Quebec. Sketch map of the eastern townships of Lower Canada, 1835.

**Jones, J.** Lithographer for maps published by the Surveyor General's Office, Melbourne, Australia. *The Parish of Nunawading in the County of Bourke*, 1853; *Township of Footscray, Parish of Cut Paw Paw*, 1854; *Storing allotments on Corio Bay Geelong*, 1855; *Country lots near Creswick Parish of Spring Hills*, 1855; *Suburban & Country lands in the Parish of Crowlands on the Wimmera River*, 1855; *The Township of Rosedale with the Agricultural Reserve...*, 1855; *Part of the parish of Yering*, 1855; *Subdivisions... in the Parish of Cassung-E-Murnong*, 1855; *Subdivisions of sections in the Parish of Conewarre, County of Grant*, 1855; *Electoral District of the Ovens*, 1855; *Diagram map shewing the worked portions of the Victorian gold fields*, 1855; *Country lands in the Parish of Bruthen...*, 1856; *The village of Bruthen on Bruthen Creek Gippsland District*, 1856; and others.

**Jones, J.** Draughtsman. *Road & Distance Map of New South Wales*, Sydney, Gibbs, Shalland & Co. 1871.

**Jones, J.W.** Australian surveyor. *Country North East of Eucla*, Adelaide 1880.

**Jones, James Alexander.** *Bogs Meath, Westmeath & Kings Co.* (2 sheets), 1811; *Map of part of the Bogs in the District of Lough Corrib in the Counties of Galway & Mayo* (4 sheets), 1814.

**Jones,** *Commander* James Felix. *Survey of the... Gulf of Cutch*, 1834; *Vestiges of Assyria*, 1855 (with J.M. Hyslop); *Surveys of Ancient Babylon* (6 sheets), 1855 (surveys used on later maps); *Enceinte of Baghdad*, 1856 (with W. Collingwood); *Sketch of the Course of the Shut ul Arab*, 1857 (with M. Green).

**Jones, James O.** *RN*. Second master aboard HMS *Retribution*. *Sevastopol Harbour*, 1854; *Black Sea Sebastopol from a Russian Survey of 1836*, London, J. Arrowsmith 1854 (both based on Russian surveys of 1836).

**Jones, John.** *A Plan of the Islands Eastward Laying from Penobscot Bay...*, 1765 MS (with Barnabas Mason).

**Jones, John.** Printer of Coventry. Produced the second edition of W. Dugdale's *The antiquities of Warwickshire illustrated...*, 1765 (first published 1656, engraved by R. Vaughan).

**Jones, John.** Instrument maker of London. See **Jones,** family.

**Jones, John** (*fl.*early-19th century). Chart and nautical instrument seller at 'North Side Old Dock, Corner of Pool Lane, Liverpool'. Liverpool seller of the charts of William Heather, Laurie & Whittle and others.

**Jones**, John Humphreys. Prepared and entered a plan for the Australian Federal Capital Design Competition, published Melbourne, Department of Home Affairs 1912.

**Jones**, Joseph [& Son]. Publishers of 47 Broad Street. *Plan of the City of Hereford*, 1858.

**Jones**, Mathew. Assistant surveyor, later 'Acting Surveyor General' on the Gold Coast. *Gold Coast Colony Open Spaces Accra...*, 1886 MS; *Plan shewing Boundaries of Government land. Axim*, 1890.

**Jones**, Phineas. American surveyor. Survey of Atkins Bay in the Kennebeck River, 1731, used by Thomas Johnston of Boston, 1753; surveys of the Kennebeck and Sagadahock Rivers were amongst those used by Thomas Johnston as sources for his map of 1754.

**Jones**, Richard. Engineer. *Plans and Profile of the Fortifications near Black River on the Moschetto Shore*, 1751 MS; *A Prospect Sections and Plan of a Fort Design'd for the Bay of Honduras, River Belease*, 1755 MS (inset on a map of the Bay of Honduras); *A correct draft of the Harbours of Port Royal and Kingston* [Jamaica], 1756 (engraved J. Mynde).

**Jones**, *Lieutenant Colonel* Robert Owen, RE. Ordnance Survey Warwickshire, 1885, 1888.

**Jones**, S. *A Plan of the Intended Harbour and Wet Dock in Saint Nicholas Bay in the Isle of Thanet...*, London 1810 (surveyed by S. Jones, engraved by V. Woodthorpe).

**Jones**, S.L. Draughtsman. *Map of Hillsdale County, Michigan*, 1857 (with Samuel Geil); *Map of Branch County, Michigan* (4 sheets), Philadelphia 1858 (with S. Geil); *Jackson County* (4 sheets), 1858 (with S. Geil); *Macomb and St. Clair Counties, Michigan*, Philadelphia 1859 (with S. Geil); *Munro County, New York* (4 sheets), Philadelphia, Geil, Harley & Siverd 1859 (with S.Geil); *Map of Kalamazoo Co., Michigan* (4 sheets), Philadelphia, Geil & Harley 1861 (surveys by I. Gross, engraved by Worley & Bracher); *Map of the State of Michigan*, 1864.

**Jones**, Samuel. See **Jones**, family.

**Jones**, Stephen. History of Poland (with map), London 1795.

**Jones**, T. Engraver. Worked for the publisher Henry Gregory, 1787.

**Jones**, T.W. *The traveller's directory: or, a pocket companion showing the course of the main road from Philadelphia to New York*, Philadelphia, M. Carey 1802, 1804 (with S.S. Moore).

**Jones**, Thomas, *junior*. *Map of the Parish of Streatham* (4 sheets), London, T. Jones Jr. c.1840.

**Jones**, Thomas. Surveyor of 62 Charing Cross, London. Cippenham Liberty [Buckinghamshire], 1841 MS.

**Jones**, Thomas. Copper globe in relief, Chicago 1894; *Jones' Model of the Earth*, 1897.

**Jones**, W. Cited as the author of *Nova Totius Angliae Delineatio: Or, A Geographical Historical Epitome of England, &c.* (with maps by Robert Morden), announced June 1698.

**Jones**, W. *Map of the County of Haldimand, Canada West* (4 sheets), Toronto 1863.

**Jones**, W. W. *Jones's Street Map of Newport* [Monmouthshire], Newport, W. Jones 1885.

**Jones**, *Captain* W.A. Campaign Map of Nebraska & Wyoming, Washington 1872-1874.

**Jones**, William. *A True Description of the Citie of Rochell* [La Rochelle], London, Thomas Jenner c.1650.

**Jones**, William (1675-1749). Mathematician from Anglesey, Fellow of the Royal Society (1712), friend of E. Halley and I. Newton. Edited some of Newton's texts. Two maps in *Y Bibl* [Welsh language Bible], Cambridge, S.P.C.K. 1746 are noted as being 'a gift from William Jones, FRS'.

**Jones**, William. Draughtsman for Thomas Telford and James Douglas's *General Plan... for the further improvement of the port of London*, 1800.

**Jones, William.** Globe and instrument maker of London. See **Jones**, family.

**Jones, Wyndham C.** Civil engineer. *Map of Oil Regions of Butler County, Pennsylvania,* Titusville 1874.

**Jones & Co.** Publishers. Classical Atlas, 1830.

**Jones & Smith** [Jones, Smith & Bye; Jones, Smith & Co.; Smith & Jones; Smith, Jones & Bye] (*fl.*from 1799). Engravers of Pentonville, London. Also known as Jones, Smith & Co., Beaufort Buildings, Strand. Engravers for [Charles] *Smith's New English Atlas*, London, C. Smith 1801, 1804, 1808, 1818, 1820, 1821, 1832, 1834-1835; Aaron Arrowsmith's *Chart of the West Indies and Spanish Dominions in North America*, 1803; Richard Phillips' *The Environs of London within twelve Miles*, 1803; 4 charts in Dessiou's *Le Petit Neptune Français*, London, W. Faden 1805 (3rd edition).

**Jong, Didrik de.** *Atlas van de Zeehavens der Bataafsche Republiek* (31 plates), Amsterdam, E. Maaskamp 1805 (with Mathias de Sallieth).

**Jong, Elisabeth de** (*fl.*1688-1697). Bookseller and publisher of The Hague, widow of Willem van Lind. Issued second edition of Johannes Leupenius' *Het Hooge Heemraadschap vande Crimpenre Warrd*, 1696 (first published 1683).

**Jonge,** family.
• **Jonge** [Jonghe], **Clement de.** (1624-1677). German art seller and map illuminator working at the sign of the 'Gekroonde Konskaert' on the Kalverstraat, Amsterdam. He also republished the works of others and compiled atlases from the works of a variety of mapmakers. He was succeeded by his son, Jacobus. Atlantic Ocean, 1664; *Brasilia* (9 sheets), 1664 (taken from a Blaeu map of 1647); *Nova Totius Terrarum Orbis Geographica Ac Hydrographica Tabula...*, 1664 (state III of Jodocus Hondius Jr.'s world map, *c.*1625, engraved by F. Huys); republished Michiel Colijn's *Hydrographica, planéque Nova Indiae Occidentalis...*, *c.*1665 (first published *c.*1622); *Icones precipuarum urbium totius Europae* (with town views), 1675; *Tabula Atlantis*, 1675.

•• **Jonge, Jacobus de** (*fl.*1677-1687). Succeeded his father in the business on the Kalverstraat.

**Jonge, Nicolai.** *Synopsis geographicae universalis*, Copenhagen, F.C. Pelt 1768.

**Jongelinx, Johannes Baptista.** Dutch engraver. Polder van Oorderen, 1723.

**Jongensweeshius, R.K.** Atlas compiler of Tilburg. *Geographische Atlas*, Amsterdam, F.H.J. Becker 1889; *Teekenatlas van Nederland in 13 Kaartjes*, Tilburg 1897 (blind maps); *Teekenatlas van Europa en de Werelddeelen*, 1897 (blind maps).

**Jongh, A. de.** *Lake Shore Louisville and South Western Railway*, 1870.

**Jongh, G.J.J. de.** *Route-Atlas van de Rotterdamsche Lloyd* (25 maps), 's-Gravenhage, Mouton & Co. 1925 and later.

**Jongh, J.W. de.** *Geschiedenisatlasje voor De Lagere School*, Groningen and 's-Gravenhage, J.B. Wolters 1927 (with A.G. van Poelje).

**Jongh, Jacobus** [Jacob] **de** (*fl.*1700-1727). Book and printseller of The Hague. Sold Willem van Swieten and Johannes Sandifort's *Kaerte... van den eijlande Oudt en Nieuw Roosenburg...* (8 sheets), 1727.

**Jonghaus, Gustav** (1807-1870). Printer and publisher of Darmstadt. Partner in Jonghaus & Venator.
• **Jonghaus & Venator.** Publisher of Darmstadt. *Urbs Roma antiqua*, 1845; *Hessen*, 1847; L.W. Ewald's *Hand-Atlas der allgemeinen Erdkunde..*, 1852-1858 and later editions; W. Eder's *Handbuch der allgemeinen Erdkunde*, 1860; *Hand-Atlas*, 1861.

**Jonghe, Clement de.** See **Jonge**, Clement de.

**Jonker, H.J.W.** Geological maps of parts of Java, 1872; Timor, 1873.

**Jönsson, J.** Swedish geologist. *Geologiska karta öfver Farsta och Gustafsberg*, 1887, 1890.

**Jöntzen, W.** Lithographer of Bremen. Europa und Nord-Amerika, 1849.

**Joop,** G. *Stockholm från Mosebacke*, 1876.

**Joosten,** Hendrik. De Kleyne wonderlijke Werelt, 1651.

**Jopp,** John. Manuscript survey of the Poona District of British India, 1824.

**Jordaens,** L. Engraver of Amsterdam. 24 plates for *Theatrum praecipuarum urbium Brabantiae*, 1660; 36 plates for Nicolaes Visscher's *Zelandiae*.

**Jordan,** Carl. German cartographer. *Reliefkarte Harzburg-Brocken*, 1896; *Bad Harzburg und Umgebung*, 1897.

**Jordan,** Claude. Voyages historiques de l'Europe, Amsterdam 1718.

**Jordan.** James B. (1838-1915). Geologist in the Mining Records Office (part of the Geological Survey) from 1858. *Stanford's Geological Map of London Shewing the Superficial Deposits*, 1870; geological model of London, c.1874 (with William Whitaker), used to compile a geological version of *Stanford's Library Map of London and its Suburbs* (24 sheets), 1877; Geological Sections of the British Isles, 1879.

**Jordan** [Jordanus; Jorden], Marcus [Mark] (1521-1595). Danish cartographer, chronicler and professor of mathematics at Copenhagen from 1550. *Gemene beschrivinge des Jodischen Landes Canaan*, c.1550 (woodcut, one example only known, at Coburg); map of Denmark, Copenhagen, Hans Vingaard 1552 (no longer extant); *Holsatiae descrip[tio]...* 1559 (woodcut, later used by de Jode, 1578, 1593, and Ortelius 1579-1594); *Descriptio Holsatiae...*, Krempe 1580 (engraved by M. Jordan); *Hodoeporicon Divi Pauli Apostoli*, 1572; *Danorum marca, vel Cimbricum, aut Daniae Regnum... M.D.LXXXV* 1585 (used by Georg Braun and Frans Hogenberg in the *Civitates orbis terrarum* IV 1588); *Cimbricae Chersonesi nunc Iutiae descriptio* [peninsula of Jutland], Ortelius c.1590 (in *Theatrum*).
**Ref.** KARROW, R.W. *Mapmakers of the sixteenth century and their maps* (1993) pp.324-326; REUMANN, K. *Karte des Marcus Jordanus uber die Herzogtumer Schleswig und Holstein...* (Kiel 1988) [in German]; MEURER, P.H. 'Die Palästina-Karte Mark Jordens' in *Speculum Orbis* (Bad Neustadt a.d. Saale: Pfaehler) 3, Jg.1 (1987) pp.9-12 [in German]; NÖRLUND, Niels E. *Danmarks kortlaegning; en historisk fremstilling udgivet met stöttet af Carlsbergfondet* (Copenhagen 1943) pp.24-30 [in Danish].

**Jordan,** Timothy. Bookseller at the Golden Lyon in Fleet Street, London, successor to Thomas Taylor. Himself succeeded by Thomas Bakewell in c.1731. Continued to sell Thomas Taylor's *England Exactly Described...*, 1729-1731 (a reworking of *Speed's Maps Epitomised*).

**Jordan,** Wilhelm (1842-1899). German geodesist, professor in Hanover. *Höhen-Karte von Baden und Württemberg*, 1871; *Karte der... Lybischen Wüste*, 1874.

**Jorden,** Mark. See **Jordan,** Marcus.

**Jordison,** J. Lithographer of Middlesbrough-on-Tees. Survey of the River Tees from Stockton to the sea, Tees Navigation Co. 1848 (surveyed by James Johnston).

**Jorge,** Juan. See **Juan** y Santacilia, Jorge.

**Jörgensen,** Gotfred. Civil engineer and draughtsman in the Department of Land and Works, Victoria, B.C. *Map of the Central Portion of British Columbia*, Victoria, Chief Commissioner of Land & Works 1892; *Map of the Western Part of British Columbia*, 1892; Northern Coast of British Columbia, 1893; *Sectional Map of the northern portion of Vancouver Island*, 1893; *Map of the Province of British Columbia...*, Montreal, Sabiston Lith. & Pub. Co. 1895 (with insets of Canada and the North Atlantic); South Eastern Vancouver Island, 1895.

**Jorio,** *Chanoine* André de. *Plan de la Ville de Naples*, 1826; *Plan de Pompei*, 1829.

**Jose,** Amaro. *The Lands of Cazembe*, 1873.

**Josenhans,** Josef (1812-1884). German theologian and missionary. *Atlas der Evangelischen Missions-Gesellschaft zu Basel*, 1857.

**Joseph,** Charles. Grand Trunk Road to Sutledge [India], 1851; *A new and improved Map of the first portion of the Grand Trunk Road from Calcutta and Benares*, Calcutta 1855; *Map of that part of India which lies between Calcutta and Lahore*, 1857.

**Joslin**, Gilman [& Son]. Globemakers of Boston, Massachusetts, 1839-1907. **Ref.** WARNER, D.J. *Rittenhouse* Vol.2, No.1 (1987) pp.100-103; 'Gilman Joslin', *Annals, Massachusetts Charitable Mechanic Association*, (Boston 1892) pp.449-450.

• **Joslin**, Gilman (1804-.c.1886). Inventor, globemaker and bookseller of Boston. Trained as a wood turner and looking glass maker, worked for globemaker Josiah Loring from c.1837 and succeeded him in 1839. Made globes for Sylvester Bliss and H.B. Nims as well as publishing globes in his own name. He also revised and republished globes by Charles Copley. *Joslin's Six Inch Terrestrial Globe...*, 1839 (engraved by William Annin); *Joslin's Six Inch Celestial Globe...*, 1839 and later updates (engraved by William Annin); *Joslin's Ten Inch Terrestrial Globe*, c.1839 and many later issues (engraved by William Annin); *Joslin's Ten Inch Celestial Globe*, c.1839 (engraved by William Annin); *New Solar Telluric Globe*, 1852; pair of 16-inch globes, 1869; *Joslin's Nine and a Half Inch Terrestrial Globe*, 1869; manufactured and published a new edition of George W. Boynton's 12-inch terrestrial globe as *Joslin's Terrestrial globe*, c.1870.

•• **Joslin**, William B. Son of Gilman Joslin, joined his father in 1874 to form Gilman Joslin & Son.

•• **Gilman Joslin & Son** (fl.1874-1907). 36-inch terrestrial globe, 1874; *Joslin's Ten Inch Terrestrial Globe*, c.1890s; *New Improved Globe*, c.1890 (revised version of Charles Copley's globe of 1852).

**Jouan**, —. Publisher of Antwerp. *Plan de la Ville et Citadelle d'Anvers*, c.1805.

**Jouan**, Henri (b.1821). French sailor and geographer. Contributed to Pacific charts published by the Dépôt de la Marine, 1860s.

**Jouanne**, —. 'Conducteur des Ponts et Chaussées'. *Plan de Nantes*, Nantes, Forest 1832, T. Veloppé 1868, 1877, 1904.

**Jouanny**, L.A. (fl.1867-1872). Worked with P.V. Jouanny on maps and atlases of Peru.

**Jouanny**, P.V. *Atlas del Perú* (13 maps), Gotha, J. Perthes 1867 (with L.A. Jouanny); Plano de Lima, 1872 (with L.A. Jouanny).

**Joubert**, Louis Martin Roch (1749-1786). Publisher and engraver of Lyon, worked with the veuve [widow] Daudet. *Gouvernement Général des Provinces du Lyonnais, Forez et Beaujolais*, 1767; *Plan Géometral de la ville de Lyon* (2 sheets), 1784.

**Jouet**, John. *A Chart of the Northern Approaches to Liverpool*, 1851.

**Joulain**, —. Worked on 3 sheets of the Cassini *Carte de France*, including Sens (sheet 46), 1752; Richelieu (sheet 66), 1756, 1758, 1759 (with Renault); Loches (sheet 30), 1759, 1760 (with others).

**Joulia**, E. French 'Administrateur des colonies'. Surveyed the upper Cavally in 1906; updated Hostain and d'Ollone's charts of the Cavally and Nuon Rivers [Ivory Coast]. **Ref.** BROC, N. *Dictionnaire illustré des explorateurs Français du XIXe siècle: I Afrique* (1988) p.182 [in French].

**Joumar**, —. *Plan de la Ville du Mans*, 1855.

**Jourdan**, —. Worked on at least 3 sheets of the Cassini *Carte de France*, including Tonnerre (sheet 82), 1752-1754 (with Noblesse); Chalon sur Saône (sheet 85), 1757-1759; Lyon (sheet 87), 1758-1762 (with La Court).

**Jourdan**, Adolphe. *Environs d'Alger*, 1884.

**Jourdan**, E. Printer. Noël and Vivien's *Carte de l'Empire Ottoman...*, Paris, Giraldon Bovinet & Co. 1825 (engraved by Giraldon-Bovinet, also published in London by J. Wyld and Cary, in Mannheim by Artaria & Fontaine, in Milan by P. & J. Vallardi, in Amsterdam by Boulton & Son and in Glasgow, by Brunin).

**Jourdan**, E.C. *Atlas Historico de laGuerra do Paraguay...*, Rio de Janeiro, Lithographia Imperial de Eduardo Rensburg 1871.

**Jourdan**, Justin. *Atlas-Guide historique et descriptif des Pyrenées de l'une à l'autre mer*, (12 maps), Paris 1875.

**Journeaux**, — *l'ainé*. 'M[archan]d d'Estampes Hôtel des Monnaies A Paris'. *Nouveau plan routier de la Ville et Faubourgs de Paris divisé en Douze Mairies...*, 1808, 1811.

**Joutel**, Henri (1640-1735). French officer from Rouen, active in North America.

Personal assistant to La Salle on his colonising mission to the Mississippi in 1684, ending up in Texas. Following the death of La Salle in 1687, Joutel made his way north to Quebec and then back to Rouen, where he compiled an account of the expedition entitled *Journal historique du dernier voyage qui feu M. de la Sale fit dans le golfe de Mexique pour trouver l'embouchure le cours de la rivière de Missisipi*, 1713 (with a map *Carte nouvelle de la Louisiane et de la Rivière de Mississipi* [sic], Paris, Estienne Robinot 1713), English edition entitled *A Journal of the Last Voyage Perform'd by Monsr. de la Sale to the Gulph of Mexico to Find out the Mouth of the Missisipi River*, 1714. **Ref.** WALDMAN & WEXLER Who was who in world exploration (1992) pp.353-354.

**Jouvency**, —. Marine surveyor. Plans and charts and for Dépôt de la Marine 1792-1807, for A.R.J. Bruny d'Entrecasteaux's *Atlas du voyage*, 1807.

**Jouvenel**, J.B. Engraver. Bruxelles, 1810.

**Jouy**, H. Atlas universel, 1834; Géographie populaire, Paris 1834.

**Jovanovits**, M. One of the international team of boundary commissioners who signed maps on behalf of the Servian Boundary Commission: *...la frontière Serbo-Turque selon article 36 du traité de Berlin...* 1878 (11 sheets plus title), 1879; also *Plan des environs de Prepolatz dressé par le topograph Russe M. Filimov*, Belgrade 1879.

**Jove**, P. See **Giovio**, Paolo.

**Jovio**, D. Andrea de. Napoli e contorni, 1819; Pozzuoli, 1820; Ville de Naples, 1826; Pompei, 1829; Napoli, 1835. **N.B.** Compare with André de **Jorio.**

**Jovius**, Paulus. See **Giovio**, Paolo.

**Joyce**, Henry [H.J.] Surveyor. *Urbis Galviae totius conatiae in Regno Hiberniae clarissimae Metropolis* [Galway], [n.d.] (dedicated to Charles II, ruled 1660-1685).

**Joyce**, Patrick Weston. Revised George Philip's *Handy Atlas of the Counties of Ireland*, 1881.

**Jozsef**, K.S. Heves-Szolusk geolog., Budapest 1868.

**Juan**, Alvaro. Spanish seaman of Almeria. Made an unusual pocket globe, post 1863. **Ref.** KROGT, P. van der The Map Collector 33 (1985) pp.29-31.

**Juan y Santacilia** [Juan], Jorge. (1717-1773). Spanish naval captain, undertook voyages to South America with Antonio de Ulloa, 1735-1744 on behalf of the Spanish king. *Plan de la Baia Ciudad de Portobello*, 1736; *Plano de la Bahía de Concepción de Chile...*, 1744; *Relacion Historica del Viaje a la America Meridional...*, Madrid, A. Marin 1748, French edition c.1752, German edition 1753 (with Antonio de Ulloa, material was re-used by John Callender, 1766, and John Pinkerton, c.1808).

**Juan de la Cosa**. See **Cosa**, Juan de la.

**Jubert**, —. *Carte des Environs de Mardick et de Dunkerque*, 1752; *Ville et Port de Toulon*, 1752.

**Jubrien** [Chalonnois], Jean (1569-1641). French surveyor. *Théâtre géographique du Royaume de France*, 1621; *Carte du Pais de Retelois*, Le Clerc 1621 (engraved by H. Picart, used by M. Tavernier 1624 and later, W. Blaeu 1631, J. Boisseau 1642, P. Mariette 1650, also Hondius & Blaeu); *Carte du Pais de Nivernois*, 1621; *Carte du pays et diocèse de Rheims*, 1623 (used by Willem Blaeu 1631 and later, A. de Fer 1646, and P. Mariette 1650).

**Judaeus** [Judaeis], Gerard de. See **Jode**, Gerard de.

**Judah**, Theodore Dehone. Civil engineer, chief engineer of the Sacramento Valley Railroad. *Map of the Sacramento Valley Railroad from the City of Sacramento To the crossing of American River at Negro Bar*, Sac. Co., San Francisco, B.F. Butler's Lith. 1854; *Map of the villages of Bellevue, Niagara Falls and Elgin*, Buffalo, Compton & Gibson 1854.

**Judd**, James. *The Shilling General Atlas* (8 maps), London 1854; *Indestructible Atlas of Modern Geography*, 1855; *The General & Historical Atlas*, 1855.

**Judd**, Philu E. (*d*.1825). Government inspector of the road from Detroit to the Miami Rapids. *Map of Michigan with part of the adjoining states*, *c*.1824 (engraved by J.O. Lewis); *Map of the United States Road from Ohio to Detroit*, 1825 (with A. Edwards and S. Vance, drawn by J. Farmer).

**Judd & Co.** Lithographers of 63 Carter Lane, Doctors Commons, London. Effingham Wilson's *The Pocket Guide and Diamond Map of London*, *c*.1884; *Map of Western Australia...*, 1890; Major J.H. Ewart's *Map of the Pokra Kingdom* [Nigeria], 1891; *Gibraltar. Enlarged from Chart No.144*, 1892; *Royal Commission on metropolitan water supply. Sketch map coloured to shew the districts supplied by each metropolitan water company...*, 1893-1894.

**Judd & Glass.** Phoenix Printing Works, London. Overland Railway through British North America, 1868.

**Juegel.** See **Jügel**, Carl Christian.

**Juel**, Rasmus. Faroe Islands, 1709-1710.

**Juettner**, Joseph. See **Jüttner**, Joseph.

**Jügel** [Juegel], Carl [Karl] Christian (1783-1869). Engraver, printer and publisher in Frankfurt am Main. *Post- und Reise Karte von Deutschland* (4 sheets), 1832; *Illustrations to the Hand-Book for Travellers on the Continent contained in a Series of Maps of the most frequented Roads through Holland, Belgium and Germany* [parallel title in German], 1842 (maps drawn and engraved by J. Lehnhardt, sold also by John Murray in London, Amyot in Paris and C. Muquardt in Brussels); *Uebersichts-Karte der Haupt-Post-Strassen & Eisenbahnen durch Deutschland* [title also in French], *c*.1844; *Grundriss von Frankfurt*, 1844; *Eisenbahn-Atlas*, 1846.

**Jügel**, Friedrich (*d*.1833). Engraver of Berlin. *Der Kremlin*, 1810.

**Jugge**, Richard. English bookseller and publisher 'at the signe of the Byble', London. Maps in Bibles, all in several editions, including: Acts of the Apostles, 1552; Eastern Mediterranean, 1552; Holy Land, 1552; one of several sellers of Richard Eden's *The Decades of the Newe Worlde*, 1555; Canaan, 1568; Daniel's Dream, 1568; Eden, 1568; route of the Exodus, 1568; Land of Promise, 1574; Richard Eden's *The History of Travayle in the West and East Indies...*, 1577 (augmented edition of *The Decades* completed by Richard Willes). **Ref.** DELANO-SMITH, C. & INGRAM, E.M. *Maps in Bibles 1500-1600* (1991); DELANO-SMITH, C. *The Map Collector* 39 (1987) pp.2-14.

**Juigne**, R. Printer of Paris. For *Atlas historique...*, 1818 (by Emmanuel Las Casas).

**Jukai** (*b*.1297). Japanese Buddhist priest. Map of the Indies, 1364 MS.

**Jukes**, Francis. Aquatint engraver of London. Henry Pelham's *A Plan of Boston in New England and its Environs... with the military works constructed in the years 1775 and 1776*, London, 1777.

**Jukes**, Joseph Beete. *Geological Map of Canada*, Montreal 1865 (with J.W. Dawson and James Robb); *Geological Map of Ireland*, 1867, 1878.

**Jukes-Brown**, Alfred Joseph. *Geological map of the neighbourhood of Cambridge*, London 1875; *Geological map of Barbados*, 1890 (with J.B. Harrison, lithographed by Malby & Sons).

**Julien**, *Lieutenant* Emile. French explorer in Central Africa, 1892-1910. *Du haut Oubangui vers le Chari par le bassin de la rivière Kotto*, 1897; *De Ouango à Mobaye par les pays Nsakara et Bougbou*, 1901. **Ref.** BROC, N. *Dictionnaire illustré des explorateurs Français du XIXe siècle: I Afrique* (1988) pp.182-183 [in French].

**Julien**, *sieur* Roch-Joseph (*fl*.1751-1780). Military geographer, publisher and mapseller 'rue du Chaume à l'hôtel de Soubise à Paris' (1752), also quai des Augustins. From 1758, he was one of the distributors of the Cassini *Carte de France*; he also collaborated with Homann's Heirs, John Rocque and Sayer & Bennett. *Atlas géographique et militaire de la France*, 1751 and later; *Bohème*, 1758; *Carte Générale d'Allemagne*, 1758; *Carte Générale de L'Electorat de Saxe*, 1758; *Nouveau Théâtre de la Guerre ou Atlas topographique et militaire...*, Paris 1758; *Théâtre du Monde*, 1762, 1768; sold Robert de Vaugondy maps, 1763; Courtalon's *Atlas élémentaire de l'Empire d'Allemagne*, 1774 (with A.

Boudet); *Partie septentrionale Des Possessions Angloise en Amerique*, 1778.

**Julius**, F. *Karte oder geographische Darstellung der Berg-, Hütten-, Salz und andere metallurgischen Werke der angrenzenden Länder des Kgr. Westphalen* [mining map of Westphalia], 1813 MS.

**Julius**, Friedrich. *Karte von Asien*, 1813; *Charte von dem Harz Gebirge und einem Theile der umliegenden Gegenden*, 1817, Braunschweig, H.C.W. Berghaus 1844.

**Jullien**, A. *Carte Générale répresentant le tracé du chemin de fer de Paris à Orléans...*, 1842 (with A. Donnet).

**Jullien**, Amédée. *La Nièvre*, 1883; *La Nièvre et le Nivernais*, 1884.

**Jumelle**, Count — de. *Carte... du Canal de St Denis à Paris*, c.1726.

**Junboll**, Th.G.J. (1802-1861). Geographical Lexicon.

**Junctinus**, F. Commentaria in sphaeram, 1577-1578.

**Jung**, —. *Vue de Sebastopol...*, Paris, c.1855 (with Gobaut, printed by Kaeppelin).

**Jung**, family. Cartographers from Rothenburg ob der Tauber, Franconia.
• **Jung**, Johann Georg, the *elder* (1583-1641). Painter, glass-cutter and cartographer. *Franconia*, 1638; *Würzburg*, 1639; *Totius Germaniae novum itinerarium*, Nuremberg 1641 (all with son Georg Conrad Jung).
•• **Jung**, Johann Georg, the *younger* (1607-1648).
•• **Jung**, Georg Conrad (1612-1691). Son of Johann Georg the elder. *Franconia*, 1638; *Würzburg*, 1639 MS (later engraved on copper); *Totius Germaniae novum itinerarium*, 1641 (all with his father); administrative map of the Duchy of Brandenburg-Ansbach, 1670.

**Jung**, Carl Emile (*b*.1836). Lexikon der Handelsgeographie, 1882.

**Jung**, Georg. German politician and cartographer. *Atlas zur Geschichte des Alten Bundes* [historical atlas], Berlin 1859.

**Jung**, J.M. Austrian draughtsman. *Österreichische Niederlande*, 1785; *Reise-Karte von Iglau*, before 1800.

**Jung**, Johann Adam. Publisher of the second German edition of Allain Manesson-Mallet's 'Description' entitled *Beschreibung des gantzen Welt-Kreises...* (5 volumes), Frankfurt am Main 1719.

**Junghuhn**, Franz Wilhelm (1809-1864). German explorer. *Atlas zur Reise durch Java*, 1845.

**Jungmann**, Carl. Engraver for Carl Ferdinand Weiland, 1848; Heinrich Kiepert, 185l; Adolf Stieler, 1860.

**Jungmeister**, A. Jn. Atlas publisher of St Petersburg, 1845.

**Jungwirth**, Franz Xaver. Bavarian engraver. *Region München*, c.1768.

**Junker**, —. *Karte der Stadt Baden und ihrer Umgebungen...*, Vienna 1830

**Junker**, Keresztély [Christoph] (1757-1841). Hungarian engraver with a workshop in Vienna, born in Pozsony [now Bratislava, Slovakia], died Vienna. Engraved for Korabinszky among others. Unter Steyermark, 1789; *Atlas von Inner Oesterreich...* (12 maps), Graz, Franz Xaver Miller c.1789-1797 (calculations by J. Liesganig, drawn by J.K. Kindermann); Klagenfurter Kreis, 1790; *Innerkrain oder Adelsberger Kreis* [Yugoslavia], Graetz 1795; *Stadt Washington*, 1796; *Karte des Canals von Wien bis Raab* (4 sheets), 1797; title for *Magyar Atlas*, 1802; *Riesen-Gebirg*, 1804.

**Junker**, *Doctor* Wilhelm Johann (1840-1892). German explorer. Travelled in Africa from 1873, travelled the White Nile from 1876, later explored Uganda and Tanzania, ending up in Zanzibar in 1886. Nord- und Central Afrika, 1880; *Reisen in Afrika*, 1889-1891; *Zentral-Afrika* (4 sheets) Gotha, Justus Perthes 1889; Buganda, 1891. **Ref.** WALDMAN & WEXLER Who was who in world exploration (1992) pp.354-355.

**Junsai**, Fujita. See **Fujita** Junsai.

**Jurany**, Wilhelm. Publisher of Leipzig. J. Lelewel's *Atlas zur Geschichte Polens*, 1846.

**Jurien-Lagravière**, Jean Pierre Edmond (1812-1892). *Carte... Canal de San Pietro*, 1844; *Carte Particulière de la Baie de Palmas*, 1845; *Carte... de la Côte... de Sardaigne*, 1846-1854.

**Jurrius**, J. *Blinde Schoolkaart van Nederlandsch Indië*, 1869.

**Justel**, Henri. *Recueil de divers voyages*, Paris, A. Cellier 1684.

**Justinianus**, Augustinus. See **Giustiniani**, Agostino.

**Just-Moers**, —. See **Moers**, Justus.

**Justo**, Jacob. See **Zaddik**, Yaakov ben Abraham.

**Juta**, J.C. & Co. Publishers. Cape Town. *Map of South Africa*, 1866; *Enlarged Map of South Africa* (4 sheets), 1891; *Excelsior Atlas*, 1899.

**Jüttner** [Juettner], *Oberleutnant* Joseph [Josef] (1775-1848). Mathematician and military cartographer in the Austro-Hungarian army. Later worked in Prague as a globemaker and publisher. Wrote a treatise on the use of globes. *Grundriss des... Haupstadt Prag* (2 sheets), 1811-1815; 12-inch terrestrial globe, Prague 1822 (with Franz Lettany); 12-inch celestial globe, 1824; *Ringkugel...* [12-inch armillary sphere], 1828; *Erdkugel* [12-inch terrestrial globe], Prague 1840 (a re-issue of the 1822 Jüttner / Lettany globe); pair of 24-inch terrestrial and celestial globes, 1836-1839, the terrestrial globe was re-issued 1846.

**Juvet**, Louis Paul (1838-1930). New York globe maker of Swiss origin. *30 Inch Globe...*, 1879; *Time Globe*, 1879.

**Juvigny**, Charles de. Siege of Budapest, Greischen 1686.

**Juzo**, Kondo. See **Kondō**, Juzo.